Neuroscience of Maladaptation

Justin R. Yates

Neuroscience of Maladaptation

Springer

Justin R. Yates
Department of Psychological Science
Northern Kentucky University
Highland Heights, KY, USA

ISBN 978-3-032-15601-3 ISBN 978-3-032-15602-0 (eBook)
https://doi.org/10.1007/978-3-032-15602-0

This Springer imprint is published by the registered company Springer Nature Switzerland AG
The registered company address is: Gewerbestrasse 11, 6330 Cham, Switzerland

This book is dedicated to my wife, Katy, and my sons, Landon and Lucas.

In Memoriam
In loving memory of Dr. Jeff Smith: colleague, mentor, friend.

Preface

I matriculated to Georgia College & State University as a psychology major, but like many students entering college, I did not know what I wanted to do with my life after obtaining my degree. I was, however, interested in attending graduate school, just so I could delay becoming a "real" adult. After taking abnormal psychology as a sophomore, I contemplated going into psychiatry. Although interested in graduate school, I was not enamored with the idea of going to medical school. My career path became much clearer during my junior year when I enrolled in physiological psychology. I immediately became fascinated by the brain and the neurobiological mechanisms that control different behaviors. As a senior, I had the opportunity to conduct animal research with my physiological psychology professor, something I had never considered doing before going to college. Ultimately, I attended the University of Kentucky and earned a Ph.D. in Experimental Psychology, where I specialized in behavioral neuroscience and psychopharmacology. I continued performing animal research as I worked in a lab that focused on addiction.

Studying addiction was eye opening. Growing up, I held negative views toward individuals with an addiction. I thought these individuals deserved little pity as they were the cause of their problems. As I learned more about addiction, I realized how wrong I was. Addiction is a biomedical disease that has a strong neurobiological basis. Now, I am a college professor, continuing my research on the neurobehavioral mechanisms of addiction. While my focus is on drug addiction, I am interested in other problematic behaviors that interfere with daily functioning, as well as the neurobiological processes that lead to various diseases and psychiatric conditions. These interests motivated me to write a book on the neuroscience of maladaptation. I want readers to gain an appreciation for how complex our nervous system is and how disruptions of various neural systems can lead individuals to engage in actions that do not benefit them or can actively cause harm to themselves and/or to others.

While the focus of this book is on neuroscience, I want to emphasize that many of the conditions discussed in this book are influenced by more than neurobiological alterations. Take addiction for example. Cognitive, behavioral, and sociocultural factors also contribute to the initiation of and the continued use of substances. My intention is not to oversimplify highly complex psychiatric disorders. Instead, I want to highlight how aberrant neural functioning can contribute to maladaptive cognitive-behavioral processes.

Highland Heights, KY, USA Justin R. Yates

Acknowledgments There is not enough space in this book to express my everlasting appreciation and love for my wife Katy and my sons Landon and Lucas for their love, support, and patience as I juggled writing this book with serving as department chair, overseeing a research laboratory, and trying to be a present husband and father.

I want to thank my parents Rita and Ray for reading drafts of each chapter and providing their continuous encouragement during the writing process.

I finally want to acknowledge two professors that influenced the trajectory of my career by inspiring me to go into the field of behavioral neuroscience and psychopharmacology: Dr. Walt Isaac and Dr. Mike Bardo, my mentors at Georgia College & State University and the University of Kentucky, respectively.

Competing Interests The author has no competing interests to declare that are relevant to the content of this manuscript.

Organization of Book

This book is organized in four parts. Part I includes Chaps. 1 and 2, which serve as a kind of prologue to this book. Chapter 1 introduces the reader to the construct of maladaptation and how it can be measured. This chapter also discusses the role of animal research in helping us elucidate the neuromechanisms of maladaptation. Chapter 2 provides an overview of neuroscience principles. Organization of the nervous system and neuroscience techniques are reviewed. This chapter is critical for those that have little background in neuroscience.

Part II covers aberrant cognitive-affective processes. Chapter 3 discusses perceptual disturbances. Individuals will learn about conditions that prevent individuals from recognizing what is in their environment, and they will learn how individuals can perceive stimuli that do not physically exist. Chapter 4 is isolated to dysfunctional pain perception; whereas some individuals experience chronic pain, other individuals do not have the ability to feel pain. Chapter 5 highlights attentional deficits, Chap. 6 delves into memory impairments, and Chap. 7 spotlights maladaptive decision-making processes. Chapter 8 illuminates how stress can become detrimental and how aberrant stress responses are linked to anxiety disorders. Chapter 9 continues this discussion, but attention shifts from anxiety disorders to mood disorders.

Part III focuses on maladaptive behavioral processes. Chapter 10 acts as an introduction to this section as it examines how the motor system can become impaired. As such, this chapter features several movement-related disorders. Chapter 11 details maladaptive habit formation, and Chap. 12 centers on motor impulsivity (i.e., doing things without first thinking about the consequences of those actions). Chapter 13 delves into problematic eating patterns, and Chap. 14 concentrates on sleep disorders. Chapters 15 and 16 are related to maladaptive social behaviors. Chapter 15 discusses several problematic social behaviors, such as social withdrawal, lack of empathy, and aggression. Chapter 16 is more specific as it covers maladaptive sexual reproductive behaviors.

Part IV includes one chapter, which serves as an epilogue to the book. Chapter 17 explores how neuroscience techniques can be used to treat various forms of maladaptation that are discussed throughout the book.

My goal in writing this book was to make it accessible to a wide audience that is not an expert in neuroscience. To this end, I try to use a more conversational tone throughout the book and begin each chapter with a personal story that relates to the topic of the chapter. I even sent each chapter to my parents, who have no training in neuroscience, to see if they could understand the content. To help guide one's reading, each chapter includes learning objectives, detailing the most important concepts covered in the chapter. Each chapter also includes several reading comprehension checks. This allows individuals to test their understanding of what they just read. An Experiment Spotlight is included in each chapter, in which one recently published study is described in more detail. Several questions are included to get the reader to think critically about the featured study. For 16 of the 17 chapters, individuals can read the entire article if they want to learn more. If using an eBook, readers can click on the link provided in the chapter to access the article. If reading the print version, you can look up the article using the information provided in the references section. Finally, an appendix is included that provides the etymology (i.e., the origin) of numerous terms that you will encounter in this book.

Contents

About the Author

Dr. Justin R. Yates is a professor and current chair in the Department of Psychological Science at Northern Kentucky University (NKU). He received his PhD in experimental psychology from the University of Kentucky in 2014. At NKU, Dr. Yates has focused on the glutamatergic basis of impulsive and risky decision making, as well as addictive-like behaviors in rats. He has mentored over 60 undergraduate students and one postdoctoral fellow, many of whom have gone on to graduate, professional, and medical school. Dr. Yates has published over three dozen research articles, and he has received over $1 million in external grant funding to support his research. His research has been recognized at the university, the regional, and the national levels, including three early career awards from the American Psychological Association. He is a Fellow of the American Psychological Association (Division 25) and the Midwestern Psychological Association.

Dr. Yates has taught courses in biopsychology, learning and behavior, cognitive processes, and research methods. In 2022, Dr. Yates published his first textbook: *Determinants of Addiction: Biological, Behavioral, Cognitive, and Sociocultural Determinants.* He was inspired to write this book after developing a special topics course on the neurobehavioral mechanisms of addiction.

List of Figures

Chapter 2

Chapter 3

Chapter 4

Chapter 5

Chapter 6

Chapter 7

Chapter 8

Chapter 9

Chapter 10

Chapter 11

Chapter 12

Chapter 13

Chapter 14

Chapter 15

Chapter 16

Chapter 17

Chapter 1
Introduction to Maladaptation

1 Introduction

If it weren't for the last minute, nothing would get done—Rita Mae Brown

The quote above perfectly captures how I operated as a high-school student, waiting until the "last minute" (sometimes literally) to complete a homework assignment or to begin studying for a test. Even when applying to colleges, I waited until the last possible moment to submit applications. I applied to two colleges only, to the chagrin of my mother. Now, I am better at managing my time, but accomplishing things in a timely fashion can be difficult with increasing work demands and familial responsibilities. I am currently a department chair trying to keep up with my research program while helping raise two children.

I feel confident in making the claim that most of us have put off doing something, whether folding laundry, studying for an exam (I see this a lot as a college professor), or something more significant like paying taxes or creating a living will. For some reading this book, procrastination may be a way of life. Procrastination is just one form of **maladaptation**; for the purpose of this book, maladaptation will be defined as the inadequate adjustment to changes in the environment. From an evolutionary perspective, organisms evolve over generations to increase their odds of survival. This is the foundation of Charles Darwin's theory of *natural selection* or what we often term "survival of the fittest".

Importantly, adaptations that once served a species well can become maladaptive later in time. Let's continue using procrastination as an example. When our earliest ancestors lived in an environment characterized by unpredictability, procrastination could have been an important survival strategy. Conservation of energy was more important when individuals needed to search for food sources or when a minor injury by today's standards may have been life threatening. As humans have evolved and have formed societies that are more predictable, procrastination can become more

J. Yates, *Neuroscience of Maladaptation*, https://doi.org/10.1007/978-3-032-15602-0_1

problematic as we have multiple responsibilities that are more time sensitive (e.g., school and work deadlines).

Procrastination can lead to difficulties for individuals, but this book will focus primarily on more severe forms of maladaptation that can negatively impact one's daily activities or interactions with other individuals. The purpose of this chapter is to introduce the ways in which researchers and clinicians quantify maladaptation in individuals. Because animal research plays a pivotal role in shaping our understanding of the neurobiological underpinnings of maladaptive processes, this chapter will also include a general discussion of the use of animals in research.

2 Learning Objectives

By the end of this chapter, you should be able to ...

1. Define maladaptation and provide examples of how it can manifest in individuals.
2. Explain how clinical assessments, self-report measures, and behavioral observations can be used to quantify maladaptive behaviors.
3. Compare and contrast the advantages and disadvantages of the different methods used to assess maladaptation.
4. Critically evaluate the importance of reliability and validity in measuring maladaptive behaviors.
5. Appreciate the need for animal research to better understand the underlying neuromechanisms of maladaptive processes.

3 General Ways of Quantifying Maladaptation

Waiting an extra day to fold laundry or to take out the trash most likely has little impact on us. How long does someone need to hold off from taking out the trash for this procrastination to be considered problematic? As defined by the American Psychological Association (n.d.), maladaptation consists of "biological traits or behavioral patterns [that] are *detrimental*, counterproductive, or otherwise interfere with optimal functioning in various domains". The term "detrimental" is important here. Procrastination, when done sparingly, is not detrimental to an individual. However, chronic procrastination can be quite detrimental. If an individual continuously lets trash accumulate in their home, this can pose serious health hazards. A student that consistently waits to start studying for exams is at risk of failing their exams and the course. Likewise, being chronically late to work can lead to one's termination.

When defined as a psychological phenomenon, maladaptation can be observed in physical behaviors, as well as cognitive-affective processes. While some forms of maladaptation may not significantly affect an individual, such as waiting to fold laundry, other forms can be harmful to an individual (e.g., substance misuse, self-injury) and/or to other individuals (e.g., driving a vehicle while intoxicated, engaging in condomless sex with multiple partners).

Maladaptation is a defining feature of many psychiatric disorders. *The Diagnostic and Statistical Manual of Mental Disorders (DSM-5)* (American Psychiatric Association, 2013) and the *International Classification of Diseases (ICD-11)* (World Health Organization, 2019) provide detailed descriptions of the psychiatric conditions recognized by doctors (note, the *ICD-11* also includes physical conditions). The first way to quantify maladaptation will center on methods used by therapists to better understand maladaptive behaviors in their clients.

When measuring a construct like maladaptation, two important concepts need to be considered: **reliability** and **validity**. Think of reliability as consistency and validity as accuracy (see Fig. 1). For reliability, the goal is to ensure that the method being used to quantify maladaptation can be consistently applied to either the same person across two or more time points and/or can be applied to multiple individuals. Furthermore, if one therapist is using a method to screen for a maladaptive behavior, another therapist should be able to do the same thing as the first therapist and get the same result. For validity, the question being asked is "does this measure of maladaptation *actually measure* what we want it to measure?" For reliability and validity to make more sense, let's apply them to the major ways we can assess maladaptive actions.

Comprehension Check 1 From a psychological viewpoint, what makes something maladaptive as opposed to inconvenient?

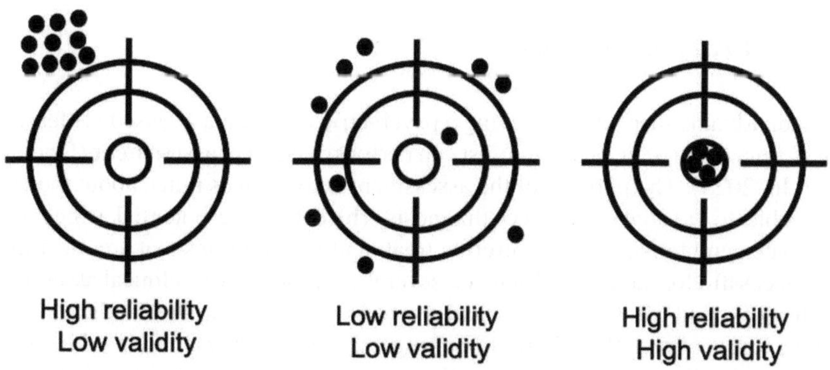

Fig. 1 The relationship between reliability and validity can be visualized using something like a target. A measure can produce consistent results, but it may not accurately measure what one wants to measure (left target). Some measures are neither reliable nor valid (center target). Measures need to be both reliable and valid (right target). Image created by author using Microsoft PowerPoint's stock images and by adding the black circles

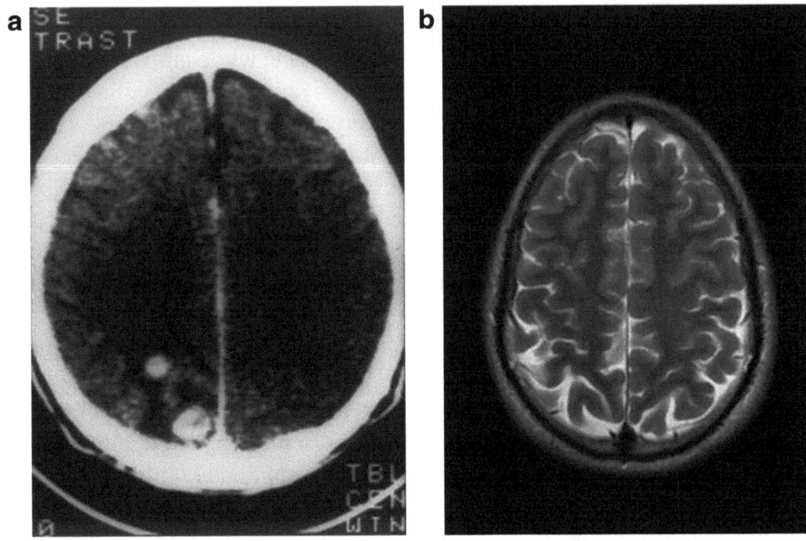

Fig. 2 (**a**) A CT scan of the brain. *Source* https://visualsonline.cancer.gov/details.cfm?imageid=2295 by the National Cancer Institute. (**b**) An MRI image of the brain. *Source* https://commons.wikimedia.org/wiki/File:MRI_Brain_T2_Axial_(17).jpg from 511KeV. CC BY-SA 4.0 (https://creativecommons.org/licenses/by-sa/4.0/deed.en), via Wikimedia Commons. *Note* These images show a *horizontal* section of the brain. The anterior portion of the brain is shown at the top of each image

> *For something to be considered maladaptive, it needs to be detrimental in such a way that it interferes with daily functioning.*

3.1 Clinical Assessments

If an individual is suspected of having a psychiatric disorder, a licensed professional such as a psychiatrist or a psychologist can perform a **clinical assessment** (Bridley & Daffin Jr., 2023). The purpose of the assessment is to learn as much about the client as possible to (a) make a correct diagnosis, (b) determine potential triggers that exacerbate symptoms, and (c) prescribe treatment interventions that are most likely to be successful for the client. There are several components of a clinical assessment. The therapist will often begin the assessment by interviewing the candidate, either asking a series of pre-drafted questions, engaging in a conversation in which topics come up naturally, or using a combination of structured and unstructured questions. Following the interview, the therapist completes a *mental status examination*. Think of this as a snapshot of the client's current mental state. During the interview, the therapist will note how the client looks, what their mood is like, how cognizant the client is of their surroundings, and how well the client functions cognitively.

The therapist may recommend that the client see their primary doctor for a physical examination to determine if a physical condition is causing their symptoms.

The therapist also has a cache of psychological tests that they can administer to the client. Commonly used by therapists, **personality inventories** include a series of questions that aim to reveal the client's self-reported behaviors, beliefs, and attitudes. The NEO Personality Inventory is one of the most commonly used assessments (McCrae & John, 1992). Consisting of 240 *likert-scale* questions (e.g., responses are measured on a scale from 1 to 5), the NEO Personality Inventory measures five major personality constructs: (1) conscientiousness, (2) agreeableness, (3) neuroticism, (4) openness, and (5) extraversion (use the acronym CANOE to remember these traits). Throughout this book, you will encounter multiple personality inventories that are used to measure maladaptive tendencies.

Let's use the NEO Personality Inventory to illustrate different kinds of reliability and validity. Specifically, I will focus on the construct neuroticism. Individuals that score high in neuroticism exhibit more anxiety, are more prone to worrying, and are more likely to experience mood swings. If neuroticism reflects a personality trait, then an individual that completes the NEO one day should have a similar score for neuroticism when the NEO is completed a month later. This is *test-retest reliability*. Furthermore, questions in the NEO should have high *internal consistency*; that is, if neuroticism is proposed to include facets such as anxiety, anger, self-consciousness, and vulnerability to stress, individuals' responses to questions related to these facets should be related to each other. *Interrater reliability* refers to the consistency at which researchers/clinicians interpret the results of an assessment. For something like the NEO, there is already a scoring rubric that is provided. This rubric details how individuals should classify individuals' personality traits based on their responses in the NEO. Interrater reliability becomes more important when assessing responses to open-ended interview questions.

Due to the numerous types of validity, I will focus on just a few here. Two forms of validity are highly similar: *content validity* and *construct validity*. If performing a clinical assessment, the therapist needs to ensure that the questions allow them to make inferences about the client's maladaptation. Content validity simply refers to the idea that one's measure represents all features of a variable being measured. Construct validity is a specific form of content validity in which the variable being measured is a construct (i.e., a concept that cannot be physically seen but is inferred by observable behaviors). Years of research have gone into ensuring that the NEO has high construct validity. This is why the NEO is one of the most widely used, if not the most used, personality inventory.

When completing an assessment, one goal is to make predictions about an individual's behavior. Knowing that an individual scores high in something like neuroticism is not necessarily informative on its own. Instead, we want to know if individuals that score high in neuroticism show fundamental differences compared to individuals that score low in neuroticism. This is where *predictive validity* comes into play. For example, research shows that individuals high in neuroticism are more likely to engage in maladaptive behaviors like illicit substance use (Kang, 2022) and self-harm (Hafferty et al., 2019). Neuroticism is also associated with procrastination (Markiewicz, 2017).

In addition to personality inventories, therapists can work directly with the client to modify problematic behaviors. As the name suggests, **applied behavior analysis (ABA)** (formally referred to as behavior modification) examines observable behaviors as opposed to personality constructs (Cooper et al., 2020). Specifically, a therapist using ABA is interested in determining the *antecedents* (i.e., triggers) and *consequences* of the client's behavior. In the case of an individual that chronically procrastinates, a potential antecedent is anxiety. Imagine a student that is assigned an oral presentation. The student may feel anxious about speaking in front of their peers. When the student avoids working on the presentation, they feel less stress as they are not currently focused on the thought of public speaking. This reduction in anxiety is reinforcing to the student, which further increases the likelihood that they continue avoiding the assignment.

Comprehension Check 2 What are the key components of a clinical assessment, and how do they contribute to understanding and addressing a client's maladaptive behaviors?

The key components of a clinical assessment are the interview, mental status examination, and psychological tests. Interviews gather information about the client's history, symptoms, and functioning. Mental status exams assess the client's current mental state. Psychological tests provide standardized measures of personality traits, cognitive abilities, and emotional states. Together, these components help clinicians understand the client's condition and develop appropriate treatment plans.

3.2 Self-Report Data

Obviously, maladaptation is not limited to those with a psychiatric disorder. Many studies rely on participants to complete a questionnaire, whether in person or online, related to some form of maladaptation. Personality inventories such as the NEO can certainly be used in such research. The types of reliability and validity discussed above apply here as well. I will introduce a couple of other types of validity in this section.

One questionnaire that measures maladaptive decision making, including procrastination, is the Melbourne Decision Making Questionnaire (MDMQ) (Mann et al., 1997). Individuals completing the MDMQ read 22 statements and rate the extent to which the statements apply to them. For example, one statement directly related to procrastination is "I often put off making decisions until the last minute". Procrastination can also be measured with the Procrastination Scale developed by Lay (1986), found in the appropriately named research article "At last, my research article on procrastination". This scale consists of 20 questions measured on a likert scale.

As procrastination can be measured in different questionnaires, so can other constructs such as neuroticism. In addition to the NEO, neuroticism can be assessed with the Eysenck Personality Questionnaire (Eysenck & Eysenck, 1975), the Zukerman-Kuhlman Personality Questionnaire (Zuckerman et al., 1993), and the Temperament and Character Inventory (Cloninger, 1994). The Minnesota Multiphasic Personality Inventory (MMPI) measures eight clinical scales, with "dysfunctional negative emotions" including items related to maladaptive anxiety, anger, and irritability. If two or more questionnaires purportedly measure the same construct, such as neuroticism, individuals should provide similar responses across them. If so, these questionnaires have high *convergent validity*. Not surprisingly, neuroticism-like traits are correlated across personality measures that include neuroticism as a personality trait (Zuckerman & Cloninger, 1996).

In addition to convergent validity, researchers want to ensure that measures have high *divergent (discriminant) validity*. If neuroticism is associated with anxiety, then individuals that score high in this trait should score low in something like extraversion, which is defined by sociability. Indeed, neuroticism and extraversion are often negatively correlated with one another (i.e., as scores in neuroticism increase, scores in extraversion decrease) (Pang et al., 2016; Uliaszek et al., 2010).

The advantage of self-report questionnaires is that researchers can gather data from numerous individuals in a short period of time. This greatly contrasts with the time needed for a therapist to conduct a clinical assessment in a single individual. There are disadvantages to self-report questionnaires. First, participants may not feel comfortable disclosing information that may make them feel embarrassed. Even though questionnaires are conducted *anonymously*, individuals may not want to disclose that they engage in certain maladaptive behaviors. This is known as *social desirability bias*. Additionally, participants may select the same response option for all questions (e.g., "strongly agree"), which decreases the validity of the survey's results.

Comprehension Check 3 What are the primary advantages and disadvantages of using self-report questionnaires to measure maladaptive behaviors?

The primary advantages of self-report questionnaires are their efficiency and cost-effectiveness, allowing researchers to gather data from large samples quickly. However, they are limited by potential biases such as social desirability bias and the inability to directly observe behaviors. Additionally, self-report questionnaires might not capture the full complexity of maladaptive behaviors.

3.3 Cognitive-Behavioral Paradigms

Because questionnaires can be limited by response bias, one way to measure forms of maladaptation is to have participants complete a procedure in a laboratory or a naturalistic setting. Some of these tasks can be completed on a computer from the comfort of one's home. Performance in one of these cognitive-behavioral tasks can be correlated with self-report measures, a form of convergent validity. See the Experiment Spotlight below for an example of how something like procrastination can be measured beyond a survey:

Experiment Spotlight: Investigating Behavioral Patterns of Procrastinators in a Wiki-based Activity

Due to copyright, the full article cannot be accessed with a link. The following link will take you to the abstract of this article: https://link.springer.com/article/10.1007/s10639-023-11893-4#article-info

Garg and Rajendran (2023) used 50 undergraduate Computer Science & Engineering students enrolled in a Java programming course. At the beginning of the semester, students completed the Tuckman Procrastination Scale (see Tuckman, 1991) and received a score between 16 and 64, with higher scores indicating greater tendencies to procrastinate. In this experiment, students that had a score between 16 and 28 were considered low procrastinators. Scores between 29 and 36 indicated moderate procrastinators, and students with scores greater than 36 were classified as high procrastinators. During the semester, students were given a group assignment (5–6 students per group) in which they had to create a Wiki Page (or pages) showing the evolution of Java. The Wiki page(s) could include a combination of text, images, diagrams, and hyperlinks to other pages.

The measured variables included the number of times the Wiki page was viewed, the number of times the Wiki page was updated, and the number of times the course module was viewed, among others. Garg and Rajendran also measured how often these activities were completed on a day-by-day basis to determine if low procrastinators completed aspects of the project sooner than moderate and high procrastinators. Overall, low-procrastinating students viewed the course module and the Wiki page more frequently and updated the Wiki page more often compared to moderate and high procrastinating students. Across each group, activity related to the Wiki page did not increase until approximately 4 days before the project deadline. However, the low procrastination group had greater course activity 3–4 days before the deadline compared to the other groups.

Questions to Consider:

1. In addition to procrastination, what data could the researchers have collected to better determine the relationship between procrastination and performance on the Wiki project?
2. Look at panel **b** of Fig. 3. What is one potential problem with presenting the data like this?
3. Based on the results of the current study, what are some potential interventions/strategies to help students reduce procrastination and improve academic performance?

The different types of reliability and validity discussed above apply to cognitive-behavioral experiments as well. There is one more validity that needs to be covered, and it is the most important kind: *internal validity*. Regardless of what type of assessment is being used to study something like procrastination, neuroticism, or any other form of maladaptation, researchers need to ensure that there is not some random factor that can better explain the results of an assessment. For example, in the Experiment Spotlight above, the researchers are interested in determining if high procrastination alters participation in a group-based project. How can the researchers be confident that their findings are not influenced by something else? That is, the high procrastinating students enrolled in the programming course may just have low motivation to earn a high grade in the class.

Comprehension Check 4 Read the following scenario: A researcher develops a new questionnaire to measure procrastination and administers it to a group of college students in an introductory psychology class. The researcher finds that procrastination scores on this new questionnaire are highly correlated with the Tuckman Procrastination Scale. Should the researcher be concerned that the questionnaire that they created is highly similar to an already existing questionnaire? Why or why not?

*No. This example illustrates convergent validity. There are many reasons why a researcher may develop a new questionnaire that measures a construct such as procrastination. The questions in the new questionnaire may be more appropriate for a specific demographic (*e.g.*, college students, children,* etc.*) or the new questionnaire may be significantly shorter than an existing questionnaire. Showing that the new questionnaire is correlated with an established questionnaire such as the Tuckman Procrastination Scale means that the researcher's new questionnaire adequately measures procrastination.*

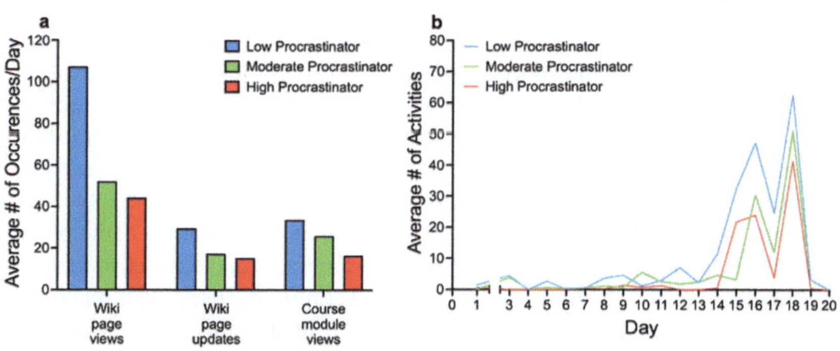

Fig. 3 Figure for Experiment Spotlight. *Note* In panel **b**, Day 2 is not included as this was a major holiday. Figure created by author using GraphPad Prism 10. Data presented in panels **a** and **b** derived from Tables 3 and 5 of Garg and Rajendran (2023), respectively

3.4 Physiological Tests

As the focus of this book is on the *neuroscience* of maladaptation, I want to highlight a few physiological tests that are commonly performed to determine if an individual's symptoms are caused by brain damage. I assume most of you reading this book have some knowledge of x-rays, which allow one to examine tissue and bone structures inside the body. A **computerized tomography** (**CT**) (also known as a **computerized axial tomography** (**CAT**)) scan uses a series of x-ray images taken from different angles to construct a more detailed image of the body, including the brain, compared to a single x-ray image (Fig. 2a).

A more detailed image of the body can be created by using **magnetic resonance imaging** (**MRI**) (Fig. 2b). Instead of using a series of x-ray images, an MRI machine uses magnets to align protons in an individual's body before a radio wave is emitted, causing the protons to spin. The radio wave is then turned off, and a machine detects how much energy is released from the protons as they return to their normal position. Because the number of protons differs across organs and structures within the body, this allows the computer to detect something like a lesion or a tumor. This technique also allows one to differentiate structures of the brain.

Imaging techniques like CAT and MRI are useful for determining if structural changes have occurred to the brain, but they provide no information concerning *functional* changes. There are several methods one can use to measure brain activity, but I will explain those in the next chapter as Chap. 2 delves into basic principles of neuroscience.

3.5 Using Animal Models

Because some forms of maladaptation are harmful, clinical studies assessing the neuromechanisms of maladaptive processes can be limited to comparing individuals with a psychiatric condition to healthy controls. Because we come from different backgrounds and go through different experiences, determining how neurobiological processes control maladaptation is confounded by many factors such as our upbringing, our social interactions, where we live, and our diet, to name a few. To better control for potential confounds, we can test animals in measures of maladaptation. Rodents, specifically mice and rats, are most commonly used in animal research, although research is conducted with non-human primates and avian species. As an animal researcher myself, I have great control over most aspects of a rat's life. Rats are housed in a temperature-and humidity-controlled room, in which the lights turn off and on at the same times each day. They live in the same type of cage, eat the same type of food, and have access to the same type of toy in their cage.

Throughout this book, you will learn about the ways we can use animals to model certain aspects of psychiatric disorders or model specific maladaptive behaviors. For example, my research focuses on substance use disorders and maladaptive decision making. Even though animal research provides opportunities to test hypotheses that are not feasible in humans, there are guidelines and laws dictating how we treat animal subjects.

Each university or college that conducts animal research must have an Institutional Animal Care and Use Committee (IACUC). The IACUC reviews all research activities conducted with vertebrate animals like rodents and birds. Researchers submit an application to the IACUC that details the (a) purpose of the study, (b) justification for using live animals instead of something like a computer simulation or cell cultures, (c) justification for the number of animals requested, (d) experimental procedures used, (e) potential adverse events that can occur, and (f) living conditions of the animals. The researcher also needs to explain how the study does not needlessly duplicate research that has already been conducted. As the current IACUC chair at my institution, I am responsible for conducting semi-annual inspections of all animal housing rooms and laboratory testing spaces, and I must submit a report to the Office of Laboratory Animal Welfare (OLAW), part of the National Institutes of Health (NIH). The NIH is a major funder for researchers conducting biomedical research. I am currently funded by the NIH.

OLAW is a government agency that is responsible for ensuring that federally funded institutions performing animal research adhere to the Public Health Service (PHS) Policy on the Humane Care and Use of Laboratory Animals. These policies can be found in *The Guide for the Humane Care and Use of Laboratory Animals* (currently in its eighth edition), often referred to as *The Guide.* One major doctrine discussed by *The Guide* is the 3 R's of research (Russell & Burch, 1959):

- *Replacement*: using a computer simulation or cell culture models in place of live animals when possible; alternatively, use the lowest order species of a live animal as possible (e.g., use invertebrate animals or rodents instead of larger mammals).
- *Reduction*: using as few animals as possible.
- *Refinement*: ensuring that experimental procedures minimize pain and distress to animals (e.g., providing pain management if using a procedure that causes pain and/or performing a surgical procedure on an animal).

There are many other *R*'s of research that are important for ensuring (a) the humane treatment of animals and (b) the veracity of claims being made by researchers. Many of the *R*'s below are applicable to both animal and human research studies.

- *Relevance*: scientific question needs to address some important medical and/or social issue.
- *Redundancy avoidance*: as mentioned above when discussing the IACUC, this *R* emphasizes the need to avoid needless duplication of an already completed experiment.
- *Regulations*: researchers need to follow all mandated federal laws and institution-wide procedures.
- *Respect*: just as researchers need to respect the rights and dignity of human research participants, preclinical researchers need to respect each animal's worth;

this involves housing animals in clean environments and handling animals appropriately to minimize stress as much as possible.

- *Responsibility*: researchers must maintain the highest of professional standards, meaning that they avoid doing things like fabricating (i.e., making up) data (sadly, this has happened in both clinical and preclinical research).
- *Reproducibility*: studies need to be designed such that they have high validity and reliability; methods need to be discussed in such a way that another researcher can copy the experiment step-by-step.
- *Robustness*: researchers use an appropriate number of research subjects that are determined before the study begins; researchers use appropriate statistical techniques to analyze their data.

The United States Department of Agriculture (USDA) enforces the Animal Welfare Act (AWA). The AWA dictates the minimum care that needs to be provided to most warm-blooded species, with the exception of rats, mice, and birds bred for research purposes. As other agencies such as the PHS oversee animal research with species like rodents and birds, the USDA is more concerned with large animal breeds like livestock. As a vast majority of preclinical studies assessing neurobiological underpinnings of maladaptive behaviors come from rodents, the USDA and the AWA will not be mentioned again in this book.

Comprehension Check 5 When researchers submit an IACUC application, they must describe their most recent *literature review*. That is, they need to explain when they last searched journal articles/books related to their proposed study. The researcher needs to indicate what database(s) they used to search their articles (e.g., Google Scholar, PubMed, PsycINFO, etc.), and they need to provide the words they used to search for articles. The researcher then needs to identify the number of sources that were found and then describe if and how these studies relate to the proposed study. Which *R* of research is directly related to comparing published studies (if they exist) to the study the researcher wants to conduct?

Redundancy avoidance. Researchers are required to discuss their most recent literature review because this helps ensure that the researcher is not conducting the same study that has already been done.

3.5.1 General Methods for Quantifying Maladaptation in Animals

There are numerous ways researchers can measure maladaptive actions in animals. Some maladaptive behaviors can simply be observed in an animal's home cage (i.e., where the animal lives when not being tested). For example, in Chap. 11, you will learn how individuals can engage in repetitious behaviors that interfere with other actions that need to be performed. These types of repetitive behaviors can be observed in lower order species.

Maladaptive behaviors are often measured by testing animals in some type of *behavioral task* outside of their home cage. One commonly used apparatus to test behavior is the **operant conditioning chamber**. Developed by the famous behavioral psychologist B.F. Skinner, the operant conditioning chamber is a box in which an animal is placed. Figure 4 shows a modern-day operant conditioning chamber used to test rats.

Operant conditioning is a form of learning in which an organism learns to associate consequences with its behaviors. For example, a rodent can be trained to press a lever (see right side of operant chamber in Fig. 4) or stick its nose in an *aperture* (see left side of operant chamber in Fig. 4) to receive a food pellet that is dispensed in a food tray (located between the two levers in Fig. 4). When the rodent receives food for completing a response on some *manipulandum* like a lever, it learns to repeat this behavior. The food acts as a **reinforcer**, increasing the likelihood that the rodent performs the same response as before. To be more specific, the food acts as a *positive* reinforcer. The word positive means that something pleasurable was added to the animal's environment. Just like the animal in the operant conditioning chamber, many of our behaviors are controlled by positive reinforcers. We continue going to work because money is added to our bank account.

The operant conditioning chamber can be altered, such that the floor is connected to an electric generator. Animals will learn to respond on a manipulandum to turn the shock generator off, thus saving themselves from the shock. Removing the unpleasant shock is reinforcing. In this situation, the rodent's behavior is controlled by *negative* reinforcement. The word negative means that something has been removed from the environment. The result is the same though. The animal will learn to press a lever/stick its nose in an aperture. As a parent with young children, I have personally

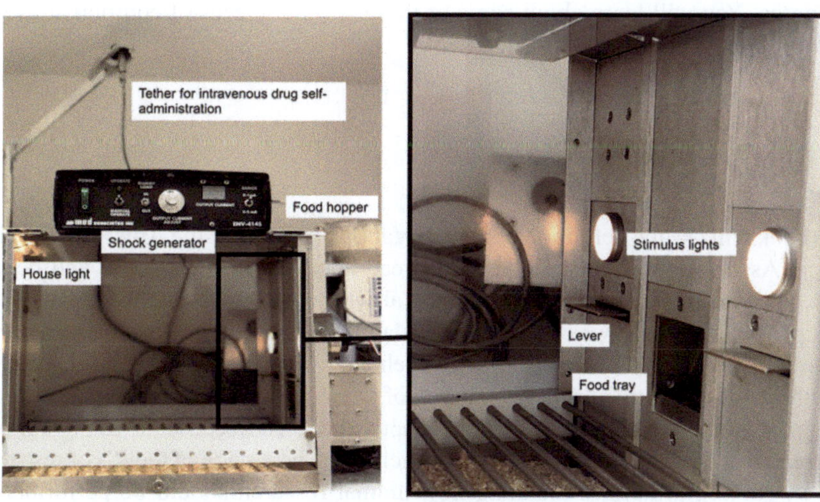

Fig. 4 An image of a modern-day operant conditioning chamber. Several major components of the operant chamber are labeled. Image was taken by the author

encountered the power of negative reinforcement on multiple occasions. As toddlers, my sons would throw temper tantrums (just like any normal toddler). If a certain action stopped my child's tantrum, I was more likely to repeat that action in the future.

The counterpart to reinforcers is **punishers**. These decrease the likelihood that a behavior is emitted in the future. Imagine if a rat is shocked each time it presses a lever. The rat will learn to stop pressing the lever. The shock acts as a positive punisher; something aversive is added to the environment, which causes a decrease in the behavior. Being reprimanded at work acts as a positive punisher. A supervisor may add something unpleasant to an employee's environment (by reprimanding them) in an effort to decrease what they deem to be a problematic behavior.

We can also withhold something that the animal finds reinforcing in order to punish a behavior. A rodent can be trained to press a lever to earn food. Eventually, we can stop giving the rodent food when it presses the lever. This is known as **extinction**, and it acts as a negative punisher. A real-life example of negative punishment is grounding a teenager for doing poorly in school. By removing access to things that the teenager enjoys (e.g., going out with friends), the goal is to decrease the likelihood that they perform poorly in school in the future.

One major advantage of the modern-day operant conditioning chamber is that it can be modified easily to suit the researcher's needs. Lights and tone generators can be added or removed. The chamber can be set up such that animals earn access to solid food like food pellets or liquid reinforcement like water or sucrose, or both. This flexibility allows researchers to study numerous types of behaviors.

Beyond the operant conditioning chamber, there are numerous pieces of equipment that have been created to measure different behaviors, allowing researchers to make inferences about perception, memory, decision making, and emotional-like responses. You will learn about some of these specific pieces of equipment as you read through this book.

4 Chapter Summary

This book will use the definition of maladaptation provided by the American Psychological Association. Therefore, you will read about various cognitive and behavioral processes that can become detrimental when performed in excess. There are multiple ways in which we can assess maladaptation, including clinical assessments, self-report questionnaires, and cognitive-behavioral paradigms. Animal research is valuable for increasing our understanding of the neurobiological processes that can become dysfunctional, thus driving maladaptive actions. Because the use of animals in research is a divisive topic, this chapter includes a discussion of the *R*'s of research, that is, guiding principles dictating how animal research should be performed to ensure the ethical treatment of animal subjects.

Answers to Experiment Spotlight Questions

1. This study did not report *how* students performed on their projects. If each type of procrastinator earned a similar grade on the project, this finding would suggest that procrastination does not negatively impact students. Another piece of information that could be beneficial is how students felt about their learning. Do high procrastinators report feeling like they learned less or that they got less from the course compared to other types of students?
2. This figure lumps things like Wiki page views, course module views, and Wiki page updates together. Being able to compare something like Wiki page updates is more informative as this is the major component of the project. Do low procrastinators significantly differ from the other groups at Day 15, or is the increased number of total activities influenced by something else, such as Wiki page views? Showing a day-by-day comparison across groups for individual metrics is more informative than summing separate measures together.
3. As a professor, one course I teach is Research Methods in Psychology. In this course, students design and execute their own experiment. This project culminates with an oral presentation and the submission of a research paper. To prevent students from waiting until the last minute to start their project, they complete multiple assignments throughout the semester that are related to their project. For example, one assignment requires that they find a specific number of sources and summarize each one. Another assignment involves completing research ethics training. Their research paper is submitted in sections to ensure that students are working on it throughout the second half of the semester instead of entirely during the final week of classes. By breaking large projects into smaller components, this can better help students stay on task.

Glossary

Applied behavior analysis (ABA) therapeutic approach that uses principles of learning to modify one's behavior.

Clinical assessment systematic evaluation of a person's psychological, biological, cognitive-behavioral, and sociocultural factors to understand their problems and to develop appropriate treatment plans.

Computerized tomography (CT) imaging technique that uses a series of X-rays to create images of the inside of the body; also known as computerized axial tomography (CAT)

Extinction process of withholding reinforcement for a previously reinforced behavior to decrease its occurrence.

Magnetic resonance imaging (MRI) imaging technique that uses magnetic fields and radio waves to create images of the inside of the body.

Maladaptation inadequate adjustment to changes in the environment; biological traits or behavioral patterns that are detrimental, counterproductive, or otherwise interfere with optimal functioning in various domains.

Operant conditioning chamber laboratory apparatus used to study animal behavior by allowing the animal to respond on a manipulandum, which then causes something to occur (e.g., illumination of a light, delivery of food, presentation of a foot shock, etc.).

Personality inventory type of questionnaire designed to assess an individual's personality traits and characteristics.

Punisher a stimulus that decreases the likelihood of a behavior being repeated.

Reinforcer a stimulus that increases the likelihood of a behavior being repeated.

Reliability consistency of a test or measurement, either across time or across different individuals.

Validity extent to which a test or measurement accurately measures what it is intended to measure.

References

American Psychiatric Association. (2013). *Diagnostic and statistical manual of mental disorders* (5th ed.). doi:10.1176/appi.books.9780890425596.

American Psychological Association. (n.d.). APA dictionary of psychology. https://dictionary.apa.org/maladaptation

Bridley, A., & Daffin Jr., L. W. (2023). *Essentials of abnormal psychology*. LibreTexts. https://socialsci.libretexts.org/Bookshelves/Psychology/Psychological_Disorders/Essentials_of_Abnormal_Psychology_(Bridley_and_Daffin)

Cloninger, C. R. (1994). *The temperament and character inventory (TCI): A guide to its development and use.* Center for Psychobiology of Personality, Washington University.

Cooper, J. O., Heron, T. E., & Heward, W. L. (2020). *Applied behavior analysis.* Pearson.

Eysenck, H. J., & Eysenck, S. B. G. (1975). Manual of the Eysenck personality questionnaire. *Hodder and Stoughton.*

Garg, A., & Rajendran, R. (2023). Investigating behavioral patterns of procrastinators in a wiki-based activity. *Education and Information Technologies, 28*, 16705–16721. https://doi.org/10.1007/s10639-023-11893-4

Hafferty, J. D., Navrady, L. B., Adams, M. J., Howard, D. M., Campbell, A. I., Whalley, H. C., Lawrie, S. M., Nicodemus, K. K., Porteous, D. J., Deary, I. J., & McIntosh, A. M. (2019). The role of neuroticism in self-harm and suicidal ideation: Results from two UK population-based cohorts. *Social Psychiatry and Psychiatric Epidemiology, 54*(12), 1505–1518. https://doi.org/10.1007/s00127-019-01725-7

Kang, W. (2022). Big five personality traits predict illegal drug use in young people. *Acta Psychologica, 231*, 103794. https://doi.org/10.1016/j.actpsy.2022.103794

Lay, C. (1986). At last, my research article on procrastination. *Journal of Research in Personality, 20*, 474–495. https://doi.org/10.1016/0092-6566(86)90127-3

Mann, L., Burnett, P., Radford, M., & Ford, S. (1997). The Melbourne decision making questionnaire: An instrument for measuring patterns for coping with decisional conflict. *Journal of Behavioral Decision Making, 10*(1), 1–19. https://doi.org/10.1002/(SICI)1099-0771(199703)10:1<1::AID-BDM242>3.0.CO;2-X

Markiewicz, K. (2017). Mediators of relationships between procrastination and neuroticism. *Acta Neuropsychologica, 15*(3), 303–313. https://doi.org/10.5604/01.3001.0010.6095

McCrae, R. R., & John, O. P. (1992). An introduction to the five-factor model and its applications. *Journal of Personality, 60*(2), 175–215. https://doi.org/10.1111/j.1467-6494.1992.tb00970.x

Pang, Y., Cui, Q., Wang, Y., Chen, Y., Wang, X., Han, S., Zhang, Z., Lu, G., & Chen, H. (2016). Extraversion and neuroticism related to the resting-state effective connectivity of amygdala. *Scientific Reports, 6*, 35484. https://doi.org/10.1038/srep35484

Russell, W. M. S., & Burch, R. L. (1959). *The principles of humane experimental technique*. Methuen.

Tuckman, B. W. (1991). The development and concurrent validity of the procrastination scale. *Educational and Psychological Measurement, 51*(2), 473–480. https://doi.org/10.1177/0013164491512022

Uliaszek, A. A., Zinbarg, R. E., Mineka, S., Craske, M. G., Sutton, J. M., Griffith, J. W., Rose, R., Waters, A., & Hammen, C. (2010). The role of neuroticism and extraversion in the stress-anxiety and stress-depression relationships. *Anxiety, Stress, and Coping, 23*(4), 363–381. https://doi.org/10.1080/10615800903377264

World Health Organization. (2019). *International statistical classification of diseases and related health problems* (11th ed.). https://icd.who.int/

Zuckerman, M., & Cloninger, C. R. (1996). Relationships between Cloningers, Zuckermans, and Eysencks dimensions of personality. *Personality and Individual Differences, 21*(2), 283–285. https://doi.org/10.1016/0191-8869(96)00042-6

Zuckerman, M., Kuhlman, D. M., Joireman, J., Teta, P., & Kraft, M. (1993). A comparison of three structural models for personality: The big three, the big five, and the alternative five. *Journal of Personality and Social Psychology, 65*(4), 757–768. https://doi.org/10.1037/0022-3514.65.4.757

Chapter 2
Basic Neuroscience Principles

1 Introduction

Although I waited until the final minute to submit applications to colleges, I was accepted to a wonderful school: Georgia College & State University, a small liberal arts college in the center of the state (not to be confused with Georgia State University in Atlanta). I matriculated as a psychology major with a minor in history. History was my first love, but I was unsure what I would do with a bachelor's degree in that field. As an incoming freshman, I had no desire to teach. Even as a psychology major, I did not know what I wanted to do with my life. Not until my junior year did I discover what I wanted to do with my psychology degree. I took a course called Physiological Psychology, and I immediately became enamored with neuroscience and psychopharmacology research. I ended up serving as a research assistant for the professor that taught that class, Dr. Walt Isaac. Dr. Isacc recommended that I attend the same graduate program that he had attended years earlier, the University of Kentucky. Now as a professor at Northern Kentucky University, I continue conducting research on the underlying neuromechanisms of addiction and maladaptive decision making.

While Chap. 1 introduced the concept of maladaptation, the purpose of Chap. 2 is to discuss basic neuroscience principles that are necessary for understanding the neuroscience of maladaptation. If you already have a strong neuroscience background, this chapter can either be skipped or can be used as a refresher. For those with little background in neuroscience, this chapter will enable you to better understand what will be covered in subsequent chapters.

2 Learning Objectives

By the end of this chapter, you should be able to …

1. Explain the process of neurotransmission.

J. Yates, *Neuroscience of Maladaptation*, https://doi.org/10.1007/978-3-032-15602-0_2

2. Identify major brain structures and explain their functions.
3. Differentiate various neuroscience techniques.

3 Overview of the Nervous System

Whether making sense of the world around us, moving our hand to swat away an annoying fly, or experiencing fear, our **nervous system** controls everything that we do. Essentially, our nervous system helps make us who we are. To understand how the nervous system works, you need to know what **neurotransmitters** are. These are chemicals that act as messengers within the nervous system. There are numerous neurotransmitters in the mammalian nervous system, which control various aspects of behavior and cognition. In subsequent chapters, you will learn how dysregulation of certain neurotransmitter systems contributes to maladaptation.

How exactly do neurotransmitters act as messengers in the nervous system? To answer this question, we need to look at one type of cell that makes up the nervous system.

3.1 Neurons: The Basic Building Block of the Nervous System

The nervous system is composed of multiple specialized cells that have distinct, but complementary, roles. For now, I will focus on **neurons**; I will come back to the other cell types later in this chapter. Neurons communicate with one another by transmitting chemicals like neurotransmitters or *hormones*. While neurotransmitters determine if the next neuron will send a message to another neuron, hormones travel to an organ (e.g., stomach) to influence its activity.

3.1.1 Primary Structures of the Neuron

There are multiple types of neurons, each with its own physical and functional characteristics (see Fig. 1 of a schematic of one type of neuron). All neurons, like most cells in the human body, have a cell body that contains the nucleus, where genetic information is stored (see Fig. 1, top left inset), and organelles, substructures that help keep the neuron alive (e.g., mitochondria). Neurons also contain **dendrites**, branch-like projections that contain **receptors**. The neurotransmitter that is released from one neuron will attach itself to receptors located on dendrites of a neighboring neuron. Specifically, neurotransmitters act as **agonists** at their respective receptors. As you will learn later, certain medications/drugs act as agonists at specific receptors while others act as **antagonists** (e.g., block the receptor).

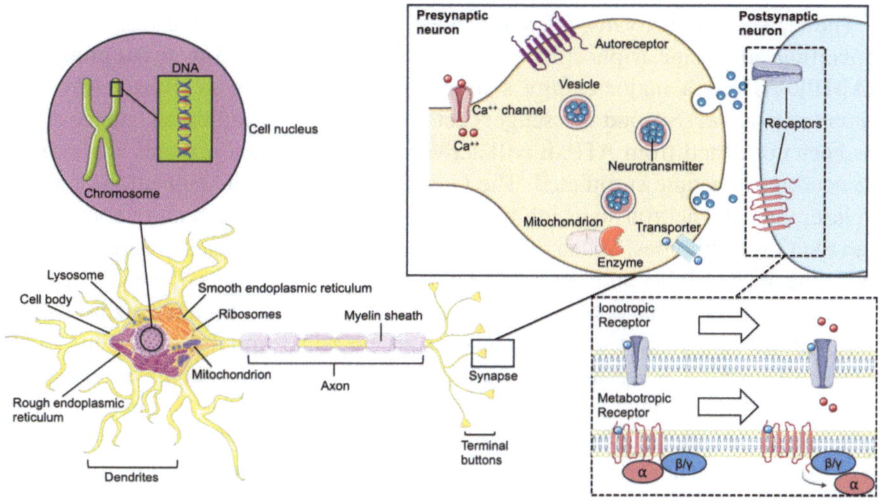

Fig. 1 Schematic of a neuron. The inside of the cell body contains organelles that are necessary for the neuron's survival. In the image, several organelles are labeled. The mitochondrion provides energy to the cell. Ribosomes are important for synthesizing proteins and attach themselves to the rough endoplasmic reticulum. The smooth endoplasmic reticulum is involved in the synthesis of lipids (fats). Lysosomes break down waste materials and foreign invaders. The cell nucleus contains chromosomes, which house our DNA (deoxyribonucleic acid) (top left inset). Dendrites contain ionotropic and/or metabotropic receptors that are activated by neurotransmitters (bottom right inset). *Note* The blue sphere indicates a neurotransmitter while the red spheres represent ions. The top right inset shows a schematic of the synapse. Image created by the author using Servier Medical Art (https://smart.servier.com/; CC BY 4.0, https://creativecommons.org/licenses/by/4.0/) and Microsoft PowerPoint

There are two major receptor types: *ionotropic* and *metabotropic* (see Fig. 1, bottom right inset) (Barrera & Edwardson, 2008; Roth, 2019). An ionotropic receptor is composed of subunits (four or five depending on the specific receptor) that surround an ion channel (hence the name ionotropic). When the neurotransmitter activates the receptor, this causes the ion channel to open, allowing ions to enter the neuron. Some ionotropic receptors allow sodium (Na^+) to enter the cell while others allow chloride (Cl^-) to enter. Notice that Na^+ is positively charged and Cl^- is negatively charged. Keep this in mind as the type of ion entering the neuron influences what it will do (or not do).

Metabotropic receptors get their name because they alter metabolic processes inside the neuron. These receptors are very different from ionotropic receptors. Not only do they look significantly different from ionotropic receptors, but they function in a distinct manner. Instead of having an ion channel embedded in the receptor, metabotropic receptors activate a *G protein*. The G protein dissociates from the receptor to interact with other proteins inside the neuron. There are different types of G proteins; depending on which one is activated, different events will occur in the neuron.

The G_s subtype activates adenyl cyclase, an important enzyme responsible for converting adenosine triphosphate (ATP) into **cyclic adenosine monophosphate (cAMP)**. ATP is a major energy source of cells, including neurons. cAMP is a *second messenger*. Second messengers act kind of like a "middleman". Once cAMP has been converted from ATP, it will activate other important proteins. This causes the neuron to become stimulated. The G_i subtype prevents the activation of adenyl cyclase; thus, the neuron stays inhibited. The G_q subtype is much more complicated than the G_s/G_i subtypes. When a G_q protein is activated, it causes different proteins inside the neuron to become activated. I will spare you the details of the G_q signaling pathway. Just know that activation of G_q-coupled receptors stimulates neurons.

The **axon** is a projection that transmits an electrical impulse to the end of the neuron; this electrical signal is what allows a neuron to send a chemical message to the next neuron (see next section for more details). Many axons are insulated by a fatty substance called the myelin sheath. The significance of the myelin sheath will be discussed in the next section. At the end of the axon is the **terminal button**, which is where chemical messengers are released.

Despite sharing features such as a cell body, dendrites, and axons, different neuron types can look drastically different from one another. As you read through this book, you will be introduced to the various neuron types and will learn where they are primarily located and how they function.

Comprehension Check 1 Why do you think there are ionotropic and metabotropic receptors in our nervous system? In other words, what is the advantage of having two different receptor types?

Ionotropic receptors allow us to process information and react quickly to stimuli as the agonist allows ions to enter the neuron directly through the receptor. Our visual and auditory systems, perception of pain, and muscle contractions are largely controlled by ionotropic receptors. Conversely, metabotropic receptors are used to control many continuous processes like heart rate, digestion, and inhibition of purposeless movements

The Action Potential

There are positively and negatively charged ions located inside and outside of neurons. Na^+ ions, which are positively charged, are found in large numbers outside of the neuron. Potassium (K^+) ions are primarily located inside the neuron. The inside of the neuron also contains a large number of *anions* in the form of negatively charged proteins and nucleic acids. This causes the inside of the neuron to be more negatively charged compared to the outside of the neuron. This is known as the neuron's **resting membrane potential** (see Fig. 2, top left inset). At this point, the neuron is *polarized*, meaning there is an electrical charge difference across the cell membrane.

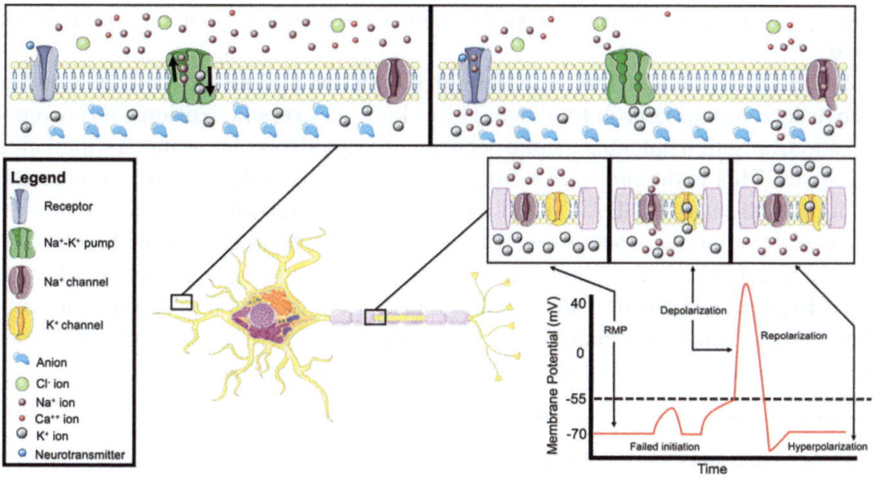

Fig. 2 The neuron at resting membrane potential (top left inset) and during the action potential (top right inset). The axon inset shows the flow of sodium (Na^+) and potassium (K^+) ions in and out of the neuron during the action potential. The graph below this inset shows the change in voltage that occurs throughout the action potential. Image created by the author using Servier Medical Art (https://smart.servier.com/; CC BY 4.0, https://creativecommons.org/licenses/by/4.0/) and Microsoft PowerPoint

If a neuron is sufficiently stimulated, this will cause positively charged ions, primarily Na^+, to enter the neuron. The inside of the neuron becomes less negatively charged compared to the outside of the neuron. This is known as **depolarization** (remember, during the resting membrane potential, the neuron is polarized). When depolarization occurs, voltage-gated Na^+ channels open, allowing even more Na^+ into the neuron. This results in a reversal of the neuron's voltage (~ +40 mV) (Fig. 2, top right inset).

When enough excitation occurs, the neuron generates an **action potential** (Kress & Mennerick, 2009). An electrical current propagates down the length of the axon, opening additional voltage-gated Na^+ channels. In a myelinated axon, there are regularly spaced gaps in which there is no myelin sheath (known as *nodes of Ranvier*). The action potential "jumps" from one node to next, a process known as **saltatory conduction**. Saltatory conduction allows the action potential to travel down the axon at a faster rate. The analogy I always use to explain saltatory conduction is skipping a rock on water. Skipping a rock allows it to cover more distance in a shorter period of time compared to attempting to push it across the surface of the water. The importance of myelin sheath becomes obvious if you are familiar with multiple sclerosis, a disorder in which the immune system damages one's myelin sheath. Because the action potential can no longer propagate efficiently down the axon, neuronal communication becomes disrupted. This leads to the long list of symptoms associated with multiple sclerosis: sensory and motor disturbances, visual and speech impairments, fatigue, gastrointestinal issues, sexual dysfunction, cognitive problems, and altered mood.

As the action potential travels the length of the axon, K^+ ions are removed from the axon to decrease the neuron's voltage. This is known as **repolarization**. During repolarization, an action potential cannot occur again, regardless of how much stimulation it receives. This is the *absolute refractory period.* This is due to Na^+ ion channels becoming desensitized. Even when the neuron has returned to resting membrane potential, K^+ continues to flow out of K^+ ion channels. Now, the inside of the neuron is more negatively charged compared to its resting membrane potential, which is called **hyperpolarization**. During hyperpolarization, an action potential can occur if it receives enough stimulation; this is the *relative refractory period*. Figure 2 (see axon inset and graph below) shows the steps of the action potential.

Comprehension Check 2 In Figure 2, what does a failed initiation mean? What results in a failed initiation?

A failed initiation is when there is insufficient stimulation from a neighboring neuron to cause an action potential. There are a few reasons this can occur. First, too few neighboring neurons release neurotransmitters to stimulate the neuron of interest (this is known as spatial summation*). Second, the neighboring neurons do not send neurotransmitters in a timely enough fashion (*temporal summation*). Third, the neuron could receive both excitatory and inhibitory neurotransmitters from neighboring neurons. The inhibitory neurotransmitters can negate the effects of the excitatory neurotransmitters*

I do want to highlight that the traditional view of action potentials as being purely electrical events has been challenged. The mechanical pulse model posits that mechanical waves propagate down the axon, causing physical deformations to occur in the neuron (Heimburg, 2025). Going through the details of the mechanical pulse model is beyond the scope of this book, but I encourage individuals to read the article cited in this paragraph.

3.1.2 Exocytosis

Once the action potential reaches the terminal button, the first thing that happens is calcium (Ca^{++}) channels embedded in the terminal button open. This allows an influx of Ca^{++} into the terminal button. The increased Ca^{++} forces **vesicles** that contain neurotransmitters/hormones to migrate from the interior of the terminal button to the outer edge. When the vesicle binds to the outer membrane of the terminal button, it opens, releasing its contents into the **synapse**, the space between neurons. Neurotransmitters cross the synapse and bind to receptors located on surrounding neurons. The process by which chemicals are released from the terminal button is called **exocytosis** (see Fig. 1, top right inset) (Wu et al., 2014).

After the neurotransmitter has been released into the synapse and has bound to its corresponding receptor, something needs to happen to this neurotransmitter to prevent it from overstimulating the receptor. For some neurotransmitter systems, the neurotransmitter can be taken back into the *presynaptic neuron* (i.e., the neuron that released the neurotransmitter) through a *transporter*. This process is called **reuptake**. This allows the excess neurotransmitter to be repackaged into vesicles for future use. Another way to control neurotransmitter levels is through enzymatic degradation. Certain enzymes selectively metabolize neurotransmitters. The final way in which neurotransmitter levels can be regulated is by **autoreceptor** activation. Autoreceptors are located on the presynaptic neuron; when the neurotransmitter in the synapse binds to an autoreceptor, the synthesis and release of additional neurotransmitter is limited. This acts as a *negative feedback loop* (see Fig. 1, top right inset).

Comprehension Check 3 Describe the stages of neurotransmission. In other words, how do neurons communicate with each other?

When a neuron is sufficiently stimulated, it generates an action potential that travels down the length of the axon. Ca^{++} *then enters the terminal button, forcing vesicles to bind to the presynaptic membrane. This causes neurotransmitter to be released into the synapse, where it binds to receptors located on dendrites of neighboring neurons. If the neurotransmitter released is excitatory, the neighboring neuron becomes depolarized, increasing the likelihood that it fires an action potential.*

3.2 *Organization of the Nervous System*

The nervous system is subdivided into two major systems. The *central nervous system (CNS)* includes the brain and the spinal cord, and the *peripheral nervous system (PNS)* includes everything else in our body. I will begin with a discussion of the PNS.

3.2.1 Peripheral Nervous System (PNS)

The PNS is composed of the somatic nervous system and the autonomic nervous system, with the former controlling voluntary motor actions and the latter regulating functions we cannot control consciously (Murtazina & Adameyko, 2023). For example, our somatic nervous system allows us to grab a pencil and to use it correctly while our autonomic nervous system controls processes such as breathing, heart rate, and blood pressure. Imagine how unmanageable our lives would become if we had to consciously tell our heart to pump blood through our body or to command our lungs to take in oxygen and to expel carbon dioxide.

The autonomic nervous system can be subdivided further into the parasympathetic and the sympathetic nervous systems. When we are at rest, our parasympathetic nervous system is active. Processes such as heart rate, blood pressure, breathing,

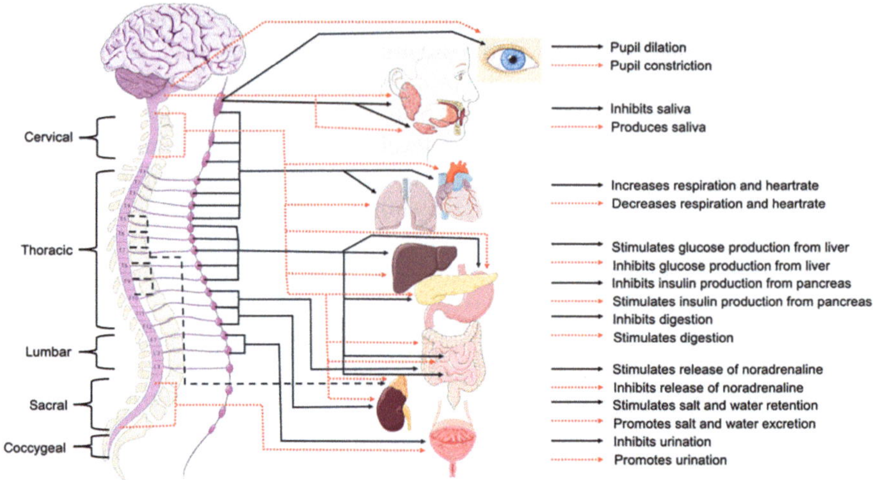

Fig. 3 Schematic showing how the sympathetic nervous system (denoted by solid black arrows) and the parasympathetic nervous system (denoted by dashed red arrows) control various organs: eyes, salivary glands, lungs, heart, stomach and intestines, pancreas, liver, adrenal glands (which sit on top of each kidney), and bladder. Image created by the author using Servier Medical Art (https://smart.servier.com/; CC BY 4.0, https://creativecommons.org/licenses/by/4.0/) and Microsoft PowerPoint

and digestion occur as normal. As such, the parasympathetic system is sometimes colloquially called the "rest and digest" system. The autonomic nervous system becomes active when we encounter a stressor or something we perceive as a potential threat. Our heart rate, blood pressure, and breathing increase while digestion comes to a halt. This is known as the "fight-or-flight" response. This is an adaptive mechanism that helps prepare us to respond to this threat, either by "fighting" it or "fleeing" from it. As you will learn in Chap. 8, a dysregulated sympathetic nervous system is implicated in certain types of maladaptation. Figure 3 shows a comparison between the parasympathetic and the sympathetic nervous systems.

3.2.2 Central Nervous System (CNS)

Before discussing the CNS, I need to cover a couple of key concepts. First, there are specific terms neuroscientists use to discuss the location of one part of the nervous system relative to another region. This makes navigating the nervous system easier. For example, if brain region A is in front of brain region B, we say that A is *anterior* to B. We could also say that B is *posterior* to (behind) A. We can also use the terms *rostral* and *caudal* in place of anterior and posterior, respectively. If brain region A is above brain region B, we say that A is *dorsal* to B. Conversely, we state that B is *ventral* to (below) A. Sometimes the terms *superior* and *inferior* are used to denote regions that are above and below other structures, respectively. The term *medial* is used to denote that we are getting closer to the middle of the brain, whereas the term *lateral* is used when discussing regions further away from the midline of the brain. Being able to recognize these terms is useful as multiple brain regions are named based on their position relative to other brain regions.

In addition to direction, an important distinction needs to be made between *gray matter* and *white matter*. Gray matter is a collection of cell bodies of neurons whereas white matter consists of myelinated axons. In the CNS, a collection of cell bodies is called a *nucleus*, not to be confused with the nucleus found within the cell body; a collection of axons is called a *tract*. Different terms are used to denote a collection of cell bodies (*ganglion*) and axons (*nerve*) in the PNS.

Sensory information traveling from the PNS to the brain first enters the spinal cord, which is protected by the spinal column. Specifically, sensory information enters the spinal cord through the dorsal root of a spinal nerve. One thing I need to note is that the directionality terms described above are slightly altered for the spinal cord in biped animals such as humans. This is because our nervous system bends at a 90° angle. The term dorsal means "on the back", so the dorsal root is closest to our back. Conversely, the ventral root is closer to our stomach (ventral is a term of or relating to the abdomen) and transmits motor information from the brain to the PNS. This is what allows us to move. Many of you are familiar with the paralysis that results from damage to the spinal cord. This occurs because of damage to the ventral root. When an individual suffers a traumatic injury to the spinal cord, both the dorsal and the ventral root are damaged, leading to paralysis and loss of sensation. However, there can be situations in which just the dorsal root is damaged. For example, a herniated disk can compress the dorsal root, leading to multiple symptoms such as pain, tingling sensations in the extremities, and/or loss of feeling (Jang & Garraway, 2024). Similarly, the ventral root can be selectively damaged. While sensory systems are left intact, individuals may have difficulty moving certain body parts that receive information from the affected ventral root (Pinchi et al., 2019).

Comprehension Check 4 In the next section, you will learn about the frontal cortex. The frontal cortex has multiple subregions. One such subregion is known as the ventromedial prefrontal cortex (vmPFC). Another region is called the dorsolateral prefrontal cortex (dlPFC). Based on their names, where do you think these two parts of the brain are located relative to one another?

The vmPFC and the dlPFC are located diagonally to each other. The dlPFC is located closer to the edge of the brain and is located above the vmPFC. The vmPFC is closer to the midline of the brain.

Brain

Our brains are divided into three major areas: the forebrain (also known as the *prosencephalon*), the midbrain (*mesencephalon*), and the hindbrain (*rhombencephalon*). The italicized words are derived from the Greek words *enkephalos* ("brain") and *proso* ("forward"), *mesos* ("middle"), and *rhombos* ("rhombus"). Taken together, prosencephalon and mesencephalon make sense. Why is the hindbrain called the rhombencephalon if *rhombos* does not mean "behind" or "back"? To answer this, I need to describe the *ventricles* of the brain, which are cavities filled with cerebrospinal fluid (CSF). There are four such ventricles in the brain. The fourth ventricle, found in the hindbrain, is shaped like a rhombus.

This section is intended to simply introduce the major divisions of the brain. I will not describe individual brain structures in detail, yet. This is because you will learn more about many of these structures and how they influence maladaptive behaviors throughout the book. Just as multiple structures are named based on their position relative to other structures, there are numerous brain structures that are named after Greek or Latin words. The appendix of this book provides the etymology of each brain structure featured in each chapter.

Forebrain

The forebrain is divided into two distinct regions: the *telencephalon* and the *diencephalon* (Fig. 4a-c). In many chapters of this book, you will learn about the contribution of the telencephalon and the diencephalon to maladaptive behaviors. The telencephalon includes the cerebral cortex, the outermost layer of the brain and what most of us think of when we hear the word "brain". The cerebral cortex is composed of *gyri* and *sulci*, which are the ridges and grooves that give it its folded appearance. The cerebral cortex is subdivided into four cortices (or lobes): frontal cortex, parietal cortex, temporal cortex, and occipital cortex.

If we peel away the cerebral cortex, we will encounter several different regions of the telencephalon. The limbic system, located medial to the temporal cortex, consists of brain structures that are important for learning, memory, emotion, and motivation. The limbic structures that will be featured most prominently in this book are the amygdala, the cingulate cortex, and the hippocampus. The telencephalon also contains the basal ganglia, which are largely responsible for controlling voluntary movements. The major structures of the diencephalon are the thalamus and the hypothalamus.

Midbrain

The midbrain contains the *tectum* and the *tegmentum* (Fig. 4d). The tectum houses structures that are involved in vision and audition (superior colliculus and inferior colliculus, respectively). The tegmentum contains structures that are involved in movement, pain perception, and reward (see inset of Fig. 4d). Notable structures of the tegmentum include the substantia nigra and the ventral tegmental area (VTA). You will read more about the VTA in the Experiment Spotlight in the present chapter.

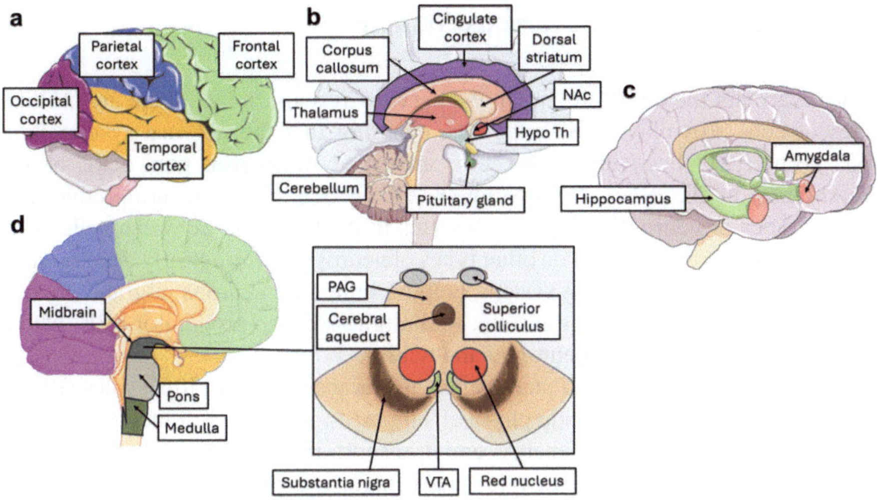

Fig. 4 Schematic showing the location of various brain structures: (**a**) cerebral cortex, (**b**) and (**c**) subcortical structures that make up the telencephalon and the diencephalon, (**d**) structures of the mensencephalon and the rhombencephalon. Image created by the author using Servier Medical Art (https://smart.servier.com/; CC BY 4.0, https://creativecommons.org/licenses/by/4.0/) and Microsoft PowerPoint. *Note* Hypo Th = hypothalamus, PAG = periaqueductal gray, VTA = ventral tegmental area

Hindbrain

There are two major subdivisions of the hindbrain: the *metencephalon* and the *myelencephalon* (Fig. 4d). The metencephalon consists of the pons and the cerebellum. The pons is one major area of the brain where information from the left side of the brain cross to the right side and vice versa. The cerebellum is primarily known for its role in motor coordination. However, the cerebellum is involved in associative learning, aspects of language, and timing. You will encounter the cerebellum multiple times in this book. The myelencephalon contains the medulla oblongata, located just above the spinal cord. Damage to the medulla prevents autonomic functions like heart rate and breathing from occurring. As such, severe damage to the medulla is often fatal; it is not surprising that this structure is tucked away at the base of our brain.

Comprehension Check 5 Why is the term basal ganglia a misnomer?

Ganglia are collections of cell bodies in the PNS . As the basal ganglia are located in the CNS, the correct name for this brain region should be basal nuclei.

3.3 Other Cells of the Nervous System

Neurons are not the only cell type in the nervous system. In fact, there are multiple types of cells that play important roles, ensuring that the nervous system functions properly. Collectively, these cells are called **neuroglia**, glial cells, or glia (Verkhratsky et al., 2019). The myelin sheath surrounding axons are produced by specific types of neuroglia (oligodendrocytes in the CNS and Schwaan cells in the PNS)! Neurons are dependent on other types of neuroglia. During early development, radial glia act as a scaffolding system, helping immature neurons migrate to where they need to go in the nervous system. Radial glia are also progenitor cells, having the ability to produce other neuroglia and neurons.

Astrocytes, named for their star-shaped appearance, have multiple roles (Ruparelia et al., 2023). They provide physical support to neurons, and they can provide neurons with glycogen for energy and transport nutrients from blood vessels. Speaking of blood vessels, astrocytes are a major component of the **blood-brain barrier (BBB)**, a semi-permeable barrier that protects the brain from toxins and pathogens while allowing in things such as oxygen, glucose, amino acids (building blocks of proteins), and certain vitamins. Astrocytes are somewhat like neurons in the sense that they can release neurotransmitters and are involved in reuptake of specific neurotransmitters. They also regulate ion levels that are outside neurons.

I will highlight a couple of additional neuroglia here. **Microglia** are immune cells that protect the nervous system in the event a toxin or pathogen enters the brain (Lannes et al., 2017). They generate an immune response such as inflammation, and they engage in *phagocytosis*; this means that microglia "eat" debris and pathogens found in the nervous system. Like radial glia, microglia are involved in neurogenesis. One important role of microglia is *synaptic pruning*, a process by which the number of synapses between neurons is reduced. When we are younger, the number of synapses we have increases substantially until adolescence. At that point, the number of synapses begins to decrease. As you will learn, there is considerable interest in understanding how dysfunctional microglia contribute to numerous psychiatric conditions.

Ependymal cells are found in the spinal cord and in the ventricles of the brain (Xie & Li, 2024). Ependymal cells produce CSF, which fills the ventricles and surrounds the spinal cord and the brain. CSF acts as a cushion to protect the CNS.

4 Techniques for Studying the Neuroscience of Maladaptation

How do we know what each brain does or how neurotransmitters influence specific behaviors? There are many tools that researchers can use to answer these questions. A discussion of these neuroscience techniques is crucial as the information you will learn throughout this book comes from research using these methods.

4.1 Case Studies

Before the development of neuroscience techniques like computed axial tomography (CAT) scans and magnetic resonance imaging (MRI) that you learned about in Chap. 1, as well as the techniques you will learn about below, our understanding of how abnormal brain functioning contributes to maladaptive behaviors came from **case studies**. Specifically, an individual with damage to part of the brain, whether due to a stroke or mechanical damage, was examined for an extended period to understand how such damage altered them. Students are often taught that research studies need to have large sample sizes to generate results that are meaningful. While larger sample sizes tend to be better than smaller sample sizes, we cannot discount the influence case studies have had on the field of neuroscience. Throughout this book, you will encounter several case studies.

4.2 Brain Imaging

In addition to CAT scans and MRIs, one technique that can be used is **diffusion tensor imaging (DTI)** (Fig. 5a) (Alexander et al., 2007). DTI measures the diffusion of water molecules. As myelinated axons are bundled together, forming a pathway between two or more brain regions, water diffuses in the direction of these axons. Techniques like MRI and DTI are useful for determining structural differences between individuals with a psychiatric disorder and control individuals. As I mentioned in Chap. 1, structural imaging does not provide information about functional differences in the brain.

4.3 Measuring Brain Activity

One of the simplest ways of measuring cortical activity is with the **electroencephalogram (EEG)** (Fig. 5b) (Chaddad et al., 2023). A cap with electrodes is placed on an individual's head, which takes advantage of the fact that neurons use an electrical signal (action potential) to stimulate release of neurotransmitters. Relatedly, **magnetoencephalography (MEG)** examines neuronal activity, but does so by measuring magnetic fields produced by electrical currents in the brain (Fig. 5c) (Yang et al., 2024). MEG is better at pinpointing where activity is occurring, but it is more cost prohibitive compared to EEG. EEG and MEG are primarily limited to measuring cortical activity.

If one wants to determine if subcortical areas are active, different approaches are often used. One popular imaging technique is **functional MRI (fMRI)** (Glover, 2011). fMRI combines MRI by measuring blood flow in the brain. Brain regions that are more active will have greater blood flow, specifically oxygenated blood

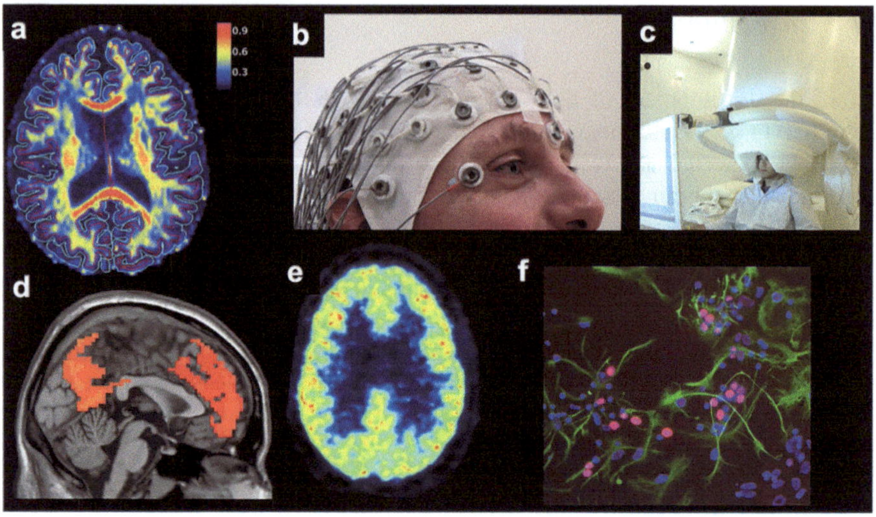

Fig. 5 Various neuroscience techniques used to image the brain and/or measure neural activity. (**a**) Results of a diffusion tensor imaging (DTI) experiment; red-shaded areas indicate greater white matter concentrations. *Source* modified from Figure 1 of Stock et al. (2020), https://doi.org/10.3389/fphys.2020.00116, CC BY 4.0, https://creativecommons.org/licenses/by/4.0/). (**b**) Individual wearing a cap of electrodes, which are used to measure cortical activity. *Source* https://www.flickr.com/photos/tim_uk/8135755109/ from Tim Sheerman-Chase. CC BY 2.0 (https://creativecommons.org/licenses/by/2.0/deed.en), via Flickr. (**c**) Individual sitting in a magnetoencephalography (MEG) machine. *Source* https://commons.wikimedia.org/wiki/File:NIMH_MEG.jpg from Oaktree b, via Wikimedia Commons. Image is in the public domain as it originally came from the National Institute of Mental Health. (**d**) A representative image generated from functional magnetic resonance (fMRI). *Source* modified from Figure 1 of Lv et al. (2018), https://doi.org/10.3174/ajnr.A5527, by permission of the American Journal of Neuroradiology (open access). (**e**) Representative positron emission tomography image. *Source* Figure 1 of Cervenka et al. (2022), https://doi.org/10.1038/s41398-022-01990-2, CC BY 4.0 (https://creativecommons.org/licenses/by/4.0/), via Springer Nature. (**f**) cFos expression of active neurons (pink circles). *Source* https://commons.wikimedia.org/wiki/File:CFos-expression-in-stimulated-neurons.jpg from Gerry Shaw. CC BY-SA 4.0 (https://creativecommons.org/licenses/by-sa/4.0/deed.en), via Wikimedia Commons

(Fig. 6d). Researchers often use fMRI to measure **functional connectivity** between brain regions (Rogers et al., 2007). Although other techniques such as EEG and MEG can be used to measure functional connectivity, fMRI is most often used in functional connectivity studies. Functional connectivity can be best compared to correlation. Positive functional connectivity means that two or more brain regions become activated at the same time when an individual is at rest or when they are performing a task. Conversely, negative functional connectivity indicates that as one brain region becomes active, a separate brain region becomes less active. Even if brain regions are not structurally connected (i.e., neurons do not travel from one brain region to another), they can still be functionally connected. Functional connectivity is a term that you will consistently encounter as you read through this book.

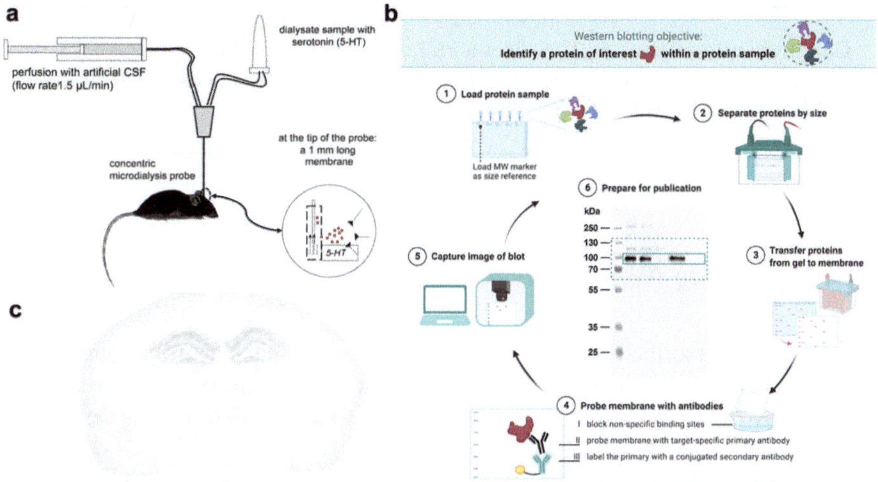

Fig. 6 A few neuroscience techniques that are used with animal subjects. (**a**) Schematic showing how microdialysis is performed. *Source* Figure 1 of Gardier (2013), https://doi.org/10.3389/fphar.2013.00098. CC BY 3.0 (https://creativecommons.org/licenses/by/3.0/), via Frontiers. (**b**) Simplified schematic showing how Western blots are performed. *Source* Figure 1 of Kroon et al. (2022). https://doi.org/10.1371/journal.pbio.3001783. CC BY 4.0 (https://creativecommons.org/licenses/by/4.0/), via PLOS. (**c**) Image of radiolabeled MK-801 binding to a type of glutamatergic receptor in the hippocampus and cortical areas of a rat. *Source* Author's own work

Instead of measuring oxygenated blood flow, a **positron emission tomography (PET) scan** tracks where in the brain a radioactive substance (tracer) travels (Shukla & Kumar, 2006). Commonly, radioactive glucose is used because the brain utilizes glucose for energy (Fig. 5e). PET imaging also allows one to measure receptor/transporter levels in the brain as radioactive glucose can be substituted with a radioactive receptor agonist/antagonist or transporter blocker.

In preclinical research, general brain activity is inferred from measuring cFos expression, an immediate early gene (see Fig. 5f) (Chung, 2015). Immediate early genes are rapidly and transiently expressed in cells in response to a wide range of stimuli. Because all neurons have a cell nucleus that contains genetic information, researchers can easily measure cFos expression. Increased cFos expression in a brain region suggests greater activity. First, the animal is euthanized, and its brain is taken for *immunohistochemistry*. The researcher slices the brain into sections and then exposes these sections to an antibody that binds to the cFos protein. This is called a *primary antibody*. Because visualizing where the primary antibody has bound to cFos can be difficult, a *secondary antibody* is added, which binds to the primary antibody. For example, the secondary antibody can be a fluorescent tag (think of using a highlighter to identify a passage in your book you want to easily come back to); this makes detecting the primary antibody easier.

Comprehension Check 6 How does a PET scan differ from fMRI?

A PET scan measures the amount of a radioactive substance in the brain whereas fMRI measures blood flow to a region of interest.

4.4 Measuring Neurotransmitter Levels

The techniques described above provide information about the neurobiological bases of maladaptation at the macro level. If one wants to determine the neurochemical mechanisms (i.e., neurotransmitters) that control maladaptive actions, alternative techniques are needed.

Magnetic resonance spectrometry (MRS) is used in conjunction with MRI to measure specific chemicals in the CNS, albeit what can be detected with MRS is limited (Tognarelli et al., 2015). As MRS can only detect small molecules, this technique is limited to identifying just a few neurotransmitters. However, MRS can be used to detect some precursor chemicals that are used to make other types of neurotransmitters. One advantage of MRS is that it can be used in both humans and in animals.

In animals, there are several methods for measuring neurotransmitter levels in the brain. **Microdialysis** involves the collection of fluid from a specific region of the brain (Fig. 2.6a) (Shippenberg & Thompson, 2001). A guide cannula is inserted into a region of interest. A probe can then be inserted into the guide cannula. This probe is connected to two lines, one that slowly perfuses something like artificial CSF into the brain. At the end of the probe is a semi-permeable membrane. As artificial CSF exits the probe, this forces fluid from inside the brain to go through the probe and out a second line connected to the probe. Microdialysis samples are placed in a special machine, which can measure the levels of a specific neurotransmitter in the sample.

There are several *electroanalytic* methods that can be used to quantify neurotransmitter levels in a brain region. For simplicity, I will discuss two here: **amperometry** and **voltammetry** (Rafiee et al., 2024). There are several variations of amperometry and voltammetry, but discussing each type is way beyond the scope of this book. Each method involves inserting a microelectrode into the brain, and a current is applied. There are typically two or three electrodes. The recording (or working) electrode is inserted into the brain and detects changes in current due to *oxidation* (loss of electrons) or *reduction* (gain of electrons). The reference electrode is placed in a separate brain region, usually one that is not of interest to the researcher. The reference electrode is important because it maintains a stable potential; changes in current detected by the recording electrode are compared to the reference electrode. A stimulating electrode can be implanted in a third brain region that has a connection with the region associated with the recording electrode. This allows one to determine if stimulating this electrode causes changes in neurotransmitter release where the recording electrode is located. In amperometry, the electrode is held at a constant potential, which allows for detection of a single neurotransmitter. In voltammetry, the potential changes, increasing and decreasing during the recording. This allows a researcher to measure more than one neurotransmitter in the same animal during the same recording.

Comprehension Check 7 What is one major advantage of using a technique like voltammetry compared to something like microdialysis?

Because voltammetry measures changes in current, this allows researchers to measure changes in neurotransmitter levels at a much faster rate compared to waiting for dialysate to be collected.

4.5 Quantifying mRNA and Protein Expression

Messenger ribonucleic acid (mRNA) contains instructions for converting deoxyribonucleic acid (DNA) into protein. Measuring mRNA and/or protein levels allows one to determine if maladaptation is associated with elevated or reduced levels of something like a receptor or a transporter. There are several methods for detecting mRNA expression in the nervous system.

Northern blotting uses *gel electrophoresis* to separate RNA molecules based on their size (Trayhurn, 1996). As the RNA molecules are separated, black bands will appear on a gel inside an apparatus. To determine which bands are associated with the mRNA of interest, fragments of RNA called *probes* are placed on the gel, which then bind to the mRNA of interest. A fluorescent tag can be added to the probe to increase the ability to detect the mRNA.

Reverse transcription polymerase chain reaction (RT-PCR) involves making millions to billions of copies of a DNA sample via thermal cycling (exposing DNA sample to cycles of heating and cooling) (Artika et al., 2022). This is necessary because the amount of DNA in some regions is so low that it is nearly impossible to quantify. The first step in RT-PCR is to transcribe RNA back to *complementary DNA (cDNA)* by an enzyme called reverse transcriptase. cDNA is exposed to thermal cycling because RNA would be damaged during this process. To isolate a gene of interest, a primer is added with DNA polymerase and nucleotides (building blocks of DNA) in a solution. This solution is what undergoes thermal cycling.

In situ hybridization is a technique in which tissue is exposed to a probe containing a complementary strand of the RNA sequence of interest (Veselinyová et al., 2021). This causes hybridization to occur; that is, the single-stranded probe will form a pair with the single-stranded RNA in the tissue. Like Northern blotting, a tag is added to the probe, which allows the researcher to visualize and quantify mRNA expression.

The primary method for measuring protein expression is Western blotting (Fig. 6b) (Kurien & Scofield, 2015). The method for Western blotting is very similar to Northern blotting. Instead of separating mRNA by its size, proteins are separated. After the proteins have been separated by weight, the gel is placed in a transfer apparatus with a membrane. Using primary and secondary antibodies, the researcher can isolate the protein of interest. This is necessary because some proteins have similar molecular weights.

Comprehension Check 8 What is the difference between a Northern blot and a Western blot?

Northern blots measure mRNA expression. Western blots measure protein expression.

4.6 Autoradiography

In vitro autoradiography uses *radiolabeled* (i.e., radioactive) **ligands** to measure the density of a receptor or transporter in the nervous system (Fig. 6c) (Kuhar, 2001). The tissue of interest is placed in a container with a film; as the radioactive material begins to decay, it leaves an invisible image on the film. The film is submerged in a developing reagent, which causes the receptor binding to appear on the film. The film is then placed in a fixative reagent to prevent the film from further developing once it is exposed to light. Darker areas indicate higher density of the radiolabeled ligand. I performed *in vitro* autoradiography experiments as part of my dissertation. In fact, the image in Fig. 6c comes from my dissertation work.

4.7 Altering Neural Activity

So far, the techniques I have covered allow researchers to measure brain activity, neurotransmitter release, or mRNA/protein levels. While important, these techniques are more correlative in nature and do not necessarily allow one to state that changes in any of these areas lead to maladaptive behavior. There are approaches that enable one to directly alter brain activity. First, direct current stimulation, most commonly in the form of **transcranial direct current stimulation (tDCS)**, uses low-level electrical currents to alter the resting membrane potential of neurons (Thair et al., 2017). If an *anode* is used, neurons are more likely to fire an action potential; conversely, a *cathode* increases hyperpolarization, thus decreasing the likelihood an action potential occurs. Going back to the discussion of voltammetry, anodes and cathodes are both electrodes. Anodes lead to oxidation, and cathodes lead to reduction.

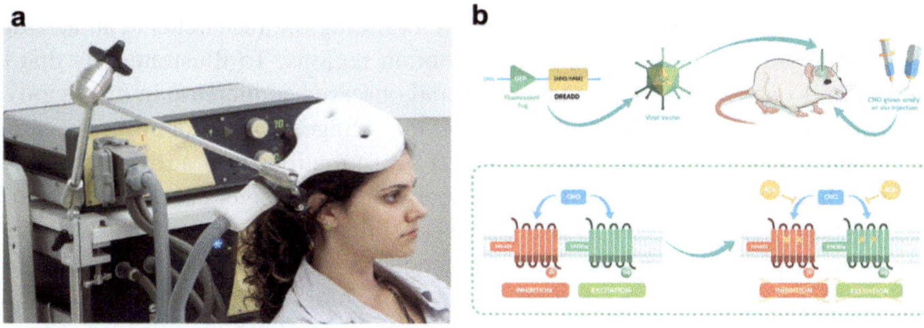

Fig. 7 (**a**) An individual demonstrating how transcranial magnetic stimulation (TMS) works. *Source* https://commons.wikimedia.org/wiki/File:Neuro-ms.png from Baburov. CC BY-SA 4.0 (https://creativecommons.org/licenses/by-sa/4.0/deed.en), via Wikimedia Commons. (**b**) A schematic of chemogenetics. *Source* Figure 1 of Chapter 3.4 of Ju et al. (2018), https://ecampusontario.pressbooks.pub/neurosciencecdn/chapter/3-4-chemogenetic-methods-to-examine-the-brain-and-behaviour/. CC BY 4.0 (https://creativecommons.org/licenses/by/4.0/), via eCampusOntario

Instead of using an electric current to modulate neuronal activity, **transcranial magnetic stimulation** (**TMS**) alters neural activity through the use of a magnet (Chail et al., 2018). An electromagnetic coil is placed over an individual's scalp. Introduction of a magnetic pulse produces an electric current in the brain (Fig. 7a). Like tDCS, TMS can be used to either depolarize or hyperpolarize neurons.

A crude way to alter neural activity in animals is to **lesion** a specific brain region (Vaidya et al., 2019). Quinolinic acid can be infused in a brain region, causing *excitotoxicity* to occur, which damages neurons. Alternatively, the drugs muscimol and baclofen can be directly infused into a region of interest. This causes temporary inactivation of the brain region. The advantage of inactivation studies is that baclofen and muscimol injections do not cause permanent damage to the brain, unlike quinolinic acid. Related to muscimol/baclofen infusions, researchers can directly inject a drug that targets a specific receptor, transporter, or enzyme into a brain region of interest.

Today, there are more sophisticated techniques for isolating the contribution of specific brain regions to behaviors of interest. These techniques involve insertion of viral vectors that infect neurons in a brain region. The first technique I want to highlight is **chemogenetics** (Fig. 7b) (Campbell & Merchant, 2018). A chemically engineered receptor packaged in a virus is inserted into neurons. For example, *designer receptors exclusively activated by a designer drug (DREADD)* only respond to an otherwise pharmacologically inert drug. That is, these receptors will not respond to naturally occurring neurotransmitters. This allows the researcher to activate these receptors when they specifically want to by injecting the animal with the designer drug.

Optogenetics is similar to chemogenetics in that a viral vector is infused into the brain (Emiliana et al., 2022). A viral vector causes the neuron to express light-sensitive ion channels. An animal can be connected to an optic cable that emits a specific color. Depending on what type of virus is inserted into the brain, the light may activate or inhibit neurons in this brain region.

In addition to altering activity of a single brain region, researchers can activate or inhibit a pathway between two or more brain regions. To illustrate why this is important: if one lesions the frontal cortex and observes an increase in maladaptive behavior, they will not be able to determine if the change in behavior occurs because the frontal cortex is no longer able to receive information from a different region or if the change occurs because the frontal cortex can no longer send messages to a different part of the brain. By focusing on a specific pathway, researchers can better answer these questions. This is where optogenetics/chemogenetics is advantageous. A researcher can inject two different viral vectors into different parts of the brain within the same pathway. The first viral vector contains the light-sensitive ion channel or the DREADD, and the second viral vector contains a *tracer*. The tracer inserts itself into the neuron via the terminal buttons and works its way to the cell body.

Comprehension Check 9 Imagine a researcher wants to determine if inactivating the pathway between two brain regions affects an animal's fear response. Of the techniques that you have learned about so far, which one would you use?

Using something like optogenetics or chemogenetics would allow you to insert two viral vectors, one in brain region A and one in brain region B. This would ensure that only neurons in this pathway express light-sensitive ion channels or DREADDs.

4.8 Genetic Manipulations

Above, I described how the immediate early gene cFos can be measured to infer neuronal activity. Researchers can also directly alter the genetic makeup of animals to study how a specific protein/enzyme in the brain regulates physiological processes and behaviors. There are several ways that researchers can genetically alter an animal such as a mouse (Shakweer et al., 2023). One method involves *pronuclear injection* of DNA into an immature egg cell of a female that has not been fertilized by a male's sperm (called an oocyte). This creates a **transgenic animal** that can be bred with a similar transgenic animal to create offspring that have a specific gene of interest. Another method involves creating **knockout/knock-in models** that show either decreased or increased expression of a gene, respectively. *Embryonic stem cells* can be modified and then injected into a blastocyst, a structure that forms following fertilization of an egg. You will learn more about embryonic stem cells in Chap. 17.

If a researcher wants to determine if gene expression in a specific brain region controls certain processes, a viral vector containing **small interfering RNA (siRNA)** can be infused into the brain. This prevents the translation of mRNA into protein. To increase mRNA expression, a viral vector containing DNA can be injected into the brain region of interest.

Genetically modified animal models can be used with other techniques described above (e.g., behavioral pharmacology, optogenetics) to provide even more insights on the underlying neurobiology of specific behaviors. The following Experiment Spotlight is just one example.

Experiment Spotlight: Cannabinoid CB1 Receptors Are Expressed in a Subset of Dopamine Neurons and Underlie Cannabinoid-Induced Aversion, Hypoactivity, and Anxiolytic Effects in Mice

The full article can be accessed with this link:

https://www.jneurosci.org/content/43/3/373.long

The VTA is located in the midbrain and is an important mediator of reward learning. Specifically, the VTA sends dopaminergic neurons to the nucleus accumbens. When we encounter stimuli that have been repeatedly paired with rewards (e.g., the smell of a cake baking), dopaminergic neurons in the VTA become active to these stimuli instead of the reward itself. In a sense, dopaminergic neurons in the VTA serve to tell us that something rewarding is on its way. In addition to dopamine, *endocannabinoids* also regulate aspects of reward. In the current study, Han et al. (2023) used a variety of techniques covered in this chapter to examine how cannabinoid receptors located in dopaminergic neurons in the VTA control several behaviors in transgenic mice.

The mice used in the experiment are called DAT-Cre, CB1-flox, and DA-CB1-KO. Let's break this down. DAT stands for dopamine transporter. This protein takes up excess dopamine from the synapse. Cre is short for Cre recombinase, which acts like a molecular pair of scissors and glue. Cre recombinase flanks a gene or DNA segment that the researcher is interested in. If the gene/DNA segment is not necessary for survival, it can be deleted once Cre recombinase "cuts" it. Alternatively, researchers can insert a new DNA sequence where the original was cut by Cre recombinase. Put these together, a DAT-Cre mouse is one in which Cre recombinase is located in neurons that express DAT only. As DAT is located on dopaminergic neurons, researchers can ensure that Cre recombinase is present in dopamine neurons only.

To understand CB1-flox mice, I need to discuss a couple of concepts. First, CB1 refers to the CB_1 receptor. Endocannabinoids bind to this receptor to exert their effects in the nervous system. There is another type of cannabinoid receptor (CB_2), but it is primarily isolated to the peripheral nervous system. Second, the term "flox" is a portmanteau of "flank" and "loxP", with loxP being a portmanteau itself, combining "locus of X-over (crossing over)" and "plasmid". In the case of a CB1-flox mouse, loxP sites are inserted on both ends of the gene that encodes the CB_1 receptor (*CNR1* gene). These loxP sites are important because they allow Cre recombinase to cut the *CNR1* gene. Think of loxP sites as flags.

Researchers can then breed DAT-Cre mice with CB1-flox mice to create a conditional knockout model. When a DAT-Cre mouse mates with a CB1-flox mouse, the offspring has dopaminergic neurons containing both Cre recombinase and the *CNR1* gene that has been flanked by loxP sites. This causes the *CNR1* gene to become "cut" from the DNA segment. Now, these mice lack CB_1 receptors, but only in dopaminergic neurons.

Incorporating an optogenetic approach, Han et al. infused a viral vector containing a light-sensitive ion channel into the VTA of mice. Mice were trained in a special chamber to press a lever to turn on a fiber optic connected to a probe that extends into the VTA. Stimulating the VTA is reinforcing to animals and is known as optical intracranial self-stimulation. Compared to DAT-Cre mice, DA-CB1-KO mice respond less to receive optical stimulation of the VTA. Han et al. also found that cannabinoids decrease optical intracranial self-stimulation in DAT-Cre mice, an effect that was largely absent in DA-CB1-KO mice.

Different mice were also used for a microdialysis experiment, in which dopamine levels were measured from the nucleus accumbens. DA-CB1-KO mice showed higher baseline levels of dopamine release in the nucleus accumbens compared to DAT-Cre mice. Similar to the optical intracranial self-stimulation data, administration of Δ9-tetrahydrocannabinol (THC), a cannabinoid best known for being the psychoactive compound in cannabis, decreases dopamine levels in the nucleus accumbens in DAT-Cre, but not DA-CB1-KO, mice.

In addition to the findings above, Han et al. found that DA-CB1-KO mice are more hyperactive compared to both DAT-Cre and CB1-flox mice and are less sensitive to the hypolocomotor effects of THC. While THC produces anxiolytic-like effects in DAT-Cre mice, it does not significantly alter anxiety-like behavior in DA-CB1-KO mice.

Questions to Consider:

1. How does the use of conditional knockout mice (DA-CB1-KO) strengthen the conclusion that CB1 receptors in dopaminergic neurons specifically mediate the observed effects of cannabinoids on reward-related behaviors and dopamine release?
2. Based on the findings of this study, propose a potential mechanism by which CB1 receptors in dopaminergic neurons influence reward-related behaviors, such as motivation and locomotor activity.
3. Considering the role of the endocannabinoid system in various physiological and psychological processes, what are the potential implications of these findings for understanding and treating conditions such as addiction, anxiety, and movement disorders?

5 Chapter Summary

This chapter introduced the basic principles of neuroscience, provided an overview of the nervous system and the role of neurons as its fundamental building blocks. The steps involved in neurotransmission were detailed as understanding this process is fundamental for understanding how certain forms of maladaptation occur. Various brain structures were briefly introduced; many of these structures will be discussed in greater detail throughout the rest of the book. Because numerous neuroscience methods are used to elucidate the neuroscience underlying processes like maladaptation, the final part of this chapter highlighted many of these techniques, ranging from brain imaging to manipulating neural circuits. Having a basic understanding of these techniques will better enable the reader to understand how brain regions and neurotransmitter systems contribute to maladaptation.

Answers to Experiment Spotlight Questions

1. By selectively eliminating CB1 receptors only in dopamine neurons, the researchers can isolate the role of these receptors in the observed behaviors. If the effects of cannabinoids were observed to be significantly altered or absent in DA-CB1-KO mice compared to control groups, it provides strong evidence for the specific involvement of CB1 receptors in dopaminergic neurons in mediating those effects. This approach helps to rule out the possibility that the observed effects are due to CB1 receptors located in other brain regions.
2. A potential mechanism by which CB1 receptors in dopaminergic neurons influence reward-related behaviors is by modulating dopamine release. The study found that DA-CB1-KO mice exhibited higher baseline dopamine levels in the nucleus accumbens compared to control mice. This suggests that CB1 receptors normally act to inhibit dopamine release. Cannabinoids, by activating CB1 receptors, could further decrease dopamine release, contributing to their effects on reward-related behaviors. Additionally, CB1 receptors might influence the firing rate of dopamine neurons, which could also affect dopamine release and subsequent reward processing. Further research is needed to fully elucidate the precise mechanisms involved.
3. The findings of this study have significant implications for understanding and treating conditions such as addiction, anxiety, and movement disorders. The involvement of the endocannabinoid system in reward processing suggests that targeting CB1 receptors could be a potential therapeutic strategy for addiction. Additionally, the observed effects of cannabinoids on locomotor activity and anxiety-like behavior raise the possibility of developing cannabinoid-based treatments for movement disorders and anxiety disorders. However, it is essential to carefully consider the potential side effects and misuse liability of such treatments. Future research should focus on identifying specific subtypes of CB1 receptors or developing more selective cannabinoid compounds to minimize adverse effects while maximizing therapeutic benefits.

Glossary

Action potential electrical signal transmitted along the membrane (typically, the axon) of a neuron, culminating in the release of neurotransmitter/hormone from the terminal button of the neuron.

Agonist chemical that binds to and activates a receptor, triggering a biological response similar to that produced by a neurotransmitter.

Amperometry technique that involves the measurement of an electric current to detect and quantify chemicals in a solution, such as neurotransmitters in a brain region.

Antagonist chemical that binds to a receptor, preventing the naturally occurring neurotransmitter from triggering its biological effects.

Autoreceptor a type of receptor located on a neuron that responds to the neurotransmitter/hormone released by the same cell, primarily acting as a negative feedback loop to limit the further release of neurotransmitter/hormone.

Axon a long projection of a neuron that conducts electrical impulses away from the cell body to the terminal buttons.

Blood-brain barrier (BBB) highly selective, semi-permeable membrane that protects the brain from harmful substances in the blood while allowing essential nutrients and gases to pass through.

Case study an-depth examination of a specific person, group, or event within a real-world context.

Cyclic adenosine monophosphate (cAMP) an intracellular second messenger that plays a vital role in many biological processes by relaying signals from outside the cell to the inside, often activating protein kinases.

Chemogenetics technique that uses genetically engineered receptors and specifically designed drugs to control and study the activity of specific cells, most often neurons, in living organisms.

Dendrite a branching extension of a neuron that receives incoming signals from other neurons and transmits these signals toward the cell body.

Depolarization decrease in the electrical potential across a cell membrane, typically caused by an influx of sodium ions, leading to the generation of an action potential.

Diffusion tensor imaging (DTI) a technique used in conjunction with magnetic resonance imaging to measure the diffusion of water molecules to visualize and analyze the structure and connectivity of white matter fibers in the brain.

Electroencephalogram (EEG) technique in which electrodes placed on an individual's scalp detect electrical activity in cortical regions of the brain.

Exocytosis process by which a neuron releases a neurotransmitter/hormone from the synaptic vesicle to the extracellular space.

Functional connectivity a statistical concept that describes the temporal correlation of neurophysiological activity between different brain regions.

Functional magnetic resonance imaging (fMRI) technique in which changes in blood flow are measured in the brain, allowing one to infer brain function.

Hyperpolarization Increase in the negative charge inside the neuron, making it less likely to generate an action potential.

Knock-in model genetically engineered organism where a specific DNA sequence is inserted or replaced within a targeted gene.

Knockout model genetically engineered organism where a specific gene has been inactivated or deleted.

Lesion area of the brain that has been damaged.

Ligand molecule or ion that binds to a receptor or to another molecule.

Magnetic resonance spectroscopy (MRS) technique that uses magnetic fields to measure and analyze the chemical composition of tissue.

Magnetoencephalography (MEG) technique that measures magnetic fields produced by electrical activity in the brain.

Microdialysis technique in which extracellular fluid is collected from the brain, thus allowing for the quantification of neurotransmitter/metabolite concentrations.

Microglia type of glial cell that acts as the primary immune cell of the central nervous system, protecting the brain and spinal cord by clearing away dead cells and destroying pathogens.

Neuroglia specialized cells that support and protect neurons.

Neuron specialized cell that transmits electrical signals, allowing communication between cells in the nervous system.

Neurotransmitter chemical messenger that transmits signals between neurons, allowing communication between cells in the nervous system.

Nervous system complex system of various types of cells that communicate with each other to control physiological, motor, cognitive, and emotional processes.

Optogenetics technique that uses light to control the activity of specific neurons in living organisms by introducing light-sensitive proteins into neurons.

Positron emission tomography (PET) scan technique that measures metabolic activity in tissue and organs by detecting gamma rays emitted from a radioactive tracer.

Receptor protein located on or within a cell to which a ligand binds, thus allowing for certain biological processes to occur.

Repolarization restoration of the resting membrane potential in a neuron by causing the efflux of potassium ions.

Resting membrane potential the electrical charge difference across a neuron's membrane when it is not generating an action potential.

Reuptake process by which neurotransmitters are transported back into the presynaptic neuron after release.

Salatatory conduction rapid propagation of an action potential along a myelinated axon by "jumping" from one node of Ranvier to the next.

Small interfering RNA (siRNA) short, double-stranded RNA molecule that disrupts gene expression by binding to and degrading messenger RNA, thus preventing protein synthesis.

Synapse junction between two neurons where neurotransmitters are released from the presynaptic neuron and bind to receptors located on the postsynaptic neuron

Terminal button end of an axon where neurotransmitters are stored.

Transcranial direct current stimulation (tDCS) technique in which a low-intensity electrical current is applied to one's scalp, thus modulating brain activity.

Transcranial magnetic stimulation (TMS) technique in which magnetic fields are applied above one's head to induce electrical currents in the brain, thus modulating brain activity.

Transgenic animal organism that has been genetically modified by introducing DNA from another species into its genome.

Vesicle small, membrane-bound sac that stores substances like neurotransmitters and hormones; found in the terminal buttons of neurons.

Voltammetry technique that measures the current produced by a chemical species as the applied potential is varied; allows for the quantification of specific neurotransmitters.

References

Alexander, A. L., Lee, J. E., Lazar, M., & Field, A. S. (2007). Diffusion tensor imaging of the brain. *Neurotherapeutics, 4*(3), 316–329. https://doi.org/10.1016/j.nurt.2007.05.011

Artika, I. M., Dewi, Y. P., Nainggolan, I. M., Siregar, J. E., & Antonjaya, U. (2022). Real-time polymerase chain reaction: Current techniques, applications, and role in COVID-19 diagnosis. *Genes, 13*(12), 2387. https://doi.org/10.3390/genes13122387

Barrera, N. P., & Edwardson, J. M. (2008). The subunit arrangement and assembly of ionotropic receptors. *Trends in Neurosciences, 31*(11), 569–576. https://doi.org/10.1016/j.tins.2008.08.001

Campbell, E. J., & Marchant, N. J. (2018). The use of chemogenetics in behavioural neuroscience: Receptor variants, targeting approaches and caveats. *British Journal of Pharmacology, 175*(7), 994–1003. https://doi.org/10.1111/bph.14146

Cervenka, S., Frick, A., Bodén, R., & Lubberink, M. (2022). Application of positron emission tomography in psychiatry-methodological developments and future directions. *Translational Psychiatry, 12*(1), 248. https://doi.org/10.1038/s41398-022-01990-2

Chaddad, A., Wu, Y., Kateb, R., & Bouridane, A. (2023). Electroencephalography signal processing: A comprehensive review and analysis of methods and techniques. *Sensors, 23*(14), 6434. https://doi.org/10.3390/s23146434

Chail, A., Saini, R. K., Bhat, P. S., Srivastava, K., & Chauhan, V. (2018). Transcranial magnetic stimulation: A review of its evolution and current applications. *Industrial Psychiatry Journal, 27*(2), 172–180. https://doi.org/10.4103/ipj.ipj_88_18

Chung L. (2015). A brief introduction to the transduction of neural activity into Fos signal. *Development & Reproduction, 19*(2), 61–67. https://doi.org/10.12717/DR.2015.19.2.061

Emiliani, V., Entcheva, E., Hedrich, R., Hegemann, P., Konrad, K. R., Lüscher, C., Mahn, M., Pan, Z. H., Sims, R. R., Vierock, J., & Yizhar, O. (2022). Optogenetics for light control of biological systems. *Nature Reviews. Methods Primers, 2*, 55. https://doi.org/10.1038/s43586-022-00136-4

Gardier, A. M. (2013). Antidepressant activity: contribution of brain microdialysis in knock-out mice to the understanding of BDNF/5-HT transporter/5-HT autoreceptor interactions. *Frontiers in Pharmacology, 4*, 98. https://doi.org/10.3389/fphar.2013.00098

Glover, G. H. (2011). Overview of functional magnetic resonance imaging. *Neurosurgery Clinics of North America, 22*(2), 133–vii. https://doi.org/10.1016/j.nec.2010.11.001

Han, X., Liang, Y., Hempel, B., Jordan, C. J., Shen, H., Bi, G. H., Li, J., & Xi, Z. X. (2023). Cannabinoid CB1 receptors are expressed in a subset of dopamine neurons and underlie cannabinoid-induced aversion, hypoactivity, and anxiolytic effects in mice. *The Journal of Neuroscience, 43*(3), 373–385. https://doi.org/10.1523/JNEUROSCI.1493-22.2022

Heimburg, T. (2025). The mechanical properties of nerves, the size of the action potential, and consequences for the brain. *Chemistry and Physics of Lipids, 267*, 105461. https://doi.org/10.1016/j.chemphyslip.2024.105461

Jang, K., & Garraway, S. M. (2024). A review of dorsal root ganglia and primary sensory neuron plasticity mediating inflammatory and chronic neuropathic pain. *Neurobiology of Pain, 15*, 100151. https://doi.org/10.1016/j.ynpai.2024.100151

Ju, W., Santos, A., Freeman, A., & Daniele, E. (2018). *Neuroscience: Canadian 1st edition*. e Campus Ontario.

Kress, G. J., & Mennerick, S. (2009). Action potential initiation and propagation: Upstream influences on neurotransmission. *Neuroscience, 158*(1), 211–222. https://doi.org/10.1016/j.neuroscience.2008.03.021

Kroon, C., Breuer, L., Jones, L., An, J., Akan, A., Mohamed Ali, E. A., Busch, F., Fislage, M., Ghosh, B., Hellrigel-Holderbaum, M., Kazezian, V., Koppold, A., Moreira Restrepo, C. A., Riedel, N., Scherschinski, L., Urrutia Gonzalez, F. R., & Weissgerber, T. L. (2022). Blind spots on western blots: Assessment of common problems in western blot figures and methods reporting with recommendations to improve them. *PLoS Biology, 20*(9), e3001783. https://doi.org/10.1371/journal.pbio.3001783

Kuhar, M. J. (2001). In vitro autoradiography. *Current Protocols in Pharmacology, 00*, 8.1.1–8.1.9. https://doi.org/10.1002/0471141755.ph0801s00

Kurien, B. T., & Scofield, R. H. (2015). Western blotting: An introduction. *Methods in Molecular Biology, 1312*, 17–30. https://doi.org/10.1007/978-1-4939-2694-7_5

Lannes, N., Eppler, E., Etemad, S., Yotovski, P., & Filgueira, L. (2017). Microglia at center stage: A comprehensive review about the versatile and unique residential macrophages of the central nervous system. *Oncotarget, 8*(69), 114393–114413. https://doi.org/10.18632/oncotarget.23106

Lv, H., Wang, Z., Tong, E., Williams, L. M., Zaharchuk, G., Zeineh, M., Goldstein-Piekarski, A. N., Ball, T. M., Liao, C., & Wintermark, M. (2018). Resting-state functional MRI: Everything that nonexperts have always wanted to know. *American Journal of Neuroradiology, 39*(8), 1390–1399. https://doi.org/10.3174/ajnr.A5527

Murtazina, A., & Adameyko, I. (2023). The peripheral nervous system. *Development, 150*(9), dev201164. https://doi.org/10.1242/dev.201164

Pinchi, E., Frati, A., Cantatore, S., D'Errico, S., Russa, R., Maiese, A., Palmieri, M., Pesce, A., Viola, R. V., Frati, P., & Fineschi, V. (2019). Acute spinal cord injury: A systematic review investigating miRNA families involved. *International Journal of Molecular Sciences, 20*(8), 1841. https://doi.org/10.3390/ijms20081841

Rafiee, M., Abrams, D. J., Cardinale, L., Goss, Z., Romero-Arenas, A., & Stahl, S. S. (2024). Cyclic voltammetry and chronoamperometry: Mechanistic tools for organic electrosynthesis. *Chemical Society Reviews, 53*(2), 566–585. https://doi.org/10.1039/d2cs00706a

Rogers, B. P., Morgan, V. L., Newton, A. T., & Gore, J. C. (2007). Assessing functional connectivity in the human brain by fMRI. *Magnetic Resonance Imaging, 25*(10), 1347–1357. https://doi.org/10.1016/j.mri.2007.03.007

Roth, B. L. (2019). Molecular pharmacology of metabotropic receptors targeted by neuropsychiatric drugs. *Nature Structural & Molecular Biology, 26*(7), 535–544. https://doi.org/10.1038/s41594-019-0252-8

Rupareliya, V. P., Singh, A. A., Butt, A. M. A H., & Kumar, H. (2023). The molecular soldiers of the CNS: Astrocytes, a comprehensive review on their roles and molecular signatures. *European Journal of Pharmacology, 959*, 176048. https://doi.org/10.1016/j.ejphar.2023.176048

Shakweer, W. M. E., Krivoruchko, A. Y., Dessouki, S. M., & Khattab, A. A. (2023). A review of transgenic animal techniques and their applications. *Journal of Genetic Engineering & Biotechnology, 21*(1), 55. https://doi.org/10.1186/s43141-023-00502-z

Shippenberg, T. S., & Thompson, A. C. (2001). Overview of microdialysis. *Current Protocols in Neuroscience, Chapter, 7*(Unit7), 1. https://doi.org/10.1002/0471142301.ns0701s00

Shukla, A. K., & Kumar, U. (2006). Positron emission tomography: An overview. *Journal of Medical Physics, 31*(1), 13–21. https://doi.org/10.4103/0971-6203.25665

Stock, B., Shrestha, M., Seiler, A., Foerch, C., Hattingen, E., Steinmetz, H., Deichmann, R., Wagner, M., & Gracien, R. M. (2020). Distribution of cortical diffusion tensor imaging changes in multiple sclerosis. *Frontiers in Physiology, 11*, 116. https://doi.org/10.3389/fphys.2020.00116

Thair, H., Holloway, A. L., Newport, R., & Smith, A. D. (2017). Transcranial direct current stimulation (tDCS): A beginner's guide for design and implementation. *Frontiers in Neuroscience, 11*, 641. https://doi.org/10.3389/fnins.2017.00641

Tognarelli, J. M., Dawood, M., Shariff, M. I., Grover, V. P., Crossey, M. M., Cox, I. J., Taylor-Robinson, S. D., & McPhail, M. J. (2015). Magnetic resonance spectroscopy: Principles and techniques: Lessons for clinicians. *Journal of Clinical and Experimental Hepatology, 5*(4), 320–328. https://doi.org/10.1016/j.jceh.2015.10.006

Trayhurn, P. (1996). Northern blotting. *Proceedings of the Nutrition Society, 55*, 583–589. https://doi.org/10.1079/PNS19960051

Vaidya, A. R., Pujara, M. S., Petrides, M., Murray, E. A., & Fellows, L. K. (2019). Lesion studies in contemporary neuroscience. *Trends in Cognitive Sciences, 23*(8), 653–671. https://doi.org/10.1016/j.tics.2019.05.009

Verkhratsky, A., Ho, M. S., Zorec, R., & Parpura, V. (2019). The concept of neuroglia. *Advances in Experimental Medicine and Biology, 1175*, 1–13. https://doi.org/10.1007/978-981-13-9913-8_1

Veselinyová, D., Mašlanková, J., Kalinová, K., Mičková, H., Mareková, M., & Rabajdová, M. (2021). Selected in situ hybridization methods: Principles and application. *Molecules, 26*(13), 3874. https://doi.org/10.3390/molecules26133874

Wu, L. G., Hamid, E., Shin, W., & Chiang, H. C. (2014). Exocytosis and endocytosis: Modes, functions, and coupling mechanisms. *Annual Review of Physiology, 76*, 301–331. https://doi.org/10.1146/annurev-physiol-021113-170305

Xie, S., & Li, F. (2024). Ependymal cells: Roles in central nervous system infections and therapeutic application. *Journal of Neuroinflammation, 21*(1), 255. https://doi.org/10.1186/s12974-024-03240-2

Yang, Y., Luo, S., Wang, W., Gao, X., Yao, X., & Wu, T. (2024). From bench to bedside: Overview of magnetoencephalography in basic principle, signal processing, source localization and clinical applications. *NeuroImage. Clinical, 42*, 103608. https://doi.org/10.1016/j.nicl.2024.103608

Chapter 3
Perceptual Disturbances

1 Introduction

The screams from the first-floor apartment kept me and my wife awake. The individual that lived directly below us was yelling for help, claiming that someone was trying to kill them. Suddenly, we could hear a door from downstairs slam shut. My wife creeped over to our front door to look through the peephole. To her (and my) surprise, my wife could see our neighbor trying to enter our apartment. During the next few months, our sleep was interrupted by cries for help, our neighbor convinced that someone had broken into their apartment. Even though my wife and I knew that no one other than our neighbor was in that apartment, their shrieks kept us awake.

Sadly, our neighbor suffered from **schizophrenia**, a debilitating psychiatric condition characterized by numerous disturbances in cognition. Specifically, our neighbor experienced auditory **hallucinations**, often claiming that they could hear someone say that they were in their apartment. Even though schizophrenia has numerous symptoms, this chapter will discuss just one type of disturbance observed in individuals with this disease: distortions in **perception**.

Hallucinations are not the only type of perceptual disturbance that individuals can experience. Whereas hallucinations involve the presence of something that is not there, there are situations in which individuals have difficulty recognizing something that is there. This chapter will focus on perceptual distortions that cannot be explained by mechanical damage to an organ like the eye. You will first learn how perception *normally* works before reading about some perceptual disturbances that can negatively impact one's life.

2 Learning Objectives

By the end of this chapter, you should be able to …

J. Yates, *Neuroscience of Maladaptation*, https://doi.org/10.1007/978-3-032-15602-0_3

1. Differentiate sensation from perception.
2. Describe the methods for measuring perception in a laboratory setting.
3. Outline the neural pathways involved in each sensory system, and explain how other parts of the brain integrate sensory information.
4. Explain how damage to different parts of the brain affects perception.
5. Discuss how dysregulation of several neurotransmitter systems can lead to perceptual disturbances, particularly hallucinations.

3 Sensation vs. Perception

To understand why perceptual distortions occur, you first need to understand *how* perception works. Perception arises from *sensation.* Think of sensation as a purely biological process and perception as a purely cognitive process. Integration of sensation and perception allows us to recognize and understand things in our environment at any given moment (or to recognize that we do not recognize something that we have been presented). You may have learned previously that we have "five senses": vision, audition (hearing), gustation (taste), olfaction (smell), and somatosensation (touch). In addition to these five senses, we have three additional senses: vestibular (balance and spatial orientation), proprioception (body position, posture, and movement), and interoception (internal bodily states).

Sensation begins when a receptor on a sensory neuron is stimulated. Let's use vision as an example. When light enters the pupil, it hits the back portion of the inner eye called the *retina.* The retina contains millions of specialized cells known as rods and cones. When these cells are stimulated, they transmit messages to other types of cells in the eye (Fig. 1a). Ultimately, information from the retina is transmitted to the occipital cortex, which then transfers information to other parts of the brain for further interpretation. This is when perception begins to occur. Figure. 1b shows two major paths involved in the perception of visual information. You will learn more about the neuromechanisms of perception below, but first, I want to discuss how perception can be measured in a laboratory setting.

3.1 Measuring Perception in a Laboratory Setting

If we want to determine the neural processes involved in perception, we can use one of the neuroscience techniques you learned about in the previous chapter to measure brain activity. Because perception is a subjective experience, researchers need to use the neuroscience techniques in conjunction with laboratory procedures that expose individuals to certain types of stimuli. There is an entire branch of psychology called **psychophysics** that deals with the relationship between physical stimuli and mental processes like perception. Studies measuring perception often have participants do at least one of the following:

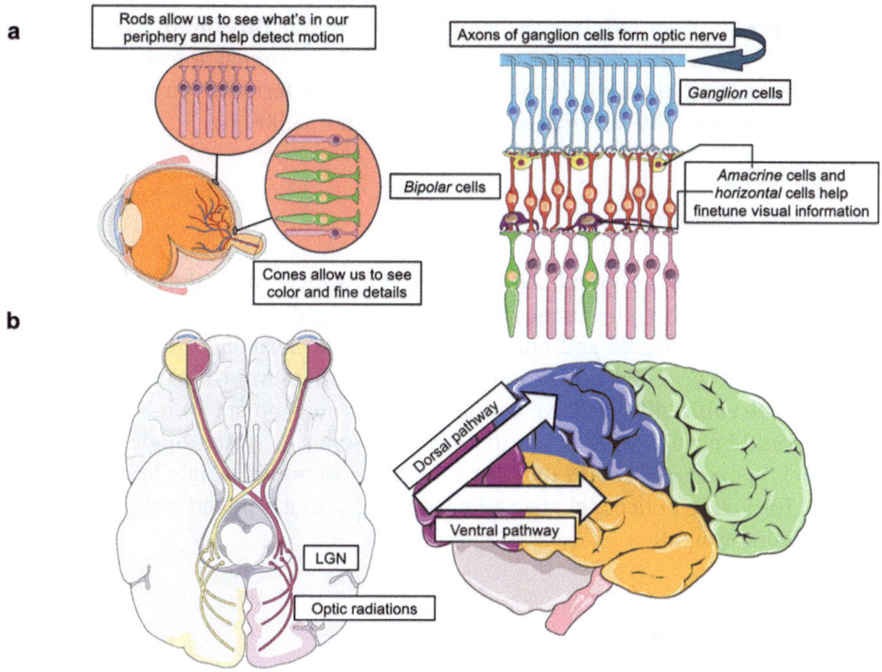

Fig. 1 Schematic of the cells located in the retina (**a**) and the pathway of the optic nerve/tract (**b**). Once visual information reaches the occipital cortex, this information is transferred to both the parietal and the temporal cortices of the brain for further processing. The connection between the occipital cortex and the parietal cortex is known as the dorsal ("where") pathway. The connection between the occipital cortex and the temporal cortex is known as the ventral ("what") pathway. *Note* LGN = lateral geniculate nucleus (in thalamus). Image created by the author using Servier Medical Art (https://smart.servier.com/; CC BY 4.0, https://creativecommons.org/licenses/by/4.0/) and Microsoft PowerPoint

1) detect if a stimulus is present or not (detection)
2) determine if two stimuli differ from one another (**discrimination**)
3) identify an unknown stimulus from a set of stimuli (identification)
4) identify the intensity/magnitude of a stimulus (**scaling**) (Hatzfeld, et al., 2017)

3.1.1 Detection

Psychophysical studies often determine **absolute thresholds**; that is, what is the minimum *intensity* needed for an individual to detect the presence of a stimulus at least 50% of the time. For example, a researcher can place headphones on a participant and play different tones. The researcher can determine an individual's lowest detectable sound frequency; you most likely know a sound's frequency as *pitch* (measured in hertz or Hz). In humans, the average absolute threshold for pitch is approximately 20 Hz, although mine is closer to 150 Hz.

The opposite of an absolute threshold is a **terminal threshold**, the maximum intensity of a stimulus that can be perceived. The terminal threshold for pitch in humans is approximately 20,000 Hz; however, the terminal threshold tends to decrease as we age. My terminal threshold is around 12,000 Hz.

3.1.2 Discrimination

As a kid, you may have completed worksheets in which you had to identify which image was unlike the other images. For very young children, they may be given an image of three squares and one triangle. Whereas absolute thresholds are used to measure detection, discrimination studies often measure **difference thresholds** (or **just-noticeable differences**). Pretend I give you 5-pound weight. How much weight would I need to add for you to notice a difference…just a single pound or an additional 5 pounds? What if I initially gave you a 50-pound weight? Less weight is required for you to notice a change when starting with a 5-pound weight compared to a 50-pound weight.

3.1.3 Identification

At first glance, identification seems like the same thing as detection. However, there is an important distinction between the two. Detection simply refers to our ability to notice some stimulus. Identification refers to our ability to detect if a specific stimulus is present among other types of stimuli. Signal detection tasks are often used to determine how well an individual can identify if a *target* stimulus is present among *distractor* stimuli.

The number of distractor stimuli can be adjusted across individual trials of signal detection tasks. Not surprisingly, when more distractors are present, detecting the target stimulus becomes more difficult. Imagine you have to find your friend. If you walk into a room, and your friend is just one of two individuals in that room, finding your friend will be easy. Conversely, trying to find your friend at a concert with thousands of other people is going to be more challenging.

On an individual trial of a signal detection task, there are four possible outcomes:

1) Hit—the individual correctly identifies the presence of the target stimulus
2) Miss—the individual fails to identify the presence of the target stimulus
3) Correct rejection—the individual correctly identifies that the target stimulus is not present
4) False alarm—the individual reports the presence of the target stimulus when it was not present

Signal detection theory is an important framework used to understand how well individuals can differentiate relevant stimuli (targets in signal detection tasks) from extraneous stimuli (distractors; also called *noise*). Signal detection has real-world implications. When driving, failing to detect a stimulus, like a child running into the road, can lead to catastrophe. In a criminal case, an eyewitness can mistakenly claim that an innocent person was the perpetrator of a crime. Some of the perceptual disturbances you will learn about in this chapter can be framed in the context of signal detection. In some conditions, individuals cannot perceive certain stimuli (miss). In other conditions, individuals report perceiving something even when no stimulus is present (e.g., hallucinations) (false alarm).

3.1.4 Scaling

Scaling is closely related to discrimination. Scaling involves differentiating two or more stimuli. Instead of merely identifying if there is a difference between stimuli, individuals are required to quantify the intensity of a stimulus. There are various ways to measure scaling, with many of them involving individuals making a discrimination. A simple way to measure scaling is to present a series of stimuli to an individual and have them rate how intense the stimulus is (e.g., on a scale of 1–10). For example, individuals can hear a series of tones, and they have to identify how loud they think each tone is.

3.2 Measuring Perception in Animals

Discrimination tasks are often used to measured perception in animals. Let's use pigeons as an example. Pigeons, like humans, are primarily visual animals that can differentiate numerous colors. A pigeon can be placed in an operant conditioning chamber and then trained to peck at a key light that is illuminated green to receive access to food. If the key light is illuminated red, responding on the key does nothing. What do you think the pigeon does in this task? If you guessed that it responds when the light is green but barely responds when the light is red, you would be correct. The pigeon has learned to discriminate the green light from the red light. Rodents do not have the visual acuity of avian species like pigeons, but they can learn to discriminate the presence of a light from the absence of light. They can also discriminate colors, although they are partially color blind as they cannot detect red. Animals can also be trained to respond when a tone is presented.

Comprehension Check 1 Explain the difference between sensation and perception, and provide an example of how these processes work together using the visual system.

Sensation is a purely biological process whereas perception is a purely cognitive process. However, sensation leads to perception. In the visual system, sensation begins when light enters the pupil and hits specialized cells. This information is then transmitted to the occipital cortex. Perception occurs when the occipital cortex processes this information and transfers it to other parts of the brain for further interpretation, allowing us to recognize and understand what we see.

4 Neuroanatomical Substrates of Perception

Each sensory system has its own pathway in the brain. These pathways will be briefly described below. Additionally, I will discuss some perceptual abnormalities that can occur following damage to parts of these pathways. Most of the evidence provided in this section comes from case studies where individuals have experienced damage to a specific part of the brain.

4.1 Vision

Axons from the optic nerve of each eye travel to each hemisphere of the brain. The optic chiasm is where axons from the nasal side of each optic nerve cross to the contralateral (opposite) side of the brain (see Fig. 1b). Most axons of the optic nerve travel to the lateral geniculate nucleus (LGN) of the thalamus, with some axons terminating in the superior colliculus (part of the tectum). Information from the LGN travels to the primary visual cortex (also known as V1) in the occipital cortex via the optic radiations.

The parietal cortex and the temporal cortex receive information from the occipital cortex. The dorsal pathway (occipital-parietal regions) is often colloquially referred to as the "where" pathway while the ventral pathway (occipital-temporal regions) is nicknamed the "what" pathway. Structures in the "where" pathway identify the location of a visual stimulus; this allows us to navigate our space and to coordinate our movements to perform a task (e.g., reaching out to grab a cup on a table). The "what" pathway enables us to differentiate the numerous shapes and colors that make up complex figures. In short, the brain regions in this pathway allow us to recognize a coffee cup from a spoon and to identify our friends and family members from a crowd of strangers (Kravitz et al., 2013). Figure. 2 shows each pathway in more detail by highlighting specific brain structures.

4.1.1 Cortical Blindness and Blindsight

The primary visual cortex is critical for processing incoming visual information and transmitting this information to both dorsal and ventral pathways. Damage to the primary visual cortex leads to "cortical blindness"; any visual information traveling to the damaged part of the brain will not be perceived by the individual (Saionz et al.,

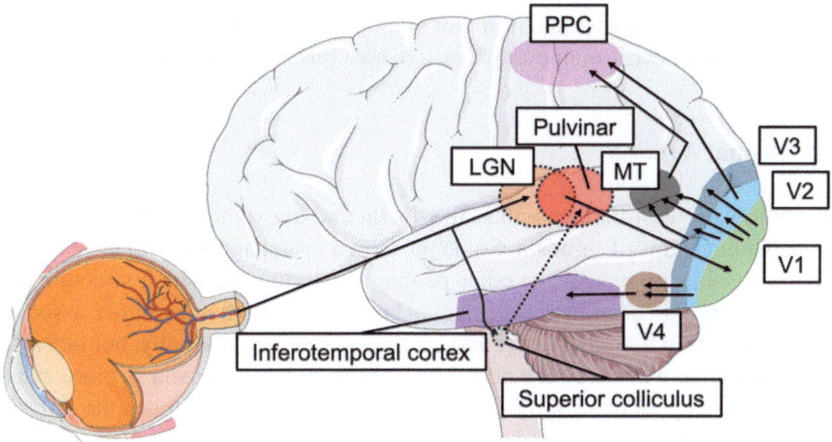

Fig. 2 A schematic showing the visual pathways in more detail. *Note* LGN = lateral geniculate nucleus of the thalamus, MT = middle temporal area, PPC = posterior parietal cortex, V1 = primary visual cortex, V2 = secondary visual cortex, V3 = third visual area, V4 = visual area 4. Image created by the author using Servier Medical Art (https://smart.servier.com/; CC BY 4.0, https://creativecommons.org/licenses/by/4.0/) and Microsoft PowerPoint

2022). However, an unusual phenomenon is observed in some individuals with a damaged primary visual cortex: **blindsight**. Even if someone is cortically blind, they can detect the presence of stimuli (Ajina & Bridge, 2016).

In one study, a patient known as SBR, who had bilateral damage to the primary visual cortex, was asked to detect the presence of a *Gabor patch* presented on the screen. A Gabor patch is a series of black and white stripes within a circular window. The contrast between the white and the black lines can be adjusted. Control individuals (i.e., those with no damage to the visual cortex) detected the presence of the Gabor patch on every trial, regardless of the contrast between the white and the black lines. Despite having bilateral damage to the primary visual cortex, SBR was able to detect Gabor patches with high contrast over 70% of the time; when contrast was low, SBR's performance decreased to below 60%. In comparison, individuals with unilateral damage to the primary visual cortex were able to detect high-contrast Gabor patches a little less than 90% of the time, but they still had difficulty detecting the presence of low-contrast images (~65% accuracy) (Ajina & Bridge, 2019).

The prevailing view of blindsight is that there is a dissociation between conscious awareness and automatic processing of information. However, some have argued against this notion, instead claiming that blindsight is merely a form of degraded vision (Phillips, 2020). Regardless of this debate, research on blindsight has enabled us to understand that other cortical areas process visual information. From the primary visual cortex, visual information is sent to the secondary visual cortex (V2). The secondary visual cortex is known to respond to differences in color, spatial frequency, patterns, and object orientation (Anzai et al., 2007; Hegdé & Van Hessen, 2000; Hubel & Livingstone, 1987; Roe & T'so, 1999; Tao et al., 2012). Given that the secondary visual cortex is spared in individuals like SBR, this may account for how individuals can detect stimuli even if they are not consciously aware of their presence.

4.1.2 Visual Agnosia

Damage to various structures within the dorsal and the ventral visual pathways can lead to distinct forms of **agnosia**. Someone with a visual agnosia will have difficulty recognizing objects, images, or other people, depending on which structure in the visual pathway has been damaged. Agnosia can be classified as either associative or apperceptive. In associative agnosia, perception occurs normally, but individuals cannot recognize what they are perceiving. Apperceptive agnosia is more extreme as individuals experience a failure in perception. Figure 3. illustrates the difference between associative agnosia and apperceptive agnosia.

Prosopagnosia (also known as "face blindness") is a type of agnosia in which individuals have difficulty perceiving faces, including one's own face. This condition is often associated with damage to the fusiform gyrus, part of the ventral pathway. The fusiform gyrus contains the fusiform face area (FFA), which forms a network with the occipital face area and the posterior temporal sulcus. To be more specific,

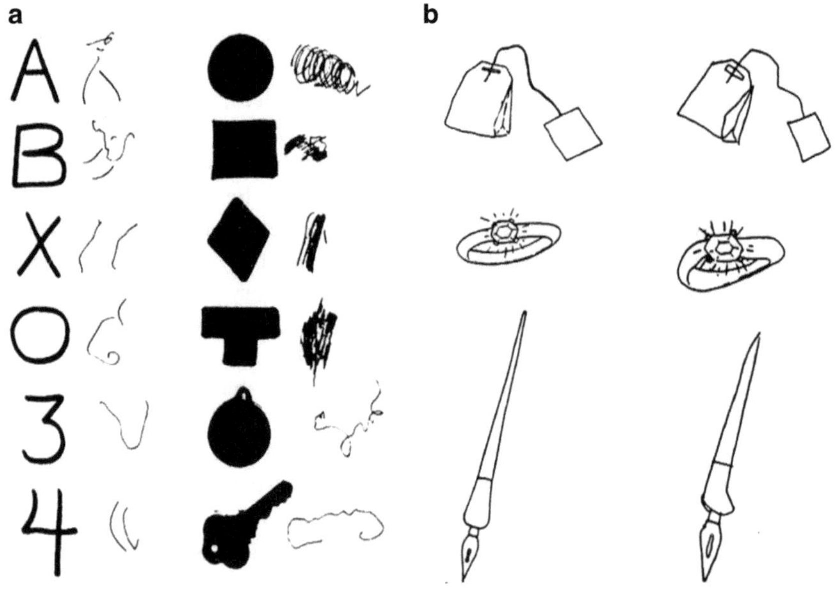

Fig. 3 Individuals with apperceptive agnosia experience a complete breakdown in perception. When asked to draw a copy of what they see, they are unable to do so (**a**). In contrast, individuals with associative agnosia can copy what they see, but they have difficulty identifying the object that they are copying (**b**). For example, these individuals can copy a teabag, but they cannot identify that what they are drawing is a teabag. *Source* Figure 1 of Baugh et al. (2017), https://doi.org/10.1016/B978-0-12-809324-5.00253-4, by permission from Elsevier

damage to the fusiform gyrus is more likely to lead to apperceptive prosopagnosia whereas damage to the anterior temporal face area is more likely to lead to associative agnosia (Corrow et al., 2016).

Look at the image in Fig. 4. How would you describe this image? Understanding what is happening in this image is not difficult. Even if you do not know which city is being depicted (Chicago, Illinois), you will be able to say that the image depicts a city skyline. You will also recognize that the city is positioned next to a body of water. Individuals with **simultanagnosia** will have a much more difficult time deciphering this image. Simultanagnosia is the inability to perceive an image as a whole (Dalrymple et al., 2013). Instead of recognizing the image as a city skyline, the individual may state that they are looking at a building, or they may say that the picture shows a lake. Damage to part of the dorsal pathway, particularly in the parietal-occipital and the middle occipital regions and parietal cortex, is associated with simultanagnosia. Damage to the frontal eye field (FEF), located in the middle frontal area of the brain, is also associated with simultanagnosia (Chechlacz et al., 2012).

The dorsal pathway also contains a brain structure that is important for detecting motion: the middle temporal area (MT, also known as V5). Damage to this part of the brain leads to a condition known as **akinetopsia** (aka, "motion blindness") (Van Swol et al., 2024). Imagine what your life would be like if you perceived the world as a series of photographs. This makes simple actions we take for granted much more

Fig. 4 Can you identify what is being shown in this picture? *Source* https://www.pexels.com/photo/photo-of-city-buildings-near-river-2535631/ from Amit Thakral, via Pexels (https://www.pexels.com/license/)

difficult (e.g., pouring a glass of water) and more complex behaviors impossible (e.g., driving a vehicle).

Comprehension Check 2 Describe the differences between the dorsal and ventral visual pathways in the brain, including their functions and the types of perceptual abnormalities that can occur if these pathways are damaged.

The dorsal pathway helps us identify where an object is located, which enables us to coordinate our movements accordingly. The ventral pathway allows us to identify what an object is. Damage to either pathway can lead to a form of agnosia. Damage to the dorsal pathway can lead to simultanagnosia, which is the inability to see images as a whole. Damage to the ventral pathway can lead to prosopagnosia, in which we cannot identify faces.

4.2 Audition

Instead of converting light waves into electrical signals, the cochlea (derived from Greek word meaning "snail") in the inner ear converts sound waves into electrical signals that are transmitted through the cochlear nerve. This nerve travels through several brain structures, including the cochlear nucleus in the brainstem, the inferior colliculus, and the medial geniculate nucleus (MGN) of the thalamus, before reaching the primary auditory cortex located in the temporal lobe. Like vision, there is a dorsal and a ventral pathway for audition. The ventral pathway (temporal-frontal) is involved in recognizing specific sounds ("what" pathway); the dorsal pathway (temporal-parietal) integrates sounds with sensorimotor actions ("where" pathway) (Rauschecker, 2012) (see Fig. 5 for a schematic of the auditory system).

Like the visual system, individuals can develop various forms of auditory agnosia (Miceli & Caccia, 2023). Generalized auditory agnosia is marked by difficulty perceiving sounds emitted from a wide range of stimuli. This form of agnosia contrasts with more specific types of auditory agnosia. Some individuals may have difficulty perceiving auditory speech while others may show perceptual deficits for environmental sounds. **Amusia**, also called tone deafness, is a type of auditory agnosia that is specific for music. For example, individuals with amusia cannot determine if a piece of music is out of tune.

Damage to parts of the primary auditory cortex and the secondary auditory cortex is associated with auditory agnosia. General auditory agnosia, which is the most common form of auditory agnosia, often results from bilateral damage to the temporal lobe. Verbal auditory agnosia is more associated with damage to the superior temporal cortex (STC). Disruptions in environmental sound perception may arise from damage to the posterior region of the superior gyrus and to the middle temporal gyrus (MTG) (Slevc & Shell, 2015).

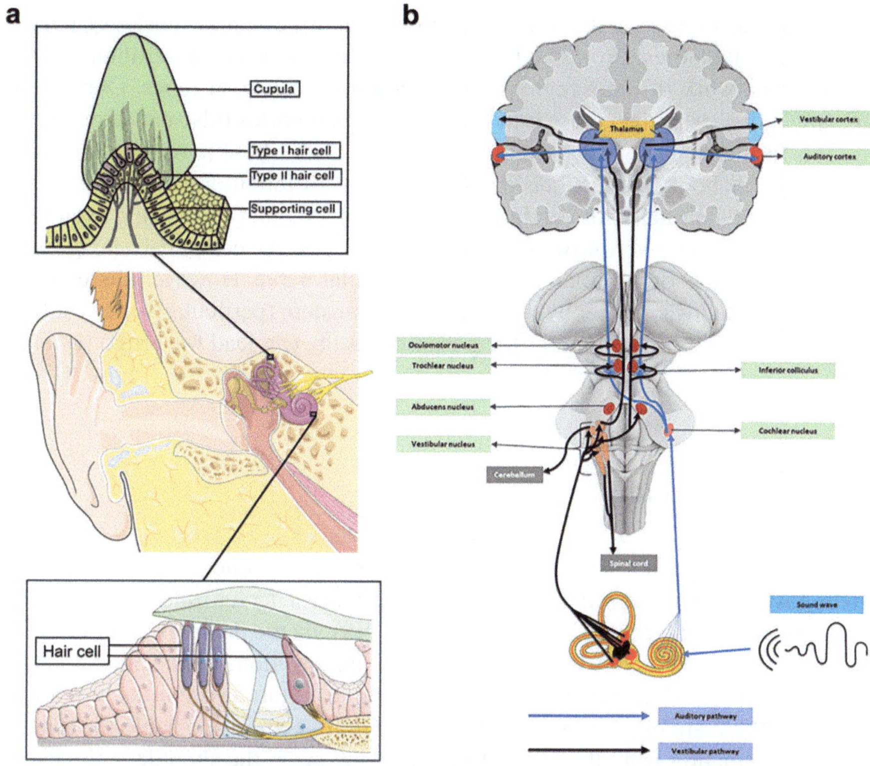

Fig. 5 (**a**) Both the auditory and the vestibular systems rely on stimulation of hair cells. Hair cells for the auditory system are located in the cochlea, a snail shell-shaped structure found in the inner ear. Hair cells for the vestibular system are found in the semicircular canals near the cochlea. Axons from hair cells form the vestibulocochlear nerve. *Source* Servier Medical Art (https://smart.servier.com/; CC BY 4.0, https://creativecommons.org/licenses/by/4.0/), with the inset of the internal structure of the semicircular canals coming from Figure 2 of Jiang et al. (2021), https://link.springer.com/article/10.1007/s00204-021-03067-3. CC BY 4.0 https://creativecommons.org/licenses/by/4.0/, via Springer Nature. (**b**) A schematic showing the auditory and the vestibular pathways in the brain. Note that the pathways in the brain are different for the auditory and the vestibular systems. *Source* Figure 1 of Kaliyappan et al. (2022), https://doi.org/10.3389/fneur.2022.850337. CC BY 4.0 (https://creativecommons.org/licenses/by/4.0/), via Frontiers

4.3 Vestibulation

The vestibular system, like the auditory system, originates in the inner ear (refer back to Fig. 5). The semicircular canals are located next to the cochlea. These canals are filled with *perilymph*, a type of fluid, which moves when we move. There are special cells that detect this motion, and this information is sent to the brain by the vestibular nerve. The vestibular nerve and the cochlear nerve are connected; hence, these nerves are collectively called the vestibulocochlear nerve. The vestibular nerve travels to the vestibular nuclei in the brainstem. These nuclei have connections with multiple brain regions, including the cerebellum, eye muscles, and the ventral posterior nucleus (VPN) of the thalamus. From the thalamus, neurons travel to the vestibular cortex, which is a network of brain regions in the temporal-parietal cortices. One notable structure in the vestibular cortex is the posterior insular cortex (also called the insula; the term insula will be used throughout the rest of the book). As you will learn throughout this book, the insula is involved in many different processes.

You may be familiar with **vertigo**, a sensation of spinning or swaying even when someone is stationary. Vertigo most commonly results from mechanical damage to the semicircular canals or inflammation of the vestibular nerve. However, disruption to the vestibular pathway, either by damage to the brainstem (particularly the vestibular nuclei), part of the cerebellum, or the posterior insula, can lead to vertigo (Li et al., 2023).

4.4 Somatosensation

For now, I am going to focus on somatosensation that does not result in pain. Pain will be covered in the next chapter. Recall that peripheral nerves transmitting sensory information enter the central nervous system (CNS) via the dorsal root of the spinal cord. One thing you have not learned yet is the organization of the spinal cord and its 31 pairs of spinal nerves (see right side of Fig. 6). Starting from the base of the brainstem, there are eight *cervical* nerves (C1-C8) that are responsible for sending sensory information from the back of the head, the neck, the shoulders, the hands, and parts of the arms (although the C1 nerve acts more as a motor neuron than a sensory neuron). Twelve *thoracic* nerves (T1-T12) are responsible for sending sensory information from the chest, the upper abdomen, and parts of the back to the CNS. The five *lumbar* nerves (L1–L5) provide sensory inputs for the lower abdomen, the groin, the thighs, the upper foot, and the outer part of the legs. The five *sacral* nerves (S1–S5) also transmit sensory information from the lower body, including the pelvic region, the groin (particularly the genital areas), parts of the leg, and the buttocks. The *coccygeal* nerve is responsible for just one region: the coccyx (tailbone).

What about sensory information for the head? Instead of traveling through spinal nerves, several *cranial nerves* transmit sensory information to the brain. There are 12 total cranial nerves. While this is the first time mentioning the term cranial nerve, you already learned about a couple of them above. The optic nerve is cranial nerve II, and the vestibulocochlear nerve is cranial nerve VIII. Specific to somatosensation of the head, the trigeminal nerve (cranial nerve V) sends sensory information from the face, the teeth, and the sinuses to the brain.

Tactile stimulation below the neck travels through the peripheral nervous system (PNS) and through the spinal cord and the brainstem. All cervical nerves and the first six thoracic nerves travel to the cuneate nucleus of the medulla whereas all nerves below the sixth thoracic nerve transmit information to the gracile nucleus of the medulla. Sensory information is then sent to the VPN before traveling to the primary

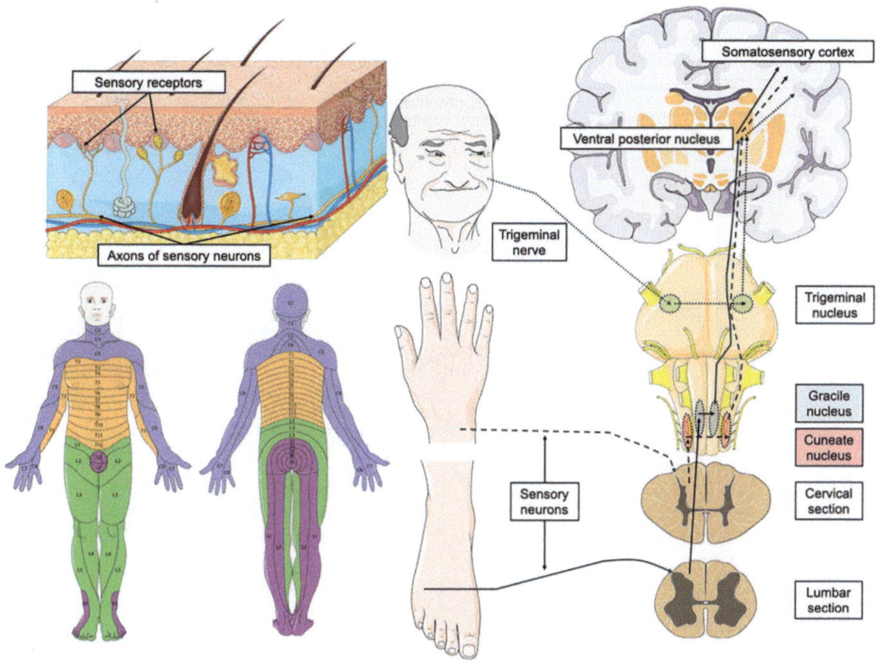

Fig. 6 Schematic detailing the somatosensory system. There are special types of receptors under the skin that detect touch and pressure (top left image). Sensory neurons transmit this information to the central nervous system (CNS). With the exception of the face and head, somatosensory information from the peripheral nervous system (PNS) is sent to the CNS via spinal nerves. The bottom left image shows a *dermatome*. A dermatome shows which parts of the body are served by various spinal nerves. For example, sensory information from the upper neck enters the spinal cord at the level of the second cervical vertebra. The trigeminal nerve, one of 12 cranial nerves, directly transmits somatosensory information originating from the face and head to the brain via the pons. Image created by the author using Servier Medical Art (https://smart.servier.com/; CC BY 4.0, https://creativecommons.org/licenses/by/4.0/) and Microsoft PowerPoint

somatosensory cortex (the most anterior portion of the parietal cortex). Instead of entering the brain through the brainstem, the trigeminal nerve enters through the pons before following the same path as spinal nerves.

Asterognosis is a type of agnosia in which individuals cannot recognize objects by touch alone (Facci et al., 2024). If you blindfolded someone with asterognosis and gave them something like sand, they would have difficulty telling you that they were touching sand. Lesions to the primary somatosensory cortex have been observed in individuals with asterognosis, with additional lesions observed in the secondary somatosensory cortex and part of the posterior parietal cortex (PPC) in those with more severe asterognosis (Bohlhalter et al., 2002).

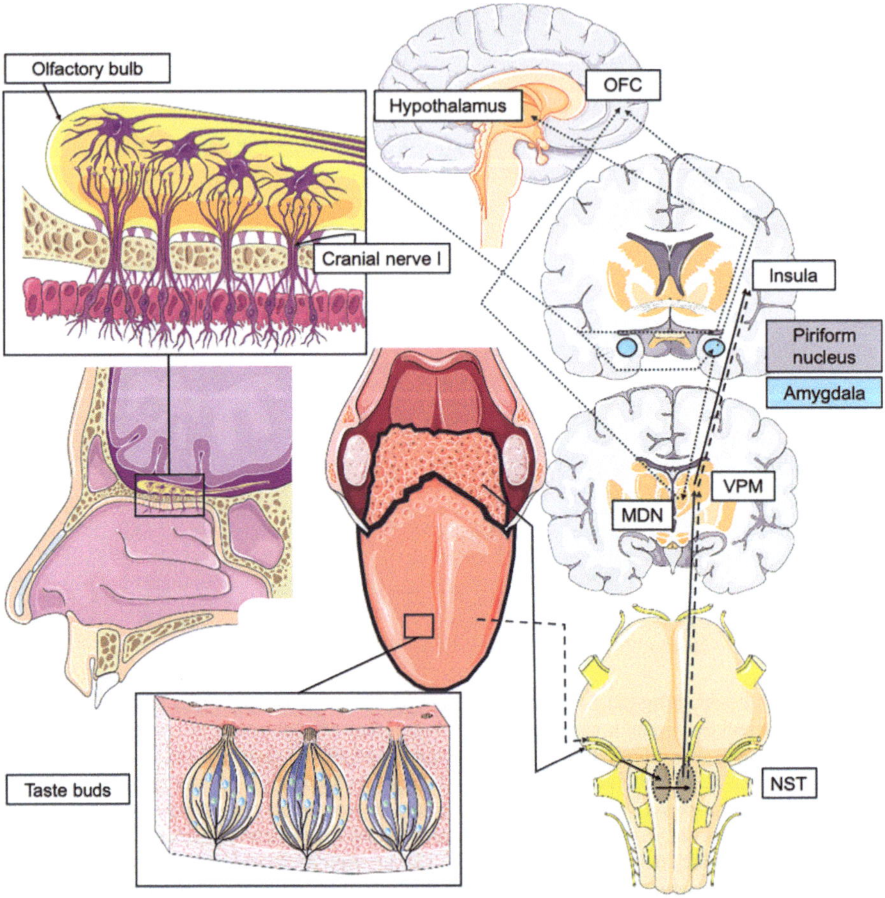

Fig. 7 A simplified schematic showing the olfactory and the gustatory pathways. *Note* MDN = mediodorsal nucleus, NST = nucleus of the solitary tract, OFC = orbitofrontal cortex, VPM = ventral posterior medial nucleus. Image created by the author using Servier Medical Art (https://smart.servier.com/; CC BY 4.0, https://creativecommons.org/licenses/by/4.0/) and Microsoft PowerPoint

5 Chemical Senses: Gustation and Olfaction

Because receptors in our tongue and in our nose respond to chemical molecules as opposed to physical stimuli like light/sound waves or mechanical pressure on the skin, they are collectively known as the chemical senses. Taste and olfaction interact with each other in an important way—they collectively influence our perception of *flavor*. I guarantee all of you have experienced what eating is like when you have a cold or have a severely stuffy nose. Even though your taste receptors are not affected by a cold, your inability to smell makes food taste bland.

Taste is interesting as two separate nerves transmit gustatory information to the brain. The facial nerve (cranial nerve VII) is responsible for the anterior two-thirds of the tongue while the glossopharyngeal nerve (cranial nerve IX) regulates taste from the posterior one-third of the tongue. Both nerves then travel to the nucleus of the solitary tract (NST). Neurons from this region project to the VPN. Finally, information regarding taste travels to the gustatory cortex, composed of the anterior insula (I told you that you would learn more about the insula) and the frontal operculum.

Olfaction is unique because neurons do not travel to the thalamus before reaching the primary olfactory cortex. Instead, the olfactory nerve (cranial nerve I) projects to the amygdala and the prepiriform cortex (primary olfactory cortex), located in the temporal cortex. As part of the limbic system, the amygdala is more known for its role in emotional memories, particularly fear and aggression. The amygdala will be covered more extensively in Chaps. 8 and 9. Given that the olfactory nerve sends direct projections to the amygdala may explain why certain odors can trigger strong memories of an event. For example, a specific perfume may elicit memories of one's grandmother. We will return to this idea in Chap. 9. Neurons from the amygdala and the prepiriform cortex then travel to the mediodorsal nucleus (MDN) of the thalamus and to the hypothalamus. Neurons originating in the thalamus transmit information to the orbitofrontal cortex (OFC), and neurons originating in the hypothalamus transmit information to the lateral posterior OFC. Additionally, the insula acts as a secondary olfactory processing center (Boot et al., 2024). This may account for why taste and olfaction are highly linked to one another. Figure. 3 shows the gustatory and the olfactory pathways in the brain.

Individuals can develop gustatory or olfactory agnosia, in which they have difficulty recognizing familiar tastes or odors, respectively, following head trauma. Presumably, gustatory agnosia results from damage to the anterior insula/frontal operculum; however, other regions may be involved. In one case study, a 39-year-old woman had part of her left anteromedial temporal lobe, which contains the amygdala, removed. Interestingly, following her surgery, the woman was no longer able to recognize different tastes (Small et al., 2005). Olfactory agnosia has been reported in an individual that had to have her right amygdala and hippocampal region removed due to epilepsy. Her agnosia was limited to odors presented to her right nostril (Lehrner et al., 1997).

Comprehension Check 3 Imagine an individual experiences damage to cranial nerve VIII. What are some perceptual disturbances this individual may encounter?

Because cranial nerve VIII (vestibulocochlear nerve) transmits information regarding audition and vestibulation, an individual can experience several issues. Hearing loss can occur, and individuals can experience vertigo. Another auditory disturbance that can occur that has not been covered is tinnitus (ringing noise).

5.1 Thalamus as a Common Mediator of Perception

Notice, the thalamus is used by each major sensory system. The thalamus is often nicknamed the "relay station" of the brain. Although "relay station" is a fitting term, it oversimplifies what the thalamus does. Instead of acting as a mere stopping point for sensory systems, the thalamus has a more active role, known to filter sensory information and to regulate motor functions, our sleep-wake cycle, and aspects of consciousness (Bosch-Bouju et al., 2013; Gent et al., 2018; Whyte et al., 2024).

Thalamic damage can cause a wide range of problems for an individual. Directly related to what has been covered in this chapter so far, damage to the thalamus is associated with perceptual distortions. Beyond perception, dysfunction of the thalamus is linked to insomnia, fatigue, cognitive impairments, and pain (Grossman & Inglese, 2016). As such, the thalamus will be covered in multiple chapters of this book.

5.2 Importance of Association Areas in Perception

The thalamus is not the only structure that helps process sensory information. The cerebral cortex contains several **association areas**. These regions are involved in *higher-order* brain functions like memory, decision making, and language. Throughout this book, you will learn how these higher-order functions can become disrupted (see Chaps. 6, 7, and 15 for examples).

Above, you learned about the neural pathways of various sensory systems. In real life, our sensory systems often work in tandem. Take a minute to stop what you are doing and to think about which sensory systems are engaged at this very moment. If you have been physically reading this book, your visual system has been active. However, your visual system is not working by itself. Your auditory system is engaged. You may have a screen reader that enables you to listen to the contents of this book, or you may be "hearing" the words as you silently read to yourself. You may also hear the sounds of an air conditioning unit, cars driving by, or roommates walking around. Your somatosensory system is also working at this moment. Even if you are not holding a physical copy of the book, your body is sending messages to your brain about the clothes you are wearing or the blanket that is covering part of your body.

As your brain receives multiple pieces of sensory information, association areas will become activated to help put all of this information together. Your brain is trying to make sense of what is happening. For example, imagine your favorite restaurant. What does it look like on the inside? What does the food smell and taste like? What sounds do you typically hear when you're there? What do the seats feel like when you sit on them? All of your sensory systems are being engaged at once in this environment. While the thalamus acts as a "relay station", association areas are like the glue that binds these sensations together.

For now, I want to briefly highlight one study that showed how association areas become engaged. Rats were trained to stick their nose into an aperture that varied in width. Manipulating the size of the aperture created a different tactile sensation. Unsurprisingly, neurons in the primary somatosensory cortex were activated when rats stuck their nose into the aperture. Additionally, neurons in the primary visual cortex were engaged because rats use this system to help direct their movement to the aperture. The researchers of this study also found that neurons in two other parts of the brain were highly active during task performance: the prefrontal cortex (PFC) and the PPC (which was mentioned in passing earlier) (Deolindo et al., 2018).

The PFC is the most anterior portion of the brain and is implicated in many higher-order functions, including the three listed above: memory, decision making, and language. The PPC is located between the primary visual cortex and the primary somatosensory cortex and helps integrate visual and somatosensory information. This area is crucially important for our spatial awareness. As the rat performs the task of inserting its nose into the aperture, the PPC helps inform the rat of where it is inside the testing chamber; that way, the rat knows where the aperture is relative to its current position. The PFC helps store this information, which allows the rat to remember how to perform the task the next day.

Comprehension Check 4 Explain how the thalamus and association areas contribute to perception?

Both the thalamus and association areas integrate information from multiple sensory systems. Because we are bombarded with different types of stimuli at any given moment, integrating sensory information better allows for us to understand what is happening around us and how to respond accordingly. For example, the sound of a dog barking in isolation may not elicit a strong response from you. However, if you are in front of a dog that is barking and baring its teeth, your brain's ability to integrate both pieces of information will allow you to identify the dog as a potential threat. In the future, if you encounter this same dog, you will know to keep your distance.

6 Neurochemical Substrates of Perception

Our sensory systems depend heavily on a balance between excitatory and inhibitory neurotransmission. Recall from the previous chapter that neurotransmitters are special chemicals that are transmitted between neurons. The first two neurotransmitters I want to introduce are glutamate and GABA (gamma-aminobutyric acid), which are the major excitatory and inhibitory neurotransmitters in the nervous system. Approximately 80–90% of all neurons release glutamate (McKenna, 2007). However, only 20% of neurons are GABAergic (Sahara et al., 2012). If the number of glutamatergic neurons significantly outnumbers GABAergic neurons, how does the nervous system maintain a balance between excitation and inhibition?

First, GABAergic neurons can inhibit multiple glutamatergic neurons simultaneously. The second point is more convoluted, requiring me to explain glutamatergic and GABAergic receptors in more detail. This will also provide an opportunity to reinforce concepts that were introduced in Chap. 2. I will start with GABAergic receptors. There are two main types of GABA receptors: $GABA_A$ and $GABA_B$. $GABA_A$ receptors are ionotropic, composed of five subunits that surround an ion channel. When these receptors are activated, the ion channel opens, allowing negatively charged chloride (Cl^-) ions to enter the neuron. This hyperpolarizes the neuron, making it less likely that it will fire an action potential. $GABA_B$ receptors are metabotropic. The end result is the same as $GABA_A$ receptors, but the way these receptors hyperpolarize the neuron is different. When a $GABA_B$ receptor is activated, this causes a G_i protein to dissociate from the receptor. As an inhibitory G protein, this leads to the inhibition of adenyl cyclase.

Glutamatergic receptors are much more complicated than GABAergic receptors. There are three major ionotropic receptor subtypes: N-methyl-D-aspartate (NMDA), α-amino-3-hydroxy-5-methyl-4-isoxazolepropionic acid (AMPA), and kainate. AMPA and kainate receptors allow positively charged sodium (Na^+) ions into the neuron. NMDA receptors allow both Na^+ and calcium (Ca^{++}) ions into the neuron. As such, all three ionotropic glutamate receptors are excitatory. There are also three classes of metabotropic glutamate receptors: Group I, Group II, and Group III. Group I metabotropic glutamate receptors ($mGluR_1$ and $mGluR_5$) are also excitatory. Stimulation of these receptors causes a G_q protein to break away from the receptor. G_q proteins do not interact with adenyl cyclase. Instead, they stimulate phospholipase C, which kickstarts a chain reaction inside the neuron. I will not go into full detail about the G_q signaling pathway. You will learn more about this pathway later in this book (Chap. 9). This reaction allows Ca^{++} to enter the cell, thus depolarizing it. Group II ($mGluR_2$, $mGluR_3$, and $mGluR_4$) and Group III ($mGluR_6$, $mGluR_7$, and $mGluR_8$) have the same mechanism of action as $GABA_B$ receptors, meaning they are inhibitory! Group II and Group III metabotropic glutamate receptors also act as autoreceptors. These receptors are located on the presynaptic neuron and act as a negative feedback loop. Thus, when these receptors are activated on a presynaptic glutamatergic neuron, this causes the neuron to release less glutamate.

Glutamatergic activity is critical for almost each sensory system. For example, when sound waves stimulate the ear drum, this causes specialized cells in your inner ear to release glutamate. This stimulates neurons, allowing information to travel through the vestibulocochlear nerve to the brain. The one major exception is gustation, in which a molecule called adenosine triphosphate (ATP) is released from neurons originating from the tongue. This is interesting because ATP is primarily used as an energy source for most cells in the body.

To prevent sensory overload, inhibitory neurotransmitters like GABA help control neuronal responses to stimuli, thus enabling us to discriminate one light from another or different types of touch (Schmidt-Wilcke et al., 2018). Think of it this way—we are constantly bombarded by stimuli in our environment. As I write this paragraph, I can hear a low humming sound coming from somewhere within my office building and can hear my colleague speaking to someone else from down the hall; the fluorescent light from my office mixes with the little sunlight coming from a small window that takes up the top 10% of one of my walls, and I can differentiate how the keys of my laptop feel compared to the edge of my desk pressed up against my forearms. Without GABAergic inputs, things would just blend together, creating a perceptual mess.

GABA is certainly not the only neurotransmitter that helps regulate perception. For now, I am going to focus on one specific neurotransmitter: serotonin (5-hydroxytryptamine (5-HT)). Serotonin is primarily found in our gastrointestinal tract, with only 1–2% located in the CNS. In the CNS, serotonergic cell bodies are isolated to the raphe nuclei, which are located in the brainstem. Serotonergic neurons project to multiple brain regions such as the cerebellum, the basal ganglia, the limbic system, the thalamus, and the cerebral cortex, including those that are involved in perception (e.g., primary visual cortex, primary auditory cortex).

When released from a presynaptic neuron, serotonin can bind to multiple 5-HT receptor types, labeled 5-HT_1, 5-HT_2, 5-HT_3, 5-HT_4, 5-HT_5, 5-HT_6, and 5-HT_7, with some of these receptor types further subdivided (e.g., 5-HT_{2A}, 5-HT_{2C}). Table 1 provides a breakdown of each 5-HT receptor. You do not need to worry about memorizing each 5-HT receptor type. Instead, I want you to appreciate the complexity of this system. For now, just know that in cortical regions, 5-HT receptors are located on both glutamatergic and GABAergic neurons. Evidence indicates that serotonin functions to decrease responses to sensory information (Jacob & Nienborg, 2018). As you will soon see, this chapter will focus primarily on one of these serotonergic receptors: the 5-HT_{2A} receptor. You will encounter serotonin numerous times as you go through each chapter of this book. Although serotonin is primarily isolated to the gastrointestinal tract, it plays a significant role in various cognitive-affective processes.

Comprehension Check 5 Explain how the balance between excitatory and inhibitory neurotransmission aids in perception.

As the major excitatory neurotransmitter, glutamate is largely responsible for transmitting sensory information to the brain for processing. Without GABA, our brain would become overwhelmed with sensory inputs. This would make prioritizing relevant stimuli from background "noise" extremely difficult. GABA helps us discriminate stimuli from one another. Other neurotransmitters also help mediate perception. Serotonin can bind to receptors that are excitatory or inhibitory, which helps further finetune our perception

Table 1 Comparison of Serotonin Receptors

Receptor	Type	Effects on Receptor Potential	Localization in Brain (Mengod et al., 2010)
5-HT_1	G_i-coupled	↓	
5-HT_{1A}			Hippocampus, cerebral cortex, dorsal raphe nuclei, entorhinal cortex
5-HT_{1B}			Basal ganglia, substantia nigra, dorsal and median raphe nuclei, parafascular nucleus of the thalamus
5-HT_{1D}			Basal ganglia, cerebral cortex, substantia nigra, hippocampus
5-HT_{1E}			Entorhinal cortex, dorsal striatum, amygdala, hypothalamus
5-HT_{1F}			CA1-CA3 regions of hippocampus, cerebral cortex, dorsal raphe nuclei
5-HT_2	G_q-coupled	↑	
5-HT_{2A}			Cerebral cortex, dorsal striatum, nucleus accumbens, olfactory tubercle, hippocampus
5-HT_{2B}			Hypothalamus, cerebellum, amygdala
5-HT_{2C}			Choroid plexus, cerebral cortex, nucleus accumbens, hippocampus, amygdala, dorsal striatum, substantia nigra
5-HT_3	Ionotropic	↑	Brainstem, hippocampus, amygdala, cerebral cortex
5-HT_4	G_s-coupled	↑	Basal ganglia, hippocampus, amygdala, olfactory tubercle, nucleus accumbens, substantia nigra, cerebral cortex
5-HT_5	G_i-coupled	↓	Cerebral cortex, hippocampus, amygdala, hypothalamus, basal ganglia
5-HT_6	G_s-coupled	↑	Dorsal striatum, nucleus accumbens, olfactory tubercle, choroid plexus, hippocampus, cerebral cortex, thalamus, hypothalamus, substantia nigra
5-HT_7	G_s-coupled	↑	Thalamus, hippocampus, hypothalamus, cerebral cortex, amygdala

7 Hallucinations

I doubt I need to convince anyone reading this book of the importance of perception. Imagine the difficulty you would have if you suddenly went blind or deaf. Although these are extreme examples, most of you have experienced how loss of perception can be frustrating. As discussed earlier in this chapter, food often tastes bland when we are sick. This was a common complaint reported by individuals affected by the COVID-19 virus (Hannum et al., 2023). While losing perceptual abilities can be challenging, some individuals are prone to perceiving stimuli that are not present, leading to difficulties for the individual.

Hallucinations are highly prevalent in schizophrenia and related schizoaffective disorders (e.g., symptoms of schizophrenia and mood disorders; you will learn more about mood disorders in Chap. 9). Schizophrenia is not the only psychiatric condition associated with hallucinations. There are several disorders that you will learn about throughout this book in which individuals can experience hallucinations (Waters & Fernyhough, 2017).

Although Hollywood movies often depict individuals with schizophrenia experiencing visual hallucinations (e.g., *A Beautiful Mind*), auditory hallucinations are the most common type of hallucination, with approximately 75% of individuals with schizophrenia and 46–57% of individuals with other psychiatric conditions experiencing them (Waters & Fernyhough, 2017). Visual hallucinations occur in just 27% of individuals with schizophrenia (Waters et al., 2014), with individuals experiencing the perception of light stimuli appearing and disappearing quickly, objects appearing smaller or larger, and objects appearing to change orientation (Silverstein & Lai, 2021). Hallucinations across multiple modalities is more common (53% of individuals with schizophrenia) than unimodal hallucinations (27%) (Lim et al., 2016). In other words, a large percentage of individuals with schizophrenia experience hallucinations for more than one sensory modality.

Hallucinations are not isolated to individuals with a psychiatric disorder. Even individuals without a psychiatric disorder can experience hallucinations (Linscott & van Os, 2013). One recent study reported that a high proportion of individuals identified experiencing auditory (29.4%), visual (21.5%), tactile (19.9%), and olfactory (17.3%) hallucinations, with the most common hallucinations including hearing voices, seeing shadows, being touched, and smelling fire (Linszen et al., 2022). Although not as widely studied, gustatory hallucinations can occur (Mitchell et al., 2017; Mueser et al., 1990).

You may have encountered something like the following situation during your life. I know when I was younger, I could feel the sensation of someone giving me a kiss on my forehead when I was in bed. Even though I opened my eyes as soon as I felt the sensation of lips pressed on my head, I did not see anyone in my room. What may have happened is that I experienced a *hypnagogic* hallucination, a false perception that occurs as we fall asleep. Individuals can also experience these hallucinations as they awaken (known as *hypnopompic* hallucinations). These hallucinations generally do not cause any distress to the individual, but some hypnagogic/hypnopompic hallucinations can be terrifying. In one case report, a young woman began experiencing hypnopompic hallucinations following a car accident; she reported being fully awake but not being able to move and that she could hear footsteps and could feel a presence in the room (McCarty & Chesson Jr., 2009). Chapter 14 provides a formal definition for hypnagogic and hypnopompic hallucinations.

7.1 Neuromechanisms Underlying Hallucinations

Increased activation of regions in the occipitotemporal cortex ("what" pathway) is observed in individuals experiencing a visual hallucination, including the lingual gyrus and the cuneus (Zmigrod et al., 2016) (see Fig. 8). These regions are involved in visual memories (Gilboa et al., 2004), enabling us to compare a stimulus in our view to things we have stored in memory. This helps us recognize objects in our environment. Hyperactivity of regions like the lingual gyrus and the cuneus may lead to a false alarm, erroneously concluding a stimulus is in our visual field when it is not.

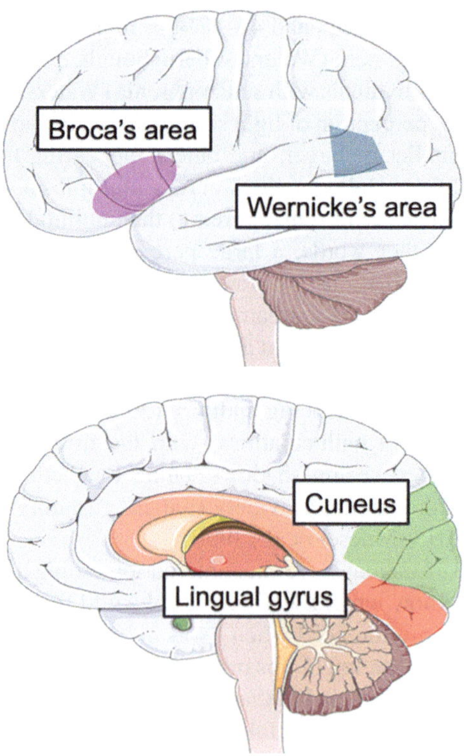

Fig. 8 During an auditory hallucination, several brain regions involved in audition are activated, including Broca's area and Wernicke's area. During visual hallucinations, areas such as the lingual gyrus and the cuneus become activated. Image created by the author using Servier Medical Art (https://smart.servier.com/; CC BY 4.0, https://creativecommons.org/licenses/by/4.0/) and Microsoft PowerPoint

Auditory hallucinations are associated with activation of the superior temporal gyrus (STG), which contains the primary auditory cortex and Wernicke's area (Zmigrod et al., 2016). Wernicke's area is primarily isolated to our dominant hemisphere (left for most individuals) and makes up the posterior region of the STG. This area is critically important for speech, with original hypotheses arguing that this region controls the comprehension of speech (Harpaz et al., 2009); however, some argue that this region is important for speech production instead of language comprehension (Binder, 2015). Related to Wernicke's area is Broca's area, which is also primarily located in the left hemisphere of the brain, and is involved in speech production, specifically coordinating linguistic information across multiple brain regions (Flinker et al., 2015) (you will learn more about Wernicke's area and Broca's area in Chap. 15). Yet, individuals experiencing hallucinations often experience increased activation of this region in the right hemisphere (Zmigrod et al., 2016). The Experiment Spotlight below focuses on the relationship between brain volume in language-associated areas and susceptibility to hallucinations.

Experiment Spotlight: Exploring the Relationship Between Hallucination Proneness and Brain Morphology

The full article can be accessed with this link: https://doi.org/10.1016/j.neuroimage.2024.120942

In this study (Alotaibi et al., 2024), individuals first completed an online survey in which they answered questions from the *Launay-Slade Hallucination Scale* (Modified Version) (LSHS) and were required to identify if they have ever been diagnosed with auditory verbal hallucinations by a healthcare professional. If individuals indicated that they had received such a diagnosis, they were excluded from the rest of the experiment.

Seventy-five individuals were invited to undergo an MRI (magnetic resonance imaging) scan and to complete two additional questionnaires: the *Dissociative Experience Scale* and the *Persecution and Deservedness Scale.* Several participants were excluded from the experiment because they had metal implants that would make going through an MRI scanner dangerous (n = 6) or had an abnormal brain scan (n = 1).

The results of this study showed that participants who were more prone to hallucinations had larger brain areas related to language, like the STG and Broca's area. Paradoxically, these results contrast with what is typically observed in individuals diagnosed with a disorder like schizophrenia. That is, clinical populations that experience hallucinations often have *decreased* brain volume in these regions.

Questions to Consider:

1. Why do you think the researchers chose to exclude individuals with a prior diagnosis of auditory verbal hallucinations from the study? How might including these individuals have affected the results?
2. The study found that a non-clinical population (students) prone to hallucinations had larger brain areas related to language, while clinical populations (e.g., individuals with schizophrenia) often show decreased brain volume in these regions. What are some possible explanations for this paradoxical finding?
3. How might the findings of this study influence future research or clinical practices related to hallucinations? What additional questions or hypotheses could

be explored to further understand the relationship between brain structure and hallucination proneness?

Like visual and auditory hallucinations, increased activity in regions involved in processing of other sensory information is implicated in other types of hallucinations. For example, stimulating the medial orbitofrontal cortex (mOFC), which contains the piriform cortex, or the olfactory sulcus, containing the olfactory bulb and olfactory tract, evokes olfactory perceptions in healthy individuals (Bérard et al., 2021; Kumar et al., 2012). Increased activation is observed in the insula of individuals that experience gustatory hallucinations (Levy & Henkin, 2004), and somatosensory hallucinations are linked to a hyperactive primary somatosensory cortex (Michael et al., 2022).

In short, brain regions involved in sensory processing become activated even in the absence of an external stimulus, leading to a hallucination. Why do these regions become activated in the absence of a stimulus? To answer this question, I need to discuss how dysregulation of several neurotransmitter systems can cause hallucinations.

7.1.1 Serotonin Hypothesis of Hallucinations

Dysregulated serotoninergic activity in cortical regions is largely implicated in hallucinations. In particular, the 5-HT_{2A} receptor is a major determinant of perceptual disturbances. This receptor is located on glutamatergic and GABAergic neurons in cortical areas that process sensory information (Charnay & Léger, 2010). Recall that the 5-HT_{2A} receptor is excitatory. Increased activation of 5-HT_{2A} receptors, particularly on glutamatergic neurons, may account for the hyperactivity observed in cortical regions during a hallucination. In fact, increased 5-HT_{2A} receptor binding is observed in the visual ventral pathway of individuals experiencing visual hallucinations (Ballanger et al., 2010).

Additional evidence supporting the serotonergic hypothesis of hallucinations comes from the effects of certain drugs. **Psychedelic hallucinogens** such as lysergic diethylamide (LSD; better known as "acid"), psilocybin ("magic mushrooms"), and mescaline (peyote) act as 5-HT_{2A} agonists. Individuals that use psychedelic hallucinogens experience intense alterations to perception, including hallucinations. Psychedelic-induced hallucinations can be blocked with a 5-HT_{2A} antagonist (Kometer et al., 2013). Conversely, atypical antipsychotic drugs, commonly used to treat schizophrenia and related disorders, have several mechanisms of action, including acting as 5-HT_{2A} antagonists, as well as 5-HT_{1A} agonists. Remember, the 5-HT_{1A} receptor is inhibitory.

Hallucinations are a rare side effect of selective serotonin reuptake inhibitors (SSRIs) and serotonin and norepinephrine reuptake inhibitors (SNRIs), classes of medications used to treat mood and anxiety disorders (Capaldi 2nd & Carr, 2010; Tripp & Golden, 2009). You will learn more about these drugs in Chaps. 8 and 9. In short, these medications cause serotonin levels to increase in the synapse, allowing for increased stimulation of 5-HT_{2A} receptors. However, because SSRIs and SNRIs merely increase serotonin levels in the synapse, this can cause serotonin to bind to both excitatory (5-HT_{2A}) and inhibitory (5-HT_{1A}) serotonin receptors. This may explain why hallucinations are a rare side effect as opposed to a common occurrence, which is a good thing. If medications used to treat psychiatric conditions commonly caused hallucinations, they would be rarely prescribed, if at all.

7.1.2 Glutamate Hypothesis of Hallucinations

Linking hallucinations exclusively to serotonin grossly oversimplifies the neurochemical bases of such perceptual abnormalities. Above, I mentioned that 5-HT_{2A} receptors are located on glutamatergic neurons in cortical areas. If 5-HT_{2A} receptors are upregulated (i.e., there are more receptors than normal) or too much serotonin is being released into the synapse, this can cause overexcitation of glutamatergic neurons, leading to hallucinations. This suggests that there is an important interaction between serotonin and glutamate.

Yet, a different hypothesis argues that *hypofunctional* NMDA receptors located on cortical GABAergic receptors leads to hallucinations. If GABA receptors are not sufficiently stimulated, they cannot inhibit other cortical areas of the brain, which can then lead to hallucinations (Cohen et al., 2015).

One major piece of evidence supporting the hypoactive glutamate hypothesis of hallucinations comes from **dissociative hallucinogens**. Drugs like ketamine ("Special K") and phencyclidine (PCP; "angel dust") alter perception and produce hallucinations, but they also cause individuals to *dissociate*, a feeling of detachment from oneself or the environment (think of observing oneself from above). This is what distinguishes these hallucinogens from the psychedelic hallucinogens described above. Dissociative hallucinogens act as NMDA channel blockers. That is, when the ion channel opens after the receptor has been stimulated, drugs like ketamine and PCP enter the channel, but they do not enter the neuron. They block the ion channel, preventing positively charged ions from entering. In short, dissociative hallucinogens act as NMDA receptor antagonists.

7.1.3 Dopamine Hypothesis of Hallucinations

One final neurotransmitter I want to highlight in this chapter is dopamine. Dopamine and serotonin are chemical cousins, belonging to a class of neurotransmitters called *monoamines* (you will be introduced to another major monoamine in Chap. 5). Instead of originating from the brainstem, dopaminergic cell bodies originate from several regions, most notably the substantia nigra and the ventral tegmental area (VTA) (both located in the tegmentum). For now, I will focus on VTA-projecting dopaminergic neurons, as they travel to limbic structures and to cortical areas.

Dopamine binds to two different receptor types: D_1-like receptors and D_2-like receptors. D_1-like receptors, consisting of the D_1 and the D_5 receptors, are excitatory as they are G_s-coupled. Stimulation of these receptors leads to activation of adenyl cyclase. D_2-like receptors, consisting of D_2, D_3, and D_4 receptors, are G_i-coupled and inhibitory. Dopamine receptors are located on glutamatergic neurons in cortical areas, particularly the PFC.

The dopamine hypothesis was the first one proposed to explain the neurochemical basis of schizophrenia, and thus hallucinations. Originally, the idea was that hyperactive mesolimbic and mesocortical pathways are associated with the symptoms of schizophrenia (Iversen & Iversen, 2007). Above, you learned that atypical antipsychotic drugs interact with 5-HT receptors. What I did not tell you is that these medications also act as dopamine D_2-like receptor antagonists. I also did not tell you that atypical antipsychotic drugs are the "newer" antipsychotic drug class (new is a relative term here as the first atypical antipsychotic was synthesized in 1958, although it was not approved by the Food and Drug Administration until 1990). Instead of saying that a drug class is "newer" or "older", they are classified as first-generation, second-generation, third-generation, and so on. Atypical antipsychotics constitute the second generation of antipsychotic medications.

The first-generation antipsychotic drugs (also known as "typical" antipsychotics) have little **affinity** for 5-HT receptors, instead acting primarily as dopamine D_2-like receptor antagonists. The finding that typical antipsychotics ameliorated symptoms like hallucinations helped pave the way for the first dopamine hypothesis. Unfortunately, typical antipsychotic drugs are not that effective at treating other symptoms of schizophrenia. Even worse, they produce horrible side effects. You will learn about these side effects in Chap. 10. With the emergence of atypical antipsychotic drugs, the dopamine hypothesis of schizophrenia was modified. Instead of saying that dopamine is hyperactive everywhere, the hypothesis now states that a hyperactive mesolimbic dopamine system underlies symptoms like hallucinations while a hypoactive dopaminergic system controls other cognitive disturbances like blunted affect (i.e., decreased emotional responses), memory impairments, etc. (da Silva Alves et al., 2008).

How exactly does a hyperactive mesolimbic dopamine system contribute to hallucinations? Dopamine in the ventral striatum, particularly in the nucleus accumbens (NAc), is important for determining the relevance of incoming stimuli. The ventral striatum sends information to part of the basal ganglia, which projects neurons to the MDN. You learned about this nucleus of the thalamus in the context of olfaction. Beyond processing olfactory information, the MDN is involved in more complex cognitive functions. This part of the thalamus has connections with the PFC, the limbic system, and the temporal cortex (Georgescu et al., 2020). A hyperactive ventral striatum may assign false signals as being relevant and then communicate this information to the thalamus. The thalamus then relays this information to other parts of the brain. As the temporal cortex is where the primary auditory cortex is located, this can account for why individuals with schizophrenia often experience auditory hallucinations

One final piece of evidence supporting the dopamine hypothesis of hallucinations comes from **psychostimulant** drugs like methamphetamine and cocaine. Psychostimulants, when taken at high concentrations, can cause hallucinations and other symptoms associated with schizophrenia (Curran et al., 2004). Psychostimulants increase dopamine levels in the brain, particularly in the ventral striatum. Similar to individuals with schizophrenia, auditory hallucinations are more common than other types of hallucinations following long-term stimulant use (Srisurapanont et al., 2003).

7.1.4 Integrating the Three Hypotheses of Hallucinations

Figure 9 shows how dysregulation of serotonin, glutamate, and dopamine can collectively cause hallucinations. As discussed above, hyperactive serotonergic receptors on glutamatergic neurons in areas like the occipital cortex can lead to increased activity in areas like the cuneus and the lingual gyrus, giving rise to visual hallucinations. In the PFC, deficient NMDA receptor activation on GABAergic neurons does not allow for inhibition of 5-HT_{2A}-containing glutamatergic neurons. This leads to overstimulation of dopaminergic neurons in the VTA. This then causes excess dopamine release in the NAc, which subsequently causes hyperactivity in the dorsal striatum and the thalamus. This can lead to false detection of auditory/visual stimuli.

Comprehension Check 6 Explain how serotonin, glutamate, and dopamine interact to influence hallucinations.

As glutamatergic neurons are found throughout the cerebral cortex, a hyperactive serotonergic system (specifically, upregulation of 5-HT_{2A} receptors) can lead to overexcitation of glutamatergic neurons. Glutamatergic neurons from the PFC depolarize dopaminergic neurons in the VTA, which then causes excess dopamine release in the ventral striatum. Glutamatergic neurons also project directly to the ventral striatum, which can further augment dopaminergic activity. Glutamatergic transmission between the PFC and other cortical areas like the primary visual cortex occurs.

8 Chapter Summary

This chapter discusses the neuromechanisms underlying perception. Specifically, this chapter covers the neuroanatomical pathways involved in various sensory systems and discusses how damage to parts of each pathway can lead to perceptual disturbances like cortical blindness and various agnosias. Additional brain regions involved in perception are highlighted, including the thalamus and association areas. The neurochemical substrates of perception are discussed, focusing on the balance between excitatory and inhibitory neurotransmitters. This chapter concludes by

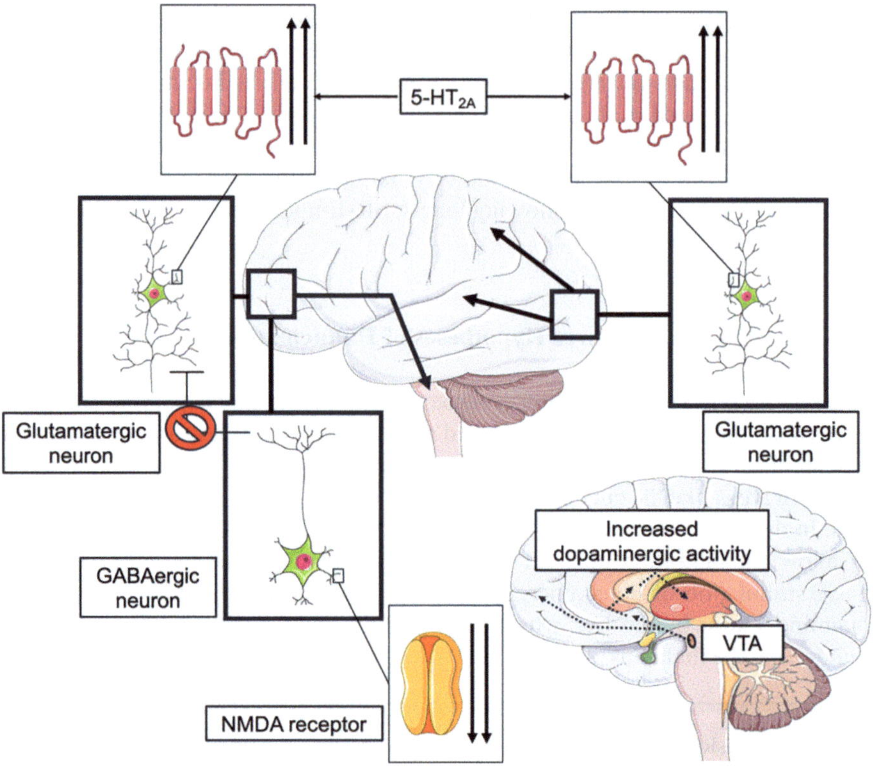

Fig. 9 Schematic showing how dysregulation of serotonin, glutamate, and dopamine can give rise to visual and/or auditory hallucinations. See text for details. Image created by the author using Servier Medical Art (https://smart.servier.com/; CC BY 4.0, https://creativecommons.org/licenses/by/4.0/) and Microsoft PowerPoint

exploring the neurotransmitters implicated in hallucinations, presenting hypotheses involving serotonin, glutamate, and dopamine, as well as how interactions between these neurotransmitter systems may contribute to perceptual distortions.

Answers to Experiment Spotlight Questions

1. The researchers did not want to include individuals with a clinical diagnosis of auditory verbal hallucinations because they wanted to isolate hallucination proneness from potential confounding variables related to a clinical condition such as medication usage or other treatments.
2. One possible explanation is that in non-clinical populations, larger brain volumes in language-related areas might reflect a form of neural plasticity or enhanced connectivity that predisposes individuals to hallucinations without leading to a clinical condition. In contrast, in clinical populations, the reduction in brain volume might be a result of neurodegenerative processes or the impact of the disorder itself.

Another explanation could be that the increased brain volume in non-clinical individuals represents a compensatory mechanism that allows them to experience hallucinations without significant functional impairment, whereas in clinical populations, this mechanism might be overwhelmed or dysfunctional.

3. The findings suggest that brain structure plays a complex role in hallucination proneness, which could influence future research to explore the protective or adaptive aspects of increased brain volume in non-clinical populations. Clinically, this might lead to new approaches in understanding and treating hallucinations by focusing on enhancing or preserving brain volume in specific areas. Future research could investigate whether similar patterns are observed in other non-clinical populations, the longitudinal changes in brain volume related to hallucination proneness, and the functional connectivity between these brain regions. Additionally, studies could explore the genetic and environmental factors that contribute to these structural differences and their impact on hallucination experiences.

Glossary

Absolute threshold minimum intensity of a stimulus needed for detection 50% of the time.

Affinity the strength of the attraction between a drug molecule and its target receptor or binding site within the body.

Agnosia a neurological disorder characterized by the inability to recognize familiar objects, persons, sounds, shapes, or smells while the specific sense is not defective.

Akinetopsia a neurological condition in which individuals cannot perceive the motion of an object or another person; colloquially referred to as "motion blindness".

Amusia a neurological disorder characterized by the inability to recognize, process, or produce music; colloquially referred to as "tone deafness".

Association area region of the cerebral cortex involved in higher-order brain functions such as learning, memory, thinking, and understanding.

Asterognosis the inability to recognize objects by touch alone, despite intact sensory function; also known as tactile agnosia.

Blindsight a phenomenon in which an individual can respond to a visual stimulus despite not being able to see it.

Difference threshold the minimum amount by which a stimulus must be changed in order for a difference to be detectable a certain proportion of time (typically proportion of 0.5). Also known as a just-noticeable difference.

Discrimination the ability of an animal to distinguish two or more stimuli from one another.

Dissociative hallucinogen a class of psychoactive substances that disrupt the integration of sensory and cognitive information, leading to feelings of detachment from oneself and the environment, hallucinations, and altered states of consciousness.
Hallucination perception of a stimulus that is not present.
Perception process of interpreting sensory information to form a mental representation of the world.
Prosopagnosia a neurological condition that impairs the ability to recognize familiar faces, including one's own; colloquially referred to as "face blindness".
Psychedelic hallucinogen a class of psychoactive substances that alter perception, mood, and cognitive processes, producing profound changes in sensory experiences and consciousness.
Psychophysics a branch of psychology that focuses on studying the relationship between physical stimuli and mental processes.
Psychostimulant a class of drugs that increase alertness, attention, and energy by stimulating the central nervous system, often leading to increased heart rate, blood pressure, and a sense of euphoria.
Scaling the process of quantifying the relationship between physical stimuli and their perceived psychological magnitudes.
Schizophrenia mental disorder characterized by perceptual distortions, emotional disturbances, disorganized thoughts, and impaired social functioning.
Signal detection theory framework used to measure how well individuals can differentiate between relevant stimuli from extraneous stimuli ("noise").
Terminal threshold the maximum intensity of a stimulus beyond which no further increase in sensation is perceived.
Vertigo sensation of spinning or swaying, even when stationary.

References

Ajina, S., & Bridge, H. (2016). Blindsight and unconscious vision: What they teach us about the human visual system. *The Neuroscientist, 23*(5), 529–541. https://doi.org/10.1177/1073858416673817

Ajina, S., & Bridge, H. (2019). Subcortical pathways to extrastriate visual cortex underlie residual vision following bilateral damage to V1. *Neuropsychologia, 128*, 140–149. https://doi.org/10.1016/j.neuropsychologia.2018.01.007

Alotaibi, A. M., Alosaimi, M. H., Alshammari, N. S., Orfali, R. S., Alwatban, A. Z., Alsharif, R. A., Meyer, G. F., & Bentall, R. P. (2024). Exploring the relationship between hallucination proneness and brain morphology. *Neuro Image, 304*, 120942. https://doi.org/10.1016/j.neuroimage.2024.120942

Anzai, A., Peng, X., & Van Essen, D. C. (2007). Neurons in monkey visual area V2 encode combinations of orientations. *Nature Neuroscience, 10*(10), 1313–1321. https://doi.org/10.1038/nn1975

Ballanger, B., Strafella, A. P., van Eimeren, T., Zurowski, M., Rusjan, P. M., Houle, S., & Fox, S. H. (2010). Serotonin 2A receptors and visual hallucinations in Parkinson disease. *Archives of Neurology, 67*(4), 416–421. https://doi.org/10.1001/archneurol.2010.35

Baugh, L. A., Desanghere, L., & Marotta, J. J. (2017). Agnosia. *Reference Module in Neuroscience and Biobehavioral Psychology*. doi:https://doi.org/10.1016/B978-0-12-809324-5.00253-4

Bérard, N., Landis, B. N., Legrand, L., Tyrand, R., Grouiller, F., Vulliémoz, S., Momjian, S., & Boëx, C. (2021). Electrical stimulation of the medial orbitofrontal cortex in humans elicits pleasant olfactory perceptions. *Epilepsy & Behavior, 114*(Pt A), 107559. https://doi.org/10.1016/j.yebeh.2020.107559

Binder, J. R. (2015). The Wernicke area: Modern evidence and a reinterpretation. *Neurology, 85*(24), 2170–2175. https://doi.org/10.1212/WNL.0000000000002219

Bohlhalter, S., Fretz, C., & Weder, B. (2002). Hierarchical versus parallel processing in tactile object recognition: A behavioural-neuroanatomical study of aperceptive tactile agnosia. *Brain, 125*(Pt 11), 2537–2548. https://doi.org/10.103/brain/awf245

Boot, E., Levy, A., Gaeta, G., Gunasekara, N., Parkkinen, E., Kontaris, E., Jacquot, M., & Tachtsidis, I. (2024). fNIRS a novel neuroimaging tool to investigate olfaction, olfactory imagery, and crossmodal interactions: A systematic review. *Frontiers in Neuroscience, 18*, 1266664. https://doi.org/10.3389/fnins.2024.1266664

Bosch-Bouju, C., Hyland, B. I., & Parr-Brownlie, L. C. (2013). Motor thalamus integration of cortical, cerebellar and basal ganglia information: Implications for normal and parkinsonian conditions. *Frontiers in Computational Neuroscience, 7*, 163. https://doi.org/10.3389/fncom.2013.00163

Capaldi, V. F., & Carr, R. B. (2010). Citalopram-induced hallucinations and delusions in a young adult. *General Hospital Psychiatry, 32*(6), 648.e1–648.e6483. https://doi.org/10.1016/j.genhosppsych.2010.07.008

Charnay, Y., & Léger, L. (2010). Brain serotonergic circuitries. *Dialogues in Clinical Neuroscience, 12*(4), 471–487. https://doi.org/10.31887/DCNS.2010.12.4/ycharnay

Chechlacz, M., Rotshtein, P., Hansen, P. C., Riddoch, J. M., Deb, S., & Humphreys, G. W. (2012). The neural underpinings of simultanagnosia: Disconnecting the visuospatial attention network. *Journal of Cognitive Neuroscience, 24*(3), 718–735. https://doi.org/10.1162/jocn_a_00159

Cohen, S. M., Tsien, R. W., Goff, D. C., & Halassa, M. M. (2015). The impact of NMDA receptor hypofunction on GABAergic neurons in the pathophysiology of schizophrenia. *Schizophrenia Research, 167*(1–3), 98–107. https://doi.org/10.1016/j.schres.2014.12.026

Corrow, S. L., Dalrymple, K. A., & Barton, J. J. (2016). Prosopagnoia: Current perspectives. *Eye and Brain, 8*, 165–175. https://doi.org/10.2147/EB.S92838

Curran, C., Byrappa, N., & McBride, A. (2004). Stimulant psychosis: Systematic review. *The British Journal of Psychiatry, 185*, 196–204. https://doi.org/10.1192/bjp.185.3.196

da Silva Alves, F., Figee, M., van Amelsvoort, T., Veltman, D., & de Haan, L. (2008). The revised dopamine hypothesis of schizophrenia: Evidence from pharmacological MRI studies with atypical antipsychotic medication. *Psychopharmacology Bulletin, 41*(1), 121–132.

Dalrymple, K. A., Barton, J. J., & Kingstone, A. (2013). A world unglued: Simultanagnosia as a spatial restriction of attention. *Frontiers in Human Neuroscience, 7*, 145. https://doi.org/10.3389/fnhum.2013.00145

Denaro, S., D'Aprile, S., Alberghina, C., Pavone, A. M., Torrisi, F., Giallongo, S., Longhitano, L., Mannino, G., Lo Furno, D., Zappalà, A., Giuffrida, R., Tibullo, D., Li Volti, G., Vicario, N., & Parenti, R. (2022). Neurotrophic and immunomodulatory effects of olfactory ensheathing cells as a strategy for neuroprotection and regeneration. *Frontiers in Immunology, 13*, 1098212. https://doi.org/10.3389/fimmu.2022.1098212

Deolindo, C. S., Kunicki, A. C. B., da Silva, M. I., Lima Brasil, F., & Moioli, R. C. (2018). Neuronal assemblies evidence distributed interactions within a tactile discrimination task in rats. *Frontiers in Neural Circuits, 11*, 114. https://doi.org/10.3389/fncir.2017.00114

Facci, L., Basilico, S., Sellitto, M., Gelosa, G., Gandola, M., & Bottini, G. (2024). Unilateral tactile agnosia as an onset symptom of corticobasal syndrome. *Frontiers in Human Neuroscience, 18*, 1401578. https://doi.org/10.3389/fnhum.2024.1401578

Flinker, A., Korzeniewska, A., Shestyuk, A. Y., Franaszczuk, P. J., Dronkers, N. F., Knight, R. T., & Crone, N. E. (2015). Redefining the role of Broca's area in speech. *Proceedings of the National*

Academy of Sciences of the United States of America, 112(9), 2871–2875. https://doi.org/10.1073/pnas.1414491112

Garrido-Palomino, I., Fryer, S., Giles, D., González-Rosa, J. J., & España-Romero, V. (2020). Attentional differences as a function of rock climbing performance. *Frontiers in Psychology, 11*, 1550. https://doi.org/10.3389/fpsyg.2020.01550

Gent, T. C., Bassetti, C., & Adamantidis, A. R. (2018). Sleep-wake control and the thalamus. *Current Opinion in Neurobiology, 52*, 188–197. https://doi.org/10.1016/j.conb.2018.08.002

Georgescu, I. A., Popa, D., & Zagrean, L. (2020). The anatomical and functional heterogeneity of the mediodorsal thalamus. *Brain Sciences, 10*(9), 624. https://doi.org/10.3390/brainsci10090624

Gilboa, A., Winocur, G., Grady, C. L., Hevenor, S. J., & Moscovitch, M. (2004). Remembering our past: Functional neuroanatomy of recollection of recent and very remote personal events. *Cerebral Cortex, 14*(11), 1214–1225. https://doi.org/10.1093/cercor/bhh082

Grossman, E. J., & Inglese, M. (2016). The role of thalamic damage in mild traumatic brain injury. *Journal of Neurotrauma, 33*(2), 163–167. https://doi.org/10.1089/neu.2015.3965

Hannum, M. E., Koch, R. J., Ramirez, V. A., Marks, S. S., Toskala, A. K., Herriman, R. D., Lin, C., Joseph, P. V., & Reed, D. R. (2023). Taste loss as a distinct symptom of COVID-19: A systematic review and meta-analysis. *Chemical Senses, 48*, bjad043. https://doi.org/10.1093/chemse/bjad043

Harpaz, Y., Levkovitz, Y., & Lavidor, M. (2009). Lexical ambiguity resolution in Wernicke's area and its right homologue. *Cortex, 45*(9), 1097–1103. https://doi.org/10.1016/j.cortex.2009.01.002

Hatzfeld, C., Kühner, M., Söllner, S., Khanh, T. Q., & Kupnik, M. (2017). Human perception measures for product design and development–A tutorial to measurement methods and analysis. *Multimodal Technologies and Interaction, 1*(4), 28. https://doi.org/10.3390/mti1040028

Hegdé, J., & Van Essen, D. C. (2000). Selectivity for complex shapes in primate visual area V2. *The Journal of Neuroscience, 20*(5), RC61. https://doi.org/10.1523/JNEUROSCI.20-05-j0001.2000

Hubel, D. H., & Livingstone, M. S. (1987). Segregation of form, color, and stereopsis in primate area 18. *The Journal of Neuroscience, 7*(11), 3378–3415. https://doi.org/10.1523/JNEUROSCI.07-11-03378.1987

Iversen, S. D., & Iversen, L. L. (2007). Dopamine: 50 years in perspective. *Trends in Neurosciences, 30*(5), 188–193. https://doi.org/10.1016/j.tins.2007.03.002

Jacob, S. N., & Nienborg, H. (2018). Monoaminergic neuromodulation of sensory processing. *Frontiers in Neural Circuits, 12*, 51. https://doi.org/10.3389/fncir.2018.00051

Jiang, L., Zheng, Z., & He, Y. (2021). Progress in protecting vestibular hair cells. *Archives of Toxicology, 95*(8), 2613–2623. https://doi.org/10.1007/s00204-021-03067-3

Kaliyappan, K., Chen, Y. C., & Krishnan Muthaiah, V. P. (2022). Vestibular cochlear manifestations in COVID-19 cases. *Frontiers in Neurology, 13*, 850337. https://doi.org/10.3389/fneur.2022.850337

Kometer, M., Schmidt, A., Jäncke, L., & Vollenweider, F. X. (2013). Activation of serotonin 2A receptors underlies the psilocybin-induced effects on α oscillations, N170 visual-evoked potentials, and visual hallucinations. *The Journal of Neuroscience, 33*(25), 10544–10551. https://doi.org/10.1523/JNEUROSCI.3007-12.2013

Kravitz, D. J., Saleem, K. S., Baker, C. I., Ungerleider, L. G., & Mishkin, M. (2013). The ventral visual pathway: An expanded neural framework for the processing of object quality. *Trends in Cognitive Sciences, 17*(1), 26–49. https://doi.org/10.1016/j.tics.2012.10.011

Kumar, G., Juhász, C., Sood, S., & Asano, E. (2012). Olfactory hallucinations elicited by electrical stimulation via subdural electrodes: Effects of direct stimulation of olfactory bulb and tract. *Epilepsy & Behavior, 24*(2), 264–268. https://doi.org/10.1016/j.yebeh.2012.03.027

Lech, R. K., & Suchan, B. (2014). Involvement of the human medial temporal lobe in a visual discrimination task. *Behavioural Brain Research, 268*, 22–30. https://doi.org/10.1016/j.bbr.2014.03.030

Lehrner, J., Baumgartner, C., Serles, W., Olbrich, A., Pataraia, E., Bacher, J., Aull, S., & Deecke, L. (1997). Olfactory prodromal symptoms and unilateral olfactory dysfunction are associated in patients with right mesial temporal lobe epilepsy. *Epilepsia, 38*(9), 1042–1044. https://doi.org/10.1111/j.1528-1157.1997.tb01490.x

Levy, L. M., & Henkin, R. I. (2004). Brain gamma-aminobutyric acid levels are decreased in patients with phantageusia and phantosmia demonstrated by magnetic resonance spectroscopy. *Journal of Computer Assisted Tomography, 28*(6), 721–727. https://doi.org/10.1097/00004728-200411000-00001

Li, Y., Qi, L., Schaper, F. L. W. V. J., Wu, D., Friedrich, M., Du, J., Yu, T., Wang, Q., Wang, X., Wang, D., Jin, G., Liu, A., Fan, C., Wang, Y., Fox, M. D., & Ren, L. (2023). A vertigo network derived from human brain lesions and brain stimulation. *Brain. Communications, 5*(2), fcad071. https://doi.org/10.1093/braincomms/fcad071

Lim, A., Hoek, H. W., Deen, M. L., Blom, J. D., & GROUP Investigators. (2016). Prevalence and classification of hallucinations in multiple sensory modalities in schizophrenia spectrum disorders. *Schizophrenia Research, 176*(2–3), 493–499. https://doi.org/10.1016/j.schres.2016.06.010

Linscott, R. J., & van Os, J. (2013). An updated and conservative systematic review and meta-analysis of epidemiological evidence on psychotic experiences in children and adults: On the pathway from proneness to persistence to dimensional expression across mental disorders. *Psychological Medicine, 43*(6), 1133–1149. https://doi.org/10.1017/S0033291712001626

Linszen, M. M. J., de Boer, J. N., Schutte, M. J. L., Begemann, M. J. H., de Vries, J., Koops, S., Blom, R. E., Bohlken, M. M., Heringa, S. M., Blom, J. D., & Sommer, I. E. C. (2022). Occurrence and phenomenology of hallucinations in the general population: A large online survey. *Schizophrenia, 8*(1), 41. https://doi.org/10.1038/s41537-022-00229-9

McCarty, D. E., & Chesson, A. L., Jr. (2009). A case of sleep paralysis with hypnopompic hallucinations. Recurrent isolated sleep paralysis associated with hypnopompic hallucinations, precipitated by behaviorally induced insufficient sleep syndrome. *Journal of Clinical Sleep Medicine, 5*(1), 83–84.

McKenna, M. C. (2007). The glutamate-glutamine cycle is not stoichiometric: Fates of glutamate in brain. *Journal of Neuroscience Research, 85*(15), 3347–3358. https://doi.org/10.1002/jnr.21444

Mengod, G., Cortés, R., Vilaró, M. T., & Hoyer, D. (2010). Distribution of 5-HT receptors in the central nervous system. In C. Müller & B. Jacobs (Eds.), *Handbook of behavioral neurobiology of serotonin* (pp. 123–138). Elsevier. https://doi.org/10.1016/B978-0-12-374634-4.00006-X

Miceli, G., & Caccia, A. (2023). The auditory agnosias: A short review of neurofunctional evidence. *Current Neurology and Neuroscience Reports, 23*(11), 671–679. https://doi.org/10.1007/s11910-023-01302-1

Michael, G. A., Salgues, S., Plancher, G., & Duran, G. (2022). Cues to body related distortions and hallucinations? Spontaneous sensations correlate with EEG oscillatory activity recorded at rest in the somatosensory cortices. *Psychiatry Research. Neuroimaging, 324*, 111506. https://doi.org/10.1016/j.pscychresns.2022.111506

Mitchell, C. A. A., Maybery, M. T., Russell-Smith, S. N., Collerton, D., Gignac, G. E., & Waters, F. (2017). The structure and measurement of unusual sensory experiences in different modalities: The Multi-Modality Unusual Sensory Experiences Questionnaire (MUSEQ). *Frontiers in Psychology, 8*, 1363. https://doi.org/10.3389/fpsyg.2017.01363

Mueser, K. T., Bellack, A. S., & Brady, E. U. (1990). Hallucinations in schizophrenia. *Acta Psychiatrica Scandinavica, 82*(1), 26–29. https://doi.org/10.1111/j.1600-0447.1990.tb01350.x

Phillips, I. (2021). Blindsight is qualitatively degraded conscious vision. *Psychological Review, 128*(3), 558–584. https://doi.org/10.1037/rev0000254

Rauschecker, J. P. (2012). Ventral and dorsal streams in the evolution of speech and language. *Frontiers in Evolutionary Neuroscience, 4*, 7. https://doi.org/10.3389/fnevo.2012.00007

Roe, A. W., & Ts'o, D. Y. (1999). Specificity of color connectivity between primate V1 and V2. *Journal of Neurophysiology, 82*(5), 2719–2730. https://doi.org/10.1152/jn.1999.82.5.2719

Sahara, S., Yanagawa, Y., O'Leary, D. D., & Stevens, C. F. (2012). The fraction of cortical GABAergic neurons is constant from near the start of cortical neurogenesis to adulthood. *The Journal of Neuroscience, 32*(14), 4755–4761. https://doi.org/10.1523/JNEUROSCI.6412-11.2012

Saionz, E. L., Busza, A., & Huxlin, K. R. (2022). Rehabilitation of visual perception in cortical blindness. *Handbook of Clinical Neurology, 184*, 357–373. https://doi.org/10.1016/B978-0-12-819410-2.00030-8

Schmidt-Wilcke, T., Fuchs, E., Funke, K., Vlachos, A., Müller-Dahlhaus, F., Puts, N. A. J., Harris, R. E., & Edden, R. A. E. (2018). GABA-From inhibition to cognition: Emerging concepts. *The Neuroscientist, 24*(5), 501–515. https://doi.org/10.1177/1073858417734530

Silverstein, S. M., & Lai, A. (2021). The phenomenology and neurobiology of visual distortions and hallucinations in schizophrenia: An update. *Frontiers in Psychiatry, 12*, 684720. https://doi.org/10.3389/fpsyt.2021.684720

Slevc, L. R., & Shell, A. R. (2015). Auditory agnosia. *Handbook of Clinical Neurology, 129*, 573–587. https://doi.org/10.1016/B978-0-444-62630-1.00032-9

Small, D. M., Bernasconi, N., Bernasconi, A., Sziklas, V., & Jones-Gotman, M. (2005). Gustatory agnosia. *Neurology, 64*(2), 311–317. https://doi.org/10.1212/01.WNL.0000149515.77718.35

Srisurapanont, M., Ali, R., Marsden, J., Sunga, A., Wada, K., & Monteiro, M. (2003). Psychotic symptoms in methamphetamine psychotic in-patients. *The International Journal of Neuropsychopharmacology, 6*(4), 347–352. https://doi.org/10.1017/S1461145703003675

Tao, X., Zhang, B., Smith, E. L., 3rd, Nishimoto, S., Ohzawa, I., & Chino, Y. M. (2012). Local sensitivity to stimulus orientation and spatial frequency within the receptive fields of neurons in visual area 2 of macaque monkeys. *Journal of Neurophysiology, 107*(4), 1094–1110. https://doi.org/10.1152/jn.00640.2011

Tripp, A. C., & Golden, S. A. (2009). Transient complex visual hallucinations with venlafaxine treatment: A case report. *Primary Care Companion to the Journal of Clinical Psychiatry, 11*(2), 83. https://doi.org/10.4088/pcc.08l00647

Van Swol, J. M., Thompson, E. B., Joffe, J. A., Nguyen, S. A., & Berman, E. L. (2024). Akinetopsia: A systematic review. *Journal of Neuro-Ophthalmology, 44*(3), e483–e488. https://doi.org/10.1097/WNO.0000000000002032

Waters, F., & Fernyhough, C. (2017). Hallucinations: A systematic review of points of similarity and difference across diagnostic classes. *Schizophrenia Bulletin, 43*(1), 32–43. https://doi.org/10.1093/schbul/sbw132

Waters, F., Collerton, D., Ffytche, D. H., Jardri, R., Pins, D., Dudley, R., Blom, J. D., Mosimann, U. P., Eperjesi, F., Ford, S., & Larøi, F. (2014). Visual hallucinations in the psychosis spectrum and comparative information from neurodegenerative disorders and eye disease. *Schizophrenia Bulletin, 40 Suppl 4*(Suppl 4), S233–S245. doi:https://doi.org/10.1093/schbul/sbu036

Whyte, C. J., Redinbaugh, M. J., Shine, J. M., & Saalmann, Y. B. (2024). Thalamic contributions to the state and contents of consciousness. *Neuron, 112*(10), 1611–1625. https://doi.org/10.1016/j.neuron.2024.04.019

Zmigrod, L., Garrison, J. R., Carr, J., & Simons, J. S. (2016). The neural mechanisms of hallucinations: A quantitative meta-analysis of neuroimaging studies. *Neuroscience and Biobehavioral Reviews, 69*, 113–123. https://doi.org/10.1016/j.neubiorev.2016.05.037

Chapter 4
Aberrant Pain Perception

1 Introduction

A burning sensation radiated through both shoulders and up the back of my neck. It was the middle of the night, and lying flat on my back was excruciating. I felt as if I had two torn rotator cuffs. Lying on either side did not alleviate the pain. Eventually, I went downstairs and sat upright on my couch, fighting back tears as the pain continued to intensify. This was not the first time I had experienced a burning sensation in my shoulders; however, this was the worst occurrence. For over 20 years, I have suffered from this burning sensation in my shoulder, although it is primarily isolated to my left shoulder. I cannot isolate what triggers this pain as it seemingly occurs at random. Months or years can pass between each episode, but when it appears, it can be debilitating.

I first experienced this pain when I was a junior in high school. I ended up sitting upright in the living room recliner, frustrated I could not fall asleep. As the pain worsened, I had to wake my mom up. Unsure of what was happening to me, my parents took me to the emergency room. With an IV in my wrist, I could feel something flow into my vein. As this substance made its way through my circulatory system, I felt a warm sensation, almost unpleasantly warm. Within 30 s, I felt relaxed, and the pain started to dampen. I became giddy, laughing at something my mom said to my dad while they sat by my bed.

When discussing the major sensory systems, I intentionally did not discuss sensation for pain. Specifically, this chapter will focus on **nociception**, the sensory system involved in encoding *noxious* stimuli, including pain. As annoying as pain is, it serves a vital role in our survival. The searing pain you experience when you burn yourself is your body's way of telling you to move away from the stimulus that caused the pain; failure to do so can lead to permanent damage or worse. Unfortunately, pain perception can become dysregulated. Some individuals experience chronic pain with no known cause while others cannot perceive pain.

J. Yates, *Neuroscience of Maladaptation*, https://doi.org/10.1007/978-3-032-15602-0_4

2 Learning Objectives

By the end of this chapter, you should be able to …

1. Define nociception and differentiate it from other forms of somatosensation.
2. Explain the peripheral and the central mechanisms that control pain perception.
3. Describe the neurobiological mechanisms involved in attenuating pain perception.
4. Explain the potential reasons why conditions such as chronic pain and insensitivity to pain occur.

3 Nociception

Although nociception is a form of somatosensation, the neural mechanisms controlling pain perception are quite distinct from detecting something like someone brushing against our skin or feeling pressure on a limb.

3.1 Peripheral Mechanisms of Pain

Our peripheral nervous system (PNS) contains specialized axons that terminate in our dermis (a middle layer of skin) called *free nerve endings* that contain *nociceptors* (Fig. 1). Some free nerve endings are thin and contain no myelination (called C fibers). Other axons are much larger and contain myelination (called Aδ fibers) (Markenson, 1996). Recall from Chap. 2 that myelination allows action potentials to quickly propagate down the length of the axon. When you receive a shot at the doctor's office, you may notice that the initial penetration of the needle results in a sharp, stinging sensation. This is then followed by a mild "achy" sensation. The immediate pain you experience results from sensory information traveling through Aδ fibers while the mild, longer-lasting pain is due to slower transmission of sensory information through C fibers. When free nerve endings are activated, nerves transmit pain information toward the central nervous system (CNS), primarily using the neurotransmitter glutamate for interneuronal communication.

When we experience mechanical damage to our body (e.g., skin tear or bone break), a type of Na^+ ion channel located on the free nerve ending is activated (Persson et al., 2010). This causes an action potential to occur, transmitting information to the spinal cord. Before discussing what happens in the spinal cord, I want to explain what else happens at the site of an injury.

First, **prostaglandins**, hormone-like *lipids*, are produced, effectively signaling that something is wrong. Lipids are compounds that are insoluble in water (think fats and oils). Hormones are similar to neurotransmitters, except that they travel through the bloodstream as opposed to between neurons. Prostaglandins are largely

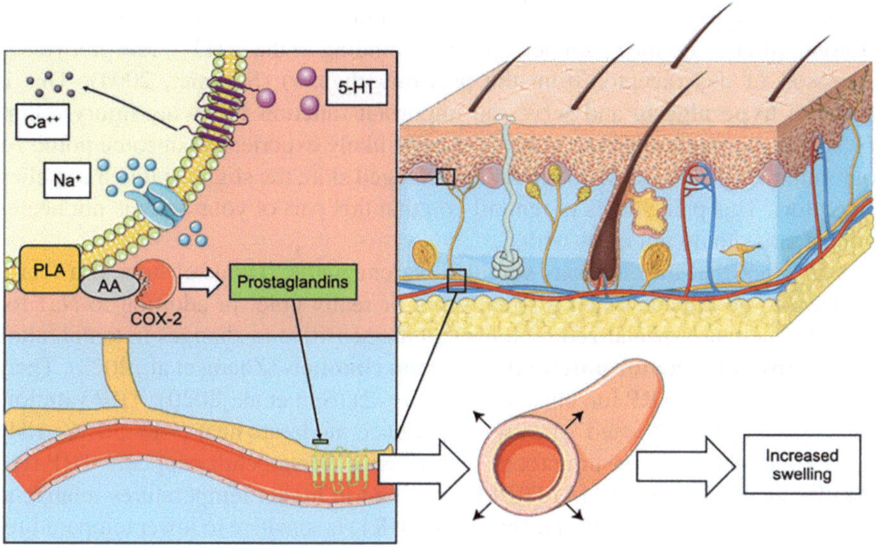

Fig. 1 Peripheral mechanisms controlling pain. Phospholipase A (PLA), which is located in the phospholipid layer of the cell membrane, first activates arachidonic acid (AA). The enzyme cyclooxygenase-2 (COX-2) converts AA into prostaglandins. Prostaglandins bind to prostaglandin receptors, which are metabotropic. In the illustration, the prostaglandin receptor is located on a blood vessel. Activation of this receptor causes the blood vessel to dilate. This causes swelling that is often associated with mechanical damage to the skin. Sodium (Na^+) enters Na^+ ion channels. This depolarizes the neuron, making it more likely to send a message to neighboring neurons. This process allows information regarding pain to travel to the brain. Additionally, serotonin (5-HT) can bind to excitatory 5-HT receptors, like the 5-HT_{2A}. Because this receptor is coupled to a G_q protein, this causes calcium (Ca^{++}) levels to increase inside the neuron, further depolarizing it. This process helps increase neuronal transmission, thus amplifying the pain signal. Image created by the author using Servier Medical Art (https://smart.servier.com/; CC BY 4.0, https://creativecommons.org/licenses/by/4.0/) and Microsoft PowerPoint

responsible for inflammation we experience following an injury. They also help kickstart a fever to help kill any viruses and/or bacteria that have infiltrated our body (Ito et al., 2001).

Prostaglandins involved with inflammation are produced from something called *arachidonic acid* by the enzyme **cyclooxygenase-2** (**COX-2**). Arachidonic acid is a polyunsaturated omega-6 fatty acid. You may be familiar with the term omega fatty acid, as this form of fat is abundant in certain types of food. Omega-6 fatty acids are found in nuts, seeds, vegetable oils, eggs, chicken, and grains. The term arachidonic is derived from the Latin word *arachis*, meaning "peanut". Ironically, peanuts do not contain much arachidonic acid; however, it received its name due to its chemical structure being similar to arachidic acid, which is found in peanuts. Arachidonic acid is found in the *phospholipids* of cell membranes. These phospholipids create a lipid bilayer, which helps provide mechanical strength to a cell. This protects the cell from rupturing. When an injury occurs, arachidonic acid is activated by phospholipase A.

Like prostaglandins, serotonin acts as a pain modulator. At the site of the injury, serotonin often magnifies pain perception by binding to the 5-HT_{2A} receptor (recall discussion of this receptor from the previous chapter) (Sommer, 2004). This is known as **hyperalgesia** and serves an important function following injury. Think of a sunburn, something most of us have most likely experienced at some point. As your immune system works to repair the damaged skin, the slightest touch can elicit discomfort. This pain serves to remind you that this part of your body is not healed and that you should not mess with it.

Pain is not the only noxious stimulus we encounter. Think of how you feel if you have to go outside when it's really hot or really cold. In addition to Na^+ ion channels, there are specialized proteins that are sensitive to changes in temperature called **transient receptor potential (TRP) ion channels** (Zhang et al., 2023). There are several types of TRP ion channels (see Fig. 2) (Sun et al., 2020). TRP vanilloid type 1 (TRPV1) is activated when we experience moderate heat (~40–50 °C; 104–122 °F) and when we consume foods with capsaicin (e.g., chili peppers). TRP type 2 (TRPV2) and TRP type M3 (TRPM3) respond to higher temperatures relative to TRPV1 while the cool-menthol receptor 1 (CMR1) is sensitive to lower temperatures (< 30 °C; 86 °F). CMR1, as the name suggests, detects the presence of menthol (think of peppermint).

Notice that TRP ion channels respond to certain chemicals. Chemicals are another type of noxious stimulus. This is why capsaicin is used in pepper spray as a form of personal protection and as a crowd deterrent. Mustard oil and wasabi activate TRP, subfamily A, member 1 (TRPA1). If either mustard oil or wasabi gets into one's eyes, the individual will experience a burning sensation because of activation of this receptor.

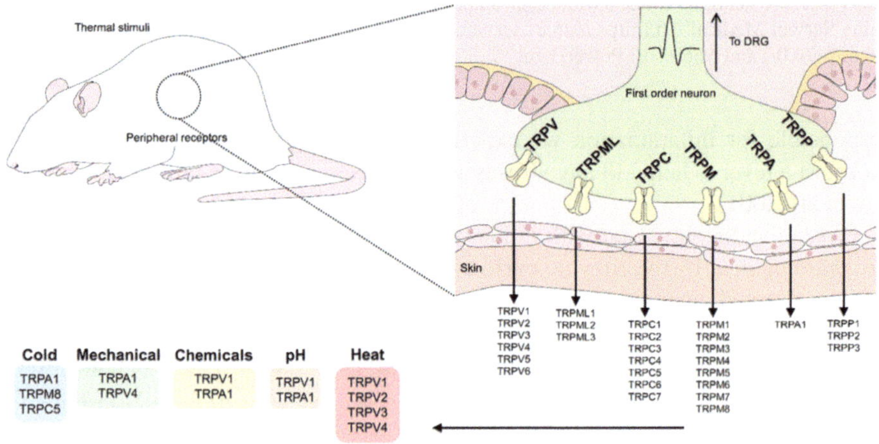

Fig. 2 A representation of the families and subfamilies of TRP ion channels. *Note* DRG = dorsal root ganglion. *Source* Figure 1 of Lezama-García et al. (2022), https://doi.org/10.3390/ani12010106. CC BY 4.0 (https://creativecommons.org/licenses/by/4.0/), via MDPI

Comprehension Check 1 Describe the role of prostaglandins in the peripheral mechanisms of pain. How can prostaglandins be blocked to limit pain transmission?

Prostaglandins are hormone-like lipids that are produced at the site of tissue damage. They contribute to the inflammatory response and can also initiate a fever to help the body fight off infection. Prostaglandins are produced from arachidonic acid by the enzyme cyclooxygenase-2 (COX-2). As you will read about below, there are medications that block COX-2. This prevents the conversion of arachidonic acid to prostaglandins.

3.2 Central Mechanisms of Pain

Figure 3 shows the pain pathways of the CNS. Neurons in the spinal cord release the neurotransmitter glutamate, as well as *substance P* (Humes et al., 2024). Substance P is a **neuropeptide**, not a neurotransmitter (the P stands for peptide) (see inset of Fig. 2). Discussing the difference between neurotransmitter and neuropeptides is not important at this moment. Just know that glutamate is the primary neurotransmitter released when we experience pain. Substance P is released when we experience intense pain or chronic pain. Think of substance P as an amplifier, similar to how serotonin can modulate pain in the PNS. If you stub your toe, glutamatergic activity will increase, allowing for this pain signal to reach your brain. If you break your foot, both glutamate and substance P will be released, magnifying the pain you experience.

As in the PNS, increasing the intensity of the pain signal in the spinal cord is important for preventing us from causing more harm to ourselves. If you stub your toe, you can still walk around even if you experience slight discomfort. Walking on a broken foot can lead to additional damage to the bones in your foot. As such, attempting to put any weight on your broken foot will cause more glutamate and substance P to be released in your spinal cord. This is your body's way of telling you to stay off your foot.

Whereas axons transmitting information about somatosensation do not cross to the contralateral side of the body until reaching the medulla, axons transmitting information about nociception immediately cross to the right side of the spinal cord and vice versa. Eventually, axons travel to the ventral posterolateral and the ventral posteromedial nuclei of the thalamus. Neurons from these thalamic nuclei project to different parts of the cerebral cortex.

First, the primary somatosensory cortex processes information such as pain intensity (i.e., how much something hurts), pain localization (i.e., where it hurts), and the type of pain experienced (e.g., burning, stabbing, etc.) (Vierck et al., 2013). The secondary somatosensory cortex helps integrate information about the painful stimulus with the other senses (Bretas et al., 2020). For example, if you place your hand on a hot stovetop, you will experience a burning sensation; you may also see that the burner is red. Your secondary somatosensory cortex pieces this information together.

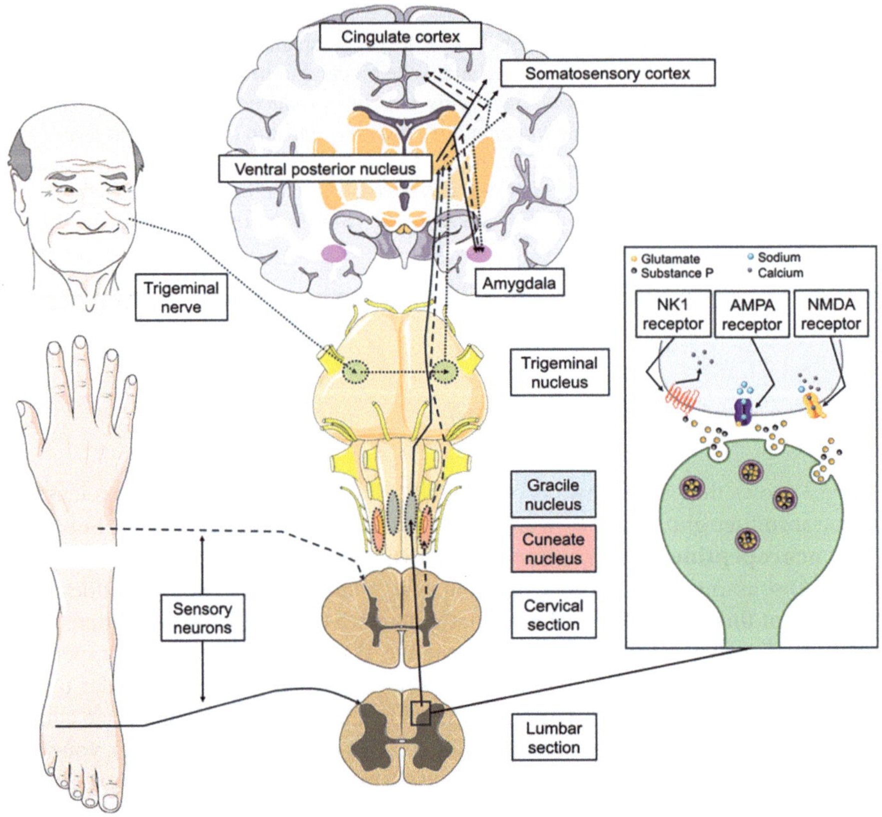

Fig. 3 A schematic showing how pain is transmitted from the peripheral nervous system (PNS) to the central nervous system (CNS). Glutamate is the primary neurotransmitter released in the spinal cord (see inset); however, substance P can be released to signal more intense pain. Substance P binds to a G protein-coupled receptor called the neurokinin receptor 1 (NK1). Activation of this receptor activates intra-signaling pathways. One result is further increased calcium (Ca^{++}) levels. Image created by the author using Servier Medical Art (https://smart.servier.com/; CC BY 4.0, https://creativecommons.org/licenses/by/4.0/) and Microsoft PowerPoint

This is important because you will learn to be careful around a stove in which one of the burners is red.

The secondary somatosensory cortex also integrates information about pain with the emotional response to pain; the insula also helps integrate sensory information about pain with its emotional aspects (Orenius et al., 2017). Experiencing physical pain elicits emotional responses. This is advantageous as experiencing a negative emotional state is not fun. While the secondary somatosensory cortex and the insula help integrate information about the painful stimulus with the cognitive-affective aspects of pain, there are other brain regions that are responsible for generating such emotional responses to pain.

One prominent brain region associated with pain-induced emotional responses is the anterior cingulate cortex (ACC) (Bliss et al., 2016). Instead of processing information about pain itself, the ACC determines how distressing or unpleasant the painful stimulus feels to us. For most of us, getting injured elicits a negative emotional response. Going back to the example of stubbing one's toe, this experience can elicit anger and frustration.

The second brain region associated with emotional responses to pain is the amygdala (Neugebauer et al., 2020). The amygdala is an important limbic structure that is involved in emotional memories and fear. Imagine a young child sticks their finger in an electrical outlet. Not only will this experience lead to pain, but the sudden, unexpected shock will most likely scare the child. This fear response is largely driven by the amygdala. Because the amygdala processes emotional memories, it will encode the events that led up to the child's fear. Now, when the child sees an outlet, they will most likely avoid it as they have learned the painful consequences associated with sticking their finger into it.

Like the PNS, prostaglandins in the CNS are important for pain perception. When we are exposed to noxious stimuli, prostaglandins (specifically prostaglandin E2) are released in the spinal cord, an effect that appears to be influenced by substance P. Prostaglandins are released from the postsynaptic neuron and bind to a G protein-coupled receptor on the presynaptic neuron, causing the further release of glutamate. Collectively, increased activity of prostaglandins, substance P, and glutamate lead to **central sensitization** (Latremoliere & Woolf, 2009; Seybold, 2009). Central sensitization is very similar to hyperalgesia I described above. In hyperalgesia, individuals experience increased pain to stimuli that normally elicit pain. In central sensitization, individuals experience both hyperalgesia and **allodynia**, pain to stimuli that normally do not elicit pain (e.g., someone gently brushing their hand on an individual can elicit pain).

3.3 Mechanisms for Managing Pain

As a teenager, I did not have the training in behavioral pharmacology I have now. I do not know the exact drug that I received that night over 20 years ago, but I am positive I received an **opioid**. Opioids are powerful *analgesics* and can be derived naturally from the opium poppy plant or can be fully synthesized in a laboratory. Opioids are also found in our body; these are called *endogenous* opioids (endogenous just means "originating within an organism"). Not surprisingly, endogenous opioids help attenuate pain responses. Being able to turn down the pain signal is important as long-term pain is detrimental, as you are about to read about below.

There are three major types of endogenous opioids: endorphins, enkephalins, and dynorphins. Endorphins are derived from the protein *proopiomelanocortin (POMC)*. Enkephalins and dynorphins come from proenkephalin and prodynorphin, respectively. Each endogenous opioid binds to a different opioid receptor: mu, delta, and kappa. Each of these receptors is metabotropic and inhibitory (G_i-coupled). Analgesia is primarily controlled by the binding of endorphins to mu opioid receptors (Trescot et al., 2008).

After the primary somatosensory cortex has received pain signals, the brain will begin to send messages back down toward the spinal cord. In the midbrain, the periaqueductal gray (PAG) projects neurons to part of the medulla, which then causes the release of endogenous opioids in the spinal cord. Because opioids bind to inhibitory receptors, this causes neurons transmitting pain signals to become less active (see Fig. 4).

Opioids are not the only neurotransmitter that inhibits pain signals. Endocannabinoids are endogenous **cannabinoids**. The most famous *exogenous* cannabinoid is

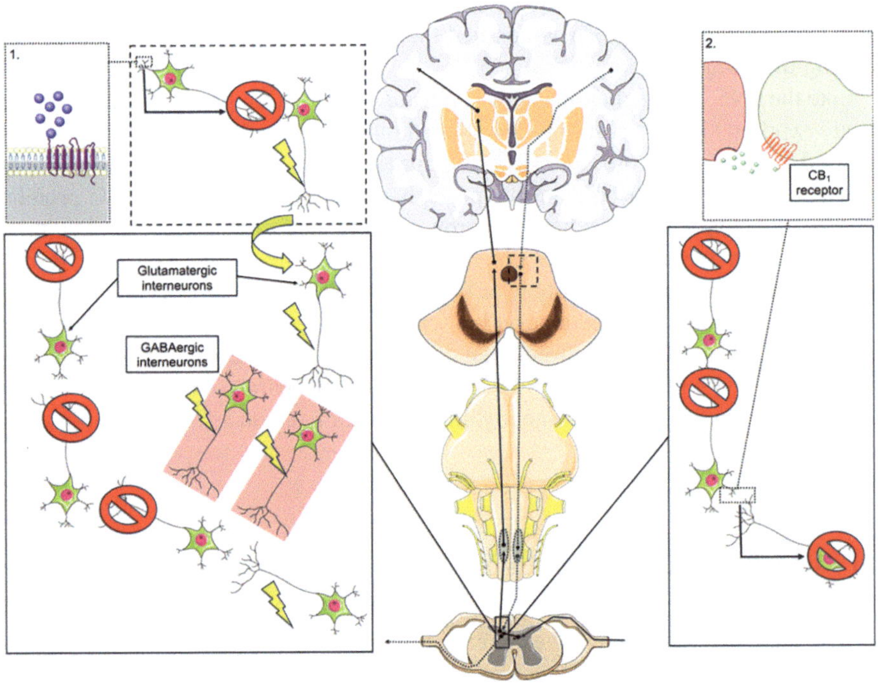

Fig. 4 There are several mechanisms by which pain transmission can be attenuated. **1** Endogenous opioids in the periaqueductal gray (PAG) bind to opioid receptors located on inhibitory GABAergic (gamma-aminobutyric acid) neurons (see top left insets). Because opioid receptors are inhibitory, this causes glutamatergic neurons in the descending pathway of the spinal cord to become disinhibited. Glutamatergic neurons then stimulate GABAergic neurons in the spinal cord. This causes glutamatergic neurons in the ascending pathway to become inhibited (see bottom left inset). **2** Endocannabinoids are released from dendrites of postsynaptic neurons in the spinal cord, which bind to cannabinoid (CB) receptors on the terminal button of glutamatergic neurons entering the spinal cord. This causes further inhibition of the ascending pathway (see right insets). Image created by the author using Servier Medical Art (https://smart.servier.com/; CC BY 4.0, https://creativecommons.org/licenses/by/4.0/) and Microsoft PowerPoint

Δ-9 tetrahydrocannabinol (THC), which is derived from the cannabis plant. Individuals reading this may be more familiar with the terms "marijuana" or "weed". The most well-known endocannabinoids are anandamide and 2-arachidonoylglycerol (2-AG). Both endocannabinoids and exogenous cannabinoids bind to two different cannabinoid receptor subtypes: CB_1 and CB_2. These receptors are metabotropic and coupled to a G_i protein. Unlike other neurotransmitter systems that you have learned about, the cannabinoid system relies on *retrograde neurotransmission*. That is, endocannabinoids are released from the cell body of the postsynaptic neuron and bind to CB receptors located on the terminal button of the presynaptic neuron. This allows cannabinoids to inhibit the activity of presynaptic neurons.

Experiment Spotlight: Combined Endocannabinoid and Cyclooxygenase Inhibition Additively Attenuates Post-Surgical Pain

The full article can be accessed with this link: https://doi.org/10.1089/can.2024.0088.

I will provide a comprehensive summary of the methods and the results of this study (Rodriguez et al., 2025) below.

Male and female C57BL/6 J mice (one of the most widely used *inbred* strains in research) underwent a surgical procedure in which an incision was made in the left hind paw. Control animals were anesthetized, but no incision was made. Twenty-four hours following surgery, mice were tested in several tasks.

First, mice were given a thermal preference test. The test apparatus consists of two hot/cold plates; these plates create a temperature differential in the apparatus. The researchers measured where mice spent their time in the apparatus. This test allows one to determine thermal sensitivity.

Next, mice were subjected to a wire-climbing test. Mice were placed in a small plastic cylinder that contained a wire mesh that wrapped around 75% of the cylinder. The researchers measured how much time mice spent climbing. If mice are in pain, they will not spend as much time climbing.

Somewhat related to the wire-climbing test is the inverted grid grip strength test. Mice were placed on a grid floor. The floor was turned upside down, and the researchers measured how long the mice could maintain their grip on the grid floor. The grid floor was suspended above padding to prevent the mice from being injured.

Finally, mice were tested for mechanical allodynia. A researcher used three *von Frey filaments* to stimulate the plantar surface of each hind paw. von Frey filaments are nylon fibers that come in different diameters and differ in stiffness. The three von Frey filaments used in the current study had a gauge of 0.16–6. The researchers measured how quickly the mice moved their paw in response to tactile stimulation.

The researchers found that hind paw incision significantly increased mechanical allodynia. That is, mice that had received an incision to their hind paw moved their paw at a much faster rate following stimulation with von Frey filaments compared to control mice. This indicates that these mice are more sensitive to mechanical stimulation. However, hind paw incision did not affect thermal preference, climbing behavior, or grip strength.

At this point, the researchers wanted to determine the contribution of arachidonic acid and cannabinoids on mechanical allodynia. Remember, COX-2 is derived from arachidonic acid. To target arachidonic acid levels, the researchers injected mice with either JZL184 or MJN110, which act as inhibitors of the enzyme monoacylglycerol lipase (MAGL). MAGL converts 2-AG, which you read about above, into arachidonic acid. Notice the direct relationship between endocannabinoids and COX-2. Inhibiting MAGL decreased mechanical pain sensitivity as mice took longer to withdraw their paw following von Frey filament stimulation.

Researchers also treated some mice with diclofenac, which inhibits COX-2, and other mice with a drug known as LY2828360, which acts as an agonist at CB_2 receptors. These drugs, like the MAGL inhibitors, decreased pain sensitivity. The researchers also showed that antagonizing CB_2 receptors with SR144528 blocks the anti-nociceptive effects of JZL184.

The final result I want to highlight is that combining JZL184 and diclofenac decreased pain sensitivity to a greater extent compared to when either drug was administered alone.

Questions to Consider:

1. Even though this study was conducted with mice, what are some real-world implications of the present results?
2. In addition to decreasing pain to a greater extent when combined, what is another advantage of combining a MAGL inhibitor and diclofenac?
3. This experiment used hind paw incision to model postsurgical pain. What is one potential limitation to examining the effects of MAGL inhibitors, a CB_2 receptor agonist, and diclofenac on postoperative pain only?

When you have experienced pain and/or a fever, you probably have taken something like acetaminophen (Tylenol) or ibuprofen (Advil; Motrin). These medications are *non-steroidal anti-inflammatory drugs (NSAIDs).* Based on their name, can you guess how these medications exert their anti-pain effects? You should be able to discern that these medications help reduce inflammation. Specifically, NSAIDs act as COX-2 inhibitors. They prevent the conversion of arachidonic acid to prostaglandins by shutting down the activity of COX-2. In the Experiment Spotlight above, the researchers used the NSAID diclofenac.

Serotoninergic activity in the spinal cord is also an important mediator of pain (Hao et al., 2023). While serotonin acts to increase pain transmission in the PNS, it often has the opposite role in the CNS. Because serotonin can bind to inhibitory receptors (e.g., $5\text{-}HT_{1A}$) in addition to excitatory receptors, it can reduce the rate of neural transmission in the spinal cord. There is increasing interest in using psychedelic hallucinogens like lysergic acid diethylamide (LSD) to treat chronic pain (Robinson et al., 2024). Recall from the previous chapter that psychedelic hallucinogens primarily act as $5\text{-}HT_{2A}$ agonists.

Comprehension Check 2 Compare and contrast the peripheral mechanisms and the central mechanisms of pain transmission.

Comparison: Both the PNS and the CNS rely heavily on glutamate to transmit information regarding pain. Prostaglandins are present in both the PNS and the CNS and are important mediators of inflammation. Additionally, serotonin acts as a modulator of pain neurotransmission in both PNS and CNS.

Differences: In the PNS, pain perception begins when nociceptors on free nerve endings are stimulated. In the CNS, free nerve endings are not present; instead, messages from peripheral neurons are transmitted to neurons in the spinal cord. While serotonin activity in the PNS primarily intensifies pain transmission, it often acts to inhibit pain perception in the CNS.

4 Neuropathic Pain

As mentioned in the Introduction, pain is important because it signals that something is wrong. Unfortunately, some individuals live with *neuropathic pain*, in which they experience **dysesthesia** (pain with no physical cause) and/or allodynia. I am just one individual that experiences dysesthesia.

Neuropathic pain typically results from damage to peripheral nerves that transmit sensory information to the spinal cord (*peripheral neuropathy*) (England & Asbury, 2004). Medical conditions like diabetes and multiple sclerosis can also lead to peripheral neuropathy. *Trigeminal neuralgia* is a debilitating disorder that causes severe pain in the face and is often caused by neurovascular compression of the trigeminal nerve (Gambeta et al., 2020). For some individuals, neuropathic pain results from a maladaptive response that is independent of physical damage to our neurons. A couple of examples will be discussed below.

4.1 Phantom Limb Pain

When my mother was younger, she was involved in an industrial accident; her pointer and middle fingers on her left hand were crushed. The damage was so severe that her fingers had to be amputated above the first knuckle. Even though this accident occurred decades ago, my mom still experiences pain where her fingers should be. My mother is not alone as some individuals that have received an amputation experience **phantom limb pain**. I need to note that not everyone experiences a sensation of pain. Instead, they may report feeling like the missing limb is still attached.

Why does phantom limb pain occur? There are several factors proposed to explain why this phenomenon occurs (Subedi & Grossberg, 2011) (Fig. 5). First, **neuromas** form in nerves leading from the missing limb/appendage. These neuromas have increased Na^+ ion channels, meaning the severed neuron becomes more easily stimulated. This hyperexcitability leads to increased pain transmission to the CNS.

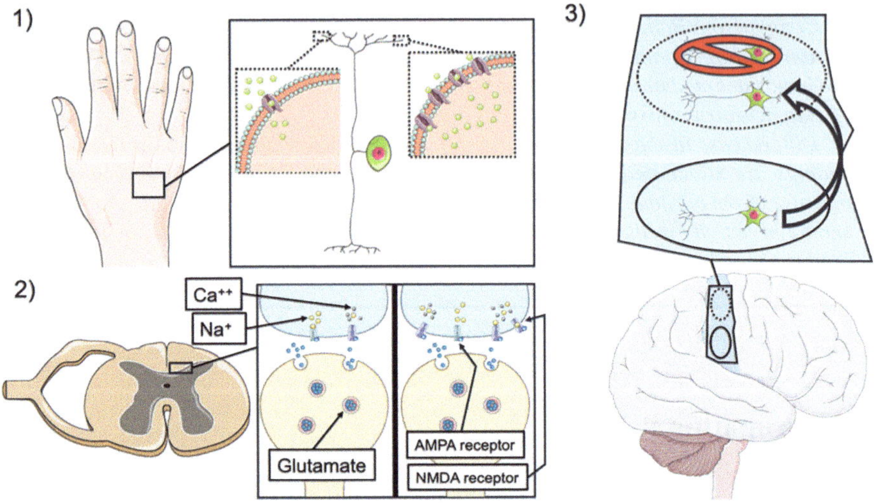

Fig. 5 There are several potential mechanisms that account for phantom limb pain. **1** A neuroma can form on a sensory neuron, which leads to increased sodium (Na^+) ion channels. Increased Na^+ influx causes sensory neurons to become hyperactive, transmitting pain signals about the missing limb. **2** In the spinal cord, neurons become hyperactive. This can be due to increased glutamate receptor, specifically N-methyl-D-aspartate (NMDA) receptor, expression. This allows more Na^+ and calcium (Ca^{++}) into the neuron, causing greater excitation and increased pain transmission. **3** Neural plasticity can occur, in which neurons from one part of the somatosensory cortex take over another part of the somatosensory cortex that no longer receives messages about a specific part of the body (indicated by the dashed circle). Image created by the author using Servier Medical Art (https://smart.servier.com/; CC BY 4.0, https://creativecommons.org/licenses/by/4.0/) and Microsoft PowerPoint

Second, neurons in the dorsal root of the spinal cord can undergo compensatory changes that make them hyperactive. Increased glutamatergic activity at N-methyl-D-aspartate (NMDA) receptors is observed in the dorsal horn of the spinal cord. Neurons containing NMDA receptors are more easily excited, causing pain messages to be transmitted through the spinal cord to the brain.

Third, the somatosensory cortex is organized in such a way that specific regions in this cortex are devoted to regions of our body. Larger areas of the somatosensory cortex are devoted to parts of the body that rely heavily on touch, such as our fingers and tongue. If the somatosensory cortex no longer receives messages from neurons associated with the missing limb/appendage, cortical reorganization can occur. Neurons from other parts of the somatosensory cortex can "take over" the unused regions of this brain structure. This is an example of **neural plasticity**. Typically, neural plasticity is advantageous. In fact, learning requires *synaptic plasticity*, changes in the strength between neurons (synaptic plasticity will be formally defined in Chap. 6). In this case, neural plasticity is detrimental as individuals experience pain emanating from a missing appendage.

4.2 Fibromyalgia

In addition to phantom limb pain, my mother suffers from **fibromyalgia**, a medical disorder primarily characterized by chronic pain. Individuals with fibromyalgia can also experience other symptoms, including allodynia, fatigue, sleep disturbances, and cognitive dysfunction ("fibrofog"). At the beginning of this section, I described a random burning sensation I experience. Although never officially diagnosed with fibromyalgia, I live with multiple symptoms associated with this disorder.

The exact cause of fibromyalgia is unknown, although genetic factors seemingly contribute to fibromyalgia. Genes that have received some attention include those that encode for the serotonin transporter, TRPV2 ion channel, *myelin transcription factor*, and *neurexin 3* (D'Agnelli et al., 2019). A discussion of how mutations to these genes may underlie fibromyalgia will be provided below.

The serotonin transporter removes excess serotonin from the synapse. As such, dysregulated serotonin transporter activity can lead to abnormal serotonin levels in the PNS and/or the CNS. Decreased expression of and/or activity of serotonin transporters in the PNS will result in excess serotonin levels. Remember, serotonin often acts to potentiate pain in the PNS. Indeed, individuals with fibromyalgia have lower serotonin transporter function in the PNS (Bazzichi et al., 2006). This means that these individuals have more serotonin in the PNS, which allows for increased stimulation of 5-HT_{2A} receptors.

As you learned earlier, TRPV2 ion channels respond to higher temperatures compared to TRPV1 ion channels. Specifically, these receptors respond to *noxious heat* (above 52 °C), as well as to mechanical stimulation and certain chemicals, including cannabidiol, better known as CBD (Oda et al., 2021). Mutations to the *TRPV2* gene may lead to TRPV2 ion channels that are hyperactive, leading to transmission of pain signals even in the absence of pain-producing stimuli.

Myelin transcription factor, as the name suggests, is involved in myelin gene transcription. This transcription factor influences oligodendrocyte progenitor cell proliferation and differentiation. Oligodendrocytes are a type of cell that form the myelin sheath of neurons in the CNS. A mutation in this gene may interfere with the production of myelin sheath, causing disruptions in action potentials. This can potentially lead to aberrant pain signaling in the CNS.

Neurexin 3 belongs to a class of proteins called neurexins. Neurexins are *presynaptic cell adhesion molecules*, helping form excitatory and inhibitory synapses in the nervous system. Think of neurexins as a tether that helps keep the presynaptic neuron's terminal button close to the dendrites of the postsynaptic neuron. If neurexin 3 does not function properly, the presynaptic terminal button will not be tethered as closely to the postsynaptic neuron, causing disruptions in neuronal communication. Potentially, this de-tethering of neurons can prevent inhibitory neurotransmitters from attenuating pain messages in the CNS (e.g., endocannabinoids may have more difficulty traveling from the postsynaptic neuron to the presynaptic neuron).

5 Congenital Insensitivity to Pain (CIP)

I have lost track of the number of times I have experienced pain during my life: getting a papercut, stepping on my son's Legos, wrecking a dirt bike, recovering from surgery, and being in a car crash, to name a few examples. As annoying as pain can be, it is a necessary mechanism that alerts us to a problem. If an individual breaks a bone, the resulting pain notifies them that something is wrong and that they should stop trying to use the limb or appendage that has been damaged.

While some individuals experience pain with no known cause, others cannot experience pain. Congenital insensitivity to pain (CIP) refers to a series of genetic disorders that cause an individual to feel no pain. In some forms of CIP, a mutation in the **neurotrophic tyrosine kinase receptor type 1** gene (Rodríguez-Blanque et al., 2024) prevents *nerve growth factor (NGF)* from stimulating the neuron. NGF is crucial for the development and survival of neurons. For example, individuals with the specific CIP **congenital insensitivity to pain with anhidrosis (CIPA)** do not have enough neurotrophic tyrosine kinase receptors, resulting in insufficient NGF activity. This causes many sensory neurons to die. This is what prevents individuals from perceiving pain or temperature. Additionally, sympathetic neurons do not properly develop, leading to *anhidrosis* (absence of sweat production).

Other forms of CIP result from a mutation in a gene that encodes for Na^+ ion channels located on sensory neurons. Although sensory neurons develop normally, Na^+ ion channels are not sufficiently activated to cause the transmission of pain signals to the brain (Drissi et al., 2020).

Being able to feel no pain may sound great, but CIP-related disorders are extremely dangerous. For example, approximately 20% of children with CIPA die before the age of 3 due to excessive fever (Rosemberg et al., 1994). Normally, when we have a serious infection, the release of prostaglandins leads to inflammation and pain. Infants and young children with CIP do not cry when hungry, and they engage in forms of self-mutilation such as biting fingers. They are at increased risk of bone fractures; as they show no pain response, the child's guardians may not suspect that any injury has occurred. This puts the child at risk of developing an infection like osteomyelitis (bacterial infection of bone), which can be fatal if untreated.

Compounding the issues described above, a large percentage of individuals with mutations to the neurotrophic tyrosine kinase receptor type 1 gene, like CIPA, have intellectual disabilities. The absence of neurotrophic tyrosine kinase receptors does not affect sensory and sympathetic neurons only. Other neurons are negatively impacted by this genetic mutation. Neurons in the brain can fail to develop properly, resulting in cognitive dysfunction. Due to these cognitive impairments, individuals may have difficulty refraining from engaging in self-injurious behavior, even if someone explains why that type of behavior is dangerous.

6 Chapter Summary

Although pain is unpleasant, without it, we would be at risk for serious injury and even death. This chapter explored nociception, the sensory system responsible for encoding noxious stimuli, including pain. The peripheral mechanisms of pain, including the role of free nerve endings, prostaglandins, and specialized proteins called TRP ion channels, were first discussed. Next, the central mechanisms of pain were covered, including the role of glutamate, substance P, and various brain regions such as the somatosensory cortex, the insula, the ACC, and the amygdala. Several maladaptive pain conditions were discussed in detail. Neuropathic pain can arise from damage to peripheral nerves. Multiple potential mechanisms may account for conditions such as phantom limb pain and fibromyalgia. CIP, a rare genetic disorder that prevents individuals from feeling pain, was also detailed.

Answers to Experiment Spotlight Questions

1. These results indicate that combining drugs that block COX-2 and MAGL can help alleviate pain resulting from surgery. This is important because traditional analgesics (e.g., opioids) have high misuse potential and can lead to overdose.
2. After reading the experiment spotlight, you learned that diclofenac is an NSAID like acetaminophen and ibuprofen. Repeated administration of NSAIDs is associated with several adverse health effects (e.g., liver damage). If an NSAID can be combined with something like an MAGL inhibitor, the dose can be lowered, thus reducing the likelihood of side effects.
3. While developing alternative treatments for postoperative pain is important, this animal model does not fully capture the human experience of pain. This study does not address how treatment with these drugs can alter chronic pain resulting from something like nerve damage.

Glossary

Allodynia sensation of pain to a stimulus that normally does not elicit pain.

Cannabinoid chemical compound that is found in the body, as well as in external sources such as the cannabis plant.

Central sensitization a condition where the central nervous system amplifies sensory input, leading to hypersensitivity to pain and other stimuli.

Congenital insensitivity to pain with anhidrosis (CIPA) a disorder resulting from a genetic mutation in which individuals experience little to no pain and produce little to no sweat.

Cyclooxygenase-2 an enzyme that plays a key role in the production of prostaglandins, which contribute to inflammation and pain.

Dysesthesia unpleasant, abnormal sense of touch, often resulting in a perception of pain.

Fibromyalgia medical condition in which individuals experience chronic pain and other symptoms such as hypersensitivity to touch, fatigue, sleep disturbances, and cognitive dysfunction.

Hyperalgesia increased sensitivity to pain or an extreme response to pain.

Neural plasticity the brain's ability to reorganize itself by forming new neural connections throughout life.

Neuroma a growth or tumor, typically benign, of a nerve.

Neuropeptide a small protein-like molecule produced by neurons that acts as chemical messengers within the nervous system.

Neurotrophic tyrosine kinase receptor type 1 protein that acts as a receptor for nerve growth factor, which is crucial for neuron development and survival.

Nociception the neural process of encoding noxious stimuli.

Opioid a class of drugs that act on the nervous system to relieve pain and can be derived from the opium poppy plant or synthesized in a laboratory.

Phantom limb pain a phenomenon in which individuals experience pain emanating from a missing limb or appendage.

Prostaglandin a hormone-like lipid; there are various prostaglandins that serve multiple purposes.

Transient receptor potential ion channel family of receptors that detects changes in temperature as well as the presence of certain chemicals like capsaicin and menthol.

References

Bazzichi, L., Giannaccini, G., Betti, L., Mascia, G., Fabbrini, L., Italiani, P., De Feo, F., Giuliano, T., Giacomelli, C., Rossi, A., Lucacchini, A., & Bombardieri, S. (2006). Alteration of serotonin transporter density and activity in fibromyalgia. *Arthritis Research & Therapy, 8*(4), R99. https://doi.org/10.1186/ar1982

Bliss, T. V., Collingridge, G. L., Kaang, B. K., & Zhuo, M. (2016). Synaptic plasticity in the anterior cingulate cortex in acute and chronic pain. *Nature Reviews. Neuroscience, 17*(8), 485–496. https://doi.org/10.1038/nrn.2016.68

Bretas, R. V., Taoka, M., Suzuki, H., & Iriki, A. (2020). Secondary somatosensory cortex of primates: Beyond body maps, toward conscious self-in-the-world maps. *Experimental Brain Research, 238*(2), 259–272. https://doi.org/10.1007/s00221-020-05727-9

D'Agnelli, S., Arendt-Nielsen, L., Gerra, M. C., Zatorri, K., Boggiani, L., Baciarello, M., & Bignami, E. (2019). Fibromyalgia: Genetics and epigenetics insights may provide the basis for the development of diagnostic biomarkers. *Molecular Pain, 15*. https://doi.org/10.1177/1744806918819944

Drissi, I., Woods, W. A., & Woods, C. G. (2020). Understanding the genetic basis of congenital insensitivity to pain. *British Medical Bulletin, 133*(1), 65–78. https://doi.org/10.1093/bmb/ldaa003

England, J. D., & Asbury, A. K. (2004). Peripheral neuropathy. *Lancet, 363*(9427), 2151–2161. https://doi.org/10.1016/S0140-6736(04)16508-2

Gambeta, E., Chichorro, J. G., & Zamponi, G. W. (2020). Trigeminal neuralgia: An overview from pathophysiology to pharmacological treatments. *Molecular Pain, 16*, 1744806920901890. https://doi.org/10.1177/1744806920901890

Hao, S., Shi, W., Liu, W., Chen, Q. Y., & Zhuo, M. (2023). Multiple modulatory roles of serotonin in chronic pain and injury-related anxiety. *Frontiers in Synaptic Neuroscience, 15*, 1122381. https://doi.org/10.3389/fnsyn.2023.1122381

Humes, C., Sic, A., & Knezevic, N. N. (2024). Substance P's impact on chronic pain and psychiatric conditions—A narrative review. *International Journal of Molecular Sciences, 25*(11), 5905. https://doi.org/10.3390/ijms25115905

Ito, S., Okuda-Ashitaka, E., & Minami, T. (2001). Central and peripheral roles of prostaglandins in pain and their interactions with novel neuropeptides nociceptin and nocistatin. *Neuroscience Research, 41*(4), 299–332. https://doi.org/10.1016/s0168-0102(01)00289-9

Latremoliere, A., & Woolf, C. J. (2009). Central sensitization: A generator of pain hypersensitivity by central neural plasticity. *The Journal of Pain, 10*(9), 895–926. https://doi.org/10.1016/j.jpain.2009.06.012

Lezama-García, K., Mota-Rojas, D., Pereira, A. M. F., Martínez-Burnes, J., Ghezzi, M., Domínguez, A., Gómez, J., de Mira Geraldo, A., Lendez, P., Hernández-Ávalos, I., Falcón, I., Olmos-Hernández, A., & Wang, D. (2022). Transient receptor potential (TRP) and thermoregulation in animals: Structural biology and neurophysiological aspects. *Animals, 12*(1), 106. https://doi.org/10.3390/ani12010106

Markenson, J. A. (1996). Mechanisms of chronic pain. *The American Journal of Medicine, 101*(1A), 6S–18S. https://doi.org/10.1016/s0002-9343(96)00133-7

Neugebauer, V., Mazzitelli, M., Cragg, B., Ji, G., Navratilova, E., & Porreca, F. (2020). Amygdala, neuropeptides, and chronic pain-related affective behaviors. *Neuropharmacology, 170*, 108052. https://doi.org/10.1016/j.neuropharm.2020.108052

Oda, M., Fujiwara, Y., Ishizaki, Y., & Shibasaki, K. (2021). Oxidation sensitizes TRPV2 to chemical and heat stimuli, but not mechanical stimulation. *Biochemistry and Biophysics Reports, 28*, 101173. https://doi.org/10.1016/j.bbrep.2021.101173

Orenius, T. I., Raij, T. T., Nuortimo, A., Näätänen, P., Lipsanen, J., & Karlsson, H. (2017). The interaction of emotion and pain in the insula and secondary somatosensory cortex. *Neuroscience, 349*, 185–194. https://doi.org/10.1016/j.neuroscience.2017.02.047

Persson, A. K., Black, J. A., Gasser, A., Cheng, X., Fischer, T. Z., & Waxman, S. G. (2010). Sodium-calcium exchanger and multiple sodium channel isoforms in intra-epidermal nerve terminals. *Molecular Pain, 6*, 84. https://doi.org/10.1186/1744-8069-6-84

Robinson, C. L., Fonseca, A. C. G., Diejomaoh, E. M., D'Souza, R. S., Schatman, M. E., Orhurhu, V., & Emerick, T. (2024). Scoping review: The role of psychedelics in the management of chronic pain. *Journal of Pain Research, 17*, 965–973. https://doi.org/10.2147/JPR.S439348

Rodríguez-Blanque, R., Nielsen, L. M., Piqueras-Sola, B., Sánchez-García, J. C., Cortés-Martín, C., Reinoso-Cobo, A., & Cortés-Martín, J. (2024). A systematic review of congenital insensitivity to pain, a rare disease. *Journal of Personalized Medicine, 14*(6), 570. https://doi.org/10.3390/jpm14060570

Rodriguez, C. E. B., Vanegas, S. O., Reck, A. M., Schrom, Y., & Kinsey, S. G. (2025). Combined endocannabinoid and cyclooxygenase inhibition additively attenuates post-surgical pain. *Cannabis and Cannabinoid Research, 10*(1), e102–e111. https://doi.org/10.1089/can.2024.0088

Rosemberg, S., Marie, S. K., & Kliemann, S. (1994). Congenital insensitivity to pain with anhidrosis (hereditary sensory and autonomic neuropathy type IV). *Pediatric Neurology, 11*(1), 50–56. https://doi.org/10.1016/0887-8994(94)90091-4

Seybold, V. S. (2009). The role of peptides in central sensitization. *Handbook of Experimental Pharmacology, 194*, 451–491. https://doi.org/10.1007/978-3-540-79090-7_13

Sommer, C. (2004). Serotonin in pain and analgesia: Actions in the periphery. *Molecular Neurobiology, 30*(2), 117–125. https://doi.org/10.1385/MN:30:2:117

Subedi, B., & Grossberg, G. T. (2011). Phantom limb pain: Mechanisms and treatment approaches. *Pain Research and Treatment, 2011*, 864605. https://doi.org/10.1155/2011/864605

Sun, Z. C., Ma, S. B., Chu, W. G., Jia, D., & Luo, C. (2020). Canonical transient receptor potential (TRPC) channels in nociception and pathological pain. *Neural Plasticity, 2020*, 3764193. https://doi.org/10.1155/2020/3764193

Trescot, A. M., Datta, S., Lee, M., & Hansen, H. (2008). Opioid pharmacology. *Pain Physician, 11*(2 Suppl.), S133–S153.

Vierck, C. J., Whitsel, B. L., Favorov, O. V., Brown, A. W., & Tommerdahl, M. (2013). Role of primary somatosensory cortex in the coding of pain. *Pain, 154*(3), 334–344. https://doi.org/10.1016/j.pain.2012.10.021

Zhang, M., Ma, Y., Ye, X., Zhang, N., Pan, L., & Wang, B. (2023). TRP (transient receptor potential) ion channel family: Structures, biological functions and therapeutic interventions for diseases. *Signal Transduction and Targeted Therapy, 8*(1), 261. https://doi.org/10.1038/s41392-023-01464-x

Chapter 5
Maladaptive Attentional Processes

1 Introduction

"We've been having a full-blown conversation with you", my wife says after finally getting my attention. I hate to admit that this is a common occurrence as I am one of the worst multitaskers I know. If I am watching a video on my phone or checking an email, I often miss parts of conversations, even when someone is directly speaking to me. Once I know someone is speaking to me, I typically put down my device or look away from my laptop. My inattention does not cause serious problems as I am able to hold a conversation once I shift my focus to the individual. I am also able to complete my work in a timely fashion. For other individuals, attentional deficits are more severe and can be maladaptive. Children may have difficulty paying attention to their teacher and/or may not be able to complete homework assignments, thus negatively impacting their grade. Similar issues can be observed in adults; they may have a hard time following instructions at work and/or completing mundane tasks. This can lead to the individual's termination from work.

This chapter will first discuss the various types of attention we routinely use. One thing you will learn is that perception and attention are closely intertwined. This makes sense as one must perceive something to pay attention to it. You will also learn about one psychiatric condition in which attentional deficits are a defining feature. The neurotransmitters and brain regions involved in attention will be introduced, and you will learn how dysfunction to specific neurotransmitter systems and brain circuits contribute to attentional deficits.

2 Learning Objectives

By the end of this chapter, you should be able to …

J. Yates, *Neuroscience of Maladaptation*, https://doi.org/10.1007/978-3-032-15602-0_5

1. Differentiate the types of attention and describe how attention can be measured in humans and animals.
2. Identify the symptoms of the inattentive subtype of attention-deficit/hyperactivity disorder (ADHD) and explain how this form of ADHD can be modeled in animals.
3. Discuss how specific neurotransmitter systems and brain regions mediate attentional processes and how dysfunction of certain neural circuits leads to attentional deficits.

3 Types of Attention

When having a conversation with another individual, you need to focus on what the person is saying. You may look at this individual and listen intently to what they have to say. This is an example of **top-down attention**, which involves using our prior knowledge and expectations to dictate where and how to allocate our attention. Throughout your life, you have learned about social etiquette and the importance of communication. You also expect that another individual listens to what you have to say during a conversation; as such, you listen when the other person speaks.

There are several types of top-down attention we use daily. **Sustained attention** involves focusing on a stimulus or an event for an extended duration. Directly related to sustained attention is **selective attention**, in which individuals must sustain their attention on one specific stimulus/event while ignoring other things. For example, when having a conversation with another individual, we must engage in both sustained attention and selective attention. We want to focus on what the other individual has to say (sustained attention) while focusing less on other things going on around us (e.g., sounds from a TV, people walking by, vibration of a cell phone, etc.).

The opposite of selective attention is **divided attention**, also known as multitasking. Here, attention is split between two or more stimuli/activities. In the introduction to this chapter, I mentioned that I am terrible at multitasking. I am not alone as research consistently shows deficits in cognition when we switch between tasks (Rogers & Monsell, 1995). One dangerous example of divided attention is driving a vehicle while text messaging, although research shows that having a conversation with a passenger while driving can also be dangerous (Theofilatos et al., 2018).

Imagine that during a conversation, you and your friend suddenly see a bright light that briefly flashes in the sky. This flash may cause a pause in your conversation as you and your friend look at the sky, trying to make sense of what you just perceived. This is an example of **bottom-up attention**. In this example, our intention was not to look at the sky, but our attention was suddenly drawn to it due to the presentation of a *salient* (i.e., highly noticeable) stimulus. Unlike top-down attention, there is little conscious effort that goes into bottom-up attention.

3.1 Measuring Attention in Humans

Figure 1 depicts various laboratory tests used to measure attention. Sustained attention and selective attention are often measured with a *continuous performance task (CPT)* (Beck et al., 1956). There are several variations of the CPT, but the premise is similar across variations. Individuals are required to respond (typically by pressing a key on a keyboard) when they see or hear a particular stimulus. In one version of the CPT, individuals respond any time they see a specific stimulus, such as the letter "X". If individuals encounter a different stimulus, they must refrain from responding. In a widely used variant of the CPT, individuals respond when they see any letter other than the letter "X" (Conners, 1995). There are also auditory versions of the CPT (Keith, 1994).

Selective attention is also measured with visual search tasks (Treisman, 1982). Think of the *Where's Waldo* books, in which you seek out a specific character or characters. Visual search tasks require individuals to identify a target on a screen. In one version of a visual search task I have my students complete, they must detect a green circle that appears with other stimuli, including blue circles, blue squares, and/or green squares. As more stimuli appear on the screen, individuals need more time

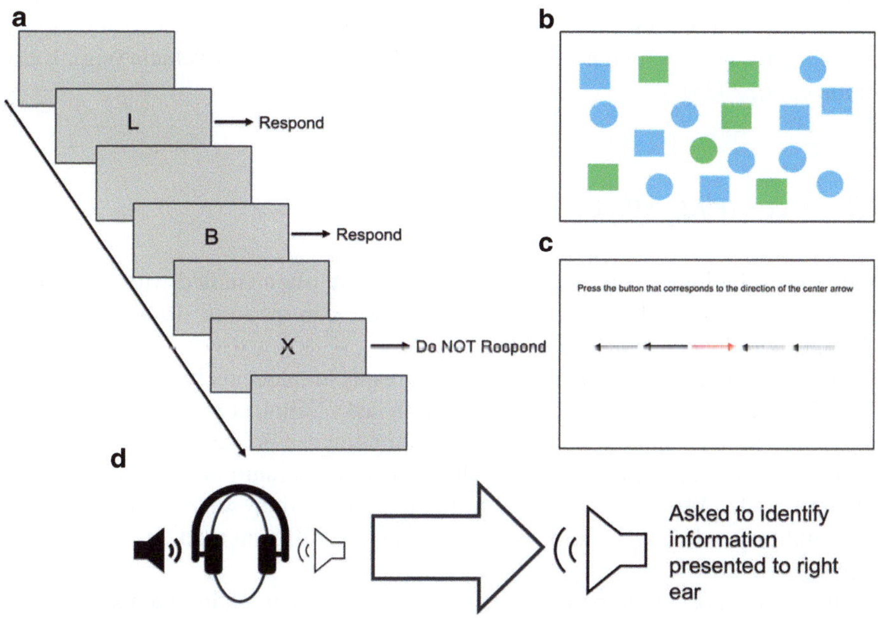

Fig. 1 Example tasks used to measure different aspects of attention in humans. **a** The continuous performance task (CPT) measures sustained attention. In this version, individuals respond each time they see any letter other than "X". **b** Visual search tasks and **c** flanker tasks measure selective attention. **d** The dichotic listening task measures divided attention. Image was created by the author using Microsoft PowerPoint

to find the target as they need to ignore the distracting stimuli and focus on finding the green circle.

Flanker tasks are like visual search tasks in that individuals must ignore distracting stimuli. Participants are instructed to respond on a keyboard when presented with a target stimulus, typically an arrow pointing to the left or to the right or a specific letter (Eriksen & Eriksen, 1974). When arrows are used, the target arrow is in the center of the screen and is flanked by arrows that point in either the same direction or in the opposite direction as the center arrow. When letters are used, the target letter is presented in the center and is flanked by distractor letters.

Divided attention can be measured numerous ways, but the goal is similar across tasks. Individuals need to perform two different actions simultaneously. This is analogous to the driving and texting example I used above. I will highlight a few examples of how divided attention can be measured in a laboratory. In the *dichotic listening task* (Broadbent, 1952), participants wear headphones and listen to two different auditory stimuli simultaneously. They are then asked to recall information regarding what they heard in one ear. Divided attention can also be measured by having individuals listen to information while viewing different stimuli (Broadbent, 1956; Moisala et al., 2015). In other tasks, individuals are asked to perform an action such as typing a string of digits on a keyboard with one hand while controlling a joystick with the other hand to track a stimulus on a screen (e.g., Farmer et al., 2018). To better mimic real-world situations, driving simulators can be used to measure divided attention by making participants complete a distractor task while driving a vehicle (e.g., Ezzati Amini et al., 2023).

3.2 Measuring Attention in Animals

As the neurobiological mechanisms of attention are often studied with primates or rodents, I will highlight a few ways in which we can measure sustained/selective attention and divided attention in these species.

Primates can be tested in similar procedures as humans, including variations of the CPT (Blackman et al., 2016), visual search tasks (Klink et al., 2023), and flanker tasks (Hassett & Hampton, 2022). Divided attention can be measured several ways. Subjects can be required to track two different visual stimuli on a computer screen (O'Neill et al., 1999) or be required to focus on stimuli presented in one modality (e.g., auditory) while ignoring stimuli presented in a different modality (e.g., visual) (Rinne et al., 2017).

One of the most common methods for measuring attention in rodents is the five-choice serial reaction time task (5-CSRTT) (Carli et al., 1983) (see Fig. 2). You will learn more about this task in Chap. 12. For now, I will briefly describe how the 5-CSRTT works. In this task, animals are placed in a special operant conditioning chamber that has five apertures. At the beginning of each trial, animals must wait a certain amount of time, typically 5 s, before a light in one of the apertures is illuminated. Animals need to stick their nose into the illuminated aperture within a certain amount of time to receive reinforcement. Animals must pay attention to the apertures to determine which one has been illuminated to correctly respond. My colleague Dr. Brittany Smith uses a touchscreen variant of the 5-CSRTT in mice (e.g., Smith et al., 2022).

There are certainly other ways we can measure attention in rodents. In some studies, animals are required to respond on one manipulandum (e.g., left lever) when a specific light is illuminated. When the light is extinguished, animals are required to respond on a different manipulandum (McGaughy & Sarter, 1995). Rodents can even perform this task using a touchscreen (Wicks et al., 2017). Similar tasks as described above can be used to measure selective attention, except a distracting stimulus (e.g., odor cue, flashing light) is presented as well (Gendle et al., 2004).

Comprehension Check 1 What is the key difference between top-down and bottom-up attention?

Top-down attention is driven by our prior knowledge and expectations and relies on conscious effort to occur. Bottom-up attention is driven by salient external stimuli that automatically capture our attention with little conscious effort.

4 Overview of Attentional Deficits

Much of our understanding about the neuroscience of attentional deficits comes from studies with individuals diagnosed with **attention-deficit/hyperactivity disorder (ADHD)**, one of the most common *neurodevelopmental disorders.* Currently, there are approximately 140 million children that have received a diagnosis of ADHD, and over 366 million adults are symptomatic for ADHD (Song et al., 2021). ADHD is not a single disorder as individuals can be diagnosed with one of three subtypes of ADHD: inattentive, hyperactive/impulsive, or combined. This chapter will focus on individuals diagnosed with the inattentive subtype of ADHD only. You will learn more about the hyperactive/impulsive subtype of ADHD in Chap. 12.

According to the *Diagnostic and Statistical Manual of Mental Disorders (DSM-5)* (American Psychiatric Association, 2013), a child up to the age of 17 must meet six of the following criteria over the span of 6 months to be diagnosed with the inattentive subtype of ADHD:

a. Often fails to give close attention to details or makes careless mistakes in schoolwork, at work, or during other activities (e.g., overlooks or misses details, work is inaccurate).
b. Often has difficulty sustaining attention in tasks or play activities (e.g., has difficulty remaining focused during lectures, conversations, or lengthy reading).
c. Often does not seem to listen when spoken to directly (e.g., mind seems elsewhere, even in the absence of any obvious distraction).

Fig. 2 Example apparatus used to measure performance in the 5-choice serial reaction time task (5-CSRTT). In this version, rats respond on a touchscreen. Notice that one of five squares is illuminated while the others remain dark. The rat needs to touch the illuminated grid to receive reinforcement. If the rat touches a dark grid, this is considered an incorrect response and indicates the rat is not paying attention. *Source* Figure 1 of Hambrecht-Wiedbusch et al. (2019), https://doi.org/10.3389/fnbeh.2019.00076. CC BY 4.0 (https://creativecommons.org/licenses/by/4.0/), via Frontiers

d. Often does not follow through on instructions and fails to finish schoolwork, chores, or duties in the workplace (e.g., starts tasks but quickly loses focus and is easily sidetracked).
e. Often has difficulty organizing tasks and activities (e.g., difficulty managing sequential tasks; difficulty keeping materials and belongings in order; messy, disorganized work; has poor time management; fails to meet deadlines).
f. Often avoids, dislikes, or is reluctant to engage in tasks that require sustained mental effort (e.g., schoolwork or homework; for older adolescents and adults, preparing reports, completing forms, reviewing lengthy papers).
g. Often loses things necessary for tasks or activities (e.g., school materials, pencils, books, tools, wallets, keys, paperwork, eyeglasses, mobile telephones).
h. Is often easily distracted by extraneous stimuli (for older adolescents and adults, may include unrelated thoughts).
i. Is often forgetful in daily activities (e.g., doing chores, running errands; for older adolescents and adults, returning calls, paying bills, keeping appointments). (p. 59).

For individuals over 17 years of age, only five symptoms need to be present for a period of 6 months to be diagnosed with the inattentive subtype. In addition to the criteria presented above, the *DSM-5* requires the following for a diagnosis of ADHD:

a. Several inattentive…symptoms were present prior to age 12 years.
b. Several inattentive…symptoms are present in two or more settings (e.g., at home, school, or work; with friends or relatives; in other activities).
c. There is clear evidence that the symptoms interfere with, or reduce the quality of, social, academic, or occupational functioning.
d. The symptoms…are not better explained by another mental disorder… (p. 60).

While attentional deficits are a defining feature of the inattentive subtype of ADHD, I do want to note that there are other psychiatric conditions in which inattention is observed. In Chap. 3, you learned about perceptual distortions linked to schizophrenia. Individuals with schizophrenia often suffer from cognitive disturbances, including attentional deficits (Carter et al., 2010). Throughout this book, you will encounter additional psychiatric conditions that include attentional deficits as a feature. For simplicity, most of the discussion surrounding the neuromechanisms of attentional deficits will come from individuals with the inattentive subtype of ADHD and animal models of ADHD (see below).

Comprehension Check 2 Why do children up to 17 years of age need to exhibit six criteria over 6 months to be diagnosed with the inattentive subtype of ADHD whereas individuals over 17 years of age need to meet five criteria only?

A lower threshold is used for older individuals because the presence of five symptoms in an older individual can cause greater problems compared to a younger individual displaying six symptoms. For example, an 18-year-old can be in both school and work. Therefore, experiencing symptoms such as failing to give close attention to details, having difficulty sustaining attention, and failing to follow instructions can lead to low grades in school and possible termination from work.

4.1 Animal Models of ADHD

To better understand the pathogenesis of ADHD, certain rodent strains that display excessive inattention are tested. The spontaneously hypertensive rat (SHR) is the most used animal model of ADHD today, but it was not originally bred for this purpose. SHRs are an inbred rat strain, derived from the Wistar-Kyoto (WKY) rat strain, which is also an inbred strain stemming from Wistar rats. Beginning in the 1960s, WKY siblings with high blood pressure were selectively bred. SHRs show impaired performance in measures of sustained/selective attention (De Bruin et al., 2003; Sagvolden, 2006) and divided attention (Jentsch, 2005).

Interestingly, there are several substrains of SHRs and WKY rats. SHRs/WKY rats bred from the company Charles River are known as SHR/NCrl and WKY/NCrl, respectively. Another substrain comes from a company formally known as Harlan Industries (SHR/NHsd and WKY/NHsd). SHR/NCrl and WKY/NCrl show equivalent deficits in sustained attention, but WKY/NCrl rats do not show the same level of hyperactivity as SHR/NCrl (Sagvolden et al., 2008). As such, the WKY/NCrl may be a better model of the inattentive subtype of ADHD, with the SHR/NCrl strain serving as a model of the *combined subtype* of ADHD (combination of inattentive and hyperactive/impulsive symptoms). This chapter will discuss some of the research findings observed in WKY/NCrl rats and SHR/NCrl rats relative to control strains. In Chap. 12, you will learn more about SHR/NCrl rats and other animal models used to mimic the hyperactive/impulsive subtype of ADHD.

5 Neuroscience of Maladaptive Attentional Processes

Our attention relies on several neurotransmitter systems and neural circuits. I will provide an overview of the neurotransmitters involved in attention, followed by a discussion of the major brain pathways that regulate distinct aspects of attention. In the sections below, I will provide evidence of how dysfunction of these systems gives rise to attentional deficits.

5.1 Neurotransmitter Systems Involved in Attention

Given that perception and attention are closely intertwined, some of the neurotransmitters you learned about in Chap. 3 also mediate attentional processes. You will also be introduced to several neurotransmitters that are important for attention.

5.1.1 Glutamate and GABA

As the major excitatory and inhibitory neurotransmitters in the brain, respectively, glutamate and gamma-aminobutyric acid (GABA) play pivotal roles in attention. Glutamate is theorized to drive top-down attentional processes, enabling individuals to engage in sustained/selective attention while GABA inhibits bottom-up attention, allowing individuals to focus on something that is immediately relevant (e.g., a coworker that is currently speaking) (Kondo et al., 2023).

You may have experienced firsthand how altering glutamate and GABA levels affects your attention. Alcohol (ethanol) blocks glutamate N-methyl-D-aspartate (NMDA) receptors and potentiates $GABA_A$ receptors. As such, alcohol acts as a depressant, slowing one's nervous system. If you have ever consumed alcohol, you may have noticed that maintaining your focus on a conversation can become challenging. This is because of alcohol's inhibitory effect on NMDA receptors and potentiating effect on $GABA_A$ receptors. Laboratory studies have consistently shown that alcohol impairs performance in attention-based tasks (e.g., Dougherty et al., 2000; Hoyer et al., 2007; Wester et al., 2010). Other drugs that block NMDA receptors, like ketamine and phencyclidine (PCP), negatively affect attentional processes in both humans (Rosenbaum et al., 1959; Zhornitsky et al., 2022) and animals (Neill et al., 2010).

Although GABA inhibits bottom-up attention to help us focus on relevant stimuli, increasing GABAergic transmission with *benzodiazepines* impairs attention (Buffett-Jerrott & Stewart, 2002), with attentional deficits reported in individuals that have recently abstained from using benzodiazepines (Crowe & Stranks, 2018). Benzodiazepines are also known as sedative-hypnotic drugs, historically used as anxiolytics, sedatives, and muscle relaxants. These drugs bind to a specific site on the $GABA_A$ receptor. Benzodiazepines do not directly activate $GABA_A$ receptors; instead, they increase the frequency at which GABA binds to the receptor. Essentially, benzodiazepines cause too much inhibition to occur in the brain, which impairs attention.

5.1.2 Norepinephrine and Dopamine

In the previous two chapters, you learned about the serotonergic system and its involvement in hallucinations and pain perception. Serotonin's "chemical cousins" dopamine and norepinephrine are involved in attention (Burk et al., 2018). Like serotonin, dopamine and norepinephrine are monoamines. Serotonin is an indoleamine while dopamine and norepinephrine are catecholamines. Not only are dopamine and norepinephrine catecholamines, but norepinephrine is directly synthesized from dopamine.

Recall that dopaminergic neurons originate from one of several regions of the brain and stimulate either excitatory D_1-like or inhibitory D_2-like receptors, which are both metabotropic (i.e., coupled to a G protein). You have already learned about a couple of dopaminergic pathways—dopamine neurons originating from the ventral tegmental area (VTA) project to the nucleus accumbens (NAc) (mesolimbic pathway) and to the cerebral cortex (mesocortical pathway). Often, these pathways are combined into the mesocorticolimbic pathway (Speranza et al., 2021). Dopaminergic neurons also originate from the substantia nigra and travel to the basal ganglia (nigrostriatal pathway). Going back to the animal models of ADHD I discussed earlier, WKY/NCrl and SHR/NCrl have increased mRNA expression of the dopamine transporter compared to WKY/NHsd in theVTA and the substantia nigra (Roessner et al., 2010). These results indicate that too much dopamine is taken back up into the presynaptic neuron, preventing dopamine from stimulating other brain regions involved in attention (more on that later in this chapter).

Noradrenergic neurons originate from the locus coeruleus, located in the brainstem. Like the dopaminergic system, noradrenergic neurons travel to cortical areas (Berridge & Spencer, 2016). Norepinephrine binds to one of three major receptor types, with each one being G protein-coupled: alpha $(\alpha)_1$, α_2, and beta (β). Noradrenergic α_1 and β receptors are excitatory, with the former coupled to a G_q protein and the latter coupled to a G_s protein (like D_1-like receptors). The α_2 receptor is inhibitory, coupled to a G_i protein (similar to D_2-like receptors). Animal research has shown that noradrenergic transmission from the locus coeruleus is important for attention as chemogenetic inhibition of noradrenergic neurons in this region impairs attention as assessed in the 5-CSRTT (Fitzpatrick et al., 2019).

The relevance of catecholamines to attention is best demonstrated by the fact that drugs that increase dopamine and/or norepinephrine levels in the brain are the primary treatments for ADHD (Groom & Cortese, 2022). The two most common prescription medications are Adderall (combination of *d*- and *l*-amphetamine salts) and Ritalin (methylphenidate). Both medications are psychostimulant drugs that exert their effects by targeting dopamine and norepinephrine transporters, albeit in different ways. Methylphenidate blocks each transporter type, preventing the reuptake of dopamine and norepinephrine from the synapse. Amphetamine causes each transporter to work in reverse, causing intracellular dopamine and norepinephrine to travel into the synapse; amphetamine also prevents the enzyme monoamine oxidase from metabolizing dopamine and norepinephrine. Like individuals with ADHD, long-term treatment with *l*-amphetamine improves deficits in sustained attention in SHR/NCrls (Sagvolden, 2011).

Even in non-ADHD individuals, amphetamine and methylphenidate improve sustained attention (MacQueen et al., 2018; Rapoport & Inoff-Germain, 2002). However, one problem is prescription stimulant misuse, particularly in college students. In one study, such stimulant misuse was identified in 8% of college students (Wong et al., 2022). Even though ADHD medications can improve aspects of attention in non-ADHD individuals, elevated dopamine and norepinephrine levels can be detrimental. Recall from Chap. 3 that stimulants can produce symptoms that mimic schizophrenia, which is attributed to increased dopamine levels in the mesolimbic pathway. Chemogenetic activation of dopaminergic neurons in the VTA and the substantia nigra also impairs attention in rats (Boekhoudt et al., 2017).

Beyond amphetamine and methylphenidate, nonstimulant drugs are used to treat ADHD. Strattera (atomoxetine) is similar to methylphenidate, but it only blocks the norepinephrine transporter. Guanfacine and clonidine are agonists at a type of noradrenergic receptor called the α_{2a} receptor. Interestingly, the α_{2a} receptor is a G_i-coupled metabotropic receptor, which is inhibitory. Intuitively, increasing activity of an inhibitory receptor subtype would worsen attention. To understand how guanfacine and clonidine can be used to treat ADHD, we need to look more closely inside the neuron.

α_{2a} receptors are often located on glutamatergic neurons. When we are in a calm, but attentive state, moderate norepinephrine levels bind to α_{2a} receptors. This action regulates the actions of calcium (Ca^{++}) that enter the neuron through glutamatergic NMDA receptors. Remember, NMDA receptors are ionotropic, allowing sodium (Na^+) and Ca^{++} to enter a neuron. When Ca^{++} enters the neuron, it triggers a series of events inside the cell. When you learned about neurons in Chap. 2, you learned that positively charged ions like Na^+ and Ca^{++} depolarize the neuron, increasing the likelihood it sends a message to another neuron. However, when NMDA receptors are not overly stimulated, small amounts of Ca^{++} enter the neuron, which causes a different sequence of events to occur in the neuron. Now, potassium (K^+) channels open, causing K^+ to exit the neuron, which hyperpolarizes (i.e., inhibits) the neuron. By stimulating α_{2a} receptors, norepinephrine helps reduce how much K^+ leaves the neuron. The end result is to limit hyperpolarization, thus helping increase neurotransmission (Arnsten, 2020).

One interesting finding is that exercise transiently improves selective attention, at least in women. This may occur because exercise activates our stress system, resulting in norepinephrine release (called adrenaline when released from the adrenal glands). You will learn more about the stress system in Chap. 8. The improved attention may result from greater connectivity between the locus coeruleus and other regions involved in attention (Mather et al., 2020).

Comprehension Check 3 How do medications for ADHD help reduce attentional deficits?

The primary ADHD treatments work to increase dopamine and norepinephrine levels in the brain by inhibiting the norepinephrine and/or dopamine transporter. This allows norepinephrine and/or dopamine levels to increase in the synapse, thus stimulating their respective receptors. Other ADHD medications work by stimulating the inhibitory noradrenergic α_{2a}*receptor. Stimulation of this receptor subtype reduces the* *amount of K^+ that exits glutamatergic neurons, preventing them from becoming too hypoactive.*

5.1.3 Acetylcholine

Another neurotransmitter involved in attention is acetylcholine. Cholinergic neurons originate from several brain regions. Projections emanating from the pedunculopontine tegmental nucleus and the laterodorsal pontine tegmentum (both located in the pons) travel to the brainstem, the cerebellum, the thalamus, and midbrain structures. The basal forebrain, located ventrally to the striatum, is the primary cholinergic output of the brain. Cholinergic neurons originating from the basal forebrain terminate in the cerebral cortex and limbic structures (Perez-Lloret & Barrantes, 2016).

Acetylcholine can bind to two different major receptor types: ionotropic nicotinic receptors and metabotropic muscarinic receptors. Nicotinic receptors get their name because nicotine, the primary psychoactive chemical in tobacco products, is the primary exogenous ligand that binds to these receptors (acetylcholine is the endogenous ligand that binds to nicotinic receptors). When nicotinic receptors are activated, an ion channel opens and allows positively charged Na^+ ions into the neuron. Muscarinic receptors are metabotropic and are both excitatory (G_q-coupled M_1, M_3, and M_5 subtypes) and inhibitory (G_i-coupled M_2 and M_4 subtypes).

Cholinergic neurons travel to the VTA and stimulate nicotinic receptors located on dopaminergic neurons. This means that acetylcholine modulates dopaminergic activity in the mesocorticolimbic pathway. As such, increasing acetylcholine levels should lead to improved attention. One anecdotal piece of evidence to support acetylcholine as a "cognitive enhancer" comes from the finding that cigarettes often lead to self-reported increases in attention (Campos et al., 2016). Additionally, a large percentage of individuals withschizophrenia (~70–80%) smokes cigarettes (Winterer, 2010). In Chap. 3, I focused on perceptual disturbances associated with schizophrenia. Individuals with schizophrenia experience a wide range of cognitive impairments, including attentional deficits (Dutterer et al., 2023; Morris et al., 2013). One hypothesis is that individuals with schizophrenia smoke to ameliorate cognitive dysfunction, although others have argued that smoking is a risk factor for the development of schizophrenia (Ding & Hu, 2021).

Because cigarettes contain thousands of chemicals, laboratory studies are useful for elucidating the effects of nicotine on cognition. To this end, nicotine administration increases alerting attention and orienting attention (Valentine & Sofuoglu, 2018), and nicotine improves performance in a 5-CSRTT in rodents and primates (Day et al., 2007; Spinelli et al., 2006). Beyond nicotine, other drugs that stimulate nicotinic receptors improve sustained attention (Kolisnyk et al., 2015). Conversely, blocking nicotinic receptors or muscarinic receptors impairs performance in the 5-CSRTT (Balachandran et al., 2018; Burk et al., 2018; Remmelink et al., 2017). Animal research has also demonstrated that mice lacking a specific subunit of the nicotinic receptor (called α_7) have difficulty sustaining attention in the 5-CSRTT (Hoyle et al., 2006; Kolisnyk et al., 2015).

5.1.4 Endocannabinoids

Endocannabinoids bind to inhibitory cannabinoid receptors (CB_1 and CB_2). Like alcohol, some individuals reading this book may personally have experienced attentional deficits following cannabis use. Cannabis affects multiple cognitive processes, including attention (Urits et al., 2021), with such deficits observed in animals as well (Moore et al., 2023). As cannabis becomes legal/decriminalized, there are concerns that traffic accidents will become more common. Indeed, a recent study found that cannabis was the most detected substance in injured drivers, followed by alcohol (Brubacher et al., 2025).

Cannabinoids alter attention by modulating activity of various neurotransmitter systems. CB receptors are located on dopamine neurons in the VTA (Han et al., 2023; Zhang et al., 2017) and on cortical glutamatergic and GABAergic neurons (Hill et al., 2007). Cannabinoid receptors that are located on GABAergic and glutamatergic neurons are important for regulatingdopamine release in the mesocorticolimbic pathway (see Fig. 3). Endocannabinoids also modulate activity in the pedunculopontine tegmental nucleus, one of the primary cholinergic centers (Kovács et al., 2017).

While cannabis tends to slow reaction time and impair attention span, there is some evidence that cannabis use can improve attentional control in certain groups of individuals. Read the Experiment Spotlight below to learn more.

Experiment Spotlight: Using Task-fMRI to Explore the Relationship Between Lifetime Cannabis Use and Cognitive Control in Individuals with First-Episode Schizophrenia

The full article can be accessed with this link: https://pmc.ncbi.nlm.nih.gov/articles/PMC11317632/.

In Chap. 3, you learned about the dopamine, the serotonin, and the glutamate hypotheses of schizophrenia, particularly how alterations to these neurotransmitter systems lead to hallucinations. Cannabinoids have also been linked to schizophrenia, as long-term cannabis use has been implicated in schizophrenia symptomatology in some individuals. However, individuals with schizophrenia are more likely to use cannabis, just as they are more likely to smoke cigarettes. In the present experiment (Lesh et al., 2024), individuals were tested in a variation of the CPT (AX-CPT) while undergoing functional magnetic resonance imaging (fMRI). In the AX-CPT, individuals need to respond when they see the letter "X", but only when the letter "A" precedes it.

Participants included individuals within the first year of the first episode of psychosis (FES), with some of these individuals having a history of cannabis use (FES+CAN; $n = 48$) and some having no history of cannabis use (FES-CAN; $n = 28$). Healthy controls with no history of cannabis use were included (CON; $n = 59$).

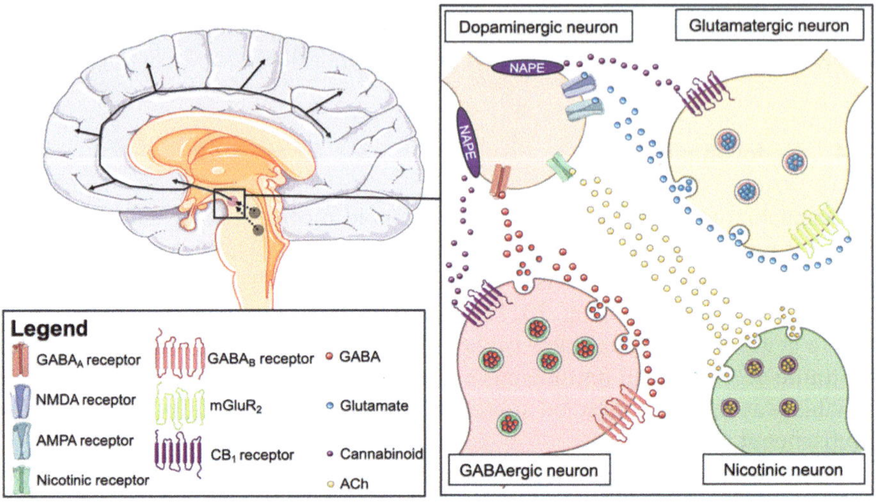

Fig. 3 A simplified schematic of a synapse in the ventral tegmental area (VTA). Cholinergic neurons originating from nuclei in the brainstem (dashed arrows) travel to the VTA, where they release acetylcholine that binds to nicotinic receptors located on dopaminergic neurons. Both GABAergic and glutamatergic neurons are located in the VTA, which help regulate the activity of dopaminergic neurons. Endocannabinoids like anandamide and 2-arachidonylglycerol (2-AG) can modulate the activity of GABAergic and glutamatergic neurons. *Note* ACh = acetylcholine, AMPA = α-amino-3-hydroxy-5-methyl-4-isoxazolepropionic acid, CB_1 = cannabinoid receptor 1, GABA = gamma-aminobutyric acid, $mGluR_2$ = metabotropic glutamate receptor 2, NAPE = N-arachidonoyl phosphatidylethanolamine (phospholipid that is critical for the synthesis of endocannabinoids), NMDA = N-methyl-D-aspartate. Image created by the author using Servier Medical Art (https://smart.servier.com/; CC BY 4.0, https://creativecommons.org/licenses/by/4.0/) and Microsoft PowerPoint

As expected, individuals without a history of psychosis (CON) performed better in the AX-CPT compared to both FES+CAN and FES-CAN groups. Interestingly, the FES+CAN group performed better than the FES-CAN group. The CON group also showed greater activation in part of the prefrontal cortex (PFC). The FES+CAN group showed a somewhat similar level of activation in the PFC compared to the CON group. The CON group also showed greater activation of brain regions in the frontal and parietal cortices during difficult trials compared to easy trials. The FES+CAN group showed less widespread activation while the FES-CAN showed no significant changes in brain activity when comparing harder trials to easier trials.

Questions to Consider:

1. One finding of the experiment is that individuals in the FES+CAN group were more likely to have a history of tobacco use compared to individuals in the FES-CAN group. How can this finding influence the interpretations of the study?
2. This experiment used one healthy control group. How could including a CON+CAN group and a CON-CAN group improve the results of the present experiment?

3. The researchers discuss some theories to explain the results of their study: (a) individuals with psychosis that use cannabis may have higher social functioning and intelligence quotient (IQ) scores prior to onset of psychosis; (b) cannabidiol (CBD) has anti-inflammatory effects, which could ameliorate some of the negative symptoms of schizophrenia; (c) individuals who use cannabis and develop schizophrenia may be less vulnerable to psychosis. Based on what you have learned in Chap. 3 and in this chapter, how could cannabis use affect dopaminergic signaling, thus improving cognitive functioning?

Comprehension Check 4 How do acetylcholine and endocannabinoids modulate dopaminergic activity in the brain?

Cholinergic neurons originating from the brainstem stimulate nicotinic receptors located on dopaminergic neurons in the VTA. This causes depolarization of dopaminergic neurons, leading to increased dopamine release in cortical and limbic regions of the brain. Cannabinoid receptors are also located on intra-VTA dopaminergic neurons; however, they act to inhibit dopaminergic activity. This can account for the cognitive-enhancing effects of nicotine products and the deleterious effects of cannabinoids on attention.

5.2 *Attention Brain Circuits*

Earlier in this chapter you learned about top-down and bottom-up attention. There are multiple neurocircuits that influence both types of attention (see Fig. 4). Before diving into these circuits, I first need to discuss a network that is highly active when we are at rest (i.e., not engaged in attentive processes): the **default mode network (DMN)**. You will encounter this network several times as you read through this book as dysregulation of this network is proposed to contribute to various psychiatric conditions. Although I did not discuss the DMN when covering schizophrenia, individuals with schizophrenia show aberrant activity in brain regions that make up this circuit (Hu et al., 2017).

5.2.1 Default Mode Network (DMN)

The DMN is most active when we are not focusing on things going on around us, such as when we daydream or let our mind wander. Several brain regions that you have not learned about make up the DMN, including the dorsomedial PFC (dmPFC), the posterior cingulate cortex (PCC), the precuneus, and the angular gyrus.

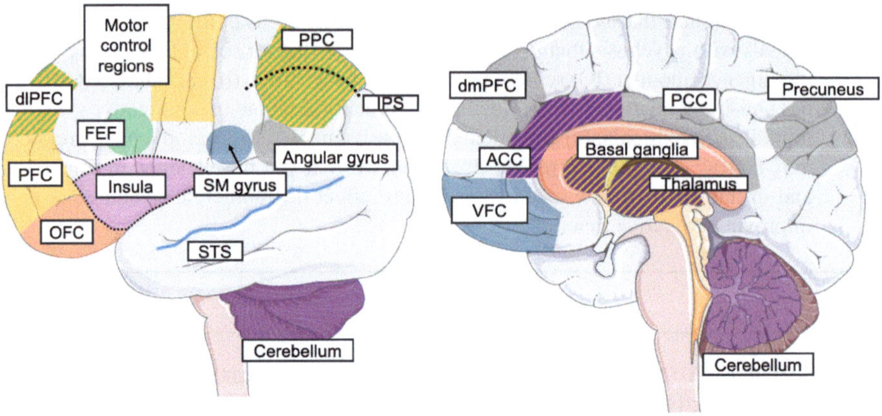

Fig. 4 A schematic showing the different attention-related networks in the brain. Brain regions associated with the default mode network (DMN) are highlighted in gray and include the dmPFC (dorsomedial prefrontal cortex), the PCC (posterior cingulate cortex), and the precuneus. Brain regions associated with the dorsal attention network (DAN) are highlighted in green and include the dlPFC (dorsolateral prefrontal cortex), the FEF (frontal eye field), the PPC (posterior parietal cortex), and the IPS (intraparietal sulcus) (indicated by dashed line). Brain regions associated with the ventral attention network (VAN) are highlighted in blue and include the SM (supramarginal) gyrus, the STS (superior temporal sulcus), and the VFC (ventral frontal cortex). Brain regions associated with the salience network are highlighted in pink and include the anterior insula and the ACC (anterior cingulate cortex). Structures associated with the cortico-basal ganglia-thalamo-cortical (CBGTC) loop are highlighted in orange and include the PFC (prefrontal cortex), the basal ganglia, the thalamus, and regions associated with motor functions; the dlPFC and the parietal cortex are also involved in this loop. Structures associated with the cerebello-thalamo-striatal-cortical (CTSC) loop are highlighted in purple and include the cerebellum and the OFC (orbitofrontal cortex); additionally, the ACC, the basal ganglia, and the thalamus are part of this circuit. Image created by the author using Servier Medical Art (https://smart.servier.com/; CC BY 4.0, https://creativecommons.org/licenses/by/4.0/) and Microsoft PowerPoint

Individuals with ADHD have increased resting-state functional connectivity within the DMN (McCarthy et al., 2013). That is, the DMN is too active in individuals with ADHD, which may account for attentional deficits in these individuals. When we need to pay attention to something, the DMN should become less active as different brain networks (to be discussed below) become active. As you will read below, the PFC is important for attentional control. For an individual with ADHD, brain regions that make up the DMN exert too much control over the PFC, which prevents it from engaging attentional networks, leading to attentional deficits.

ADHD medications help reduce the hyperactivity observed in the DMN (Silberstein et al., 2016), which may be influenced by noradrenergic stimulation of α_1 receptors in the dmPFC. For example, injection of an α_1 receptor antagonist prevents methylphenidate-induced improvements in sustained attention (Berridge et al., 2012), and intra-dmPFC administration of an α_1 receptor agonist improves such attention (Berridge & Spencer, 2016). Furthermore, WKY/NCrl rats show decreased glutamate-induced release of norepinephrine in the PFC compared to SHR/NCrl and

Sprague Dawley rats (an outbred strain) (Howells et al., 2009). By increasing noradrenergic neurotransmission in the PFC, this region may be able to surpass inputs from other regions of the DMN, allowing it to activate other brain structures that regulate attentional processes.

Beyond ADHD, I earlier mentioned that cannabis is notorious for impairing attention. Cannabis-induced impairments in attention are directly correlated with reduced deactivation of the DMN (Bossong et al., 2013). This means that cannabis prevents the switch from DMN to other neural circuits that enable attention. Let's turn our attention (pun intended) to these attention circuits.

5.2.2 Dorsal Attention Network (DAN)

The **dorsal attention network (DAN)** consists of several brain regions in the frontal and parietal cortices that help regulate top-down attention. I will first describe evidence linking the frontal cortex to top-down attention, followed by a discussion of how the parietal cortex mediates such attention.

Frontal Cortical Control of Attention

The dorsolateral PFC (dlPFC) is critical for sustained and selective attention (Kane & Engle, 2002). Intermittent theta burst stimulation, a form of transcranial magnetic stimulation (TMS) that you learned about in Chap. 2, applied to the left dlPFC improves attention (Dong et al., 2023). Conversely, using TMS to inactivate the dlPFC disrupts divided attention (Johnson et al., 2007), and pharmacological inactivation of dlPFC impairs the ability of monkeys to perform a visual search task (Iba & Sawaguchi, 2003) and impairs selective attention (Suzuki & Gottlieb, 2013).

At the neurochemical level, dopamine release is observed in frontal regions in individuals performing a sustained attention task (Aalto et al., 2005). Remember, dopaminergic neurons originating from the VTA project to cortical areas, including the PFC. In rodents, the medial PFC (mPFC) has received considerable attention as it is homologous to the human dlPFC (Uylings et al., 2003). Stimulation of dopamine D_1-like receptors in mPFC improves visual attention while blocking these receptors impairs attention (Chudasama & Robbins, 2004; Granon et al., 2000). However, intra-mPFC D_2-like receptors do not appear to regulate attention (Granon et al., 2000). As D_1-like receptors are excitatory, these results are consistent with the findings described above that increased activation of the DAN is critical for top-down attention.

Increased stimulation of dopamine D_1-like receptors in the PFC may be one major reason why psychostimulant drugs like amphetamine and methylphenidate improve symptoms of ADHD. As dopaminergic transmission originating from the VTA appears to be hypoactive in individuals with ADHD, there is insufficient dopamine to stimulate D_1-like receptors in the brain.

While glutamate increases norepinephrine release, activation of D_1-like receptors increases glutamatergic activity in the PFC (Seamans et al., 2001). D_1-like receptors are highly expressed on a type of excitatory neuron called a **pyramidalneuron** (the name comes from the pyramid-like shape of the cell body). Insufficient activation of D_1-like receptors leads to hypoactive glutamate levels, which can underlie attentional deficits. In fact, these deficits are observed following intra-mPFC blockade of glutamate NMDA receptors (Mirjana et al., 2004; Murphy et al., 2005). Adolescents treated with methylphenidate have greater glutamate levels in the right dlPFC, providing additional support that attentional deficits result from insufficient stimulation of the DAN (Bakhshi et al., 2023).

To complicate matters a little, cholinergic transmission in the PFC is important for attention. Acetylcholine levels increase in mPFC of rats completing the 5-CSRTT (Dalley et al., 2001; Passetti et al., 2000) or a sustained attention task (Arnold et al., 2002). Disrupting cholinergic transmission impairs attention (Burk et al., 2018). Like dopamine, acetylcholine is an important modulator of NMDA receptor activity (Cools & Arnsten, 2022). Collectively, increased activity in frontal regions like the dlPFC helps us focus our attention, with deficient activation of these regions being associated with inattention.

Beyond the PFC, another region implicated in top-down attention is the frontal eye field (FEF), which receives direct inputs from the dlPFC (Pierrot-Deseilligny et al., 2005). The FEF is activated when individuals focus their attention on a stimulus (Astafiev et al., 2003). In Chap. 3, I mentioned that damage to the FEF is associated with simultanagnosia, a condition in which individuals cannot see a picture as a whole. This is consistent with the finding that inactivation of the FEF leads to deficits in spatial attention (Bollimunta et al., 2018). I also want to highlight that one hypothesis is that simultanagnosia reflects a "spatial restriction" of attention (Dalrymple et al., 2013).

Neurons projecting from the FEF to visual cortices express higher concentrations of dopamine D_1-like receptors compared to D_2-like receptors (Mueller et al., 2020). Furthermore, D_1-like-containing neurons in the FEF are known to control neuronal responses in part of the visual cortex (Noudoost & Moore, 2011). This is consistent with what I discussed earlier that D_1-like receptors appear to be more important for attention compared to D_2-like receptors.

Similar to the dlPFC, noradrenergic and cholinergic activity in the FEF contribute to attentional processes. Each noradrenergic receptor subtype is highly expressed on pyramidal neurons in the FEF (Lee et al., 2020), and muscarinic receptors in the FEF are implicated in attention (Dasilva et al., 2019).

Parietal Cortical Control of Attention

Parietal regions involved in attention include the intraparietal sulcus (IPS) and the posterior parietal cortex (PPC). Early evidence linking parietal regions to attention comes from case studies in which individuals with damage to the parietal lobe experience **hemispatial neglect**. For example, an individual with damage to the right parietal lobe may fail to notice stimuli that are presented to the left visual field. Even when asked to draw an image, individuals with parietal damage will not be able to draw the entire stimulus (Chen & Goedert, 2012).

Like the FEF, the IPS is activated when individuals focus their attention on a stimulus (Astafiev et al., 2003). Pharmacological inactivation of this region with GABAergic receptor agonists disrupts saccadic eye movements in response to incoming sensory stimuli (Li et al., 1999), although impairments in selective attention are not as pronounced following inactivation of this region compared to inactivation of the dlPFC (Suzuki & Gottlieb, 2013). Also, cholinergic activity in the IPS appears to be important for orienting and reorienting attention (Davidson & Marrocco, 2000; Thiel et al., 2005). Cholinergic activity within frontoparietal regions is also important for alerting attention (Ljubojevic et al., 2018).

Collectively, these data indicate that the IPS is critical for *spatial attention*. If someone begins speaking to you, you most likely turn your head toward the speaker. Individuals with localized lesions to the IPS struggle to orient their attention to specific stimuli (Gillebert et al., 2011). Dopaminergic activity in the IPS is particularly important for spatial attention (van Kempen et al., 2022). Individuals with ADHD show impaired spatial attention; in fact, one study found that college students with ADHD were more likely to miss a target stimulus that was shown to the left side of their body compared to the right side (Jones et al., 2008).

While the IPS is important for spatial attention, the PPC is implicated in selective attention (Behrmann et al., 2004) as well as general arousal (Malhotra et al., 2009). Increased GABA concentration in the PPC is associated with *attentional blink*, a phenomenon in which detection of one stimulus is difficult if it is presented immediately after a separate target (Kihara et al., 2016). A real-life example of attentional blink can be observed when driving. Seeing a car pulled over to the side of the road may prevent a driver from seeing a deer dart in front of their car. Enhanced attentional blink is observed in individuals with ADHD (Amador-Campos et al., 2015).

Dopamine activity in the PPC, like the IPS, is important for attention; yet, too much dopamine in the PPC can be problematic. This is the same concept I brought up earlier when discussing the contributions of dopamine to attention. Specific to the PPC, mice with reduced dopamine transporters have difficulty performing an attention-based task, and they show decreased activity in the PPC (Cybulska-Klosowicz et al., 2017). Somewhat related, transcranial direct current stimulation (tDCS) applied to the PPC improves orienting attention in children with ADHD, but it hinders selective attention (Salehinejad et al., 2020). Additionally, individuals with schizophrenia have a hyperactive parietal cortex (Wible et al., 2009), which may be attributed to elevated dopamine levels observed in these individuals.

Comprehension Check 5 How do the DMN and the DAN work together to control attention?

These systems are often viewed as being "anti-correlated"; that is, they work in opposition to one another. However, both systems are important for attention. The

DMN is most active when we are in a state of rest. The DAN is largely involved with top-down attention. If the DMN is hyperactive, the DAN is not able to become sufficiently activated, which can impair attentional control. Think of a car: being able to accelerate and to brake is important. Likewise, switching from a restful state (i.e., DMN activation) to a more alert state (DAN activation) allows us to focus on stimuli that are currently relevant.

5.2.3 Ventral Attention Network (VAN)

In contrast to the DAN, the **ventral attention network (VAN)** consists of brain regions that mediate bottom-up processing. The temporoparietal junction (TPJ) spans both temporal and parietal cortices, with the supramarginal (SM) gyrus located in the parietal cortex and the superior temporal sulcus (STS) located in the temporal cortex. The posterior portion of the SM gyrus allows us to direct our attention to an unexpected stimulus (think of a bright flash of light that you were not anticipating to see) with the anterior part of this region playing an important role in target detection (Patel et al., 2019). For example, disruption of the SM gyrus with repetitive TMS impairs individuals' divided attention in a dichotic listening task (Bareham et al., 2018). Because the SM gyrus is not able to function properly, individuals have a hard time identifying information that was presented in the opposite ear.

The middle region of the STS is involved in spatial attention; the posterior region of the STS is also involved in spatial attention, but its involvement is more specific to attention directed toward social cues (Ramezanpour & Fallah, 2022). You will learn more about the STS in Chap. 15 when I discuss maladaptive social interactions.

The ventral frontal cortex (VFC), like the SM gyrus, becomes activated when we are presented with an unexpected stimulus (Vossel et al., 2014). This region of the brain is also sensitive to salient stimuli. Have you ever noticed how paying attention to something can become difficult when you are hot? There is a reason for this phenomenon. Research has shown that individuals exposed to a temperature of 50 °C (122 °F) show decreased activity in the DAN and increased activity in the VFC (Qian et al., 2020). These results suggest that activation of the VFC is a negative; however, bottom-up attention is still important. Increased inattention is correlated with reduced volume of the ventromedial PFC (vmPFC) (Albaugh et al., 2017).

The importance of bottom-up attention is apparent when we use driving as an example. If an animal unexpectedly runs into the road, brain regions in the VAN become active. Failure to attend to this stimulus can be problematic, especially if the animal is large.

5.2.4 Salience Network

A couple of additional brain regions are involved in the detection and evaluation of highly salient stimuli: the anterior cingulate cortex (ACC) and the anterior insula (Schimmelpfennig et al., 2023). These regions are involved in other aspects of attention; for example, the ACC is activated during divided attention (Loose et al., 2003), and the anterior insula guides interoceptive attention, that is, attention to internal bodily functions (Wang et al., 2019).

Whereas the VAN is primarily responsible for orienting us toward a salient stimulus, the salience network allows us to assign value to such a stimulus. If there is a bright flash in the sky, the VAN becomes activated, and you turn your head toward the direction where the flash occurred. Brain regions in the salience network are then activated. This is valuable because you need to determine if the flash is coming from a random firework or if there is lightning in the distance, indicating an approaching storm.

Researchers can use an *oddball task* to measure attention to salient stimuli. In this task, participants are shown a series of identical stimuli (e.g., blue square). Eventually, a random stimulus, the "oddball" will be presented (e.g., red oval). The ACC and the anterior insula are activated when the oddball stimulus is presented (Molnar-Szakacs & Uddin, 2022), but these regions are not as active in individuals with the inattentive subtype of ADHD (Orinstein & Stevens, 2014). Despite the altered neural responses to an "oddball" stimulus, individuals with ADHD tend to perform similarly in the oddball task compared to healthy controls. Less activation may occur in individuals with ADHD because they have a hyperactive salience network at baseline (e.g., when not performing a cognitive task). Individuals that score higher in a self-report questionnaire related to ADHD symptoms have greater connectivity between the dorsal ACC (dACC) and the anterior insula (Janes et al., 2018). Similar hyperactivity is observed in children with ADHD (Damiani et al., 2021). A hyperactive salience network may account for why individuals with ADHD become easily distracted.

Comprehension Check 6 What is the relationship between the VAN and the salience network?

The VAN and the salience network are closely related and are both critical mediators of bottom-up attention. Both networks are activated by presentation of salient stimuli. There is an important distinction between these networks. The VAN is important for orienting/reorienting us to salient stimuli while the salience network allows us to assign value to a salient stimulus (e.g., this network enables us to determine how important a salient stimulus is).

5.2.5 Cortico-Basal Ganglia-Thalamo-Cortical (CBGTC) Loop

As the name suggests, the **cortico-basal ganglia-thalamo-cortical (CBGTC) loop** is a neural circuit consisting of the cerebral cortex (the cortico and the cortical parts of the loop), the basal ganglia, and the thalamus (Alexander et al., 1986). I have already discussed the contributions of cortical areas to attention, so this section will focus primarily on the basal ganglia and the thalamus.

When I briefly introduced the basal ganglia in Chap. 2, I wrote that this region is largely involved in motor functions. You will learn much more about the basal ganglia in Chaps. 10 and 11. For now, just know that the basal ganglia consist of several structures: the globus pallidus, the caudate nucleus, and the putamen. Collectively, the caudate nucleus and the putamen make up the dorsal striatum. The putamen and the globus pallidus make up the lentiform nucleus.

The basal ganglia and the thalamus are important for *sensory gating* (Redinbaugh & Saalmann, 2024). In the context of attention, the basal ganglia and the thalamus help select which stimuli are currently relevant and transmit those stimuli to other parts of the brain. Relatedly, the basal ganglia are heavily involved in attention shifting (van Schouwenburg et al., 2010). Let's use driving as an example. When driving down a highway, your attention is constantly shifting. To ensure that traffic has not come to a stop, you will focus your attention further down the highway. You will then shift your focus closer to your car's current position to make sure there are no upcoming potholes or other obstructions. You will also check your side and rearview mirrors to check for something like an emergency vehicle approaching. Your attention will also be directed to the sides of your vehicle to ensure that another vehicle is not merging where you are.

Inattention is negatively correlated with dopamine transporter availability in the caudate nucleus (Itagaki et al., 2024). This means there are fewer dopamine transporters to remove excess dopamine from the synapse. Additionally, WKY/NCrl rats have greater evoked dopamine release compared to SHR/NCrl rats in the dorsal striatum (Miller et al., 2012). These findings seem counterintuitive as I have discussed that a hypoactive dopaminergic system is a biomarker for ADHD. Remember, too much dopamine can be problematic. The basal ganglia have dopamine D_1-like and D_2-like receptors, meaning that excess dopamine can bind to both receptor subtypes, thus preventing them from engaging in selective gating.

If dopamine levels in the striatum are too elevated, increased binding to both D_1-like and D_2-like receptors can lead to a situation in which this part of the brain is insufficiently stimulated. Indeed, individuals with ADHD have lower levels of glutamate and glutamine (precursor to glutamate) (you will often see the combination of glutamate and glutamine abbreviated as Glx) in the basal ganglia; furthermore, in adults that are not currently on medication for ADHD, lower basal ganglia Glx concentrations are associated with inattention (Maltezos et al., 2014). Collectively, these results suggest that decreased activity of the basal ganglia gives rise to attentional deficits.

The inattentive subtype of ADHD is associated with morphological changes to the thalamus, specifically an enlarged thalamus (Fu et al., 2021). However, other research has shown that the pulvinar (a specific nucleus of the thalamus) is smaller in children with ADHD compared to controls (Ivanov et al., 2010). The pulvinar is especially important for selective gating, particularly for visual attention (Purushothaman et al., 2012). A smaller pulvinar may inhibit the flow of visual stimuli to other cortical areas for processing, thus interfering with selective attention. Beyond morphological changes, decreased thalamic activity is observed in individuals with ADHD (Källstrand et al., 2024).

Notice that hypoactivity of the basal ganglia and the thalamus is implicated in attentional deficits. Earlier, you learned that hypoactivity of the frontal cortex and parts of the parietal cortex are linked to inattention. These parts of the cortex make up the cortico and the cortical parts of the CBGTC loop. Overall, attentional deficits appear to arise because the CBGTC loop is hypoactive, thus preventing individuals from effectively attending to relevant stimuli (e.g., missing important announcements made during class).

5.2.6 Cerebello-Thalamo-Striato-Cortical (CTSC) Loop

Closely related to the CBGTC loop is the **cerebello-thalamo-striato-cortical (CTSC) loop**. The major difference between these pathways is the inclusion of the cerebellum in the latter. Like the basal ganglia, the cerebellum is primarily associated with motor control. Yet, this region is involved in higher-order cognitive processes, including attention. One major function of the cerebellum is *predictive coding* (Schmahmann, 2010).

Let's return to the example of driving I used earlier when discussing the basal ganglia. If you leave for school or work at the same time each day and take the same route to school/work, you have a general sense of what traffic is going to be like. If you must leave during rush hour, you can predict that there will be heavy traffic. You also know where that traffic is going to be. You also have a feel for how your vehicle handles when you drive. That is, you know how hard you need to push your foot down on the accelerator to go the speed you want, and you know how hard you need to press the brake to stop on time. Your ability to do these things is influenced by cerebellar activity.

How is predictive coding related to attention? Predictive coding and sensory gating are closely related. Sensory gating helps direct the flow of information from one part of the brain (e.g., basal ganglia/thalamus) to other parts of the brain such as the frontal cortex for further processing. Predictive coding, like sensory gating, acts as a filter. Imagine you are driving on the interstate during non-rush hour time. If you see heavy traffic stopped on the interstate ahead of you, this will cause a large "prediction error" to occur as this event is not expected given the time of day. If a lower prediction error is generated, you may not notice the discrepancy quickly enough to stop before crashing into another car. The lower prediction error can occur because one is not paying attention, instead focusing on something else like texting while driving.

Predictive coding can also be applied to something as simple as a conversation. Depending on who you around, you can predict what types of conversations you will have. If an individual says something completely unexpected (e.g., your friend tells you they have a terminal illness), this will capture your attention due to the large prediction error that occurs. Poor predictive coding can lead to situations in which an individual struggles to attend to a conversation.

We know the cerebellum is important for attention as morphological and functional changes are observed in the cerebellum of individuals with ADHD (Cundari et al., 2023), and individuals with damage to the cerebellum perform worse in measures of attention compared to controls (Reumers et al., 2025). Conversely, stimulating the cerebellum improves attention (Esterman et al., 2017).

The cerebellum receives excitatory inputs from various brain regions, including parts of the cerebral cortex. However, many neurons in the cerebellum are inhibitory, with the **Purkinje cell** being one of the most unique neurons in the central nervous system (CNS). This neuron has a distinct appearance, marked by numerous dendrites that look like a treetop. Despite the abundance of inhibitory neurons in the cerebellum, parts of the cerebellum, known as the deep cerebellar nuclei, project glutamatergic neurons to the thalamus. Thus, a dysregulated cerebellum leads to insufficient activity of the thalamus. The result is poor predictive coding and poor sensory gating, which leads to impaired attentional processing.

Comprehension Check 7 How do the CBGTC and the CTSC loops mediate attention?

These neural circuits are important for sensory gating and predictive coding, respectively. Both neural circuits include similar brain structures, such as the cerebral cortex, the striatum, and the thalamus. Collectively, these circuits allow us to make predictions about incoming stimuli and to filter relevant stimuli from irrelevant stimuli.

6 Chapter Summary

This chapter began by defining different types of attention and methods for measuring attention in both humans and animals. An overview of the inattentive subtype of ADHD was provided, and a couple of animal models of ADHD were introduced: the WKY/NCrl as a model of the inattentive subtype and the SHR/NCrl as a model of the combined subtype. This chapter delved into the neuroscience of maladaptive attentional processes, outlining the neurotransmitter systems involved in attention, with a major focus on dopamine and norepinephrine. Furthermore, the brain circuits involved in attention were covered. Overall, hyperactivity of the DMN and the salience network and hypoactivity of circuits such as the DAN, the CBGTC loop, and the CTSC loop are associated with attentional deficits.

Answers to Experiment Spotlight Questions

1. As you learned earlier in this chapter, tobacco contains nicotine, which binds to nicotinic acetylcholine receptors. Stimulating nicotinic receptors improves attention. Thus, the finding that cannabis use in individuals with FES improves attentional control may be influenced more by their history of tobacco smoking instead of cannabis use.
2. To better determine if cannabis use selectively improves attentional control in individuals with FES, comparing performance between FES+CAN, FES-CAN, CON+CAN, and CON-CAN is important. As you have learned, cannabis is typically associated with impaired cognitive processing. Having CON+CAN and CON-CAN groups would enable the researchers to determine if long-term cannabis use worsens attention in individuals without a history of psychosis. If long-term cannabis use is associated with improved AX-CPT performance in both FES and CON individuals, this may suggest that some other factor beyond cannabis is leading to improved performance (e.g., co-occurring tobacco use).
3. In Chap. 3, you learned that excess dopamine in the limbic system is implicated in hallucinations. Individuals with schizophrenia have a hyperactive dopaminergic system in the limbic system. In this chapter, you learned that cannabinoid receptors, which are inhibitory, are located in the VTA. Long-term cannabis use, like tobacco use could act as a form of self-medication to better regulate dopamine levels. Cannabis can inhibit glutamatergic neurons that excite dopaminergic neurons. This would limit dopamine levels in the mesolimbic pathway, alleviating symptoms such as hallucinations. However, tobacco use helps stimulate dopaminergic transmission in the brain, particularly in the mesocorticolimbic pathway. Thus, individuals with schizophrenia may be more likely to alternate between tobacco products and cannabis as a way to achieve optimal dopamine levels. If dopamine levels are too high, individuals experience symptoms of schizophrenia. If dopamine levels are too low, individuals experience cognitive deficits.

Glossary

Bottom-up attention a form of attention that occurs automatically and is driven by salient stimuli in the environment.

Cerebello-thalamo-striato-cortical (CTSC) loop a neural pathway involving the cerebellum, thalamus, striatum, and cortex that plays critical role in motor control and cognitive processes, including attention.

Cortico-basal ganglia-thalamo-cortical (CBGTC) loop a neural circuit involving the cortex, basal ganglia, and thalamus that plays a critical role in motor control and cognitive processes, including attention.

Default mode network (DMN) a network of brain regions that are most active when a person is at rest and is not focused on the external world.

Dorsal attention network (DAN) a network of frontal-parietal brain regions, including the intraparietal sulcus and the frontal eye field, that regulate top-down attentional processes.

Hemispatial neglect neuropsychological condition in which individuals fail to attend to or perceive stimuli located in one half of the visual field; typically results from damage to the contralateral side of the parietal lobe.

Purkinje cell a large, distinctive inhibitory neuron located in the cerebellar cortex.

Pyramidal neuron most common excitatory neuron found in the brain; named for its triangular-shaped cell body.

Top-down attention a form of attention that requires conscious effort and is driven by our knowledge and prior experiences.

Ventral attention network (VAN) a network consisting primarily of the ventral frontal cortex and the temporoparietal junction that is involved in bottom-up attentional processes.

References

Aalto, S., Brück, A., Laine, M., Någren, K., & Rinne, J. O. (2005). Frontal and temporal dopamine release during working memory and attention tasks in healthy humans: A positron emission tomography study using the high-affinity dopamine D2 receptor ligand [11C]FLB 457. *The Journal of Neuroscience, 25*(10), 2471–2477. https://doi.org/10.1523/JNEUROSCI.2097-04.2005

Albaugh, M. D., Orr, C., Chaarani, B., Althoff, R. R., Allgaier, N., D'Alberto, N., Hudson, K., Mackey, S., Spechler, P. A., Banaschewski, T., Brühl, R., Bokde, A. L. W., Bromberg, U., Büchel, C., Cattrell, A., Conrod, P. J., Desrivières, S., Flor, H., Frouin, V., Gallinat, J., … Potter, A. S. (2017). Inattention and reaction time variability are linked to ventromedial prefrontal volume in adolescents. *Biological Psychiatry, 82*(9), 660–668. https://doi.org/10.1016/j.biopsych.2017.01.003

Alexander, G. E., DeLong, M. R., & Strick, P. L. (1986). Parallel organization of functionally segregated circuits linking basal ganglia and cortex. *Annual Review of Neuroscience, 9*, 357–381. https://doi.org/10.1146/annurev.ne.09.030186.002041

Amador-Campos, J. A., Aznar-Casanova, J. A., Bezerra, I., Torro-Alves, N., & Sánchez, M. M. (2015). Attentional blink in children with attention deficit hyperactivity disorder. *Brazilian Journal of Psychiatry, 37*(2), 133–138. https://doi.org/10.1590/1516-4446-2014-1415

American Psychiatric Association. (2013). *Diagnostic and statistical manual of mental disorders* (5th ed.). https://doi.org/10.1176/appi.books.9780890425596

Arnold, H. M., Burk, J. A., Hodgson, E. M., Sarter, M., & Bruno, J. P. (2002). Differential cortical acetylcholine release in rats performing a sustained attention task versus behavioral control tasks that do not explicitly tax attention. *Neuroscience, 114*(2), 451–460. https://doi.org/10.1016/s0306-4522(02)00292-0

Arnsten, A. F. T. (2020). Guanfacine's mechanism of action in treating prefrontal cortical disorders: Successful translation across species. *Neurobiology of Learning and Memory, 176*, 107327. https://doi.org/10.1016/j.nlm.2020.107327

Astafiev, S. V., Shulman, G. L., Stanley, C. M., Snyder, A. Z., Van Essen, D. C., & Corbetta, M. (2003). Functional organization of human intraparietal and frontal cortex for attending, looking, and pointing. *The Journal of Neuroscience, 23*(11), 4689–4699. https://doi.org/10.1523/JNEUROSCI.23-11-04689.2003

Bakhshi, S., Tehrani-Doost, M., & Batouli, S. A. H. (2023). Fronto-cerebellar neurometabolite alterations after methylphenidate in children and adolescents with ADHD: A proton magnetic resonance spectroscopy study. *Journal of Attention Disorders, 27*(4), 410–422. https://doi.org/10.1177/10870547221146238

Balachandran, R. C., Sieg, M. L., Tran, C. T. Q., Clancy, B. M., Beaudin, S. A., & Eubig, P. A. (2018). Cholinergic and dopaminergic interactions alter attention and response inhibition in Long-Evans rats performing the 5-choice serial reaction time task. *Pharmacology, Biochemistry, and Behavior, 175*, 160–173. https://doi.org/10.1016/j.pbb.2018.10.006

Bareham, C. A., Georgieva, S. D., Kamke, M. R., Lloyd, D., Bekinschtein, T. A., & Mattingley, J. B. (2018). Role of the right inferior parietal cortex in auditory selective attention: An rTMS study. *Cortex, 99*, 30–38. https://doi.org/10.1016/j.cortex.2017.10.003

Beck, L. H., Bransome, E. D., Jr., Mirsky, A. F., Rosvold, H. E., & Sarason, I. (1956). A continuous performance test of brain damage. *Journal of Consulting Psychology, 20*(5), 343–350. https://doi.org/10.1037/h0043220

Behrmann, M., Geng, J. J., & Shomstein, S. (2004). Parietal cortex and attention. *Current Opinion in Neurobiology, 14*(2), 212–217. https://doi.org/10.1016/j.conb.2004.03.012

Berridge, C. W., Shumsky, J. S., Andrzejewski, M. E., McGaughy, J. A., Spencer, R. C., Devilbiss, D. M., & Waterhouse, B. D. (2012). Differential sensitivity to psychostimulants across prefrontal cognitive tasks: Differential involvement of noradrenergic α_1 - and α_2-receptors. *Biological Psychiatry, 71*(5), 467–473. https://doi.org/10.1016/j.biopsych.2011.07.022

Berridge, C. W., & Spencer, R. C. (2016). Differential cognitive actions of norepinephrine a2 and a1 receptor signaling in the prefrontal cortex. *Brain Research, 1641*(Pt B), 189–196. https://doi.org/10.1016/j.brainres.2015.11.024

Blackman, R. K., Crowe, D. A., DeNicola, A. L., Sakellaridi, S., MacDonald, A. W., 3rd., & Chafee, M. V. (2016). Monkey prefrontal neurons reflect logical operations for cognitive control in a variant of the AX continuous performance task (AX-CPT). *The Journal of Neuroscience, 36*(14), 4067–4079. https://doi.org/10.1523/JNEUROSCI.3578-15.2016

Boekhoudt, L., Voets, E. S., Flores-Dourojeanni, J. P., Luijendijk, M. C., Vanderschuren, L. J., & Adan, R. A. (2017). Chemogenetic activation of midbrain dopamine neurons affects attention, but not impulsivity, in the five-choice serial reaction time task in rats. *Neuropsychopharmacology, 42*(6), 1315–1325. https://doi.org/10.1038/npp.2016.235

Bollimunta, A., Bogadhi, A. R., & Krauzlis, R. J. (2018). Comparing frontal eye field and superior colliculus contributions to covert spatial attention. *Nature Communications, 9*(1), 3553. https://doi.org/10.1038/s41467-018-06042-2

Bossong, M. G., Jansma, J. M., van Hell, H. H., Jager, G., Kahn, R. S., & Ramsey, N. F. (2013). Default mode network in the effects of Δ9-Tetrahydrocannabinol (THC) on human executive function. *PLoS ONE, 8*(7), e70074. https://doi.org/10.1371/journal.pone.0070074

Broadbent, D. E. (1952). Failures of attention in selective listening. *Journal of Experimental Psychology, 44*(6), 428–433. https://doi.org/10.1037/h0057163

Broadbent, D. E. (1956). Successive responses to simultaneous stimuli. *Quarterly Journal of Experimental Psychology, 8*, 145–152. https://doi.org/10.1080/17470215608416814

Brubacher, J. R., Erdelyi, S., Chan, H., Simmons, S., Atkinson, P., Besserer, F., Clarke, D. B., Davis, P., Daoust, R., Émond, M., Eppler, J., Lee, J. S., MacPherson, A., Magee, K., Mercier, E., Ohle, R., Parsons, M., Rao, J., Rowe, B. H., Taylor, J., … Wishart, I. (2025). Prevalence of impairing substance use in injured drivers. *JAMA Network Open, 8*(4), e256379. https://doi.org/10.1001/jamanetworkopen.2025.6379

Buffett-Jerrott, S. E., & Stewart, S. H. (2002). Cognitive and sedative effects of benzodiazepine use. *Current Pharmaceutical Design, 8*(1), 45–58. https://doi.org/10.2174/1381612023396654

Burk, J. A., Blumenthal, S. A., & Maness, E. B. (2018). Neuropharmacology of attention. *European Journal of Pharmacology, 835*, 162–168. https://doi.org/10.1016/j.ejphar.2018.08.008

Campos, M. W., Serebrisky, D., & Castaldelli-Maia, J. M. (2016). Smoking and cognition. *Current Drug Abuse Reviews, 9*(2), 76–79. https://doi.org/10.2174/1874473709666160803101633

Carli, M., Robbins, T. W., Evenden, J. L., & Everitt, B. J. (1983). Effects of lesions to ascending noradrenergic neurones on performance of a 5-choice serial reaction task in rats; implications for theories of dorsal noradrenergic bundle function based on selective attention and arousal. *Behavioural Brain Research, 9*(3), 361–380. https://doi.org/10.1016/0166-4328(83)90138-9

Carter, J. D., Bizzell, J., Kim, C., Bellion, C., Carpenter, K. L., Dichter, G., & Belger, A. (2010). Attention deficits in schizophrenia–preliminary evidence of dissociable transient and sustained deficits. *Schizophrenia Research, 122*(1–3), 104–112. https://doi.org/10.1016/j.schres.2010.03.019

Chen, P., & Goedert, K. M. (2012). Clock drawing in spatial neglect: A comprehensive analysis of clock perimeter, placement, and accuracy. *Journal of Neuropsychology, 6*(2), 270–289. https://doi.org/10.1111/j.1748-6653.2012.02028.x

Chudasama, Y., & Robbins, T. W. (2004). Dopaminergic modulation of visual attention and working memory in the rodent prefrontal cortex. *Neuropsychopharmacology, 29*(9), 1628–1636. https://doi.org/10.1038/sj.npp.1300490

Conners, C. K. (1995). *Conners' continuous performance test computer program: User's manual.* Multi-Health Systems.

Cools, R., & Arnsten, A. F. T. (2022). Neuromodulation of prefrontal cortex cognitive function in primates: The powerful roles of monoamines and acetylcholine. *Neuropsychopharmacology, 47*(1), 309–328. https://doi.org/10.1038/s41386-021-01100-8

Crowe, S. F., & Stranks, E. K. (2018). The residual medium and long-term cognitive effects of benzodiazepine use: An updated meta-analysis. *Archives of Clinical Neuropsychology, 33*(7), 901–911. https://doi.org/10.1093/arclin/acx120

Cundari, M., Vestberg, S., Gustafsson, P., Gorcenco, S., & Rasmussen, A. (2023). Neurocognitive and cerebellar function in ADHD, autism and spinocerebellar ataxia. *Frontiers in Systems Neuroscience, 17*, 1168666. https://doi.org/10.3389/fnsys.2023.1168666

Cybulska-Klosowicz, A., Laczkowska, M., Zakrzewska, R., & Kaliszewska, A. (2017). Attentional deficits and altered neuronal activation in medial prefrontal and posterior parietal cortices in mice with reduced dopamine transporter levels. *Molecular and Cellular Neurosciences, 85*, 82–92. https://doi.org/10.1016/j.mcn.2017.09.004

Dalley, J. W., McGaughy, J., O'Connell, M. T., Cardinal, R. N., Levita, L., & Robbins, T. W. (2001). Distinct changes in cortical acetylcholine and noradrenaline efflux during contingent and noncontingent performance of a visual attentional task. *The Journal ofNeuroscience, 21*(13), 4908–4914. https://doi.org/10.1523/JNEUROSCI.21-13-04908.2001

Dalrymple, K. A., Barton, J. J., & Kingstone, A. (2013). A world unglued: Simultanagnosia as a spatial restriction of attention. *Frontiers in Human Neuroscience, 7*, 145. https://doi.org/10.3389/fnhum.2013.00145

Damiani, S., Tarchi, L., Scalabrini, A., Marini, S., Provenzani, U., Rocchetti, M., Oliva, F., & Politi, P. (2021). Beneath the surface: Hyper-connectivity between caudate and salience regions in ADHD fMRI at rest. *European Child & Adolescent Psychiatry, 30*(4), 619–631. https://doi.org/10.1007/s00787-020-01545-0

Dasilva, M., Brandt, C., Gotthardt, S., Gieselmann, M. A., Distler, C., & Thiele, A. (2019). Cell class-specific modulation of attentional signals by acetylcholine in macaque frontal eye field. *Proceedings of the National Academy of Sciences of the United States of America, 116*(40), 20180–20189. https://doi.org/10.1073/pnas.1905413116

Davidson, M. C., & Marrocco, R. T. (2000). Local infusion of scopolamine into intraparietal cortex slows covert orienting in rhesus monkeys. *Journal ofNeurophysiology, 83*(3), 1536–1549. https://doi.org/10.1152/jn.2000.83.3.1536

Day, M., Pan, J. B., Buckley, M. J., Cronin, E., Hollingsworth, P. R., Hirst, W. D., Navarra, R., Sullivan, J. P., Decker, M. W., & Fox, G. B. (2007). Differential effects of ciproxifan and nicotine on impulsivity and attention measures in the 5-choice serial reaction time test. *Biochemical Pharmacology, 73*(8), 1123–1134. https://doi.org/10.1016/j.bcp.2006.12.004

De Bruin, N. M., Kiliaan, A. J., De Wilde, M. C., & Broersen, L. M. (2003). Combined uridine and choline administration improves cognitive deficits in spontaneously hypertensive rats. *Neurobiology of Learning and Memory, 80*(1), 63–79. https://doi.org/10.1016/s1074-7427(03)00024-8

Ding, J. B., & Hu, K. (2021). Cigarette smoking and schizophrenia: Etiology, clinical, pharmacological, and treatment implications. *Schizophrenia Research and Treatment, 2021*, 7698030. https://doi.org/10.1155/2021/7698030

Dong, L., Chen, W. C., Su, H., Wang, M. L., Du, C., Jiang, X. R., Mei, S. F., Chen, S. J., Liu, X. J., & Liu, X. B. (2023). Intermittent theta burst stimulation to the left dorsolateral prefrontal cortex improves cognitive function in polydrug use disorder patients: A randomized controlled trial. *Frontiers in Psychiatry, 14*, 1156149. https://doi.org/10.3389/fpsyt.2023.1156149

Dougherty, D. M., Marsh, D. M., Moeller, F. G., Chokshi, R. V., & Rosen, V. C. (2000). Effects of moderate and high doses of alcohol on attention, impulsivity, discriminability, and response bias in immediate and delayed memory task performance. *Alcoholism: Clinical and Experimental Research, 24*(11), 1702–1711.

Dutterer, J., Bansal, S., Robinson, B., & Gold, J. M. (2023). Sustained attention deficits in schizophrenia: Effect of memory load on the Identical Pairs Continuous Performance Test. *Schizophrenia Research. Cognition, 33*, 100288. https://doi.org/10.1016/j.scog.2023.100288

Eriksen, B. A., & Eriksen, C. W. (1974). Effects of noise letters upon the identification of a target letter in a nonsearch task. *Perception & Psychophysics, 16*(1), 143–149. https://doi.org/10.3758/BF03203267

Esterman, M., Thai, M., Okabe, H., DeGutis, J., Saad, E., Laganiere, S. E., & Halko, M. A. (2017). Network-targeted cerebellar transcranial magnetic stimulation improves attentional control. *NeuroImage, 156*, 190–198. https://doi.org/10.1016/j.neuroimage.2017.05.011

Ezzati Amini, R., Al Haddad, C., Batabyal, D., Gkena, I., De Vos, B., Cuenen, A., Brijs, T., & Antoniou, C. (2023). Driver distraction and in-vehicle interventions: A driving simulator study on visual attention and driving performance. *Accident Analysis and Prevention, 191*, 107195. https://doi.org/10.1016/j.aap.2023.107195

Farmer, G. D., Janssen, C. P., Nguyen, A. T., & Brumby, D. P. (2018). Dividing attention between tasks: Testing whether explicit payoff functions elicit optimal dual-task performance. *Cognitive Science, 42*(3), 820–849. https://doi.org/10.1111/cogs.12513

Fitzpatrick, C. M., Runegaard, A. H., Christiansen, S. H., Hansen, N. W., Jørgensen, S. H., McGirr, J. C., de Diego Ajenjo, A., Sørensen, A. T., Perrier, J. F., Petersen, A., Gether, U., Woldbye, D. P. D., & Andreasen, J. T. (2019). Differential effects of chemogenetic inhibition of dopamine and norepinephrine neurons in the mouse 5-choice serial reaction time task. *Progress in Neuro-Psychopharmacology & Biological Psychiatry, 90*, 264–276. https://doi.org/10.1016/j.pnpbp.2018.12.004

Fu, C., Chen, S., Qian, A., Zhou, R., Zhou, J., Li, J., Cheng, J., Yang, C., Zhao, K., & Wang, M. (2021). Larger thalamus correlated with inattentive severity in the inattentive subtype of ADHD without comorbidity. *Psychiatry Research, 304*, 114079. https://doi.org/10.1016/j.psychres.2021.114079

Gendle, M. H., White, T. L., Strawderman, M., Mactutus, C. F., Booze, R. M., Levitsky, D. A., & Strupp, B. J. (2004). Enduring effects of prenatal cocaine exposure on selective attention and reactivity to errors: Evidence from an animal model. *Behavioral Neuroscience, 118*(2), 290–297. https://doi.org/10.1037/0735-7044.118.2.290

Gillebert, C. R., Mantini, D., Thijs, V., Sunaert, S., Dupont, P., & Vandenberghe, R. (2011). Lesion evidence for the critical role of the intraparietal sulcus in spatial attention. *Brain, 134*(Pt 6), 1694–1709. https://doi.org/10.1093/brain/awr085

Granon, S., Passetti, F., Thomas, K. L., Dalley, J. W., Everitt, B. J., & Robbins, T. W. (2000). Enhanced and impaired attentional performance after infusion of D1 dopaminergic receptor agents into rat prefrontal cortex. *The Journal of Neuroscience, 20*(3), 1208–1215. https://doi.org/10.1523/JNEUROSCI.20-03-01208.2000

Groom, M. J., & Cortese, S. (2022). Current pharmacological treatments for ADHD. *Current Topics in Behavioral Neurosciences, 57*, 19–50. https://doi.org/10.1007/7854_2022_330

Han, X., Liang, Y., Hempel, B., Jordan, C. J., Shen, H., Bi, G. H., Li, J., & Xi, Z. X. (2023). Cannabinoid CB1 receptors are expressed in a subset of dopamine neurons and underlie cannabinoid-induced aversion, hypoactivity, and anxiolytic effects in mice. *The Journal of Neuroscience, 43*(3), 373–385. https://doi.org/10.1523/JNEUROSCI.1493-22.2022

Hassett, T. C., & Hampton, R. R. (2022). Control of attention in rhesus monkeys measured using a flanker task. *Attention, Perception & Psychophysics, 84*(7), 2155–2166. https://doi.org/10.3758/s13414-022-02452-z

Hambrecht-Wiedbusch, V. S., LaTendresse, K. A., Avidan, M. S., Nelson, A. G., Phyle, M., Ajluni, R. E., & Mashour, G. A. (2019). General anesthesia does not have persistent effects on attention in rodents. *Frontiers in Behavioral Neuroscience, 13*, 76. https://doi.org/10.3389/fnbeh.2019.00076

Hill, E. L., Gallopin, T., Férézou, I., Cauli, B., Rossier, J., Schweitzer, P., & Lambolez, B. (2007). Functional CB1 receptors are broadly expressed in neocortical GABAergic and glutamatergic neurons. *Journal of Neurophysiology, 97*(4), 2580–2589. https://doi.org/10.1152/jn.00603.2006

Howells, F. M., Bindewald, L., & Russell, V. A. (2009). Cross-fostering does not alter the neurochemistry or behavior of spontaneously hypertensive rats. *Behavioral and Brain Functions, 5*, 24. https://doi.org/10.1186/1744-9081-5-24

Hoyer, W. J., Semenec, S. C., & Buchler, N. E. (2007). Acute alcohol intoxication impairs controlled search across the visual field. *Journal of Studies on Alcohol and Drugs, 68*(5), 748–758. https://doi.org/10.15288/jsad.2007.68.748

Hoyle, E., Genn, R. F., Fernandes, C., & Stolerman, I. P. (2006). Impaired performance of alpha7 nicotinic receptor knockout mice in the five-choice serial reaction time task. *Psychopharmacology (Berl), 189*(2), 211–223. https://doi.org/10.1007/s00213-006-0549-2

Hu, M. L., Zong, X. F., Mann, J. J., Zheng, J. J., Liao, Y. H., Li, Z. C., He, Y., Chen, X. G., & Tang, J. S. (2017). A review of the functional and anatomical default mode network in schizophrenia. *Neuroscience Bulletin, 33*(1), 73–84. https://doi.org/10.1007/s12264-016-0090-1

Iba, M., & Sawaguchi, T. (2003). Involvement of the dorsolateral prefrontal cortex of monkeys in visuospatial target selection. *Journal of Neurophysiology, 89*(1), 587–599. https://doi.org/10.1152/jn.00148.2002

Itagaki, S., Ohnishi, T., Toda, W., Sato, A., Matsumoto, J., Ito, H., Ishii, S., Yamakuni, R., Miura, I., & Yabe, H. (2024). Reduced dopamine transporter availability in drug-naive adult attention-deficit/hyperactivity disorder. *PCN Reports, 3*(1), e177. https://doi.org/10.1002/pcn5.177

Ivanov, I., Bansal, R., Hao, X., Zhu, H., Kellendonk, C., Miller, L., Sanchez-Pena, J., Miller, A. M., Chakravarty, M. M., Klahr, K., Durkin, K., Greenhill, L. L., & Peterson, B. S. (2010). Morphological abnormalities of the thalamus in youths with attention deficit hyperactivity disorder. *The American Journal of Psychiatry, 167*(4), 397–408. https://doi.org/10.1176/appi.ajp.2009.09030398

Janes, A. C., Gilman, J. M., Frederick, B. B., Radoman, M., Pachas, G., Fava, M., & Evins, A. E. (2018). Salience network coupling is linked to both tobacco smoking and symptoms of attention deficit hyperactivity disorder (ADHD). *Drug and Alcohol Dependence, 182*, 93–97. https://doi.org/10.1016/j.drugalcdep.2017.11.005

Jentsch, J. D. (2005). Impaired visuospatial divided attention in the spontaneously hypertensive rat. *Behavioural Brain Research, 157*(2), 323–330. https://doi.org/10.1016/j.bbr.2004.07.011

Johnson, J. A., Strafella, A. P., & Zatorre, R. J. (2007). The role of the dorsolateral prefrontal cortex in bimodal divided attention: Two transcranial magnetic stimulation studies. *Journal of Cognitive Neuroscience, 19*(6), 907–920. https://doi.org/10.1162/jocn.2007.19.6.907

Jones, K. E., Craver-Lemley, C., & Barrett, A. M. (2008). Asymmetrical visual-spatial attention in college students diagnosed with ADD/ADHD. *Cognitive and Behavioral Neurology, 21*(3), 176–178. https://doi.org/10.1097/WNN.0b013e318185e6a9

Källstrand, J., Niklasson, K., Lindvall, M., & Claesdotter-Knutsson, E. (2024). Reduced thalamic activity in ADHD under ABR forward masking conditions. *Applied Neuropsychology. Child, 13*(3), 222–228. https://doi.org/10.1080/21622965.2022.2155520

Kane, M. J., & Engle, R. W. (2002). The role of prefrontal cortex in working-memory capacity, executive attention, and general fluid intelligence: An individual-differences perspective. *Psychonomic Bulletin & Review, 9*(4), 637–671. https://doi.org/10.3758/bf03196323

Keith, R. W. (1994). *The auditory continuous performance test.* Psychological Corp.

Kihara, K., Kondo, H. M., & Kawahara, J. I. (2016). Differential contributions of GABA concentration in frontal and parietal regions to individual differences in attentional blink. *The Journal of Neuroscience, 36*(34), 8895–8901. https://doi.org/10.1523/JNEUROSCI.0764-16.2016

Klink, P. C., Teeuwen, R. R. M., Lorteije, J. A. M., & Roelfsema, P. R. (2023). Inversion of pop-out for a distracting feature dimension in monkey visual cortex. *Proceedings of the National Academy of Sciences of the United States of America, 120*(9), e2210839120. https://doi.org/10.1073/pnas.2210839120

Kolisnyk, B., Al-Onaizi, M. A., Prado, V. F., & Prado, M. A. (2015). α7 nicotinic ACh receptor-deficient mice exhibit sustained attention impairments that are reversed by β2 nicotinic ACh receptor activation. *British Journal of Pharmacology, 172*(20), 4919–4931. https://doi.org/10.1111/bph.13260

Kondo, H. M., Terashima, H., Kihara, K., Kochiyama, T., Shimada, Y., & Kawahara, J. I. (2023). Prefrontal GABA and glutamate-glutamine levels affect sustained attention. *Cerebral Cortex, 33*(19), 10441–10452. https://doi.org/10.1093/cercor/bhad294

Kovács, A., Bordás, C., Bíró, T., Hegyi, Z., Antal, M., Szücs, P., & Pál, B. (2017). Direct presynaptic and indirect astrocyte-mediated mechanisms both contribute to endocannabinoid signaling in the pedunculopontine nucleus of mice. *Brain Structure & Function, 222*(1), 247–266. https://doi.org/10.1007/s00429-016-1214-0

Lee, M., Mueller, A., & Moore, T. (2020). Differences in noradrenaline receptor expression across different neuronal subtypes in macaque frontal eye field. *Frontiers in Neuroanatomy, 14*, 574130. https://doi.org/10.3389/fnana.2020.574130

Lesh, T. A., Rhilinger, J., Brower, R., Mawla, A. M., Ragland, J. D., Niendam, T. A., & Carter, C. S. (2024). Using task-fMRI to explore the relationship between lifetime cannabis use and cognitive control in individuals with first-episode schizophrenia. *Schizophrenia Bulletin Open, 5*(1), sgae016. https://doi.org/10.1093/schizbullopen/sgae016

Li, C. S., Mazzoni, P., & Andersen, R. A. (1999). Effect of reversible inactivation of macaque lateral intraparietal area on visual and memory saccades. *Journal of Neurophysiology, 81*(4), 1827–1838. https://doi.org/10.1152/jn.1999.81.4.1827

Ljubojevic, V., Luu, P., Gill, P. R., Beckett, L. A., Takehara-Nishiuchi, K., & De Rosa, E. (2018). Cholinergic modulation of frontoparietal cortical network dynamics supporting supramodal attention. *The Journal of Neuroscience, 38*(16), 3988–4005. https://doi.org/10.1523/JNEUROSCI.2350-17.2018

Loose, R., Kaufmann, C., Auer, D. P., & Lange, K. W. (2003). Human prefrontal and sensory cortical activity during divided attention tasks. *Human Brain Mapping, 18*(4), 249–259. https://doi.org/10.1002/hbm.10082

MacQueen, D. A., Minassian, A., Kenton, J. A., Geyer, M. A., Perry, W., Brigman, J. L., & Young, J. W. (2018). Amphetamine improves mouse and human attention in the 5-choice continuous performance test. *Neuropharmacology, 138*, 87–96. https://doi.org/10.1016/j.neuropharm.2018.05.034

Malhotra, P., Coulthard, E. J., & Husain, M. (2009). Role of right posterior parietal cortex in maintaining attention to spatial locations over time. *Brain, 132*(Pt 3), 645–660. https://doi.org/10.1093/brain/awn350

Maltezos, S., Horder, J., Coghlan, S., Skirrow, C., O'Gorman, R., Lavender, T. J., Mendez, M. A., Mehta, M., Daly, E., Xenitidis, K., Paliokosta, E., Spain, D., Pitts, M., Asherson, P., Lythgoe, D. J., Barker, G. J., & Murphy, D. G. (2014). Glutamate/glutamine and neuronal integrity in adults with ADHD: A proton MRS study. *Translational Psychiatry, 4*(3), e373. https://doi.org/10.1038/tp.2014.11

Mather, M., Huang, R., Clewett, D., Nielsen, S. E., Velasco, R., Tu, K., Han, S., & Kennedy, B. L. (2020). Isometric exercise facilitates attention to salient events in women via the noradrenergic system. *NeuroImage, 210*, 116560. https://doi.org/10.1016/j.neuroimage.2020.116560

McCarthy, H., Skokauskas, N., Mulligan, A., Donohoe, G., Mullins, D., Kelly, J., Johnson, K., Fagan, A., Gill, M., Meaney, J., & Frodl, T. (2013). Attention network hypoconnectivity with default and affective network hyperconnectivity in adults diagnosed with attention-deficit/hyperactivity disorder in childhood. *JAMA Psychiatry, 70*(12), 1329–1337. https://doi.org/10.1001/jamapsychiatry.2013.2174

McGaughy, J., & Sarter, M. (1995). Behavioral vigilance in rats: Task validation and effects of age, amphetamine, and benzodiazepine receptor ligands. *Psychopharmacology (Berl), 117*(3), 340–357. https://doi.org/10.1007/BF02246109

Miller, E. M., Pomerleau, F., Huettl, P., Russell, V. A., Gerhardt, G. A., & Glaser, P. E. (2012). The spontaneously hypertensive and Wistar Kyoto rat models of ADHD exhibit subregional differences in dopamine release and uptake in the striatum and nucleus accumbens. *Neuropharmacology, 63*(8), 1327–1334. https://doi.org/10.1016/j.neuropharm.2012.08.020

Mirjana, C., Baviera, M., Invernizzi, R. W., & Balducci, C. (2004). The serotonin 5-HT2A receptors antagonist M100907 prevents impairment in attentional performance by NMDA receptor blockade in the rat prefrontal cortex. *Neuropsychopharmacology, 29*(9), 1637–1647. https://doi.org/10.1038/sj.npp.1300479

Moisala, M., Salmela, V., Salo, E., Carlson, S., Vuontela, V., Salonen, O., & Alho, K. (2015). Brain activity during divided and selective attention to auditory and visual sentence comprehension tasks. *Frontiers in Human Neuroscience, 9*, 86. https://doi.org/10.3389/fnhum.2015.00086

Molnar-Szakacs, I., & Uddin, L. Q. (2022). Anterior insula as a gatekeeper of executive control. *Neuroscience and Biobehavioral Reviews, 139*, 104736. https://doi.org/10.1016/j.neubiorev.2022.104736

Moore, C. F., Davis, C. M., Sempio, C., Klawitter, J., Christians, U., & Weerts, E. M. (2023). Effects of oral Δ9-tetrahydrocannabinol and cannabidiol combinations on a sustained attention task in rats. *Experimental and Clinical Psychopharmacology, 31*(5), 881–887. https://doi.org/10.1037/pha0000635

Morris, R., Griffiths, O., Le Pelley, M. E., & Weickert, T. W. (2013). Attention to irrelevant cues is related to positive symptoms in schizophrenia. *Schizophrenia Bulletin, 39*(3), 575–582. https://doi.org/10.1093/schbul/sbr192

Mueller, A., Krock, R. M., Shepard, S., & Moore, T. (2020). Dopamine receptor expression among local and visual cortex-projecting frontal eye field neurons. *Cerebral Cortex, 30*(1), 148–164. https://doi.org/10.1093/cercor/bhz078

Murphy, E. R., Dalley, J. W., & Robbins, T. W. (2005). Local glutamate receptor antagonism in the rat prefrontal cortex disrupts response inhibition in a visuospatial attentional task. *Psychopharmacology (Berl), 179*(1), 99–107. https://doi.org/10.1007/s00213-004-2068-3

Neill, J. C., Barnes, S., Cook, S., Grayson, B., Idris, N. F., McLean, S. L., Snigdha, S., Rajagopal, L., & Harte, M. K. (2010). Animal models of cognitive dysfunction and negative symptoms of schizophrenia: Focus on NMDA receptor antagonism. *Pharmacology & Therapeutics, 128*(3), 419–432. https://doi.org/10.1016/j.pharmthera.2010.07.004

Noudoost, B., & Moore, T. (2011). Control of visual cortical signals by prefrontal dopamine. *Nature, 474*(7351), 372–375. https://doi.org/10.1038/nature09995

O'Neill, J., Fitten, L. J., Siembieda, D. W., Crawford, K. C., Halgren, E., Fisher, A., & Refai, D. (1999). Divided attention-enhancing effects of AF102B and THA in aging monkeys. *Psychopharmacology (Berl), 143*(2), 123–130. https://doi.org/10.1007/s002130050927

Orinstein, A. J., & Stevens, M. C. (2014). Brain activity in predominantly-inattentive subtype attention-deficit/hyperactivity disorder during an auditory oddball attention task. *Psychiatry Research, 223*(2), 121–128. https://doi.org/10.1016/j.pscychresns.2014.05.012

Passetti, F., Dalley, J. W., O'Connell, M. T., Everitt, B. J., & Robbins, T. W. (2000). Increased acetylcholine release in the rat medial prefrontal cortex during performance of a visual attentional task. *The European Journal of Neuroscience, 12*(8), 3051–3058. https://doi.org/10.1046/j.1460-9568.2000.00183.x

Patel, G. H., Sestieri, C., & Corbetta, M. (2019). The evolution of the temporoparietal junction and posterior superior temporal sulcus. *Cortex, 118*, 38–50. https://doi.org/10.1016/j.cortex.2019.01.026

Perez-Lloret, S., & Barrantes, F. J. (2016). Deficits in cholinergic neurotransmission and their clinical correlates in Parkinson's disease. *NPJ Parkinson's Disease, 2*, 16001. https://doi.org/10.1038/npjparkd.2016.1

Pierrot-Deseilligny, C.h., Müri, R. M., Nyffeler, T., & Milea, D. (2005). The role of the human dorsolateral prefrontal cortex in ocular motor behavior. *Annals of the New York Academy of Sciences, 1039*, 239–251. https://doi.org/10.1196/annals.1325.023

Purushothaman, G., Marion, R., Li, K., & Casagrande, V. A. (2012). Gating and control of primary visual cortex by pulvinar. *Nature Neuroscience, 15*(6), 905–912. https://doi.org/10.1038/nn.3106

Qian, S., Yan, S., Zhou, C., Shi, Z., Wang, Z., Xiong, Y., & Zhou, Y. (2020). Resting-state brain activity predicts selective attention deficits during hyperthermia exposure. *International Journal of Hyperthermia, 37*(1), 220–230. https://doi.org/10.1080/02656736.2020.1735536

Ramezanpour, H., & Fallah, M. (2022). The role of temporal cortex in the control of attention. *Current Research in Neurobiology, 3*, 100038. https://doi.org/10.1016/j.crneur.2022.100038

Rapoport, J. L., & Inoff-Germain, G. (2002). Responses to methylphenidate in Attention-Deficit/Hyperactivity Disorder and normal children: Update 2002. *Journal of Attention Disorders, 6*(Suppl. 1), S57–S60. https://doi.org/10.1177/070674370200601s07

Redinbaugh, M. J., & Saalmann, Y. B. (2024). Contributions of basal ganglia circuits to perception, attention, and consciousness. *Journal of Cognitive Neuroscience, 36*(8), 1620–1642. https://doi.org/10.1162/jocn_a_02177

Remmelink, E., Chau, U., Smit, A. B., Verhage, M., & Loos, M. (2017). A one-week 5-choice serial reaction time task to measure impulsivity and attention in adult and adolescent mice. *Scientific Reports, 7*, 42519. https://doi.org/10.1038/srep42519

Reumers, S. F. I., Bongaerts, F. L. P., de Leeuw, F. E., van de Warrenburg, B. P. C., Schutter, D. J. L. G., & Kessels, R. P. C. (2025). Cognition in cerebellar disorders: What's in the profile? A systematic review and meta-analysis. *Journal of Neurology, 272*(3), 250. https://doi.org/10.1007/s00415-025-12967-8

Rinne, T., Muers, R. S., Salo, E., Slater, H., & Petkov, C. I. (2017). Functional imaging of audio-visual selective attention in monkeys and humans: How do lapses in monkey performance affect cross-species correspondences? *Cerebral Cortex, 27*(6), 3471–3484. https://doi.org/10.1093/cercor/bhx092

Roessner, V., Sagvolden, T., Dasbanerjee, T., Middleton, F. A., Faraone, S. V., Walaas, S. I., Becker, A., Rothenberger, A., & Bock, N. (2010). Methylphenidate normalizes elevated dopamine transporter densities in an animal model of the attention-deficit/hyperactivity disorder combined type, but not to the same extent in one of the attention-deficit/hyperactivity disorder inattentive type. *Neuroscience, 167*(4), 1183–1191. https://doi.org/10.1016/j.neuroscience.2010.02.073

Rogers, R. D., & Monsell, S. (1995). Costs of a predictible switch between simple cognitive tasks. *Journal of Experimental Psychology: General, 124*(2), 207–231. https://doi.org/10.1037/0096-3445.124.2.207

Rosenbaum, G., Cohen, B. D., Luby, E. D., Gottlieb, J. S., & Yelen, D. (1959). Comparison of sernyl with other drugs: Simulation of schizophrenic performance with sernyl, LSD-25, and amobarbital (amytal) sodium; I. Attention, motor function, and proprioception. *Archives of General Psychiatry, 1*, 651–656. https://doi.org/10.1001/archpsyc.1959.03590060113013

Sagvolden, T. (2006). The alpha-2A adrenoceptor agonist guanfacine improves sustained attention and reduces overactivity and impulsiveness in an animal model of Attention-Deficit/

Hyperactivity Disorder (ADHD). *Behavioral and Brain Functions, 2*, 41. https://doi.org/10.1186/1744-9081-2-41
Sagvolden, T. (2011). Impulsiveness, overactivity, and poorer sustained attention improve by chronic treatment with low doses of l-amphetamine in an animal model of Attention-Deficit/Hyperactivity Disorder (ADHD). *Behavioral and Brain Functions, 7*, 6. https://doi.org/10.1186/1744-9081-7-6
Sagvolden, T., Dasbanerjee, T., Zhang-James, Y., Middleton, F., & Faraone, S. (2008). Behavioral and genetic evidence for a novel animal model of Attention-Deficit/Hyperactivity Disorder Predominantly Inattentive Subtype. *Behavioral and Brain Functions, 4*, 56. https://doi.org/10.1186/1744-9081-4-56
Salehinejad, M. A., Ghayerin, E., Nejati, V., Yavari, F., & Nitsche, M. A. (2020). Domain-specific involvement of the right posterior parietal cortex in attention network and attentional control of ADHD: A randomized, cross-over, sham-controlled tDCS study. *Neuroscience, 444*, 149–159. https://doi.org/10.1016/j.neuroscience.2020.07.037
Schimmelpfennig, J., Topczewski, J., Zajkowski, W., & Jankowiak-Siuda, K. (2023). The role of the salience network in cognitive and affective deficits. *Frontiers in Human Neuroscience, 17*, 1133367. https://doi.org/10.3389/fnhum.2023.1133367
Schmahmann, J. D. (2010). The role of the cerebellum in cognition and emotion: Personal reflections since 1982 on the dysmetria of thought hypothesis, and its historical evolution from theory to therapy. *Neuropsychology Review, 20*(3), 236–260. https://doi.org/10.1007/s11065-010-9142-x
Seamans, J. K., Durstewitz, D., Christie, B. R., Stevens, C. F., & Sejnowski, T. J. (2001). Dopamine D1/D5 receptor modulation of excitatory synaptic inputs to layer V prefrontal cortex neurons. *Proceedings of the National Academy of Sciences of the United States of America, 98*(1), 301–306. https://doi.org/10.1073/pnas.98.1.301
Silberstein, R. B., Pipingas, A., Farrow, M., Levy, F., & Stough, C. K. (2016). Dopaminergic modulation of default mode network brain functional connectivity in attention deficit hyperactivity disorder. *Brain and Behavior, 6*(12), e00582. https://doi.org/10.1002/brb3.582
Smith, B. L., Guzman, T. A., Brendle, A. H., Laaker, C. J., Ford, A., Hiltz, A. R., Zhao, J., Setchell, K. D. R., & Reyes, T. M. (2022). Perinatal morphine exposure leads to sex-dependent executive function deficits and microglial changes in mice. *eNeuro, 9*(5), ENEURO.0238-22.2022. https://doi.org/10.1523/ENEURO.0238-22.2022
Song, P., Zha, M., Yang, Q., Zhang, Y., Li, X., & Rudan, I. (2021). The prevalence of adult attention-deficit hyperactivity disorder: A global systematic review and meta-analysis. *Journal of Global Health, 11*, 04009. https://doi.org/10.7189/jogh.11.04009
Speranza, L., di Porzio, U., Viggiano, D., de Donato, A., & Volpicelli, F. (2021). Dopamine: The neuromodulator of long-term synaptic plasticity, reward and movement control. *Cells, 10*(4), 735. https://doi.org/10.3390/cells10040735
Spinelli, S., Ballard, T., Feldon, J., Higgins, G. A., & Pryce, C. R. (2006). Enhancing effects of nicotine and impairing effects of scopolamine on distinct aspects of performance in computerized attention and working memory tasks in marmoset monkeys. *Neuropharmacology, 51*(2), 238–250. https://doi.org/10.1016/j.neuropharm.2006.03.012
Suzuki, M., & Gottlieb, J. (2013). Distinct neural mechanisms of distractor suppression in the frontal and parietal lobe. *Nature Neuroscience, 16*(1), 98–104. https://doi.org/10.1038/nn.3282
Theofilatos, A., Ziakopoulos, A., Papadimitriou, E., & Yannis, G. (2018). How many crashes are caused by driver interaction with passengers? A meta-analysis approach. *Journal of Safety Research, 65*, 11–20. https://doi.org/10.1016/j.jsr.2018.02.001
Thiel, C. M., Zilles, K., & Fink, G. R. (2005). Nicotine modulates reorienting of visuospatial attention and neural activity in human parietal cortex. *Neuropsychopharmacology, 30*(4), 810–820. https://doi.org/10.1038/sj.npp.1300633
Treisman, A. (1982). Perceptual grouping and attention in visual search for features and for objects. *Journal of Experimental Psychology: Human Perception and Performance, 8*(2), 194–214. https://doi.org/10.1037/0096-1523.8.2.194

Urits, I., Charipova, K., Gress, K., Li, N., Berger, A. A., Cornett, E. M., Kassem, H., Ngo, A. L., Kaye, A. D., & Viswanath, O. (2021). Adverse effects of recreational and medical cannabis. *Psychopharmacology Bulletin, 51*(1), 94–109.

Uylings, H. B., Groenewegen, H. J., & Kolb, B. (2003). Do rats have a prefrontal cortex? *Behavioural Brain Research, 146*(1–2), 3–17. https://doi.org/10.1016/j.bbr.2003.09.028

Valentine, G., & Sofuoglu, M. (2018). Cognitive effects of nicotine: Recent progress. *Current Neuropharmacology, 16*(4), 403–414. https://doi.org/10.2174/1570159X15666171103152136

van Kempen, J., Brandt, C., Distler, C., Bellgrove, M. A., & Thiele, A. (2022). Dopamine influences attentional rate modulation in Macaque posterior parietal cortex. *Scientific Reports, 12*(1), 6914. https://doi.org/10.1038/s41598-022-10634-w

van Schouwenburg, M. R., den Ouden, H. E., & Cools, R. (2010). The human basal ganglia modulate frontal-posterior connectivity during attention shifting. *The Journal of Neuroscience, 30*(29), 9910–9918. https://doi.org/10.1523/JNEUROSCI.1111-10.2010

Vossel, S., Geng, J. J., & Fink, G. R. (2014). Dorsal and ventral attention systems: Distinct neural circuits but collaborative roles. *The Neuroscientist, 20*(2), 150–159. https://doi.org/10.1177/1073858413494269

Wang, X., Wu, Q., Egan, L., Gu, X., Liu, P., Gu, H., Yang, Y., Luo, J., Wu, Y., Gao, Z., & Fan, J. (2019). Anterior insular cortex plays a critical role in interoceptive attention. *eLife, 8*, e42265. https://doi.org/10.7554/eLife.42265

Wester, A. E., Verster, J. C., Volkerts, E. R., Böcker, K. B., & Kenemans, J. L. (2010). Effects of alcohol on attention orienting and dual-task performance during simulated driving: An event-related potential study. *Journal of Psychopharmacology, 24*(9), 1333–1348. https://doi.org/10.1177/0269881109348168

Wible, C. G., Preus, A. P., & Hashimoto, R. (2009). A cognitive neuroscience view of schizophrenic symptoms: Abnormal activation of a system for social perception and communication. *Brain Imaging and Behavior, 3*(1), 85–110. https://doi.org/10.1007/s11682-008-9052-1

Wicks, B., Waxler, D. E., White, K. M., Duncan, N., Bergmann, J., Cole, R. D., Parikh, V., & Bangasser, D. A. (2017). Method for testing sustained attention in touchscreen operant chambers in rats. *Journal of Neuroscience Methods, 277*, 30–37. https://doi.org/10.1016/j.jneumeth.2016.12.003

Winterer, G. (2010). Why do patients with schizophrenia smoke? *Current Opinion in Psychiatry, 23*(2), 112–119. https://doi.org/10.1097/YCO.0b013e3283366643

Wong, S. H. M., Stevens, C., Liu, C. H., & Chen, J. A. (2022). Prevalence and Correlates of Prescription stimulant misuse among US college students: Results from a national survey. *The Journal of Clinical Psychiatry, 84*(1), 22m14420. https://doi.org/10.4088/JCP.22m14420

Zhang, H. Y., Gao, M., Shen, H., Bi, G. H., Yang, H. J., Liu, Q. R., Wu, J., Gardner, E. L., Bonci, A., & Xi, Z. X. (2017). Expression of functional cannabinoid CB_2 receptor in VTA dopamine neurons in rats. *Addiction Biology, 22*(3), 752–765. https://doi.org/10.1111/adb.12367

Zhornitsky, S., Tourjman, V., Pelletier, J., Assaf, R., Li, C. R., & Potvin, S. (2022). Acute effects of ketamine and esketamine on cognition in healthy subjects: A meta-analysis. *Progress in Neuro-Psychopharmacology & Biological Psychiatry, 118*, 110575. https://doi.org/10.1016/j.pnpbp.2022.110575

Chapter 6
Maladaptive Memories

1 Introduction

"Where is it?" I recheck the crack between the two couch cushions. Once again, the remote to the TV has magically vanished. No one in the house, including myself, can remember who had the remote last and where they placed it. My oldest son swears he didn't use it. My youngest child chimes in with a "Me not know where the remote went". I pace our living room, get on my hands and knees to check under the couch and the ottoman, and pull the cushion off our IKEA chair. I even check the kitchen to see if the remote was placed on the counter. Misplacing an item is something most (if not all) of us have experienced. While frustrating, the consequences of failing to remember where we placed something are generally innocuous. Similarly, many of us experience the frustrating "tip-of-the-tongue" phenomenon, struggling to come up with the right expression, or we have difficulty remembering the name of an acquaintance we recently met.

The memory lapses mentioned above are quite normal and do not severely impact our lives. This chapter will delve into maladaptive memory processes. Some individuals may struggle to remember large portions of their life and/or cannot form new memories. However, other individuals may have a hyperactive memory system, often "reliving" memories of traumatic events. These memory disturbances can be quite disruptive to an individual's physical and/or psychological well-being.

2 Learning Objectives

By the end of this chapter, you should be able to …

1. Compare and contrast the modal model of memory with the working memory model.

J. Yates, *Neuroscience of Maladaptation*, https://doi.org/10.1007/978-3-032-15602-0_6

2. Describe the ways in which we can measure memory processes in a laboratory setting.
3. Discuss the brain structures involved in the various forms of memory.
4. Explain the cellular mechanisms of learning and memory.
5. Identify disorders characterized by loss of memory and disorders characterized by maladaptive memory processes, and explain the neuromechanisms underlying these disorders.

3 Overview of Memory

Because of the complexity of memory, I first need to explain how memory is structured and how it functions. Multiple models have been proposed to explain how memory works. For simplicity, I will highlight two of the major theories that have been proposed. I will also discuss some of the ways in which different memory systems can be measured in a laboratory. This knowledge will help further your understanding of the neuromechanisms underlying dysregulated memory processes.

3.1 Modal Model of Memory

Proposed by Dr. Richard Atkinson and Dr. Richard Shiffrin, the modal model of memory (also known as the multi-store model of memory) likens memory to a computer, with an input, a processor, and an output (Fig. 1a). According to Atkinson and Shiffrin (1968), information moves from one mode of memory to another in a linear fashion. When we are presented with some event or piece of information, we briefly store this information in **sensory memory**, which lasts approximately less than 1s. Essentially, sensory memory is a form of attention. If we fail to pay attention to some stimulus, we will not be able to store this information for later use. Think of the expression "in one ear and out the other".

If we pay attention to a stimulus, information about the stimulus is transferred from sensory memory to **short-term memory (STM)**. As the name suggests, STM retains to-be-remembered information for approximately 30–60s. The amount of information that can be stored in STM ranges from 3–7 items (Miller, 1956); however, we can increase STM storage by *chunking* information. Although telephone numbers contain 10 individual units, we can store that information in STM easily. This is because we chunk telephone numbers into 3 units as opposed to 10 individual units. To keep something in STM, we engage in *maintenance rehearsal*, which consists of verbally or mentally repeating information. If we fail to transfer information out of STM, we experience *encoding failure*. We will not be able to extract this information in the future.

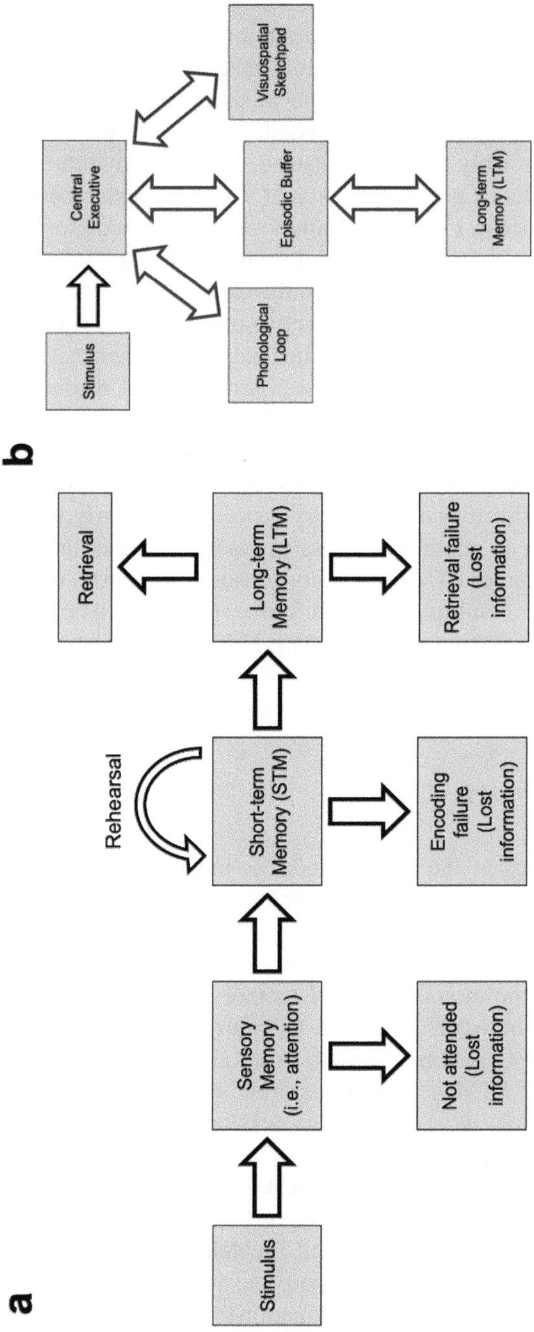

Fig. 1 Schematic representations of the modal model of memory (**a**) and the working memory model (**b**). Image created by the author using Microsoft PowerPoint

The final memory unit is **long-term memory (LTM)**. To transfer information from STM to LTM, an individual must engage in *elaborative rehearsal*. During elaborative rehearsal, an individual links new information to information already stored in LTM. In contrast to STM, the amount of information that can be stored in LTM is virtually limitless. If this is the case, why do we struggle to remember information? One popular theory is *interference* (Underwood, 1957). Here, the information that we fail to remember is not truly forgotten. Instead, other information we have learned at some point interferes or gets in the way of the to-be-remembered information. Two possibilities can occur. First, information we have learned previously can interfere with our ability to remember newer information. Second, information we have learned recently can make retrieving older information more difficult. These types of interference are called proactive and retroactive interference, respectively. We have not truly lost what we have stored in LTM. Instead, we experience a form of *retrieval failure*. Later in this chapter, you will learn about several medical conditions in which memory loss is a core symptom. You will also learn about the neurobiological changes that occur that are associated with such memory loss.

The modal model has been heavily criticized due to its oversimplification of memory. Atkinson and Shiffrin treat memory as occurring linearly, traveling from one memory storage center to another in a passive manner. Another major limitation of this model is that it assumes LTM is a unitary construct. However, evidence indicates that LTM is composed of multiple subtypes. These subtypes will be discussed in more detail below. Before discussing these subtypes, I want to describe an alternative view of memory: the **working memory** model.

3.2 *Working Memory*

Given the shortcomings of the modal model of memory, others have argued that memory is much more dynamic, with complex interactions occurring between STM and LTM. Dr. George Miller proposed the concept of working memory (Miller et al., 1960), a concept that was expanded by Dr. Alan Baddeley (Baddeley & Hitch, 1974). In contrast to the modal model of memory, working memory is dynamic, as information flows between STM and LTM in a reciprocal fashion (Fig. 1b). According to Baddeley, working memory is an alternative to STM. Whereas the modal model assumes STM is merely a storage center that briefly retains to-be-remembered information, working memory acts as both a storage center and as a processing center.

Another distinction between STM and working memory is that working memory is fractioned into several components as opposed to consisting of a unitary system. I will not go into too much detail here, but Baddeley termed these subsystems the central executive, the visuo-spatial sketchpad, and the phonological loop. The central executive is responsible for allocating information to the other subsystems. The visuo-spatial sketchpad processes information that is visual or spatial in nature ("mind's eye"). The phonological loop processes auditory or written information. Information gets transmitted to LTM via the episodic buffer. When we need to retrieve information from LTM, the episodic buffer sends the information back to the central executive. The central executive sends this information back to the phonological loop and/or visuo-spatial sketchpad for further processing, or it can send it back to LTM. This allows us to extract information from LTM to assist us as we process information in working memory.

Comprehension Check 1 The modal model of memory is criticized for its linear and passive view of memory processing. Based on the description of the working memory model, explain one specific way in which the working memory model addresses this limitation of the modal model.

In contrast to the modal model of memory, the working memory model is more dynamic, allowing for memories to be extracted from long-term storage into STM to be manipulated before being sent back to LTM. This bidirectional flow highlights a more active role of short-term/working memory in not just storing but also manipulating information with the aid of LTM.

3.3 Long-Term Memory (LTM)

LTM can be subdivided into two major subdivisions: **explicit memory** and **implicit memory**. Explicit memory (also known as declarative memory) involves the conscious recollection of personal memories (*episodic memory*) and facts/concepts that one has learned (*semantic memory*). Another type of explicit memory is *prospective memory*, which entails remembering to perform some action in the future. For example, remembering to attend a meeting that you made 2 weeks previously is a form of prospective memory.

Implicit memory (also known as nondeclarative memory) stores processes that occur unconsciously. One type of implicit memory is *procedural memory*, which encodes processes in which we engage on a routine basis. Habits are a form of procedural memory; complex activities such as driving a car are stored in procedural memory. You will learn more about habits and how they can become maladaptive in Chap. 11. *Priming* is another form of implicit memory. Exposure to a particular stimulus can unconsciously influence our responding to a subsequent stimulus. There are multiple types of priming. I will provide one example. In a word-stem completion task, individuals are shown multiple words in quick succession. Eventually, individuals are shown an incomplete word and are asked to complete the word. Imagine if individuals are shown the word stem CO___ and previously had seen the word "corn". Even though participants can complete the word stem any way they want, they often use the word "corn" because they previously had encountered that word.

Comprehension Check 2 What is the key difference between explicit memory and implicit memory? Provide a real-world example of explicit memory and implicit memory.

The key difference between these types of memory is one's conscious awareness during recall. Whereas individuals are aware that they are learning explicit memories, implicit memory occurs even though we are not aware that learning is occurring. A real-world example of explicit memory is remembering details from a movie you watched last week. An example of implicit memory is the improved speed and fluidity one exhibits when typing on a keyboard over time.

4 Measuring Memory in a Laboratory

Like attention, there are numerous laboratory tests that are used to study the various types of memory. Going through each laboratory test is beyond the scope of this book, but I will provide a general description of how memory can be measured. To measure STM/working memory, a memory span task can be used (Dempster, 1978). Individuals are presented with a sequence of either words, letters, or numbers (when numbers are used, this task is often referred to as a *digit span* task). Individuals then need to recall the sequence that they were just presented. As the session progresses, the lists increase in length. One's memory span is calculated as the longest number of sequential units that can be recalled correctly. For example, if using a digit span task (Woods et al., 2011), individuals may be first presented the digits 5 9 2, with each number being briefly presented on a computer screen. After this initial trial, the number of items to be recalled increases by one (e.g., 2 7 0 3).

Another way to measure working memory is with the *n*-back test (Kirchner, 1958). Individuals are presented with a sequence of stimuli and are asked to indicate when the current stimulus matches a stimulus from *n* positions earlier in the sequence. Imagine someone is given this sequence: X V T C **V** Y R R X V **R** T **V** T R X. In this variant of the *n*-back test, individuals must identify when the current stimulus was presented three steps earlier. Those letters are bolded in the sequence provided above.

LTM can be measured several ways, with one simple form being a free recall task (Cohen et al., 2017). Participants are given a list of to-be-remembered items (typically words), and they are asked to recall these items after a certain amount of time has passed. This task is somewhat analogous to studying for an exam that requires short-answer or essay responses. LTM can also be measured using recognition tasks. In one variation of a recognition task, participants identify if a stimulus was previously presented. Multiple-choice questions measure one's recognition.

Memory processes can be studied in animals. To study working memory, a delayed matching-to-sample (DMTS) task is used (e.g., Kangas et al., 2011). This task can be studied in multiple species, such as primates (including humans), rodents, and birds.

In rodents, a stimulus light above one lever is illuminated. After a short delay, two or more stimulus lights are illuminated. Rodents are required to remember which stimulus light was previously illuminated. Responding on the correct manipulandum leads to reinforcement. This procedure is similar in avian species, but different colored lights are often used.

In rodents, LTM is often measured with the Morris water maze and the radial arm maze. In the Morris water maze, rodents must locate a platform that is obscured (Morris, 1984). The top of the platform sits just below the surface of the water, and the water is made opaque, usually by adding low fat milk powder. The time needed to find the platform is recorded. The radial arm maze consists of eight arms that branch out from a center platform (Olton & Samuelson, 1976). Only four of the arms are baited with food. Rodents must remember which arms are baited with food, and they must remember which arms they have already explored. This second component taps more into working memory. One major benefit of the radial arm maze is that it measures both working memory and LTM.

5 Neuromechanisms of Memory

As there are multiple types of memory, one should not be surprised to learn that different forms of memory are subserved by distinct brain regions. A review of the brain regions involved in memory will be provided first before discussing the cellular mechanisms of learning and memory.

5.1 Brain Regions Involved in Memory

Much of the evidence supporting the contribution of specific brain regions to memory (see Fig. 2) originally came from case studies. In fact, much of our knowledge about memory comes from what is perhaps the most famous case study of all time. Continue reading to find out more.

5.1.1 Hippocampus

Having suffered from reoccurring seizures, Henry Molaison (known as patient H.M. until his death in 2008) had part of his brain removed to ameliorate his epilepsy. In 1953, we did not have a complete understanding of the functions of the medial temporal lobe (MTL) (the region removed from Molaison), which includes the hippocampus. Following his surgery, Molaison experienced severe **amnesia**. He could not remember many events that had happened before his surgery (retrograde amnesia). More shockingly, Henry could not form certain types of memory (anterograde amnesia).

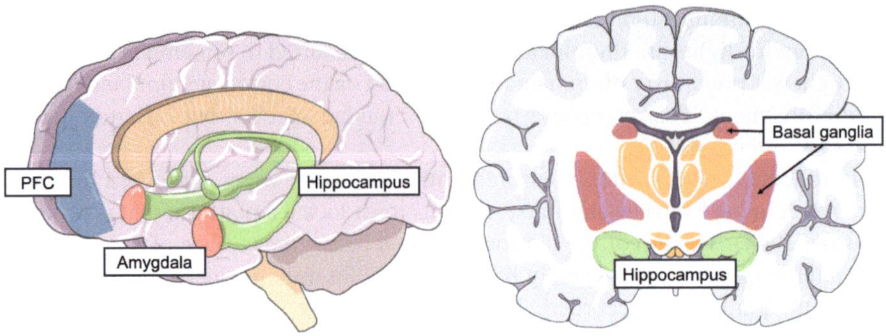

Fig. 2 Schematic of major brain regions involved in memory processes. *Note* PFC = prefrontal cortex. Image created by the author using Servier Medical Art (https://smart.servier.com/; CC BY 4.0, https://creativecommons.org/licenses/by/4.0/) and Microsoft PowerPoint

Dr. Suzanne Corkin (1937–2016) spent over four decades studying Molaison. During this time, Molaison never learned who Dr. Corkin was. He developed a sense of familiarity with her, but he would think that they were classmates. Although Molaison could not learn new explicit memories (e.g., poor performance in digit span tasks), he showed that he could form implicit memories. To illustrate, Molaison was tested in a mirror tracing task. In this procedure, he had to draw a line between two parallel lines, but he could not look down at his hand. Instead, he had to use a mirror to guide his hand. This may sound easy, but this task can be difficult for those uninitiated with it. Each day Molaison showed improvements on the task, but he always believed that he was performing the task for the first time (Corkin, 2002). In other words, he never remembered completing the task previously, but he could learn to perform the task. Dr. Corkin's work supported the idea that LTM can be fractioned into explicit and implicit subtypes.

When Dr. William Scoville performed Molaison's bilateral MTL resection (fancy way of saying that he cut this part of the brain out), he removed more than just the hippocampus. How do we know that the memory deficits experienced by Molaison were due to removal of the hippocampus and not a different structure? Techniques like functional magnetic resonance imaging (fMRI) show increased activation of the hippocampus when individuals memorize a series of words (Greicius et al., 2003). Additionally, transcranial magnetic stimulation (TMS) to the hippocampus improves working memory performance (San Agustín et al., 2023). Animal studies have shown that lesions to the hippocampus impair performance in tasks that require consolidation of STM to LTM, such as maze learning (Bolhuis et al., 1994; Kaada et al., 1961).

We know that the hippocampus is involved in explicit memory, but what exactly does it do? The hippocampus *consolidates* STM into LTM. This explains why Molaison had difficulty forming new memories. If memories are not stored in the hippocampus, why did Molaison experience retrograde amnesia? Not only is the hippocampus activated during encoding of information, but its activity increases during retrieval of information (Greicius et al., 2003). The hippocampus is critical for reconsolidation of memories after they have been recalled. Damage to the hippocampus can prevent reconsolidation of memories, leading to memory impairments.

In addition to consolidating STM into LTM, the hippocampus is involved in spatial memory (Zhao et al., 2016). One piece of evidence linking the hippocampus to spatial memory comes from a study showing that London taxi drivers have a larger posterior hippocampus compared to non-taxi drivers (Maguire et al., 2000). There is also evidence that exercise increases the size of the hippocampus in older adults, corresponding to improved spatial memory (Erickson et al., 2011). More recently, animal research has demonstrated that chemogenetic inhibition of the hippocampus impairs performance in a Morris water maze (Cinalli et al., 2023).

5.1.2 Amygdala

In addition to removing Henry Molaison's hippocampus, Dr. Scoville removed part of his amygdala. Like the hippocampus, the amygdala is important for declarative memory (Inman et al., 2018), but one of its major functions is to help consolidate memories involving emotionally arousing stimuli/events (Qasim et al., 2023). When individuals perform a DMTS task involving emotional expressions, there is increased activity of the amygdala, the hippocampus, and the orbitofrontal cortex (OFC) (LoPresti et al., 2008).

One way researchers have studied the contribution of the amygdala to emotional memories is by testing animals in a *fear conditioning* paradigm. Imagine a rat that is placed in a box that includes a floor that can be electrified. Before the rat is shocked, a tone turns on. If this process is repeated multiple times, the rat will learn to associate the tone with the shock. Now, the tone can be presented without the rat receiving a shock. What do you think the rat does when it hears the tone? Because the rat expects to be shocked when it hears the tone, the rat displays a *freezing* response. That is, the rat will stop what it's doing and will try to remain motionless. This is an evolutionary behavior that rodents (and other animals) perform to prevent detection by a predator. Essentially, we use freezing as a measure of fear in an animal like a rat. The rat learning to freeze in response to a tone previously paired with shock is a form of implicit learning known as **Pavlovian conditioning** (named for Dr. Ivan Pavlov who first described this form of learning). You will also see Pavlovian conditioning referred to as classical conditioning.

Animals that receive damage to part of the amygdala, known as the basolateral nucleus (BLA), show very little freezing when placed back in an environment associated with shock 30 days after the final conditioning session (Poulos et al., 2009). On the flip side, stimulation of the amygdala can enhance fear-like memories (Kellett & Kokkinidis, 2004). We will come back to this later in this chapter.

5.1.3 Prefrontal Cortex (PFC)

In the previous chapter, you learned how parts of the PFC mediate attentional processes. While the hippocampus consolidates STM into LTM, there is increasing evidence that memories are stored in parts of the PFC. However, the PFC is not merely a storage bin. When individuals use both working memory and LTM, PFC activity increases (Ranganath et al., 2003). This makes sense according to Baddeley's working memory model. Remember, Baddeley argued that working memory is not a storage bin; instead, individuals actively recruit LTM when performing tasks that require STM. For example, monkeys trained to perform a DMTS task show increased neuronal firing of the dorsolateral PFC (dlPFC) during the delay period (Sawaguchi & Yamane, 1999). Likewise, humans performing *n*-back tasks show increased activation of the dlPFC (Rovetti et al., 2021).

5.1.4 Basal Ganglia

Returning to Henry Molaison, we now know that his ability to form implicit memories is due to the fact his basal ganglia were left intact. In addition to its role in sensory gating (see previous chapter), the basal ganglia are involved in procedural memory. One piece of evidence supporting the role of the basal ganglia in implicit memory comes from individuals with Parkinson's disease, a neurodegenerative condition that affects dopaminergic transmission in the basal ganglia. You will learn more about Parkinson's disease in Chap. 10, but individuals with this disorder show impairments in implicit memory (van Asselen et al., 2009). Animal research also shows that lesions to the dorsal striatum disrupt procedural memory in rats (Reading et al., 1991). The basal ganglia are particularly important for habit learning, a topic that will be explored in greater detail in Chap. 11.

Comprehension Check 3 Considering the distinct functions of these brain regions, explain how damage to the hippocampus and the basal ganglia would result in different types of memory impairment. Provide an example of a specific task that would be significantly impaired by damage to each of these regions.

Damage to the hippocampus primarily impairs the formation of new explicit memories and recall of older explicit memories, whereas damage to the basal ganglia primarily affects implicit memory. A task significantly impaired by hippocampal damage would be learning the names of new acquaintances or remembering the layout of a new city. A task significantly impaired by damage to the basal ganglia would be learning to ride a bicycle or playing a musical instrument, even though the individual might still consciously remember the instructions for performing each task.

5.2 Cellular Mechanisms of Memory

So far in this book, you have learned about the neuromechanisms of maladaptive processes at a more macro level. That is, you have learned how dysfunction of brain regions and neurotransmitter systems can promote maladaptation. However, you have not learned much about the cellular mechanisms implicated in various forms of maladaptation. This is a good chapter to introduce such mechanisms as there is sufficient understanding of the cellular mechanisms controlling memory.

To understand the cellular mechanisms of memory, you need to learn more about the glutamatergic system, which is pivotal for learning and memory. As you have already learned, glutamate can bind to both ionotropic and metabotropic glutamate receptors (mGluRs), with each major receptor type being further subdivided. Recall that there are three ionotropic glutamate receptor types, with each one being composed of four subunits that form an ion channel: N-methyl-D-aspartate (NMDA) receptor, α-amino-3-hydroxy-5-methyl-4-isoxazolepropionic acid (AMPA) receptor, and kainate receptor. Differentiating AMPA receptors from kainate receptors can be difficult, so you may see AMPA receptors and kainate receptors referred to as non-NMDA receptors. There are also three groups of mGluRs, identified as Group I, Group II, and Group III. Group I mGluRs include $mGluR_1$ and $mGluR_5$ and are coupled to a G_q protein. Group II ($mGluR_2$ and $mGluR_3$) and Group III ($mGluR_4$, $mGluR_6$, $mGluR_7$, $mGluR_8$) are coupled to a G_i protein; thus, these receptors are inhibitory.

For learning to occur, neurons need to change the type of connections they have with other neurons. Remember, neurons form synapses with one another, which allows a neurotransmitter/neuropeptide to travel from one neuron to another to stimulate a receptor. The "strength" of these synapses can change, becoming either stronger or weaker. This is known as **synaptic plasticity**, with the terms *long-term potentiation (LTP)* and *long-term depression (LTD)* used to describe increased or decreased synaptic strength, respectively. LTP and LTD are directly related to **Hebbian synapses**, named for Dr. Donald Hebb. The expression "neurons that fire together, wire together" is used to describe such synapses. LTP and LTD will be described in the sections below. As you will learn, LTP and LTD are heavily dependent on glutamatergic activity.

5.2.1 Long-Term Potentiation (LTP)

LTP is composed of two major stages: early-phase LTP and late-phase LTP. Each phase is associated with activation of NMDA receptors. However, NMDA receptor activation is dependent on activity of non-NMDA receptors. When AMPA/kainate receptors are stimulated, sodium (Na^+) ions enter the neuron, depolarizing the neuron. This causes a magnesium (Mg^{++}) ion trapped in the channel of the NMDA receptor to be displaced. Now, when the NMDA receptor is stimulated, both Na^+ and calcium (Ca^{++}) enter the neuron, further depolarizing the neuron.

During early-phase LTP, Ca^{++} that enters the neuron activates **protein kinases** like Ca^{++}/calmodulin-dependent kinase II (CaMKII), leading to *phosphorylation*. Phosphorylation means that a phosphate group (PO_4^{3-}) has been added to some molecule or ion. Inside a neuron, phosphorylation can lead to different things. Specific to LTP, phosphorylation leads to the insertion of additional AMPA receptors at the postsynaptic membrane (see Fig. 3a). By increasing the number of AMPA receptors, the postsynaptic neuron is now more sensitive to glutamate that is released from the presynaptic neuron (i.e., the synapse has become strengthened).

Because additional AMPA receptors have been added, one thing you will notice is that the dendrites of neurons begin to look different. If you zoom in on a dendrite, you will notice that it has small protrusions extending out. These protrusions are known as *dendritic spines*, and they serve to increase the surface area of dendrites. During LTP, dendritic spines become larger to accommodate increased AMPA receptor expression (Fig. 3b). Researchers can determine if LTP has occurred by measuring something known as the *AMPA-to-NMDA ratio*. If this ratio is high, one can infer that LTP has occurred (Fig. 3c).

Disrupting non-NMDA receptor signaling in the hippocampus disrupts LTP (Debray et al., 1997) and memory retention in animals (Riedel et al., 1999). Older monkeys, who perform worse on working memory tasks, have reduced dendritic spine density and fewer synapses in the PFC (Dumitriu et al., 2010). This finding suggests that LTP does not occur as readily in older individuals, which may account for memory impairments observed in these individuals.

LTP is primarily influenced by glutamatergic signaling, but additional neurotransmitters can modulate LTP. Using one example, when dopamine binds to D_1-like receptors, adenyl cyclase is activated, which causes cyclic adenosine monophosphate (cAMP) levels to increase. cAMP then activates protein kinase A, which allows Ca^{++} to enter the neuron. The increased Ca^{++} levels lead to morphological and functional changes in neurons that are associated with LTP (Speranza et al., 2021).

While critical for LTP, too much neuronal excitation can be detrimental to learning. In older individuals, higher neuronal excitability is associated with *decreased* learning capacity (List et al., 2013). This is most likely due to the fact that excessive Ca^{++} levels lead to cell damage (Weber, 2012). To prevent overexcitation of neurons, another mechanism beyond LTP is required. This is where LTD becomes important.

5.2.2 Long-Term Depression (LTD)

Whereas LTP involves strengthening of synapses, LTD causes a synapse to weaken. Not only does LTD prevent excitotoxicity from occurring, but it is required for learning to occur. This is because synapses need to be "reset" via LTD, allowing them to be strengthened at a later point via LTP. Using a computer as an analogy, there is only so much information you can store on a computer. Similarly, synapses cannot become infinitely strengthened. Just as you need to delete/move old files

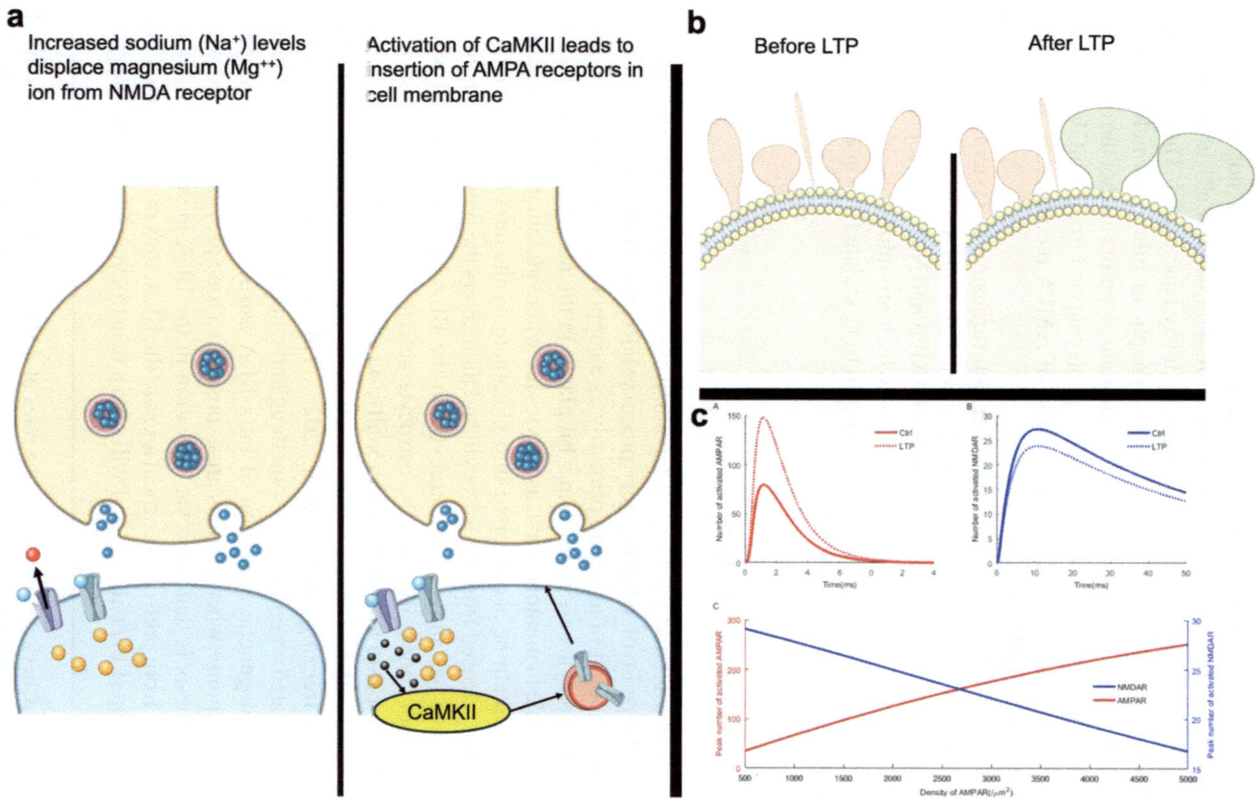

Fig. 3 (**a**) Schematic representation of long-term potentiation (LTP) and (**b**) Schematic representation of dendritic spines before and after LTP. The spines in green illustrate the morphological changes that occur to dendritic spines during LTP. *Source* Servier Medical Art (https://smart.servier.com/; CC BY 4.0, https://creativecommons.org/licenses/by/4.0/ and Microsoft PowerPoint. (**c**) A representation of how AMPA-to-NMDA receptor ratios can be measured. Before LTP, solid lines in (A) and (B) show that NMDA receptors outnumber AMPA receptors. During LTP, the dashed lines show that AMPA receptors now outnumber NMDA receptors. Therefore, the AMPA-to-NMDA receptor ratio has increased. The figure shown in (C) of panel **c** further highlights that the density of AMPA receptors increases while the density of NMDA receptors decreases during LTP. *Source* Figure 3 of Guo (2025), https://doi.org/10.1016/j.neures.2024.10.003, with permission from Elsevier

that are no longer needed, LTD works to clear space by weakening the connection between certain neurons.

LTD, like LTP, involves AMPA receptors. If AMPA receptors are insufficiently stimulated, LTD occurs instead of LTP. Because the neuron is not depolarized enough, low Ca^{++} concentrations enter the neuron. The lower Ca^{++} levels do not activate CaMKII. Instead, a different signaling pathway is activated. The end result is reduced AMPA receptor expression on the postsynaptic membrane, which weakens the connection between neurons. Ionotropic glutamate receptors are not the only ones that drive LTD. Low-frequency stimulation of postsynaptic Group I mGluRs leads to a similar result as low-frequency stimulation of AMPA receptors: reduced cell surface expression of AMPA receptors.

Similar to LTP, LTD is modulated by other neurotransmitter systems. Stimulation of cannabinoid CB receptors on excitatory neurons leads to reduced glutamate release, which leads to insufficient stimulation of AMPA and NMDA receptors. Stimulation of alpha-2 (α_2) noradrenergic receptors, which are inhibitory, also leads to LTD (Marzo et al., 2010). Dopamine is also known to modulate LTD in certain parts of the brain (Sheynikhovich et al., 2013).

5.2.3 Role of Astrocytes in Memory

The discussion of LTP and LTD focused exclusively on neurons. Yet, neurons are not the only cell type involved in memory processes. Recall from Chap. 2 that astrocytes were once thought of as acting primarily as a support system for neurons. We now know that astrocytes are more than a bit player in the nervous system, instead playing a crucial role in memory. First, astrocytes are where glutamate is synthesized. The amino acid glutamine is converted into glutamate by the enzyme glutaminase. Glutamate is then released from the astrocyte and then enters the presynaptic neuron.

Recent evidence has shown that astrocytes in the BLA become activated when mice receive a foot shock; importantly, astrocyte activity remains elevated several days following exposure to shock. Additionally, when mice are placed back in the context associated with foot shock, astrocyte activity corresponds to the initiation and the termination of freezing (Suthard et al., 2023). This finding indicates that astrocytes become activated when we learn something and when we remember something. Furthermore, chemogenetic activation of intra-BLA astrocytes increases a conditioned freezing response when animals are exposed to a tone previously paired with shock. One hypothesis is that astrocyte activation in the BLA increases activity of neurons projecting from prefrontal cortical areas to the BLA, thus increasing memory for an aversive event (Lei et al., 2022). I will return to astrocytes later in this chapter.

Comprehension Check 4 Considering the roles of AMPA and NMDA receptors in synaptic plasticity, explain how the activation of these receptors contributes differently to the strengthening (LTP) and weakening (LTD) of synaptic connections.

In LTP, the initial activation of AMPA receptors by glutamate leads to the depolarization of the postsynaptic neuron. This depolarization removes the Mg^{++} block from the NMDA receptor, allowing it to be activated by glutamate. The influx of both Na^{+} and, critically, Ca^{++} ions through the NMDA receptor triggers intracellular signaling cascades, including the activation of protein kinases. These kinases lead to the insertion of more AMPA receptors into the postsynaptic membrane and changes in dendritic spine structure, ultimately strengthening the synaptic connection and making the postsynaptic neuron more sensitive to subsequent glutamate release.

In contrast, LTD occurs when there is low-frequency stimulation, resulting in only a modest release of glutamate and thus a smaller depolarization of the postsynaptic neuron. While AMPA receptors are still activated, the depolarization is often insufficient to fully remove the Mg^{++} block from the NMDA receptor, leading to a smaller influx of Ca^{++}. This lower level of Ca^{++} influx activates different signaling pathways, which result in the internalization or removal of AMPA receptors from the postsynaptic membrane. This reduction in AMPA receptors weakens the synaptic connection, making the postsynaptic neuron less responsive to subsequent glutamate release.

6 Disorders Characterized by Memory Loss

Earlier, you learned about Henry Molaison and how removal of his hippocampus led to severe retrograde and anterograde amnesia. There are multiple diseases that cause amnesia. This section will highlight some of these diseases and why they occur.

6.1 Disorders Involving Dementia

Dementia is a debilitating condition that encompasses numerous symptoms, including memory loss. Dementia is associated with multiple diseases, such as Alzheimer's disease, frontotemporal disease (FTD), Huntington's disease, vascular dementia, dementia with Lewy bodies, Creutzfeldt-Jakob disease, and chronic traumatic encephalopathy (CTE). In short, these diseases cause widespread damage throughout the nervous system; memory loss occurs because of damage to brain structures like the hippocampus and the PFC. At the cellular level, these diseases cause extensive damage to neurons, thus disrupting processes like LTP and LTD. I want to highlight that these diseases are caused by different factors, and they differentially affect parts of the brain. I will briefly review various conditions associated with dementia below.

6.1.1 Alzheimer's Disease

Alzheimer's disease is the most widely known condition that involves dementia, and it is characterized by *plaques* and *neurofibrillary tangles* that form throughout the brain (Fig. 4). This occurs because proteins called *beta amyloid* and *tau protein* accumulate at higher-than-normal levels. Beta amyloid is created when the amyloid precursor protein is cleaved by two enzymes. Normally, beta amyloid is removed from the nervous system. In Alzheimer's disease, beta amyloid is not removed. This causes increasing protein levels to build up on cell bodies of neurons (plaques), damaging the cell body.

Tau protein is important for providing structural support to neurons; however, when tau protein levels are too high, this causes axons to get tangled (neurofibrillary tangles). This further prevents neurons from communicating with one another, leading to a wide range of impairments, including memory loss (Breijyeh & Karaman, 2020).

Above, I mentioned that beta amyloid is not adequately removed in Alzheimer's disease. A potential reason is linked to dysfunction of a protein called *apolipoprotein E* (APOE) (Belaidi et al., 2025). This protein is involved in lipid (i.e., fat) transport and neuronal repair. In the central nervous system (CNS), APOE is primarily produced by astrocytes. APOE then binds to one of several types of receptors of the low-density lipoprotein (LDL) receptor family. When APOE binds to the LDL receptor (LDLR) located on a cell like a neuron, lipids are taken back into the cell. This process is known as endocytosis and is the opposite of exocytosis that you learned about in Chap. 2. The low-density lipoprotein receptor-related protein 1 (LRP1) acts similarly to LDL receptors, but they also help remove certain ligands from the synapse, including beta amyloid and tau proteins.

I also need to mention that APOE has three major isoforms (i.e., variants): APOE2, APOE3, and APOE4. Individuals with the APOE4 isoform are more likely to develop Alzheimer's disease, whereas those with the APOE2 isoform are significantly less likely to develop Alzheimer's disease. Why is this the case? Research has demonstrated that the expression of APOE4 leads to multiple deficiencies that can promote the buildup of beta amyloid and tau proteins. First, the presence of APOE4 causes the blood-brain barrier (BBB) to weaken. As you have already learned, the BBB prevents pathogens from entering the brain from the peripheral nervous system (PNS). A weakened BBB allows more pathogens to enter the brain, which increases the risk for neuroinflammation. Another finding is that APOE4 has a poor lipidation profile. Lipidation is the process in which a lipid is added to a protein such as amyloid beta. As such, APOE4 is not as effective at collecting beta amyloid and tau, enabling them to build up to excess levels.

To further complicate matters, there are additional receptors in LDL family that are implicated in Alzheimer's disease. The very-low-density lipoprotein receptor (VLDLR) and the APOE receptor 2 have a different function compared to LDLR and LRP1. Stimulation of these receptors activates the reelin signaling pathway. This pathway is crucial for neuronal migration, dendritic growth, and synaptic plasticity. Reelin is a glycoprotein, meaning that it is a protein with at least one carbohydrate

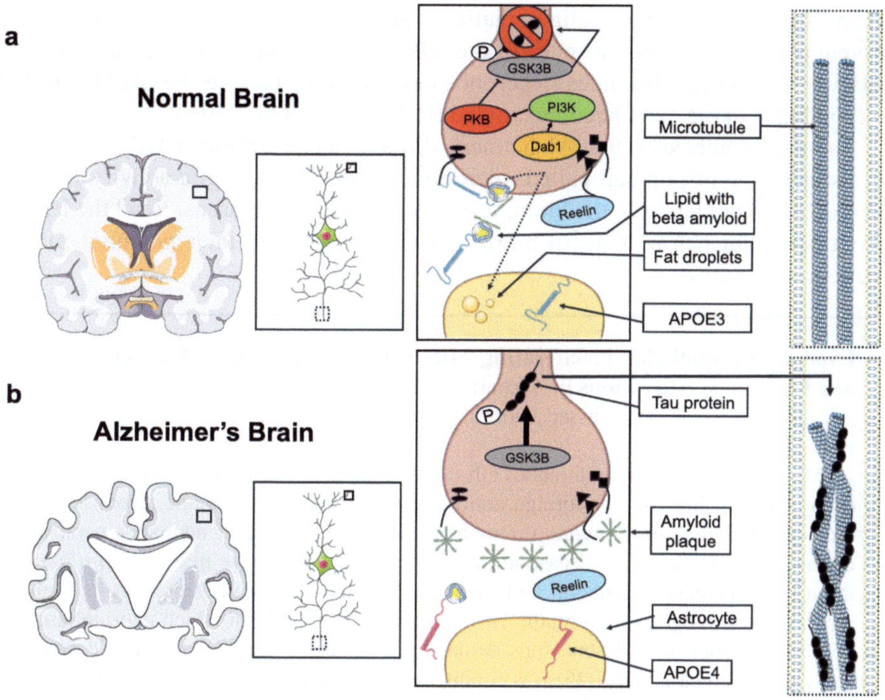

Fig. 4 A comparison of a healthy brain (**a**) and a brain affected by Alzheimer's disease (**b**). In each image, part of the cortex is highlighted, which shows a pyramidal neuron (left inset). In the middle inset, one dendrite of the pyramidal neuron is shown with part of an astrocyte. Astrocytes produce apolipoprotein E (APOE). In a normal brain, APOE3 is effective at lipidation; that is, it can add a lipid to proteins such as beta amyloid. When APOE3 binds to low-density lipoprotein receptor-related protein 1 (LRP1) (line with two solid circles), the lipid and the attached amyloid beta protein are taken up into the neuron. The lipid is broken down and transferred back to the astrocyte. Additionally, reelin can bind to very-low-density lipoprotein receptor (VLDLR) or APOE receptor 2 (denoted by the line with two triangles). This activates the reelin signaling pathway. First, Dab1 (disabled-1) activates PI3K (phosphatidylinositol 3-kinase), which then activates PKB (protein kinase B). PKB inhibits GSK3β (glycogen synthase kinase 3 beta). Inhibition of GSK3β prevents phosphorylation of tau protein. Individuals with the APOE4 allele are at increased risk of developing Alzheimer's disease. Because APOE4 is less effective at lipidation, amyloid beta levels increase in the synaptic space, in which they form clumps (plaques). Additionally, the reelin signaling pathway is less effective. Because GSK3β is not inhibited via activation of the reelin signaling pathway, hyperphosphorylation of tau proteins occurs. The right inset shows the microtubules that provide structural support to the axon. Excess accumulation of tau protein causing the microtubules inside axons to become tangled (neurofibrillary tangles). Image created by the author using Servier Medical Art (https://smart.servier.com/; CC BY 4.0, https://creativecommons.org/licenses/by/4.0/) and Microsoft PowerPoint

chain. Stimulation of the reelin signaling pathway is important for regulating tau protein phosphorylation. Presence of the APOE4 allele impairs the reelin signaling pathway; tau protein becomes hyperphosphorylated, increasing the likelihood that microtubules in the axon become tangled (Hunsberger et al., 2019).

Another hypothesis is that Alzheimer's disease can be attributed to dysregulation of another type of glia cell called microglia. Excessive microglia activity leads to increased inflammation in the brain, resulting in damaged neurons. Read the Experiment Spotlight below to learn more about microglia and their potential role in Alzheimer's disease.

Experiment Spotlight: Proliferating Microglia Exhibit Unique Transcriptional and Functional Alterations in Alzheimer's Disease

The full article can be accessed with this link: https://pmc.ncbi.nlm.nih.gov/articles/PMC12140498/

Microglia are important immune cells of the brain, kickstarting an immune response if any signs of infection or foreign entities are detected. This immune response can include inflammation. Previous research has already shown that neurodegenerative diseases like Alzheimer's disease are characterized by excessive microglia *proliferation* (i.e., increased numbers) and microglia death. This means that there is high turnover of microglia in the brain. This can create a situation in which neuroinflammation is chronically present, thus damaging other cells like neurons. The current study (Villacampa et al., 2025) was conducted to identify potential subpopulations of microglia that are associated with chronic inflammation in the brain. Numerous assays were performed. I will highlight the most relevant methods below.

First, the researchers examined the brains of a transgenic mouse type called the APP/PS1 (amyloid precursor protein/presenilin 1). These mice are used as an animal model of Alzheimer's disease as they are bred to have high beta amyloid levels. Brains were taken at either 4 months, 6 months, or 12 months of age. In addition to using APP/PS1 mice, the researchers examined postmortem tissue that was collected from individuals with Alzheimer's disease, individuals with mild cognitive impairment (but not diagnosed with Alzheimer'sdisease), and controls.

Immunohistochemistry was performed using both mouse and human brain tissue to detect Ki67-positive and DDX3-positive cells. Ki67 (antigen Kiel 67) is a protein that is associated with cellular proliferation. DDX-3 (DEAD-box RNA helicase 3) is a type of enzyme that is involved in RNA metabolism; specifically, DDX-3 aids in transcription of ribonucleic acid (RNA) into deoxyribonucleic acid (DNA), transporting RNA molecules within a cell, translating messenger RNA (mRNA) into protein, and degrading RNA molecules. Increased DDX-3 expression indicates increased microglia activation. Cells were also stained with IBA1 (ionized calcium-binding adaptor molecule 1) because this protein is largely confined to microglia. Thus, cells labeled with both Ki67 and IBA1 can be classified as proliferating microglia.

Additionally, the researchers performed fluorescence-activated cell sorting (FACS) to physically separate proliferating microglia from non-proliferating microglia. Tissue was exposed to the antibodies CD11b-PE, CD45-FITC, and Ki67. CD11b-PE is a protein that is found on the surface of microglia. CD45-FITC is found on the surface of immune cells, including microglia. In a separate experiment, mice were injected with methoxy-X04, which is a fluorescent tag that labels beta amyloid plaques, before their brains were taken. This experiment was performed to determine which proliferating microglia were actively trying to remove beta amyloid.

RNA sequencing was also performed. This technique allows one to determine the sequence of RNA molecules, and thus, isolate which genes are being expressed and how much they are being expressed. This experiment was performed to determine if proliferating microglia display *differentially expressed genes* (often abbreviated as DEGs in the literature) compared to non-proliferating microglia.

In APP/PS1 mice and individuals with Alzheimer's disease, an increase in microglia proliferation was observed, particularly in the cortex and in the hippocampus. In wildtype mice (i.e., control mice), the number of microglia decreased across age; conversely, in APP/PS1 mice, the number of microglia increased. Approximately 40% of proliferating microglia were found near beta amyloid plaques in mice that were 6 or 12 months of age, compared to approximately 10% for mice at 4 months of age. Proliferating microglia were less likely to clear beta amyloid compared to non-proliferating microglia.

RNA sequencing revealed that proliferating microglia in APP/PS1 mice show differential gene expression compared to proliferating microglia in wildtype mice and compared to non-proliferating microglia. The changes in gene expression are not apparent in APP/PS1 mice at 4 months of age, but these changes begin to emerge at 6 months of age and persist at 12 months of age. Of note, proliferating microglia express genes that are related to inflammation. In particular, DDX-3 levels were significantly higher in microglia extracted from APP/PS1 mice relative to wildtype controls at 12 months of age.

Questions to Consider:

1. How do the results better help us understand how microglia contribute to the pathology of conditions like Alzheimer's disease?
2. Based on the results of this study, what approach can be taken in an effort to slow the progression of Alzheimer's disease?
3. In the experiment, I described the immunohistochemistry experiment that was conducted to visualize proliferating and non-proliferating microglia. I also described the FACS experiment. What was the purpose of performing these two assays?

6.1.2 Frontotemporal Dementia (FTD)

As the name suggests, FTD is related to damage to the frontotemporal regions of the brain. Like Alzheimer's disease, a buildup of tau protein is observed. However, another protein is proposed to contribute to FTD: TAR DNA-binding protein 43 (TDP-43). As a DNA and an RNA binding protein, TDP-43 is important for gene expression and RNA splicing. One hypothesis is that TDP-43 can become displaced, moving from inside the nucleus to outside the nucleus. This causes TDP-43 to clump together, which damages the neuron (Kellett et al., 2025) (Fig. 5).

Another potential cause of FTD is an excessive repeat of part of the *C9orf72* gene, which encodes the protein C9orf72 (chromosome 9 open reading frame 72)

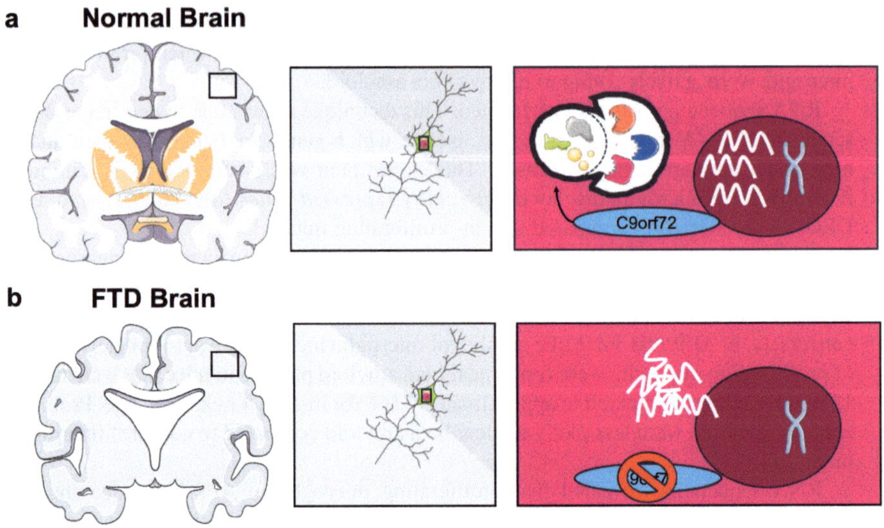

Fig. 5 A schematic illustrating the pathology associated with frontotemporal dementia (FTD). In a healthy brain (**a**), TAR DNA-binding protein 43 (TDP-43) (pictured as a white squiggly line) is found in the cell nucleus. The *C9orf72* gene encodes the C9orf72 protein, which plays a critical role in autophagy. Debris is collected and sent to a lysosome, an organelle that is responsible for breaking down waste products in the cell. In the image, you can see a lysosome binding to a vesicle that contains debris. In FTD (**b**), TDP-43 escapes the cell nucleus, where it forms clumps. Because the *C9orf72* gene does not work properly, autophagy does not occur as it should. As such, debris continues to build up inside the neuron, ultimately damaging it. Image created by the author using Servier Medical Art (https://smart.servier.com/; CC BY 4.0, https://creativecommons.org/licenses/by/4.0/) and Microsoft PowerPoint

(DeJesus-Hernandez et al., 2011). C9orf72 is proposed to assist in several processes. One of its major functions is to aid in *autophagy*, the process by which old or damaged cell parts are broken down and removed from the cell; the C9orf72 protein is also important in regulating immune responses (Pang & Hu, 2021; Smeyers et al., 2021). If the *C9orf72* gene becomes dysregulated, the C9orf72 protein can no longer remove debris from cells (Fig. 5). This leads to high levels of debris inside cells, including neurons, and inflammation. This combination then causes neurons to die.

6.1.3 Huntington's Disease

Huntington's disease is a genetic disorder caused by a mutated gene for the huntingtin protein. The huntingtin protein is hypothesized to play several important roles in neurons, such as helping transport materials into the neuron and through axons. The mutated huntingtin protein causes neurons to begin to die. Individuals with Huntington's disease and those with prodromal Huntington's disease (i.e., those that do not show overt symptoms) show impairments in implicit learning (Ghilardi et al.,

2008; van Asselen et al., 2012). This is due to damage that occurs to the basal ganglia. Eventually, the damage spreads to other parts of the brain, particularly the PFC, leading to deficits in declarative memory (Montoya et al., 2006). You will learn more about Huntington's disease in Chap. 10.

6.1.4 Vascular Dementia

Vascular dementia is caused by reduced blood flow to any part of the brain that controls memory functions (e.g., frontal cortex). One culprit of reduced blood flow is a *stroke*, which can be caused by a blood clot or from a ruptured blood vessel. High blood pressure and conditions like diabetes can also lead to reduced blood flow to the brain. Regardless of the underlying cause, neurons become damaged due to oxygen and glucose deprivation. This damage leads to disruptions in neuronal communication, resulting in various symptoms like memory impairment.

6.1.5 Dementia With Lewy Bodies

Dementia with Lewy bodies results from high deposits of a protein called alpha-synuclein that form clumps known as Lewy bodies (named for Dr. Fritz Heinrich Lewy). Normally, alpha-synuclein is found in axon terminals and helps the presynaptic neuron release neurotransmitters into the synapse. However, the alpha-synuclein protein can become misfolded. The faulty alpha-synuclein gets released into the synapse and gets taken into the postsynaptic neuron. Alpha-synuclein then clumps together, forming a Lewy body (Fig. 6). Individuals with Alzheimer's disease can also develop Lewy bodies.

Several factors have been proposed to cause misfolding of alpha-synuclein proteins (Negi et al., 2024). In addition to exposure to environmental toxins, one hypothesis argues that mutations of the *SNCA* gene (which encodes alpha-synuclein) cause the protein to malfunction, leading to misfolding. Alterations in **post-translational modification** to alpha-synuclein can increase the probability that this protein becomes misfolded. Post-translational modification, as the name implies, occurs after a protein has been translated from mRNA. After the protein has been created, additional changes can be made to its structure, which modifies what it does. You have already encountered one example of a post-translational modification earlier in this chapter. Phosphorylation alters a protein by adding a phosphate group. *Nitration* and *acetylation* are additional examples of post-translational modifications; nitration involves the addition of a nitro group (NO_2; one nitrogen atom and two oxygen atoms) while acetylation means that an acetyl group (CH3CO; CH3 is known as a methyl group and contains one carbon atom and three hydrogen atoms while CO is a carbonyl group and consists of one carbon atom and one oxygen atom) has been added to a protein (see Fig. 6).

Acetylation of alpha-synuclein is protective against the formation of Lewy bodies as the addition of an acetyl group helps slow the accumulation of alpha-synuclein

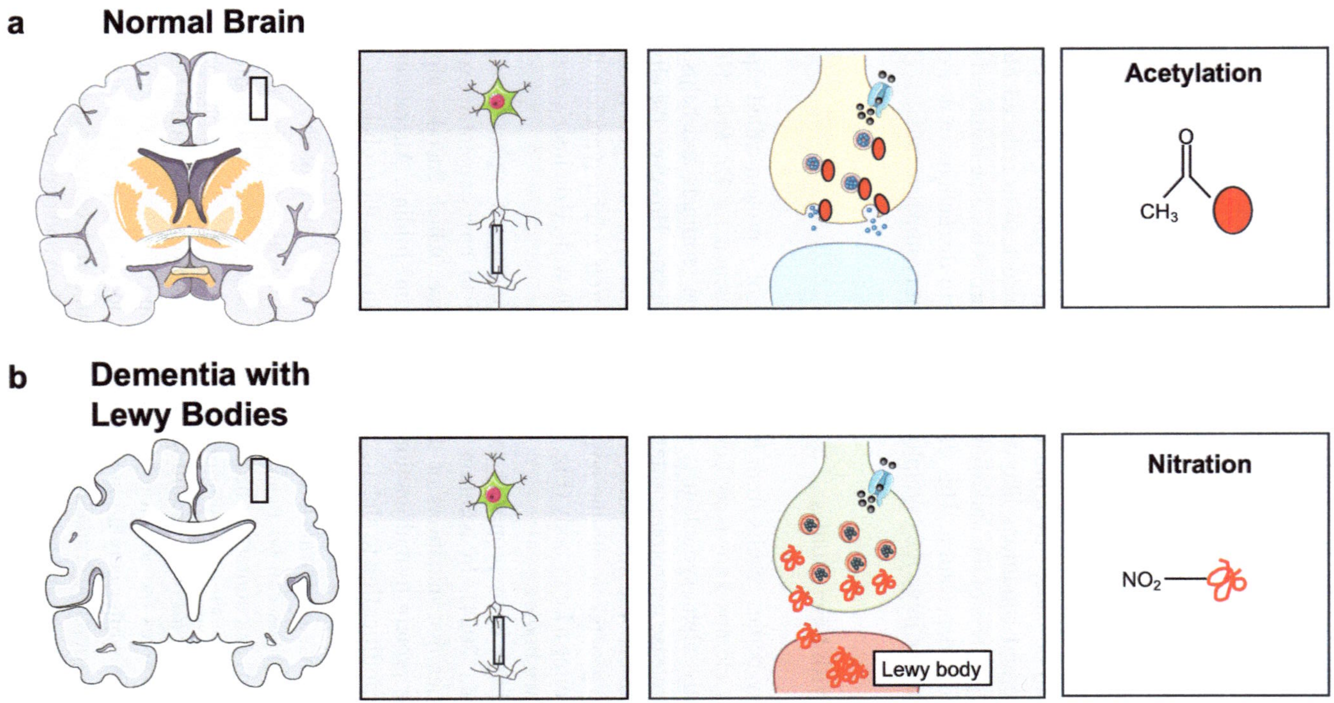

Fig. 6 A schematic comparing a healthy brain (**a**) to one affected by Lewy bodies (**b**). Alpha-synuclein normally helps the vesicle release its contents into the synaptic space. However, when the alpha-synuclein protein becomes misfolded, it is more likely to be released into the synapse and absorbed into dendrites of surrounding neurons. As the alpha-synuclein continues to clump together and with other organelles inside the cell, a Lewy body is formed. The far right insets show examples of post-translational modifications that can be made to the alpha-synuclein protein. Acetylation is protective against Lewy bodies whereas nitration is a risk factor for the formation of Lewy bodies. Image created by the author using Servier Medical Art (https://smart.servier.com/; CC BY 4.0, https://creativecommons.org/licenses/by/4.0/) and Microsoft PowerPoint

proteins. Conversely, nitration appears to be a risk factor for the development of Lewy bodies. Nitration of alpha-synuclein can lead to **oxidative stress**. I need to provide a brief chemistry lesson to explain why oxidative stress occurs and why this is problematic.

Molecules are groups of two or more atoms (e.g., nitro and acetyl groups). The molecule you may be most familiar with is H_2O, which is water. Atoms contain a nucleus that houses protons (positively charged) and neutrons (no charge). Electrons (negatively charged) surround the nucleus. Atoms typically have the same number of protons and electrons, but the number of electrons can be either higher or lower than the number of protons. These are called ions, which you have already learned about (e.g., Na^+ and Ca^{++} ions). When there are more electrons than protons, the ion is negatively charged and is called an anion. Positively charged ions are called cations. Some molecules have an uneven number of electrons and are highly unstable, meaning that they try to strip an electron from nearby molecules. These unsteady molecules are known as *free radicals*, and they can cause irreparable damage to cells.

The counterpart to free radicals is *antioxidants*, a term you may be familiar with. Vitamin C is a well-known antioxidant. Antioxidants can donate an electron to a free radical without becoming unstable. Oxidative stress occurs when free radicals outnumber antioxidants. Going back to nitration, NO_2 is a free radical. When NO_2 levels increase too much, oxidative stress occurs.

6.1.6 Creutzfeldt-Jakob Disease

Creutzfeldt-Jakob disease is caused by an abnormal protein called a *prion*. A prion is a misfolded protein that then causes other proteins to become misfolded. When discussing dementia with Lewy bodies, the misfolding of alpha-synuclein proteins is similar to the misfolding that occurs with prions. The formation of prions leads to a chain reaction-like event, in which numerous proteins become misshapen. Folded proteins no longer function as normal. As neurons contain numerous proteins (e.g., receptors, transporters, ion channels, microtubules, alpha-synuclein, etc.), prions wreak havoc on neuronal transmission. Due to the widespread damage caused by prions, individuals with Creutzfeldt-Jakob disease often die within 1 year after symptoms emerge.

Creutzfeldt-Jakob disease belongs to a group of diseases called *transmissible spongiform encephalopathies*. Even if you have never heard of Creutzfeldt-Jakob disease, you may have heard of "mad cow disease", which is formally called bovine spongiform encephalopathy. These diseases can be transmitted across animals, and Creutzfeldt-Jakob disease can result from consumption of beef from a cow with bovine spongiform encephalopathy. The term "spongiform" is used because prion diseases cause the brain to resemble a sponge as brain tissue deteriorates. "Encephalopathy" is a term that indicates that the brain no longer functions as it should. In fact, each disease that has been described in this chapter is an encephalopathy.

6.1.7 Chronic Traumatic Encephalopathy (CTE)

Originally known as "punch-drunk syndrome" and dementia pugilistica, the symptoms of CTE were first studied in boxers. CTE is now known to result from repetitive blows to the head; as such, individuals that play other contact sports like football or individuals that experience domestic violence are at risk of developing CTE (Saulle & Greenwald, 2012). One of the most extreme cases of CTE came from Chris Benoit, a professional wrestler that was described as having the brain of a 70-year-old Alzheimer's patient despite being 40 years old at the time of his death following a double-murder suicide. CTE is associated with hyperphosphorylated tau protein like in Alzheimer's disease; however, the way the tau protein is folded differs across each condition (Falcon et al., 2019). This means that the pathology of Alzheimer's disease and CTE is not identical, which emphasizes that these conditions can be differentiated.

Researchers are still trying to understand the exact reason why repeated head injuries lead to CTE. Like in Alzheimer's disease, microglia-induced neuroinflammation is associated with elevated tau protein levels observed in individuals with CTE (Cherry et al., 2016). This makes intuitive sense considering that damage to the body produces a state of inflammation (see Chap. 4). The key with CTE is that this condition is not the result of a single traumatic brain injury; instead, multiple "sub-concussive" blows to the head are a risk factor for CTE. When an individual experiences multiple hits to the head, microglia can become dysregulated, leading to a chronic state of inflammation in the brain.

Comprehension Check 5 Contrast the primary neuropathological features of Alzheimer's disease and Huntington's disease, and explain how these distinct features lead to differences in the initial types of memory impairment observed.

In Alzheimer's disease, the primary neuropathological features are the accumulation of amyloid plaques and neurofibrillary tangles that damage neurons throughout the brain, particularly in the hippocampus and PFC. This leads to impairments in explicit memory.

In contrast, Huntington's disease is characterized by a genetic mutation leading to an abnormal huntingtin protein, which causes the degeneration of neurons primarily in the basal ganglia and later in the cortex. The initial memory impairments in Huntington's disease often manifest as impairments in implicit memory.

6.2 *Dissociative Disorders*

Memory loss is not isolated to individuals with forms of dementia. I now want to briefly discuss a class of mental disorders known as dissociative disorders. As defined by the American Psychiatric Association (2024), "dissociative disorders involve problems with memory, identity, emotion, perception, behavior and sense of self". The term "dissociative" is used because individuals with these disorders often feel disconnected (dissociated) from themselves or from the world around them. There are several major dissociative disorders, with the most common one being **dissociative amnesia**. Individuals with dissociative amnesia often cannot remember a period of their life while others may not be able to remember specific events within a period of time. In very extreme cases, an individual may not know who they are. Memory loss tends to be related to a traumatic or a stressful event.

Symptoms of dissociative amnesia are negatively correlated with hippocampal volume; that is, individuals with a smaller hippocampus are more likely to experience more symptoms (Dimitrova et al., 2023). On top of having a smaller hippocampus, individuals with dissociative amnesia have a hypoactive hippocampus and PFC (Taïb et al., 2023). Expanding on this finding, a recent study found increased activation of the right anterior dlPFC and the right ventrolateral PFC (vlPFC) and decreased activation of the medial PFC (mPFC) and the hippocampus (Marsh et al., 2025). These results suggest that when individuals are exposed to stimuli associated with a traumatic/stressful event, the dlPFC and the vlPFC inhibit brain regions associated with memory retrieval. In a way, the brain is trying to "protect" the individual from having to re-experience something traumatic.

In popular culture, **dissociative identity disorder (DID)** is often used as a plot device. This is because DID is characterized by the presence of two or more distinct personalities in an individual. Unfortunately, the media often portrays DID incorrectly, with television shows and movies including characters with DID that transition to a violent persona and/or show large physical changes across personalities. Individuals with DID are rarely violent (Webermann & Brand, 2017) and tend to use multiple personalities as a coping mechanism for a traumatic experience (Krause-Utz, 2022).

Given the complexity of DID and the rarity of this disorder (observed in ~1.0–1.5% of the population; Brand et al., 2016), unraveling the neurobiological bases of this condition is extremely difficult. Concerning memory impairment associated with DID, hippocampal volumes are lower in individuals with DID compared to healthy controls (Dimitrova et al., 2023, Vermetten et al., 2006). While early work indicated that individuals with DID have a smaller amygdala (Vermetten et al., 2006), more recent work has not provided as much support for this finding (Blihar et al., 2021).

7 Disorders Characterized by Maladaptive Memory Processes

Maladaptive memories can exist beyond memory loss. Some individuals seemingly remember things, but what they remember is inaccurate or did not occur. In other cases, reactivation of certain memories can be maladaptive and cause serious psychological harm to an individual. A few prominent examples will be discussed below.

7.1 Confabulation

H.M. may be the most studied individual in cognitive neuroscience; however, the case study that receives the most attention in psychology is Phineas Gage. Gage worked for the Rutland & Burlington Railroad as a foreman. On September 13, 1848, Gage was using a 13-pound tamping rod to pack gunpowder into a hole. The gunpowder was used to blast through rocks that would have blocked the path of the track. The gunpowder ignited prematurely, causing the tamping rod to blow through Gage's head. The rod entered Gage's left cheek and damaged a large part of his left PFC before exiting his skull. Miraculously, Gage survived the accident and even remained conscious.

Although Gage survived, his life changed. In addition to personality changes resulting from damage to the PFC, Gage experienced **confabulation**, a disorder in which people unintentionally create false memories (i.e., engage in "honest lying"). Confabulation does not result from exclusive damage to the PFC. Patients with lesions to a wide range of brain regions, including the OFC and limbic structures, experience confabulation (Schnider & Ptak, 1999).

How does damage to the regions listed above contribute to confabulation? The ventromedial PFC (vmPFC) is involved in *schema representation.* Think of a schema as a blueprint that enables us to organize information about things in the world. For example, if someone asks you to describe a dog, you can most likely do this easily. This is because we have a schema for dogs that we can retrieve from memory. As we learn more about the world, we can update our schemas and store this new information in memory. Individuals with lesions to the vmPFC that experience confabulation have difficulty identifying words that belong to a specific schema (Ghosh et al., 2014).

In addition to schema representation, the vmPFC appears to be important for *reality monitoring*, evaluating if something is real or imagined. Individuals that show reduced activity of the caudal region of the vmPFC are more likely to identify a self-generated stimulus (i.e., imagined) as being real (Turner et al., 2008).

One feature of confabulation is misidentifying when events occurred. Someone may claim they went to a specific amusement park when they were a child when they really went there when they were 18. The OFC plays a pivotal role in governing temporal processing (Johnson et al., 2022). Impaired functioning of the OFC provides an account for errors in recollecting events in the correct order.

There is one brain region I have not discussed in this book that has been identified as an important mediator of memory: the mammillary bodies, located in the diencephalon (Fig. 7). The best evidence supporting the involvement of the mammillary bodies in memory comes from individuals with **Korsakoff's syndrome**. This disease is often associated with individuals that are severely dependent on alcohol. These individuals often fail to get enough thiamin (better known as vitamin B1), which results in massive damage to the mammillary bodies. Individuals with Korsakoff's syndrome experience anterograde amnesia and confabulation. Often, Korsakoff's syndrome is combined with *Wernicke's syndrome* (Wernicke-Korsakoff syndrome)

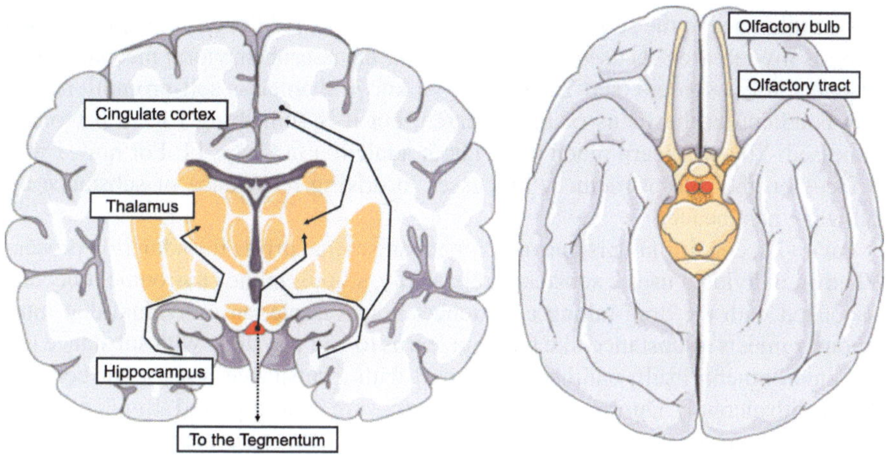

Fig. 7 Location of the mammillary bodies (highlighted in red) as viewed from a coronal section (left image) and from the ventral side of the brain (right image). The left image also shows the connections between the mammillary bodies and other structures involved in memory. Image created by the author using Servier Medical Art (https://smart.servier.com/; CC BY 4.0, https://creativecommons.org/licenses/by/4.0/) and Microsoft PowerPoint

as long-term thiamin deficiencies cause issues beyond memory impairments, such as language difficulties.

Specifically related to memory, the mammillary bodies have a direct connection with the thalamus, and they receive information from the hippocampus. The pathway between the mammillary bodies and the thalamus is important for regulating thalamo-cortical and hippocampo-cortical circuits, which influence temporal memory and *contextual memory* (Dillingham et al., 2021). Contextual memory refers to our ability to recall details associated with an event, such as the time in which the event occurred, where the event occurred, the people that were present at the time of the event, and the type of emotions elicited by the event. Damage to the mammillary bodies causes disruptions in thalamo-cortical and hippocampo-cortical circuits, which then leads to amnesia and confabulation.

7.2 Addiction

Addiction is defined by the American Psychological Association as "a state of psychological and/or physical dependence on the use of drugs or other substances, such as alcohol, or on activities or behaviors, such as sex, exercise, and gambling" (n.d.). Importantly, addiction is classified as a chronic, **relapsing** disorder (National Institute on Drug Abuse, 2011). This means that individuals are likely to resume using a substance or engaging in an addictive behavior following *abstinence*. Recurrence of substance use can be insidious as individuals may be at increased risk for overdose as their tolerance to the substance decreases over time. Addiction is deeply personal to me as my research focuses on understanding the neurobehavioral mechanisms of substance use disorders (see Yates, 2023—I know, shameless self-promotion). Not only is addiction the focus of my research, but it is something I have personally witnessed. You will learn much more about addiction in Chap. 11. For now, I want to focus on the role of memory processes in driving recurrence of substance use following abstinence.

Pavlovian conditioning is one major mechanism that drives the addiction process. When an individual uses a substance repeatedly, stimuli in the environment become associated with the drug-taking experience. For example, an individual that intravenously injects a substance like fentanyl learns to pair a syringe with substance use. The environment, itself, can become paired with a substance (e.g., a bar becomes paired with alcohol). During abstinence, re-exposure to drug-paired stimuli can elicit withdrawal symptoms and/or intense cravings by activating memories related to substance use.

Figure 8 shows neuronal changes that occur when an individual learns to associate a stimulus with drug. In this example, an individual that uses a syringe to inject a drug encodes several pieces of information. First, there is the visual stimulus of the syringe, which holds the drug to be injected. Second, there is the physical sensation of the needle piercing the skin and the drug flowing through the vein. Third, the individual experiences the physical and the psychological effects of the drug. Numerous brain regions help process the different aspects of the drug-taking experience, including visual pathways, regions of the mesocorticolimbic pathway, and the striatum. The hippocampus receives inputs from these various regions. Going back to the concept of Hebbian synapses, the connections between presynaptic neurons and hippocampal neurons become stronger. Eventually, presentation of a drug-paired stimulus alone can elicit strong responses from hippocampal neurons that previously received inputs from neurons transmitting information about the drug itself.

7.3 Posttraumatic Stress Disorder (PTSD)

Posttraumatic stress disorder (PTSD) is a mental and behavioral disorder that results from exposure to a traumatic event, such as warfare, physical/emotional abuse, domestic violence, and surviving a mass shooting, to name a few. Individuals with PTSD experience multiple symptoms, often "reliving" the traumatic event, showing distress to trauma-related cues, having disturbing thoughts and dreams about the traumatic event, and experiencing increased activity of the sympathetic nervous system's fight-or-flight response. Like addiction, you will learn more about PTSD later in this book (see Chap. 8). For now, I want to discuss the neural bases for the distorted memories that individuals with PTSD experience.

Dysregulated amygdalar function is often observed in individuals with PTSD. Remember, not only is the amygdala involved in emotional memories, but it is important for fear conditioning. We easily learn to form associations between certain stimuli

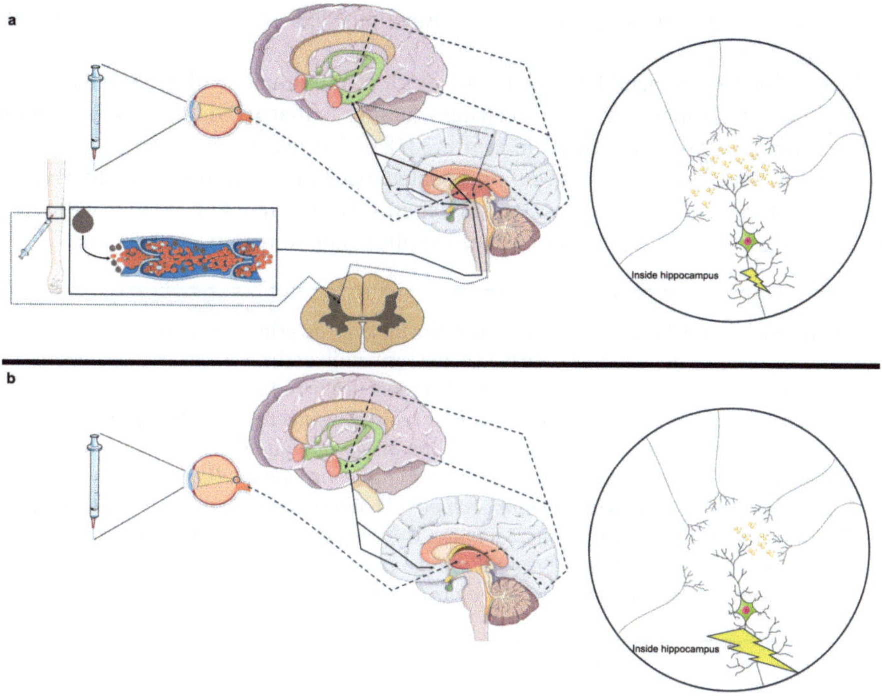

Fig. 8 An example of the neural mechanisms associated with Pavlovian conditioning as related to recurrence of substance use. (**a**) Seeing a syringe activates visual pathways (dashed arrows), and information regarding the sensation of the needle piercing the skin travels through the spinal cord before reaching the somatosensory cortex (dotted arrows). When injected intravenously, the drug travels through the blood stream, crossing the blood-brain barrier (BBB) and activating neurons in the ventral tegmental area (VTA); the mesocorticolimbic pathway and the striatum are then activated (solid arrows). Pyramidal neurons in the hippocampus receive inputs from various brain regions such as the temporal and the parietal cortices (dashed arrows), the somatosensory cortex (dotted arrow), and the PFC and striatum (solid arrows). Repeated drug administration increases the synaptic strength of pyramidal neurons. (**b**) Presentation of a drug-paired stimulus, activates the visual pathways, eventually causing neurotransmitter release in the hippocampus. Due to the formation of Hebbian synapses, pyramidal neurons become highly active. The hippocampus has direct connections with brain regions in the mesocorticolimbic pathway, which then triggers cravings for the substance. Image created by the author using Servier Medical Art (https://smart.servier.com/; CC BY 4.0, https://creativecommons.org/licenses/by/4.0/) and Microsoft PowerPoint

and emotional responses related to fear. During fear learning, LTP is largely observed in part of the amygdala. Because the amygdala becomes hyperactive following trauma, individuals with PTSD may show an increased fear response when presented with a non-threatening stimulus. For example, military veterans may mistake the sound of a car backfiring for a gunshot. This occurs because of a phenomenon known as **generalization**. Because the sound of a car backfiring resembles the sound of a gunshot, this leads to activation of the same neural pathways responsible for encoding information about gunshots and the emotional/physical responses associated with them (e.g., fear).

In addition to enhanced fear-related memories, individuals with PTSD also show general deficits in memory. One consistent finding is that PTSD is associated with reduced hippocampal volume (Xie et al., 2018). The prevailing hypothesis is that exposure to a major traumatic event or repeated exposure to trauma leads to overexcitation of hippocampal neurons, causing excitotoxicity. In Chap. 8, you will learn the exact mechanisms that lead to increased glutamate in response to trauma.

> **Comprehension Check 6** Propose a common underlying principle about how dysregulation in specific brain regions can lead to the formation or the retrieval of maladaptive memories across the seemingly different disorders discussed above.
>
> *Dysregulation of the hippocampus and/or regions connected with the hippocampus can contribute to conditions such as confabulation, addiction, and PTSD. Confabulation is associated with damage to the PFC and the mammillary bodies, which have direct connections with the hippocampus. Memory impairments associated with PTSD are attributed to hippocampal damage. Conversely, part of the addiction process, namely recurrence of substance use following abstinence, is driven by reactivation of drug memories when certain stimuli are presented.*

8 Chapter Summary

We rely extensively on memory and often take this process for granted. Being able to perform simple tasks is often dependent on a properly functioning memory system. In this chapter, you learned that memory is not a unitary construct, as it can be fractioned into different types (e.g., explicit vs. implicit memory). You also learned about the brain regions involved in different forms of memory and about the cellular mechanisms that allow for the formation of memories. Synaptic plasticity, particularly in the hippocampus, is critical for the consolidation of STM into LTM, specifically explicit memories, but it is also important for reconsolidating memories. Damage to this region can lead to both retrograde and anterograde amnesia. Memory impairments are also observed following damage to cortical areas such as the PFC. Confabulation can also result from cortical damage and damage to other brain regions such as the mammillary bodies. The basal ganglia are important for implicit memory, as damage to this system (e.g., Huntington's disease) impairs this form of learning. This chapter concluded by discussing a couple of disorders characterized, at least in part, by maladaptive memories: addiction and PTSD. You will learn more about these disorders in later chapters as aberrant memory processes make up just a portion of symptoms associated with them.

Answers to Experiment Spotlight Questions:

1. Typically, microglia are activated in response to a pathogen/foreign invader that has made its way into the brain. Microglia are normally effective at neutralizing such threats. This study shows that proliferating microglia are ineffective at removing beta amyloid, which contributes to the buildup of amyloid plaques that ultimately damage neurons. Mechanistically, this study highlights that proliferating microglia show differential gene expression, meaning that these microglia function differently compared to non-proliferating microglia. This can potentially explain why proliferating microglia are ineffective at removing beta amyloid.
2. Given that the DDX-3 gene is implicated in inflammatory responses and was upregulated in proliferating microglia, one potential therapeutic approach is to inhibit the DDX-3 protein. Blocking this protein could disrupt proliferating microglia, thus allowing existing microglia to more efficiently perform their job of removing excess beta amyloid.
3. At first glance, the immunohistochemistry and the FACS assays seem to provide the same information about proliferating and non-proliferating microglia. However, these techniques have an important distinction. In the immunohistochemistry experiment, the researchers are merely visualizing where proliferating and non-proliferating microglia are located in brain regions like the cortex and the hippocampus. FACS allows one to physically separate these cells from tissue. By separating these cells from brain tissue, the researchers could then perform the RNA sequencing experiment, thus allowing them to determine if proliferating microglia display differential gene expression compared to non-proliferating microglia.

Glossary

Amnesia a form of memory loss that can be characterized by the inability to recall past events and/or the inability to learn new information.

Confabulation the creation of fabricated, distorted, or misinterpreted memories in the absence of conscious awareness of deception, often occurring in individuals with memory disorders.

Dementia a general term for a decline in mental ability severe enough to interfere with daily life.

Dissociative amnesia a dissociative disorder characterized by the inability to recall important personal information, usually of a traumatic or stressful nature, that is too extensive to be explained by ordinary forgetfulness.

Dissociative identity disorder (DID) a dissociative disorder characterized by the presence of two or more distinct personality states, each with its own patterns of perceiving, relating to, and thinking about the environment and self.

Explicit memory the conscious, intentional recollection of factual information, past experiences, and concepts.

Generalization the tendency to respond in a similar way to different stimuli that resemble one another.

Hebbian synapse a neural connection that strengthens when the presynaptic neuron consistently contributes to the firing of the postsynaptic neuron; the expression "neurons that fire together, wire together" is often used to describe such synapses.

Implicit memory a type of memory that does not require conscious thought and is expressed through performance rather than recollection.

Korsakoff's syndrome a chronic memory disorder typically caused by a severe deficiency of thiamine (vitamin B1), often linked to chronic alcohol use, characterized by difficulties in forming new memories and recalling recent events, and sometimes including confabulation.

Long-term memory (LTM) the relatively permanent storage of information, ranging from factual knowledge to personal experiences and skills.

Oxidative stress an imbalance in the body between the production of harmful free radicals (i.e., unstable molecules) and the ability of antioxidants (type of molecule that can donate an electron) to neutralize them, potentially leading to cellular damage.

Pavlovian (classical) conditioning a learning process in which an association is made between a neutral stimulus and a biologically relevant stimulus, resulting in the neutral stimulus eliciting a learned, reflexive response.

Post-translational modification the modification of proteins after they have been translated by mRNA; these modifications often alter their physical and chemical properties, stability, activity, and localization.

Protein kinase an enzyme that modifies other proteins by chemically adding phosphate groups (phosphorylation).

Relapse the recurrence of a past condition or ailment after apparent recovery.

Sensory memory the briefest form of memory, holding sensory information for a very short period of time, typically fractions of a second to a few seconds.

Short-term memory (STM) a cognitive system that temporarily holds information for a brief period, typically seconds to minutes.

Synaptic plasticity the ability of synapses to strengthen or weaken over time in response to changes in their activity.

Working memory a cognitive system that temporarily holds and manipulates information needed for complex tasks such as learning, reasoning, and comprehension.

References

American Psychiatric Association. (2024). *What are dissociative disorders?* Retrieved November 7, 2024, from https://www.psychiatry.org/patients-families/dissociative-disorders/what-are-dissociative-disorders

American Psychological Association. (n.d.). *Substance use, abuse, and addiction.* Retrieved November 7, 2024, from https://www.apa.org/topics/substance-use-abuse-addiction#:~:text=Addiction%20is%20a%20state%20of,the%20APA%20Dictionary%20of%20Psychology

Atkinson, R. C., & Shiffrin, R. M. (1968). Chapter: Human memory: A proposed system and its control processes. In Spence, K. W., & Spence, J. T. The psychology of learning and motivation (Vol. 2, pp. 89–195). Academic Press.

Baddeley, A. D., & Hitch, G. (1974). Working memory. In G. H. Bower (Ed.), The psychology of learning and motivation: Advances in research and theory (Vol. 8, pp. 47–89). Academic Press.

Belaidi, A. A., Bush, A. I., & Ayton, S. (2025). Apolipoprotein E in Alzheimer's disease: Molecular insights and therapeutic opportunities. *Molecular Neurodegeneration, 20*(1), 47. https://doi.org/10.1186/s13024-025-00843-y

Blihar, D., Crisafio, A., Delgado, E., Buryak, M., Gonzalez, M., & Waechter, R. (2021). A meta-analysis of hippocampal and amygdala volumes in patients diagnosed with dissociative identity disorder. *Journal of Trauma & Dissociation, 22*(3), 365–377. https://doi.org/10.1080/15299732.2020.1869650

Bolhuis, J. J., Stewart, C. A., & Forrest, E. M. (1994). Retrograde amnesia and memory reactivation in rats with ibotenate lesions to the hippocampus or subiculum. *The Quarterly Journal of Experimental Psychology. B, Comparative and Physiological Psychology, 47*(2), 129–150.

Brand, B. L., Sar, V., Stavropoulos, P., Krüger, C., Korzekwa, M., Martínez-Taboas, A., & Middleton, W. (2016). Separating fact from fiction: An empirical examination of six myths about dissociative identity disorder. *Harvard Review of Psychiatry, 24*(4), 257–270. https://doi.org/10.1097/HRP.0000000000000100

Breijyeh, Z., & Karaman, R. (2020). Comprehensive review on Alzheimer's disease: Causes and treatment. *Molecules, 25*(24), 5789. https://doi.org/10.3390/molecules25245789

Cherry, J. D., Tripodis, Y., Alvarez, V. E., Huber, B., Kiernan, P. T., Daneshvar, D. H., Mez, J., Montenigro, P. H., Solomon, T. M., Alosco, M. L., Stern, R. A., McKee, A. C., & Stein, T. D. (2016). Microglial neuroinflammation contributes to tau accumulation in chronic traumatic encephalopathy. *Acta Neuropathologica Communications, 4*(1), 112. https://doi.org/10.1186/s40478-016-0382-8

Cinalli, D. A., Jr., Cohen, S. J., Calubag, M., Oz, G., Zhou, L., & Stackman, R. W., Jr. (2023). DREADD-inactivation of dorsal CA1 pyramidal neurons in mice impairs retrieval of object and spatial memories. *Hippocampus, 33*(1), 6–17. https://doi.org/10.1002/hipo.23484

Cohen, M. S., Rissman, J., Hovhannisyan, M., Castel, A. D., & Knowlton, B. J. (2017). Free recall test experience potentiates strategy-driven effects of value on memory. *Journal of Experimental Psychology. Learning, Memory, and Cognition, 43*(10), 1581–1601. https://doi.org/10.1037/xlm0000395

Corkin, S. (2002). What's new with the amnesic patient H.M.? *Nature Reviews Neuroscience, 3*(2), 153–160. https://doi.org/10.1038/nrn726

Debray, C., Diabira, D., Gaiarsa, J. L., Ben-Ari, Y., & Gozlan, H. (1997). Contributions of AMPA and GABA(a) receptors to the induction of NMDAR-dependent LTP in CA1. *Neuroscience Letters, 238*(3), 119–122. https://doi.org/10.1016/s0304-3940(97)00865-3

DeJesus-Hernandez, M., Mackenzie, I. R., Boeve, B. F., Boxer, A. L., Baker, M., Rutherford, N. J., Nicholson, A. M., Finch, N. A., Flynn, H., Adamson, J., Kouri, N., Wojtas, A., Sengdy, P., Hsiung, G. Y., Karydas, A., Seeley, W. W., Josephs, K. A., Coppola, G., Geschwind, D. H., Wszolek, Z. K., et al. (2011). Expanded GGGGCC hexanucleotide repeat in noncoding region of C9ORF72 causes chromosome 9p-linked FTD and ALS. *Neuron, 72*(2), 245–256. https://doi.org/10.1016/j.neuron.2011.09.011

Dempster, F. N. (1978). Memory span and short-term memory capacity: A developmental study. *Journal of Experimental Child Psychology, 26*(3), 419–431. https://doi.org/10.1016/0022-0965(78)90122-4

Dillingham, C. M., Milczarek, M. M., Perry, J. C., & Vann, S. D. (2021). Time to put the mammillothalamic pathway into context. *Neuroscience and Biobehavioral Reviews, 121*, 60–74. https://doi.org/10.1016/j.neubiorev.2020.11.031

Dimitrova, L. I., Dean, S. L., Schlumpf, Y. R., Vissia, E. M., Nijenhuis, E. R. S., Chatzi, V., Jäncke, L., Veltman, D. J., Chalavi, S., & Reinders, A. A. T. S. (2023). A neurostructural biomarker of dissociative amnesia: A hippocampal study in dissociative identity disorder. *Psychological Medicine, 53*(3), 805–813. https://doi.org/10.1017/S0033291721002154

Dumitriu, D., Hao, J., Hara, Y., Kaufmann, J., Janssen, W. G., Lou, W., Rapp, P. R., & Morrison, J. H. (2010). Selective changes in thin spine density and morphology in monkey prefrontal cortex correlate with aging-related cognitive impairment. *The Journal of Neuroscience, 30*(22), 7507–7515. https://doi.org/10.1523/JNEUROSCI.6410-09.2010

Erickson, K. I., Voss, M. W., Prakash, R. S., Basak, C., Szabo, A., Chaddock, L., Kim, J. S., Heo, S., Alves, H., White, S. M., Wojcicki, T. R., Mailey, E., Vieira, V. J., Martin, S. A., Pence, B. D., Woods, J. A., McAuley, E., & Kramer, A. F. (2011). Exercise training increases size of hippocampus and improves memory. *Proceedings of the National Academy of Sciences of the United States of America, 108*(7), 3017–3022. https://doi.org/10.1073/pnas.1015950108

Falcon, B., Zivanov, J., Zhang, W., Murzin, A. G., Garringer, H. J., Vidal, R., Crowther, R. A., Newell, K. L., Ghetti, B., Goedert, M., & Scheres, S. H. W. (2019). Novel tau filament fold in chronic traumatic encephalopathy encloses hydrophobic molecules. *Nature, 568*(7752), 420–423. https://doi.org/10.1038/s41586-019-1026-5

Ghilardi, M. F., Silvestri, G., Feigin, A., Mattis, P., Zgaljardic, D., Moisello, C., Crupi, D., Marinelli, L., Dirocco, A., & Eidelberg, D. (2008). Implicit and explicit aspects of sequence learning in pre-symptomatic Huntington's disease. *Parkinsonism & Related Disorders, 14*(6), 457–464. https://doi.org/10.1016/j.parkreldis.2007.11.009

Ghosh, V. E., Moscovitch, M., Melo Colella, B., & Gilboa, A. (2014). Schema representation in patients with ventromedial PFC lesions. *The Journal of Neuroscience, 34*(36), 12057–12070. https://doi.org/10.1523/JNEUROSCI.0740-14.2014

Greicius, M. D., Krasnow, B., Boyett-Anderson, J. M., Eliez, S., Schatzberg, A. F., Reiss, A. L., & Menon, V. (2003). Regional analysis of hippocampal activation during memory encoding and retrieval: fMRI study. *Hippocampus, 13*(1), 164–174. https://doi.org/10.1002/hipo.10064

Guo, Q. (2025). Quantitative analysis of the interaction between NMDA and AMPA receptors in glutamatergic synapses based on mathematical model. *Neuroscience Research, 212*, 127–135. https://doi.org/10.1016/j.neures.2024.10.003

Hunsberger, H. C., Pinky, P. D., Smith, W., Suppiramaniam, V., & Reed, M. N. (2019). The role of APOE4 in Alzheimer's disease: Strategies for future therapeutic interventions. *Neuronal Signaling, 3*(2), NS20180203. https://doi.org/10.1042/NS20180203

Inman, C. S., Manns, J. R., Bijanki, K. R., Bass, D. I., Hamann, S., Drane, D. L., Fasano, R. E., Kovach, C. K., Gross, R. E., & Willie, J. T. (2018). Direct electrical stimulation of the amygdala enhances declarative memory in humans. *Proceedings of the National Academy of Sciences of the United States of America, 115*(1), 98–103. https://doi.org/10.1073/pnas.1714058114

Johnson, E. L., Chang, W. K., Dewar, C. D., Sorensen, D., Lin, J. J., Solbakk, A. K., Endestad, T., Larsson, P. G., Ivanovic, J., Meling, T. R., Scabini, D., & Knight, R. T. (2022). Orbitofrontal cortex governs working memory for temporal order. *Current Biology, 32*(9), R410–R411. https://doi.org/10.1016/j.cub.2022.03.074

Kaada, B. R., Rasmussen, E. W., & Kveim, O. (1961). Effects of hippocampal lesions on maze learning and retention in rats. *Experimental Neurology, 3*, 333–355. https://doi.org/10.1016/0014-4886(61)90009-7

Kangas, B. D., Berry, M. S., & Branch, M. N. (2011). On the development and mechanics of delayed matching-to-sample performance. *The Journal of the Experimental Analysis of Behavior, 95*(2), 221–236. https://doi.org/10.1901/jeab.2011.95-221

Kellett, E. A., Bademosi, A. T., & Walker, A. K. (2025). Molecular mechanisms and consequences of TDP-43 phosphorylation in neurodegeneration. *Molecular Neurodegeneration, 20*(1), 53. https://doi.org/10.1186/s13024-025-00839-8

Kellett, J., & Kokkinidis, L. (2004). Extinction deficit and fear reinstatement after electrical stimulation of the amygdala: Implications for kindling-associated fear and anxiety. *Neuroscience, 127*(2), 277–287. https://doi.org/10.1016/j.neuroscience.2004.05.011

Kirchner, W. K. (1958). Age differences in short-term retention of rapidly changing information. *Journal of Experimental Psychology, 55*(4), 352–358. https://doi.org/10.1037/h0043688

Krause-Utz, A. (2022). Dissociation, trauma, and borderline personality disorder. *Borderline Personality Disorder and Emotion Dysregulation, 9*(1), 14. https://doi.org/10.1186/s40479-022-00184-y

Lei, Z., Xie, L., Li, C. H., Lam, Y. Y., Ramkrishnan, A. S., Fu, Z., Zeng, X., Liu, S., Iqbal, Z., & Li, Y. (2022). Chemogenetic activation of astrocytes in the basolateral amygdala contributes to fear memory formation by modulating the amygdala-prefrontal cortex communication. *International Journal of Molecular Sciences, 23*(11), 6092. https://doi.org/10.3390/ijms23116092

List, J., Kübke, J. C., Lindenberg, R., Külzow, N., Kerti, L., Witte, V., & Flöel, A. (2013). Relationship between excitability, plasticity and thickness of the motor cortex in older adults. *NeuroImage, 83*, 809–816. https://doi.org/10.1016/j.neuroimage.2013.07.033

LoPresti, M. L., Schon, K., Tricarico, M. D., Swisher, J. D., Celone, K. A., & Stern, C. E. (2008). Working memory for social cues recruits orbitofrontal cortex and amygdala: A functional magnetic resonance imaging study of delayed matching to sample for emotional expressions. *The Journal of Neuroscience, 28*(14), 3718–3728. https://doi.org/10.1523/JNEUROSCI.0464-08.2008

Maguire, E. A., Gadian, D. G., Johnsrude, I. S., Good, C. D., Ashburner, J., Frackowiak, R. S., & Frith, C. D. (2000). Navigation-related structural change in the hippocampi of taxi drivers. *Proceedings of the National Academy of Sciences of the United States of America, 97*(8), 4398–4403. https://doi.org/10.1073/pnas.070039597

Marsh, L. C., Apšvalka, D., Kikuchi, H., Abe, N., Kawaguchi, J., Kopelman, M. D., & Anderson, M. C. (2025). Prefrontally mediated inhibition of memory systems in dissociative amnesia. *Psychological Medicine, 54*(16), 1–9. Advance online publication. https://doi.org/10.1017/S0033291724003040

Marzo, A., Bai, J., Caboche, J., Vanhoutte, P., & Otani, S. (2010). Cellular mechanisms of long-term depression induced by noradrenaline in rat prefrontal neurons. *Neuroscience, 169*(1), 74–86. https://doi.org/10.1016/j.neuroscience.2010.04.046

Miller, G. A. (1956). The magical number seven, plus or minus two: Some limits on our capacity for processing information. *Psychological Review, 63*(2), 81–97. https://doi.org/10.1037/h0043158

Miller, G. A., Galanter, E., & Pribram, K. H. (1960). *Plans and the structure of behavior*. Holt, Rinehart and Winston, Inc.

Montoya, A., Pelletier, M., Menear, M., Duplessis, E., Richer, F., & Lepage, M. (2006). Episodic memory impairment in Huntington's disease: A meta-analysis. *Neuropsychologia, 44*(10), 1984–1994. https://doi.org/10.1016/j.neuropsychologia.2006.01.015

Morris, R. (1984). Developments of a water-maze procedure for studying spatial learning in the rat. *Journal of Neuroscience Methods, 11*(1), 47–60. https://doi.org/10.1016/0165-0270(84)90007-4

National Institute on Drug Abuse. (2011). *Drug Misuse and Addiction*. Retrieved November 7, 2024, from https://nida.nih.gov/publications/drugs-brains-behavior-science-addiction/drug-misuse-addiction

Negi, S., Khurana, N., & Duggal, N. (2024). The misfolding mystery: α-Synuclein and the pathogenesis of Parkinson's disease. *Neurochemistry International, 177*, 105760. https://doi.org/10.1016/j.neuint.2024.105760

Olton, D. S., & Samuelson, R. J. (1976). Remembrance of places passed: Spatial memory in rats. *Journal of Experimental Psychology: Animal Behavior Processes, 2*(2), 97–116. https://doi.org/10.1037/0097-7403.2.2.97

Pang, W., & Hu, F. (2021). Cellular and physiological functions of C9ORF72 and implications for ALS/FTD. *Journal of Neurochemistry, 157*(3), 334–350. https://doi.org/10.1111/jnc.15255

Poulos, A. M., Li, V., Sterlace, S. S., Tokushige, F., Ponnusamy, R., & Fanselow, M. S. (2009). Persistence of fear memory across time requires the basolateral amygdala complex. *Proceedings of the National Academy of Sciences of the United States of America, 106*(28), 11737–11741. https://doi.org/10.1073/pnas.0905257106

Qasim, S. E., Mohan, U. R., Stein, J. M., & Jacobs, J. (2023). Neuronal activity in the human amygdala and hippocampus enhances emotional memory encoding. *Nature Human Behaviour, 7*(5), 754–764. https://doi.org/10.1038/s41562-022-01502-8

Ranganath, C., Johnson, M. K., & D'Esposito, M. (2003). Prefrontal activity associated with working memory and episodic long-term memory. *Neuropsychologia, 41*(3), 378–389. https://doi.org/10.1016/s0028-3932(02)00169-0

Reading, P. J., Dunnett, S. B., & Robbins, T. W. (1991). Dissociable roles of the ventral, medial and lateral striatum on the acquisition and performance of a complex visual stimulus-response habit. *Behavioural Brain Research, 45*(2), 147–161. https://doi.org/10.1016/s0166-4328(05)80080-4

Riedel, G., Micheau, J., Lam, A. G., Roloff, E. L., Martin, S. J., Bridge, H., de Hoz, L., Poeschel, B., McCulloch, J., & Morris, R. G. (1999). Reversible neural inactivation reveals hippocampal participation in several memory processes. *Nature Neuroscience, 2*(10), 898–905. https://doi.org/10.1038/13202

Rovetti, J., Goy, H., Nurgitz, R., & Russo, F. A. (2021). Comparing verbal working memory load in auditory and visual modalities using functional near-infrared spectroscopy. *Behavioural Brain Research, 402*, 113102. https://doi.org/10.1016/j.bbr.2020.113102

San Agustín, A., Crevillén, D., Soto-León, V., Moreno, J. C., Oliviero, A., & Pons, J. L. (2023). Transcranial magnetic stimulation combined with endogenous human hippocampal and motor cortical activity enhances memory. *PLoS One, 18*(12), e0295413. https://doi.org/10.1371/journal.pone.0295413

Saulle, M., & Greenwald, B. D. (2012). Chronic traumatic encephalopathy: A review. *Rehabilitation Research and Practice, 2012*, 816069. https://doi.org/10.1155/2012/816069

Sawaguchi, T., & Yamane, I. (1999). Properties of delay-period neuronal activity in the monkey dorsolateral prefrontal cortex during a spatial delayed matching-to-sample task. *Journal of Neurophysiology, 82*(5), 2070–2080. https://doi.org/10.1152/jn.1999.82.5.2070

Schnider, A., & Ptak, R. (1999). Spontaneous confabulators fail to suppress currently irrelevant memory traces. *Nature Neuroscience, 2*(7), 677–681. https://doi.org/10.1038/10236

Sheynikhovich, D., Otani, S., & Arleo, A. (2013). Dopaminergic control of long-term depression/long-term potentiation threshold in prefrontal cortex. *The Journal of Neuroscience, 33*(34), 13914–13926. https://doi.org/10.1523/JNEUROSCI.0466-13.2013

Smeyers, J., Banchi, E.-G., & Latouche, M. (2021). C9ORF72: What it is, what it does, and why it matters. *Frontiers in Cellular Neuroscience, 15*, 661447. https://doi.org/10.3389/fncel.2021.661447

Speranza, L., di Porzio, U., Viggiano, D., de Donato, A., & Volpicelli, F. (2021). Dopamine: The neuromodulator of long-term synaptic plasticity, reward and movement control. *Cells, 10*(4), 735. https://doi.org/10.3390/cells10040735

Suthard, R. L., Senne, R. A., Buzharsky, M. D., Pyo, A. Y., Dorst, K. E., Diep, A. H., Cole, R. H., & Ramirez, S. (2023). Basolateral amygdala astrocytes are engaged by the acquisition and expression of a contextual fear memory. *The Journal of Neuroscience, 43*(27), 4997–5013. https://doi.org/10.1523/JNEUROSCI.1775-22.2023

Taïb, S., Yrondi, A., Lemesle, B., Péran, P., & Pariente, J. (2023). What are the neural correlates of dissociative amnesia? A systematic review of the functional neuroimaging literature. *Frontiers in Psychiatry, 14*, 1092826. https://doi.org/10.3389/fpsyt.2023.1092826

Turner, M. S., Simons, J. S., Gilbert, S. J., Frith, C. D., & Burgess, P. W. (2008). Distinct roles for lateral and medial rostral prefrontal cortex in source monitoring of perceived and imagined events. *Neuropsychologia, 46*(5), 1442–1453. https://doi.org/10.1016/j.neuropsychologia.2007.12.029

Underwood, B. J. (1957). Interference and forgetting. *Psychological Review, 64*(1), 49–60. https://doi.org/10.1037/h0044616

van Asselen, M., Almeida, I., Andre, R., Januário, C., Gonçalves, A. F., & Castelo-Branco, M. (2009). The role of the basal ganglia in implicit contextual learning: A study of Parkinson's disease. *Neuropsychologia, 47*(5), 1269–1273. https://doi.org/10.1016/j.neuropsychologia.2009.01.008

van Asselen, M., Almeida, I., Júlio, F., Januário, C., Campos, E. B., Simões, M., & Castelo-Branco, M. (2012). Implicit contextual learning in prodromal and early stage Huntington's disease patients. *Journal of the International Neuropsychological Society, 18*(4), 689–696. https://doi.org/10.1017/S1355617712000288

Vermetten, E., Schmahl, C., Lindner, S., Loewenstein, R. J., & Bremner, J. D. (2006). Hippocampal and amygdalar volumes in dissociative identity disorder. *The American Journal of Psychiatry, 163*(4), 630–636. https://doi.org/10.1176/ajp.2006.163.4.630

Villacampa, N., Sarlus, H., Martorell, P., Bhalla, K., Gomez-Castro, S., Vieira-Saecker, A., Slutzkin, I., Händler, K., Venegas, C., McManus, R., Ulas, T., Beyer, M., Segal, E., & Heneka, M. T. (2025). Proliferating microglia exhibit unique transcriptional and functional alterations in Alzheimer's disease. *ASN Neuro, 17*(1), 2506406. https://doi.org/10.1080/17590914.2025.2506406

Weber, J. T. (2012). Altered calcium signaling following traumatic brain injury. *Frontiers in Pharmacology, 3*, 60. https://doi.org/10.3389/fphar.2012.00060

Webermann, A. R., & Brand, B. L. (2017). Mental illness and violent behavior: The role of dissociation. *Borderline Personality Disorder and Emotion Dysregulation, 4*, 2. https://doi.org/10.1186/s40479-017-0053-9

Woods, D. L., Kishiyamaa, M. M., Lund, E. W., Herron, T. J., Edwards, B., Poliva, O., Hink, R. F., & Reed, B. (2011). Improving digit span assessment of short-term verbal memory. *Journal of Clinical and Experimental Neuropsychology, 33*(1), 101–111. https://doi.org/10.1080/13803395.2010.493149

Xie, H., Claycomb Erwin, M., Elhai, J. D., Wall, J. T., Tamburrino, M. B., Brickman, K. R., Kaminski, B., McLean, S. A., Liberzon, I., & Wang, X. (2018). Relationship of hippocampal volumes and posttraumatic stress disorder symptoms over early posttrauma periods. *Biological Psychiatry. Cognitive Neuroscience and Neuroimaging, 3*(11), 968–975. https://doi.org/10.1016/j.bpsc.2017.11.010

Yates, J. R. (2023). *Determinants of addiction: Neurobiological, behavioral, cognitive, and sociocultural factors*. Academic Press.

Zhao, R., Grunke, S. D., Keralapurath, M. M., Yetman, M. J., Lam, A., Lee, T. C., Sousounis, K., Jiang, Y., Swing, D. A., Tessarollo, L., Ji, D., & Jankowsky, J. L. (2016). Impaired recall of positional memory following chemogenetic disruption of place field stability. *Cell Reports, 16*(3), 793–804. https://doi.org/10.1016/j.celrep.2016.06.032

Chapter 7
Maladaptive Decision Making

1 Introduction

When I was in eighth grade, *Jackass: The Movie* had just been released. Based on the MTV television series, *Jackass* followed pranks and stunts performed by Johnny Knoxville, Bam Margera, Steve-O, and others. Despite being underage, my friend and I were able to see the movie (I think my friend's dad took us). Inspired by the stunts that we saw, my friend and I decided to perform our own stunt when we returned to his house. We took a Razor scooter and tied it to the back of his bicycle. While my friend rode his bike, I sat on the scooter, letting myself be pulled around the neighborhood. This was not the most exciting stunt, but we did not have many options (nor the budget of MTV). However, my friend went down a steep hill, picking up speed. I was unable to stay on the scooter and fell off, getting some nasty road rash on my backside.

Sometimes, we make bad decisions. These bad decisions may not cause significant harm. I may have received some road rash, but I was fortunate that I did not land face first on the pavement. Other decisions can have life-altering consequences. For example, investing one's life savings in an ultimately unsuccessful business can lead to financial ruin. This chapter will focus on maladaptive decision making, the process in which we make decisions that lead to unbeneficial outcomes.

2 Learning Objectives

By the end of this chapter, you should be able to …

1. Identify the major forms of maladaptive decision making that are commonly studied.
2. Discuss the contribution of several neurotransmitter systems to maladaptive decision making.

J. Yates, *Neuroscience of Maladaptation*, https://doi.org/10.1007/978-3-032-15602-0_7

3. Explain the neural circuitry that underlies suboptimal decision making, impulsive decision making, and risky decision making.

3 Overview of Decision Making

We make numerous decisions every day. Think of the questions you ask yourself nearly every single day:

Do I want to get out of bed?

What am I going to eat for breakfast? Lunch? Dinner?

What do I want to listen to on my way to work/school?

What do I want to wear today?

What do I want to do when I am finished with work/school?

There are some decisions we may make just once but can radically change the trajectory of our lives. These decisions include selecting a college to attend, choosing a life partner, having a child or children, and entering a specific career, just to name a few.

Rooted in economics, *normative* models of decision making assume we think rationally about the choices presented to us. According to **expected utility theory**, an individual bases their decision on the *expected utility* of each alternative and the respective probability of each choice. Utility refers to how useful or beneficial the outcome will be. To illustrate expected utility, I am going to give you a choice between two alternatives. Would you rather have \$50 guaranteed (100% delivery) or \$100 with a 40% chance of winning? To calculate expected utility for each alternative, one just needs to multiply the monetary amount with the probability of its delivery. For the guaranteed option, the expected utility is \$50 (\$50 × 1.0). For the uncertain option, the expected utility is \$40 (\$100 × 0.4). According to expected utility theory, the individual should always choose the more advantageous option (\$50 guaranteed).

How does this relate to real-world decision making? Let's use one of the questions from above: "Do I want to get out bed?" Suppose you have work in the morning. You can either get up and go to work, or you can call out sick and stay in bed longer. In our example, let's make a couple of assumptions. First, suppose you place high utility on sleeping (90% utility), and you place lower utility on going to work because you do not particularly enjoy your job (40% utility). Second, you must weigh the probabilities of each outcome. If you sleep in, you are most likely going to enjoy staying in bed. We will assign a probability of 0.8 that you enjoy staying in bed. If you go to work, you most likely will not enjoy your day as much, so we will set the probability of enjoyment to 0.3. The expected utilities for each outcome are 0.72 (0.9 × 0.8) and 0.12 (0.4 × 0.3), respectively. Therefore, you should stay in bed and call out of work. The question becomes: is this what you would *really* do? If you have enough sick days or vacation days built up, perhaps. In most cases, this will not occur because we need to go to work to keep our job.

The major limitation of the expected utility theory is that it often fails to account for how people truly make decisions when faced with uncertainty. *Descriptive* models of decision making try to explain how individuals actually make decisions as opposed to normative models that describe decision making under ideal circumstances. One of the most famous descriptive models of decision making is **prospect theory**, proposed by Dr. Daniel Kahneman (1934–2024) and Dr. Amos Tversky (1937–1996). Tversky and Kahneman asked participants to make a choice between two outcomes: 100% chance to win $30 or 80% chance to win $45. According to expected utility theory, individuals should choose the second option because its expected utility is higher ($36) compared to the first outcome ($30). However, 78% of the participants chose option 1.

Tversky and Kahneman also asked participants if they would rather choose a 25% chance of winning $30 or a 20% chance of winning $45. Here, expected utility theory would predict that individuals choose the second option (expected utility of $9 vs. $7.50). Indeed, more participants chose this option (58%) compared to the first option. Notice, when both outcomes are unlikely to occur, individuals select the lower probability option that leads to a higher payout. However, when faced with two outcomes that are associated with guaranteed or near-guaranteed payouts, participants select the option with lower expected utility. According to prospect theory, we make decisions based on the amount of gains or losses from what we have at the moment. We tend to be risk averse when encountering two outcomes that most likely will lead to gains (e.g., 100% chance to win $30 vs. 80% to win $45). Conversely, we are more willing to take risks when faced with two low-probability outcomes (25% chance to win $30 vs. 20% chance to win $45).

Prospect theory also claims that we are more sensitive to losses compared to gains. That is, people feel more strongly about losing $50 than winning $50. I like to propose the following scenario to my students. I ask them to imagine how they would feel if I decided to give them 10 bonus points because I was feeling generous. My students show some excitement, but the response is muted for the most part. I then ask them to imagine how they would feel if I deducted 10 points from each student. Even though this is a hypothetical situation, students show a much stronger emotional response to the thought of losing points as opposed to gaining the same number of points.

Comprehension Check 1 Your employer offers you a choice: a $5 gift card or a $5 lottery ticket. The jackpot is worth $1 billion. According to expected utility theory, which option has higher utility, and why? How can prospect theory be used to explain how people decide between these alternatives?

To answer the first question, you need to know the probability of winning the lottery. The probability of winning will depend on the specific lottery that you play, but for the major lotteries, the probability of winning is approximately 1 in 300 million (0.00000000333%). The expected utility of each alternative is $5 vs. $3.33. This means that the gift card is a better value than the lottery ticket. According to prospect theory, *individuals may choose the lottery ticket because both options are not particularly great (a $5 gift card will not get you much). In this event, we may take a "go big or go home" attitude and choose to play the lottery.*

What I want to highlight is that our decision making is often *irrational*. If we made decisions according to expected utility theory, we would not engage in maladaptive decision making. This section is going to focus on three major forms of maladaptive decision making that individuals can make. Certain clinical populations are more likely to engage in these forms of decision making, but we all have made a poor decision that falls under at least one of the subsections below.

3.1 Suboptimal Decision Making

One of the most utilized procedures for measuring decision making is the Iowa gambling task (IGT) (Fig. 1a, left side of image). Participants are presented with four decks of cards on a computer screen and can either win or lose money after selecting just one card on an individual trial. The four decks differ from one another in the distribution of rewarded cards and penalized cards, as well as in the amount of money gained or lost during a trial.

In the original IGT, Decks A and B led to delivery of $100 while Decks C and D led to delivery of $50. The catch is that the probability of losing money is highest in Deck A, and the amounts lost during a penalized trial are greatest for Deck B. Across the session, participants earn more money if they consistently select cards from Decks C and D. During the early stages of IGT performance, individuals are making decisions under *ambiguity*. That is, individuals have not quite learned the contingencies of reinforcement associated with each deck.

What I do want to highlight about the IGT is that it can measure a form of maladaptive decision making known as **suboptimal decision making**, in which individuals make less favorable outcomes due to inefficient evaluation of each option's outcome. In the IGT, consistently choosing Decks A and B leads to a suboptimal outcome as one earns less money throughout the session. In real life, individuals often make suboptimal choices. Taking out a 7-year car loan to reduce one's monthly payment is suboptimal as the individual will end up paying much more compared to taking out a loan with a shorter life. Choosing a romantic partner based on physical attraction alone can lead to an unfulfilling relationship. Some students refuse to apply to graduate schools located further from home, ultimately failing to get accepted into a school.

Suboptimal decision making can be measured in animals as well. The rat/rodent gambling task (rGT) is an animal analog of the IGT (Zeeb et al., 2009). In the rGT, rodents are tested in an operant conditioning chamber similar to the one described for the five-choice serial reaction time task (5-CSRTT; refer back to Chap. 5). There are

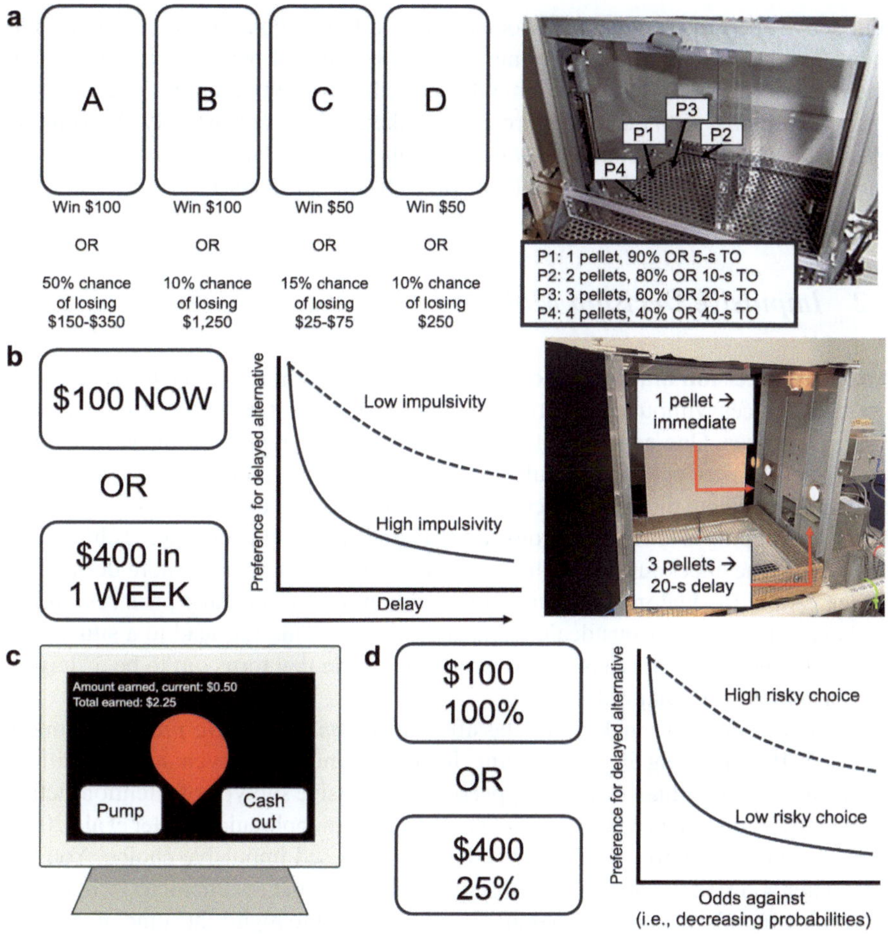

Fig. 1 Various methods for measuring maladaptive decision making. (**a**) The Iowa gambling task (IGT) and the rat/rodent gambling task (rGT) measure decision making under ambiguity and decision making under uncertainty. As such, these tasks measure suboptimal decision making and risky decision making. Image of IGT created by the author using Microsoft PowerPoint. Image of the operant chamber comes from https://commons.wikimedia.org/wiki/File:5CSRTT_for_mice.jpg from Bd008. CC BY-SA 4.0 (https://creativecommons.org/licenses/by-sa/4.0/deed.en), via Wikimedia Commons (**b**) Delay discounting is used to measure impulsive decision making. The amount of each alternative and the delay of each alternative can be adjusted such that a discounting curve can be generated (see middle panel of **b**). The schematic of delay discounting was created by the author using Microsoft PowerPoint. Image of the operant chamber comes from Figure 1 of Salcido et al. (2022), https://doi.org/10.3389/fpain.2021.793958. CC BY 4.0 (https://creativecommons.org/licenses/by/4.0/), via Frontiers. (**c**) An example of the balloon analog risk task (BART), which is used to measure risky decision making. Image created by the author using Servier Medical Art (https://smart.servier.com/; CC BY 4.0, https://creativecommons.org/licenses/by/4.0/) and Microsoft PowerPoint. (**d**) Probability discounting is similar to delay discounting, but it measures risky decision making; although not shown, animals can be tested in probability discounting. Image created by the author using Microsoft PowerPoint

four active apertures, which are associated with different contingencies of reinforcement. The magnitude of reinforcement, the probability of receiving reinforcement, and the time-out period (i.e., the time the animal must wait before they can respond again) associated with each aperture differ. Like the IGT, consistently choosing the options associated with smaller reward amounts is more advantageous as the rat can earn more food during the session (Fig. 1a, right side of image).

3.2 *Impulsive Decision Making*

Impulsive decision making (also known as impulsive choice) is the tendency to act or make choices without careful thought or consideration for the potential outcomes of such choices. One example of impulsive decision making that many of us have most likely experienced is “impulse shopping”. We may purchase an item without even thinking about the potential consequences of making such a purchase. Typically, impulse shopping is fairly innocuous; we may grab a candy bar or a bag of chips while standing in the checkout line. Other impulsive decisions can have more detrimental effects. An individual may purchase a car without performing any prior research or without comparing various dealerships and vehicles. This can lead to a situation in which the individual purchases an overpriced vehicle that turns out to be a “lemon” (i.e., a vehicle with major defects).

Impulsive decision making is classified as a *transdiagnostic process* (Amlung et al., 2019), meaning it is a defining feature of multiple psychiatric conditions. For example, individuals with the hyperactive/impulsive subtype of attention-deficit/hyperactivity disorder (ADHD) (Fox et al., 2023), schizophrenia (Weller et al., 2014), and dementia (Godefroy et al., 2023) display increased impulsive choice. You will encounter additional disorders in subsequent chapters that are characterized, at least in part, by impulsive decision making. For now, I want to highlight some of the ways such maladaptive impulsive decision making can be measured.

The Barratt Impulsiveness Scale (BIS-11) is a self-report questionnaire that asks individuals about different forms of impulsivity, including impulsive decision making (Patton et al., 1995). The most common behavioral measure of impulsive decision making in both humans and animals is delay discounting (Fig. 1b). Participants are presented with choices between smaller, immediate rewards (e.g., $100 now) and larger, delayed rewards (e.g., $1000 delivered after a delay of 1 month). The delay to the larger reward is systematically varied across trials. The participants’ choices are used to estimate a discounting rate, which reflects how quickly the value of the reward declines as the delay increases. Obviously, asking animals to choose between two hypothetical monetary rewards is not possible. Instead, an animal is trained to respond on one of two manipulanda. Eventually, the animal learns that responding on one manipulandum leads to near immediate delivery of a small food reward (e.g., one food pellet) and that responding on the other manipulandum leads to delivery of a large food reward (e.g., four food pellets) following a delay that is adjusted throughout the session.

3.3 *Risky Decision Making*

Risky decision making (also known as risky choice) involves uncertainty. Let's return to the IGT. With sufficient training, individuals should begin to learn the contingencies of reinforcement associated with each deck of cards. That is, they should learn that one deck of cards gives a bigger payout but that there is a higher chance that a large monetary loss can result if that deck is selected. At this point, each decision is based on *uncertainty* as the individual does not know if their choice will lead to a win or to a loss.

Risky decision making is not inherently maladaptive. Many of our decisions involve those in which we cannot predict the outcome. Think about applying to colleges. You can compare multiple schools, make several campus visits, and apply for various scholarships. Yet, you do not know if attending a specific school is going to benefit you long-term. I took a risk by moving out of state to attend graduate school. There were no guarantees that I would succeed in my graduate program. With that said, there are risky choices that are problematic. Driving a vehicle while intoxicated or having unprotected sex with a stranger can have catastrophic consequences.

Impulsive decision making and risky decision making often co-occur. Going back to this chapter's introduction, my friend and I engaged in both impulsive and risky decision making. We did not think about the potential consequences of emulating what we saw in *Jackass*. We also took a physical risk as there was the potential for injury. Similarly, purchasing a vehicle on a whim is both impulsive and risky as there is the potential that the car may have numerous mechanical issues.

In addition to the IGT (with sufficient training), there are other methods we can use to measure risky decision making. In the balloon analog risk task (BART), individuals inflate a virtual balloon to earn money (Fig. 1c). Participants earn more money each time they inflate the balloon, but the risk is the balloon popping, resulting in the individual losing the money they have accumulated during that trial. Probability-discounting tasks are very similar to delay-discounting procedures, except individuals make choices between small, guaranteed reinforcers and large, probabilistic reinforcers. Individuals that consistently choose the large, probabilistic reinforcer are considered to display higher levels of risky decision making.

Like delay discounting, probability discounting is used in both human and animal subjects (Fig. 1d). Another measure of risk-based decision making used in animals is the risky decision task (RDT). In the RDT, animals choose between a large food reward and a small food reward, but the large reward is paired with probabilistic delivery of foot shock (Simon et al., 2009). This task differs from other measures of risky choice as positive punishment (foot shock) is used instead of negative punishment (losing out on money/food).

Comprehension Check 2 What is the key difference between suboptimal decision making and impulsive decision making?

> *Suboptimal decision making involves misevaluation of alternatives, leading an individual to choose an option that leads to less favorable long-term outcomes. Impulsive decision making can also lead to selection of outcomes that are less favorable long-term, but this form of decision making is characterized by a lack of consideration of each alternative.*

4 Neurobiological Underpinnings of Maladaptive Decision Making

You have already learned about the mesocorticolimbic pathway, in which neurons from the ventral tegmental area (VTA) project to both the nucleus accumbens (NAc) and the prefrontal cortex (PFC). Because dopamine is a major neurotransmitter found in the PFC, its involvement in decision making has been widely studied. I will begin this section by discussing the evidence for dopaminergic control of decision making stemming from populations known to have aberrant dopaminergic signaling and from behavioral pharmacology experiments.

Dopamine is certainly not the only neurotransmitter involved in decision making. You will learn about the contribution of other major neurotransmitters to decision making. I will then discuss the neural circuitry that underlies suboptimal, impulsive, and risky decision making.

4.1 Optimal Catecholamine Levels Are Necessary for Adaptive Decision Making

There are several disorders characterized by aberrant dopamine signaling that lead to maladaptive decision making. You have already learned about ADHD. Recall that individuals with ADHD have low catecholamine levels. Substance use disorders, which I mentioned briefly in the previous chapter, are also associated with hypoactivity of the dopaminergic system as continued substance use causes dopamine levels to become depleted (Diana, 2011). In the next chapter, you will learn more about Parkinson's disease, a movement disorder that results from destruction of dopaminergic neurons originating from a different part of the brain.

Each of the disorders listed above is associated with deficits in decision making (Colautti et al., 2023; Garon et al., 2006; Kovács et al., 2017; Mallorquí-Bagué et al., 2016; Roshani et al., 2020), including increased impulsive decision making (Bickel et al., 2019; Fox et al., 2023; Mies et al., 2019; Pennisi et al., 2023) and maladaptive risky decision making (Dai et al., 2016; Drechsler et al., 2010; Humphreys & Lee, 2011; Lawrence et al., 2009). As such, dopaminergic contributions to impulsive choice have been widely studied in comparison to noradrenergic contributions. This section will focus primarily on the role of dopamine in mediating impulsive decision making.

In healthy volunteers, consumption of an amino acid cocktail designed to reduce dopamine levels causes decision-making deficits (Sevy et al., 2006), an effect that does not occur following depletion of norepinephrine levels (O'Carroll & Papps, 2003). Conversely, administering a catechol-O-methyltransferase (COMT) inhibitor (note, COMT metabolizes catecholamines like dopamine) decreases impulsive decision making (Kayser et al., 2012). A hypoactive dopaminergic state accounts for why psychostimulant drugs such as amphetamine (Adderall, Dexedrine) and methylphenidate (Ritalin, Concerta) are effective in treating conditions like ADHD. Even in rodents, methylphenidate attenuates impulsive choice (Koloski et al., 2024; Wooters & Bardo, 2011). The selective norepinephrine transporter inhibitor atomoxetine (Strattera) and an alpha-2 (α_2) receptor antagonist reduce impulsive choice in animals (Schippers et al., 2016; Sun et al., 2012). These drugs may work to ameliorate impulsive choice by increasing dopamine levels in the brain (Bymaster et al., 2002; Devoto et al., 2004).

As with other conditions covered in this book, there are situations in which too much dopamine can be detrimental. When individuals use substances like methamphetamine or fentanyl, dopamine levels increase in the mesocorticolimbic pathway; this dopamine influx can lead to maladaptive decision making. There is some evidence to support this as a noradrenergic α_2 receptor agonist, which regulates the further release of dopamine, improves IGT performance in individuals with a substance use disorder (Zhang et al., 2012).

Mice that have had the dopamine transporter "knocked down" (i.e., decreased) show impaired decision making under ambiguity, although risky decision making is similar to controls (Young et al., 2011). Fewer dopamine transporters means that dopamine stays in the synapse for longer periods. Amphetamine, which reverses the flow of dopamine, as well as norepinephrine transporters, impairs performance in the rGT (Baarendse et al., 2013; Zeeb et al., 2009), and cocaine self-administration increases impulsive choice in animals (Mendez et al., 2010). This last finding mirrors the increased delay discounting often observed in individuals with a psychostimulant use disorder (Coffey et al., 2003; Hoffman et al., 2006). Increasing dopamine neurotransmission also increases preference for risky options in healthy volunteers (Hirschbichler et al., 2022; Peters et al., 2020).

In summary, both hypoactive and hyperactive dopaminergic states contribute to maladaptive decision making. This is why psychostimulants improve aspects of decision making in individuals with low baseline catecholamine levels (e.g., those with ADHD) but exacerbate maladaptive decision making in healthy individuals (e.g., someone misusing amphetamine).

4.2 Serotonin Modulates Dopaminergic Control of Decision Making

Like dopamine, serotoninergic neurons project to the frontal cortex. As you will learn in Chap. 9, serotonin has received considerable attention for its role in impulsivity governed by strong emotional responses. Serotonin has also been implicated in impulsive decision making; a variant of the gene that encodes the serotonin transporter is associated with impulsive decision making in children with ADHD (Sonuga-Barke et al., 2011), and low serotonin levels are linked to heightened impulsive choice (Mobini et al., 2000; Schweighofer et al., 2008).

Serotonin is also involved in general decision making as assessed in IGT-type procedures. However, discrepancies are observed across human and animal studies. In humans, providing the precursor to serotonin impairs decision making under ambiguity, but not decision making under risk (Gendle & Golding, 2010). Administration of a serotonin transporter inhibitor also increases delay discounting (i.e., increases impulsive decision making) without altering risky decision making (Zorick et al., 2022).

In animals, depleting serotonin levels promotes suboptimal decision making in the rGT (Koot et al., 2012) while administration of serotonin transporter inhibitors does not alter impulsive choice (Baarendse & Vanderschuren, 2012). This finding is consistent with those showing that stimulation of 5-$HT_{1A/1B}$ receptors (which often act to reduce serotonin levels originating from the raphe nuclei) impairs rGT performance (Zeeb et al., 2009) and increases impulsive choice (Blasio et al., 2012; Korte et al., 2017; Winstanley et al., 2005). I now need to reconcile these seemingly contradictory findings.

When serotonin levels are increased (such as giving participants the precursor to serotonin), there may be overstimulation of 5-$HT_{2A/C}$ receptors, causing increased concentrations of other neurotransmitters like dopamine and glutamate. In fact, animal studies show that activation of 5-$HT_{2A/C}$ receptors increases impulsive choice (Blasio et al., 2012), with opposite results reported following 5-HT_{2C} receptor antagonism (Paterson et al., 2012). Conversely, low serotonin levels can also cause hyperactivity of the dopaminergic system as there is not enough serotonin to bind to receptors that regulate the release of dopamine. In other words, dopaminergic neurons become disinhibited.

In short, serotonin, like dopamine, needs to operate at an optimal level to promote effective decision making. Too much serotonin can directly stimulate the release of excess dopamine. Conversely, not enough serotonin can prevent the inhibition of dopaminergic neurons. The end result is the same—a failure in the serotonin system promotes a hyperactive dopaminergic system.

4.3 Acetylcholine and Maladaptive Decision Making

In Chap. 5, I discussed acetylcholine's role in attention and how individuals with schizophrenia often smoke cigarettes, possibly to improve cognitive deficits caused by the disorder. Nicotine, which binds to acetylcholine nicotinic receptors, is the primary psychoactive compound found in tobacco products. Use of tobacco products is associated with improved decision making as assessed with the IGT (Zandonai et al., 2018), and tobacco withdrawal worsens performance in the IGT (Buelow & Suhr, 2014). In contrast to the acute effects of nicotine, long-term tobacco use leads to increased impulsive decision making, although it does not affect probability discounting (Białaszek et al., 2017; Ohmura et al., 2005).

In animals, the effects of nicotine on impulsive choice are mixed. Increased impulsive choice (Kolokotroni et al., 2011; Locey & Dallery, 2009), particularly in animals with low baseline levels of impulsivity (Kayir et al., 2014), and decreased impulsive choice (Ozga & Anderson, 2018) have been observed. Additional research has shown that nicotine does not selectively alter impulsive choice, instead causing increased preference for small rewards over large rewards, regardless if a delay is enacted (Locey & Dallery, 2011). In other words, nicotine can produce suboptimal decision making. Think of it this way. If I asked you, "Would you rather have $10 right now or $100 right now?", what would you choose? Why would you choose less money if both alternatives were made available immediately with no strings attached? Nicotine can make an animal choose the smaller reward, even when the large reward is made available immediately.

Concerning risky choice, nicotine consistently increases risky choice in probability-discounting procedures (Giner et al., 2022; Mendez et al., 2012; Nolder & Anderson, 2023), but not in the rGT or the RDT (Betts et al., 2021; Kelly et al., 2025; Silveira et al., 2015). One study did report a decrease in risky choice in the RDT following nicotine administration (Mitchell et al., 2011), but nicotine decreased preference for the large food reward in the RDT, even if shock is never delivered. Once again, this finding indicates that increasing cholinergic neurotransmission can promote suboptimal choice in animals.

Collectively, these results also illustrate the importance of maintaining a homeostatic level of acetylcholine for decision making. In clinical populations (e.g., schizophrenia, individuals with tobacco dependence), nicotine can improve decision-making processes. Yet, nicotine can promote maladaptive decisions, particularly suboptimal decision making. One potential reason for this phenomenon relates to dopamine (again). Cholinergic projections from the brainstem stimulate dopaminergic neurons in the VTA. Low acetylcholine levels can cause insufficient activation of the VTA, leading to a hypo-dopaminergic state. Yet, excess cholinergic activity can cause hyperactivity of the mesocorticolimbic pathway.

4.4 Long-Term Stimulation of Endocannabinoid System Promotes Maladaptive Decision Making

The easiest way to understand how endocannabinoids affect decision making is to think of what cannabis does to individuals. Acute cannabis use increases preference for risky alternatives (Lane et al., 2005). Likewise, adolescents that use cannabis take more risks in the BART (Hanson et al., 2014). Cannabis use is also associated with real-world maladaptive risky choice, such as aggressive driving (Ortiz-Peregrina et al., 2021) and unprotected sex (Agrawal et al., 2016; Bustamante et al., 2022).

While cannabis use is associated with risky decision making, acute cannabis administration does not alter suboptimal decision making (McDonald et al., 2003; Ramaekers et al., 2006; Vadhan et al., 2007). Also, cannabidiol, a compound found in cannabis, fails to alter impulsive decision making (Hindocha et al., 2018). Acute administration of cannabinoid receptor agonists, including Δ9-tetrahydrocannabinol (THC), does not significantly alter decision making in the rGT (Ferland et al., 2018; but see Gueye et al., 2016 for modest improvements in rGT performance following CB_1 receptor activation). These findings are similar to those showing that endocannabinoid levels in individuals do not predict IGT performance (Fagundo et al., 2013).

Long-term cannabis use is associated with several deficits in decision making, as individuals that use cannabis habitually perform suboptimally in the IGT (Ferland et al., 2023), with poor decision making being correlated with severity of cannabis use disorder (Gonzalez et al., 2012). Chronic THC exposure during adolescence promotes suboptimal choice in a delay-discounting procedure (i.e., increased preference for a small reward even when the large reward is delivered immediately) (Hamidullah et al., 2021) and increased risky choice in the rGT (Ferland et al., 2023), and repeated administration of a CB_1 receptor agonist during adolescence increases impulsive choice during adulthood (Johnson et al., 2019). This last finding is interesting because acute THC reduces impulsive choice in rodents (Wiskerke et al., 2011), but a CB_1 receptor agonist increases impulsive choice in the spontaneously hyperactive rat (SHR), the animal model of ADHD you learned about in Chap. 5 (Leffa et al., 2019).

These results provide evidence that transient activation of the endocannabinoid system can increase risky decision making, but long-term disturbances of this system are required for the emergence of suboptimal decision making and impulsive choice. One potential mechanism to account for these findings is that the endocannabinoid system is important for long-term depression (LTD), one of the cellular mechanisms that controls learning and memory (see previous chapter). LTD is a form of synaptic plasticity that decreases the number of synapses in a brain region. This means that with repeated cannabis use, synaptic changes occur in parts of the brain that regulate general decision making and impulsive choices. At this point, turning our attention to the neural circuitry involved in maladaptive decision making is a good idea to better understand how dysregulation of neurotransmitter systems in specific parts of the brain can impair our decision-making capabilities.

Comprehension Check 3 Explain why psychostimulants like methylphenidate (Ritalin) improve decision-making in someone with ADHD but worsen it in a healthy individual who misuses them.

Psychostimulants like methylphenidate work by increasing dopamine (and norepinephrine) levels in the brain. In individuals with ADHD, their baseline catecholamine levels are often too low (a hypoactive state), contributing to decision-making deficits. For them, the stimulant brings dopamine levels into a more optimal range, improving functions like impulse control. However, in healthy individuals who misuse these drugs, their dopamine levels become too elevated. This can cause overstimulation of dopamine D_2 *receptors in the PFC, leading to hypoactive functioning in this region, thus leading to impaired decision making.*

4.5 Neural Circuitry Underlying Maladaptive Decision Making

One important function of the PFC is executive control. We often equate the PFC as the rational part of the brain, controlling the more primitive urges of the limbic system. The PFC is involved in working memory, which you learned about in the previous chapter. Working memory and decision making are highly linked. When we make a decision, we rely on working memory. Imagine you are with a group of friends, and you are trying to decide what to do that evening. Your group may narrow down your choices to three alternatives: (1) hanging out at one friend's apartment, (2) going to a movie theater, or (3) eating at a restaurant. As you weigh these options, your working memory system is activated. You may remember having a bad time at this one friend's apartment, so you may advocate for options 2 or 3. Alternatively, you may have heard from others that the restaurant suggested by your friend group is not that good. Thus, we rely on our memory to make decisions.

The neural circuitry involved in maladaptive decision making largely involves the PFC and its connections. As such, most of this section will focus on prefrontal cortical circuits that you have already learned about, including the mesocorticolimbic pathway and the cortico-basal ganglia-thalamo-cortical (CBGTC) loop. Given the complexity of decision making, there are numerous brain regions that contribute to this cognitive function. In this section, the neural circuitry of each major type of maladaptive decision making will be discussed independently. Yet, you will notice major overlap in the neural circuitry that governs each type of decision making covered in this chapter. Figure 2 shows the major brain regions implicated in decision making.

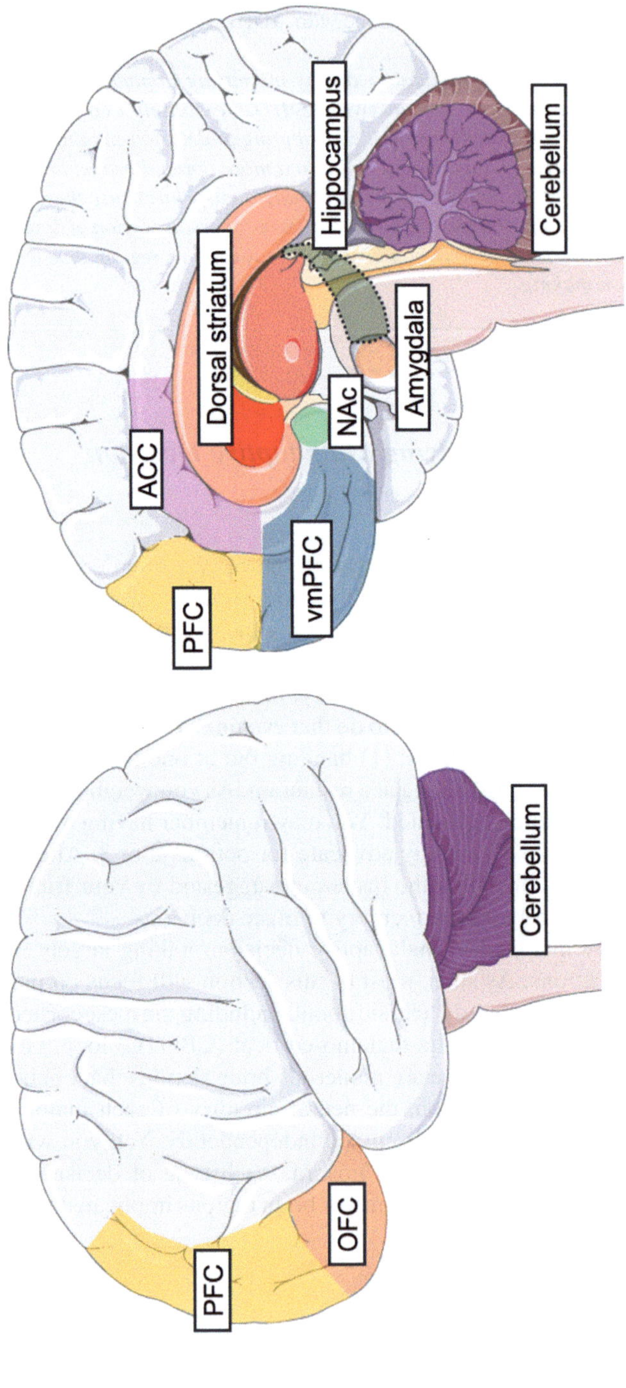

Fig. 2 Brain regions heavily implicated in maladaptive decision making. *Note* ACC = anterior cingulate cortex; NAc = nucleus accumbens; OFC = orbitofrontal cortex; PFC = prefrontal cortex; vmPFC = ventromedial PFC. Image created by the author using Servier Medical Art (https://smart.servier.com/; CC BY 4.0, https://creativecommons.org/licenses/by/4.0/) and Microsoft PowerPoint

4.5.1 Neural Circuitry of Suboptimal Decision Making

The IGT was originally designed by Dr. Antoine Bechara to measure decision making in individuals with damage to the ventromedial PFC (vmPFC). Individuals with damage to this part of the brain perform poorly on the IGT, often selecting cards from the disadvantageous decks (Bechara et al., 1994). Similar impairments in decision making are observed in rodents following damage to the PFC (Shaver et al., 2019) and following inactivation of the medial PFC (mPFC) (Zeeb et al., 2015).

The vmPFC is involved in *value-based* decision making. When we make a choice between two or more options, we often assign a subjective value to each one. Going back to the example from above in which friends are trying to decide on a group activity. When you are given the three options of spending time at a friend's apartment, going to the movies, or eating at a restaurant, you often provide a "score" for each activity based on multiple factors. You may use personal preferences, past experiences, and/or goals to help score each alternative. If there is a movie playing that you desperately want to see, this will increase the score you assign option 2. If no movie looks appealing, you will score this option more unfavorably. Individuals with vmPFC damage have difficulty engaging in this type of decision making, hence why they perform poorly in the IGT.

Functional magnetic resonance imaging (fMRI) shows that the lateral PFC (lPFC) is active during early trials of the IGT, indicating this region is important for decision making under ambiguity. There is some evidence that the right PFC is more important for decision making in the IGT, as patients with damage to the right dorsolateral PFC (dlPFC) demonstrate suboptimal performance in this task compared to individuals with damage to the left dlPFC and to controls (Clark et al., 2003). Like the lPFC, the anterior cingulate cortex (ACC) is highly active when individuals make decisions under ambiguity in the IGT (Hartstra et al., 2010) and is more active when individuals make suboptimal choices in a "Blackjack gambling task" (Hewig et al., 2009).

Cortico-Basal Ganglia-Thalamo-Cortical (CBGTC) Loop

Returning to thedopaminergic basis of suboptimal decision making, you learned that both hyperactivity and hypoactivity of the dopaminergic system can impair decision making. Let's examine why this is the case. While overexpression of dopamine D_1 receptors in the PFC causes rodents to perform worse in an rGT (Beyer et al., 2021), a hyperactive dopamine system may impair decision making by overstimulating dopamine D_2-like receptors (Fig. 3a). Whereas blocking D_1-like receptors has no effect in the rGT, stimulating D_2-like receptors impairs rGT performance while blocking these receptors has the opposite effect (Di Ciano et al., 2017; Zeeb et al., 2009). As D_2-like receptors are inhibitory and serve as autoreceptors, overstimulation of these receptors then leads to a hypoactive dopaminergic state, characterized by decreased dopamine release.

If the dopaminergic system is hypoactive, there is not enough stimulation of D_2-like receptors, which leads to a temporary spike in dopamine levels (Fig. 3b). Because

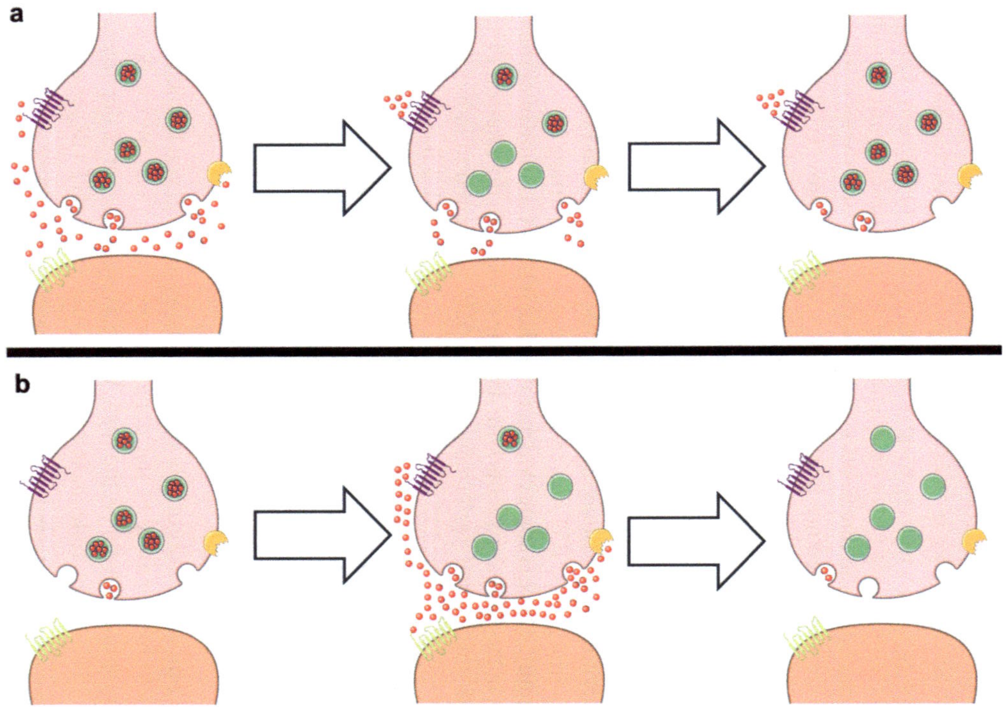

Fig. 3 Schematic showing how loss of dopamine homeostasis can lead to a hypoactive dopaminergic state. (**a**) Excess dopamine release into the synapse causes excessive activation of dopamine D_2-like receptors (located on presynaptic neuron). In addition to excess binding of these inhibitory autoreceptors, catechol-O-methyltransferase (COMT), which is bound to the presynaptic membrane, metabolizes dopamine. As dopamine continues binding to D_2-like receptors, less dopamine is released into the synapse, leading to low dopamine levels in the synapse. (b) Low dopamine release causes insufficient stimulation of D_2-like receptors. This causes a "runaway" effect where too much dopamine is released into the synapse. As COMT metabolizes the excess dopamine in the synapse, not enough dopamine reaches D_2-like receptors. Eventually, the dopamine inside the presynaptic neuron is depleted, leading to a hypoactive dopaminergic state. Image created by the author using Servier Medical Art (https://smart.servier.com/; CC BY 4.0, https://creativecommons.org/licenses/by/4.0/) and Microsoft PowerPoint

only so much dopamine can be released from a neuron, insufficient neuronal inhibition via D_2-like receptors leads to dopamine depletion, thus leading to a hypoactive dopaminergic state. Indeed, individuals with substance use disorders often have fewer dopamine D_2-like receptors in the PFC (Goldstein & Volkow, 2011), which most likely accounts for decision-making deficits often observed in these individuals.

Notice that the end result is the same. Whether the dopaminergic system is hyperactive or hypoactive, decreasing dopaminergic neurotransmission in the PFC impairs decision making. This prefrontal cortical dopamine deficiency detrimentally affects other dopaminergic regions of the brain. Activation of the PFC is important for increasing dopamine levels in the dorsal striatum of the CBGTC loop (Adrover et al., 2020). Poor decision making observed in Parkinson's disease is related to lower activation of the putamen following losses (Gescheidt et al., 2013). Likewise, rats that perform poorly in an rGT show less neuronal activity in the dorsomedial striatum (DMS) (Fitoussi et al., 2015).

There is direct evidence that a hypoactive dopaminergic system in the dorsal striatum is related to decision making deficits. While lower dopamine transporter availability is observed in the dorsal striatum of individuals with gambling disorder, dopamine transporter availability is *negatively* correlated with IGT task performance (Pettorruso et al., 2019). This means that there are more dopamine transporters to remove dopamine from the synapse in individuals that perform worse in the IGT. Administration of pramipexole, a selective dopamine D_3 receptor agonist (which is part of the D_2-like family), impairs performance in the rGT by causing overexcitation of the globus pallidus (Kubota et al., 2024). Relatedly, individuals with Parkinson's disease show less functional connectivity between the ACC and the internal capsule of the globus pallidus when completing the IGT (Gescheidt et al., 2013). Disulfiram, better known as Antabuse (used to treat alcohol use disorder), improves performance in the rGT by decreasing striatal norepinephrine levels while increasing dopamine levels (Di Ciano et al., 2018).

Hypoactivity of the dorsal striatum can impair decision making by making it difficult for individuals to select the appropriate behavioral response. The DMS is important for learning a task while the dorsolateral striatum (DLS) enables us to engage in habitual actions (i.e., those that require little conscious control). If the dorsal striatum, especially the DMS, is hypoactive, individuals will have difficulty learning to associate outcomes with specific choices. This can account for why individuals with lower dopamine levels in the dorsal striatum perform worse in the IGT. You will learn more about the DMS and the DLS and their contributions to learning in Chap. 11.

Mesocorticolimbic Pathway

Because the NAc is heavily involved in reinforcement learning, this region is important for evaluating the potential rewards of each alternative. This is a crucial element of decision making. When deciding on an activity for the evening, you and your friends are going to evaluate how entertaining each option is relative to the costs involved with each activity. The NAc is highly sensitive to reward magnitudes; that is, when evaluating potential fun vs. costs, the NAc is going to be activated more by "how much fun" the group will have performing an activity. This means that overexcitation of the NAc could lead to situations in which individuals select options that may not be the most beneficial in the long run. For example, choosing an incredibly expensive, but fun, activity may not be the best option if there is an equally exciting activity that is significantly less expensive.

Individuals with gambling disorder that show increased dopamine release in the ventral striatum perform worse in the IGT compared to healthy controls (Linnet et al., 2011b); this increased dopamine release is also positively correlated with self-reported excitement levels in these individuals (Linnet et al., 2011a). Relatedly, individuals with comorbid Parkinson's disease and impulse-control disorders have a hyperactive ventral striatum, which includes the NAc (Paz-Alonso et al., 2020) (note, you will learn about impulse-control disorders in Chap. 12). Even rats that display poor decision making in an rGT have increased dopamine metabolism in the NAc core (e.g., higher dopamine metabolites compared to dopamine) (Fitoussi et al., 2015). This could indicate that there is excessive dopamine that quickly gets metabolized. Additional animal research has demonstrated that chemogenetic inhibition of NAc-projecting dopamine neurons improves decision making in the rGT (Hynes et al., 2020). Overall, these results suggest that hyperactivity of the NAc can bias individuals toward high magnitude alternatives that may not be the most optimal option.

Hyperactivity in the NAc should not be too surprising given the low dopamine levels in the PFC. Insufficient stimulation of D_2-like receptors will lead to excess activation of the NAc (Del Arco & Mora, 2008). The most likely explanation for this effect is that stimulation of D_2-like receptors on prefrontal glutamatergic neurons is important for regulating the NAc. When D_2-like receptors are not sufficiently stimulated, glutamatergic neurons become disinhibited, leading to increased NAc activity.

Corticolimbic System

Not to be confused with the mesocorticolimbic pathway, the corticolimbic system includes the PFC, the hippocampus, and the amygdala, and it is critically important for regulating memory. I mentioned the importance of working memory to decision making above when discussing the PFC. As the hippocampus consolidates short-term memories into long-term memory, its contribution to decision making is similar to that of the PFC. For example, alcohol use disorder leads to impaired decision making, most likely due to physical alterations that occur to the brain (recall Korsakoff's syndrome from Chap. 6). In this group, impaired performance on the IGT is associated with lower gray matter volumes in the hippocampus, as well as in other regions such as the vmPFC and the dorsal ACC (dACC) (Le Berre et al., 2014). Hippocampal damage is also associated with impaired IGT performance in individuals with epilepsy (Labudda et al., 2009).

In addition to memory processes, the amygdala enables us to learn the association between stimuli and rewards. Impaired decision making in the IGT is observed in patients with amygdalar lesions (Bechara et al., 1999). Consistent findings are observed in the rGT following damage to the amygdala (Tremblay et al., 2021; Zeeb & Winstanley, 2011). Furthermore, disconnecting the pathway between the amygdala and the OFC impairs learning of the rGT (Zeeb & Winstanley, 2013). Impairing amygdalar function can increase suboptimal decision making and can prevent individuals from learning the association between certain cues that are associated with choice outcomes. In the IGT, individuals need to remember which deck of cards leads to each outcome. If this process is impaired, individuals can have difficultly using these contextual cues to guide decision making.

There is evidence that serotoninergic activity in the amygdala is important for guiding decision making; rats that display poor decision making in an rGT have lower serotonin metabolism in the basolateral nucleus (BLA) (Fitoussi et al., 2015). There are a couple of ways to interpret this finding. Lower metabolism could indicate that not enough serotonin is being metabolized, causing excess serotonin levels to increase in the synapse. The BLA contains a high number of GABAergic (gamma-aminobutyric acid) interneurons that contain the 5-HT_{2A} receptor. If this receptor is overstimulated, too much GABA release occurs in the amygdala, leading to hypoactivity in this region. However, low metabolism could indicate that there are lower baseline levels of serotonin. If less serotonin is being released, there is less of it to metabolize. In this case, a hypoactive amygdala can still result. Serotoninergic neurons in the BLA contain the 5-HT_{1A} autoreceptor. Deficient serotonin levels would result in insufficient activation of these receptors, which would lead to further increases in serotonin release; this would enable serotonin to bind to 5-HT_{2A} receptors on GABAergic neurons, thus inhibiting the amygdala.

Corticocerebellar Pathway

The dorsal striatum is not the only motor control region implicated in maladaptive decision making. The cerebellum is important for predictive coding (Chap. 5). When an individual makes a decision under conditions of ambiguity, they are effectively making a prediction about the outcomes of each alternative. In the IGT, individuals should learn that consistently choosing the high reward options is suboptimal as they are less likely to win or more likely to lose larger amounts of money. If the cerebellum does not work properly, individuals will have more difficulty predicting that selecting the high reward option is more than likely going to lead to less money across the session.

To this end, patients with cerebellar lesions perform worse on the IGT compared to healthy controls, with their performance being similar to those with damage to the frontal cortex (de Oliveira Cardoso et al., 2014). Individuals with cerebellar ataxia, like those with cerebellar lesions, perform worse on the IGT (Lai et al., 2023). You will learn more about cerebellar ataxia in Chap. 10.

Beyond those with cerebellar ataxia, the contribution of the cerebellum to decision making has been largely studied with individuals that use cannabis. Increased cannabis use is associated with impaired decision making (see above), and individuals that use cannabis have increased activity of the cerebellum and decreased activity of the PFC (Bolla et al., 2005; but see Vaidya et al., 2012 for increased activity in the vmPFC), and impaired performance in the IGT is associated with hyperactive functional connectivity between the cerebellum and regions such as basal ganglia structures, the NAc, and structures of the default mode network (DMN) (Francis et al., 2023). Recall that the DMN is most active when we refrain from engaging in cognitive effort (see Chap. 5). Overall, administration of cannabis activates CB_1 receptors located in the cerebellum, which inhibits this region, thus leading to impaired decision making by interfering with one's ability to make predictions about their choices.

Comprehension Check 4 How does hypoactivity of the PFC contribute to hyperactivity of the NAc, and how does this contribute to suboptimal decision making?

A hypoactive dopaminergic system in the PFC can lead to hyperactivity in the NAc. Specifically, insufficient stimulation of D_2-like receptors on prefrontal glutamatergic neurons causes these neurons to become disinhibited. This increased activity then leads to increased NAc activity, which biases individuals toward high magnitude but potentially suboptimal choices.

4.5.2 Neural Circuitry of Impulsive Decision Making

Similar to suboptimal decision making, aberrant prefrontal cortical activity is a major contributor of impulsive decision making. This should not be surprising given that these forms of maladaptive decision making are highly correlated (Franken et al., 2008). For example, choosing to receive \$100 immediately vs. \$1000 in 2 weeks is both impulsive and suboptimal as one can earn significantly more money by waiting a short period of time.

This section will begin by focusing on the contribution of the PFC to impulsive decision making, followed by a discussion of how another part of the frontal cortex influences impulsive decisions. Brain regions in the CBGTC loop and the limbic system will be covered as well.

Prefrontal Cortex (PFC)

Although delay discounting is positively correlated with left dlPFC activation in healthy adults (Ikegami & Sorama, 2023), a hypoactive PFC tends to be implicated in impulsive decision making (Gianotti et al., 2012), just like with suboptimal decision making. Lesions to the vmPFC lead to increased impulsive decision making in humans (Mok et al., 2021), and adults with ADHD performing a delay-discounting task show less activation of the dlPFC (Ortiz et al., 2015). Somewhat similarly, lesions to the mPFC increase impulsive choice in animals (e.g., Churchwell et al., 2009), but pharmacological inactivation of this region has no effect on delay discounting (Feja & Koch, 2014). This finding suggests that damage to excitatory neurons, as opposed to over-inhibition via activation of GABAergic receptors, modulates impulsive choice. Indeed, transcranial magnetic stimulation (TMS) of the mPFC decreases impulsive choice (Cho et al., 2015), and methylphenidate decreases impulsive choice, possibly by enhancing functional connectivity between the mPFC and the putamen (Daood et al., 2022) via increased glutamatergic activity (Di Miceli & Gronier, 2015). Conversely, inhibiting prefrontal cortical activity by administering a CB_1 agonist into PFC induces impulsive choice (Fatahi et al., 2018). As CB_1 receptors are located on glutamatergic neurons, stimulating these receptors causes inhibition of excitatory neurons, thus contributing to hypoactivity of the PFC.

You have previously learned about the dorsal attention network (DAN) and its relevance to attentional control (see Chap. 5). The frontal cortex, as well as the parietal cortex, is an important component of this network. Increased functional connectivity observed between the DAN and the insula (a key component of the salience network) predicts greater impulsive decision making in smokers (Clewett et al., 2014). A potential explanation is that increased activity of both the DAN and the insula biases these individuals to immediate rewards, accounting for why they display enhanced impulsive choice and continue to smoke despite long-term negative consequences of tobacco use.

Like suboptimal decision making, dopaminergic activity in the PFC mediates impulsive decision making, with both hypoactivity and hyperactivity of this system contributing to impulsive choices. Let's cover the evidence supporting a hypoactive dopaminergic system to impulsive decision making. Systemic blockade of D_1-like receptors, but not D_2-like receptors, increases impulsive choice (Koffarnus et al., 2011; van Gaalen et al., 2006). Individuals with a specific genotype for COMT, the enzyme that metabolizes catecholamines, display enhanced delay discounting, and this genotype leads to reduced dopamine levels in the PFC (Gianotti et al., 2012). Depleting mPFC dopamine levels induces impulsive choice in rodents (Freund et al., 2014). Droxidopa, a norepinephrine prodrug, decreases impulsive choice in rats (Dela Peña et al., 2021). One potential mechanism to explain this finding is that increasednorepinephrine levels in the PFC promote increased prefrontal cortical dopamine release.

A seemingly paradoxical finding is that both stimulation and blockade of intra-mPFC D_1-like receptors increases impulsive choice (Loos et al., 2010; Pardey et al., 2013). This is related to the idea that both hyper-and hypo-dopaminergic states can exacerbate impulsive behavior. Studies consistently show that antagonizing intra-PFC D_2-like receptors increases impulsive choice (Pardey et al., 2013; Yates et al., 2014). Blocking these autoreceptors prevents them from inhibiting dopamine release, allowing dopamine levels to rise in the synapse. Overexpression of D_1-like receptors in the PFC leads to increased impulsive choice (Sonntag et al., 2014), and high impulsive animals have increased dopamine D_1-like receptor gene expression in the mPFC. As mentioned earlier, $5\text{-}HT_{1A/B}$ agonists increase impulsive choice; the ability of these drugs to influence impulsive decision making is attributed to increased prefrontal cortical dopamine release (Korte et al., 2017).

Orbitofrontal Cortex (OFC)

In contrast to the PFC, lesions to the OFC either increase (e.g., Mobini et al., 2002) or decrease (e.g., Winstanley et al., 2004) impulsive choice. Research has even shown that subregions within the OFC differentially control impulsive choice in rats; damage to the medial OFC (mOFC) increases impulsive choice while damage to the lateral OFC (lOFC) decreases impulsive choice (Mar et al., 2011). This point is important as it highlights functional differences across structures of the frontal cortex. The discrepancies observed between the PFC and the OFC may be attributed to differences in serotoninergic activity in these regions. Increased serotonin transporter function, indicating *lower* serotonin levels, in the OFC is associated with impulsive choice in rats (Darna et al., 2015), and stimulation of intra-OFC 5-HT_{1A} receptors *decreases* impulsive choice (Yates et al., 2014), which differs from the effects of 5-HT_{1A} receptor agonists injected systemically.

There is an important distinction in what the OFC does compared to the PFC. The PFC is critical for executive functions like working memory and decision making. The OFC is more similar to the limbic system, as it is involved in reward processing and emotion regulation. In fact, the OFC and the NAc have reciprocal connections with one another. Even within the OFC, the mOFC and the lOFC have different functions, as the mOFC compares value of reward alternatives and predicts outcomes of each choice, and the lOFC helps modify our behaviors in response to unexpected outcomes and uses environmental contexts to drive certain behaviors. Damage to the mOFC, then, may impair individuals' ability to compare an immediately delivered reward that is suboptimal with a delayed reward that has a better payoff. Wouldn't damage to the lOFC also promote impulsive behavior? This is where interpreting the results of animal studies examining impulsive choice becomes difficult.

In one of the standard delay-discounting procedures, animals choose between a small, immediate reward and a large, delayed reward. At the beginning of the session, there are no delays to either reward. As the session advances, the delay to the large reward systematically increases. At the beginning of the session, animals should consistently choose the large reward as they can earn more food; this is analogous to me asking if you would rather have $10 now or $100 now. Because the lOFC is important for behavior modification, damage to this region may not lead to a decrease in impulsive choice *per se*. Instead, the animal may engage in a form of perseveration, something you will learn more about in Chap. 11. This illustrates the difficulty of dissociating certain constructs like response perseveration, suboptimal decision making, and impulsive choice.

Dorsal Striatum

Less activation of the caudate nucleus is observed in individuals with ADHD performing a delay-discounting task (Ortiz et al., 2015), and lower activity in the putamen is aputamenssociated with increased impulsive decision making in healthy volunteers (Smith et al., 2016). These results are consistent with preclinical research demonstrating that lesions to the striatum increase impulsive choice (Dunnett et al., 2012; Tedford et al., 2015). Lower white matter connections between the striatum and the PFC are also correlated with increased impulsive choice in healthy participants (Kim & Im, 2019). This suggests a hypoactive connection between the PFC and the striatum, which may account for the hypoactivity observed in the dorsal striatum.

Like the PFC, both decreased and increased dopamine levels in the dorsal striatum can lead to a hypoactive dorsal striatum. First, individuals with ADHD have increased DAT availability (Dalley et al., 2008), meaning that there is lower availability of dopamine to stimulate its constituent receptors. Relatedly, reduced impulsive choice following COMT inhibition corresponds to increased activity in the ventral putamen as there is more dopamine to bind to dopamine receptors (Kayser et al., 2012). However, increased striatal dopamine release is also observed in high impulsive animals (Moreno et al., 2021). Excess dopamine release can lead to overstimulation of dopamine D_2-like receptors located on GABAergic **medium spiny neurons (MSNs)**. Inhibition of MSNs prevents them from inhibiting other inhibitory neurons, leading to a net hypoactivity of the striatum.

Nucleus Accumbens (NAc)

Much of our knowledge on the contribution of the NAc to impulsive choice comes from animal studies as human imaging studies often focus on the ventral striatum, which contains more than the NAc (Basar et al., 2010). In contrast to what is observed following lesions of the PFC, lesions to the NAc consistently increase impulsive choice or increase suboptimal decision making in animals as assessed with delay discounting (e.g., Bezzina et al., 2007; Cardinal et al., 2001). That is, lesions to the NAc can lead to decreased preference for the large magnitude reinforcer, even when its delivery is not delayed (Cardinal et al., 2001).

Increasing dopamine levels in this region increases impulsive choice (Korte et al., 2017). However, in the NAc core, decreased D_1-like receptor binding and D_2-like receptor binding/mRNA expression is observed in high impulsive animals (Barlow et al., 2018; Jupp et al., 2013; Shen et al., 2024; Simon et al., 2013). Impulsive decision making appears to be driven more by reduced activation of D_2-like receptors compared to D_1-like receptors. In the ventral striatum, reduced binding of D_2-like receptors is associated with increased impulsive choice (Oberlin et al., 2015). This finding is similar to the induction of impulsive choice following knockdown of D_2-like receptors in the VTA (Bernosky-Smith et al., 2018). Impulsivity precipitated by D_1 overexpression in the PFC may also stem from reduced D_2 receptor expression in the NAc (Sonntag et al., 2014). These results align with the increased impulsive choice observed following damage to the NAc.

The NAc is also composed of MSNs. Like the dorsal striatum, fewer D_2-like receptors allow GABAergic neurons to become disinhibited, enabling them to stimulate other inhibitory neurons. There is another type of interneuron in the NAc that is cholinergic and excitatory. I mention this because selectively increasing dopamine D_2-like receptor expression on these cholinergic interneurons increases impulsive decision making (Cavallaro et al., 2023). This further supports the hypothesis that hypoactivity of the NAc contributes to impulsive choice. This is consistent with what I discussed earlier concerning the OFC, particularly the mOFC, and its role in impulsive decision making. In fact, mOFC neurons project to the NAc, and decreased activation of these neurons is observed in high impulsive animals (Wang et al. 2019).

Hippocampus and Amygdala

Let's modify the example from earlier of friends trying to decide on an activity to do. Now, imagine a group of friends is trying to plan a vacation. Some friends in the group want to take a trip as soon as possible, but they want to go somewhere that you have been to previously and know is not that exciting. You prefer to wait a little bit longer, so there is time to find a destination that will provide a great experience for the entire friend group. By using information stored in memory and evaluating the rewarding (and nonrewarding in this case) aspects of your previous trip, you are able to make a decision regarding immediate reinforcement compared to delayed reinforcement. Thus, a functioning hippocampus and amygdala would seem to be important for promoting choices that reflect self-control. This is the case as hippocampal/amygdalar lesions and pharmacological inactivation of these regions increase impulsive choice (Abela & Chudasama, 2013; Abela & Chudasama, 2014; Churchwell et al., 2009; Hernandez et al., 2022; Mariano et al., 2009; Winstanley et al., 2004). However, things are about to get more complicated as impulsive choice governed by the hippocampus and the amygdala does not merely come down to activation vs. inactivation.

In a delay-discounting task, impulsive choice is associated with greater activation of a network of brain regions including the hippocampus and the amygdala when choosing large, delayed rewards (Stanger et al., 2013), but elevated amygdalar activation in response to winning immediate rewards is stronger in high impulsive individuals (Ludwig et al., 2015). Essentially, when an impulsive individual chooses a large, delayed reward, there is higher activation of the hippocampus as the individual encodes information regarding their choice and the delay to reward receipt. The amygdala is important for assigning emotional valence to this response and for evaluating the reward. For someone that engages in impulsive decision making, their hippocampus may need to work harder to encode information involving events that will occur in the future. Interestingly, increased white matter volume in the hippocampus is associated with impulsive choice (Yu, 2012), indicating more myelinated axons in this region of high impulsive individuals. This suggests that these individuals need more neurons to help the hippocampus function normally.

Additionally, impulsive individuals may show enhanced amygdalar activity because waiting is aversive. Indeed, adolescents with ADHD show enhanced activation of the amygdala, as well as the dlPFC, when presented with cues that signal that a reward will be delayed (Van Dessel et al., 2018). Collectively, human studies suggest that hyperactivity of the hippocampus and the amygdala contributes to impulsive choice, which is incongruent with the findings that damage/pharmacological inactivation of these regions increase impulsive choice.

In addition to the finding that lesions to or pharmacological inactivation of the hippocampus/amygdala increase impulsive choice, preventing **neurogenesis** of hippocampal neurons promotes such decision making (Seib et al., 2021). This is where things get more complicated. The hippocampus contains many glutamatergic receptors. Theoretically, disrupting glutamatergic signaling in this region should increase impulsive choice, mirroring what is observed following pharmacological inactivation with GABA receptor agonists. Yet, deletion of N-methyl-D-aspartate (NMDA) receptors from hippocampal pyramidal neurons *reduces* delay discounting in mice (Masuda et al., 2020). Relatedly, increasing hippocampal serotonin levels by disrupting function of the serotonin transporter transiently decreases impulsive choice (Zoratto et al., 2013). The hippocampus contains a high concentration of 5-HT_1 receptors, which largely serve to inhibit glutamatergic neurons. Likewise, direct stimulation of hippocampal noradrenergic α_2 receptors reduces impulsive choice by inhibiting glutamatergic activity (Abela & Chudasama, 2014), potentially by increasing GABA release from surrounding astrocytes (Gaidin et al., 2020).

Beyond astrocyte-mediated release of GABA, the hippocampus also contains GABAergic neurons that regulate neuronal transmission. Adolescent animals display increased impulsive choice and higher CB_1 receptor expression in parts of the hippocampus compared to adults (Romero-Torres et al., 2023). CB_1 receptors are highly expressed on GABAergic interneurons in the hippocampus. This results in glutamatergic neurons becoming disinhibited.

On one hand, damaging the hippocampus increases impulsive choice, but other evidence suggests that overactivation of the hippocampus exacerbates impulsive choice. How can these seemingly contradictory findings be reconciled? In addition to being located on glutamatergic neurons, glutamate receptors are found on GABAergic neurons. Often, lesion studies are performed by infusing the NMDA receptor agonist quinolinic acid, which causes excitotoxicity, damaging the neuron. Because NMDA receptors are located on both glutamatergic and GABAergic neurons, quinolinic acid causes widespread impairments in hippocampal function (think of this as like producing a stroke in an animal). Pharmacological inactivation studies work by infusing drugs that bind to GABAergic receptors. In the hippocampus, GABAergic neurons inhibit other neurons, including glutamatergic ones. Now, glutamate is free to stimulate GABAergic neurons. Thus, overactivation of glutamate receptors in the hippocampus can lead to net inhibition of this region as GABAergic neurons become overly active. Tying everything together, increased

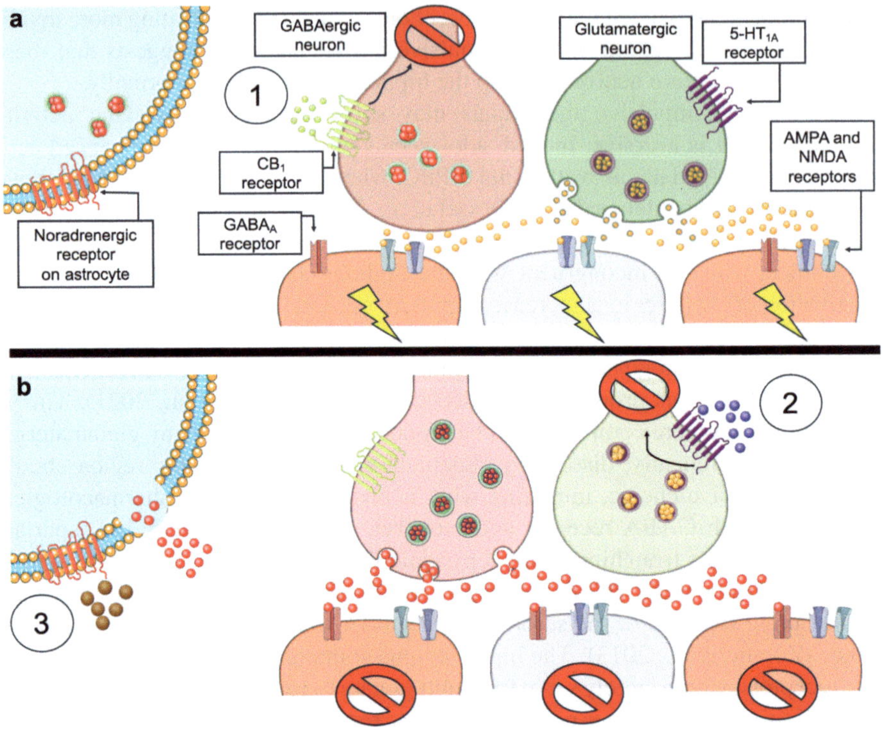

Fig. 4 (**a**) Overexcitation of glutamatergic neurons can lead to net inhibition in the hippocampus, which can exacerbate impulsive decision making. In this schematic, there is one glutamatergic neuron (in green) and one GABAergic neuron (in red) forming synapses with one glutamatergic neuron and two GABAergic neurons. GABAergic neurons can become inhibited via enhanced stimulation of cannabinoid CB_1 receptors located on presynaptic GABAergic neurons (1). Glutamate can stimulate ionotropic glutamate receptors located on postsynaptic neurons. This will excite both glutamatergic neurons and GABAergic neurons. Notice that in the image above, there are more GABAergic neurons that are stimulated. These GABAergic neurons can then inhibit other glutamatergic neurons, leading to inhibition. (**b**) Reducing excess excitation in the hippocampus can ameliorate impulsive decision making. There are a couple mechanisms by which this can be achieved. First, activation of presynaptic $5\text{-}HT_{1A}$ receptors on glutamatergic neurons can directly inhibit glutamatergic neurons (2). Second, stimulation of noradrenergic receptors on surrounding astrocytes can cause the release of GABA into the synapse (3). Because GABA binds to GABA receptors located on both glutamatergic and GABAergic receptors, this results in inhibition of both receptor types. This helps keep neuronal transmission at an optimal level. Image created by the author using Servier Medical Art (https://smart.servier.com/; CC BY 4.0, https://creativecommons.org/licenses/by/4.0/) and Microsoft PowerPoint

inhibition within the hippocampus and amygdala seems to lead to impulsive choices (Fig. 4).

Anterior Cingulate Cortex (ACC)

In the context of decision making, the ACC has a couple of important functions. First, the ACC is involved in conflict monitoring and error detection. If someone chooses to participate in an activity that ends up being unpleasant, this will produce an error signal. That is, the outcome of the activity is discrepant from what the individual had originally expected. Second, the ACC integrates information about rewards and their associated effort/risks. Ideally, when making a decision, you weigh the pros and cons of each option to determine which one is more ideal. Based on this description, you would assume that disrupting ACC activity impairs decision making. Well, this is partially correct.

Adults with ADHD show reduced ACC activation when completing a delay-discounting task (Ortiz et al., 2015). Administration of a CB_1 receptor agonist into the ACC promotes preference for a small, immediate reward over a large, delayed reward (Fatahi et al., 2018). However, the ACC is more active when individuals make impulsive decisions (Vorobyev et al., 2015), with glutamate concentrations being higher in the ACC of individuals that display greater delay discounting (Schmaal et al., 2012). Remember, the ACC has multiple functions, so increased and decreased activation of this region can promote different types of behaviors that can trigger an impulsive decision.

In the case of a hypoactive ACC, error detection processes may become impaired, making it more difficult for individuals to understand that immediate reinforcement is not always better than waiting for a better outcome. In substance use disorders, individuals continue to choose the immediate reinforcement of escaping withdrawal symptoms by using a substance; these individuals have difficulty understanding that continued use contributes to negative health and interpersonal problems. One study has even shown that chronic cannabis use prevents individuals from noticing that they had committed an error in a task, which was correlated with ACC hypoactivity (Hester et al., 2009).

When the ACC is hyperactive, individuals may be more sensitive to the effort associated with waiting longer to receive a reward relative to the magnitude of the small reward. Because the effort of waiting is weighted more heavily that receiving a less optimal reward faster, this can bias individuals to consistently choose outcomes that have short-term benefits. For example, trying to achieve abstinence requires a sustained, high-effort commitment across the lifespan. This can seem unachievable for some. Instead, individuals can take a substance to reduce withdrawal symptoms.

Habenula

I now want to briefly introduce a brain structure that has not been discussed in this book, the habenula (Fig. 5). The habenula is located in the epithalamus, which sits above the thalamus. The habenula is subdivided into lateral and medial subregions, with each subregion receiving inputs from different brain regions. Notably, the medial habenula receives direct inputs from the basal forebrain, which includes the NAc,

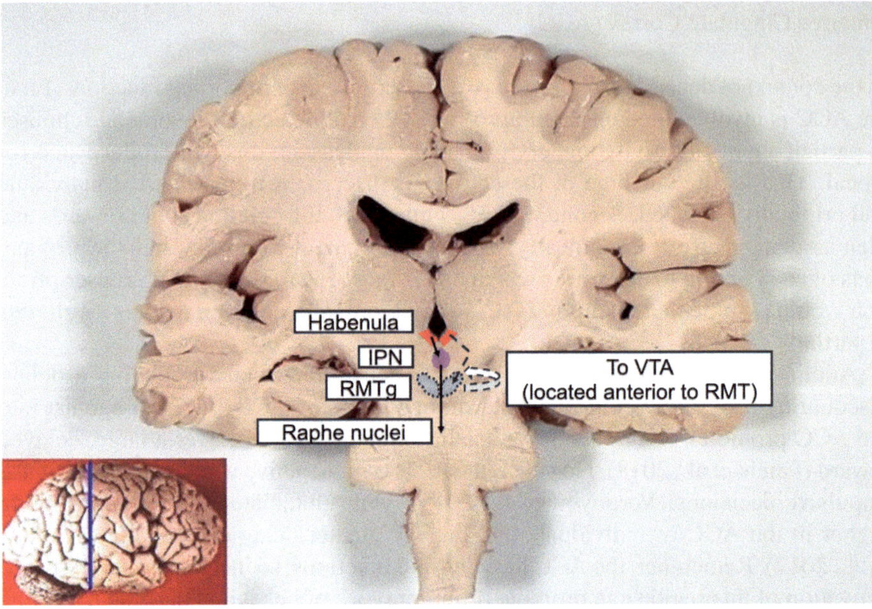

Fig. 5 Image showing location of the habenula, the interpeduncular nucleus (IPN), and the rostromedial tegmental nucleus (RMTg). Solid lines indicate projections originating from the lateral habenula. Dashed lines indicate projections originating from the medial habenula. *Source* https://commons.wikimedia.org/wiki/File:Human_brain_frontal_(coronal)_section.JPG from Dr. John Beal. CC BY 2.5 (https://creativecommons.org/licenses/by/2.5/deed.en), via Wikimedia Commons. Microsoft PowerPoint was used to label brain regions

while the lateral habenula receives inputs from the PFC and the globus pallidus (Ables et al., 2023; Boulos et al., 2017). Relevant to the present chapter, the habenula is important for encoding rewarding and aversive aspects of stimuli and is involved in value-based decision making (Hikosaka, 2010). Overall, lesions to the habenula increase impulsive choice (Lee et al., 2021), although subtle differences are observed when each specific subregion is targeted.

The medial habenula sends cholinergic neurons to the interpeduncular nucleus (IPN), located in the tegmentum. Genetically removing this cholinergic pathway from mice causes them to become more impulsive (Kobayashi et al., 2013). Increased impulsive choice most likely occurs because the IPN projects GABAergic neurons to the raphe nuclei. Without these GABAergic inputs, the raphe nuclei can become disinhibited, leading to increased serotonin release that overstimulates 5-$HT_{2A/C}$ receptors in other parts of the brain.

In contrast to the medial habenula, inactivation of the lateral habenula promotes suboptimal decision making in delay discounting as opposed to a change in impulsive choice *per se* (Baker et al., 2019). The lateral habenula sends glutamatergic neurons to the rostromedial tegmental nucleus (RMTg). The RMTg then projects GABAergic neurons to the VTA. Therefore, the lateral habenula indirectly controls dopaminergic activity. When the lateral habenula is inhibited, the RMTg does not become sufficiently activated, leading to disinhibition of the VTA. As you read earlier in this chapter, excess dopamine is related to impaired suboptimal decision making, which provides an account of why disrupting the lateral habenula leads to suboptimal choice in delay discounting as opposed to a selective change in impulsive choice.

Cerebellum

Finally, I want to briefly mention the potential role of the cerebellum to impulsive decision making. Earlier in this chapter, you learned that damage to the cerebellum impairs decision making by increasing suboptimal choice. Reduced activation of the cerebellum is also observed in adults with ADHD completing a delay-discounting task (Ortiz et al., 2015), as well as in other conditions characterized by impulsive decision making (Carlisi et al., 2017; Murphy et al., 2017). However, direct current stimulation applied to the cerebellum does not alter delay discounting (Wynn et al., 2019). There are two potential explanations. One, reduced activity observed in the cerebellum may be a byproduct of altered neuronal transmission from other regions such as the PFC. Two, Wynn et al. (2019) applied direct current stimulation to healthy volunteers, meaning that activating the cerebellum above baseline levels is insufficient to alter impulsive decision making. In sum, these findings seemingly indicate that failure in predictive coding can account for both suboptimal and impulsive decision making.

Comprehension Check 5 Lesion studies often show that damage to the hippocampus/amygdala increases impulsive choice, while human imaging studies sometimes show greater *activation* in these areas in highly impulsive individuals. How can one reconcile these discrepant findings?

Both extremes of activity (hypoactivity vs. *hyperactivity) can ultimately lead to impulsive choice due to impaired information processing within these regions. For the hippocampus and amygdala, damage or inactivation can directly disrupt their ability to perform functions crucial for self-control, such as encoding long-term information about delayed rewards or attaching appropriate emotional valence to outcomes. Conversely, hyperactivity observed in imaging studies of impulsive individuals might indicate that these regions are working harder to compensate for underlying inefficiencies, or that over-inhibition within these systems ultimately leads to a net functional hypoactivity. Overactivation of glutamate receptors in the hippocampus, for example, could lead to net inhibition by excessively stimulating GABAergic neurons. In essence, regardless of the initial cause, an imbalanced or inefficient state within these regions often leads to increased inhibition, thus resulting in impulsive decision making.*

4.5.3 Neural Circuitry of Risky Decision Making

By now, you should have a general idea of which brain regions will be covered below. Most of the research examining risky decision making has focused on prefrontal cortical areas and limbic structures.

Dissociation between Prefrontal Cortex and Orbitofrontal Cortex

Whereas the lPFC is important for decision making under ambiguity, the mOFC is more active in later trials of the IGT, suggesting that this region is more involved in decision making under risk (Hartstra et al., 2010). Lesions to the OFC consistently decrease risky choice (Abela & Chudasama, 2013; Mobini et al., 2002; Orsini et al., 2015). While depleting dopamine levels from the OFC does not alter risky decision making (Mai & Hauber, 2015), blocking D_1-like receptors in the mOFC reduces risky choice while stimulating D_2-like receptors increases risky choice (Jenni et al., 2021). How can this be the case?

If decision making is dictated by a fine balance of activation of D_1-like and D_2-like receptors, then globally depleting dopamine levels will not affect this intricate balance. This can explain why dopamine depletion fails to alter risky choice. Another consideration is that other neurotransmitters can mediate OFC activity, negating the effects of dopamine depletion. For example, brain injury-induced increases in risky choice are influenced by reduced norepinephrine transporter expression in the OFC (Knapp et al., 2024). This causes excess norepinephrine levels to increase in the OFC. Norepinephrine then influences dopamine release, which can bind to D_2-like receptors.

Activation of the mPFC is observed in individuals prior to selection of a risky alternative over a safe alternative (Matthews et al., 2004), although increased risky decision making in probability discounting is reported following lesions to the vmPFC (Mok et al., 2021). This last finding is consistent with the impaired decision making and increased impulsive choice observed in individuals with vmPFC damage. The Experiment Spotlight below details how a form of transcranial direct current stimulation (tDCS) can improve risky decision making in children with ADHD.

Experiment Spotlight: Transcranial Random Noise Stimulation (tRNS) Improves Hot and Cold Executive Functions in Children with Attention Deficit-Hyperactivity Disorder (ADHD)

The link to the article can be found here: https://www.nature.com/articles/s41598-024-57920-3

In the study by Nejati et al. (2024), 18 children (7–12 years of age) with ADHD were recruited to complete a series of tests to measure executive functions. First, the circle tracing task requires one to trace a circle with their index finger. The children were instructed to perform this task twice, once as quickly as they wanted and once as slowly as possible. What the participants did not know is that they had a 12-minute period in which they could trace the circle, meaning they needed to trace the circle very slowly during this time. Second, participants completed a go/no-go task, which measures a different form of impulsivity (see Chap. 12). Third, participants completed a 1-back test (refer back to Chap. 6 for *n*-back tests). Fourth, participants completed the Wisconsin card sorting task, which measures cognitive flexibility (specifically perseveration; see Chap. 11 for a discussion of perseveration). Finally, participants completed the BART, which is directly relevant to the present chapter.

Participants were subjected to **transcranial random noise stimulation (tRNS)**, closely related to tDCS. The major difference between these techniques is that tRNS applies an alternating electric current at random frequencies whereas tDCS applies a consistent current. In the present study, participants experienced currents that ranged from −500 to 500 μA (microamperes). Electrodes were placed over the left dlPFC and the right vmPFC. Each participant completed two sessions 1 week apart, one in which tRNS was applied and one in which they were given a *sham* session. During the sham session, the current was applied for 30 s only before being discontinued.

tRNS significantly increased the time children spent tracing the circle when instructed to trace as slowly as possible, indicating improved ongoing response inhibition. tRNS also increased accuracy on no-go trials of the go/no-go task, improved working memory as assessed with the 1-back test, and increased the number of successful pumps in the BART. That is, children given tRNS were more likely to earn money in this task compared to when they received sham stimulation.

Questions to Consider:

1. What are some advantages of using tRNS instead of tDCS?
2. In the sham condition, why is the tRNS applied for 30 s instead of never being turned on?
3. Why do you think one electrode was placed over the left dlPFC and the other over the right vmPFC?

While systemic administration of D_1-like or D_2-like receptor agonists increases risky choice, and administration of D_1-like or D_2-like receptor antagonists produces the opposite effect (St Onge & Floresco, 2009), intra-PFC D_1-like receptors are important mediators of risky choice (Münster et al., 2024). However, when these drugs are locally administered to specific PFC subregions, different findings can be observed. Whereas blockade of intra-mPFC D_1-like receptors reduces risky choice, antagonism of D_2-like receptors increases risky choice (St Onge et al., 2011). This finding highlights a subtle difference in how intra-PFC dopamine influences risky decision making and impulsive decision making. Recall that intra-PFC blockade of both D_1-like and D_2-like receptors increases impulsive choice. Thus, impulsive choice and risky choice can be correlated with each other due to aberrations in D_2-like receptor function in the PFC.

Reward-Sensitive Structures

During deliberation between a large, risky option and a small, safe option, increased activation of the NAc is observed in individuals that ultimately choose the risky alternative, mirroring what is observed in the mPFC (Matthews et al., 2004; Op de Macks et al., 2018). Just as in the PFC, dopamine depletion in the NAc has no effect on risky choice (Mai & Hauber, 2015); increasing dopaminergic activity does increase risky choice (e.g., Freels et al., 2020; St Onge et al., 2012). Dopamine release in the NAc shell is directly related to the size of a reward (Beyene et al., 2010). If an individual has excessive dopamine release in the NAc, they may overweigh the importance of a high-risk, large reward in favor of a smaller, less risky option.

Lesions to the amygdala increase risky choice (Orsini et al., 2015), but pharmacological inactivation of the amygdala induces suboptimal decision making in probability discounting (van Holstein et al., 2020). This is consistent with the finding that damage to the amygdala induces suboptimal decision making and impulsive choice. Above, I mentioned that individuals show greater activation of the mPFC and the NAc before selecting a large, risky alternative. There is some evidence to suggest that a similar increase in BLA activity occurs while animals deliberate between rewards associated with different levels of risk. In one study, optogenetic inhibition of the BLA before a rat decides between each alternative decreases risky choice; yet, if this region is inhibited while the rat receives punishment following selection of the large, risky option, they will display greater risky choice (Orsini et al., 2017). Let's use an example to illustrate what I mean here.

Suppose you are deciding on a place to eat for dinner. Imagine you are choosing between a place where you consistently eat and a place you have never visited. Damage to the amygdala can promote a form of risk aversion by propelling the individual to stick with what they know. This is because the individual cannot fully process the potential benefits and losses associated with the risky option (i.e., novel restaurant). Now, imagine a situation in which you are deciding on a place to eat. If you got sick after eating at one restaurant previously, the BLA encoded the context of the restaurant with the aversive experience of getting ill. Even if the food at this restaurant did not cause your illness, you will pair the act of getting sick to the food. This is known as *conditioned taste aversion.* If your amygdala does not function properly, you may not be able to retrieve the aversive association you had previously formed. In a way, the bond between environmental context and the feeling of being sick has been broken. This may promote the individual to try the restaurant again.

The ACC is also activated when individuals choose a high-risk alternative compared to a low-risk alternative (Christopoulos et al., 2009; Rodrigo et al., 2014; Smith et al., 2009), and functional connectivity between the OFC and the ACC is associated with risk preference (Wang et al., 2019). This is consistent with what I described above—the OFC becomes active when we make risky choices. Yet, ACC activation is higher in individuals that discontinue pumping a balloon in the BART, indicating this region is important for loss aversion (Fukunaga et al., 2012). In fact, direct stimulation of the ACC promotes risk aversion in healthy volunteers (Mattavelli et al., 2022). These results are not as contradictory as they first appear.

Given the ACC's contributions to conflict monitoring, this region becomes active when we make a high-risk decision because there is greater conflict between the potential for a large reward and the potential for a significant loss. Greater activation of the ACC does not necessarily map onto preference for the risky option. In the BART, the ACC becomes active as the individual decides to stop pumping the balloon as this is related to conflict monitoring. There is a conflict between pumping the balloon to earn additional money and the risk of the balloon popping. The finding that direct stimulation of the ACC promotes risk aversion means that this region is more important for governing risk aversion and that dysregulation of the ACC can promote maladaptive risky choices.

Motor Control Areas

A network of brain regions, including the dorsal striatum, is associated with risk-taking behavior (Rao et al., 2008; Wang et al., 2022). In a gambling task, individuals with obsessive-compulsive disorder (OCD) show greater activation of the putamen, as well as the OFC and the ACC (Moreira et al., 2020). There is some evidence that the cerebellum is involved in risky decision making as well. One study found that gray matter volume of the left cerebellum is correlated with risk tolerance as assessed with the BART (Quan et al., 2022). Furthermore, functional connectivity between the right cerebellum and the left caudate nucleus is associated with risk taking in the BART (Huo et al., 2023).

Suppose the activity you and your friends choose is to go to a casino. At the casino, you can play a more conservative game like the penny slot machine, or you can take on more risk by playing a game that requires you to spend more money. As I noted above, the NAc becomes more active when we make a risky decision. Excessive dopamine release in the NAc, particularly the core, can influence dorsal striatal activity as the core is important for establishing motor behaviors related to reinforcement (see Chap. 11 for more details).

As you continue playing a game like Blackjack, you learn the skills needed to play the game effectively such as approaching the table, speaking with the dealer, and playing the game itself. The cerebellum is also recruited due to its role in predictive coding. When you play a game that involves probabilistic wins, this inherently involves the cerebellum as it makes predictions about what you will experience as you play.

Hyperactivity of the dorsal striatum may lead to habitual risky decision making (e.g., gambling disorder). At the same time, dysfunction of the cerebellum may impair individuals' ability to predict what they will experience when they engage in a high-risk activity. Perhaps, they overestimate the probability that they will win when they place a gamble.

Comprehension Check 6 The OFC and the ACC are involved in risky decision making, albeit with some seemingly contradictory findings. How can one reconcile the idea that the OFC and ACC are active during risky choices, but OFC lesions decrease risky choice while ACC stimulation decreases risky choice?

The OFC is crucial for evaluating and potentially overweighing large potential rewards, making individuals more inclined toward risky options when it's functioning normally. When damaged, this bias toward large, risky rewards is reduced, leading to less risky behavior. The ACC's activation during high-risk decisions is attributed to its role in conflict monitoring and error detection, rather than directly promoting risk taking. When there's a high potential for both large rewards and significant losses, the ACC becomes active to process this conflict. The finding that direct ACC stimulation promotes risk aversion supports this; it suggests that a properly functioning ACC acts as a brake, signaling the potential negative consequences and prompting individuals to avoid risk. Therefore, its activity in risky situations reflects the brain's attempt to manage the risk, not necessarily to pursue it.

5 Chapter Summary

This chapter dived into the complex neural circuits associated with various forms of maladaptive decision making. You learned the difference between suboptimal, impulsive, and risky decision making, as well as the neural correlates of each type of decision making. I know this was a lot of information to unpack. Here are the main points to take from the present chapter:

1. The PFC is crucially important for decision making. Dysfunction of this region can give rise to various forms of maladaptive decision making.
2. Too much activation in the NAc is associated with each form of maladaptive decision making covered in this chapter. Excess activation of the NAc may stem from dysfunction of the PFC, which normally acts to inhibit NAc activity. Increased NAc activity may bias individuals to rewards that are delivered immediately and/or options that appear to have the "highest value".
3. Impairment of memory-associated structures such as the hippocampus and the amygdala leads to maladaptive decision making. Dysfunction of the hippocampus can make retrieving information from long-term memory difficult and make thinking about future rewards challenging. This can cause individuals to opt for immediate rewards (impulsive choice) or fail to learn that some alternatives are not as advantageous as others (suboptimal decision making). In addition to memory, the amygdala is critical for assigning value to rewards. Deficient activity of the amygdala is associated with problematic valuation of alternatives that are maladaptive.

4. The ACC is important for conflict monitoring. When we make a decision, we are in a state of conflict. Aberrant signaling in the ACC can disrupt this cognitive process, leading us to make maladaptive decisions.
5. There is some discrepancy concerning how dorsal striatal activity mediates maladaptive decision making. A hypoactive dorsal striatum is implicated in suboptimal and impulsive decision making, perhaps by interfering with one's ability to learn the association between a behavioral response and its outcome. A hyperactive dorsal striatum seems to be involved in risky choice; this could reflect that some risky behaviors can become habitual (e.g., gambling).

Answers to Experiment Spotlight Questions:

1. In tDCS, a single, continuous current is applied to a brain region of interest. This is not ideal for a couple of reasons. One, individuals can report side effects such as tingling, itching, and burning where the electrodes have been placed. Two, this technique requires individuals to consider polarity. That is, a cathode can be placed over a region to cause hyperpolarization whereas an anode leads to depolarization of neurons. Because tRNS uses currents that vary in intensity, the side effect profile of this technique is minimal. This procedure is also simpler as researchers do not need to worry about placing an anode or a cathode over the region of interest.
2. For a sham condition to work, participants cannot tell they are not receiving tRNS. By turning the machine on for 30 s, participants can hear the equipment start up as normal. This can convince the participant that they are receiving tRNS.
3. These regions (as well as their respective hemispheres) were chosen based on studies suggesting that the left dlPFC is important for "cold" executive functions. These executive functions lack an emotional component. In the present study, all but one test measured cold executive functioning. "Hot" executive functions are those that include an emotional component; this type of executive function is largely controlled by the right vmPFC. The BART was used to measure hot executive functioning.

Glossary

Expected utility theory states that we consider the expected utility (i.e., usefulness) and the probability of each option when deciding between alternatives; this is a normative model of decision making.

Impulsive decision making tendency to act or make choices without careful thought or consideration of the potential consequences.

Medium spiny neuron (MSN) type of GABAergic neuron primarily found in the dorsal and the ventral striatum.

Neurogenesis the process by which new neurons are generated.

Prospect theory posits that individuals are more risk averse when faced with expected gains but are more risk seeking when faced with expected losses; also argues that we are more sensitive to losses than to gains; this is a descriptive model of decision making.

Risky decision making choosing between options with uncertain outcomes, where the potential for both significant gains and losses exists.

Suboptimal decision making process of choosing options that lead to less favorable outcomes due to inefficient evaluation of each option's outcome.

Transcranial random noise stimulation (tRNS) a non-invasive brain stimulation technique that applies a low-intensity alternating electrical current with randomized frequencies and amplitudes to the scalp, aiming to modulate cortical excitability and brain function.

References

Abela, A. R., & Chudasama, Y. (2013). Dissociable contributions of the ventral hippocampus and orbitofrontal cortex to decision-making with a delayed or uncertain outcome. *European Journal of Neuroscience, 37*(4), 640–647. https://doi.org/10.1111/ejn.12071

Abela, A. R., & Chudasama, Y. (2014). Noradrenergic α2A-receptor stimulation in the ventral hippocampus reduces impulsive decision-making. *Psychopharmacology, 231*(3), 521–531. https://doi.org/10.1007/s00213-013-3262-y

Ables, J. L., Park, K., & Ibañez-Tallon, I. (2023). Understanding the habenula: A major node in circuits regulating emotion and motivation. *Pharmacological Research, 190*, 106734. https://doi.org/10.1016/j.phrs.2023.106734

Adrover, M. F., Shin, J. H., Quiroz, C., Ferré, S., Lemos, J. C., & Alvarez, V. A. (2020). Prefrontal cortex-driven dopamine signals in the striatum show unique spatial and pharmacological properties. *The Journal of Neuroscience, 40*(39), 7510–7522. https://doi.org/10.1523/JNEUROSCI.1327-20.2020

Agrawal, A., Few, L., Nelson, E. C., Deutsch, A., Grant, J. D., Bucholz, K. K., Madden, P. A., Heath, A. C., & Lynskey, M. T. (2016). Adolescent cannabis use and repeated voluntary unprotected sex in women. *Addiction, 111*(11), 2012–2020. https://doi.org/10.1111/add.13490

American Psychiatric Association. (2013). *Diagnostic and statistical manual of mental disorders* (5th ed.). doi:10.1176/appi.books.9780890425596.

Amlung, M., Marsden, E., Holshausen, K., Morris, V., Patel, H., Vedelago, L., Naish, K. R., Reed, D. D., & McCabe, R. E. (2019). Delay discounting as a transdiagnostic process in psychiatric disorders: A meta-analysis. *JAMA Psychiatry, 76*(11), 1176–1186. https://doi.org/10.1001/jamapsychiatry.2019.2102

Baarendse, P. J., & Vanderschuren, L. J. (2012). Dissociable effects of monoamine reuptake inhibitors on distinct forms of impulsive behavior in rats. *Psychopharmacology, 219*(2), 313–326. https://doi.org/10.1007/s00213-011-2576-x

Baarendse, P. J., Winstanley, C. A., & Vanderschuren, L. J. (2013). Simultaneous blockade of dopamine and noradrenaline reuptake promotes disadvantageous decision making in a rat gambling task. *Psychopharmacology, 225*(3), 719–731. https://doi.org/10.1007/s00213-012-2857-z

Baker, P. M., Rao, Y., Rivera, Z. M. G., Garcia, E. M., & Mizumori, S. J. Y. (2019). Selective functional interaction between the lateral habenula and hippocampus during different tests of response flexibility. *Frontiers in Molecular Neuroscience, 12*, 245. https://doi.org/10.3389/fnmol.2019.00245

Barlow, R. L., Gorges, M., Wearn, A., Niessen, H. G., Kassubek, J., Dalley, J. W., & Pekcec, A. (2018). Ventral striatal D2/3 receptor availability is associated with impulsive choice behavior as well as limbic corticostriatal connectivity. *The International Journal of Neuropsychopharmacology, 21*(7), 705–715. https://doi.org/10.1093/ijnp/pyy030

Basar, K., Sesia, T., Groenewegen, H., Steinbusch, H. W., Visser-Vandewalle, V., & Temel, Y. (2010). Nucleus accumbens and impulsivity. *Progress in Neurobiology, 92*(4), 533–557. https://doi.org/10.1016/j.pneurobio.2010.08.007

Bechara, A., Damasio, A. R., Damasio, H., & Anderson, S. W. (1994). Insensitivity to future consequences following damage to human prefrontal cortex. *Cognition, 50*(1–3), 7–15. https://doi.org/10.1016/0010-0277(94)90018-3

Bechara, A., Damasio, H., Damasio, A. R., & Lee, G. P. (1999). Different contributions of the human amygdala and ventromedial prefrontal cortex to decision-making. *The Journal of Neuroscience, 19*(13), 5473–5481. https://doi.org/10.1523/JNEUROSCI.19-13-05473.1999

Bernosky-Smith, K. A., Qiu, Y. Y., Feja, M., Lee, Y. B., Loughlin, B., Li, J. X., & Bass, C. E. (2018). Ventral tegmental area D2 receptor knockdown enhances choice impulsivity in a delay-discounting task in rats. *Behavioural Brain Research, 341*, 129–134. https://doi.org/10.1016/j.bbr.2017.12.029

Betts, G. D., Hynes, T. J., & Winstanley, C. A. (2021). Pharmacological evidence of a cholinergic contribution to elevated impulsivity and risky decision-making caused by adding win-paired cues to a rat gambling task. *Journal of Psychopharmacology, 35*(6), 701–712. https://doi.org/10.1177/0269881120972421

Beyene, M., Carelli, R. M., & Wightman, R. M. (2010). Cue-evoked dopamine release in the nucleus accumbens shell tracks reinforcer magnitude during intracranial self-stimulation. *Neuroscience, 169*(4), 1682–1688. https://doi.org/10.1016/j.neuroscience.2010.06.047

Beyer, D. K. E., Horn, L., Klinker, N., & Freund, N. (2021). Risky decision-making following prefrontal D1 receptor manipulation. *Translational Neuroscience, 12*(1), 432–443. https://doi.org/10.1515/tnsci-2020-0187

Bezzina, G., Cheung, T. H., Asgari, K., Hampson, C. L., Body, S., Bradshaw, C. M., Szabadi, E., Deakin, J. F., & Anderson, I. M. (2007). Effects of quinolinic acid-induced lesions of the nucleus accumbens core on inter-temporal choice: A quantitative analysis. *Psychopharmacology, 195*(1), 71–84. https://doi.org/10.1007/s00213-007-0882-0

Białaszek, W., Marcowski, P., & Cox, D. J. (2017). Differences in delay, but not probability discounting, in current smokers, E-cigarette users, and never smokers. *The Psychological Record, 67*(2), 223–230. https://doi.org/10.1007/s40732-017-0244-1

Bickel, W. K., Athamneh, L. N., Basso, J. C., Mellis, A. M., DeHart, W. B., Craft, W. H., & Pope, D. (2019). Excessive discounting of delayed reinforcers as a trans-disease process: Update on the state of the science. *Current Opinion in Psychology, 30*, 59–64. https://doi.org/10.1016/j.copsyc.2019.01.005

Blasio, A., Narayan, A. R., Kaminski, B. J., Steardo, L., Sabino, V., & Cottone, P. (2012). A modified adjusting delay task to assess impulsive choice between isocaloric reinforcers in non-deprived male rats: Effects of 5-HT2A/C and 5-HT1A receptor agonists. *Psychopharmacology, 219*(2), 377–386. https://doi.org/10.1007/s00213-011-2517-8

Bolla, K. I., Eldreth, D. A., Matochik, J. A., & Cadet, J. L. (2005). Neural substrates of faulty decision-making in abstinent marijuana users. *NeuroImage, 26*(2), 480–492. https://doi.org/10.1016/j.neuroimage.2005.02.012

Boulos, L. J., Darcq, E., & Kieffer, B. L. (2017). Translating the habenula-from rodents to humans. *Biological Psychiatry, 81*(4), 296–305. https://doi.org/10.1016/j.biopsych.2016.06.003

Buelow, M. T., & Suhr, J. A. (2014). Risky decision making in smoking and nonsmoking college students: Examination of Iowa gambling task performance by deck type selections. *Applied Neuropsychology. Child, 3*(1), 38–44. https://doi.org/10.1080/21622965.2012.691065

Bustamante, M. J., Palfai, T. P., Luehring-Jones, P., Maisto, S. A., & Simons, J. S. (2022). Cannabis use and sexual risk among MSM who drink: Understanding why more frequent cannabis users may engage in higher rates of condomless sex. *Drug and Alcohol Dependence, 232*, 109282. https://doi.org/10.1016/j.drugalcdep.2022.109282

Bymaster, F. P., Katner, J. S., Nelson, D. L., Hemrick-Luecke, S. K., Threlkeld, P. G., Heiligenstein, J. H., Morin, S. M., Gehlert, D. R., & Perry, K. W. (2002). Atomoxetine increases extracellular levels of norepinephrine and dopamine in prefrontal cortex of rat: A potential mechanism for efficacy in attention deficit/hyperactivity disorder. *Neuropsychopharmacology, 27*(5), 699–711. https://doi.org/10.1016/S0893-133X(02)00346-9

Cardinal, R. N., Pennicott, D. R., Sugathapala, C. L., Robbins, T. W., & Everitt, B. J. (2001). Impulsive choice induced in rats by lesions of the nucleus accumbens core. *Science, 292*(5526), 2499–2501. https://doi.org/10.1126/science.1060818

Carlisi, C. O., Norman, L., Murphy, C. M., Christakou, A., Chantiluke, K., Giampietro, V., Simmons, A., Brammer, M., Murphy, D. G., & MRC AIMS consortium, Mataix-Cols, D., & Rubia, K. (2017). Comparison of neural substrates of temporal discounting between youth with autism spectrum disorder and with obsessive-compulsive disorder. *Psychological Medicine, 47*(14), 2513–2527. https://doi.org/10.1017/S0033291717001088

Cavallaro, J., Yeisley, J., Akdoğan, B., Salazar, R. E., Floeder, J. R., Balsam, P. D., & Gallo, E. F. (2023). Dopamine D2 receptors in nucleus accumbens cholinergic interneurons increase impulsive choice. *Neuropsychopharmacology, 48*(9), 1309–1317. https://doi.org/10.1038/s41386-023-01608-1

Cho, S. S., Koshimori, Y., Aminian, K., Obeso, I., Rusjan, P., Lang, A. E., Daskalakis, Z. J., Houle, S., & Strafella, A. P. (2015). Investing in the future: Stimulation of the medial prefrontal cortex reduces discounting of delayed rewards. *Neuropsychopharmacology, 40*(3), 546–553. https://doi.org/10.1038/npp.2014.211

Christopoulos, G. I., Tobler, P. N., Bossaerts, P., Dolan, R. J., & Schultz, W. (2009). Neural correlates of value, risk, and risk aversion contributing to decision making under risk. *The Journal of Neuroscience, 29*(40), 12574–12583. https://doi.org/10.1523/JNEUROSCI.2614-09.2009

Churchwell, J. C., Morris, A. M., Heurtelou, N. M., & Kesner, R. P. (2009). Interactions between the prefrontal cortex and amygdala during delay discounting and reversal. *Behavioral Neuroscience, 123*(6), 1185–1196. https://doi.org/10.1037/a0017734

Clark, L., Manes, F., Antoun, N., Sahakian, B. J., & Robbins, T. W. (2003). The contributions of lesion laterality and lesion volume to decision-making impairment following frontal lobe damage. *Neuropsychologia, 41*(11), 1474–1483. https://doi.org/10.1016/s0028-3932(03)00081-2

Clewett, D., Luo, S., Hsu, E., Ainslie, G., Mather, M., & Monterosso, J. (2014). Increased functional coupling between the left fronto-parietal network and anterior insula predicts steeper delay discounting in smokers. *Human Brain Mapping, 35*(8), 3774–3787. https://doi.org/10.1002/hbm.22436

Coffey, S. F., Gudleski, G. D., Saladin, M. E., & Brady, K. T. (2003). Impulsivity and rapid discounting of delayed hypothetical rewards in cocaine-dependent individuals. *Experimental and Clinical Psychopharmacology, 11*(1), 18–25. https://doi.org/10.1037/1064-1297.11.1.18

Colautti, L., Iannello, P., Silveri, M. C., & Antonietti, A. (2023). Decision-making under ambiguity and risk and executive functions in Parkinson's disease patients: A scoping review of the studies investigating the Iowa gambling task and the game of dice. *Cognitive, Affective, & Behavioral Neuroscience, 23*(5), 1225–1243. https://doi.org/10.3758/s13415-023-01106-3

Dai, Z., Harrow, S. E., Song, X., Rucklidge, J. J., & Grace, R. C. (2016). Gambling, delay, and probability discounting in adults with and without ADHD. *Journal of Attention Disorders, 20*(11), 968–978. https://doi.org/10.1177/1087054713496461

Dalley, J. W., Mar, A. C., Economidou, D., & Robbins, T. W. (2008). Neurobehavioral mechanisms of impulsivity: Fronto-striatal systems and functional neurochemistry. *Pharmacology, Biochemistry, and Behavior, 90*(2), 250–260. https://doi.org/10.1016/j.pbb.2007.12.021

Daood, M., Peled-Avron, L., Ben-Hayun, R., Nevat, M., Aharon-Peretz, J., Tomer, R., & Admon, R. (2022). Fronto-striatal connectivity patterns account for the impact of methylphenidate on choice impulsivity among healthy adults. *Neuropharmacology, 216*, 109190. https://doi.org/10.1016/j.neuropharm.2022.109190

Darna, M., Chow, J. J., Yates, J. R., Charnigo, R. J., Beckmann, J. S., Bardo, M. T., & Dwoskin, L. P. (2015). Role of serotonin transporter function in rat orbitofrontal cortex in impulsive choice. *Behavioural Brain Research, 293*, 134–142. https://doi.org/10.1016/j.bbr.2015.07.025

de Oliveira Cardoso, C., Branco, L. D., Cotrena, C., Kristensen, C. H., Schneider Bakos, D. D., & Fonseca, R. P. (2014). The impact of frontal and cerebellar lesions on decision making: Evidence from the Iowa gambling task. *Frontiers in Neuroscience, 8*, 61. https://doi.org/10.3389/fnins.2014.00061

Del Arco, A., & Mora, F. (2008). Prefrontal cortex-nucleus accumbens interaction: In vivo modulation by dopamine and glutamate in the prefrontal cortex. *Pharmacology, Biochemistry, and Behavior, 90*(2), 226–235. https://doi.org/10.1016/j.pbb.2008.04.011

Dela Peña, I., Shen, G., & Shi, W. X. (2021). Droxidopa alters dopamine neuron and prefrontal cortex activity and improves attention-deficit/hyperactivity disorder-like behaviors in rats. *European Journal of Pharmacology, 892*, 173826. https://doi.org/10.1016/j.ejphar.2020.173826

Devoto, P., Flore, G., Pira, L., Longu, G., & Gessa, G. L. (2004). Alpha2-adrenoceptor mediated co-release of dopamine and noradrenaline from noradrenergic neurons in the cerebral cortex. *Journal of Neurochemistry, 88*(4), 1003–1009. https://doi.org/10.1046/j.1471-4159.2003.02239.x

Di Ciano, P., Cormick, P. M., Stefan, C., Wong, E., Kim, A., Remington, G., & Le Foll, B. (2017). The effects of buspirone on occupancy of dopamine receptors and the rat gambling task. *Psychopharmacology, 234*(22), 3309–3320. https://doi.org/10.1007/s00213-017-4715-5

Di Ciano, P., Manvich, D. F., Pushparaj, A., Gappasov, A., Hess, E. J., Weinshenker, D., & Le Foll, B. (2018). Effects of disulfiram on choice behavior in a rodent gambling task: Association with catecholamine levels. *Psychopharmacology, 235*(1), 23–35. https://doi.org/10.1007/s00213-017-4744-0

Di Miceli, M., & Gronier, B. (2015). Psychostimulants and atomoxetine alter the electrophysiological activity of prefrontal cortex neurons, interaction with catecholamine and glutamate NMDA receptors. *Psychopharmacology, 232*(12), 2191–2205. https://doi.org/10.1007/s00213-014-3849-y

Diana, M. (2011). The dopamine hypothesis of drug addiction and its potential therapeutic value. *Frontiers in Psychiatry, 2*, 64. https://doi.org/10.3389/fpsyt.2011.00064

Drechsler, R., Rizzo, P., & Steinhausen, H. C. (2010). Decision making with uncertain reinforcement in children with attention deficit/hyperactivity disorder (ADHD). *Child Neuropsychology, 16*(2), 145–161. https://doi.org/10.1080/09297040903190774

Dunnett, S. B., Heuer, A., Lelos, M., Brooks, S. P., & Rosser, A. E. (2012). Bilateral striatal lesions disrupt performance in an operant delayed reinforcement task in rats. *Brain Research Bulletin, 88*(2–3), 251–260. https://doi.org/10.1016/j.brainresbull.2011.04.002

Fagundo, A. B., de la Torre, R., Jiménez-Murcia, S., Agüera, Z., Pastor, A., Casanueva, F. F., Granero, R., Baños, R., Botella, C., Del Pino-Gutierrez, A., Fernández-Real, J. M., Fernández-García, J. C., Frühbeck, G., Gómez-Ambrosi, J., Menchón, J. M., Moragrega, I., Rodríguez, R., Tárrega, S., Tinahones, F. J., & Fernández-Aranda, F. (2013). Modulation of the endocannabinoids N-arachidonoylethanolamine (AEA) and 2-arachidonoylglycerol (2-AG) on executive functions in humans. *PLoS One, 8*(6), e66387. https://doi.org/10.1371/journal.pone.0066387

Fatahi, Z., Reisi, Z., Rainer, G., Haghparast, A., & Khani, A. (2018). Cannabinoids induce apathetic and impulsive patterns of choice through CB1 receptors and TRPV1 channels. *Neuropharmacology, 133*, 75–84. https://doi.org/10.1016/j.neuropharm.2018.01.021

Feja, M., & Koch, M. (2014). Ventral medial prefrontal cortex inactivation impairs impulse control but does not affect delay-discounting in rats. *Behavioural Brain Research, 264*, 230–239. https://doi.org/10.1016/j.bbr.2014.02.013

Ferland, J. N., Carr, M. R., Lee, A. M., Hoogeland, M. E., Winstanley, C. A., & Pattij, T. (2018). Examination of the effects of cannabinoid ligands on decision making in a rat gambling task.

Pharmacology, Biochemistry, and Behavior, 170, 87–97. https://doi.org/10.1016/j.pbb.2018.05.012

Ferland, J. N., Ellis, R. J., Betts, G., Silveira, M. M., de Firmino, J. B., Winstanley, C. A., & Hurd, Y. L. (2023). Long-term outcomes of adolescent THC exposure on translational cognitive measures in adulthood in an animal model and computational assessment of human data. *JAMA Psychiatry, 80*(1), 66–76. https://doi.org/10.1001/jamapsychiatry.2022.3915

Fitoussi, A., Le Moine, C., De Deurwaerdère, P., Laqui, M., Rivalan, M., Cador, M., & Dellu-Hagedorn, F. (2015). Prefronto-subcortical imbalance characterizes poor decision-making: Neurochemical and neural functional evidences in rats. *Brain Structure & Function, 220*(6), 3485–3496. https://doi.org/10.1007/s00429-014-0868-8

Fox, A. E., Nicholson, A. M., Singha, D., Thieret, B. A. S., Ortiz, M., & Visser, E. J. (2023). Timing and delay discounting in attention-deficit/hyperactivity disorder: A translational approach. *Developmental Psychobiology, 65*(5), e22399. https://doi.org/10.1002/dev.22399

Francis, A. N., Camprodon, J. A., & Filbey, F. (2023). Functional hyperconnectivity between corticocerebellar networks and altered decision making in young adult cannabis users: Evidence from 7T and multivariate pattern analysis. *Psychiatry Research. Neuroimaging, 331*, 111613. https://doi.org/10.1016/j.pscychresns.2023.111613

Franken, I. H., van Strien, J. W., Nijs, I., & Muris, P. (2008). Impulsivity is associated with behavioral decision-making deficits. *Psychiatry Research, 158*(2), 155–163. https://doi.org/10.1016/j.psychres.2007.06.002

Freels, T. G., Gabriel, D., Lester, D. B., & Simon, N. W. (2020). Risky decision-making predicts dopamine release dynamics in nucleus accumbens shell. *Neuropsychopharmacology, 45*(2), 266–275. https://doi.org/10.1038/s41386-019-0527-0

Freund, N., MacGillivilray, H. T., Thompson, B. S., Lukkes, J. L., Stanis, J. J., Brenhouse, H. C., & Andersen, S. L. (2014). Sex-dependent changes in ADHD-like behaviors in juvenile rats following cortical dopamine depletion. *Behavioural Brain Research, 270*, 357–363. https://doi.org/10.1016/j.bbr.2014.05.024

Fukunaga, R., Brown, J. W., & Bogg, T. (2012). Decision making in the balloon analogue risk task (BART): Anterior cingulate cortex signals loss aversion but not the infrequency of risky choices. *Cognitive, Affective, & Behavioral Neuroscience, 12*(3), 479–490. https://doi.org/10.3758/s13415-012-0102-1

Gaidin, S. G., Zinchenko, V. P., Sergeev, A. I., Teplov, I. Y., Mal'tseva, V. N., & Kosenkov, A. M. (2020). Activation of alpha-2 adrenergic receptors stimulates GABA release by astrocytes. *Glia, 68*(6), 1114–1130. https://doi.org/10.1002/glia.23763

Garon, N., Moore, C., & Waschbusch, D. A. (2006). Decision making in children with ADHD only, ADHD-anxious/depressed, and control children using a child version of the Iowa gambling task. *Journal of Attention Disorders, 9*(4), 607–619. https://doi.org/10.1177/1087054705284501

Gendle, M. H., & Golding, A. C. (2010). Oral administration of 5-hydroxytryptophan (5-HTP) impairs decision making under ambiguity but not under risk: Evidence from the Iowa gambling task. *Human Psychopharmacology, 25*(6), 491–499. https://doi.org/10.1002/hup.1139

Gescheidt, T., Mareček, R., Mikl, M., Czekóová, K., Urbánek, T., Vaníček, J., Shaw, D. J., & Bareš, M. (2013). Functional anatomy of outcome evaluation during Iowa gambling task performance in patients with Parkinson's disease: An fMRI study. *Neurological Sciences, 34*(12), 2159–2166. https://doi.org/10.1007/s10072-013-1439-0

Gianotti, L. R., Figner, B., Ebstein, R. P., & Knoch, D. (2012). Why some people discount more than others: Baseline activation in the dorsal pfc mediates the link between COMT genotype and impatient choice. *Frontiers in Neuroscience, 6*, 54. https://doi.org/10.3389/fnins.2012.00054

Giner, P., Maynez-Anchondo, L., Liley, A. E., Uribe, K. P., Frietze, G. A., Simon, N. W., & Mendez, I. A. (2022). Increased risky choice and reduced *CHRNB2* expression in adult male rats exposed to nicotine vapor. *International Journal of Molecular Sciences, 23*(3), 1231. https://doi.org/10.3390/ijms23031231

Godefroy, V., Sezer, I., Bouzigues, A., Montembeault, M., Koban, L., Plassmann, H., & Migliaccio, R. (2023). Altered delay discounting in neurodegeneration: Insight into the underlying mechanisms and perspectives for clinical applications. *Neuroscience and Biobehavioral Reviews, 146*, 105048. https://doi.org/10.1016/j.neubiorev.2023.105048

Goldstein, R. Z., & Volkow, N. D. (2011). Dysfunction of the prefrontal cortex in addiction: Neuroimaging findings and clinical implications. *Nature Reviews. Neuroscience, 12*(11), 652–669. https://doi.org/10.1038/nrn3119

Gonzalez, R., Schuster, R. M., Mermelstein, R. J., Vassileva, J., Martin, E. M., & Diviak, K. R. (2012). Performance of young adult cannabis users on neurocognitive measures of impulsive behavior and their relationship to symptoms of cannabis use disorders. *Journal of Clinical and Experimental Neuropsychology, 34*(9), 962–976. https://doi.org/10.1080/13803395.2012.703642

Gueye, A. B., Trigo, J. M., Vemuri, K. V., Makriyannis, A., & Le Foll, B. (2016). Effects of various cannabinoid ligands on choice behaviour in a rat model of gambling. *Behavioural Pharmacology, 27*(2–3 Spec Issue), 258–269. https://doi.org/10.1097/FBP.0000000000000222

Hamidullah, S., Lutelmowski, C. D., Creighton, S. D., Luciani, K. R., Frie, J. A., Winters, B. D., & Khokhar, J. Y. (2021). Effects of vapourized THC and voluntary alcohol drinking during adolescence on cognition, reward, and anxiety-like behaviours in rats. *Progress in Neuro-Psychopharmacology & Biological Psychiatry, 106*, 110141. https://doi.org/10.1016/j.pnpbp.2020.110141

Hanson, K. L., Thayer, R. E., & Tapert, S. F. (2014). Adolescent marijuana users have elevated risk-taking on the balloon analog risk task. *Journal of Psychopharmacology, 28*(11), 1080–1087. https://doi.org/10.1177/0269881114550352

Hartstra, E., Oldenburg, J. F., Van Leijenhorst, L., Rombouts, S. A., & Crone, E. A. (2010). Brain regions involved in the learning and application of reward rules in a two-deck gambling task. *Neuropsychologia, 48*(5), 1438–1446. https://doi.org/10.1016/j.neuropsychologia.2010.01.012

Hernandez, C. M., McQuail, J. A., Ten Eyck, T. W., Wheeler, A. R., Labiste, C. C., Setlow, B., & Bizon, J. L. (2022). $GABA_B$ receptors in prelimbic cortex and basolateral amygdala differentially influence intertemporal decision making and decline with age. *Neuropharmacology, 209*, 109001. https://doi.org/10.1016/j.neuropharm.2022.109001

Hester, R., Nestor, L., & Garavan, H. (2009). Impaired error awareness and anterior cingulate cortex hypoactivity in chronic cannabis users. *Neuropsychopharmacology, 34*(11), 2450–2458. https://doi.org/10.1038/npp.2009.67

Hewig, J., Straube, T., Trippe, R. H., Kretschmer, N., Hecht, H., Coles, M. G., & Miltner, W. H. (2009). Decision-making under risk: An fMRI study. *Journal of Cognitive Neuroscience, 21*(8), 1642–1652. https://doi.org/10.1162/jocn.2009.21112

Hikosaka, O. (2010). The habenula: From stress evasion to value-based decision making. *Nature Reviews. Neuroscience, 11*(7), 503–513. https://doi.org/10.1038/nrn2866

Hindocha, C., Freeman, T. P., Grabski, M., Crudgington, H., Davies, A. C., Stroud, J. B., Das, R. K., Lawn, W., Morgan, C. J. A., & Curran, H. V. (2018). The effects of cannabidiol on impulsivity and memory during abstinence in cigarette dependent smokers. *Scientific Reports, 8*(1), 7568. https://doi.org/10.1038/s41598-018-25846-2

Hirschbichler, S. T., Rothwell, J. C., & Manohar, S. G. (2022). Dopamine increases risky choice while D2 blockade shortens decision time. *Experimental Brain Research, 240*(12), 3351–3360. https://doi.org/10.1007/s00221-022-06501-9

Hoffman, W. F., Moore, M., Templin, R., McFarland, B., Hitzemann, R. J., & Mitchell, S. H. (2006). Neuropsychological function and delay discounting in methamphetamine-dependent individuals. *Psychopharmacology, 188*(2), 162–170. https://doi.org/10.1007/s00213-006-0494-0

Humphreys, K. L., & Lee, S. S. (2011). Risk taking and sensitivity to punishment in children with ADHD, ODD, ADHD+ODD, and controls. *Journal of Psychopathology and Behavioral Assessment, 33*(3), 299–307. https://doi.org/10.1007/s10862-011-9237-6

Huo, H., Lesage, E., Dong, W., Verguts, T., Seger, C. A., Diao, S., Feng, T., & Chen, Q. (2023). The neural substrates of how model-based learning affects risk taking: Functional coupling between right cerebellum and left caudate. *Brain and Cognition, 172*, 106088. https://doi.org/10.1016/j.bandc.2023.106088

Hynes, T. J., Ferland, J. M., Feng, T. L., Adams, W. K., Silveira, M. M., Tremblay, M., Chernoff, C. S., Brodie, H. G., Ebsary, S. A., Russell, B., Kaur, S., & Winstanley, C. A. (2020). Chemogenetic inhibition of dopaminergic projections to the nucleus accumbens has sexually dimorphic effects in the rat gambling task. *Behavioral Neuroscience, 134*(4), 309–322. https://doi.org/10.1037/bne0000372

Ikegami, M., & Sorama, M. (2023). Differential neural correlates in the prefrontal cortex during a delay discounting task in healthy adults: An fNIRS study. *Brain Sciences, 13*(5), 758. https://doi.org/10.3390/brainsci13050758

Jenni, N. L., Li, Y. T., & Floresco, S. B. (2021). Medial orbitofrontal cortex dopamine D_1/D_2 receptors differentially modulate distinct forms of probabilistic decision-making. *Neuropsychopharmacology, 46*(7), 1240–1251. https://doi.org/10.1038/s41386-020-00931-1

Johnson, K. R., Boomhower, S. R., & Newland, M. C. (2019). Behavioral effects of chronic WIN 55, 212-2 administration during adolescence and adulthood in mice. *Experimental and Clinical Psychopharmacology, 27*(4), 348–358. https://doi.org/10.1037/pha0000271

Jupp, B., Caprioli, D., Saigal, N., Reverte, I., Shrestha, S., Cumming, P., Everitt, B. J., Robbins, T. W., & Dalley, J. W. (2013). Dopaminergic and GABA-ergic markers of impulsivity in rats: Evidence for anatomical localisation in ventral striatum and prefrontal cortex. *European Journal of Neuroscience, 37*(9), 1519–1528. https://doi.org/10.1111/ejn.12146

Kayir, H., Semenova, S., & Markou, A. (2014). Baseline impulsive choice predicts the effects of nicotine and nicotine withdrawal on impulsivity in rats. *Progress in Neuro-Psychopharmacology & Biological Psychiatry, 48*, 6–13. https://doi.org/10.1016/j.pnpbp.2013.09.007

Kayser, A. S., Allen, D. C., Navarro-Cebrian, A., Mitchell, J. M., & Fields, H. L. (2012). Dopamine, corticostriatal connectivity, and intertemporal choice. *The Journal of Neuroscience, 32*(27), 9402–9409. https://doi.org/10.1523/JNEUROSCI.1180-12.2012

Kelly, M., Garner, M., Cooper, E. M., & Orsini, C. A. (2025). Cholinergic regulation of decision making under risk of punishment. *Neurobiology of Learning and Memory, 217*, 108018. https://doi.org/10.1016/j.nlm.2024.108018

Kim, B., & Im, H. I. (2019). The role of the dorsal striatum in choice impulsivity. *Annals of the New York Academy of Sciences, 1451*(1), 92–111. https://doi.org/10.1111/nyas.13961

Knapp, C. P., Papadopoulos, E., Loweth, J. A., Raghupathi, R., Floresco, S. B., Waterhouse, B. D., & Navarra, R. L. (2024). Perturbations in risk/reward decision making and frontal cortical catecholamine regulation induced by mild traumatic brain injury. *Behavioural Brain Research, 467*, 115002. https://doi.org/10.1016/j.bbr.2024.115002

Kobayashi, Y., Sano, Y., Vannoni, E., Goto, H., Suzuki, H., Oba, A., Kawasaki, H., Kanba, S., Lipp, H. P., Murphy, N. P., Wolfer, D. P., & Itohara, S. (2013). Genetic dissection of medial habenula-interpeduncular nucleus pathway function in mice. *Frontiers in Behavioral Neuroscience, 7*, 17. https://doi.org/10.3389/fnbeh.2013.00017

Koffarnus, M. N., Newman, A. H., Grundt, P., Rice, K. C., & Woods, J. H. (2011). Effects of selective dopaminergic compounds on a delay-discounting task. *Behavioural Pharmacology, 22*(4), 300–311. https://doi.org/10.1097/FBP.0b013e3283473bcb

Kolokotroni, K. Z., Rodgers, R. J., & Harrison, A. A. (2011). Acute nicotine increases both impulsive choice and behavioural disinhibition in rats. *Psychopharmacology, 217*(4), 455–473. https://doi.org/10.1007/s00213-011-2296-2

Koloski, M. F., Terry, A., Lee, N., & Ramanathan, D. S. (2024). Methylphenidate, but not citalopram, decreases impulsive choice in rats performing a temporal discounting task. *Frontiers in Psychiatry, 15*, 1385502. https://doi.org/10.3389/fpsyt.2024.1385502

Koot, S., Zoratto, F., Cassano, T., Colangeli, R., Laviola, G., van den Bos, R., & Adriani, W. (2012). Compromised decision-making and increased gambling proneness following dietary serotonin depletion in rats. *Neuropharmacology, 62*(4), 1640–1650. https://doi.org/10.1016/j.neuropharm.2011.11.002

Korte, S. M., Prins, J., Van den Bergh, F. S., Oosting, R. S., Dupree, R., Korte-Bouws, G. A., Westphal, K. G., Olivier, B., Denys, D. A., Garland, A., & Güntürkün, O. (2017). The 5-$HT_{1A/1B}$-receptor agonist eltoprazine increases both catecholamine release in the prefrontal cortex and dopamine release in the nucleus accumbens and decreases motivation for reward and "waiting" impulsivity, but increases "stopping" impulsivity. *European Journal of Pharmacology, 794*, 257–269. https://doi.org/10.1016/j.ejphar.2016.11.024

Kovács, I., Richman, M. J., Janka, Z., Maraz, A., & Andó, B. (2017). Decision making measured by the Iowa gambling task in alcohol use disorder and gambling disorder: A systematic review and meta-analysis. *Drug and Alcohol Dependence, 181*, 152–161. https://doi.org/10.1016/j.drugalcdep.2017.09.023

Kubota, H., Zhou, X., Zhang, X., Watanabe, H., & Nagai, T. (2024). Pramipexole hyperactivates the external globus pallidus and impairs decision-making in a mouse model of Parkinson's disease. *International Journal of Molecular Sciences, 25*(16), 8849. https://doi.org/10.3390/ijms25168849

Labudda, K., Frigge, K., Horstmann, S., Aengenendt, J., Woermann, F. G., Ebner, A., Markowitsch, H. J., & Brand, M. (2009). Decision making in patients with temporal lobe epilepsy. *Neuropsychologia, 47*(1), 50–58. https://doi.org/10.1016/j.neuropsychologia.2008.08.014

Lai, R. Y., Desai, N. A., Amlang, C. J., Lin, C. R., Chen, T. X., Minyetty, M. J., Amokrane, N., & Kuo, S. H. (2023). Gambling associated risk-taking decision in cerebellar ataxia. *Parkinsonism & Related Disorders, 107*, 105252. https://doi.org/10.1016/j.parkreldis.2022.105252

Lane, S. D., Cherek, D. R., Tcheremissine, O. V., Lieving, L. M., & Pietras, C. J. (2005). Acute marijuana effects on human risk taking. *Neuropsychopharmacology, 30*(4), 800–809. https://doi.org/10.1038/sj.npp.1300620

Lawrence, A. J., Luty, J., Bogdan, N. A., Sahakian, B. J., & Clark, L. (2009). Problem gamblers share deficits in impulsive decision-making with alcohol-dependent individuals. *Addiction, 104*(6), 1006–1015. https://doi.org/10.1111/j.1360-0443.2009.02533.x

Le Berre, A. P., Rauchs, G., La Joie, R., Mézenge, F., Boudehent, C., Vabret, F., Segobin, S., Viader, F., Allain, P., Eustache, F., Pitel, A. L., & Beaunieux, H. (2014). Impaired decision-making and brain shrinkage in alcoholism. *European Psychiatry, 29*(3), 125–133. https://doi.org/10.1016/j.eurpsy.2012.10.002

Lee, Y. A., Kim, Y. J., Lee, J. S., Lee, S., & Goto, Y. (2021). Imbalance between dopamine and serotonin caused by neonatal habenula lesion. *Behavioural Brain Research, 409*, 113316. https://doi.org/10.1016/j.bbr.2021.113316

Leffa, D. T., Ferreira, S. G., Machado, N. J., Souza, C. M., Rosa, F. D., de Carvalho, C., Kincheski, G. C., Takahashi, R. N., Porciúncula, L. O., Souza, D. O., Cunha, R. A., & Pandolfo, P. (2019). Caffeine and cannabinoid receptors modulate impulsive behavior in an animal model of attentional deficit and hyperactivity disorder. *The European Journal of Neuroscience, 49*(12), 1673–1683. https://doi.org/10.1111/ejn.14348

Linnet, J., Møller, A., Peterson, E., Gjedde, A., & Doudet, D. (2011a). Dopamine release in ventral striatum during Iowa gambling task performance is associated with increased excitement levels in pathological gambling. *Addiction, 106*(2), 383–390. https://doi.org/10.1111/j.1360-0443.2010.03126.x

Linnet, J., Møller, A., Peterson, E., Gjedde, A., & Doudet, D. (2011b). Inverse association between dopaminergic neurotransmission and Iowa gambling task performance in pathological gamblers and healthy controls. *Scandinavian Journal of Psychology, 52*(1), 28–34. https://doi.org/10.1111/j.1467-9450.2010.00837.x

Locey, M. L., & Dallery, J. (2009). Isolating behavioral mechanisms of intertemporal choice: Nicotine effects on delay discounting and amount sensitivity. *Journal of the Experimental Analysis of Behavior, 91*(2), 213–223. https://doi.org/10.1901/jeab.2009.91-213

Locey, M. L., & Dallery, J. (2011). Nicotine and the behavioral mechanisms of intertemporal choice. *Behavioural Processes, 87*(1), 18–24. https://doi.org/10.1016/j.beproc.2011.02.006

Loos, M., Pattij, T., Janssen, M. C., Counotte, D. S., Schoffelmeer, A. N., Smit, A. B., Spijker, S., & van Gaalen, M. M. (2010). Dopamine receptor D1/D5 gene expression in the medial prefrontal cortex predicts impulsive choice in rats. *Cerebral Cortex, 20*(5), 1064–1070. https://doi.org/10.1093/cercor/bhp167

Ludwig, V. U., Nüsser, C., Goschke, T., Wittfoth-Schardt, D., Wiers, C. E., Erk, S., Schott, B. H., & Walter, H. (2015). Delay discounting without decision-making: Medial prefrontal cortex and amygdala activations reflect immediacy processing and correlate with impulsivity and anxious-depressive traits. *Frontiers in Behavioral Neuroscience, 9*, 280. https://doi.org/10.3389/fnbeh.2015.00280

Mai, B., & Hauber, W. (2015). Orbitofrontal or accumbens dopamine depletion does not affect risk-based decision making in rats. *Cognitive, Affective, & Behavioral Neuroscience, 15*(3), 507–522. https://doi.org/10.3758/s13415-015-0353-8

Mallorquí-Bagué, N., Fagundo, A. B., Jimenez-Murcia, S., de la Torre, R., Baños, R. M., Botella, C., Casanueva, F. F., Crujeiras, A. B., Fernández-García, J. C., Fernández-Real, J. M., Frühbeck, G., Granero, R., Rodríguez, A., Tolosa-Sola, I., Ortega, F. J., Tinahones, F. J., Alvarez-Moya, E., Ochoa, C., Menchón, J. M., & Fernández-Aranda, F. (2016). Decision making impairment: A shared vulnerability in obesity, gambling disorder and substance use disorders? *PLoS One, 11*(9), e0163901. https://doi.org/10.1371/journal.pone.0163901

Mar, A. C., Walker, A. L., Theobald, D. E., Eagle, D. M., & Robbins, T. W. (2011). Dissociable effects of lesions to orbitofrontal cortex subregions on impulsive choice in the rat. *The Journal of Neuroscience, 31*(17), 6398–6404. https://doi.org/10.1523/JNEUROSCI.6620-10.2011

Mariano, T. Y., Bannerman, D. M., McHugh, S. B., Preston, T. J., Rudebeck, P. H., Rudebeck, S. R., Rawlins, J. N., Walton, M. E., Rushworth, M. F., Baxter, M. G., & Campbell, T. G. (2009). Impulsive choice in hippocampal but not orbitofrontal cortex-lesioned rats on a nonspatial decision-making maze task. *The European Journal of Neuroscience, 30*(3), 472–484. https://doi.org/10.1111/j.1460-9568.2009.06837.x

Masuda, A., Sano, C., Zhang, Q., Goto, H., McHugh, T. J., Fujisawa, S., & Itohara, S. (2020). The hippocampus encodes delay and value information during delay-discounting decision making. *eLife, 9*, e52466. https://doi.org/10.7554/eLife.52466

Mattavelli, G., Lo Presti, S., Tornaghi, D., & Canessa, N. (2022). High-definition transcranial direct current stimulation of the dorsal anterior cingulate cortex modulates decision-making and executive control. *Brain Structure & Function, 227*(5), 1565–1576. https://doi.org/10.1007/s00429-022-02456-3

Matthews, S. C., Simmons, A. N., Lane, S. D., & Paulus, M. P. (2004). Selective activation of the nucleus accumbens during risk-taking decision making. *Neuroreport, 15*(13), 2123–2127. https://doi.org/10.1097/00001756-200409150-00025

McDonald, J., Schleifer, L., Richards, J. B., & de Wit, H. (2003). Effects of THC on behavioral measures of impulsivity in humans. *Neuropsychopharmacology, 28*(7), 1356–1365. https://doi.org/10.1038/sj.npp.1300176

Mendez, I. A., Gilbert, R. J., Bizon, J. L., & Setlow, B. (2012). Effects of acute administration of nicotinic and muscarinic cholinergic agonists and antagonists on performance in different cost-benefit decision making tasks in rats. *Psychopharmacology, 224*(4), 489–499. https://doi.org/10.1007/s00213-012-2777-y

Mendez, I. A., Simon, N. W., Hart, N., Mitchell, M. R., Nation, J. R., Wellman, P. J., & Setlow, B. (2010). Self-administered cocaine causes long-lasting increases in impulsive choice in a delay discounting task. *Behavioral Neuroscience, 124*(4), 470–477. https://doi.org/10.1037/a0020458

Mies, G. W., de Water, E., Wiersema, J. R., & Scheres, A. (2019). Delay discounting of monetary gains and losses in adolescents with ADHD: Contribution of delay aversion to choice. *Child Neuropsychology, 25*(4), 528–547. https://doi.org/10.1080/09297049.2018.1508563

Mitchell, M. R., Vokes, C. M., Blankenship, A. L., Simon, N. W., & Setlow, B. (2011). Effects of acute administration of nicotine, amphetamine, diazepam, morphine, and ethanol on risky decision-making in rats. *Psychopharmacology, 218*(4), 703–712. https://doi.org/10.1007/s00213-011-2363-8

Mobini, S., Body, S., Ho, M. Y., Bradshaw, C. M., Szabadi, E., Deakin, J. F., & Anderson, I. M. (2002). Effects of lesions of the orbitofrontal cortex on sensitivity to delayed and probabilistic reinforcement. *Psychopharmacology, 160*(3), 290–298. https://doi.org/10.1007/s00213-001-0983-0

Mobini, S., Chiang, T. J., Ho, M. Y., Bradshaw, C. M., & Szabadi, E. (2000). Effects of central 5-hydroxytryptamine depletion on sensitivity to delayed and probabilistic reinforcement. *Psychopharmacology, 152*(4), 390–397. https://doi.org/10.1007/s002130000542

Mok, J. N. Y., Green, L., Myerson, J., Kwan, D., Kurczek, J., Ciaramelli, E., Craver, C. F., & Rosenbaum, S. R. (2021). Does ventromedial prefrontal cortex damage really increase impulsiveness? Delay and probability discounting in patients with focal lesions. *Journal of Cognitive Neuroscience, 33*(9), 1–19. Advance online publication. https://doi.org/10.1162/jocn_a_01721

Moreira, P. S., Macoveanu, J., Marques, P., Coelho, A., Magalhães, R., Siebner, H. R., Soares, J. M., Sousa, N., & Morgado, P. (2020). Altered response to risky decisions and reward in patients with obsessive–compulsive disorder. *Journal of Psychiatry & Neuroscience, 45*(2), 98–107. https://doi.org/10.1503/jpn.180226

Moreno, M., Azocar, V., Vergés, A., & Fuentealba, J. A. (2021). High impulsive choice is accompanied by an increase in dopamine release in rat dorsolateral striatum. *Behavioural Brain Research, 405*, 113199. https://doi.org/10.1016/j.bbr.2021.113199

Münster, A., Huster, J., Sommer, S., Traxler, C., Votteler, A., & Hauber, W. (2024). Enhanced risky choice in male rats elicited by the acute pharmacological stressor yohimbine involves prefrontal dopamine D1 receptor activation. *The International Journal of Neuropsychopharmacology, 27*(2), pyae006. https://doi.org/10.1093/ijnp/pyae006

Murphy, C. M., Christakou, A., Giampietro, V., Brammer, M., Daly, E. M., Ecker, C., Johnston, P., Spain, D., Robertson, D. M., MRC AIMS Consortium, Murphy, D. G., & Rubia, K. (2017). Abnormal functional activation and maturation of ventromedial prefrontal cortex and cerebellum during temporal discounting in autism spectrum disorder. *Human Brain Mapping, 38*(11), 5343–5355. https://doi.org/10.1002/hbm.23718

Nejati, V., Dehghan, M., Shahidi, S., Estaji, R., & Nitsche, M. A. (2024). Transcranial random noise stimulation (tRNS) improves hot and cold executive functions in children with attention deficit-hyperactivity disorder (ADHD). *Scientific Reports, 14*(1), 7600. https://doi.org/10.1038/s41598-024-57920-3

Nolder, K. A., & Anderson, K. G. (2023). Effects of acute and chronic nicotine administration on probability discounting. *Behavioural Pharmacology, 34*(8), 468–476. https://doi.org/10.1097/FBP.0000000000000753

Oberlin, B. G., Albrecht, D. S., Herring, C. M., Walters, J. W., Hile, K. L., Kareken, D. A., & Yoder, K. K. (2015). Monetary discounting and ventral striatal dopamine receptor availability in nontreatment-seeking alcoholics and social drinkers. *Psychopharmacology, 232*(12), 2207–2216. https://doi.org/10.1007/s00213-014-3850-5

O'Carroll, R. E., & Papps, B. P. (2003). Decision making in humans: The effect of manipulating the central noradrenergic system. *Journal of Neurology, Neurosurgery, and Psychiatry, 74*(3), 376–378. https://doi.org/10.1136/jnnp.74.3.376

Ohmura, Y., Takahashi, T., & Kitamura, N. (2005). Discounting delayed and probabilistic monetary gains and losses by smokers of cigarettes. *Psychopharmacology, 182*(4), 508–515. https://doi.org/10.1007/s00213-005-0110-8

Op de Macks, Z. A., Flannery, J. E., Peake, S. J., Flournoy, J. C., Mobasser, A., Alberti, S. L., Fisher, P. A., & Pfeifer, J. H. (2018). Novel insights from the yellow light game: Safe and risky decisions differentially impact adolescent outcome-related brain function. *NeuroImage, 181*, 568–581. https://doi.org/10.1016/j.neuroimage.2018.06.058

Orsini, C. A., Hernandez, C. M., Singhal, S., Kelly, K. B., Frazier, C. J., Bizon, J. L., & Setlow, B. (2017). Optogenetic inhibition reveals distinct roles for basolateral amygdala activity at discrete time points during risky decision making. *The Journal of Neuroscience, 37*(48), 11537–11548. https://doi.org/10.1523/JNEUROSCI.2344-17.2017

Orsini, C. A., Trotta, R. T., Bizon, J. L., & Setlow, B. (2015). Dissociable roles for the basolateral amygdala and orbitofrontal cortex in decision-making under risk of punishment. *The Journal of Neuroscience, 35*(4), 1368–1379. https://doi.org/10.1523/JNEUROSCI.3586-14.2015

Ortiz, N., Parsons, A., Whelan, R., Brennan, K., Agan, M. L., O'Connell, R., Bramham, J., & Garavan, H. (2015). Decreased frontal, striatal and cerebellar activation in adults with ADHD during an adaptive delay discounting task. *Acta Neurobiologiae Experimentalis, 75*(3), 326–338. https://doi.org/10.55782/ane-2015-2038

Ortiz-Peregrina, S., Ortiz, C., & Anera, R. G. (2021). Aggressive driving behaviours in cannabis users. The influence of consumer characteristics. *International Journal of Environmental Research and Public Health, 18*(8), 3911. https://doi.org/10.3390/ijerph18083911

Ozga, J. E., & Anderson, K. G. (2018). Reduction in delay discounting due to nicotine and its attenuation by cholinergic antagonists in Lewis and Fischer 344 rats. *Psychopharmacology, 235*(1), 155–168. https://doi.org/10.1007/s00213-017-4752-0

Pardey, M. C., Kumar, N. N., Goodchild, A. K., & Cornish, J. L. (2013). Catecholamine receptors differentially mediate impulsive choice in the medial prefrontal and orbitofrontal cortex. *Journal of Psychopharmacology, 27*(2), 203–212. https://doi.org/10.1177/0269881112465497

Paterson, N. E., Wetzler, C., Hackett, A., & Hanania, T. (2012). Impulsive action and impulsive choice are mediated by distinct neuropharmacological substrates in rat. *The International Journal of Neuropsychopharmacology, 15*(10), 1473–1487. https://doi.org/10.1017/S14611457 11001635

Patton, J. H., Stanford, M. S., & Barratt, E. S. (1995). Factor structure of the Barratt impulsiveness scale. *Journal of Clinical Psychology, 51*(6), 768–774. https://doi.org/10.1002/1097-4679(199 511)51:6<768::aid-jclp2270510607>3.0.co;2-1

Paz-Alonso, P. M., Navalpotro-Gomez, I., Boddy, P., Dacosta-Aguayo, R., Delgado-Alvarado, M., Quiroga-Varela, A., Jimenez-Urbieta, H., Carreiras, M., & Rodriguez-Oroz, M. C. (2020). Functional inhibitory control dynamics in impulse control disorders in Parkinson's disease. *Movement Disorders, 35*(2), 316–325. https://doi.org/10.1002/mds.27885

Pennisi, P., Salehinejad, M. A., Corso, A. M., Merlo, E. M., Avenanti, A., & Vicario, C. M. (2023). Delay discounting in Parkinson's disease: A systematic review and meta-analysis. *Behavioural Brain Research, 436*, 114101. https://doi.org/10.1016/j.bbr.2022.114101

Peters, J., Vega, T., Weinstein, D., Mitchell, J., & Kayser, A. (2020). Dopamine and risky decision-making in gambling disorder. *eNeuro, 7*(3), ENEURO.0461-19.2020. https://doi.org/10.1523/ENEURO.0461-19.2020

Pettorruso, M., Martinotti, G., Cocciolillo, F., De Risio, L., Cinquino, A., Di Nicola, M., Camardese, G., Migliara, G., Moccia, L., Conte, E., Janiri, L., & Di Giuda, D. (2019). Striatal presynaptic dopaminergic dysfunction in gambling disorder: A ^{123}I-FP-CIT SPECT study. *Addiction Biology, 24*(5), 1077–1086. https://doi.org/10.1111/adb.12677

Quan, P., He, L., Mao, T., Fang, Z., Deng, Y., Pan, Y., Zhang, X., Zhao, K., Lei, H., Detre, J. A., Kable, J. W., & Rao, H. (2022). Cerebellum anatomy predicts individual risk-taking behavior and risk tolerance. *NeuroImage, 254*, 119148. https://doi.org/10.1016/j.neuroimage.2022.119148

Ramaekers, J. G., Kauert, G., van Ruitenbeek, P., Theunissen, E. L., Schneider, E., & Moeller, M. R. (2006). High-potency marijuana impairs executive function and inhibitory motor control. *Neuropsychopharmacology, 31*(10), 2296–2303. https://doi.org/10.1038/sj.npp.1301068

Rao, H., Korczykowski, M., Pluta, J., Hoang, A., & Detre, J. A. (2008). Neural correlates of voluntary and involuntary risk taking in the human brain: An fMRI study of the balloon analog risk task (BART). *NeuroImage, 42*(2), 902–910. https://doi.org/10.1016/j.neuroimage.2008.05.046

Rodrigo, M. J., Padrón, I., de Vega, M., & Ferstl, E. C. (2014). Adolescents' risky decision-making activates neural networks related to social cognition and cognitive control processes. *Frontiers in Human Neuroscience, 8*, 60. https://doi.org/10.3389/fnhum.2014.00060

Romero-Torres, B. M., Alvarado-Ramírez, Y. A., Duran-Alonzo, S. R., Ruiz-Contreras, A. E., Herrera-Solis, A., Amancio-Belmont, O., Prospéro-García, O. E., & Méndez-Díaz, M. (2023). A potential role of hippocampus on impulsivity and alcohol consumption through CB1R.

Pharmacology, Biochemistry, and Behavior, 225, 173558. https://doi.org/10.1016/j.pbb.2023.173558

Roshani, F., Piri, R., Malek, A., Michel, T. M., & Vafaee, M. S. (2020). Comparison of cognitive flexibility, appropriate risk-taking and reaction time in individuals with and without adult ADHD. *Psychiatry Research, 284*, 112494. https://doi.org/10.1016/j.psychres.2019.112494

Salcido, C. A., Argenbright, C. M., Aguirre, T., Trujillo, A. D., & Fuchs, P. N. (2022). The use of an FR1 schedule operant approach-avoidance paradigm to measure the aversiveness of neuropathic and inflammatory pain. *Frontiers in Pain Research, 2*, 793958. https://doi.org/10.3389/fpain.2021.793958

Schippers, M. C., Schetters, D., De Vries, T. J., & Pattij, T. (2016). Differential effects of the pharmacological stressor yohimbine on impulsive decision making and response inhibition. *Psychopharmacology, 233*(14), 2775–2785. https://doi.org/10.1007/s00213-016-4337-3

Schmaal, L., Goudriaan, A. E., van der Meer, J., van den Brink, W., & Veltman, D. J. (2012). The association between cingulate cortex glutamate concentration and delay discounting is mediated by resting state functional connectivity. *Brain and Behavior, 2*(5), 553–562. https://doi.org/10.1002/brb3.74

Schweighofer, N., Bertin, M., Shishida, K., Okamoto, Y., Tanaka, S. C., Yamawaki, S., & Doya, K. (2008). Low-serotonin levels increase delayed reward discounting in humans. *The Journal of Neuroscience, 28*(17), 4528–4532. https://doi.org/10.1523/JNEUROSCI.4982-07.2008

Seib, D. R., Espinueva, D. F., Princz-Lebel, O., Chahley, E., Stevenson, J., O'Leary, T. P., Floresco, S. B., & Snyder, J. S. (2021). Hippocampal neurogenesis promotes preference for future rewards. *Molecular Psychiatry, 26*(11), 6317–6335. https://doi.org/10.1038/s41380-021-01165-3

Sevy, S., Hassoun, Y., Bechara, A., Yechiam, E., Napolitano, B., Burdick, K., Delman, H., & Malhotra, A. (2006). Emotion-based decision-making in healthy subjects: Short-term effects of reducing dopamine levels. *Psychopharmacology, 188*(2), 228–235. https://doi.org/10.1007/s00213-006-0450-z

Shaver, T. K., Ozga, J. E., Zhu, B., Anderson, K. G., Martens, K. M., & Vonder Haar, C. (2019). Long-term deficits in risky decision-making after traumatic brain injury on a rat analog of the Iowa gambling task. *Brain Research, 1704*, 103–113. https://doi.org/10.1016/j.brainres.2018.10.004

Shen, H., Ma, Z., Hans, E., Duan, Y., Bi, G. H., Chae, Y. C., Bonifazi, A., Battiti, F. O., Newman, A. H., Xi, Z. X., & Yang, Y. (2024). Involvement of dopamine D3 receptor in impulsive choice decision-making in male rats. *Neuropharmacology, 257*, 110051. https://doi.org/10.1016/j.neuropharm.2024.110051

Silveira, M. M., Malcolm, E., Shoaib, M., & Winstanley, C. A. (2015). Scopolamine and amphetamine produce similar decision-making deficits on a rat gambling task via independent pathways. *Behavioural Brain Research, 281*, 86–95. https://doi.org/10.1016/j.bbr.2014.12.029

Simon, N. W., Beas, B. S., Montgomery, K. S., Haberman, R. P., Bizon, J. L., & Setlow, B. (2013). Prefrontal cortical-striatal dopamine receptor mRNA expression predicts distinct forms of impulsivity. *The European Journal of Neuroscience, 37*(11), 1779–1788. https://doi.org/10.1111/ejn.12191

Simon, N. W., Gilbert, R. J., Mayse, J. D., Bizon, J. L., & Setlow, B. (2009). Balancing risk and reward: A rat model of risky decision making. *Neuropsychopharmacology, 34*(10), 2208–2217. https://doi.org/10.1038/npp.2009.48

Smith, B. W., Mitchell, D. G., Hardin, M. G., Jazbec, S., Fridberg, D., Blair, R. J., & Ernst, M. (2009). Neural substrates of reward magnitude, probability, and risk during a wheel of fortune decision-making task. *NeuroImage, 44*(2), 600–609. https://doi.org/10.1016/j.neuroimage.2008.08.016

Smith, C. T., Wallace, D. L., Dang, L. C., Aarts, E., Jagust, W. J., D'Esposito, M., & Boettiger, C. A. (2016). Modulation of impulsivity and reward sensitivity in intertemporal choice by striatal and midbrain dopamine synthesis in healthy adults. *Journal of Neurophysiology, 115*(3), 1146–1156. https://doi.org/10.1152/jn.00261.2015

Sonntag, K. C., Brenhouse, H. C., Freund, N., Thompson, B. S., Puhl, M., & Andersen, S. L. (2014). Viral over-expression of D1 dopamine receptors in the prefrontal cortex increase high-risk behaviors in adults: Comparison with adolescents. *Psychopharmacology, 231*(8), 1615–1626. https://doi.org/10.1007/s00213-013-3399-8

Sonuga-Barke, E. J., Kumsta, R., Schlotz, W., Lasky-Su, J., Marco, R., Miranda, A., Mulas, F., Oades, R. D., Banaschewski, T., Mueller, U., Andreou, P., Christiansen, H., Gabriels, I., Uebel, H., Kuntsi, J., Franke, B., Buitelaar, J., Ebstein, R., Gill, M., Anney, R., et al. (2011). A functional variant of the serotonin transporter gene (SLC6A4) moderates impulsive choice in attention-deficit/hyperactivity disorder boys and siblings. *Biological Psychiatry, 70*(3), 230–236. https://doi.org/10.1016/j.biopsych.2011.01.040

St Onge, J. R., Abhari, H., & Floresco, S. B. (2011). Dissociable contributions by prefrontal D1 and D2 receptors to risk-based decision making. *The Journal of Neuroscience, 31*(23), 8625–8633. https://doi.org/10.1523/JNEUROSCI.1020-11.2011

St Onge, J. R., Ahn, S., Phillips, A. G., & Floresco, S. B. (2012). Dynamic fluctuations in dopamine efflux in the prefrontal cortex and nucleus accumbens during risk-based decision making. *The Journal of Neuroscience, 32*(47), 16880–16891. https://doi.org/10.1523/JNEUROSCI.3807-12.2012

St Onge, J. R., & Floresco, S. B. (2009). Dopaminergic modulation of risk-based decision making. *Neuropsychopharmacology, 34*(3), 681–697. https://doi.org/10.1038/npp.2008.121

Stanger, C., Elton, A., Ryan, S. R., James, G. A., Budney, A. J., & Kilts, C. D. (2013). Neuroeconomics and adolescent substance abuse: Individual differences in neural networks and delay discounting. *Journal of the American Academy of Child and Adolescent Psychiatry, 52*(7), 747–755.e6. https://doi.org/10.1016/j.jaac.2013.04.013

Sun, H., Cocker, P. J., Zeeb, F. D., & Winstanley, C. A. (2012). Chronic atomoxetine treatment during adolescence decreases impulsive choice, but not impulsive action, in adult rats and alters markers of synaptic plasticity in the orbitofrontal cortex. *Psychopharmacology, 219*(2), 285–301. https://doi.org/10.1007/s00213-011-2419-9

Tedford, S. E., Persons, A. L., & Napier, T. C. (2015). Dopaminergic lesions of the dorsolateral striatum in rats increase delay discounting in an impulsive choice task. *PLoS One, 10*(4), e0122063. https://doi.org/10.1371/journal.pone.0122063

Tremblay, M., Adams, W. K., & Winstanley, C. A. (2021). Kindling of the basolateral or central nucleus of the amygdala increases suboptimal choice in a rat gambling task and increases motor impulsivity in risk-preferring animals. *Behavioural Brain Research, 398*, 112941. https://doi.org/10.1016/j.bbr.2020.112941

Vadhan, N. P., Hart, C. L., van Gorp, W. G., Gunderson, E. W., Haney, M., & Foltin, R. W. (2007). Acute effects of smoked marijuana on decision making, as assessed by a modified gambling task, in experienced marijuana users. *Journal of Clinical and Experimental Neuropsychology, 29*(4), 357–364. https://doi.org/10.1080/13803390600693615

Vaidya, J. G., Block, R. I., O'Leary, D. S., Ponto, L. B., Ghoneim, M. M., & Bechara, A. (2012). Effects of chronic marijuana use on brain activity during monetary decision-making. *Neuropsychopharmacology, 37*(3), 618–629. https://doi.org/10.1038/npp.2011.227

Van Dessel, J., Sonuga-Barke, E., Mies, G., Lemiere, J., Van der Oord, S., Morsink, S., & Danckaerts, M. (2018). Delay aversion in attention deficit/hyperactivity disorder is mediated by amygdala and prefrontal cortex hyper-activation. *Journal of Child Psychology and Psychiatry, 59*(8), 888–899. https://doi.org/10.1111/jcpp.12868

van Gaalen, M. M., van Koten, R., Schoffelmeer, A. N., & Vanderschuren, L. J. (2006). Critical involvement of dopaminergic neurotransmission in impulsive decision making. *Biological Psychiatry, 60*(1), 66–73. https://doi.org/10.1016/j.biopsych.2005.06.005

van Holstein, M., Mac Leod, P. E., & Floresco, S. B. (2020). Basolateral amygdala—Nucleus accumbens circuitry regulates optimal cue-guided risk/reward decision making. *Progress in Neuro-Psychopharmacology & Biological Psychiatry, 98*, 109830. https://doi.org/10.1016/j.pnpbp.2019.109830

Vorobyev, V., Kwon, M. S., Moe, D., Parkkola, R., & Hämäläinen, H. (2015). Risk-taking behavior in a computerized driving task: Brain activation correlates of decision-making, outcome, and peer influence in male adolescents. *PLoS One, 10*(6), e0129516. https://doi.org/10.1371/journal.pone.0129516

Wang, M., Chen, Z., Zhang, S., Xu, T., Zhang, R., Suo, T., & Feng, T. (2019). High self-control reduces risk preference: The role of connectivity between right orbitofrontal cortex and right anterior cingulate cortex. *Frontiers in Neuroscience, 13*, 194. https://doi.org/10.3389/fnins.2019.00194

Wang, M., Zhang, S., Suo, T., Mao, T., Wang, F., Deng, Y., Eickhoff, S., Pan, Y., Jiang, C., & Rao, H. (2022). Risk-taking in the human brain: An activation likelihood estimation meta-analysis of the balloon analog risk task (BART). *Human Brain Mapping, 43*(18), 5643–5657. https://doi.org/10.1002/hbm.26041

Wang, Z., Yue, L., Cui, C., Liu, S., Wang, X., Li, Y., & Ma, L. (2019). Top-down control of the medial orbitofrontal cortex to nucleus accumbens core pathway in decisional impulsivity. *Brain Structure & Function, 224*(7), 2437–2452. https://doi.org/10.1007/s00429-019-01913-w

Weller, R. E., Avsar, K. B., Cox, J. E., Reid, M. A., White, D. M., & Lahti, A. C. (2014). Delay discounting and task performance consistency in patients with schizophrenia. *Psychiatry Research, 215*(2), 286–293. https://doi.org/10.1016/j.psychres.2013.11.013

Winstanley, C. A., Theobald, D. E., Cardinal, R. N., & Robbins, T. W. (2004). Contrasting roles of basolateral amygdala and orbitofrontal cortex in impulsive choice. *The Journal of Neuroscience, 24*(20), 4718–4722. https://doi.org/10.1523/JNEUROSCI.5606-03.2004

Winstanley, C. A., Theobald, D. E., Dalley, J. W., & Robbins, T. W. (2005). Interactions between serotonin and dopamine in the control of impulsive choice in rats: Therapeutic implications for impulse control disorders. *Neuropsychopharmacology, 30*(4), 669–682. https://doi.org/10.1038/sj.npp.1300610

Wiskerke, J., Stoop, N., Schetters, D., Schoffelmeer, A. N., & Pattij, T. (2011). Cannabinoid CB1 receptor activation mediates the opposing effects of amphetamine on impulsive action and impulsive choice. *PLoS One, 6*(10), e25856. https://doi.org/10.1371/journal.pone.0025856

Wooters, T. E., & Bardo, M. T. (2011). Methylphenidate and fluphenazine, but not amphetamine, differentially affect impulsive choice in spontaneously hypertensive, Wistar-Kyoto and Sprague-Dawley rats. *Brain Research, 1396*, 45–53. https://doi.org/10.1016/j.brainres.2011.04.040

Wynn, S. C., Driessen, J. M. A., Glennon, J. C., Brazil, I. A., & Schutter, D. J. L. G. (2019). Cerebellar transcranial direct current stimulation improves reactive response inhibition in healthy volunteers. *Cerebellum, 18*(6), 983–988. https://doi.org/10.1007/s12311-019-01047-z

Yates, J. R. (2024). Aberrant glutamatergic systems underlying impulsive behaviors: Insights from clinical and preclinical research. *Progress in Neuro-Psychopharmacology & Biological Psychiatry, 135*, 111107. https://doi.org/10.1016/j.pnpbp.2024.111107

Yates, J. R., Perry, J. L., Meyer, A. C., Gipson, C. D., Charnigo, R., & Bardo, M. T. (2014). Role of medial prefrontal and orbitofrontal monoamine transporters and receptors in performance in an adjusting delay discounting procedure. *Brain Research, 1574*, 26–36. https://doi.org/10.1016/j.brainres.2014.06.004

Young, J. W., van Enkhuizen, J., Winstanley, C. A., & Geyer, M. A. (2011). Increased risk-taking behavior in dopamine transporter knockdown mice: Further support for a mouse model of mania. *Journal of Psychopharmacology, 25*(7), 934–943. https://doi.org/10.1177/0269881111400646

Yu, R. (2012). Regional white matter volumes correlate with delay discounting. *PLoS One, 7*(2), e32595. https://doi.org/10.1371/journal.pone.0032595

Zandonai, T., Chiamulera, C., Mancabelli, A., Falconieri, D., & Diana, M. (2018). A preliminary investigation on smokeless tobacco use and its cognitive effects among athletes. *Frontiers in Pharmacology, 9*, 216. https://doi.org/10.3389/fphar.2018.00216

Zeeb, F. D., Baarendse, P. J., Vanderschuren, L. J., & Winstanley, C. A. (2015). Inactivation of the prelimbic or infralimbic cortex impairs decision-making in the rat gambling task. *Psychopharmacology, 232*(24), 4481–4491. https://doi.org/10.1007/s00213-015-4075-y

Zeeb, F. D., Robbins, T. W., & Winstanley, C. A. (2009). Serotonergic and dopaminergic modulation of gambling behavior as assessed using a novel rat gambling task. *Neuropsychopharmacology, 34*(10), 2329–2343. https://doi.org/10.1038/npp.2009.62

Zeeb, F. D., & Winstanley, C. A. (2011). Lesions of the basolateral amygdala and orbitofrontal cortex differentially affect acquisition and performance of a rodent gambling task. *The Journal of Neuroscience, 31*(6), 2197–2204. https://doi.org/10.1523/JNEUROSCI.5597-10.2011

Zeeb, F. D., & Winstanley, C. A. (2013). Functional disconnection of the orbitofrontal cortex and basolateral amygdala impairs acquisition of a rat gambling task and disrupts animals' ability to alter decision-making behavior after reinforcer devaluation. *The Journal of Neuroscience, 33*(15), 6434–6443. https://doi.org/10.1523/JNEUROSCI.3971-12.2013

Zhang, X. L., Wang, G. B., Zhao, L. Y., Sun, L. L., Wang, J., Wu, P., Lu, L., & Shi, J. (2012). Clonidine improved laboratory-measured decision-making performance in abstinent heroin addicts. *PLoS One, 7*(1), e29084. https://doi.org/10.1371/journal.pone.0029084

Zoratto, F., Tringle, A. L., Bellenchi, G., Speranza, L., Travaglini, D., di Porzio, U., Perrone-Capano, C., Laviola, G., Dreyer, J. L., & Adriani, W. (2013). Impulsivity and home-cage activity are decreased by lentivirus-mediated silencing of serotonin transporter in the rat hippocampus. *Neuroscience Letters, 548*, 38–43. https://doi.org/10.1016/j.neulet.2013.05.076

Zorick, T., Okita, K., Renard, K. B., Mandelkern, M. A., Brody, A. L., & London, E. D. (2022). The effects of citalopram and thalamic dopamine $D_{2/3}$ receptor availability on decision-making and loss aversion in alcohol dependence. *Psychiatry Journal, 2022*, 5663274. https://doi.org/10.1155/2022/5663274

Chapter 8
Maladaptive Stress

1 Introduction

With a pounding chest, I could feel sweat pooling under my shirt. I felt nauseous as I got out of my seat and headed to the front of the classroom. My hands shook as I loaded my PowerPoint presentation on the computer. I waited for my 12th grade AP English teacher to signal the start of my presentation, which seemed like an eternity. My presentation was on tobacco use and had to be 15 min long; at the time, this was a Herculean task as I thought there was no way I could speak on a topic for 15 min. Once I was instructed to begin, I unleashed a whirlwind of words, neglecting to pause to take a breath. Luckily, I was able to finish my presentation without fainting. However, my 15-min talk lasted 12 min, and I earned a C- on the presentation.

The experience described above is not an isolated one. Most, if not all, of us experience **anxiety** at some point in our lives. I have mentioned anxiety throughout this book, but I have not provided a formal definition for this term. Anxiety is a feeling of worry, nervousness, or unease, typically about an event or something with an uncertain outcome. As a college professor, I see anxiety in my students daily. I have students that clearly understand the material but perform poorly on quizzes/exams. The primary culprit is testing anxiety.

While anxiety is a common occurrence and is transient, anxiety can become maladaptive, disrupting one's day-to-day activities. I often teach our program's research methods class, in which students conduct their own experiment and present the results to the class. During one semester, a student told me they were going to take a 0% on the presentation because they did not want to present in front of others. As the presentation was worth a considerable number of points, refusing to give a presentation had a detrimental effect on this student's grade; luckily, they still passed the course.

Directly related to anxiety is **stress**, a state of mental or emotional strain or tension resulting from adverse or demanding circumstances. Like anxiety, we all face some form of stress (called a *stressor*). As you will learn in this chapter, stress has a

J. Yates, *Neuroscience of Maladaptation*, https://doi.org/10.1007/978-3-032-15602-0_8

useful function. Unfortunately, too much stress can be problematic, causing both physical and psychological harm to an individual. This chapter will begin with a discussion of stress, followed by a discussion of anxiety disorders. You will learn about the mechanisms that lead to dysregulated stress responses, which are implicated in anxiety disorders. In fact, maladaptive stress is associated with multiple conditions. This will not be the only chapter in which you learn how stress contributes to other psychiatric conditions.

2 Learning Objectives

By the end of this chapter, you should be able to …

1. Discuss how stress is an adaptive process and how dysregulation of the stress system causes maladaptive behaviors.
2. Differentiate the various anxiety disorders.
3. Explain how animals can be used to model stress and anxiety.
4. Detail the neuromechanisms underlying anxiety and stress-related disorders.

3 Types of Stress

Dr. Hans Selye (1907–1982) was an endocrinologist responsible for pioneering research on physiological responses to stress. He proposed that there are two major types of stress we experience: *distress* and *eustress*. Distress is what many of us think when we hear the term stress. That is, distress is a negative experience, such as losing a loved one. Conversely, eustress is a positive form of stress that ultimately benefits the individual. Examples of eustress I have personally experienced are my wedding and the birth of my children, especially my first child as I had no idea what to expect as a new father. Some have argued that using the terms distress and eustress is unnecessary and that the term stress should be used to avoid ambiguity (Bienertova-Vasku et al., 2020).

Regardless of the terminology that is used, not all stress is inherently bad. Dr. Robert Yerkes (1876–1956) and Dr. John Dodson (1879–1955) attempted to describe the relationship between physiological arousal/stress and performance on cognitive-behavioral tasks (Yerkes & Dodson, 1908). If someone performs an easy task like focusing attention on a single stimulus, higher levels of arousal/stress are associated with *better* performance. However, if one performs a more complex task like dividing attention across two or more stimuli, performance is optimal when stress levels are moderate. Increased arousal/stress can *impair* performance in complex tasks.

Here is an example of how the Yerkes-Dodson law can be applied to real-life behaviors. If I play a simple video game like *Mario Kart*, having higher arousal/stress will improve my performance as I am more actively engaged in the game. Lower arousal levels can impair my performance as I may be less likely to pay attention to the game as I play; perhaps I am more focused on talking to my wife or my sons. If I play a more difficult game (the original *Contra* comes to mind), I will not make it far if my arousal levels are too low (see above). However, if I have too much stress, my performance can start to decrease as I may not be able to react in time to an enemy attack. The relationship between arousal/stress and performance is often called the **Yerkes-Dodson law**.

Stressors can be classified by how long they last or how frequently they occur (Crosswell & Lockwood, 2020). Almost all of us encounter *daily* (or near-daily) stressors, like sitting in traffic. For parents with young infants, a daily stressor may be waking up frequently during the night (the term "sleeping like a baby" is quite the misnomer). There are *acute* stressors that we encounter periodically, such as giving a presentation or having an unexpected expense. *Chronic* stressors last for a longer period compared to acute stressors. Examples of such stress include homelessness, repeated abuse, and living with a chronic illness. *Life events* are major changes that occur in one's life that affect their day-to-day activities. Life events can be positive (e.g., getting married, having a child) or negative (e.g., losing a job, losing a loved one). Some life events are classified as *traumatic life events.* Here, the individual is placed in a situation in which their well-being is severely jeopardized. Physical/sexual abuse, witnessing the death of another individual, and living through a terrorist attack are examples of traumatic life events.

4 Fear as a Form of Stress

Directly related to stress is **fear**; like anxiety, fear is something I have already mentioned in this book but have yet to provide a formal definition. Fear is an emotional response that occurs when we perceive some threat or danger. Most of us have experienced fear at some point in our life. Fear is often characterized by a feeling of dread and a strong urge to avoid or escape the stimulus that induces the fear response. Children often have a fear of the dark; as such, they may try to sleep in the same room as their parent, or they may request a nightlight for their room. Fear, like stress, is adaptive. Many individuals show fear responses to stimuli that can cause great bodily harm or death, such as venomous animals. However, like stress, fear can become dysregulated, with individuals showing intense fear to stimuli that pose no real threat to the individual. I will discuss dysregulated fear below.

5 Anxiety and Anxiety-Related Disorders

You have already learned about one major psychiatric condition that involves dysregulated stress/anxiety. Chapter 6 introduced posttraumatic stress disorder (PTSD), which is classified as a trauma- and stressor-related disorder in the *Diagnostic and*

Statistical Manual of Mental Disorders (DSM-5) (American Psychiatric Association, 2013) and as a disorder specifically associated with stress in the *International Classification of Diseases (ICD-11)* (World Health Organization, 2022). Recall that PTSD results from exposure to an extremely traumatic event (e.g., military combat; emotional, physical, sexual abuse; serious accidents; etc.). Cues associated with the traumatic event can elicit intense physiological reactions and psychological distress. As such, individuals with PTSD often engage in avoidance behavior to reduce the likelihood of experiencing these symptoms. *Acute stress disorder* is a transient form of PTSD, with symptoms lasting between 3 days and 1 month after trauma exposure.

Beyond PTSD, the *DSM-5* and the *ICD-11* have an entire section devoted to anxiety disorders. The major anxiety disorders will be briefly discussed below. For simplicity, I will use the *DSM-5* criteria when discussing anxiety disorders below.

5.1 Generalized Anxiety Disorder (GAD)

Individuals with generalized anxiety disorder (GAD) have chronic anxiety that is not attributable to a single source. These individuals excessively worry, but they do not know why they feel anxious. According to the *DSM-5*, individuals need to have three (or more) of the following six symptoms consistently across a 6-month period:

1. Restlessness or feeling keyed up or on edge.
2. Being easily fatigued.
3. Difficulty concentrating or mind going blank.
4. Irritability.
5. Muscle tension.
6. Sleep disturbance (difficulty falling or staying asleep, or restless, unsatisfying sleep) (p. 222).

For children under the age of 18, only one of the symptoms above is required.

5.2 Panic Disorder

Panic disorder is characterized by panic attacks, which are intense experiences of fear. Like GAD, there is not a specific stimulus that triggers a panic attack. For a diagnosis of panic disorder, individuals must experience 4 (or more) of the following symptoms of a panic attack:

1. Palpitations, pounding heart, or accelerated heart rate.
2. Sweating.
3. Trembling or shaking.
4. Sensations of shortness of breath or smothering.
5. Feelings of choking.

6. Chest pain or discomfort.
7. Nausea or abdominal distress.
8. Feeling dizzy, unsteady, light-headed, or faint.
9. Chills or heat sensations.
10. Paresthesias (numbness or tingling sensations).
11. Derealization (feelings of unreality) or depersonalization (being detached from oneself).
12. Fear of losing control or "going crazy".
13. Fear of dying… (p. 208)

Additionally, individuals that experience a panic attack must experience, for at least 1 month, persistent worry about the occurrence of additional panic attacks and/or significant maladaptive changes in behavior designed to avoid having panic attacks.

5.3 *Phobic Disorder*

In contrast to GAD and panic disorder, phobic disorder is characterized by intense fear of a specific object or situation. Importantly, the fear is disproportionate to the actual danger associated with the object or situation. For example, a common phobia is *arachnophobia* (fear of arachnids like spiders). Even if a spider is enclosed in a terrarium and has no way of getting to the individual, the individual will still experience intense fear. Two types of phobic disorder are included as their own psychiatric disorders in the *DSM-5*. These disorders will be discussed in the subsections below.

5.3.1 Social Anxiety Disorder

In the Introduction, I provided an example in which I experienced anxiety before giving a presentation in front of my classmates and teacher. Although I was anxious about the ordeal, I was able to get through my presentation. For others, social situations can cause extreme anxiety, fear, and distress. An individual with social anxiety disorder experiences intense fear/anxiety when placed in a social situation in which they may face possible scrutiny, such as interacting with others, eating in front of others, or giving a speech. Additionally, an individual with social anxiety disorder fears that their anxiety symptoms will be observed and negatively evaluated by others.

5.3.2 Agoraphobia

Individuals with agoraphobia have a fear they will be placed in situations in which escape is difficult and/or where help may be unavailable if the individual experiences symptoms characteristic of panic disorder (e.g., if the person is alone in a parking lot). According to the *DSM-5*, individuals with agoraphobia show fear/anxiety about two or more of the following situations:

1. Using public transportation (e.g., automobiles, buses, trains, ships, planes).
2. Being in open spaces (e.g., parking lots, marketplaces, bridges).
3. Being in enclosed places (e.g., shops, theaters, cinemas).
4. Standing in line or being in a crowd.
5. Being outside of the home alone (p. 217).

6 Stress as Antecedent to Other Maladaptive Behaviors

Stress is a common trigger for maladaptive behaviors that you have learned about in this book. One prevailing theory for schizophrenia is that individuals are genetically predisposed to develop this condition, but symptoms do not manifest until after the individual has experienced a major stressor (Betensky et al., 2008). My grandfather's symptoms did not manifest until after he returned home after serving in World War II and after experiencing a traumatic stressor. Stress also negatively impacts cognitive functions, particularly attention and memory (Kim & Kim, 2023; Liu et al., 2020). As I noted in the Introduction, I have had numerous students perform poorly on exams despite seemingly understanding the concepts covered in class. In Chap. 6, I briefly discussed addiction. Stress is a major determinant of substance use, is a predictor of addiction, and is an antecedent to recurrence of substance use (Sinha, 2008).

Comprehension Check 1 How can the Yerkes-Dodson law be used to explain why individuals with anxiety disorders often show working memory deficits?

According to the Yerkes-Dodson law, cognitive-behavioral performance is generally optimal at intermediate stress levels. Because individuals with anxiety disorders show elevated stress responses, this leads to impairments in working memory.

7 Animal Models of Stress, Anxiety, and Fear

Subjecting people to highly stressful situations is difficult due to ethical reasons. This is why we often rely on animal studies to better understand the neuromechanisms underlying stress, anxiety, and fear. There are numerous ways researchers can induce stress in an animal (Atrooz et al., 2021) (Fig. 1). Like humans, animals can be exposed to environmental stressors such as loud noises, disruption of the light–dark cycle, and food/water restriction. Animals can also be subjected to mild physical pain (typically a foot shock), physical restraint (Fig. 1a), and forced swimming.

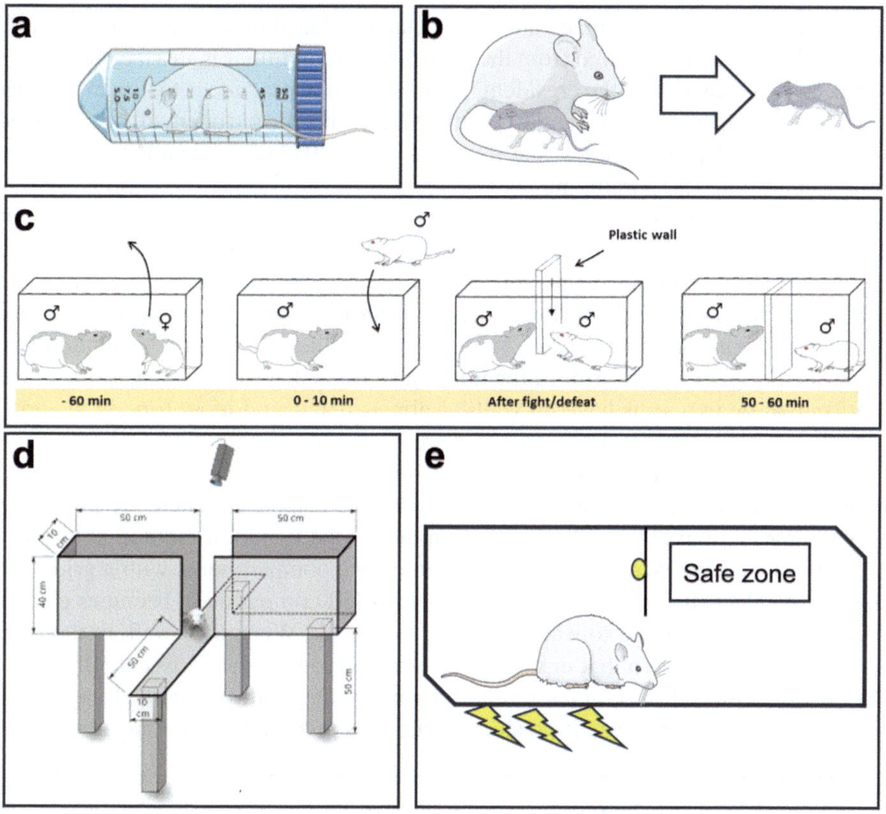

Fig. 1 Methods for inducing stress in rodents: (**a**) restraint stress, (**b**) maternal separation, (**c**) social defeat, (**d**) elevated plus maze, (**e**) shuttle box to measure conditioned avoidance. Panels (**a**), (**b**), and (**e**) were created using images from Servier Medical Art (https://smart.servier.com/; CC BY 4.0, https://creativecommons.org/licenses/by/4.0/) and Microsoft PowerPoint. Panel (**c**) is a slightly modified image of Figure 1 from Rajalingam et al. (2020), https://doi.org/10.1186/s12868-020-00574-1. CC BY 4.0 (https://creativecommons.org/licenses/by/4.0/), via Springer Nature. Panel (**d**) comes from https://commons.wikimedia.org/wiki/File:ElevatedPlusMaze.svg from Samueljohn.de. CC BY-SA 3.0 (https://creativecommons.org/licenses/by-sa/3.0/deed.en), via Wikimedia Commons

Research often takes advantage of the fact that rodents, like other mammals, are social animals. Multiple stressors involve the absence of prosocial interactions or the presence of aversive social interactions. Social isolation is a known stressor in mammals. Relatedly, *maternal separation* involves isolating young animals from the mother for extended periods of time (Fig. 1b). **Social defeat** utilizes antagonistic social interactions to induce stress in animals (Fig. 1c). The *resident-intruder paradigm* is the most common method for measuring social defeat. Because many rodent strains are generally docile, researchers house male rodents with a female for approximately 1 week before testing begins. This causes the male to become territorial, promoting it to attack any males that are placed in its cage. Before testing begins, the female is removed from the cage, and an unfamiliar male (the "intruder") is placed in the cage with the "resident" male. Common practice is to use an intruder that is smaller than the resident as this better ensures that the intruder will be subjected to social defeat by the resident. See Koolhaas et al. (2013) for a detailed description of a resident-intruder paradigm.

Anxiety/fear can be measured by examining an animal's *exploratory behavior*. Rodents can be placed in an *open field*, an arena-like apparatus with walls to prevent the animal from escaping. Some animals will explore the middle of the arena while others will stay close to the walls (a form of **thigmotaxis**). Another option is to test animals in the *elevated plus maze*. The maze looks like a "plus sign", and the maze is elevated (Fig. 1d). Two arms of the maze have walls that surround each side of the platform while two arms have no walls. Animals that spend more time in the arms with the closed walls are considered to display increased anxiety.

In Chap. 6, I briefly discussed conditioned fear in animals. Likewise, *conditioned avoidance* procedures can be used to infer fear-like responses. There are numerous ways to measure avoidance learning in animals. One common method is to place animals in a *shuttle box*, which is composed of two compartments with a grid floor that can be electrified (Fig. 1e). Imagine that the left compartment becomes electrified. The animal is going to want to escape the shock. When they walk to the right compartment, they learn that this side does not lead to foot shock. Eventually, when the animal is placed back in the left compartment, they will immediately approach the right compartment, even if no shock is administered in the left compartment.

Comprehension Check 2 What is a real-life analog to the social defeat paradigm?

Think of individuals that are bullied by another individual or by a group of individuals. Being subjected to bullying is stressful. Some reading this book may have been the victims of bullying and can speak to the anxiety they felt when they knew they had to encounter their bully or bullies at school or at work.

8 Neuromechanisms Governing Dysregulated Stress Responses

There are multiple mechanisms that regulate stress. In this section, I will present the major systems involved in stress and fear. I will also discuss how dysfunction of these systems contributes to anxiety disorders and to conditions like PTSD.

8.1 *Hypothalamic–Pituitary–Adrenal (HPA) Axis*

The primary "stress system" is the **hypothalamic–pituitary–adrenal (HPA) axis** (Fig. 2), which is part of the *neuroendocrine system.* This means that the HPA axis relies on transmission of both neurotransmitters and hormones. The stress response begins with the paraventricular nucleus (PVN) of the hypothalamus. Located in the ventral portion of the hypothalamus, near the third ventricle, the produces and releases a crucial hormone called **corticotropin-releasing hormone (CRH)** (also known as corticotropin-releasing factor or CRF).

CRH travels through the bloodstream before binding to one of two types of G protein-coupled receptors (CRHR1 and CRHR2) in the pituitary gland, located at the base of the hypothalamus. This then causes the pituitary gland to secrete and release **adrenocorticotropic hormone (ACTH)**, which travels to the adrenal glands located above each kidney.

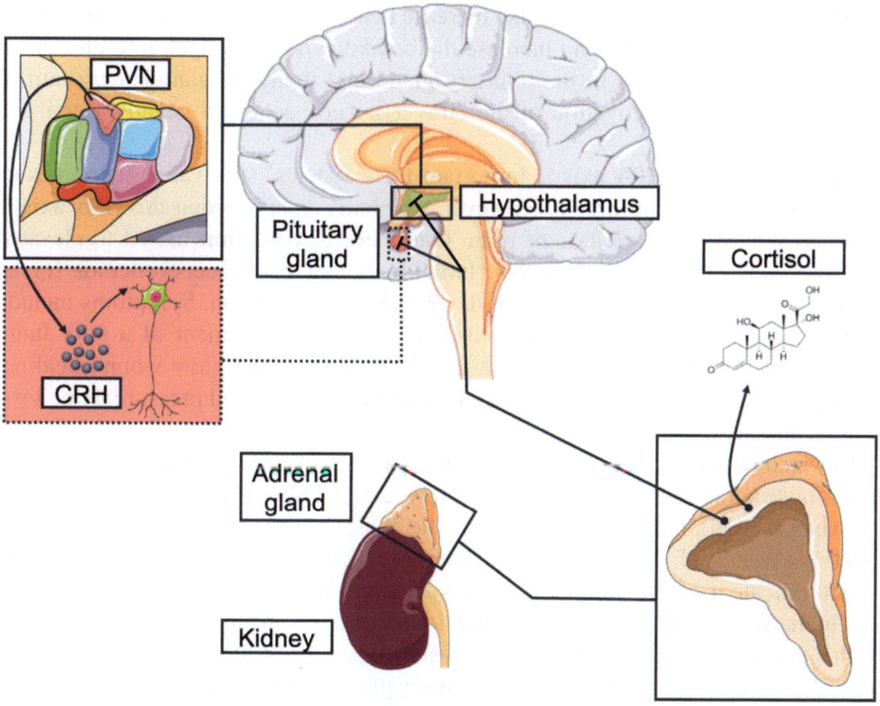

Fig. 2 A schematic of the hypothalamic–pituitary–adrenal (HPA) axis. See text for a description of the HPA axis. *Note* CRH = corticotropin-releasing hormone, PVN = paraventricular nucleus. Image created by the author using Servier Medical Art (https://smart.servier.com/; CC BY 4.0, https://creativecommons.org/licenses/by/4.0/) and Microsoft PowerPoint. The cortisol chemical structure comes from https://commons.wikimedia.org/wiki/File:Cortisol3.svg from NEUROtiker, via Wikimedia Commons (image is in the public domain)

In the adrenal gland, ACTH binds to a type of G protein-coupled receptor known as the *melanocortin 2 receptor (MC2R)*. At this point, the adrenal glands begin to secrete and release the hormone **cortisol**, as well as the hormones noradrenaline and adrenaline, into the bloodstream. Collectively, cortisol and noradrenaline/adrenaline are integral components of our "fight-or-flight" response. I will first discuss what cortisol does, followed by a discussion of the effects of noradrenaline/adrenaline on the stress response.

Cortisol belongs to a group of **steroids** known as *glucocorticoids*, which belong to a larger class of steroids called *corticosteroids*. When individuals hear the word "steroid", they often think of anabolic steroids, which are synthetic and often misused by bodybuilders/athletes. The term steroid refers to a chemical whose structure consists of four fused rings (see cortisol's chemical structure in Fig. 2). Cortisol binds to G protein-coupled glucocorticoid receptors throughout the body, causing widespread effects to occur. Some primary functions of cortisol include increasing glucose release from the liver (which provides energy), suppressing inflammation, and regulating blood pressure. Like cortisol, noradrenaline and adrenaline increase blood pressure. These hormones also increase heart rate and respiratory rate, reduce gut motility and digestion, and increase the breakdown of glycogen into glucose in the liver. Cortisol returns to the pituitary gland and thehypothalamus to stimulate receptors to stop the activation of the HPA axis (i.e., acts as a negative feedback loop).

Activation of the HPA axis is important for a species' survival as it enables us to respond to imminent threats. However, problems emerge when the HPA axis is activated for extended periods of time. One medical condition linked to elevated cortisol levels is *Cushing syndrome*. Individuals that take glucocorticoids for an extended period of time are at risk of developing this condition. Symptoms include weight gain, particularly in the face and the trunk, development of a fatty lump between the shoulders, purple stretch marks, frail skin, and slow wound healing. Other potential symptoms include muscle weakness, high blood pressure, bone loss, fatigue, and anxiety.

Elevated cortisol is also observed in individuals with anxiety disorders and other conditions characterized by increased anxiety (e.g., PTSD) (Patel et al., 2024; Sousa-Lima et al., 2019). Individuals with a specific allele of the *CRHR1* gene (encodes for CRH) are at risk of developing panic disorder (Weber et al., 2016). These findings do not indicate if excess cortisol leads to symptoms associated with these conditions or if increased cortisol levels occur when individuals begin experiencing symptoms. To this end, cortisol administration increases anxiety in healthy volunteers (Grillon et al., 2011); likewise, corticosterone (primary glucocorticoid in small mammals and birds) increases anxiety-like behavior in the elevated plus maze (Mitra & Sapolsky, 2008). These findings indicate that individuals with a hyperactive HPA axis are more prone to anxiety and anxiety/stress-related disorders.

When cortisol levels exceed a certain level, the negative feedback mechanism meant to regulate the release of further ACTH and CRH becomes overwhelmed. Imagine a bouncer at a bar trying to stop 100 patrons from entering at once as opposed to a single individual. Strategies aimed at blunting activity of the HPA axis should help ameliorate excessive stress responses. Indeed, administration of a CRH1 antagonist decreases stress responses in animals (Ayala et al., 2004), supporting this hypothesis. However, in one study with humans, a CRH1 antagonist had no effects on anxiety, instead increasing a fear response to a cue that predicted shock administration (Grillon et al., 2015). This finding suggests that cortisol may have differential effects in different parts of the brain. Although cortisol is secreted in the peripheral nervous system (PNS), glucocorticoid receptors are found in several locations in the brain, particularly the amygdala, the frontal cortex, and the hippocampus. You will learn about the specific contributions of these brain regions to stress, fear, and anxiety later. At this point, let's discuss noradrenergic contributions to maladaptive stress.

Comprehension Check 3 What is the role of the PVN in the stress response?
The PVN kickstarts the stress response by releasing CRH. CRH stimulates the pituitary gland, which then secretes ACTH, which travels to the adrenal gland.

8.1.1 Noradrenergic Contributions to Maladaptive Stress

As described, noradrenaline is released from the adrenal gland in response to ACTH stimulation. Norepinephrine also contributes to stress responses as a major neurotransmitter of the sympathetic nervous system. In Chap. 2, you learned about the autonomic nervous system, which consists of the parasympathetic and the sympathetic nervous systems. In that chapter, I mentioned that the sympathetic nervous system is highly associated with the "fight-or-flight" response. I now need to provide additional information about the sympathetic nervous system.

Neurons of the sympathetic nervous system originate in the thoracic and the lumbar regions of the spinal cord. The axons of these neurons are short as they only need to travel to the **sympathetic chain** (Fig. 3a), which is a long collection of cell bodies that runs parallel to the spinal column. The neurons projecting from the spinal cord to the sympathetic chain are called *preganglionic neurons*. Preganglionic neurons release acetylcholine, which stimulates excitatory nicotinic receptors on *postganglionic neurons*.

The cell bodies of postganglionic neurons originate in the sympathetic chain, and the axons of these neurons are much longer compared to preganglionic neurons. This is because these neurons must travel from just outside of the spinal cord to different organs. Postganglionic neurons release norepinephrine, which binds to either alpha (α) or beta (β) receptor subtypes. For example, norepinephrine binds to β_1 receptors in the heart to increase heart rate and to β_2 receptors in the lung to widen airways. These actions allow us to circulate more blood throughout our body and to allow more air into our lungs, providing more oxygen for red blood cells.

In the brain, most noradrenergic neurons originate from the locus coeruleus, located in the brainstem (Fig. 3b). Noradrenergic neurons travel to the cerebral cortex,

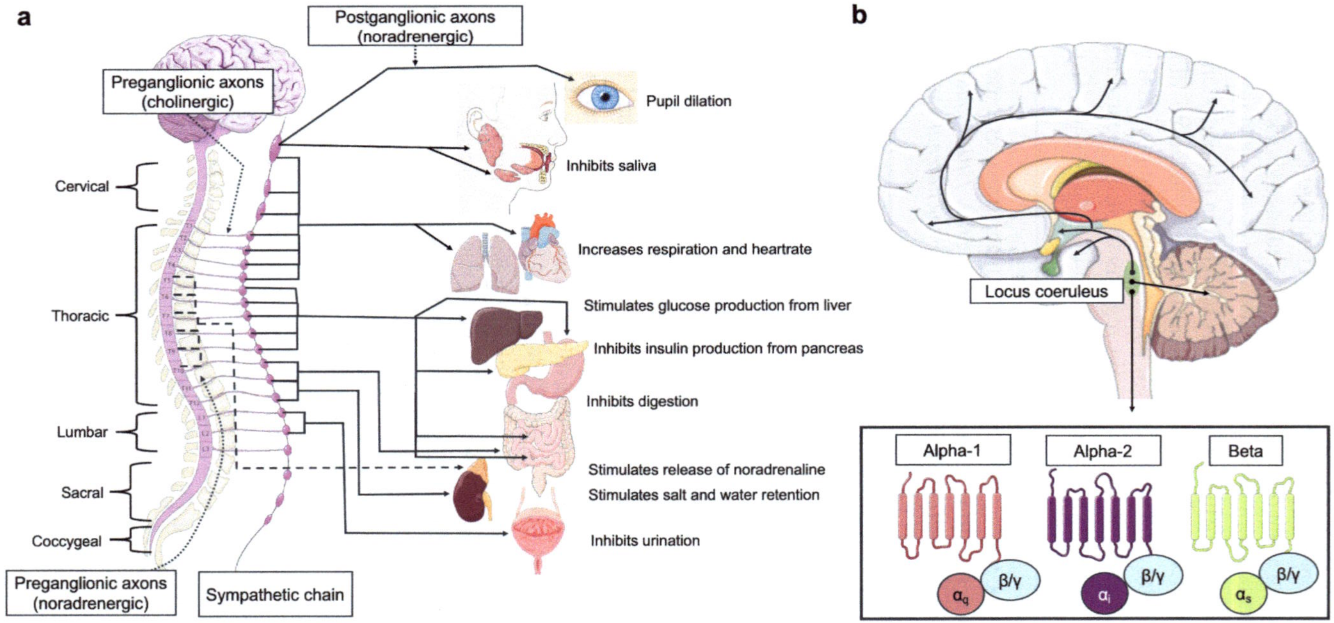

Fig. 3 (**a**) Schematic of the sympathetic nervous system. Preganglionic axons, which are primarily cholinergic, exit the spinal cord and form synapses with noradrenergic postganglionic neurons in the sympathetic chain. Noradrenergic activity influences numerous organs, preparing organisms for a "fight-or-flight" response. (**b**) Noradrenergic pathways in the brain originate from the locus coeruleus. The inset at the bottom of the figure shows the three major noradrenergic receptor subtypes. Image was created by the author using Servier Medical Art (https://smart.servier.com/; CC BY 4.0, https://creativecommons.org/licenses/by/4.0/) and Microsoft PowerPoint

the limbic system, the cerebellum, and the hypothalamus. Stress increases noradrenergic activity in the locus coeruleus (Sabban et al., 2015); norepinephrine then affects hypothalamic activity, thus contributing to activation of the HPA axis (Levy & Tasker, 2012). Anxiety-like behavior can be induced in animals by directly stimulating locus coeruleus noradrenergic neurons or by increasing norepinephrine levels in the locus coeruleus (Fentress et al., 2013; McCall et al., 2015).

In preclinical studies, one major way to induce a stress response is by administering a drug called *yohimbine* (see Chitsaz et al., 2025 for a recent example). Derived from the bark of the yohimbe tree, yohimbine is an antagonist at α_2 receptors. As α_2 receptors are inhibitory (G_i-coupled), yohimbine prevents the inhibition of neurons that contain α_2 receptors. Yohimbine can produce full panic attacks in humans (Cameron et al., 2000) and can increase self-reported "panic symptoms" (Vasa et al., 2009). Antagonism of α_2 receptors leads to increased norepinephrine levels in the locus coeruleus (Fernández-Pastor et al., 2005). This should make sense considering α_2 receptors primarily act as autoreceptors in the locus coeruleus, regulating the further release of norepinephrine. Blocking these receptors prevents this negative feedback from occurring, leading to excess norepinephrine release.

If stress and anxiety are related to hyperactivity of the noradrenergic system, decreasing activity of the locus coeruleus should reduce these symptoms. To this end, anxiolytic effects are observed when noradrenergic locus coeruleus neurons are inhibited during exposure to stress (McCall et al., 2015). Clonidine, an α_2 receptor agonist, has been used off-label to treat anxiety disorders (Hollowell et al., 2025). More commonly, anxiety disorders are treated with serotonin and norepinephrine reuptake inhibitors (SNRIs). This drug class targets both serotonin and norepinephrine transporters, thus increasing serotonin and norepinephrine levels in the brain. In fact, I have been taking duloxetine (Cymbalta) since 2021 to manage symptoms of anxiety. You will learn more about serotonin's relationship to maladaptive stress below.

If norepinephrine is associated with stress responses, why are SNRIs effective at treating anxiety disorders? Overproduction of norepinephrine can cause depletion of this neurotransmitter. This is similar to what you learned in the previous chapter concerning dopamine depletion and decision-making impairments. Indeed, corticosterone administration causes increased norepinephrine transporter expression in the locus coeruleus (Fan et al., 2014), which would not enable norepinephrine to bind to α_2 receptors to regulate the further release of norepinephrine. By preventing the reuptake of norepinephrine, norepinephrine can more easily bind to α_2 receptors, allowing for attenuation of noradrenergic signaling.

Comprehension Check 4 How can increasing norepinephrine levels lead to both anxiogenic and anxiolytic effects?

Stress leads to activation of noradrenergic neurons in the locus coeruleus. This causes the release of norepinephrine in various parts of the brain. Chronic stress can lead to depletion of norepinephrine levels. Drugs like SNRIs restore baseline *norepinephrine levels, which then bind to $\alpha_2$receptors to prevent further release of norepinephrine.*

8.2 Brain Regions Involved in Maladaptive Stress

Above, you learned that norepinephrine originates from the locus coeruleus. This region of the brainstem forms critical connections with brain regions that are highly implicated in the pathogenesis of anxiety disorders and PTSD. In this section, I will provide evidence that brain regions such as the amygdala, the frontal cortex, the hippocampus, and the cerebellum are involved in maladaptive stress responses. For now, I will focus on the influence of cortisol and/or norepinephrine in these regions. Later, I will introduce additional neurotransmitters that regulate stress responses.

8.2.1 Amygdala

The amygdala is an important limbic structure that consolidates emotional memories and processes emotional responses, particularly fear. The amygdala is particularly important for *threat learning* (think fear conditioning) (Kredlow et al., 2022). Imagine during your first day at a new job, your boss berates you for a minor issue, humiliating you in front of your coworkers. During this aversive experience, your amygdala will encode information about your boss with the negative experience of being berated. In the future, when you encounter your boss, you will experience anxiety.

Damage to the amygdala can prevent the expression of fear in individuals. In one case study, patient SM, who had bilateral amygdalar lesions, was presented with snakes and spiders, was taken to a haunted house, and was shown "emotionally evocative" films; not once did SM report experiencing any level of fear (Feinstein et al., 2011). Like patient SM, lesions to the amygdala impair fear conditioning in animals (Sun et al., 2020).

Whereas a damaged amygdala impairs fear responses, a hyperactive amygdala is implicated in psychiatric conditions marked by heightened fear, such as anxiety disorders and PTSD. Individuals with PTSD have greater amygdalar activity compared to healthy controls (Liberzon & Sripada, 2008). Additionally, altered functional connectivity is observed in individuals with anxiety disorders. For example, individuals with GAD have reduced functional connectivity between the amygdala and the cingulate cortex and medial prefrontal cortex (mPFC) and increased functional connectivity between the amygdala and the superior temporal gyrus (STG) and the insula (Wang et al., 2021). Individuals prone to elevated anxiety may identify neutral stimuli as threatening due to heightened amygdalar activity.

Not only are glucocorticoids located in the hypothalamus and the pituitary gland, but they are found throughout the brain. Social defeat stress leads to reduced expression of glucocorticoid receptors in the basolateral nucleus of the amygdala (BLA), as well as in the ventral portion of the hippocampus (Winzenried et al., 2023). However, genetic deletion of the glucocorticoid receptor from glutamatergic, but not GABAergic (gamma-aminobutyric acid), neurons in the BLA (as well as in the frontal cortex) increases hyperactivity of the HPA axis; interestingly, this results in decreased anxiety-like behavior (Hartmann et al., 2017). These results seem counterintuitive, but remember that cortisol crosses the blood-brainbarrier (BBB) to bind to glucocorticoid receptors to *inhibit* further excitation of the HPA axis. If there are fewer glucocorticoid receptors in the BLA, cortisol cannot exert its inhibitory effect on the HPA axis. This can result in overexcitation of this system, leading individuals to be in a constant stress state.

This explanation makes sense, but this does not explain how an overactive HPA axis can have paradoxical anxiolytic effects. Hartmann et al. (2017) argue that deletion of glucocorticoid receptors may decrease fear responses without necessarily altering anxiety. That is, glucocorticoid receptors on glutamatergic neurons appear to be important for the immediate processing of threatening stimuli as opposed to generating a consistent state of worry. One possibility is that intra-amygdalar glucocorticoid receptors expressed on glutamatergic neurons mediate fear learning. As such, removing these receptors could impair such learning, which is why animals show less fear-like behavior despite an overactive HPA axis.

Noradrenergic activity in the amygdala is associated with fear (Onur et al., 2009), with β receptor activation being necessary for threat detection (Díaz-Mataix et al., 2017). Blocking β receptors blunts BLA activation in response to facial expressions (Hurlemann et al., 2010), providing additional evidence that this receptor subtype is especially important for detecting potential threats. Excessiveβ receptor stimulation in the amygdala may also be a biomarker for conditions like PTSD. Direct infusion of aβ receptor agonist into the amygdala impairs extinction of fear conditioning (Dębiec et al., 2011). This means that individuals continue responding aversely to a stimulus that no longer predicts threat or to stimuli that are similar to the original fear-inducing stimulus (e.g., military veteran showing fear response to sound of car backfiring).

Norepinephrine levels are lower in the amygdala of adult rats that experienced chronic stress early in life (Dimonte et al., 2023; Luo et al., 2014). Chronic stress also leads to reduced expression of excitatory α_1 receptors in the BLA (Volk et al., 2014). These results seem contradictory to what you read above, but another major change that occurs following prolonged stress is a compensatory upregulation of norepinephrine transporters in the amygdala, as well as in other regions implicated in maladaptive stress (e.g., frontal cortex and hippocampus) (Chen et al., 2012; Fan et al., 2014). This upregulation of transporters allows for too much norepinephrine to be taken back into the presynaptic neuron. Normally, increased norepinephrine levels inhibit neuronal firing in the BLA (Buffalari & Grace, 2009; Chen & Sara, 2007), most likely by binding to α_1 receptors located on GABAergic interneurons (Braga et al., 2004). Without sufficient stimulation of these receptors, GABAergic neurons do not work as effectively, which results in a net overexcitation of amygdalar neurons.

Comprehension Check 5 Explain the seemingly contradictory effects of glucocorticoid receptors and norepinephrine on amygdalar function and their implications for anxiety and fear-related behaviors.

Glucocorticoid receptors expressed on glutamatergic neurons in the BLA mediate immediate fear learning. When these receptors are genetically deleted, the HPA axis becomes hyperactive because cortisol cannot bind to them to exert its inhibitory effect, leading to a constant stress state. However, this same deletion paradoxically decreases fear-like behaviors because glucocorticoid receptors are crucial for the immediate processing of threatening stimuli.

While excessive β receptor stimulation in the amygdala is linked to conditions like PTSD, chronic stress can lead to lower overall norepinephrine levels in the amygdala. This seems counterintuitive until you consider the compensatory changes that occur. Chronic stress causes an upregulation of norepinephrine transporters. This increased uptake of norepinephrine into the presynaptic neuron leads to a deficiency of the neurotransmitter in the synapse. This is significant because norepinephrine normally binds to α_1 receptors on GABAergic interneurons, which then inhibit other amygdalar neurons. With less norepinephrine available to activate these inhibitory interneurons, the net effect is disinhibition and overexcitation of the amygdala, contributing to heightened fear and anxiety responses.

8.2.2 Frontal Cortex

Given their role in attentional processes, including evaluating the salience of specific stimuli, and executive functions, dysfunction of the PFC and the dorsal anterior cingulate cortex (dACC) is also hypothesized to underlie anxiety disorders and PTSD (Kenwood et al., 2022). Hypoactivity of the dorsomedial PFC (dmPFC) and the dACC is observed in individuals with anxiety disorders. Due to hypoactivity in these regions, individuals fail to identify the stimulus as non-threatening. In fact, transcranial direct current stimulation (tDCS) of the frontal cortex reverses anxiety-like responses in both humans and rodents (Carnevali et al., 2020; Fang & Wang, 2021).

The ventromedial PFC (vmPFC) is crucial for inhibiting threat responses originating from the amygdala (Kredlow et al., 2022) and is more active when we view stimuli that are perceived as safe (Leschak et al., 2022). Lower functional connectivity between the vmPFC and the amygdala is associated with increased anxiety (Mehta et al., 2022), and early-life emotional abuse leads to hypoactivity of the vmPFC, as well as the orbitofrontal cortex (OFC) (Wang et al., 2013).

While a hypoactive dACC is observed in individuals with anxiety disorders, direct stimulation of the subgenual ACC (sgACC) increases heart rate and fear-like behavior in primates (Alexander et al., 2020). Also, enhanced connectivity between part of the sgACC and the BLA is important for fear-related memory encoding (Hakamata et al., 2020). Notice that the ACC is not a homogenous structure, as the dACC and the sgACC differentially contribute to stress responses. The dACC is more similar to regions like the PFC as it helps control executive functions and plays pivotal roles in conflict monitoring and error detection. The sgACC is more linked to the limbic system, receiving direct inputs from the amygdala (Dixon et al., 2017), and is involved in emotional processing. Anxiety disorders can be partially driven by a shift away from dACC control of behavior to hyperactivity of the sgACC.

Elevated cortisol levels in prefrontal cortical areas account for why stress often negatively impacts cognitive functions like working memory. For example, chronic stress produces working memory deficits in mice, with these deficits coinciding with rapid increases in corticosterone levels in the PFC (Dominguez et al., 2019). Cortisol also affects the noradrenergic system in the frontal cortex, which can promote maladaptive stress responses (Fan et al., 2014). Changes to the noradrenergic system in the frontal cortex following stress are reported in the literature. Social isolation stress leads to increased α_1 receptor protein expression in the mPFC (Kawanabe et al., 2023). Given that α_1 receptor activation causes cognitive dysfunction (Arnsten et al., 1999), this finding provides a potential mechanism accounting for stress-induced cognitive impairment often observed (Shields et al., 2016). Noradrenergic α_1 receptors are located on both glutamatergic and GABAergic neurons in the frontal cortex and are more likely to be stimulated as norepinephrine levels increase (Maletic et al., 2017). As norepinephrine levels increase in response to stress, excess stimulation of α_1 receptors, particularly on GABAergic neurons, may be why regions like the PFC become hypoactive.

In contrast to α_1 receptors, intra-mPFC infusion of an α_2 receptor agonist reduces anxiety-like behavior, but an α_2 receptor antagonist increases such behavior (Wu et al., 2019). This finding is consistent with what is reported following systemic administration of α_2 receptor ligands. Noradrenergic α_2 receptors, like α_1 receptors, are located on cortical glutamatergic and GABAergic neurons. This means that stimulation of α_2 receptors can inhibit GABAergic neurons that lead to stress-induced hypoactivity of the frontal cortex. Directly supporting this argument is the finding that α_2 receptor agonists increase norepinephrine levels in the frontal cortex (Millan et al., 2000).

Interestingly, optogenetic stimulation of β_2 receptors located on glutamatergic neurons in mPFC induces anxiety-like behavior in rodents whereas genetically silencing these receptors produces anxiolytic effects (Lei et al., 2022). Noradrenergic β_2 receptors are typically excitatory (G_s-coupled), meaning that stimulating excitatory neurotransmission can potentiate anxiety. This finding is seemingly at odds with what was discussed above. One important consideration is that β_2 receptors can couple to both G_S and G_i proteins, indicating this receptor can have both excitatory and inhibitory effects (Daaka et al., 1997). If β_2 receptors bind to G_i-couple receptors on glutamatergic neurons, this would inhibit excitatory neurotransmission, leading to hypoactivation of the frontal cortex, which is consistent with the effects of α_1 receptor activation.

8.2.3 Hippocampus

The hippocampus is an important locus for learning and memory and has reciprocal connections with the amygdala. The hippocampus helps encode information about the context associated with an aversive event. Part of the hippocampus known as the ventral subiculum receives inputs from the locus coeruleus, and norepinephrine release in this region stimulates neurons in response to a stressor (Lipski & Grace, 2013). Mild to moderate stress facilitates processes like long-term potentiation (LTP) and long-term depression (LTD), the cellular mechanisms by which learning occurs. While glutamate is important for LTP and LTD (see Chap. 6), norepinephrine levels increase in the hippocampus in response to an acute stressor, although this response decreases with a second exposure to the stressor (Hajós-Korcsok et al., 2003). This mechanism appears to be important for allowing an organism to learn about the stressor and to adapt to it. Exposure to chronic stress becomes problematic as norepinephrine levels become hypoactive (e.g., norepinephrine levels become depleted) in the hippocampus just like in the amygdala (Zhang et al., 2023). The hippocampus then helps inhibit activity of the PVN as glucocorticoid receptors are expressed in this region.

Going back to the example from earlier in this chapter—if your boss reprimands you at work in front of your peers, the hippocampus will encode more than just the experience itself; contextual cues, like your location at work where you were reprimanded, are encoded. In the future, you may feel uneasy being at work, even if your boss is not around.

Damage to the hippocampus leads to excess corticosteroid levels, including prolonged corticosteroid responses to stressors (Jankord & Herman, 2008). While anxiety disorders are associated with an enlarged hippocampus (Baksh et al., 2021), prolonged stress impairs neurogenesis of hippocampal neurons (Lucassen et al., 2015). This can also account for stress-induced working memory deficits. The enlarged hippocampus observed in anxiety disorders may reflect increased neuroinflammation resulting from chronic stress. I will return to this point at the end of this chapter. I also want to highlight that reduced functional and structural connectivity between the vmPFC and the hippocampus is observed in military veterans with increased PTSD-like symptoms (Admon et al., 2013).

Comprehension Check 6 How do the frontal cortex and the hippocampus interact with one another to influence stress responses?

The (vmPFC) is essential for inhibiting threat responses originating from the amygdala. Reduced functional and structural connectivity between the vmPFC and the hippocampus is observed in individuals with PTSD, suggesting that this impaired communication may be a key factor in the development of symptoms. This disrupted communication prevents the brain from accurately contextualizing and inhibiting fear responses, leading to an over-generalized and prolonged state of anxiety and stress. Additionally, while mild to moderate stress can facilitate hippocampal function, prolonged stress impairs hippocampal neurogenesis, which can lead to working memory deficits. This dysfunction, combined with altered frontal cortex activity, can perpetuate a cycle of maladaptive stress responses.

8.2.4 Cerebellum

Above, you learned about the amygdala and threat learning (fear conditioning). There is evidence that the cerebellum is also important for this form of learning (Doubliez et al., 2023). This makes sense given that the cerebellum regulates predictive coding processes. If your boss consistently threatens to fire you for minor errors, your cerebellum encodes this information, such that when you see your boss, it is anticipated that something aversive will happen. This helps you identify your boss as a "threat", precipitating a fear-like response. Notice that the cerebellum and the hippocampus have an overlapping function. Stress increases functional connectivity between the cerebellum and the hippocampus (Hilberdink et al., 2021).

Daily cortisol levels are associated with strengthened connectivity between the cerebellum and the amygdala (Hakamata et al., 2017). Administration of hydrocortisone and yohimbine increases functional connectivity between theamygdala and the hippocampus with the cerebellum (Rosada et al., 2024), and yohimbine increases cerebral blood flow in the cerebellum (Cameron et al., 2000). Further evidence supporting cerebellar noradrenergic activity in stress is that inhibiting the pathway between the locus coeruleus and the cerebellum prevents fear conditioning (Stanley et al., 2023).

Individuals with PTSD have greater activation of the cerebellum (Naegeli et al., 2018). Experiencing trauma can also produce morphological changes to the cerebellum to occur, such as reduced gray matter volume, as well as aberrant functional connectivity between the cerebellum and the default mode network (DMN) and the salience network (Blithikioti et al., 2022). Changes to the cerebellum are also observed in individuals with Cushing syndrome, as they have a larger cerebellum (Andela et al., 2013). As with the hippocampus, an enlarged cerebellum may reflect increased neuroinflammation.

8.3 *Other Neurotransmitters Involved in Maladaptive Stress*

While cortisol and noradrenaline/norepinephrine receive much of the spotlight concerning stress, there are numerous neurotransmitters that mediate stress responses and fear processing. Dysfunction of any of the neurotransmitters detailed below can increase symptoms of anxiety and may contribute to the etiology of anxiety and anxiety-related disorders.

8.3.1 Serotonin

As a major neuromodulator, serotonin influences activity of the brain regions involved in the pathogenesis of anxiety and anxiety-related disorders (Bauer, 2015). Stimulation of the dorsal raphe nuclei leads to increased serotonin release in the locus coeruleus (Kaehler et al., 1999), and stressful events cause serotonin levels to increase in the locus coeruleus (Kaehler et al., 2000) and increase serotoninergic activity in the amygdala (Mo et al., 2008). Serotonin is crucial for dampening locus coeruleus activity; one mechanism by which serotonin accomplishes this is by reducing glutamatergic inputs to the locus coeruleus (Aston-Jones et al., 1991). Both the inhibitory 5-HT_{1A} receptor and the excitatory 5-HT_2 receptor are involved in attenuated stress responses, but in a way that is not entirely expected.

5-HT_{1A} receptors are inhibitory and primarily act as autoreceptors. Intuitively, stimulating these receptors should decrease activity of the locus coeruleus, but that is not the case as 5-HT_{1A} agonists increase the firing rate of noradrenergic neurons (Szabo & Blier, 2001) while 5-HT_{1A} antagonists exert the opposite effect (Haddjeri et al., 1997). Conversely, stimulating the excitatory 5-HT_2 receptor decreases neuronal activity in the locus coeruleus (Szabo & Blier, 2001). 5-HT_{1A} and 5-HT_2 receptors may act to inhibit or stimulate GABAergic neurons, respectively, that inhibit noradrenergic neurons in the locus coeruleus (see Fig. 4). The role of GABA to maladaptive stress will be covered below.

An important consideration is that optimal serotonin levels are critical for regulating stress. On one hand, tryptophan depletion (resulting in less serotonin) increases

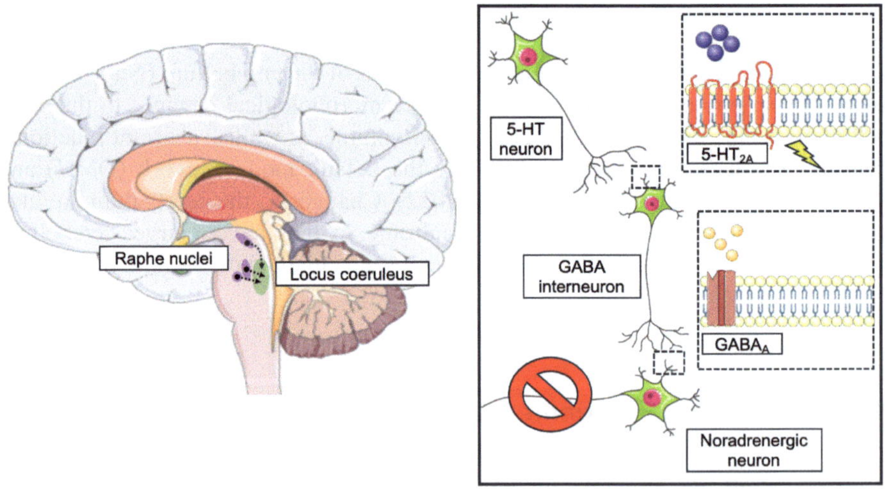

Fig. 4 Schematic showing how stimulation of serotonin 5-HT_{2A} receptors can decrease noradrenergic activity, thus ameliorating anxiety. Stimulation of 5-HT_{2A} receptors on GABAergic interneurons causes the release of GABA, which binds to GABA receptors located on noradrenergic neurons in the locus coeruleus. Image created by the author using Servier Medical Art (https://smart.servier.com/; CC BY 4.0, https://creativecommons.org/licenses/by/4.0/) and Microsoft PowerPoint

social anxiety (van Veen et al., 2009), and damage to the raphe nuclei increases anxiety-like behavior in animals (Netto et al., 2002). Also, lower serotonin levels in the hippocampus and the amygdala are observed following long-term stress (Lin et al., 2016; Zhang et al., 2023), and "high anxious" animals have lower serotonin levels in the hippocampus (File et al., 1987). Increased serotonin transporter and 5-HT_3 receptor expression are also observed following stress (Tseilikman et al., 2024). Increased serotonin transporters would lead to excess removal of serotonin from the synapse, and increased 5-HT_3 receptor expression may reflect a compensatory upregulation to account for low serotonin levels.

On the other hand, excess serotonin in the brain can worsen anxiety symptoms. Serotonin transporter knockout mice often display increased anxiety-like behavior as too much serotonin builds up in the synapse (Marazziti et al., 2015). Chronic stress has been shown to increase serotonin levels in the hippocampus (Tseilikman et al., 2024). Excessive serotonin levels may be linked to deficient 5-HT_{1A} activity as reduced 5-HT_{1A} binding is observed in regions like the amygdala and the ACC in individuals with social anxiety disorder (Lanzenberger et al., 2007).

In addition to SNRIs, another treatment option for anxiety disorders is selective serotonin reuptake inhibitors (SSRIs). This drug class was originally used for the treatment of mood disorders (see next chapter for more details), but they are now commonly prescribed for anxiety disorders. SSRIs most likely exert their therapeutic effects by normalizing serotonin levels in the brain. SSRIs, as well as the SNRI duloxetine, reduce serotonin transporter levels in the frontal cortex (Mirza et al., 2007) and increase serotonin levels in the amygdala and hippocampus (Lin et al., 2016). In short, SSRIs, like SNRIs for the noradrenergic system, help restore and maintain a balance of serotonin in the central nervous system (CNS).

8.3.2 Dopamine

A hypoactive dopaminergic system is associated with anxiety. Individuals with Parkinson's disease, which results in the death of dopaminergic neurons from the substantia nigra, often experience increased anxiety (Chen & Marsh, 2014). Bilateral dopamine lesions to the ventral tegmental area (VTA) and the substantia nigra increase anxiety-like behavior (Faggiani et al., 2018; Vieira et al., 2019), which is attenuated when the dopamine precursor L-DOPA (l-3,4-dihydroxyphenylalanine) is administered (Faggiani et al., 2018).

Both D_1-like and D_2-like receptors regulate aspects of anxiety. Dopamine originating from the VTA stimulates D_1-like receptors in the interpeduncular nucleus (IPN) which attenuates anxiety-like responses (DeGroot et al., 2020), and directly stimulating D_1-like receptors in the mPFC ameliorates anxiety-like behavior by blocking glutamatergic activity in the hippocampus (Liu et al., 2019). Reducing the number of D_1-like receptors increases anxiety-like behavior (Tong et al., 2023), and maternal separation causes a downregulation of D_1-like receptors in both the mPFC and the hippocampus (Hamdan et al., 2022).

D_2-like receptor agonists also have anxiolytic properties (Wall et al., 2003), potentially by restoring excitability of VTA neurons (Peng et al., 2021). Blocking these receptors produces anxiogenic effects (Timothy et al., 1999). Social anxiety disorder appears to be, at least partially, driven by decreased dopamine levels in frontal cortical and limbic regions; in particular, lower D_2-like receptor expression is observed in these individuals (Marazziti et al., 2015). In animals, early-life stress leads to a reduction in D_2-like receptors in the hippocampus (Hamdan et al., 2022), and removal of D_2-like receptors from hippocampal neurons potentiates anxiety-like behavior (Gulfo et al., 2023).

The findings above provide an oversimplified view of how dopamine receptors mediate anxiety. Both stimulation and blockade of D_1-like receptors in the lateral habenula increase anxiety-like behavior (Chan et al., 2017). In the previous chapter, you were introduced to the habenula; the lateral habenula innervates the rostromedial tegmental nucleus (RMTg), which then inhibits dopaminergic activity in the VTA. Stimulating D_1-like receptors in the lateral habenula leads to inhibition of the VTA, resulting in a hypoactive dopaminergic state. The lateral habenula also sends glutamatergic neurons directly to the VTA. Thus, inhibiting D_1-like receptors on these glutamatergic neurons can also result in hypoactivity of the VTA.

Also, D_2-like receptors do not always promote anxiolysis. Activating D_2-like receptors on medium spiny neurons (MSNs) in the nucleus accumbens (NAc) projecting to the ventral pallidum (VP) increases anxiety-like behavior (Correia et al., 2023). To understand why this is the case, we now need to turn our attention to another neurotransmitter system known to mediate anxiety.

Comprehension Check 7 Explain the complex relationship between serotonin and dopamine levels and anxiety. Make sure to focus on how both a deficit and an excess of serotonin can be problematic for stress regulation, and describe how a hypoactive dopaminergic system, particularly involving D_1-like and D_2-like receptors, is linked to anxiogenic behaviors.

A serotonin deficit is linked to increased anxiety, as evidenced by tryptophan depletion studies and the observation of lower serotonin levels in the hippocampus and amygdala after long-term stress. In these cases, increased serotonin transporter expression, which removes serotonin from the synapse, may be a compensatory response. Conversely, an excess of serotonin can also worsen anxiety. This is seen in serotonin transporter knockout mice, where too much serotonin accumulates in the synapse, leading to increased anxiety-like behaviors. This excess may be linked to a deficiency in the inhibitory 5-HT$_{1A}$*receptors, which are responsible for binding to serotonin and regulating its effects. Medications like SSRIs work by normalizing these serotonin levels, bringing them back into an optimal range.*

In contrast to serotonin, a hypoactive dopaminergic system is consistently associated with anxiety. Both D_1*-like and* D_2*-like receptors play a role in anxiety.* D_1*-like receptors typically have an anxiolytic effect. Early-life stress can lead to a downregulation of these receptors, which contributes to anxiety. Similarly,* D_2*-like receptors also have anxiolytic properties. Lower expression of these receptors is observed in* *individuals with social anxiety disorder, and their removal from hippocampal neurons potentiates anxiety-like behavior. This suggests that a reduction in the function or number of these dopamine receptors contributes directly to an anxious state.*

8.3.3 GABA

Notice that hyperactivity of multiple systems (HPA axis, sympathetic nervous system, limbic structures) is associated with anxiety-related disorders. One potential cause for this hyperactivity is a deficient GABAergic system. Individuals with anxiety disorders have lower GABA levels and reduced GABA receptors, and animal studies show decreased expression of GABA receptors in animals that display increased anxiety-like behavior (Arora et al., 2024; Nikolaus et al., 2010).

The amygdala relies heavily on the GABAergic system to function normally (Nuss, 2015). When we encounter a potentially threatening stimulus, the thalamus, the PFC, and the ACC relay information to the BLA. The BLA then projects glutamatergic neurons to the central nucleus of the amygdala (CeA), and the BLA activates GABAergic neurons that travel to the CeA to regulate glutamatergic inputs from the BLA to the CeA. The CeA then sends GABAergic neurons to the hypothalamus and the brainstem. This is what leads to the physical manifestations of stress, anxiety, and fear. Chronic stress leads to loss of tonic, but not phasic, $GABA_A$ receptor-mediated control of the amygdala (Liu et al., 2014). This can lead to a situation in which the CeA sends faulty signals to regions like the hypothalamus, increasing physiological/psychological responses to non-threatening stimuli.

One major treatment option for anxiety disorders is *benzodiazepines*, which target $GABA_A$ receptors (Garakani et al., 2020). The most well-known benzodiazepines include alprazolam (Xanax), diazepam (Valium), lorazepam (Ativan), and clonazepam (Klonopin). These drugs do not directly activate $GABA_A$ receptors. Instead, they potentiate $GABA_A$ receptor activity, meaning they make the receptor more sensitive to GABA stimulation.

Let's return to the finding that optogenetic stimulation of D_2-like receptors in the NAc increases anxiety-like responses. When D_2-containing MSNs are stimulated, the VP becomes inhibited. Because the VP primarily contains GABAergic neurons, inhibition of this region leads to deficient GABAergic control of other brain regions like the lateral hypothalamic area (LHA), a region involved in anxiety (Owens-French et al., 2022). Disinhibition of the LHA causes a different system to become hyperactive. I now need to introduce you to a neuropeptide that has not yet been discussed.

8.3.4 Orexin

Orexin (also known as hypocretin) is a neuropeptide that is produced in the LHA. You will learn much more about the orexinergic system in Chap. 14. For now, just know that orexin produces anxiogenic effects (Palotai et al., 2014), and hyperactivity of the orexinergic system is proposed to contribute to states of panic (Johnson et al., 2012). This is most likely due to the fact that orexin activates noradrenergic neurons in the locus coeruleus; this is supported by the finding that optogenetic stimulation of orexin neurons in the locus coeruleus increases fear conditioning while an orexin receptor antagonist infused into the locus coeruleus impairs fear conditioning (Sears et al., 2013). Similarly, blocking orexin receptors reverses anxiety-like and PTSD-like behaviors (Prajapati & Krishnamurthy, 2021; Sarfi et al., 2025).

Comprehension Check 8 How does the relationship between GABA and orexin influence anxiety?

GABA receptors in the LHA normally inhibit orexin neurons. If the GABAergic system is hypoactive, orexin neurons become disinhibited. These neurons travel to regions like the locus coeruleus, where they trigger the release of norepinephrine. This action can promote increased anxiety.

8.3.5 Endocannabinoids

I still remember one of my college experiences vividly. My roommate and his girlfriend were using cannabis. Out of nowhere, my roommate's girlfriend began to panic, claiming she could see faces melting. I had always associated cannabis with calming people and giving them the "munchies" (you will learn more about the munchies in Chap. 13). Research indicates that cannabis exerts a biphasic effect on anxiety. At low doses, cannabis is an anxiolytic, but it has anxiogenic effects at higher doses (Ahmed et al., 2022). Why is this the case?

At low concentrations, cannabinoids preferentially bind to CB receptors located on glutamatergic neurons (Rey et al., 2012). Inhibition of glutamatergic neurons can attenuate stress-induced increases in norepinephrine levels (Reyes et al., 2012). Blocking CB_1 receptors increases anxiety levels (Bergamaschi et al., 2014) and produces anxiety-like behavior in animals (Thiemann et al., 2009). In fact, the CB_1 receptor antagonist rimonabant was approved in Europe as a weight-loss drug in 2006, but it was withdrawn from the market in 2008 due to serious side effects, including increased reports of anxiety (Sam et al., 2011).

Earlier, I mentioned that tDCS to the frontal cortex reverses anxiety-like behavior in rodents. This effect is negated following CB_1 receptor antagonism or knockdown of CB_1 receptors in the amygdala (Fang & Wang, 2021). Knockdown of amygdalar CB_1 receptors also leads to increased cortisol levels in the PNS (Zhu et al., 2023).

Preventing the breakdown of endocannabinoids produces anxiolytic effects, with this effect being reversed if CB_1 receptors are removed from amygdalar astrocytes (Duan et al., 2017).

As endocannabinoid concentrations increase, CB receptors on GABAergic neurons are activated (Rey et al., 2012). As you read above, inhibiting GABAergic neurotransmission leads to anxiety. Stimulating CB_1 receptors increases prefrontal cortical norepinephrine levels (Oropeza et al., 2005), which contributes to cannabinoid-induced anxiety. My roommate's girlfriend most likely experienced severe anxiety because she used more cannabis or a more potent strain of cannabis than what she typically used, which led to excessive inhibition of the GABAergic system.

Given the misuse potential of cannabis, there is increasing interest in examining how cannabidiol (CBD), another compound derived from the cannabis plant, influences stress and anxiety. CBD exerts anxiolytic-like effects in animals (Shu et al., 2024) and has been proposed as a potential treatment for anxiety disorders (Blessing et al., 2015; Han et al., 2024).

8.3.6 Endogenous Opioids

Opioid receptors are divided into three classes: mu, kappa, and delta. While each opioid receptor controls analgesia, mu opioid receptors have historically received considerable attention due to their contribution to the addiction process. Opioids have anxiolytic effects when taken acutely (Colasanti et al., 2011). Opioid-induced anxiolysis may stem from inhibition of the noradrenergic system. For example, stimulating mu opioid receptors decreases norepinephrine release in the hippocampus (Simmons et al., 1992) while such stimulation does not alter cortisol release (Pascoe et al., 2008). Like mu opioid receptor agonists, delta opioid receptor agonists do not alter cortisol levels (Pascoe et al., 2008), but they attenuate yohimbine-induced anxiogenic effects as these receptors are located on noradrenergic neurons in the locus coeruleus (Tkaczynski et al., 2022) and on CRH-containing neurons in the amygdala (Reyes et al., 2017).

Decreased mu opioid receptor availability is observed in individuals that score higher in a self-report *anxiety* measure, notably in regions such as the amygdala, the hippocampus, the thalamus, the ventral striatum, and the insula (Nummenmaa et al., 2020). Animals with delta opioid receptor deficiency exhibit *anxiety*-like phenotypes (Nakamoto & Tokuyama, 2023). Interestingly, mice lacking the mu opioid receptor show *less* aversion to social contact following social defeat *stress* despite showing comparable increases in corticosterone levels and increased *anxiety*-like behavior (e.g., spending more time in corners of open arena) (Komatsu et al., 2011). More work is needed to better understand why deletion of the mu opioid receptor decreases social aversion, especially as long-term stress is proposed to dysregulate these receptors, leading to social withdrawal (Meier et al., 2021).

Similar to cannabinoids, excessive opioid use is associated with elevated *anxiety* (Herlinger & Lingford-Hughes, 2022). When an individual uses opioids long-term, there are compensatory changes that occur to opioid receptors (Christie, 2008). These receptors can become desensitized, meaning they do not respond as easily when a ligand binds to them. Another process that can occur is *internalization*; that is, the receptor can be removed from the cell membrane. With fewer opioid receptors available, the noradrenergic system is no longer inhibited. Due to its own adaptations, particularly noradrenergic receptor upregulation, the noradrenergic system becomes hyperactive, potentiating anxiety and stress.

If processes like desensitization and internalization of mu opioid receptors exacerbate anxiety, one would assume that blocking these receptors would also worsen anxiety. Indeed, opioid receptor antagonists increase symptoms of anxiety and cortisol levels in humans (Cohen et al., 1983; Daubenmier et al., 2014; Torpy et al., 1995). Naloxone also potentiates shock-induced norepinephrine release in the amygdala (Quirarte et al., 1998). Interestingly, the opioid receptor antagonist naloxone (Narcan) has been shown to decrease cortisol levels (Burnett et al., 1999). I need to note that naloxone is a non-selective opioid receptor antagonist, meaning that it binds to the three different opioid receptor types. Whereas selective mu and delta opioid receptor antagonists increase anxiety/stress (Khroyan et al., 2015; Perrine et al., 2006), kappa receptor antagonists have anxiolytic effects (Van't Veer & Carlezon Jr., 2013).

While mu and delta opioid receptors are associated with reinforcement learning (Pradhan et al., 2011; Valentino & Volkow, 2018), kappa receptors are highly involved in *dysphoria* (Land et al., 2008), a feeling of unease or unhappiness. Kappa receptors are located in many of the stress-related brain regions discussed in this chapter, and their activation leads to increased cortisol levels (Pascoe et al., 2008). Stress causes dynorphin levels to increase, which then bind to kappa receptors, and repeated stress leads to an upregulation of kappa receptors, particularly in the PFC (Cole et al., 2024).

Comprehension Check 9 Compare and contrast how the endocannabinoid system and the opioid system regulate anxiety.

Both systems demonstrate a biphasic effect on anxiety. Low-level activation of both systems is generally anxiolytic, while high-level or prolonged activation can become anxiogenic. A key difference lies in the specific neurotransmitter systems that endocannabinoids and opioids modulate. The endocannabinoid system exerts its anxiolytic effects by inhibiting glutamatergic activity, which in turn can attenuate stress-induced increases in norepinephrine. In contrast, the opioid system primarily exerts its anxiolytic effects by inhibiting the noradrenergic system directly.

8.4 How Our Immune System Influences Maladaptive Stress

The last thing I want to cover in this chapter is the influence of our immune system on maladaptive stress. You learned a little about the immune system in Chap. 4 when I discussed pain perception. The immune system's purpose is to protect us from pathogens such as bacteria and viruses. When our immune system detects that a pathogen has infiltrated the body, it sets off a chain reaction of events, leading to inflammation. **Cytokines** are major contributors to inflammation and are produced by various cells of the immune system, including *macrophages* and *T lymphocytes* (often called T cells). Cytokines act as an attractor, bringing additional immune cells to the site of injury/damage. Chronic stress leads to increased pro-inflammatory markers in serum (Zhang et al., 2023). Cytokines cause physical changes to occur to the amygdala and increase its activity in response to social threats (Łoś & Waszkiewicz, 2021). Read the Experiment Spotlight below to learn more about cytokines and their relationship to anxiety and alcohol use disorder.

Experiment Spotlight: Circulating Immune and Endocrine Markers in Currently Drinking and Abstinent Individuals with Alcohol Use Disorder and Controls
The link to the article can be found here: https://pmc.ncbi.nlm.nih.gov/articles/PMC12046569/.

The purpose of the study conducted by Tyler et al. (2025) was to compare immune and endocrine biomarkers between individuals with alcohol use disorder ($n = 9$), individuals currently abstaining from alcohol use (for at least 6 weeks) ($n = 10$), and individuals that have never met the criteria for alcohol use disorder (i.e., controls) ($n = 12$). I will focus on the immune biomarkers only when discussing this study. You will learn more about endocrine biomarkers in Chap. 13. The immune biomarkers measured were different cytokines, including various *interleukins (IL)* such as IL-1RA, IL-6, IL-8, IL-10, IL-18, *chemokine (C–C motif) ligand 2 (CCL2)*, and *tumor necrosis factor-alpha (TNF-α)*.

Individuals abstinent from alcohol consumption showed elevated IL-8 levels compared to healthy controls and to individuals with alcohol use disorder. Additionally, when all participants were combined into a single analysis, correlation analyses revealed that IL-18 and TNF-α levels were associated with measures of alcohol use and dependence and anxiety. In individuals with current alcohol use disorder, TNF-α levels were positively correlated with measures of alcohol dependence, but were negatively correlated with the number of past heavy drinking days. In abstinent individuals, IL-18 levels were positively correlated with the severity of alcohol use and dependence while they were negatively correlated with anxiety.

Questions to Consider:

1. What is one potential reason why IL-8 levels were higher in the abstinent group compared to the healthy control group and to the group of individuals that currently consume alcohol?

2. TNF-α levels were negatively correlated with the number of heavy drinking days in individuals that have alcohol use disorder. What can account for this negative correlation?
3. How do the results of the present study support the hypothesis that the immune system contributes to anxiety disorders?

In the CNS, microglia serve as the primary immune cell. When microglia detect the presence of a pathogen, they release cytokines and can directly engage in **phagocytosis**. That is, microglia can engulf pathogens, other cells, or debris. Phagocytosis helps clear the CNS of materials that can harm the brain. However, chronic stress causes excessive activation of microglia, particularly in the PFC, the hippocampus, and the amygdala (Schramm & Waisman, 2022; Zhang et al., 2023). Increased proliferation of microglia increases neuroinflammation in these regions, which further potentiates maladaptive responses to stress.

Reducing hyperactivity of the immune system is a potential target for treating anxiety and anxiety-like disorders. Optogenetic stimulation of noradrenergic neurons projecting from the locus coeruleus to the hippocampus improves anxiety-like behavior by reducing cytokine production and microglia activation (Zou et al., 2025). Additional research shows that inhibitingmicroglia activation ameliorates anxiety-like behavior (Li et al., 2021; Wang et al., 2024). Increasing endocannabinoid levels and stimulating delta opioid receptors ameliorate anxiety-like behavior by inhibiting neuroinflammation (Kasatkina et al., 2021; Xu et al., 2023).

9 Chapter Summary

This chapter explored the multifaceted nature of stress and how its dysregulation can lead to anxiety and anxiety-like disorders such as PTSD. The HPA axis and the noradrenergic system are significant contributors to the stress response, but there are multiple brain regions implicated in maladaptive stress, particularly the amygdala, the frontal cortex, and the hippocampus. Likewise, multiple neurotransmitters systems beyond norepinephrine influence stress responses, including serotonin, dopamine, GABA, orexin, endocannabinoids, and endogenous opioids. Stress also activates the immune system, and a hyperactive immune response can cause neuroinflammation in brain regions implicated in stress by activating microglia and promoting the release of pro-inflammatory cytokines.

Answers to Experiment Spotlight Questions

1. Maintaining abstinence is stressful. Due to the numerous environmental stimuli that become paired with alcohol consumption, individuals trying to achieve sobriety are bombarded with reminders of their previous alcohol use. This stress can account for the elevated IL-8 levels observed in these individuals.

2. In individuals with an alcohol use disorder, withdrawal is a highly stressful event. Withdrawal from alcohol can even be fatal if severe enough. If individuals have had more heavy drinking days, this means they are less likely to experience withdrawal symptoms. This can lead to a decrease in the stress response.
3. When all participants were analyzed together, cytokine levels (specifically IL-18 and TNF-α) were positively associated with measures of anxiety. This is consistent with the hypothesis that heightened immune activity can trigger feelings of anxiety.

Glossary

Adrenocorticotropic hormone (ACTH) a hormone produced by the pituitary gland that stimulates the adrenal glands to release cortisol.

Anxiety a feeling of worry, nervousness, or unease, typically about an event or something with an uncertain outcome.

Corticotropin-releasing hormone (CRH) a hormone released by the hypothalamus that stimulates the pituitary gland to release adrenocorticotropic hormone (ACTH), initiating the stress response. Also called as corticotropin-releasing factor (CRF).

Cortisol a steroid hormone produced by the adrenal glands that helps the body respond to stress by increasing glucose availability and suppressing the immune system.

Cytokine a small protein that acts as a messenger molecule, regulating immune responses and cell communication.

Fear an emotional response experienced when perceiving a threat or danger, characterized by a feeling of dread and often a strong urge to avoid or escape.

Hypothalamic–pituitary–adrenal (HPA) axis a neuroendocrine system that regulates the body's response to stress.

Phagocytosis a cellular process where a cell engulfs pathogens, other cells, or debris, allowing lysosomes to break down the ingested material.

Social defeat the experience of being repeatedly overpowered or subordinated in social interactions.

Steroid a class of chemical compounds characterized by a four-ring structure.

Stress a state of mental or emotional strain or tension resulting from adverse or demanding circumstances.

Sympathetic chain a paired bundle of nerve fibers running along each side of the vertebral column, serving as a key component of the sympathetic nervous system and facilitating the distribution of its signals throughout the body.

Thigmotaxis a behavioral response to tactile stimuli, typically referring to an organism's movement toward or away from physical contact with surfaces or objects (e.g., rodents tend to stick close to walls, avoiding open spaces).

Yerkes-Dodson law hypothesis that performance increases with physiological or mental arousal (stress) up to a point, after which it decreases.

References

Admon, R., Leykin, D., Lubin, G., Engert, V., Andrews, J., Pruessner, J., & Hendler, T. (2013). Stress-induced reduction in hippocampal volume and connectivity with the ventromedial prefrontal cortex are related to maladaptive responses to stressful military service. *Human Brain Mapping, 34*(11), 2808–2816. https://doi.org/10.1002/hbm.22100

Ahmed, M., Boileau, I., Le Foll, B., Carvalho, A. F., & Kloiber, S. (2022). The endocannabinoid system in social anxiety disorder: From pathophysiology to novel therapeutics. *Brazilian Journal of Psychiatry, 44*(1), 81–93. https://doi.org/10.1590/1516-4446-2021-1926

Alexander, L., Wood, C. M., Gaskin, P. L. R., Sawiak, S. J., Fryer, T. D., Hong, Y. T., McIver, L., Clarke, H. F., & Roberts, A. C. (2020). Over-activation of primate subgenual cingulate cortex enhances the cardiovascular, behavioral and neural responses to threat. *Nature Communications, 11*(1), 5386. https://doi.org/10.1038/s41467-020-19167-0

American Psychiatric Association. (2013). *Diagnostic and statistical manual of mental disorders* (5th ed.). https://doi.org/10.1176/appi.books.9780890425596

Andela, C. D., van der Werff, S. J., Pannekoek, J. N., van den Berg, S. M., Meijer, O. C., van Buchem, M. A., Rombouts, S. A., van der Mast, R. C., Romijn, J. A., Tiemensma, J., Biermasz, N. R., van der Wee, N. J., & Pereira, A. M. (2013). Smaller grey matter volumes in the anterior cingulate cortex and greater cerebellar volumes in patients with long-term remission of Cushing's disease: A case-control study. *European Journal of Endocrinology, 169*(6), 811–819. https://doi.org/10.1530/EJE-13-0471

Arnsten, A. F., Mathew, R., Ubriani, R., Taylor, J. R., & Li, B. M. (1999). Alpha-1 noradrenergic receptor stimulation impairs prefrontal cortical cognitive function. *Biological Psychiatry, 45*(1), 26–31. https://doi.org/10.1016/s0006-3223(98)00296-0

Arora, I., Mal, P., Arora, P., Paul, A., & Kumar, M. (2024). GABAergic implications in anxiety and related disorders. *Biochemical and Biophysical Research Communications, 724*, 150218. https://doi.org/10.1016/j.bbrc.2024.150218

Aston-Jones, G., Akaoka, H., Charléty, P., & Chouvet, G. (1991). Serotonin selectively attenuates glutamate-evoked activation of noradrenergic locus coeruleus neurons. *The Journal of Neuroscience, 11*(3), 760–769. https://doi.org/10.1523/JNEUROSCI.11-03-00760.1991

Atrooz, F., Alkadhi, K. A., & Salim, S. (2021). Understanding stress: Insights from rodent models. *Current Research in Neurobiology, 2*, 100013. https://doi.org/10.1016/j.crneur.2021.100013

Ayala, A. R., Pushkas, J., Higley, J. D., Ronsaville, D., Gold, P. W., Chrousos, G. P., Pacak, K., Calis, K. A., Gerald, M., Lindell, S., Rice, K. C., & Cizza, G. (2004). Behavioral, adrenal, and sympathetic responses to long-term administration of an oral corticotropin-releasing hormone receptor antagonist in a primate stress paradigm. *The Journal of Clinical Endocrinology and Metabolism, 89*(11), 5729–5737. https://doi.org/10.1210/jc.2003-032170

Baksh, R. A., Ritchie, C. W., Terrera, G. M., Norton, J., Raymont, V., & Ritchie, K. (2021). The association between anxiety disorders and hippocampal volume in older adults. *Psychology and Aging, 36*(2), 288–297. https://doi.org/10.1037/pag0000597

Bauer, E. P. (2015). Serotonin in fear conditioning processes. *Behavioural Brain Research, 277*, 68–77. https://doi.org/10.1016/j.bbr.2014.07.028

Bergamaschi, M. M., Queiroz, R. H., Chagas, M. H., Linares, I. M., Arrais, K. C., de Oliveira, D. C., Queiroz, M. E., Nardi, A. E., Huestis, M. A., Hallak, J. E., Zuardi, A. W., Moreira, F. A., & Crippa, J. A. (2014). Rimonabant effects on anxiety induced by simulated public speaking in healthy humans: A preliminary report. *Human Psychopharmacology, 29*(1), 94–99. https://doi.org/10.1002/hup.2374

Betensky, J. D., Robinson, D. G., Gunduz-Bruce, H., Sevy, S., Lencz, T., Kane, J. M., Malhotra, A. K., Miller, R., McCormack, J., Bilder, R. M., & Szeszko, P. R. (2008). Patterns of stress in schizophrenia. *Psychiatry Research, 160*(1), 38–46. https://doi.org/10.1016/j.psychres.2007.06.001

Bienertova-Vasku, J., Lenart, P., & Scheringer, M. (2020). Eustress and distress: Neither good nor bad, but rather the same? *BioEssays, 42*(7), 1900238. https://doi.org/10.1002/bies.201900238

Blessing, E. M., Steenkamp, M. M., Manzanares, J., & Marmar, C. R. (2015). Cannabidiol as a potential treatment for anxiety disorders. *Neurotherapeutics, 12*(4), 825–836. https://doi.org/10.1007/s13311-015-0387-1

Blithikioti, C., Nuño, L., Guell, X., Pascual-Diaz, S., Gual, A., Balcells-Olivero, M, & Miquel, L. (2022). The cerebellum and psychological trauma: A systematic review of neuroimaging studies. *Neurobiology of Stress, 17*, 100429. https://doi.org/10.1016/j.ynstr.2022.100429

Braga, M. F., Aroniadou-Anderjaska, V., Manion, S. T., Hough, C. J., & Li, H. (2004). Stress impairs alpha(1A) adrenoceptor-mediated noradrenergic facilitation of GABAergic transmission in the basolateral amygdala. *Neuropsychopharmacology, 29*(1), 45–58. https://doi.org/10.1038/sj.npp.1300297

Buffalari, D. M., & Grace, A. A. (2009). Chronic cold stress increases excitatory effects of norepinephrine on spontaneous and evoked activity of basolateral amygdala neurons. *The International Journal of Neuropsychopharmacology, 12*(1), 95–107. https://doi.org/10.1017/S1461145708009140

Burnett, F. E., Scott, L. V., Weaver, M. G., Medbak, S. H., & Dinan, T. G. (1999). The effect of naloxone on adrenocorticotropin and cortisol release: Evidence for a reduced response in depression. *Journal of Affective Disorders, 53*(3), 263–268. https://doi.org/10.1016/s0165-0327(98)00127-x

Cameron, O. G., Zubieta, J. K., Grunhaus, L., & Minoshima, S. (2000). Effects of yohimbine on cerebral blood flow, symptoms, and physiological functions in humans. *Psychosomatic Medicine, 62*(4), 549–559. https://doi.org/10.1097/00006842-200007000-00014

Carnevali, L., Pattini, E., Sgoifo, A., & Ottaviani, C. (2020). Effects of prefrontal transcranial direct current stimulation on autonomic and neuroendocrine responses to psychosocial stress in healthy humans. *Stress, 23*(1), 26–36. https://doi.org/10.1080/10253890.2019.1625884

Chan, J., Ni, Y., Zhang, P., Zhang, J., & Chen, Y. (2017). D1-like dopamine receptor dysfunction in the lateral habenula nucleus increased anxiety-like behavior in rat. *Neuroscience, 340*, 542–550. https://doi.org/10.1016/j.neuroscience.2016.11.005

Chen, F. J., & Sara, S. J. (2007). Locus coeruleus activation by foot shock or electrical stimulation inhibits amygdala neurons. *Neuroscience, 144*(2), 472–481. https://doi.org/10.1016/j.neuroscience.2006.09.037

Chen, J. J., & Marsh, L. (2014). Anxiety in Parkinson's disease: Identification and management. *Therapeutic Advances in Neurological Disorders, 7*(1), 52–59. https://doi.org/10.1177/1756285613495723

Chen, P., Fan, Y., Li, Y., Sun, Z., Bissette, G., & Zhu, M. Y. (2012). Chronic social defeat up-regulates expression of norepinephrine transporter in rat brains. *Neurochemistry International, 60*(1), 9–20. https://doi.org/10.1016/j.neuint.2011.11.003

Chitsaz, A., Ebrahimi-Ghiri, M., Zarrindast, M. R., & Khakpai, F. (2025). Synergistic interaction between clonidine and ACPA on the modulation of anxiety-like behaviors in non-acute restraint stress and acute restraint stress conditions. *Brain Research, 1847*, 149304. https://doi.org/10.1016/j.brainres.2024.149304

Christie, M. J. (2008). Cellular neuroadaptations to chronic opioids: Tolerance, withdrawal and addiction. *British Journal of Pharmacology, 154*(2), 384–396. https://doi.org/10.1038/bjp.2008.100

Cohen, M. R., Cohen, R. M., Pickar, D., Weingartner, H., & Murphy, D. L. (1983). High-dose naloxone infusions in normals. Dose-dependent behavioral, hormonal, and physiological responses. *Archives of General Psychiatry, 40*(6), 613–619. https://doi.org/10.1001/archpsyc.1983.04390010023003

Colasanti, A., Rabiner, E. A., Lingford-Hughes, A., & Nutt, D. J. (2011). Opioids and anxiety. *Journal of Psychopharmacology, 25*(11), 1415–1433. https://doi.org/10.1177/0269881110367726

Cole, R. H., Moussawi, K., & Joffe, M. E. (2024). Opioid modulation of prefrontal cortex cells and circuits. *Neuropharmacology, 248*, 109891. https://doi.org/10.1016/j.neuropharm.2024.109891

Correia, R., Coimbra, B., Domingues, A. V., Wezik, M., Vieitas-Gaspar, N., Gaspar, R., Sousa, N., Pinto, L., Rodrigues, A. J., & Soares-Cunha, C. (2023). Involvement of nucleus accumbens D2-medium spiny neurons projecting to the ventral pallidum in anxiety-like behaviour. *Journal of Psychiatry & Neuroscience, 48*(4), E267–E284. https://doi.org/10.1503/jpn.220111

Crosswell, A. D., & Lockwood, K. G. (2020). Best practices for stress measurement: How to measure psychological stress in health research. *Health Psychology Open, 7*(2), 2055102920933072. https://doi.org/10.1177/2055102920933072

Daaka, Y., Luttrell, L. M., & Lefkowitz, R. J. (1997). Switching of the coupling of the beta2-adrenergic receptor to different G proteins by protein kinase A. *Nature, 390*(6655), 88–91. https://doi.org/10.1038/36362

Daubenmier, J., Lustig, R. H., Hecht, F. M., Kristeller, J., Woolley, J., Adam, T., Dallman, M., & Epel, E. (2014). A new biomarker of hedonic eating? A preliminary investigation of cortisol and nausea responses to acute opioid blockade. *Appetite, 74*, 92–100. https://doi.org/10.1016/j.appet.2013.11.014

Dębiec, J., Bush, D. E., & LeDoux, J. E. (2011). Noradrenergic enhancement of reconsolidation in the amygdala impairs extinction of conditioned fear in rats—A possible mechanism for the persistence of traumatic memories in PTSD. *Depression and Anxiety, 28*(3), 186–193. https://doi.org/10.1002/da.20803

DeGroot, S. R., Zhao-Shea, R., Chung, L., Klenowski, P. M., Sun, F., Molas, S., Gardner, P. D., Li, Y., & Tapper, A. R. (2020). Midbrain dopamine controls anxiety-like behavior by engaging unique interpeduncular nucleus microcircuitry. *Biological Psychiatry, 88*(11), 855–866. https://doi.org/10.1016/j.biopsych.2020.06.018

Díaz-Mataix, L., Piper, W. T., Schiff, H. C., Roberts, C. H., Campese, V. D., Sears, R. M., & LeDoux, J. E. (2017). Characterization of the amplificatory effect of norepinephrine in the acquisition of Pavlovian threat associations. *Learning & Memory, 24*(9), 432–439. https://doi.org/10.1101/lm.044412.116

Dimonte, S., Sikora, V., Bove, M., Morgese, M. G., Tucci, P., Schiavone, S., & Trabace, L. (2023). Social isolation from early life induces anxiety-like behaviors in adult rats: Relation to neuroendocrine and neurochemical dysfunctions. *Biomedicine & Pharmacotherapy, 158*, 114181. https://doi.org/10.1016/j.biopha.2022.114181

Dixon, M. L., Thiruchselvam, R., Todd, R., & Christoff, K. (2017). Emotion and the prefrontal cortex: An integrative review. *Psychological Bulletin, 143*(10), 1033–1081. https://doi.org/10.1037/bul0000096

Dominguez, G., Henkous, N., Prevot, T., David, V., Guillou, J. L., Belzung, C., Mons, N., & Béracochéa, D. (2019). Sustained corticosterone rise in the prefrontal cortex is a key factor for chronic stress-induced working memory deficits in mice. *Neurobiology of Stress, 10*, 100161. https://doi.org/10.1016/j.ynstr.2019.100161

Doubliez, A., Nio, E., Senovilla-Sanz, F., Spatharioti, V., Apps, R., Timmann, D., & Lawrenson, C. L. (2023). The cerebellum and fear extinction: Evidence from rodent and human studies. *Frontiers in Systems Neuroscience, 17*, 1166166. https://doi.org/10.3389/fnsys.2023.1166166

Duan, T., Gu, N., Wang, Y., Wang, F., Zhu, J., Fang, Y., Shen, Y., Han, J., & Zhang, X. (2017). Fatty acid amide hydrolase inhibitors produce rapid anti-anxiety responses through amygdala long-term depression in male rodents. *Journal of Psychiatry & Neuroscience, 42*(4), 230–241. https://doi.org/10.1503/jpn.160116

Fernández-Pastor, B., Mateo, Y., Gómez-Urquijo, S., & Javier Meana, J. (2005). Characterization of noradrenaline release in the locus coeruleus of freely moving awake rats by in vivo microdialysis. *Psychopharmacology (Berl), 180*(3), 570–579. https://doi.org/10.1007/s00213-005-2181-y

Garakani, A., Murrough, J. W., Freire, R. C., Thom, R. P., Larkin, K., Buono, F. D., & Iosifescu, D. V. (2020). Pharmacotherapy of anxiety disorders: Current and emerging treatment options. *Frontiers in Psychiatry, 11*, 595584. https://doi.org/10.3389/fpsyt.2020.595584

Gulfo, M. C., Lebowitz, J. J., Ramos, C., Hwang, D. W., Nasrallah, K., & Castillo, P. E. (2023). Dopamine D2 receptors in hilar mossy cells regulate excitatory transmission and hippocampal function. *Proceedings of the National Academy of Sciences of the United States of America, 120*(50), e2307509120. https://doi.org/10.1073/pnas.2307509120

Haddjeri, N., de Montigny, C., & Blier, P. (1997). Modulation of the firing activity of noradrenergic neurones in the rat locus coeruleus by the 5-hydroxtryptamine system. *British Journal of Pharmacology, 120*(5), 865–875. https://doi.org/10.1038/sj.bjp.0700968

Hajós-Korcsok, E., Robinson, D. D., Yu, J. H., Fitch, C. S., Walker, E., & Merchant, K. M. (2003). Rapid habituation of hippocampal serotonin and norepinephrine release and anxiety-related behaviors, but not plasma corticosterone levels, to repeated footshock stress in rats. *Pharmacology, Biochemistry, and Behavior, 74*(3), 609–616. https://doi.org/10.1016/s0091-3057(02)01047-x

Hakamata, Y., Komi, S., Moriguchi, Y., Izawa, S., Motomura, Y., Sato, E., Mizukami, S., Kim, Y., Hanakawa, T., Inoue, Y., & Tagaya, H. (2017). Amygdala-centred functional connectivity affects daily cortisol concentrations: A putative link with anxiety. *Scientific Reports, 7*(1), 8313. https://doi.org/10.1038/s41598-017-08918-7

Hakamata, Y., Mizukami, S., Izawa, S., Moriguchi, Y., Hori, H., Kim, Y., Hanakawa, T., Inoue, Y., & Tagaya, H. (2020). Basolateral amygdala connectivity with subgenual anterior cingulate cortex represents enhanced fear-related memory encoding in anxious humans. *Biological Psychiatry. Cognitive Neuroscience and Neuroimaging, 5*(3), 301–310. https://doi.org/10.1016/j.bpsc.2019.11.008

Hamdan, J. N., Sierra-Fonseca, J. A., Flores, R. J., Saucedo, S., Jr., Miranda-Arango, M., O'Dell, L. E., & Gosselink, K. L. (2022). Early-life adversity increases anxiety-like behavior and modifies synaptic protein expression in a region-specific manner. *Frontiers in Behavioral Neuroscience, 16*, 1008556. https://doi.org/10.3389/fnbeh.2022.1008556

Han, K., Wang, J. Y., Wang, P. Y., & Peng, Y. C. (2024). Therapeutic potential of cannabidiol (CBD) in anxiety disorders: A systematic review and meta-analysis. *Psychiatry Research, 339*, 116049. https://doi.org/10.1016/j.psychres.2024.116049

Hartmann, J., Dedic, N., Pöhlmann, M. L., Häusl, A., Karst, H., Engelhardt, C., Westerholz, S., Wagner, K. V., Labermaier, C., Hoeijmakers, L., Kertokarijo, M., Chen, A., Joëls, M., Deussing, J. M., & Schmidt, M. V. (2017). Forebrain glutamatergic, but not GABAergic, neurons mediate anxiogenic effects of the glucocorticoid receptor. *Molecular Psychiatry, 22*(3), 466–475. https://doi.org/10.1038/mp.2016.87

Herlinger, K., & Lingford-Hughes, A. (2022). Opioid use disorder and the brain: A clinical perspective. *Addiction, 117*(2), 495–505. https://doi.org/10.1111/add.15636

Hilberdink, C. E., van Zuiden, M., Schrantee, A., Korosi, A., Kaiser, A., Zhutovsky, P., Ginty, A. T., Ensink, J. B. M., Lindauer, R. J. L., Vrijkotte, T. G. M., & de Rooij, S. R. (2021). Dysregulated functional brain connectivity in response to acute social-evaluative stress in adolescents with PTSD symptoms. *European Journal of Psychotraumatology, 12*(1), 1880727. https://doi.org/10.1080/20008198.2021.1880727

Hurlemann, R., Walter, H., Rehme, A. K., Kukolja, J., Santoro, S. C., Schmidt, C., Schnell, K., Musshoff, F., Keysers, C., Maier, W., Kendrick, K. M., & Onur, O. A. (2010). Human amygdala reactivity is diminished by the β-noradrenergic antagonist propranolol. *Psychological Medicine, 40*(11), 1839–1848. https://doi.org/10.1017/S0033291709992376

Faggiani, E., Naudet, F., Janssen, M. L. F., Temel, Y., & Benazzouz, A. (2018). Serotonergic neurons mediate the anxiolytic effect of l-DOPA: Neuronal correlates in the amygdala. *Neurobiology of Disease, 110*, 20–28. https://doi.org/10.1016/j.nbd.2017.11.001

Fan, Y., Chen, P., Li, Y., Cui, K., Noel, D. M., Cummins, E. D., Peterson, D. J., Brown, R. W., & Zhu, M. Y. (2014). Corticosterone administration up-regulated expression of norepinephrine

transporter and dopamine β-hydroxylase in rat locus coeruleus and its terminal regions. *Journal of Neurochemistry, 128*(3), 445–458. https://doi.org/10.1111/jnc.12459
Fang, G., & Wang, Y. (2021). Transcranial direct current stimulation (tDCS) produce anti-anxiety response in acute stress exposure rats via activation of amygdala CB1R. *Behavioural Brain Research, 400*, 113050. https://doi.org/10.1016/j.bbr.2020.113050
Feinstein, J. S., Adolphs, R., Damasio, A., & Tranel, D. (2011). The human amygdala and the induction and experience of fear. *Current Biology, 21*(1), 34–38. https://doi.org/10.1016/j.cub.2010.11.042
Fentress, H. M., Klar, R., Krueger, J. J., Sabb, T., Redmon, S. N., Wallace, N. M., Shirey-Rice, J. K., & Hahn, M. K. (2013). Norepinephrine transporter heterozygous knockout mice exhibit altered transport and behavior. *Genes, Brain, and Behavior, 12*(8), 749–759. https://doi.org/10.1111/gbb.12084
File, S. E., Curle, P. F., Baldwin, H. A., & Neal, M. J. (1987). Anxiety in the rat is associated with decreased release of 5-HT and glycine from the hippocampus. *Neuroscience Letters, 83*(3), 318–322. https://doi.org/10.1016/0304-3940(87)90107-8
Grillon, C., Hale, E., Lieberman, L., Davis, A., Pine, D. S., & Ernst, M. (2015). The CRH1 antagonist GSK561679 increases human fear but not anxiety as assessed by startle. *Neuropsychopharmacology, 40*(5), 1064–1071. https://doi.org/10.1038/npp.2014.316
Grillon, C., Heller, R., Hirschhorn, E., Kling, M. A., Pine, D. S., Schulkin, J., & Vythilingam, M. (2011). Acute hydrocortisone treatment increases anxiety but not fear in healthy volunteers: A fear-potentiated startle study. *Biological Psychiatry, 69*(6), 549–555. https://doi.org/10.1016/j.biopsych.2010.12.013
Hollowell, S., Aneja, E., & Schwartz, T. (2025). Exploring the potential benefits of clonidine for anxiety disorders. *CNS Spectrums*, 1–15. Advance online publication. https://doi.org/10.1017/S1092852925100266
Jankord, R., & Herman, J. P. (2008). Limbic regulation of hypothalamo-pituitary-adrenocortical function during acute and chronic stress. *Annals of the New York Academy of Sciences, 1148*, 64–73. https://doi.org/10.1196/annals.1410.012
Johnson, P. L., Molosh, A., Fitz, S. D., Truitt, W. A., & Shekhar, A. (2012). Orexin, stress, and anxiety/panic states. *Progress in Brain Research, 198*, 133–161. https://doi.org/10.1016/B978-0-444-59489-1.00009-4
Kaehler, S. T., Singewald, N., & Philippu, A. (1999). Dependence of serotonin release in the locus coeruleus on dorsal raphe neuronal activity. *Naunyn-Schmiedeberg's Archives of Pharmacology, 359*(5), 386–393. https://doi.org/10.1007/pl00005365
Kaehler, S. T., Singewald, N., Sinner, C., Thurnher, C., & Philippu, A. (2000). Conditioned fear and inescapable shock modify the release of serotonin in the locus coeruleus. *Brain Research, 859*(2), 249–254. https://doi.org/10.1016/s0006-8993(00)01967-3
Kasatkina, L. A., Rittchen, S., & Sturm, E. M. (2021). Neuroprotective and immunomodulatory action of the endocannabinoid system under neuroinflammation. *International Journal of Molecular Sciences, 22*(11), 5431. https://doi.org/10.3390/ijms22115431
Kawanabe, S., Mori, M., Harada, H., Murata, Y., Ohe, K., & Enjoji, M. (2023). Upregulations of α_1 adrenergic receptors and noradrenaline synthases in the medial prefrontal cortex are associated with emotional and cognitive dysregulation induced by post-weaning social isolation in male rats. *Neuroscience Letters, 797*, 137071. https://doi.org/10.1016/j.neulet.2023.137071
Kenwood, M. M., Kalin, N. H., & Barbas, H. (2022). The prefrontal cortex, pathological anxiety, and anxiety disorders. *Neuropsychopharmacology, 47*(1), 260–275. https://doi.org/10.1038/s41386-021-01109-z
Khroyan, T. V., Wu, J., Polgar, W. E., Cami-Kobeci, G., Fotaki, N., Husbands, S. M., & Toll, L. (2015). BU08073 a buprenorphine analogue with partial agonist activity at μ-receptors in vitro but long-lasting opioid antagonist activity in vivo in mice. *British Journal of Pharmacology, 172*(2), 668–680. https://doi.org/10.1111/bph.12796
Kim, E. J., & Kim, J. J. (2023). Neurocognitive effects of stress: A metaparadigm perspective. *Molecular Psychiatry, 28*(7), 2750–2763. https://doi.org/10.1038/s41380-023-01986-4

Komatsu, H., Ohara, A., Sasaki, K., Abe, H., Hattori, H., Hall, F. S., Uhl, G. R., & Sora, I. (2011). Decreased response to social defeat stress in μ-opioid-receptor knockout mice. *Pharmacology, Biochemistry, and Behavior, 99*(4), 676–682. https://doi.org/10.1016/j.pbb.2011.06.008

Koolhaas, J. M., Coppens, C. M., de Boer, S. F., Buwalda, B., Meerlo, P., & Timmermans, P. J. (2013). The resident-intruder paradigm: A standardized test for aggression, violence and social stress. *Journal of Visualized Experiments, 77*, e4367. https://doi.org/10.3791/4367

Kredlow, M. A., Fenster, R. J., Laurent, E. S., Ressler, K. J., & Phelps, E. A. (2022). Prefrontal cortex amygdala and threat processing: Implications for PTSD. *Neuropsychopharmacology 47*(1), 247–259. https://doi.org/10.1038/s41386-021-01155-7

Land, B. B., Bruchas, M. R., Lemos, J. C., Xu, M., Melief, E. J., & Chavkin, C. (2008). The dysphoric component of stress is encoded by activation of the dynorphin kappa-opioid system. *The Journal of Neuroscience, 28*(2), 407–414. https://doi.org/10.1523/JNEUROSCI.4458-07.2008

Lanzenberger, R. R., Mitterhauser, M., Spindelegger, C., Wadsak, W., Klein, N., Mien, L. K., Holik, A., Attarbaschi, T., Mossaheb, N., Sacher, J., Geiss-Granadia, T., Kletter, K., Kasper, S., & Tauscher, J. (2007). Reduced serotonin-1A receptor binding in social anxiety disorder. *Biological Psychiatry, 61*(9), 1081–1089. https://doi.org/10.1016/j.biopsych.2006.05.022

Lei, Z., Lam, Y., Li, C., Fu, Z., Ramkrishnan, A. S., Liu, S., & Li, Y. (2022). β2-adrenoceptors in the medial prefrontal cortex excitatory neurons regulate anxiety-like behavior in mice. *International Journal of Molecular Sciences, 23*(10), 5578. https://doi.org/10.3390/ijms23105578

Leschak, C. J., Hornstein, E. A., Byrne Haltom, K. E., Johnson, K. L., Breen, E. C., Irwin, M. R., & Eisenberger, N. I. (2022). Ventromedial prefrontal cortex activity differentiates sick from healthy faces: Associations with inflammatory responses and disease avoidance motivation. *Brain, Behavior, and Immunity, 100*, 48–54. https://doi.org/10.1016/j.bbi.2021.11.011

Levy, B. H., & Tasker, J. G. (2012). Synaptic regulation of the hypothalamic-pituitary-adrenal axis and its modulation by glucocorticoids and stress. *Frontiers in Cellular Neuroscience, 6*, 24. https://doi.org/10.3389/fncel.2012.00024

Li, S., Liao, Y., Dong, Y., Li, X., Li, J., Cheng, Y., Cheng, J., & Yuan, Z. (2021). Microglial deletion and inhibition alleviate behavior of post-traumatic stress disorder in mice. *Journal of Neuroinflammation, 18*(1), 7. https://doi.org/10.1186/s12974-020-02069-9

Liberzon, I., & Sripada, C. S. (2008). The functional neuroanatomy of PTSD: A critical review. *Progress in Brain Research, 167*, 151–169. https://doi.org/10.1016/S0079-6123(07)67011-3

Lin, C. C., Tung, C. S., & Liu, Y. P. (2016). Escitalopram reversed the traumatic stress-induced depressed and anxiety-like symptoms but not the deficits of fear memory. *Psychopharmacology (Berl), 233*(7), 1135–1146. https://doi.org/10.1007/s00213-015-4194-5

Lipski, W. J., & Grace, A. A. (2013). Footshock-induced responses in ventral subiculum neurons are mediated by locus coeruleus noradrenergic afferents. *European Neuropsychopharmacology, 23*(10), 1320–1328. https://doi.org/10.1016/j.euroneuro.2012.10.007

Liu, L., Zhang, L., Wang, T., & Chen, L. (2019). Dopamine D_1 receptor in the medial prefrontal cortex mediates anxiety-like behaviors induced by blocking glutamatergic activity of the ventral hippocampus in rats. *Brain Research, 1704*, 59–67. https://doi.org/10.1016/j.brainres.2018.09.024

Liu, Q., Liu, Y., Leng, X., Han, J., Xia, F., & Chen, H. (2020). Impact of chronic stress on attention control: Evidence from behavioral and event-related potential analyses. *Neuroscience Bulletin, 36*(11), 1395–1410. https://doi.org/10.1007/s12264-020-00549-9

Liu, Z. P., Song, C., Wang, M., He, Y., Xu, X. B., Pan, H. Q., Chen, W. B., Peng, W. J., & Pan, B. X. (2014). Chronic stress impairs GABAergic control of amygdala through suppressing the tonic GABAA receptor currents. *Molecular Brain, 7*, 32. https://doi.org/10.1186/1756-6606-7-32

Łoś, K., & Waszkiewicz, N. (2021). Biological markers in anxiety disorders. *Journal of Clinical Medicine, 10*(8), 1744. https://doi.org/10.3390/jcm10081744

Lucassen, P. J., Oomen, C. A., Naninck, E. F., Fitzsimons, C. P., van Dam, A. M., Czeh, B., & Korosi, A. (2015). Regulation of adult neurogenesis and plasticity by (early) stress, glucocorticoids, and inflammation. *Cold Spring Harbor Perspectives in Biology, 7*(9), a021303. https://doi.org/10.1101/cshperspect.a021303

Luo, X. M., Yuan, S. N., Guan, X. T., Xie, X., Shao, F., & Wang, W. W. (2014). Juvenile stress affects anxiety-like behavior and limbic monoamines in adult rats. *Physiology & Behavior, 135*, 7–16. https://doi.org/10.1016/j.physbeh.2014.05.035

Maletic, V., Eramo, A., Gwin, K., Offord, S. J., & Duffy, R. A. (2017). The role of norepinephrine and its α-adrenergic receptors in the pathophysiology and treatment of major depressive disorder and schizophrenia: A systematic review. *Frontiers in Psychiatry, 8*, 42. https://doi.org/10.3389/fpsyt.2017.00042

Marazziti, D., Abelli, M., Baroni, S., Carpita, B., Ramacciotti, C. E., & Dell'Osso, L. (2015). Neurobiological correlates of social anxiety disorder: An update. *CNS Spectrums, 20*(2), 100–111. https://doi.org/10.1017/S109285291400008X

McCall, J. G., Al-Hasani, R., Siuda, E. R., Hong, D. Y., Norris, A. J., Ford, C. P., & Bruchas, M. R. (2015). CRH engagement of the locus coeruleus noradrenergic system mediates stress-induced anxiety. *Neuron, 87*(3), 605–620. https://doi.org/10.1016/j.neuron.2015.07.002

Mehta, N. D., Stevens, J. S., Li, Z., Fani, N., Gillespie, C. F., Ravi, M., Michopoulos, V., & Felger, J. C. (2022). Inflammation, amygdala-ventromedial prefrontal functional connectivity and symptoms of anxiety and PTSD in African American women recruited from an inner-city hospital: Preliminary results. *Brain, Behavior, and Immunity, 105*, 122–130. https://doi.org/10.1016/j.bbi.2022.06.013

Meier, I. M., van Honk, J., Bos, P. A., & Terburg, D. (2021). A mu-opioid feedback model of human social behavior. *Neuroscience and Biobehavioral Reviews, 121*, 250–258. https://doi.org/10.1016/j.neubiorev.2020.12.013

Millan, M. J., Lejeune, F., Gobert, A., Brocco, M., Auclair, A., Bosc, C., Rivet, J. M., Lacoste, J. M., Cordi, A., & Dekeyne, A. (2000). S18616, a highly potent spiroimidazoline agonist at alpha(2)-adrenoceptors: II. Influence on monoaminergic transmission, motor function, and anxiety in comparison with dexmedetomidine and clonidine. *The Journal of Pharmacology and Experimental Therapeutics, 295*(3), 1206–1222.

Mirza, N. R., Nielsen, E. Ø., & Troelsen, K. B. (2007). Serotonin transporter density and anxiolytic-like effects of antidepressants in mice. *Progress in Neuro-Psychopharmacology & Biological Psychiatry, 31*(4), 858–866. https://doi.org/10.1016/j.pnpbp.2007.01.020

Mitra, R., & Sapolsky, R. M. (2008). Acute corticosterone treatment is sufficient to induce anxiety and amygdaloid dendritic hypertrophy. *Proceedings of the National Academy of Sciences of the United States of America, 105*(14), 5573–5578. https://doi.org/10.1073/pnas.0705615105

Mo, B., Feng, N., Renner, K., & Forster, G. (2008). Restraint stress increases serotonin release in the central nucleus of the amygdala via activation of corticotropin-releasing factor receptors. *Brain Research Bulletin, 76*(5), 493–498. https://doi.org/10.1016/j.brainresbull.2008.02.011

Naegeli, C., Zeffiro, T., Piccirelli, M., Jaillard, A., Weilenmann, A., Hassanpour, K., Schick, M., Rufer, M., Orr, S. P., & Mueller-Pfeiffer, C. (2018). Locus coeruleus activity mediates hyperresponsiveness in posttraumatic stress disorder. *Biological Psychiatry, 83*(3), 254–262. https://doi.org/10.1016/j.biopsych.2017.08.021

Nakamoto, K., & Tokuyama, S. (2023). Stress-induced changes in the endogenous opioid system cause dysfunction of pain and emotion regulation. *International Journal of Molecular Sciences, 24*(14), 11713. https://doi.org/10.3390/ijms241411713

Netto, S. M., Silveira, R., Coimbra, N. C., Joca, S. R., & Guimarães, F. S. (2002). Anxiogenic effect of median raphe nucleus lesion in stressed rats. *Progress in Neuro-Psychopharmacology & Biological Psychiatry, 26*(6), 1135–1141. https://doi.org/10.1016/s0278-5846(02)00248-8

Nikolaus, S., Antke, C., Beu, M., & Müller, H. W. (2010). Cortical GABA, striatal dopamine and midbrain serotonin as the key players in compulsive and anxiety disorders—Results from in vivo imaging studies. *Reviews in the Neurosciences, 21*(2), 119–139. https://doi.org/10.1515/revneuro.2010.21.2.119

Nummenmaa, L., Karjalainen, T., Isojärvi, J., Kantonen, T., Tuisku, J., Kaasinen, V., Joutsa, J., Nuutila, P., Kalliokoski, K., Hirvonen, J., Hietala, J., & Rinne, J. (2020). Lowered endogenous mu-opioid receptor availability in subclinical depression and anxiety. *Neuropsychopharmacology, 45*(11), 1953–1959. https://doi.org/10.1038/s41386-020-0725-9

Nuss, P. (2015). Anxiety disorders and GABA neurotransmission: A disturbance of modulation. *Neuropsychiatric Disease and Treatment, 11*, 165–175. https://doi.org/10.2147/NDT.S58841

Onur, O. A., Walter, H., Schlaepfer, T. E., Rehme, A. K., Schmidt, C., Keysers, C., Maier, W., & Hurlemann, R. (2009). Noradrenergic enhancement of amygdala responses to fear. *Social Cognitive and Affective Neuroscience, 4*(2), 119–126. https://doi.org/10.1093/scan/nsn049

Oropeza, V. C., Page, M. E., & Van Bockstaele, E. J. (2005). Systemic administration of WIN 55,212–2 increases norepinephrine release in the rat frontal cortex. *Brain Research, 1046*(1–2), 45–54. https://doi.org/10.1016/j.brainres.2005.03.036

Owens-French, J., Li, S. B., Francois, M., Leigh Townsend, R., Daniel, M., Soulier, H., Turner, A., de Lecea, L., Münzberg, H., Morrison, C., & Qualls-Creekmore, E. (2022). Lateral hypothalamic galanin neurons are activated by stress and blunt anxiety-like behavior in mice. *Behavioural Brain Research, 423*, 113773. https://doi.org/10.1016/j.bbr.2022.113773

Palotai, M., Telegdy, G., & Jászberényi, M. (2014). Orexin A-induced anxiety-like behavior is mediated through GABA-ergic, α- and β-adrenergic neurotransmissions in mice. *Peptides, 57*, 129–134. https://doi.org/10.1016/j.peptides.2014.05.003

Pascoe, J. E., Williams, K. L., Mukhopadhyay, P., Rice, K. C., Woods, J. H., & Ko, M. C. (2008). Effects of mu, kappa, and delta opioid receptor agonists on the function of hypothalamic-pituitary-adrenal axis in monkeys. *Psychoneuroendocrinology, 33*(4), 478–486. https://doi.org/10.1016/j.psyneuen.2008.01.006

Patel, V. K., Vaishnaw, A., Shirbhate, E., Kore, R., Singh, V., Veerasamy, R., & Rajak, H. (2024). Cortisol as a target for treating mental disorders: A promising avenue for therapy. *Mini Reviews in Medicinal Chemistry, 24*(6), 588–600. https://doi.org/10.2174/0113895575262104230928042150

Peng, B., Xu, Q., Liu, J., Guo, S., Borgland, S. L., & Liu, S. (2021). Corticosterone attenuates reward-seeking behavior and increases anxiety via D2 receptor signaling in ventral tegmental area dopamine neurons. *The Journal of Neuroscience, 41*(7), 1566–1581. https://doi.org/10.1523/JNEUROSCI.2533-20.2020

Perrine, S. A., Hoshaw, B. A., & Unterwald, E. M. (2006). Delta opioid receptor ligands modulate anxiety-like behaviors in the rat. *British Journal of Pharmacology, 147*(8), 864–872. https://doi.org/10.1038/sj.bjp.0706686

Pradhan, A. A., Befort, K., Nozaki, C., Gavériaux-Ruff, C., & Kieffer, B. L. (2011). The delta opioid receptor: An evolving target for the treatment of brain disorders. *Trends in Pharmacological Sciences, 32*(10), 581–590. https://doi.org/10.1016/j.tips.2011.06.008

Prajapati, S. K., & Krishnamurthy, S. (2021). Non-selective orexin-receptor antagonist attenuates stress-re-stress-induced core PTSD-like symptoms in rats: Behavioural and neurochemical analyses. *Behavioural Brain Research, 399*, 113015. https://doi.org/10.1016/j.bbr.2020.113015

Quirarte, G. L., Galvez, R., Roozendaal, B., & McGaugh, J. L. (1998). Norepinephrine release in the amygdala in response to footshock and opioid peptidergic drugs. *Brain Research, 808*(2), 134–140. https://doi.org/10.1016/s0006-8993(98)00795-1

Rajalingam, D., Nymoen, I., Jacobsen, D. P., Eriksen, M. B., Dissen, E., Nielsen, M. B., Einarsen, S. V., & Gjerstad, J. (2020). Repeated social defeat promotes persistent inflammatory changes in splenic myeloid cells; decreased expression of β-arrestin-2 (ARRB2) and increased expression of interleukin-6 (IL-6). *BMC Neuroscience, 21*(1), 25. https://doi.org/10.1186/s12868-020-00574-4

Rey, A. A., Purrio, M., Viveros, M. P., & Lutz, B. (2012). Biphasic effects of cannabinoids in anxiety responses: CB1 and GABA(B) receptors in the balance of GABAergic and glutamatergic neurotransmission. *Neuropsychopharmacology, 37*(12), 2624–2634. https://doi.org/10.1038/npp.2012.123

Reyes, B. A. S., Kravets, J. L., Connelly, K. L., Unterwald, E. M., & Van Bockstaele, E. J. (2017). Localization of the delta opioid receptor and corticotropin-releasing factor in the amygdalar complex: Role in anxiety. *Brain Structure & Function, 222*(2), 1007–1026. https://doi.org/10.1007/s00429-016-1261-6

Reyes, B. A., Szot, P., Sikkema, C., Cathel, A. M., Kirby, L. G., & Van Bockstaele, E. J. (2012). Stress-induced sensitization of cortical adrenergic receptors following a history of cannabinoid exposure. *Experimental Neurology, 236*(2), 327–335. https://doi.org/10.1016/j.expneurol.2012.05.016

Rosada, C., Lipka, R., Metz, S., Otte, C., Heekeren, H., & Wingenfeld, K. (2024). Effects of stress-related neuromodulators on amygdala and hippocampus resting state functional connectivity. *Journal of Psychopharmacology, 38*(7), 604–614. https://doi.org/10.1177/02698811241260972

Sabban, E. L., Laukova, M., Alaluf, L. G., Olsson, E., & Serova, L. I. (2015). Locus coeruleus response to single-prolonged stress and early intervention with intranasal neuropeptide Y. *Journal of Neurochemistry, 135*(5), 975–986. https://doi.org/10.1111/jnc.13347

Sam, A. H., Salem, V., & Ghatei, M. A. (2011). Rimonabant: From RIO to ban. *Journal of Obesity, 2011*, 432607. https://doi.org/10.1155/2011/432607

Sarfi, M., Elahdadi Salmani, M., Lashkarbolouki, T., & Goudarzi, I. (2025). Divergent effects of noradrenergic activation and orexin receptor 1 blockade on hippocampal structure, anxiety-like behavior, and social interaction following chronic stress. *Pharmacology, Biochemistry, and Behavior, 250*, 173997. https://doi.org/10.1016/j.pbb.2025.173997

Schramm, E., & Waisman, A. (2022). Microglia as central protagonists in the chronic stress response. *Neurology: Neuroimmunology & Neuroinflammation, 9*(6), e200023. https://doi.org/10.1212/NXI.0000000000200023

Sears, R. M., Fink, A. E., Wigestrand, M. B., Farb, C. R., de Lecea, L., & Ledoux, J. E. (2013). Orexin/hypocretin system modulates amygdala-dependent threat learning through the locus coeruleus. *Proceedings of the National Academy of Sciences of the United States of America, 110*(50), 20260–20265. https://doi.org/10.1073/pnas.1320325110

Shields, G. S., Sazma, M. A., & Yonelinas, A. P. (2016). The effects of acute stress on core executive functions: A meta-analysis and comparison with cortisol. *Neuroscience and Biobehavioral Reviews, 68*, 651–668. https://doi.org/10.1016/j.neubiorev.2016.06.038

Shu, G., He, Y., Suo, J., Wu, C., Gong, X., Xiang, Y., Yang, W., Cheng, J., Wang, Y., Chen, W., & Shen, J. (2024). Cannabidiol exhibits anxiolytic-like effects and antipsychotic-like effects in mice models. *Neuroscience Letters, 826*, 137723. https://doi.org/10.1016/j.neulet.2024.137723

Simmons, M. L., Wagner, J. J., Caudle, R. M., & Chavkin, C. (1992). Endogenous opioid regulation of norepinephrine release in guinea pig hippocampus. *Neuroscience Letters, 141*(1), 84–88. https://doi.org/10.1016/0304-3940(92)90340-d

Sinha, R. (2008). Chronic stress, drug use, and vulnerability to addiction. *Annals of the New York Academy of Sciences, 1141*, 105–130. https://doi.org/10.1196/annals.1441.030

Sousa-Lima, J., Moreira, P. S., Raposo-Lima, C., Sousa, N., & Morgado, P. (2019). Relationship between obsessive compulsive disorder and cortisol: Systematic review and meta-analysis. *European Neuropsychopharmacology, 29*(11), 1185–1198. https://doi.org/10.1016/j.euroneuro.2019.09.001

Stanley, A. T., Post, M. R., Lacefield, C., Sulzer, D., & Miniaci, M. C. (2023). Norepinephrine release in the cerebellum contributes to aversive learning. *Nature Communications, 14*(1), 4852. https://doi.org/10.1038/s41467-023-40548-8

Sun, Y., Gooch, H., & Sah, P. (2020). Fear conditioning and the basolateral amygdala. *F1000Research, 9*, F1000 Faculty Rev-53. https://doi.org/10.12688/f1000research.21201.1

Szabo, S. T., & Blier, P. (2001). Functional and pharmacological characterization of the modulatory role of serotonin on the firing activity of locus coeruleus norepinephrine neurons. *Brain Research, 922*(1), 9–20. https://doi.org/10.1016/s0006-8993(01)03121-3

Thiemann, G., Watt, C. A., Ledent, C., Molleman, A., & Hasenöhrl, R. U. (2009). Modulation of anxiety by acute blockade and genetic deletion of the CB(1) cannabinoid receptor in mice together with biogenic amine changes in the forebrain. *Behavioural Brain Research, 200*(1), 60–67. https://doi.org/10.1016/j.bbr.2008.12.035

Timothy, C., Costall, B., & Smythe, J. W. (1999). Effects of SCH23390 and raclopride on anxiety-like behavior in rats tested in the black-white box. *Pharmacology, Biochemistry, and Behavior, 62*(2), 323–327. https://doi.org/10.1016/s0091-3057(98)00157-9

Tkaczynski, J. A., Borodovitsyna, O., & Chandler, D. J. (2022). Delta opioid receptors and enkephalinergic signaling within locus coeruleus promote stress resilience. *Brain Sciences, 12*(7), 860. https://doi.org/10.3390/brainsci12070860

Tong, Q., Cui, X., Xu, H., Zhang, X., Hu, S., Huang, F., & Xiao, L. (2023). D1 receptor-expressing neurons in ventral tegmental area alleviate mouse anxiety-like behaviors via glutamatergic projection to lateral septum. *Molecular Psychiatry, 28*(2), 625–638. https://doi.org/10.1038/s41380-022-01809-y

Torpy, D. J., Grice, J. E., Hockings, G. I., Crosbie, G. V., Walters, M. M., & Jackson, R. V. (1995). Effect of sodium valproate on naloxone-stimulated ACTH and cortisol release in humans. *Clinical and Experimental Pharmacology & Physiology, 22*(6–7), 441–443. https://doi.org/10.1111/j.1440-1681.1995.tb02037.x

Tseilikman, V. E., Tseilikman, O. B., Karpenko, M. N., Traktirov, D. S., Obukhova, D. A., Shatilov, V. A., Zhukov, M. S., Manuilov, G. V., Yegorov, O. N., Aristov, M. R., Lipatov, I. A., Buksha, I. A., Epitashvili, A. E., Pashkov, A. A., & Novak, J. (2024). Unraveling the serotonergic mechanism of stress-related anxiety: Focus on co-treatment with resveratrol and selective serotonin reuptake inhibitors. *Biomedicines, 12*(11), 2455. https://doi.org/10.3390/biomedicines12112455

Tyler, R. E., Vizioli, C., Barb, J. J., Farokhnia, M., & Leggio, L. (2025). Circulating immune and endocrine markers in currently drinking and abstinent individuals with alcohol use disorder and controls. *Addiction Biology, 30*(5), e70039. https://doi.org/10.1111/adb.70039

Valentino, R. J., & Volkow, N. D. (2018). Untangling the complexity of opioid receptor function. *Neuropsychopharmacology, 43*(13), 2514–2520. https://doi.org/10.1038/s41386-018-0225-3

Van't Veer, A., & Carlezon, W. A., Jr. (2013). Role of kappa-opioid receptors in stress and anxiety-related behavior. *Psychopharmacology (Berl), 229*(3), 435–452. https://doi.org/10.1007/s00213-013-3195-5

van Veen, J. F., van Vliet, I. M., de Rijk, R. H., van Pelt, J., Mertens, B., Fekkes, D., & Zitman, F. G. (2009). Tryptophan depletion affects the autonomic stress response in generalized social anxiety disorder. *Psychoneuroendocrinology, 34*(10), 1590–1594. https://doi.org/10.1016/j.psyneuen.2009.05.007

Vasa, R. A., Pine, D. S., Masten, C. L., Vythilingam, M., Collin, C., Charney, D. S., Neumeister, A., Mogg, K., Bradley, B. P., Bruck, M., & Monk, C. S. (2009). Effects of yohimbine and hydrocortisone on panic symptoms, autonomic responses, and attention to threat in healthy adults. *Psychopharmacology (Berl), 204*(3), 445–455. https://doi.org/10.1007/s00213-009-1475-x

Vieira, J. C. F., Bassani, T. B., Santiago, R. M., de O Guaita, G., Zanoveli, J. M., da Cunha, C., & Vital, M. A. B. F. (2019). Anxiety-like behavior induced by 6-OHDA animal model of Parkinson's disease may be related to a dysregulation of neurotransmitter systems in brain areas related to anxiety. *Behavioural Brain Research, 371*, 111981. https://doi.org/10.1016/j.bbr.2019.111981

Volk, N., Paul, E. D., Haramati, S., Eitan, C., Fields, B. K. K., Zwang, R., Gil, S., Lowry, C. A., & Chen, A. (2014). MicroRNA-19b associates with Ago2 in the amygdala following chronic stress and regulates the adrenergic receptor beta 1. *The Journal of Neuroscience, 34*(45), 15070–15082. https://doi.org/10.1523/JNEUROSCI.0855-14.2014

Wall, P. M., Blanchard, R. J., Yang, M., & Blanchard, D. C. (2003). Infralimbic D2 receptor influences on anxiety-like behavior and active memory/attention in CD-1 mice. *Progress in Neuro-Psychopharmacology & Biological Psychiatry, 27*(3), 395–410. https://doi.org/10.1016/S0278-5846(02)00356-1

Wang, C. Y., Jiang, S. Y., Liao, S. M., Tian-Liu, Wu, Q. S., Pan, H. Q., Wei-Nie, Zhang, W. H., Pan, B. X., & Liu, W. Z. (2024). Dimethyl fumarate ameliorates chronic stress-induced anxiety-like behaviors by decreasing neuroinflammation and neuronal activity in the amygdala. *International Immunopharmacology, 137*, 112414. https://doi.org/10.1016/j.intimp.2024.112414

Wang, M., Cao, L., Li, H., Xiao, H., Ma, Y., Liu, S., Zhu, H., Yuan, M., Qiu, C., & Huang, X. (2021). Dysfunction of resting-state functional connectivity of amygdala subregions in drug-naïve patients with generalized anxiety disorder. *Frontiers in Psychiatry, 12*, 758978. https://doi.org/10.3389/fpsyt.2021.758978

Wang, L., Paul, N., Stanton, S. J., Greeson, J. M., & Smoski, M. J. (2013). Loss of sustained activity in the ventromedial prefrontal cortex in response to repeated stress in individuals with early-life emotional abuse: Implications for depression vulnerability. *Frontiers in Psychology, 4*, 320. https://doi.org/10.3389/fpsyg.2013.00320

Weber, H., Richter, J., Straube, B., Lueken, U., Domschke, K., Schartner, C., Klauke, B., Baumann, C., Pané-Farré, C., Jacob, C. P., Scholz, C. J., Zwanzger, P., Lang, T., Fehm, L., Jansen, A., Konrad, C., Fydrich, T., Wittmann, A., Pfleiderer, B., Ströhle, A., … Reif, A. (2016). Allelic variation in CRHR1 predisposes to panic disorder: evidence for biased fear processing. *Molecular Psychiatry, 21*(6), 813–822. https://doi.org/10.1038/mp.2015.125

Winzenried, E. T., Everett, A. C., Saito, E. R., Miller, R. M., Johnson, T., Neal, E., Boyce, Z., Smith, C., Jensen, C., Kimball, S., Brantley, A., Melendez, G., Moffat, D., Davis, E., Aponik, L., Crofts, T., Dabney, B., & Edwards, J. G. (2023). Effects of a true prophylactic treatment on hippocampal and amygdala synaptic plasticity and gene expression in a rodent chronic stress model of social defeat. *International Journal of Molecular Sciences, 24*(13), 11193. https://doi.org/10.3390/ijms241311193

World Health Organization. (2022). *ICD-11: International classification of diseases* (11th revision). https://icd.who.int/

Wu, Z., Wang, T., Li, L., Hui, Y., Zhang, Q., & Yuan, H. (2019). Activation and blockade of α_2-adrenoceptors in the prelimbic cortex regulate anxiety-like behaviors in hemiparkinsonian rats. *Biochemical and Biophysical Research Communications, 519*(4), 697–704. https://doi.org/10.1016/j.bbrc.2019.09.038

Xu, Y., Chen, R., Zhi, F., Sheng, S., Khiati, L., Yang, Y., Peng, Y., & Xia, Y. (2023). δ-opioid receptor, microglia and neuroinflammation. *Aging and Disease, 14*(3), 778–793. https://doi.org/10.14336/AD.2022.0912

Yerkes, R. M., & Dodson, J. D. (1908). The relation of strength of stimulus to rapidity of habit formation. *Journal of Comparative Neurology & Psychology, 18*, 459–482. https://doi.org/10.1002/cne.920180503

Zhang, C., Liu, B., Pawluski, J., Steinbusch, H. W. M., Kirthana Kunikullaya, U., & Song, C. (2023). The effect of chronic stress on behaviors, inflammation and lymphocyte subtypes in male and female rats. *Behavioural Brain Research, 439*, 114220. https://doi.org/10.1016/j.bbr.2022.114220

Zhu, L., Zheng, D., Li, R., Shen, C. J., Cai, R., Lyu, C., Tang, B., Sun, H., Wang, X., Ding, Y., Xu, B., Jia, G., Li, X., Gao, L., & Li, X. M. (2023). Induction of anxiety-like phenotypes by knockdown of cannabinoid type-1 receptors in the amygdala of marmosets. *Neuroscience Bulletin, 39*(11), 1669–1682. https://doi.org/10.1007/s12264-023-01081-2

Zou, H., Pu, W., Zhou, J., Li, J., Ma, L., Wang, S., Liu, C., Mou, J., Liu, X., Yu, T., Wei, Y., Xie, H., & Cao, S. (2025). Noradrenergic locus coeruleus-CA3 activation alleviates neuropathic pain and anxiety- and depression-like behaviors by suppressing microglial neuroinflammation in SNI mice. *CNS Neuroscience & Therapeutics, 31*(3), e70360. https://doi.org/10.1111/cns.70360

Chapter 9
Mood Dysregulation

1 Introduction

During my junior year of high school, I developed mononucleosis (mono) and had to miss 2 months of school. Upon hearing the news that I would need to be home-schooled, I was disappointed. I enjoyed my first semester as a junior, and I generally liked school. I was also working at the time, so missing work (and the whopping $6.75 per hour I made) was also upsetting. At the same time, I was excited about the idea of not having to wake up early to get to school on time or having to sit through classes all day. Because I often worked immediately after school, I was happy to get a break from getting home at 10:30 pm. For the first couple of weeks of being quarantined at home, I enjoyed the freedom of staying up later during the week and waking up after 9 am. I vividly remember playing *Grand Theft Auto: Vice City* on the Playstation 2 throughout the night. However, the longer my social isolation lasted, the more I noticed my **mood** change. I was unhappy and had difficulty sleeping. I was eventually prescribed zolpidem (Ambien) to help me sleep. I was fortunate that I was able to return to school by mid-March, and my mood returned to normal.

Sadly, over 300 million individuals are estimated to live with a *mood disorder* (World Health Organization, 2022). Just as I experienced worsening mood during my social isolation, depression increased during the worldwide COVID-19 pandemic that forced individuals to maintain social distancing (Zhu et al., 2023). This chapter will focus on the neuroscience of mood dysregulation. First, an overview of the various mood disorders will be provided. I will also discuss how researchers can model mood disorders in animals before detailing the neuromechanisms underlying these disorders. In addition to mood disorders, I will return to the construct of impulsivity to explain how individuals can engage in impulsive behaviors when in a heightened emotional state. ***Trigger warning:*** this chapter will include a discussion of self-injurious behavior, including suicide.

J. Yates, *Neuroscience of Maladaptation*, https://doi.org/10.1007/978-3-032-15602-0_9

2 Learning Objectives

By the end of this chapter, you should be able to …

1. Differentiate the various mood disorders.
2. Identify the ways depressive-like behavior can be measured in animals.
3. Discuss how dysfunction to the limbic-cortical-striatal-pallidal-thalamic (LCSPT) pathway and the default mode network (DMN) contributes to mood dysregulation.
4. Explain how aberrations to monoaminergic, glutamatergic, and GABAergic systems promote maladaptive emotional processing.
5. Describe how neuroinflammation is related to mood disorders.
6. Explain the neuromechanisms that underlie forms of affective impulsivity, such as negative urgency and self-injurious behavior.

3 Overview of Mood Disorders

Mood disorders are mental health conditions in which individuals experience significant and prolonged changes to their mood. Most individuals are familiar with the two most prominent mood disorders: major depressive disorder (MDD) (often simply referred to as depression) and bipolar disorder. These disorders will be discussed below, followed by a brief description of other mood disorders.

3.1 Major Depressive Disorder (MDD)

MDD is the most common depressive disorder recognized by the *Diagnostic and Statistical Manual for Mental Disorders (DSM-5)* (American Psychiatric Association, 2013) and the *International Classification of Diseases (ICD-11)* (World Health Organization, 2019). According to the *DSM-5*, a diagnosis of MDD can be made when an individual has experienced five or more of the following symptoms consistently over a 2-week period:

1. Depressed mood most of the day, nearly every day, as indicated by either subjective report (e.g., feels sad, empty, hopeless) or observation made by others (e.g., appears tearful). (**Note:** In children and adolescents, can be irritable mood.)
2. Markedly diminished interest or pleasure in all, or almost all, activities most of the day, nearly every day (as indicated by either subjective account or observation).
3. Significant weight loss when not dieting or weight gain (e.g., a change of more than 5% of body weight in a month), or decrease or increase in appetite nearly every day…
4. Insomnia or hypersomnia nearly every day.

5. Psychomotor agitation or retardation nearly every day (observable by others, not merely subjective feelings of restlessness or being slowed down).
6. Fatigue or loss of energy nearly every day.
7. Feelings of worthlessness or excessive or inappropriate guilt (which may be delusional) nearly every day (not merely self-reproach or guilt about being sick).
8. Diminished ability to think or concentrate, or indecisiveness, nearly every day (either by subjective account or as observed by others).
9. Recurrent thoughts of death (not just fear of dying), recurrent suicidal ideation without a specific plan, or a suicide attempt or a specific plan for committing suicide (American Psychiatric Association, 2013, p. 160).

Note, one of the symptoms must be either depressed mood or diminished interest in activities for a diagnosis of MDD to be made.

Notice that some of the symptoms listed above deal with alterations in cognitive-behavioral functions. MDD is more than just being sad. Individuals with MDD often show impaired attentional and memory functions (Baune et al., 2014; Culpepper, 2015), increased impulsivity (Pulcu et al., 2014; Takahashi et al., 2008), and maladaptive risky decision making (Lu et al., 2024b). When you learn about the neural underpinnings of mood disorders, you will learn why deficits in executive functioning are often observed in individuals living with conditions like MDD.

3.2 Bipolar Disorder

Bipolar disorder is not included with depressive disorders like MDD in the *DSM-5* or the *ICD-11*, instead being placed in its own category (Bipolar and Related Disorders). Additionally, bipolar disorder can be separated into two separate conditions: bipolar I disorder and bipolar II disorder. Bipolar I disorder is characterized by swings between major depressive episodes and *manic* episodes. During a manic episode, an individual has a persistently elevated mood and has increased energy that lasts for at least 1 week. Three or more of the following symptoms must be present to a significant degree for a diagnosis of bipolar disorder to be made:

1. Inflated self-esteem or grandiosity.
2. Decreased need for sleep (e.g., feels rested after only 3 h of sleep).
3. More talkative than usual or pressure to keep talking.
4. Flight of ideas or subjective experience that thoughts are racing.
5. Distractibility (i.e., attention too easily drawn to unimportant or irrelevant external stimuli), as reported or observed.
6. Increase in goal-directed activity (either socially, at work or school, or sexually) or psychomotor agitation (i.e., purposeless non-goal-directed activity).
7. Excessive involvement in activities that have a high potential for painful consequences (e.g., engaging in unrestrained buying sprees, sexual indiscretions, or foolish business investments) (American Psychiatric Association, 2013, p. 124).

Although not required for a diagnosis of bipolar I disorder, individuals may experience hypomania. In the *DSM-5*, the symptoms of a hypomanic episode are identical to a manic episode; however, an individual needs to show symptoms for at least 4 consecutive days as opposed to 1 week. Furthermore, to be considered a hypomanic episode, the episode cannot be severe enough to impair an individual's social or occupational functioning or to necessitate hospitalization. Experiencing hypomania instead of mania is the key difference between bipolar I disorder and bipolar II disorder. The criteria for a major depressive episode are identical to those described above for MDD for both bipolar I and bipolar II disorders.

Like MDD, individuals with bipolar disorder have symptoms in domains beyond mood dysregulation. These individuals show deficits in executive function, attention, and memory (Lima et al., 2018) and display enhanced impulsivity (Ahn et al., 2011; Gilbert et al., 2011; Huang et al., 2024; but see Wang et al., 2020). During an episode of mania, individuals can experience symptoms that mirror those in schizophrenia, particularly paranoia and hallucinations (Kim et al., 2025).

3.3 Other Mood Disorders

While the current chapter will focus primarily on MDD and bipolar disorder, I want to highlight a few additional mood disorders. Persistent depressive disorder is similar to MDD, except that individuals need to experience a depressed mood for most of the day for more days than not for at least 2 years (only 1 year for children and adolescents). Individuals also need to experience two or more of the following symptoms:

1. Poor appetite or overeating.
2. Insomnia or hypersomnia.
3. Low energy or fatigue.
4. Low self-esteem.
5. Poor concentration or difficulty making decisions.
6. Feelings of hopelessness (American Psychiatric Association, 2013, p. 168).

Disruptive mood dysregulation disorder (DMDD) is characterized by chronic and severe persistent irritability, most commonly in the form of temper outbursts and an angry mood in between each outburst. Cyclothymic disorder is similar to bipolar II disorder, except individuals experience hypomanic and depressive symptoms that do not meet the threshold for hypomania or major depression. To be diagnosed with cyclothymic disorder, individuals must experience symptoms of hypomania/depression for at least half the time during a 2-year period (only 1 year for children and adolescents), and they cannot be without symptoms for more than 2 months at a time. Finally, found in the Schizophrenia Spectrum and Other Psychotic Disorders category of the *DSM-5* and the Schizophrenia and Other Primary Psychotic Disorders in the *ICD-11*, schizoaffective disorders are mood disorders that feature elements observed in conditions like schizophrenia (e.g., hallucinations).

Comprehension Check 1 How is bipolar I disorder differentiated from bipolar II disorder?

Individuals with bipolar II disorder experience hypomania as opposed to mania.

4 Using Animal Models to Study the Neurobiology of Mood Disorders

In the previous chapter, you learned about the various ways in which researchers can induce stress in animals, such as restraint, forced swimming, exposure to painful stimuli, social isolation, and social defeat. Now, I am going to discuss how stress paradigms can be used to measure depressive-like behavior in animals. First, I need to discuss the seminal work of Dr. Martin Seligman.

4.1 *Learned Helplessness*

Dr. Seligman first conducted an experiment with three groups of dogs tested in two phases (Seligman & Maier, 1967) (see Fig. 1a). In phase 1, all dogs were placed in a harness that prevented them from jumping off the ground. The dogs were placed on a floor that could be electrified. Dogs in Group 1 were never exposed to shock during phase 1. Dogs in Groups 2 and 3 were subjected to random shocks. Whereas dogs in Group 2 could press a panel to turn the shock off, dogs in Group 3 could not. During phase 2 of the experiment, dogs were removed from their harness, and they could avoid shock by stepping over a barrier to get off the electrified part of the floor. Dogs in Groups 1 and 2 quickly learned to escape the shock. However, most dogs in Group 3 never learned to escape the shock. Instead, they laid on the floor and continued to receive intermittent shocks.

Notice that in Dr. Seligman's work, dogs in Group 3 experienced *unavoidable* shock. Animals that experience stressors that seem inescapable develop depressive-like symptoms. This is analogous to the depression individuals may experience when faced with a difficult circumstance that seemingly has no end in sight (e.g., long-term financial struggle). Indeed, chronic stress is a major antecedent to depressive symptoms (Anisman & Zacharko, 1982; Tafet & Nemeroff, 2016).

Dr. Seligman developed the concept of **learned helplessness**, a psychological phenomenon in which an individual stops attempting to escape or avoid negative situations, even when escape is possible, due to chronic exposure to uncontrollable stressors. Learned helplessness is an important psychological mechanism accounting for conditions like MDD (Miller et al., 1975; Pryce et al., 2011).

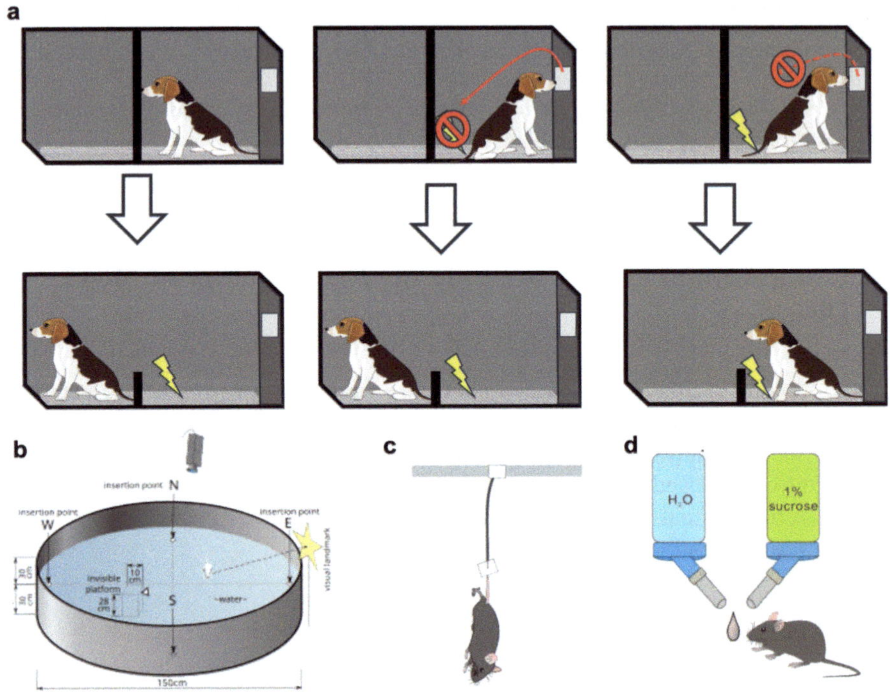

Fig. 1 Methods for measuring depressive-like behaviors in animals. (**a**) A depiction of the learned helplessness experiment conducted by Seligman and Maier (1967). See the text for a description of this experiment. Image created by the author using Servier Medical Art (https://smart.servier.com/; CC BY 4.0, https://creativecommons.org/licenses/by/4.0/) and Microsoft PowerPoint. (**b**) The forced swim test. *Source* https://commons.wikimedia.org/wiki/File:MorrisWaterMaze.svg from Samueljohn.de. CC BY-SA 3.0 (https://creativecommons.org/licenses/by-sa/3.0/deed.en), via Wikimedia Commons. (**c**) The tail suspension test. (**d**) The sucrose preference test; in this specific version, rodents can consume tap water from the bottle on the left or sucrose water from the bottle on the right. Decreased consumption of the sucrose water is interpreted as anhedonia-like behavior. Panels (**c**) and (**d**) are modified from Figs. 6 and 7, respectively, of Zhao et al. (2021), https://doi.org/. CC BY 4.0 (https://creativecommons.org/licenses/by/4.0/), via Frontiers

In addition to exposing animals to unavoidable shocks, researchers can utilize methods such as social defeat, the forced swim test (also called Porsolt test), and the tail suspension test, to name a few. See the previous chapter for a description of the social defeat paradigm. In the forced swim test, animals are placed in a Morris water maze or in a small container filled with water. There is no way for the animal to escape this environment, and they are forced to stay afloat. The tail suspension test involves holding a rodent in place by its tail. This is extremely aversive. I equate it to pulling on someone's ear. Similar to Seligman's dogs, rodents that display learned helplessness will not work as hard to stay afloat or will not work as hard to escape when suspended by their tail.

4.2 Sucrose Preference Test

Anhedonia, the inability to derive pleasure from activities that are normally enjoyable, is a defining feature of MDD (Cooper et al., 2018) and can be modeled in rodents using the *sucrose preference test* (Liu et al., 2018). Rodents find sucrose to be highly palatable. Normally, if given a choice between water and a sucrose solution, rodents will consume high levels of the sucrose solution. Exposure to chronic stress decreases consumption of sucrose (Pothion et al., 2004; Remus et al., 2015; Willner et al., 1987).

5 Neurocircuitry of Mood Disorders

There is large overlap in the neuromechanisms underlying mood disorders and anxiety disorders. This provides an explanation as to why mood disorders are often comorbid with anxiety disorders (Kalin, 2020). The hypothalamic–pituitary–adrenal axis (HPA) was covered in the previous chapter, but hyperactivity of this system is hypothesized to be a mechanism for conditions like MDD (Mikulska et al., 2021). This section will focus on a couple of major neural circuits that control different aspects of mood.

5.1 Limbic-Cortical-Striatal-Pallidal-Thalamic (LCSPT) Pathway

Overall, the limbic-cortical-striatal-pallidal-thalamic (LCSPT) pathway processes emotionally salient stimuli/events, enables us to respond accordingly, and stores information regarding emotional context (Drevets et al., 2008). I will use an example from the previous chapter to illustrate how the LCSPT pathway normally functions. When learning about stress, I had you imagine a scenario in which your boss berates you in front of your coworkers.

When your boss begins to publicly reprimand you, your hippocampus and amygdala (part of the limbic system) become active. The amygdala receives inputs from sensory areas, such as the sight of your boss and the way their voice sounds while they yell at you. As the amygdala and the hippocampus have a reciprocal connection, the hippocampus receives information from the amygdala and then begins working with the prefrontal cortex (PFC) to extract memories related to the situation in which you find yourself. From what you have learned throughout your life, you know that someone yelling at you is a bad thing.

At this point, the HPA axis has been activated, initiating a stress response (see previous chapter). Your PFC will also begin evaluating the situation in more detail. This will allow you to assess how angry your boss is at the moment and if there is anything you can do or say to mitigate the situation. The anterior cingulate cortex (ACC) performs a risk–benefit analysis of each potential action you can take during the interaction with your boss. For example, one potential action you could take is yelling back at your boss. You will most likely conclude that this is a suboptimal decision as it will make your current situation worse.

The striatum receives inputs from the frontal cortex/ACC and helps determine the motor actions you should take. If the best course of action is to remain still and listen to your boss, the striatum will select this motor action. The pallidum (globus pallidus) is an important component of the basal ganglia, transmitting information to the thalamus, which then transmits information to motor cortical areas of the brain. The thalamus also integrates information from other components of the LCSPT pathway, particularly the amygdala, to refine the motor actions you should take.

The LCSPT pathway consists of two different pathways that originate in the frontal cortex: the orbitofrontal cortex (OFC) and the medial prefrontal cortex (mPFC) (Fig. 2). I will further describe these pathways below and will discuss evidence linking dysregulation of these circuits to mood disorders.

5.1.1 Orbital Prefrontal Pathway

The orbital prefrontal pathway includes connections between the medial OFC (mOFC) and the lateral OFC (lOFC) with regions associated with perception, such as the insula, the inferior temporal cortex (IT), and the thalamus, and with memory, such as the amygdala and the hippocampus (note, the IT is also involved in memory). This pathway allows us to process affective characteristics of stimuli, including reward, aversion, and relative value (Drevets, 2008). The OFC is crucial for evaluating the emotional value of external stimuli. More specifically, the mOFC represents reward value while the lOFC processes punishment and reward omission (Rolls et al., 2020).

Let's use a different example to highlight how the orbital prefrontal pathway works. For many, smelling fresh-baked cookies elicits a positive reaction. First, olfactory information travels from the olfactory bulb to the olfactory tract, which then projects to the primary olfactory cortex (piriform cortex), as well as to the OFC, the amygdala, and the entorhinal cortex. The olfactory system is unique in that the olfactory tract does not travel to the thalamus first. Instead, the thalamus receives olfactory inputs from the primary olfactory cortex and the hippocampus. The thalamus sends information back to the OFC and sends information to the insula.

The OFC is instrumental in determining if the smell of baking cookies should elicit a positive or a negative response. The OFC works with the limbic system to compare what we are currently experiencing to what we have experienced in the past, which helps influence the emotion we should feel at that moment. We often associate the smell of cookies with the taste of cookies, which is generally reinforcing. Even if one does not like the taste of cookies, they may still have a positive reaction in response to the scent of cookies due to interactions they have had with a loved one who liked to bake cookies when they visited. Because this is a pleasant experience, the smell of cookies causes the mOFC to become active.

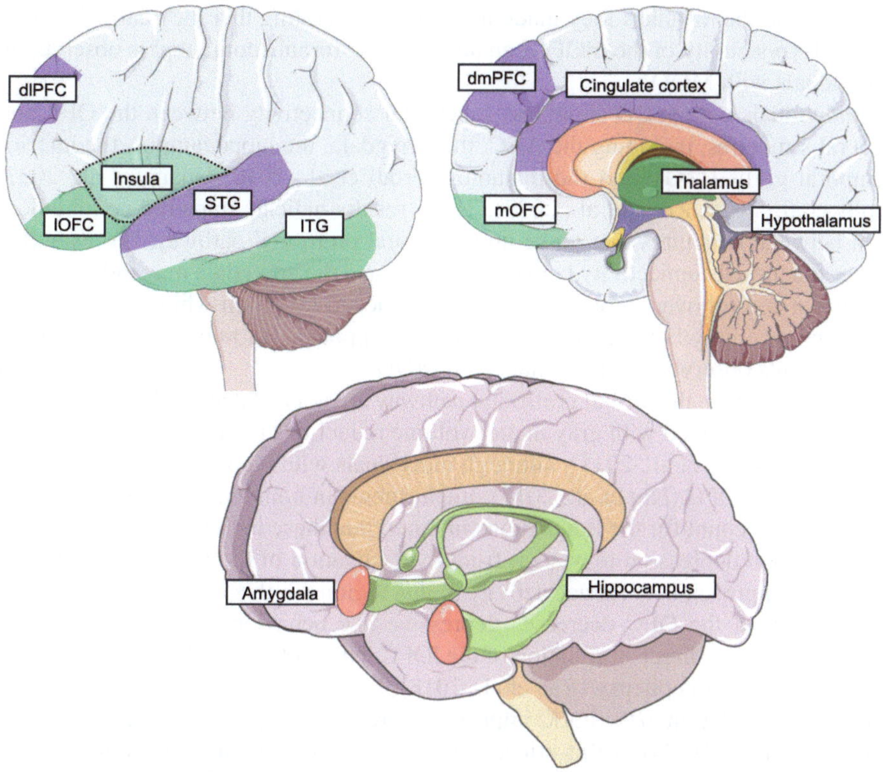

Fig. 2 Orbital prefrontal and medial prefrontal pathways of the limbic-cortical-striatal-pallidal-thalamic (LCSPT) pathway. Regions highlighted in green make up the orbital prefrontal pathway, and regions highlighted in purple make up the medial prefrontal pathway. The amygdala and the hippocampus are involved in both pathways. *Note* dlPFC = dorsolateral prefrontal cortex; dmPFC = dorsomedial prefrontal cortex; ITG = inferior temporal gyrus; lOFC = lateral orbitofrontal cortex; mOFC = medial orbitofrontal cortex; PAG = periaqueductal gray; STG = superior temporal gyrus. Image created by the author using Servier Medical Art (https://smart.servier.com/; CC BY 4.0, https://creativecommons.org/licenses/by/4.0/) and Microsoft PowerPoint

Just as smells can elicit positive reactions, there are odors that elicit strong negative reactions, such as a decomposing animal. From an evolutionary perspective, this makes sense as the aversive smell could act as a warning. Encountering a dead animal could indicate the presence of a nearby predator, or the animal could have died from a disease. The amygdala and the lOFC are important for processing the aversive properties of stimuli.

As discussed by Rolls et al. (2020), depressive symptoms are linked to hyperactivity of the lOFC and hypoactivity of the mOFC. This leads individuals to become overly sensitive to punishment and insensitive to rewarding stimuli. For example, a college student with depression can become hyperfocused on potentially failing an upcoming exam. Even if the student passes the exam with a respectable B, they will ruminate on the mistakes they made instead of celebrating that they earned a good grade. Hypoactivity of the mOFC can also account foranhedonia that is observed in individuals with MDD.

Individuals with MDD show less functional connectivity between the OFC and several structures, including the ACC, the amygdala, the hippocampus, the inferior temporal gyrus (ITG), and the thalamus (Frodl et al., 2010; Huang et al., 2023; Rolls et al., 2020; Zhou et al., 2023). These results indicate impaired communication between structures that make up the orbital prefrontal pathway. For example, lower functional connectivity between the OFC and the amygdala may indicate poor top-down control over emotional responses to negative stimuli while reduced functional connectivity between the OFC and the hippocampus may underlie memory impairments observed in individuals with MDD.

Impairments to the orbital prefrontal pathway are not isolated to MDD. Bipolar II disorder is associated with gray matter volume reductions in the mOFC, the IT, and the insula (Narita et al., 2011), although individuals with bipolar disorder also have a thinner left lOFC (Niu et al., 2017). Impairments in both lOFC and mOFC could potentially account for the fluctuations in mood that these individuals experience.

Given that depression may stem from overactivation of the lOFC, one potential treatment approach is to reduce activity in this region. To this end, pharmacological inactivation of the OFC decreases depressive-like behavior in animals (Kuniishi et al., 2017). Yet, direct stimulation of the lOFC, but not the mOFC, improves mood in individuals with epilepsy (Rao et al., 2018). More work is needed to determine if electrical stimulation to the lOFC improves depressive symptoms in individuals with conditions like MDD or if this effect is isolated to those with damage to the lOFC.

Although not part of the orbital prefrontal pathway, I want to highlight that reduced functional connectivity between the mOFC and the cerebellum is observed in individuals with MDD (Wang et al., 2018). In the last chapter, you learned that hyperactivity of the cerebellum is linked to anxiety. If the mOFC is ineffective at attenuating cerebellar activity, increased anxiety could give rise to depressive symptoms. Given the cerebellum is involved in numerous cognitive functions, this finding can also provide an account for some of the cognitive deficits observed in individuals with MDD, including attentional deficits and impulsivity (Guo et al., 2024; Keller et al., 2019).

Depressive symptoms are also linked to dysfunction of the IT. Not only is the IT involved in visual object recognition and visual working memory (Conway, 2018; Ranganath, 2006), but it is especially sensitive to emotional facial expressions (Rolls, 2007). Furthermore, the IT has strong connections with the OFC and the amygdala (Rolls, 2023). Individuals with MDD and bipolar disorder have reduced cortical thickness of the IT (Niu et al., 2017). In Chap. 6, you learned about various forms of dementia. Individuals with dementia have higher depression rates (Kitching, 2015), and depression has been linked to the development of dementia (Hakim, 2022). Reduced gray matter volume in the IT is observed in individuals with dementia with Lewy bodies (Querry et al., 2024). Collectively, these results indicate that disrupting the flow of emotionally relevant information to regions like the OFC and the amygdala impairs emotion processing in individuals, which contributes to depressive symptoms.

5.1.2 Medial Prefrontal Pathway

The medial prefrontal pathway includes the dorsomedial PFC (dmPFC) and the dorsolateral PFC (dlPFC), the cingulate cortex, the limbic system, the superior temporal gyrus (STG), the hypothalamus, and the periaqueductal gray (PAG). As discussed by Drevets et al. (2008), this pathway is known as the visceromotor system and is involved in *visceral responses*. These are instinctive, involuntary reactions we display when we encounter certain stimuli or situations. For example, images of extreme violence can elicit a visceral response such as nausea and sweating. The dmPFC is also important for emotion introspection and self-reflection (Herwig et al., 2010; Schilbach et al., 2012), and both the dmPFC and the dlPFC regulate aspects of decision making and impulse control.

Individuals with mood disorders show structural abnormalities such as reduced gray matter volume in the PFC, the ACC, and the hippocampus/parahippocampus (Duman et al., 2019; Grieve et al., 2013; Lacerda et al., 2004; Merz et al., 2018; Narita et al., 2011; Nery et al., 2009; Schmaal et al., 2017; Yang et al., 2022). As these regions are important for working memory, this finding can account for the working memory deficits that are frequently associated with mood disorders (Ha et al., 2012; Kriesche et al., 2023). Specific to bipolar disorder, reduced gray matter volume in the ventromedial PFC (vmPFC) and the ACC may provide an additional account for why mood fluctuates between mania/hypomania and depression.

Although individuals with MDD have reduced gray matter volume in the PFC, they show greater hyperactivity in the dlPFC, as well as in the vmPFC (Lin et al., 2021), and they show increased functional connectivity between the dlPFC and the OFC (Frodl et al., 2010). Hyperactivity of the dlPFC and the vmPFC may explain the decision-making deficits that are often associated with MDD (e.g., Rinaldi et al., 2020; Takahashi et al., 2008). Given that the vmPFC is important for reward processing (Rehbein et al., 2023), hyperactivity of this region can lead to anhedonia by causing too much inhibition of the nucleus accumbens (NAc), a brain region often nicknamed our "reward center".

In contrast to the dlPFC/vmPFC, hypoactivity of the dmPFC is observed in individuals with MDD (Nixon et al., 2013). A hypoactive dmPFC may not allow individuals with MDD to engage in adaptive reappraisal of negative experiences (i.e., use self-reflection to change how one interprets a situation), which contributes to negative emotions in these individuals. Let's return to the work example I referenced earlier. Getting reprimanded is not a fun experience, and having this occur in a public setting is embarrassing (and unprofessional on your boss' part). An individual would have the right to be frustrated by this situation, but they have the ability to reframe this situation in a positive way. The individual can think "although my boss went about this in the wrong manner, I now know what I need to avoid doing in the future".

Individuals that do not have the ability to use adaptive reappraisal will ruminate on the mistake they made that led to their reprimand.

Comprehension Check 2 How does dysfunction of the orbital prefrontal and the medial prefrontal pathways of the LCSPT pathway contribute to the cognitive and emotional symptoms of mood disorders?

Dysfunction of the orbital prefrontal pathway is largely associated with emotional dysregulation. Hypoactivity of the mOFC is associated with anhedonia, and depressive symptoms are linked to hyperactivity of the lOFC, a region that processes negative stimuli. Dysregulation of the medial prefrontal pathway is correlated with cognitive disturbances that can contribute to depressed mood. Hyperactivity of the dlPFC and the vmPFC leads to decision-making deficits, and hypoactivity of the dmPFC can impair adaptative reappraisal. As such, individuals are more prone to rumination, excessively thinking about negative events.

5.2 Default Mode Network (DMN)

The default mode network (DMN) was first introduced in Chap. 5 and includes the mPFC, the posterior cingulate cortex (PCC), the precuneus, and the angular gyrus. The DMN is most active when we are resting and not focused on a specific task. The DMN is hyperactive in individuals with MDD (Duman et al., 2019). This can account for the increased rumination often observed in these individuals (see above about hypoactivity of the dmPFC) (Zhou et al., 2020). Individuals with MDD also display increased connectivity between the DMN and corticolimbic regions (Zhang et al., 2025), which can further explain why individuals with mood disorders are more prone to rumination. Antidepressant medications reduce functional connectivity within the DMN, providing further support that dysfunction of this neural circuit is involved in mood dysregulation (Yan et al., 2019).

Altered functional connectivity within the DMN and between DMN structures and other brain regions has been observed in individuals with bipolar disorder. When experiencing a depressive episode, these individuals show increased functional connectivity between the PCC and the lOFC and between the dmPFC and sensorimotor, visual, and salience network regions; furthermore, pessimism is positively correlated with altered functional connectivity between the PCC and the precuneus (Yang et al., 2021). As described above, thelOFC is sensitive to punishment, and the dmPFC is involved in emotion introspection. The PCC helps process emotionally relevant stimuli. Taken together, these results further support the idea that hyperactivity of the DMN drives rumination of negative events, which can lead to learned helplessness. Related to this hypothesis is the observed hypoconnectivity between the DMN and the PFC of individuals with bipolar disorder (Zhang et al., 2025). The PFC is an important regulator of the DMN, working to suppress its activity when we engage in cognitive processing. If the PFC is less effective at suppressing the DMN, individuals may be more likely to ruminate.

Relative to healthy controls and those with MDD, individuals with bipolar disorder show greater functional connectivity between the DMN and regions that make up the dorsal attention network (DAN) (see Chap. 5) (Rai et al., 2021). As the DMN and the DAN are "anti-correlated" (i.e., when DAN activity increases, DMN activity should decrease), increased functional connectivity between these networks provides an account of attention deficits in individuals with bipolar disorder. This finding may also explain why individuals with bipolar disorder show lower perceived abilities in the areas of attention and planning (Rosso et al., 2025).

The mPFC is subdivided into the prelimbic and the infralimbic subregions. The prelimbic cortex is important for promoting goal-directed behaviors and learning new associations. The infralimbic cortex suppresses learned behaviors, enabling us to learn new associations. For example, during fear conditioning, the prelimbic cortex allows us to learn to associate an environmental stimulus with a negative emotional response. Imagine an individual is bitten by a dog; this individual's prelimbic cortex will form an association between dogs and pain, which can promote fear responses to dogs. Over time, this individual can learn that not all dogs will bite an individual. Learning this new association is dependent on infralimbic activity. Prelimbic development during adolescence is critical for mood regulation. Adolescent rats that receive lesions to the prelimbic mPFC show increased learned helplessness during adulthood, an effect that does not manifest when lesions occur during adulthood (Uliana et al., 2020). These results suggest that abnormal prelimbic development can increase susceptibility to depressive-like symptoms later in life.

Comprehension Check 3 In the passage above, you learned that the loss of anti-correlation between the DMN and the DAN is associated with attentional deficits. How do you think the breakdown in anti-correlation between these two networks influences depressive symptoms?

Normally, when the DAN is active, the DMN is suppressed and vice versa. If these regions are no longer anti-correlated, this means that there is inadequate switching between the DMN and the DAN. As the DMN is involved in introspection, individuals that get stuck in a DMN-dominated state may engage in more rumination, reflecting on negative experiences more than they should.

6 Neurotransmitter Systems Implicated in Major Depressive Disorder

The neurotransmitters implicated in depression are highly similar to those included in the previous chapter. While glucocorticoids and norepinephrine received most of the spotlight in the last chapter, serotonin's role in depression has received considerable attention. As such, I will begin by discussing the contribution of serotonin to MDD.

6.1 Serotonin

One major piece of evidence supporting serotonin hypothesis of depression comes from the pharmacodynamic profile of antidepressant drugs (see Fig. 3). Hillhouse and Porter (2015) provide a history of the development of antidepressant drugs. In the 1950s, a drug known as reserpine (extracted from Indian snakeroot) was used to treat hypertensive vascular disease. However, patients began developing depressive symptoms during treatment, which was reversed when individuals stopped taking reserpine. Reserpine inhibits the vesicular monoamine transporter (VMAT-2); this prevents monoamines like serotonin from reentering the vesicle after being taken back up into the presynaptic terminal. This then allows monoamine oxidase (MAO) to metabolize monoamines, causing their levels to decrease.

During the time reserpine was being used, a serendipitous finding in a lab paved the way for the development of antidepressant drugs. Isonicotinyl hydrazide (isoniazid) was developed to treat tuberculosis. In the early 1950s, a derivative of isoniazid was synthesized: isopropyl-isonicotinyl hydrazide (iproniazid) (Fox & Gibas, 1953). Reports of individuals experiencing elevated mood and improved sleep began emerging. Eventually, iproniazid was tested in a group of individuals with MDD and was shown to be effective at reducing depressive symptoms. Iproniazid is classified as an MAO inhibitor (MAOI). As the name indicates, MAOIs prevent MAO from metabolizing monoamines. This allows monoamine levels to increase.

In the 1950s, another drug class was identified that demonstrated antidepressant effects. Imipramine was originally synthesized as a potential antipsychotic drug (recall antipsychotic medications from Chap. 3). Although imipramine was ineffective at treating schizophrenia, it improved depressive symptoms. Imipramine became the first *tricyclic antidepressant*. Unlike MAOIs, tricyclics do not target MAO. Instead, these drugs have several mechanisms of action, including inhibiting serotonin and norepinephrine transporters and blocking noradrenergic alpha (α) receptors. The name tricyclic comes from the chemical structure of drugs in this class. Each tricyclic antidepressant drug has a chemical structure that consists of three benzene rings (this is where *tri* in tricyclic comes from), which are hexagonal shaped and contain carbon atoms at each corner. The term "cyclic" in chemistry refers to compounds with ring structures. Thus, tricyclic antidepressants are compounds

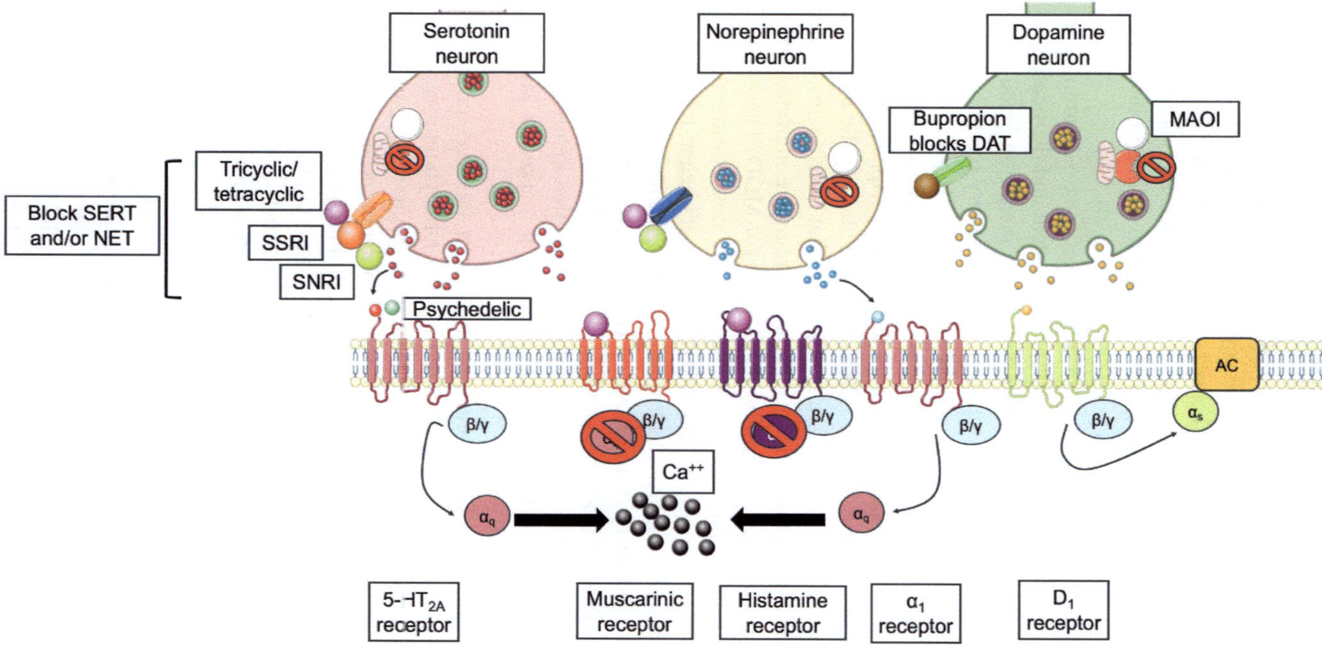

Fig. 3 Schematic showing how various antidepressant drugs target monoaminergic systems. *Note* AC = adenyl cyclase, DAT = dopamine transporter, MAOI = monoamine oxidase inhibitor; NET = norepinephrine transporter; SERT = serotonin transporter. Image created by the author using Servier Medical Art (https://smart.servier.com/; CC BY 4.0, https://creativecommons.org/licenses/by/4.0/) and Microsoft PowerPoint

made up of three benzene rings. In the 1970s, tetracyclic antidepressants were introduced. These drugs are similar to tricyclics, except their chemical structure consists of four benzene rings.

While MAOIs and tricyclic/tetracyclic antidepressant drugs alleviate depressive symptoms, their side effect profiles can be severe. MAOIs are linked to "wine and cheese syndrome". Certain foods, like cheese and certain alcoholic beverages, contain high levels of *tyramine*, a type of monoamine. If individuals eat foods high in tyramine while taking an MAOI, the excess tyramine can lead to a dangerously high blood pressure.

Tricyclic/tetracyclic antidepressants do not target just serotonin and norepinephrine. These drugs also target the cholinergic system and the neurotransmitter histamine, causing a wide range of side effects. Cholinergic-induced side effects include drug mouth, blurred vision, constipation, and urinary retention. Histamine-related side effects include drowsiness and weight gain. If you have ever taken diphenhydramine (Benadryl), you may have experienced drowsiness; this is because diphenhydramine targets histamine receptors in both the peripheral nervous system (PNS) and the central nervous system (CNS). Tricyclic/tetracyclic antidepressants also target histamine receptors in the brain, which is what leads to drowsiness.

Today, many individuals take selective serotonin reuptake inhibitors (SSRIs) or serotonin/norepinephrine reuptake inhibitors (SNRIs) for MDD. These medications were also alluded to in the previous chapter as they are now commonly prescribed for anxiety disorders. Beginning in the 1970s, researchers began first synthesizing drugs that could block the reuptake of serotonin. In 1987, the first SSRI was approved for use by the Food and Drug Administration (FDA): fluoxetine (Prozac). The first SNRI, venlafaxine (Effexor), became available in 1993. The effectiveness of SSRIs and SNRIs coincides with gene association studies that often identify alterations to the *SLC6A4* gene (serotonin transporter) as a potential marker of MDD (Flint & Kendler, 2014; Lohoff, 2010).

The beneficial effects of antidepressant drugs can be reversed by blocking serotonin synthesis. Individuals administered p-chlorophenylalanine, a tryptophan hydroxylase inhibitor, show little improvement in depressive symptoms following antidepressant treatment. However, this effect is not observed when individuals are given α-methyl-p-tyrosine, which inhibits catecholamine (e.g., norepinephrine) synthesis. Interestingly, giving unmedicated individuals p-chlorophenylalanine does not make their depression worse. These results indicate that serotonin levels need to surpass a minimum threshold level to be effective and that mood disorders may not result merely from low serotonin levels (Hamon & Blier, 2013). The contribution of other neurotransmitters to MDD will be discussed below.

Around the time that the first SSRIs were being synthesized, research provided additional support for the serotonin hypothesis of depression by showing that individuals that attempt suicide have lower levels of the serotonin metabolite 5-hydroxyindoleacetic acid (5-HIAA) in cerebrospinal fluid (CSF) (Asberg et al., 1976). Additional studies have reported decreased serotonin and/or 5-HIAA levels in serum and CSF of those that have died of suicide or have contemplated suicide (e.g., Achtyes et al., 2020; Alvarez et al., 1999; Dogan et al., 2016; Lengvenyte et al., 2023; Sun et al., 2020).

Beyond traditional antidepressant drugs, there is growing interest in using psychedelic hallucinogens for the treatment of MDD (Vinarcsik et al., 2025; Wingert et al., 2025). Recall from Chap. 3 that psychedelic hallucinogens like psilocybin ("magic mushrooms"), lysergic diethylamide (LSD; "acid"), and peyote exert their effects by stimulating 5-HT_{2A} receptors. Psychedelic hallucinogens and non-hallucinogenic 5-HT_{2A} receptor agonists attenuate depressive-like behavior in animals (Sekssaoui et al., 2024). Read the Experiment Spotlight below to learn more about the use of psilocybin in an animal model of depression.

Experiment Spotlight: Experiment Spotlight: Unraveling Psilocybin's Therapeutic Potential: Behavioral and Neuroplasticity Insights in Wistar-Kyoto and Wistar Male Rat Models of Treatment-Resistant Depression

The link to the article can be found here: https://pmc.ncbi.nlm.nih.gov/articles/PMC12226618/

Kolasa et al. (2024) tested how psilocybin affects behavioral responses and neuroplasticity in Wistar Han and Wistar-Kyoto (WKY) rats. You previously learned about WKY rats in Chap. 5 as some have argued that this strain models the inattentive subtype of attention-deficit/hyperactivity disorder (ADHD). WKY rats have also been proposed to model treatment-resistant depression. Rats were injected with a single dose of psilocybin (0.3 mg/kg) and then were tested in three different behavioral tasks over the span of 23 days. Rats were euthanized 1 day after the final behavioral test, and brain tissue and blood were collected for biochemical assays.

Rats were tested in (a) a novel object recognition task 4 h and 6 days after psilocybin administration; (b) a social interaction test 2 days and 8 days after psilocybin administration; and (c) a forced swim test 9 and 23 days after psilocybin administration. The novel object recognition task measures cognitive functioning, particularly recognition memory. The social interaction test measures motivation to engage in social behaviors. As you have already learned, the forced swim test is a method for inducing stress and can be used to measure learned helplessness-like behavior. Therefore, I will focus on the results of the forced swim test.

Following euthanasia, the PFC was removed for further testing (to be discussed below), and the researchers collected blood to analyze **brain-derived neurotrophic factor (BDNF)**. I have not discussed BDNF in this book, but this protein plays a critical role in neuroplasticity and promotes neuronal growth and survival. Disruption of BDNF signaling is linked to conditions like MDD (Yang et al., 2020), as well as numerous disorders you have already read about in this book (e.g., Gao et al., 2022; Gören, 2016; Tsai, 2017).

Quantitative reverse transcription polymerase chain reaction (qRT-PCR) was used to measure intra-PFC RNA (ribonucleic acid) of the GluA1 and the GluA2 subunits of the glutamate AMPA (α-amino-3-hydroxy-5-methyl-4-isoxazolepropionic acid) receptor, the *activity-regulated cytoskeleton-associated (ARC) protein, neurotrophin-3* and its corresponding receptor (neurotrophic tyrosine receptor kinase 3), *neurotrophin-4, nerve growth factor (NGF)* receptor, *cytidine monophosphate (CMP)-N-acetylneuraminate-poly-alpha-2,8-sialyltransferase* and *alpha-N-acetylneuraminide alpha-2,8-sialyltransferase*, and *postsynaptic density 95 (PSD-95).* Western blotting was also used to measure ARC protein expression. You already have learned about AMPA receptors, but I need to take a detour to explain what everything else does.

The **postsynaptic density (PSD)** is a network of proteins found primarily in dendrites of excitatory neurons (Fig. 4). You already know that ionotropic glutamate receptors like AMPA receptors are excitatory and are critical for neuroplasticity (think back to long-term potentiation (LTP) that you learned about in Chap. 6). PSD-95 is known as a *scaffolding protein.* Scaffolding proteins help keep other proteins "tethered" together. They are more than molecular anchors. For example, PSD-95 is important for trafficking AMPA receptors to the cell membrane, which is critical for synaptic plasticity. ARC is also involved in AMPA receptor trafficking, but it often has the opposite role of PSD-95, as its activation can lead to the internalization of AMPA receptors. Neurotrophin-3 and neurotrophin-4, like BDNF, are NGFs. These proteins directly influence the structure of the PSD by facilitating the transport of scaffolding proteins and by phosphorylating NMDA (N-methyl-D-aspartate) receptor subunits.

CMP-N-acetylneuraminate-poly-alpha-2,8-sialyltransferase and alpha-N-acetylneuraminide alpha-2,8-sialyltransferase are enzymes that synthesize *polysialic acids.* To prevent overcomplicating things, I want to mention that these enzymes are important for cell adhesion. Cell adhesion ensures the structural integrity of the synapse, allowing for efficient communication between neurons.

Kolasa et al. (2025) found the following:

1. WKY rats showed increased immobility time in the forced swim test (i.e., showed increased depressive-like behavior) compared to Wistar Han rats.
2. Psilocybin decreased immobility in the forced swim test for WKY rats when tested 9 days after treatment, but not after 23 days of treatment. Conversely, Wistar Han rats showed less immobility when tested 23 days after psilocybin treatment, but not when tested 9 days following treatment.
3. BDNF levels did not differ between WKY rats and Wistar Han rats, and psilocybin failed to alter BDNF levels in either strain.
4. WKY rats had decreased mRNA and protein expression of ARC, which was normalized following psilocybin treatment.
5. WKY rats also had reduced mRNA expression of the GluA1 subunit of the AMPA receptor and CMP-N-acetylneuraminate-poly-alpha-2,8-sialyltransferase, with psilocybin failing to normalize mRNA expression.
6. WKY rats had increased mRNA expression of neurotrophic tyrosine receptor kinase 1, NGF, and neurotrophin-4.

Questions to Consider:

1. The summary highlights that psilocybin treatment had different timing effects on immobility in the forced swim test for WKY rats versus Wistar Han rats. What might these different temporal responses suggest about the underlying biological mechanisms of psilocybin in these two rat strains? Given that WKY rats are a model for treatment-resistant depression, what are the potential clinical implications of these findings?

2. The study found that psilocybin normalized the reduced expression of ARC mRNA and protein in WKY rats, but it failed to normalize the reduced GluA1 mRNA expression. How do these specific molecular changes (or lack thereof) relate to the observed decrease in depressive-like behavior in the forced swim test?
3. Psilocybin failed to alter BDNF levels in either rat strain, despite BDNF's critical role in neuroplasticity and its link to depression. This finding contrasts with the significant changes observed in other neuroplasticity-related proteins like ARC. What are the possible reasons for this seemingly contradictory result?

Given that stress impairs 5-HT_{2A}-mediated GABAergic (gamma-aminobutyric acid) signaling in the basolateral nucleus of the amygdala (BLA) (Jiang et al., 2009), 5-HT_{2A} receptor agonists may, at least partially, ameliorate depressive symptoms by increasingGABAergic transmission in the amygdala. Furthermore, 5-HT_{2A} receptor agonists decrease resting-state functional connectivity between the amygdala and the ACC while increasing functional connectivity between the amygdala and the lPFC (Miranda, 2025).

In addition to 5-HT_{2A} receptor agonists, stimulation of 5-HT_{1A} receptors leads to antidepressant-like effects (e.g., Głuch-Lutwin et al., 2023; Hui et al., 2014; Zhou et al., 2014). Notice that stimulation of both excitatory and inhibitory serotonin receptors ameliorates depressive-like behaviors. The 5-HT_{1A} receptor is primarily located in the limbic system, and most neurons in this region are glutamatergic; as such, 5-HT_{1A} receptors serve to inhibit these glutamatergic neurons, leading to decreased activity in brain structures such as the amygdala (Glikmann-Johnston et al., 2015; Sengupta et al., 2017).

Comprehension Check 4 Compare and contrast the different classes of antidepressant drugs mentioned (MAOIs, tricyclic/tetracyclic antidepressants, and SSRIs/SNRIs). Your answer should focus on their mechanisms of action and the side effects associated with each class, explaining how these side effects are a direct result of their differing pharmacological profiles.

Each of these drug classes results in elevated serotonin and/or norepinephrine levels. While MAOIs inhibit MAO, the enzyme that metabolizes monoamines, tricyclics/tetracyclics and SSRIs/SNRIs prevent the reuptake of serotonin/norepinephrine. However, tricyclic/tetracyclic drugs have actions on other neurotransmitter systems. The differential mechanisms of action lead to varying side effect profiles across drug classes. Because MAOIs prevent the metabolism of all monoamines, individuals cannot eat certain foods that contain tryptamine as this can cause dangerously high serotonin levels to build up. Tricyclic/tetracyclic drugs, due to their actions on multiple neurotransmitter systems, cause multiple side effects, including drowsiness and dry mouth. The side effect profile of MAOIs and tricyclics/tetracyclics is a major reason why SSRIs and SNRIs are the primary treatment for MDD. However, SSRIs and SNRIs do have side effects, particularly sexual dysfunction.

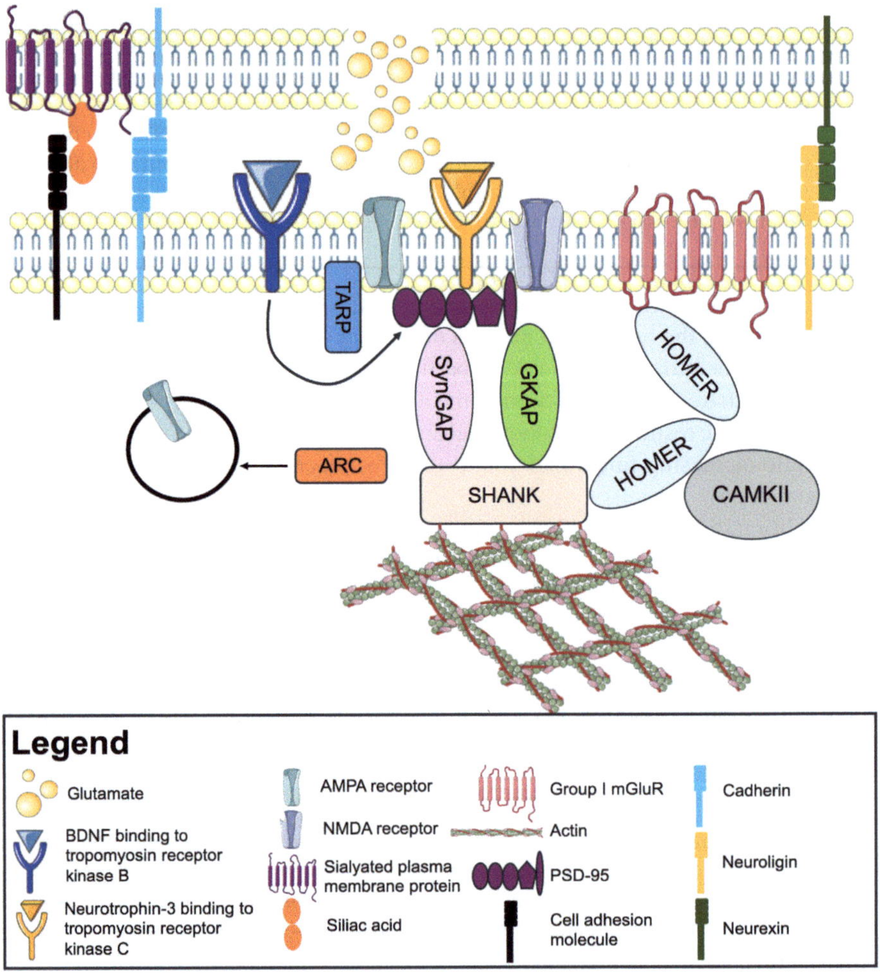

Fig. 4 A simplified schematic of the postsynaptic density (PSD). The schematic depicts an excitatory synapse. Glutamate can bind to excitatory AMPA and NMDA receptors, as well metabotropic glutamate receptors (mGluRs) that are either excitatory or inhibitory. The inside of the postsynaptic neuron contains multiple types of scaffolding proteins, including PSD-95, Homer proteins, SHANK (SRC homology 3 and multiple ankyrin repeat domains), GKAP (guanylate kinase-associated protein), and SynGAP (synaptic Ras GTPase-activating protein Actin filament provides structural support to neurons, and TARPs (transmembrane AMPA receptor regulatory protein) help transport and anchor AMPA receptors to cell membrane. ARC (activity-regulated cytoskeleton-associated) proteins contribute to long-term potentiation (LTP) and long-term depression (LTD). One major function is to internalize AMPA receptors (depicted in the image above). Brain-derived neurotrophic factor (BDNF) binds to tropomyosin receptor kinase B, which influences PSD-95 activity. Neurotrophins also bind to tropomyosin receptor kinases, which influence various cellular processes. Various cell adhesion molecules help anchor the presynaptic neuron to the postsynaptic neuron. Siliac acid also binds to certain cell adhesion molecules and to sialyated plasma membrane proteins. Image created by the author using Servier Medical Art (https://smart.servier.com/; CC BY 4.0, https://creativecommons.org/licenses/by/4.0/) and Microsoft PowerPoint

6.2 *Norepinephrine*

SNRIs are used to treat both anxiety disorders and MDD. This makes sense as noradrenergic neurons originating from the locus coeruleus project to regions involved in emotion regulation, including the limbic system, the frontal cortex, and the hypothalamus. While stress is associated with increased noradrenergic activity, depletion of norepinephrine gives rise to symptoms of anxiety and depression (e.g., Qian et al., 2023; Wang et al., 2025). Individuals with MDD also have lower norepinephrine levels in the brain (Lambert et al., 2000). Similar to what was discussed concerning serotonergic alterations in MDD, individuals that have died by suicide show perturbations to the noradrenergic system, particularly decreased norepinephrine transporter binding (Chandley & Ordway, 2012; Moret & Briley, 2011).

Earlier, I mentioned that α-methyl-p-tyrosine fails to block antidepressant drug effects; yet, administration of α-methyl-p-tyrosine to individuals with MDD can exacerbate depressive symptoms without altering depression in healthy controls (Berman et al., 1999; Hasler et al., 2008). Another thing I need to mention is that α-methyl-p-tyrosine can precipitate depressive symptoms in individuals that are treated with norepinephrine reuptake inhibitors, but not SSRIs (Miller et al., 1996). In healthy individuals, inhibiting the synthesis of norepinephrine most likely does not induce depressive symptoms as serotonin levels are unaffected, and these individuals most likely have enough norepinephrine stored in neurons that α-methyl-p-tyrosine administration does not detrimentally affect the noradrenergic system. Conversely, individuals with MDD have low neuronal norepinephrine levels; thus, preventing the synthesis of norepinephrine will render SNRIs ineffective as preventing the reuptake of norepinephrine only works if there are sufficient norepinephrine levels in the synapse.

A hypoactive noradrenergic system may stem from increased sensitivity of α_2 receptors and/or upregulation of these receptors (Cottingham & Wang, 2012). While acute stress leads to desensitized α_2 receptors (Freedman et al., 1990), a compensatory upregulation/sensitization of α_2 receptors occurs during chronic stress, which leads to low noradrenergic levels. As discussed by Cottingham and Wang (2012), antidepressant drugs, regardless of their mechanism of action, often cause downregulation of α_2 receptors. In the previous chapter, you learned that the α_2 receptor antagonist yohimbine induces stress and anxiety-like behavior in animals. However, yohimbine potentiates the antidepressant-like effects of SSRIs/SNRIs in animals (Dhir & Kulkarni, 2007). Changes to α_2 receptor expression and sensitization provide a mechanism to explain why anxiety disorders often precede mood disorders (Kessler & Wang, 2008).

Comprehension Check 5 How can one reconcile the finding that α-methyl-p-tyrosine does not block the effects of antidepressants, yet can exacerbate depressive symptoms in individuals with MDD?

Long-term stress leads to sensitized and upregulated α_2 receptors, which are inhibitory. Both SSRIs and SNRIs downregulate these receptors, but α-methyl-p-tyrosine selectively inhibits catecholamine synthesis. Thus, there is not enough norepinephrine to stimulate excitatory α_1 receptors. When an individual takes an SSRI, the influence of serotonin can offset the reduced norepinephrine levels induced by α-methyl-p-tyrosine.

6.3 Dopamine

Reduced dopamine levels in the mesocorticolimbic pathway are hypothesized to contribute to anhedonia. Not only do individuals with Parkinson's disease experience increased anxiety (see previous chapter), but they are more likely to experience anhedonia and depression (Lintel et al., 2021; Nagayama et al., 2017; Poewe, 2008). Anhedonia is also commonly observed in individuals with substance use disorders (Koob, 2022). Individuals with MDD have reduced dopamine transporter availability in the ventral tegmental area (VTA) and the striatum (Mizuno et al., 2023; Pizzagalli et al., 2019), most likely reflecting a compensatory downregulation due to hypoactive dopamine levels in the mesocorticolimbic pathway.

In 1989, the FDA approved bupropion (Wellbutrin) for the treatment of MDD. Unlike the other antidepressant drugs described above, bupropion does not target serotonin or norepinephrine. Instead, bupropion acts primarily as a dopamine transporter inhibitor (see Fig. 3). Evidence indicates that bupropion may not be effective as a long-term treatment for MDD; relative to placebo, bupropion does not significantly decrease 6-month relapse rates (Kishi et al., 2023). In addition to bupropion, atypical antipsychotic medications have been prescribed to individuals with MDD. These drugs target both serotonin 5-HT_{2A} receptors and dopamine D_2-like receptors. Like bupropion, these medications can help treat acute depression, but long-term effectiveness is limited (Jha & Mathew, 2023).

Preclinical studies have demonstrated dopaminergic alterations that occur following long-term stress that lead to depressive-like behavior. Dopamine levels are lower in prefrontal cortical areas, including the ACC, of animals that display increased learned helplessness (Kolata et al., 2018). Depleting dopamine levels from the NAc promotes learned helplessness (Bergamini et al., 2016). Decreased dopamine levels in the mesocorticolimbic pathway do not allow for sufficient stimulation of dopamine D_1-like receptors in the PFC. Decreased D_1-like receptor stimulation is linked to depressive-like behavior; when these receptors are chronically stimulated, there is a corresponding downregulation of this receptor subtype in the PFC, which corresponds to learned helplessness (Gambarana et al., 1995). If D_1-containing neurons in the mPFC are directly stimulated, mice show less learned helplessness (Wu et al., 2021).

Dopamine D_2-like receptors also mediate depressive-like behaviors, at least in animals. Genetic deletion of the D_2 receptor increases learned helplessness (Hranilovic et al., 2008), as does blockade of these receptors with haloperidol (a typical antipsychotic medication) (Petty et al., 1994). However, clinical studies are more mixed. Individuals with MDD following traumatic brain injury have decreased binding of D_2-like receptors in the dorsal striatum (Jolly et al., 2019), but other work has shown increased D_2-like receptor availability in several brain regions (Peciña et al., 2017; Shah et al., 1997) while other studies have observed no differences in D_2-like receptor binding (de Kwaasteniet et al., 2014). In fact, a meta-analysis of 43 studies found that D_2/D_3 receptor availability does not significantly differ between individuals with MDD and controls (Mizuno et al., 2023).

Comprehension Check 6 What mechanism(s) can account for the finding that bupropion has limited long-term effectiveness in treating MDD?

Although bupropion increases dopamine levels, which can help ameliorate anhedonia, individuals with MDD have reduced dopamine transporter expression and downregulated dopamine D_1-like receptor expression. Bupropion-induced increases in dopamine levels may not provide enough long-term stimulation of the dopaminergic system due to deficient D_1-like receptor expression.

6.4 Glutamate

While serotonin levels are often lower in individuals with MDD, they have increased blood plasma glutamate levels (Inoshita et al., 2018; Ogawa et al., 2018; Umehara et al., 2017) but reduced glutamine/glutamate levels in CSF (Frye et al., 2007; Levine et al., 2000). Despite having lower CSF glutamate levels, glutamate levels are positively correlated with suicidal ideation in individuals with MDD (Garakani et al., 2013). Additionally, quinolinic acid, an NMDA receptor agonist, is elevated in individuals that have previously attempted suicide, with quinolinic acid levels positively correlating with suicidal ideation (Erhardt et al., 2013).

In the brain, lower glutamate levels in the PFC are observed in individuals with first-episode depression (Draganov et al., 2020), and decreased glutamate is observed in the dlPFC, the mPFC, and the amygdala of individuals diagnosed with MDD (Ritter et al., 2022; Shirayama et al., 2017). Glutamate levels in the dlPFC are also inversely correlated with duration of depressive symptoms in children and adolescents with MDD (Caetano et al., 2005). Individuals with MDD have decreased GluN2A and GluN2B protein expression in the PFC (Feyissa et al., 2009), and downregulated NMDA receptor density in the superior temporal cortex (STC) (Nudmamud-Thanoi & Reynolds, 2004).

The discrepancy in glutamine/glutamate ratios in PNS and CNS for individuals with MDD may result from a weakened blood–brain barrier (BBB). A dysfunctional

BBB is observed in individuals with MDD (Wu et al., 2022). Glutamate released in the CNS may "leak" through a weakened BBB into the PNS. The net result would be elevated glutamate levels in the PNS and lower glutamate levels in CSF.

One piece of support for the glutamate hypothesis of depression comes from the finding that ketamine acts as a rapid antidepressant. Ketamine is an antagonist at NMDA receptors; specifically, it blocks the ion channel of these receptors. One potential mechanism for ketamine's antidepressant effects is improving dopaminergic dysfunction in areas like the NAc and the hippocampus (Belujon & Grace, 2014). However, the fact that ketamine acts as an antidepressant does not provide full support for the glutamate hypothesis of depression. Beyond ketamine, other NMDA receptor antagonists have shown little promise as antidepressants (Hillhouse & Porter, 2015). Additionally, while the affinity for ketamine at NMDA receptors is high, it acts as an agonist at dopamine D_2-like receptors and the 5-HT_{2A} receptor (Kapur & Seeman, 2002), and it acts as a dopamine and norepinephrine transporter inhibitor (Nishimura et al., 1998). Despite this caveat, the NMDA receptor antagonist dextromethorphan has recently been combined with bupropion as a treatment for MDD (known as AXS-05). So far, results have been promising, as AXS-05 is well tolerated by patients, and antidepressant effects occur quickly (Akbar et al., 2023).

6.5 GABA

One common finding is that MDD is associated with a hypoactive GABAergic system (Luscher et al., 2023). Individuals with MDD have lower GABA levels in CSF (Fee et al., 2017) and throughout the brain (Duman et al., 2019). Chronic stress reduces GABAergic neurons in the OFC (Varga et al., 2017). Additionally, deletion of GAD67 (glutamic acid decarboxylase), the enzyme that synthesizes GABA, increases learned helplessness in animals (Kolata et al., 2018). However, in the mPFC, chemogenetic inhibition of GABAergic interneurons decreases depressive-like behavior (Fogaça et al., 2021). This finding seems paradoxical, so I need to explain how inhibiting GABAergic neurons can ameliorate depressive symptoms when low GABA levels are proposed to underlie conditions like MDD. In individuals with low baseline GABA levels, inhibitory GABAergic interneurons in the mPFC are less likely to be inhibited by other GABAergic neurons. These GABAergic interneurons inhibit excitatory pyramidal neurons, leading to hypoactive connections between the mPFC and other brain regions like the hippocampus. As discussed above, a hypoactive dmPFC is associated with MDD. By inhibiting GABAergic interneurons, the balance between excitation and inhibition is restored. Glutamatergic neurons originating from the mPFC can activate neurons in other parts of the brain.

Comprehension Check 7 How can hypoactivity of both glutamate and GABA lead to depressive symptoms?

Hypoactivity of the GABAergic system can lead to a net hypoactivity of the glutamatergic system. Because GABAergic neurons contain GABAergic and glutamatergic receptors, low GABA levels could cause certain GABAergic neurons to become disinhibited. They are then able to inhibit neighboring glutamatergic neurons.

7 Neurochemical Underpinnings of Bipolar Disorder

In contrast to MDD, increased dopamine levels appear to drive bipolar disorder, particularly mania. Antipsychotic drugs are often prescribed to manage manic symptoms (Nierenberg et al., 2023). You already know that antipsychotic drugs act as dopamine D_2-like receptor antagonists, as well as serotonin 5-HT receptor antagonists, and are used to treat schizophrenia. During mania, individuals with bipolar disorder can experience symptoms that mirror those found in schizophrenia (e.g., paranoia, hallucinations). Certain antipsychotic medications have antidepressant effects in individuals with bipolar disorder, with the hypothesis being that these drugs can target monoaminergic transporters, thus increasing norepinephrine and serotonin levels (Rybakowski, 2023).

While antipsychotics are often prescribed to those with bipolar disorder, the primary treatment for this disorder is lithium, which is considered a *mood stabilizer* (Nierenberg et al., 2023). You may be familiar with lithium if you have learned about the periodic table of elements (lithium is the third chemical element). Additional mood stabilizers include valproate and lamotrigine (these drugs are also used to treat epilepsy). Overall, mood stabilizers decrease dopaminergic and glutamatergic activity, which helps ameliorate manic symptoms, and increase serotonergic and GABAergic activity, which is associated with decreased depressive symptoms (Szałach et al., 2023). The ways in which mood stabilizers influence neurotransmission are complex. Below, I will describe a couple of these mechanisms. I will focus my discussion on lithium's effects as they have been most studied.

Instead of acting as a receptor agonist/antagonist or enzyme inhibitor, lithium modifies second messenger systems. In Chap. 2, I introduced the concept of second messengers when discussing G protein-coupled metabotropic receptors. You learned a little about cyclic adenosine monophosphate (cAMP), which is activated by G_S-coupled receptors and inhibited by G_i-coupled receptors. At the time, I did not discuss what G_q-coupled receptors do. Now is the time to discuss the intracellular signaling pathway associated with this receptor subtype. When a G_q-coupled receptor is stimulated, the G protein activates the enzyme phospholipase C, which then cleaves phosphatidylinositol 4,5-bisphosphate (PIP2) into inositol 1,4,5-trisphosphate (IP3) and diacylglycerol (DAG). IP3 binds to receptors located on the endoplasmic reticulum, an important organelle, which causes calcium (Ca^{++}) levels to increase in the neuron. DAG activates protein kinase C. Collectively, activation of IP3 and DAG leads to neuronal excitation. Lithium disrupts this signaling pathway by decreasing inositol levels (Brown & Tracy, 2013) (Fig. 5).

Lithium also inhibits *glycogen synthase kinase-3 (GSK-3)*, an enzyme that is involved in several signaling pathways. Crucially, GSK-3 regulates glycogen synthesis. Glycogen is a type of carbohydrate that serves as a form of energy storage. In fact, you will learn more about glycogen in Chap. 13. GSK-3 is also involved in cell survival and differentiation, and it influences neuronal excitability. One hypothesis argues that lithium alleviates depressive symptoms by increasing neuronal excitability via GSK-3-induced inhibition whereas it decreases manic symptoms by blocking potassium (K^+) ion channels (Sato, 2021).

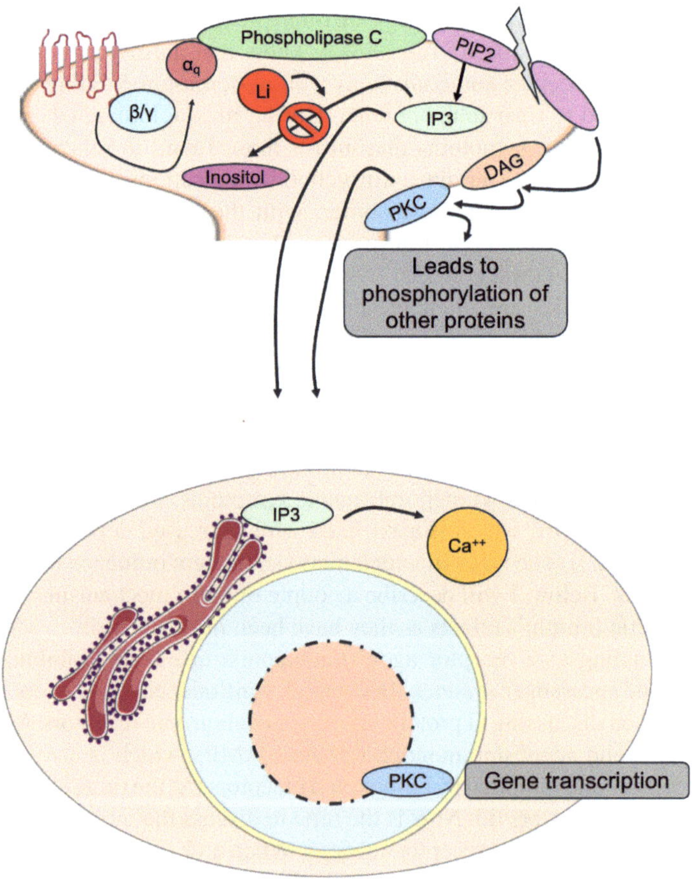

Fig. 5 One proposed mechanism by which lithium exerts its therapeutic effects for the treatment of bipolar disorder. Lithium (Li) disrupts the conversion of IP3 (1,4,5-trisphosphate) into inositol. *Note* DAG = diacylglycerol; PIP2 = phosphatidylinositol 4,5-bisphosphate; PKC = protein kinase C. Image created by the author using Servier Medical Art (https://smart.servier.com/; CC BY 4.0, https://creativecommons.org/licenses/by/4.0/) and Microsoft PowerPoint

8 Neuroinflammation and Mood Disorders

Given the high comorbidity of anxiety disorders and mood disorders, one should not be surprised to see that neuroinflammation is linked to conditions like MDD and bipolar disorder. To provide a brief recap from the previous chapter—stress produces an immune response, characterized by increased release of phagocytes, lymphocytes, and cytokines. Chronic stress leads to a long-lasting immune response, which can cause physical damage to brain structures.

Individuals with depressive disorders have increased levels of peripheral cytokine levels (Cyrino et al., 2021; Réus et al., 2023), which can then enter the brain via the *circumventricular organs*, brain structures situated near the third and the fourth ventricle. The BBB is weaker near the circumventricular organs. Increased cytokine activity in the brain can induce symptoms of MDD by overstimulating the *kynurenine* pathway. The enzyme indoleamine 2,3-dioxygenase (IDO) metabolizes tryptophan (the precursor of serotonin) to kynurenine. This means that cytokines cause too much tryptophan to be metabolized before it can be converted into serotonin (Réus et al., 2023; Zhang et al., 2021).

Increased kynurenine leads to another issue that contributes to the etiology of depressive symptoms in MDD and bipolar disorder. Kynurenine is metabolized into either kynurenic acid or quinolinic acid. These metabolites exert opposite effects on glutamatergic NMDA receptors. Earlier in this chapter, I mentioned that quinolinic acid levels are elevated in individuals that have attempted suicide; this metabolite is an NMDA receptor agonist. Overstimulation of NMDA receptors with quinolinic acid leads to excitotoxicity, leading to damaged glutamatergic neurons. Kynurenic acid, conversely, is an antagonist at NMDA receptors. Even though quinolinic acid and kynurenic acid have opposing actions, they both lead to reduced glutamatergic signaling. This coincides with the hypoactive glutamatergic system often observed in individuals with MDD.

Cytokine activity leads to formation of **reactive oxygen species (ROS),** chemically reactive molecules that can cause significant damage to cellular structures. One such structure that is highly sensitive to ROS is the mitochondrion, the energy source of cells. Mitochondrial dysfunction is impaired in MDD and bipolar disorder (Cyrino et al., 2021; Song et al., 2023). If mitochondria are damaged, neurons cannot generate adenosine triphosphate (ATP). This leads to cellular damage, which provides an account for reduced gray matter volume observed in multiple brain regions in those with depressive disorders.

As cytokines are implicated in depressive disorders, disrupting cytokine signaling should be effective in ameliorating depressive symptoms. In addition to its actions on intracellular signaling pathways, lithium reduces pro-inflammatory markers in the brain (Damri & Agam, 2024). SSRIs and SNRIs also have anti-inflammatory effects (Dionisie et al., 2021). Studies have even determined if non-steroidal anti-inflammatory drugs (NSAIDs) (refer back to Chap. 4) can be used as an adjunctive treatment for depression; unfortunately, results have not provided much support for their effectiveness in treating depressive disorders (Gong et al., 2025; Rosenblat et al.,

2016). Remember, NSAIDs block the enzyme cyclooxygenase 2 (COX-2), which is responsible for producing prostaglandins, particularly in the PNS. NSAIDs do not affect cytokines in the CNS as much as in the PNS; this is most likely why these medications are not viable treatment options for mood disorders.

Comprehension Check 8 Above, you read that neuroinflammation is implicated in both MDD and bipolar disorder. Given what you learned about lithium's mechanisms of action, how do you think lithium works to reduce neuroinflammation?

Lithium's actions on the GSK-3 signaling pathway explain how this medication can reduce neuroinflammation in the brain. GSK-3 activation increases neuronal excitability. When neurons are damaged due to excitotoxicity, this can promote neuroinflammation. By regulating the GSK-3 pathway, lithium prevents overexcitation of neurons.

9 Affective Impulsivity

Chapter 7 covered impulsive decision making. I now need to introduce an additional form of impulsivity, **affective impulsivity**. This form of impulsivity is driven by strong emotional responses. If an individual experiences intense sadness, they may be more likely to engage in impulsive behaviors.

9.1 Urgency

A term that is often used to describe affective impulsivity is *urgency* (Cyders & Smith, 2008). Note, urgency can occur when an individual experiences intense positive emotions. For example, when I was a graduate student at the University of Kentucky, the men's basketball team won the NCAA national championship. There were many individuals that engaged in a form of *positive* urgency, which manifested as excessive alcohol consumption and setting a couch on fire (this occurred one street over from my apartment). In Chap. 15, you will learn about different forms of aggression. As you will learn, one form of aggression reflects affective impulsivity as individuals engage in this behavior without thinking about it.

Negative urgency is associated with greater activation of the amygdala and reduced functional connectivity between the amygdala and the PFC (Steele et al., 2021) and is correlated with reduced ventrolateral PFC (vlPFC) activity when individuals try to perform cognitive tasks (Teh et al., 2025). These results indicate that the PFC cannot exert as much control over emotional processing centers of the brain. This predisposes individuals to engage in an action without considering its consequences.

Higher negative urgency is associated with greater activation of the OFC and the amygdala when individuals view negative images (Cyders et al., 2015). Given that thelOFC and the amygdala are both involved in the processing of punishment, individuals may display increased negative urgency because they are overly sensitive to punishment. Increased activity in the OFC may be related to altered serotoninergic activity in this region. In one of my earlier studies, my colleagues and I found increased serotonin transporter function in the OFC of rats that displayed enhanced urgency-like behavior (Yates et al., 2015). Increased transporter function would indicate lower serotonin levels in the frontal cortex. Going back to Chap. 7, low serotonin levels can promote impulsive decision making by disinhibiting dopaminergic neurons.

Negative urgency is positively correlated with resting-state functional connectivity between the dorsal ACC (dACC) and the insula (Um et al., 2020). As these structures make up part of the salience network, individuals may be more primed to detect negative stimuli. Resting-state functional connectivity between the vmPFC and the dorsal striatum is also associated with negative urgency (Leenaerts et al., 2024). The vmPFC exerts top-down control over the striatum, which is important for motor actions. If individuals are more sensitive to negative stimuli, increased connectivity between the vmPFC and the striatum can lead to a situation in which an individual impulsively responds when they encounter something aversive.

Individuals that experience more negative urgency have greater dopamine release in the vmPFC when exposed to stress (Leenaerts et al., 2024). The vmPFC has an indirect effect on motor behaviors as it has a direct connection with the subthalamic nucleus (STN), an important structure that influences activity of the basal ganglia (Mulder et al., 2014). Increased vmPFC activity could then cause hyperactivity of motor control centers, which manifests as mood-based impulsivity.

In contrast to the vmPFC, dopamine may be hypoactive in the NAc as increased dopamine transporter function is observed in this region of rats that display increased urgency-like behavior (Yates et al., 2015). This means that dopamine is cleared too quickly from the synapse, resulting in low intra-NAc dopamine levels. In Chap. 7, I discussed that increased dopamine levels in the NAc are associated with impulsive decision making. One important consideration is that increased dopamine transporter function may not necessarily lead to a hypoactive dopaminergic state in the NAc. Research indicates that dopamine D_2-like receptor binding in the striatum is negatively correlated with negative urgency, at least in individuals with gambling disorder (Clark et al., 2012). As D_2-like receptors regulate the release of dopamine from the presynaptic neuron, lower D_2-like receptor binding can result in excessive dopamine release into the synapse. Thus, increased dopamine transporter function may be a compensatory response to clear excess dopamine from the synapse.

9.2 Self-Injurious Behavior

Directly related to mood disorders, one serious form of affective impulsivity is self-injurious behavior (SIB) (Gvion et al., 2015), with suicide being the ultimate form of SIB. In this section, I will highlight some of the findings concerning the neurobiological bases of non-suicidal self-injury (NSSI). I mentioned that postmortem examination of brains from individuals that died of suicide revealed disturbances in serotoninergic and noradrenergic biomarkers. Directly related to this finding, reducing serotonin levels via tryptophan depletion increases SIB in humans (McCloskey et al., 2009).

One method for inducing SIB in animals is administering the drug pemoline. Pemoline has a somewhat similar mechanism of action as amphetamine and was originally developed to treat ADHD. Pemoline was ultimately discontinued due to concerns over liver toxicity. While pemoline increases dopamine levels, chronic treatment causes dopamine depletion from the presynaptic neuron. Animals treated with pemoline show increased SIB, such as self-biting (Mueller & Nyhan, 1982).

Another method for inducing SIB in animals is to lesion dopaminergic neurons with 6-hydroxydopamine (6-OHDA) (Stodgell et al., 1998). In fact, 6-OHDA lesions are hypothesized to serve as an animal model of *Lesch-Nyhan syndrome*, a genetic disorder that results in a mutation of a gene that encodes the enzyme hypoxanthine–guanine phosphoribosyltransferase (HGPRT). Individuals with this mutation have excessive buildup of uric acid. If you are familiar with *gout*, you have probably learned how uric acid leads to arthritic symptoms associated with this condition. For those with Lesch-Nyhan syndrome, uric acid affects more than joints; damage occurs to parts of the brain, particularly the basal ganglia (which relies heavily on dopaminergic signaling). This causes motor and cognitive disturbances, including SIB (Visser et al., 2000).

Alterations in the endogenous opioid system have also been observed in individuals that engage in NSSI. A polymorphism in the gene that encodes the kappa opioid receptor is associated with NSSI, and individuals that engage in NSSI have increased kappa opioid receptor expression and higher dynorphin levels (Liza et al., 2024). This finding is consistent with the finding that the kappa opioid receptor is involved in dysphoria (see previous chapter). NSSI may be a response to intense, negative emotions. I will come back to this below.

In addition to neurochemical disturbances, reduced gray matter volume is observed in regions such as the amygdala, the ACC, the insula, and the putamen of individuals that engage in NSSI (Liu et al., 2025; Pang et al., 2024; Won et al., 2023), and they have a hypofunctioning PFC (Koenig et al., 2021). When individuals with a history of NSSI are subjected to cold pain, they show greater activation of several brain regions, including the amygdala, the parahippocampal formation, and the OFC (Osuch et al., 2014). Similarly, when presented with emotionally salient images, individuals that engage in NSSI show stronger activity in the amygdala, the hippocampus, and the ACC, and they show greater activity in the OFC when viewing images directly related to NSSI (Plener et al., 2012). Notice that altered brain activity of individuals that engage in NSSI is similar to what was discussed for those that display increased negative urgency. Indeed, negative urgency is a major risk factor for NSSI (Peckham et al., 2020).

Adolescents that engage in NSSI have reduced functional connectivity between the amygdala and the ACC (Auerbach et al., 2021; Osuch et al., 2014; but see Westlund Schreiner et al., 2017) and within frontotemporal regions (Lu et al., 2024a). Reduced dACC activity is observed when individuals that engage in NSSI commit errors in a go/no-go task (Malejko et al., 2022). Similarly, deactivation of the dlPFC occurs when individuals who engage in NSSI are given cognitively challenging tasks (Zahid et al., 2022). Overall, brain regions involved in executive functioning are impaired in individuals that engage in NSSI. There is evidence that impulsivity, also marked by impaired functioning of prefrontal cortical areas, is a predictor of NSSI (He et al., 2025).

Reduced functional connectivity is observed between the putamen and temporoparietal junction (TPJ) (Liu et al., 2025), as well as between the NAc and the cerebellum (Chen et al., 2023). Recall that the TPJ makes up part of the ventral attention network (VAN). The TPJ is important for attentional control and self-awareness (Doricchi, 2022). As you have previously learned, the cerebellum is involved in higher-order cognitive functions in addition to motor control. The putamen, as part of the basal ganglia, is involved in more than just motor control. Like the NAc, the putamen is involved in reward processing (Muranishi et al., 2011). Overall, reduced functional connectivity between the putamen and the TPJ and the NAc and the cerebellum may promote NSSI as individuals are not able to effectively manage intense, negative emotional experiences (Wang et al., 2022).

NSSI has been conceptualized as an addiction (Blasco-Fontecilla et al., 2016). Addictive behaviors become habitual actions that require little conscious effort; likewise, NSSI is proposed to reflect a form of habit formation (read Chap. 11 to learn more about habit formation). At a neurobiological level, NSSI is associated with increased functional connectivity between the amygdala and the supplementary motor area (SMA) (Westlund Schreiner et al., 2017). When individuals encounter a negative stimulus, the amygdala causes the SMA to become active, which prepares motor sequences. In this case, the motor sequence includes actions related to self-injury.

Comprehension Check 9 What are the neurobiological and neurochemical parallels between negative urgency and self-injurious behavior (SIB).

Both negative urgency and SIB are associated with hypoactivity of the PFC, which is responsible for executive functions and impulse control, and hypoactivity of the serotoninergic system. As negative urgency and SIB can be considered affective forms of impulsivity, deficient serotonin may be the reason why the PFC is underactive, which then promotes emotionally driven impulsive behaviors.

10 Chapter Summary

This chapter explored the neurobiological underpinnings of mood disorders like MDD and bipolar disorder. Dysregulation of the LCSPT pathway and the DMN contributes to emotional dysregulation. Specifically, altered functioning of the LCSPT can promote enhanced sensitivity to punishment and ineffective adaptive reappraisal of negative situations. Hyperactivity of the DMN can drive increased rumination. Hypoactivity of several neurotransmitter systems is implicated in mood disorders, primarily monoamines, glutamate, and GABA. Collectively, aberrant neurotransmission can account for increased anhedonia, rumination, and impulsivity observed in individuals with mood disorders. This chapter concluded by discussing the neural correlates of affective impulsivity, including SIB. Given the relationship between mood disorders and impulsivity, similar mechanisms are observed for mood disorders, negative urgency, and SIB.

Answers to Experiment Spotlight Questions

1. WKY rats had significantly lower ARC mRNA and protein expression compared to Wistar Han rats. By stimulating 5-HT_{2A} receptors, psilocybin produces a rapid increase in ARC expression, which may account for the decreased immobility time in the forced swim test observed 9 days after treatment. One challenge in interpreting the results is that ARC mRNA/protein expression was not measured 9 days after psilocybin treatment; therefore, there is no way of knowing if ARC expression was further elevated at Day 9 relative to Day 23. In other words, while ARC expression was elevated 23 days following psilocybin treatment, ARC expression could have been further elevated at Day 9 and then decreased somewhat over time. For Wistar Han rats, the decreased immobility time observed at Day 23 may indicate that different mechanisms are involved in the antidepressant-like effects of psilocybin. The only change observed for Wistar Han rats was decreased PSD-95 expression.
2. These results indicate that LTP does not necessarily need to be altered to reduce depressive-like behavior. ARC is important for internalizing AMPA receptors, which is related to LTD. Both LTP and LTD are required for learning to occur as synapses need to be reset via LTD for future learning to occur. By increasing ARC expression, psilocybin may exert its therapeutic effects by regulating the balance of LTP and LTD.
3. One major factor that can account for the results is that the biochemical analysis was performed 24 days after psilocybin administration, which may have been too late to detect a change in BDNF levels. The article notes that BDNF mRNA levels often increase within a couple of hours but may decrease significantly within 24 h, suggesting the study's timing missed the peak expression period.

Glossary

Affective impulsivity the tendency to act rashly or without forethought when experiencing strong emotions.

Anhedonia the diminished ability to experience pleasure from activities that are usually enjoyable.

Brain-derived neurotrophic factor (BDNF) a protein that supports the survival, growth, and differentiation of neurons and synapses in the brain, functioning as a key modulator of neuroplasticity.

Learned helplessness a psychological condition in which an individual, after repeated exposure to uncontrollable aversive stimuli, ceases to attempt to escape or avoid negative situations, even when escape is possible.

Mood an emotional state or disposition, often prolonged or pervasive, that influences an individual's perception of the world.

Postsynaptic density (PSD) a protein-dense specialization found on the postsynaptic membrane of a chemical synapse, where it organizes neurotransmitter receptors and other signaling molecules to ensure efficient synaptic transmission.

Reactive oxygen species (ROS) chemically reactive molecules containing oxygen, formed as a natural byproduct of oxygen metabolism, that can cause significant damage to cellular structures.

References

Achtyes, E., Keaton, S. A., Smart, L., Burmeister, A. R., Heilman, P. L., Krzyzanowski, S., Nagalla, M., Guillemin, G. J., Escobar Galvis, M. L., Lim, C. K., Muzik, M., Postolache, T. T., Leach, R., & Brundin, L. (2020). Inflammation and kynurenine pathway dysregulation in post-partum women with severe and suicidal depression. *Brain, Behavior, and Immunity, 83*, 239–247. https://doi.org/10.1016/j.bbi.2019.10.017

Ahn, W. Y., Rass, O., Fridberg, D. J., Bishara, A. J., Forsyth, J. K., Breier, A., Busemeyer, J. R., Hetrick, W. P., Bolbecker, A. R., & O'Donnell, B. F. (2011). Temporal discounting of rewards in patients with bipolar disorder and schizophrenia. *Journal of Abnormal Psychology, 120*(4), 911–921. https://doi.org/10.1037/a0023333

Akbar, D., Rhee, T. G., Ceban, F., Ho, R., Teopiz, K. M., Cao, B., Subramaniapillai, M., Kwan, A. T. H., Rosenblat, J. D., & McIntyre, R. S. (2023). Dextromethorphan-bupropion for the treatment of depression: A systematic review of efficacy and safety in clinical trials. *CNS Drugs, 37*(10), 867–881. https://doi.org/10.1007/s40263-023-01032-5

Alvarez, J. C., Cremniter, D., Lesieur, P., Gregoire, A., Gilton, A., Macquin-Mavier, I., Jarreau, C., & Spreux-Varoquaux, O. (1999). Low blood cholesterol and low platelet serotonin levels in violent suicide attempters. *Biological Psychiatry, 45*(8), 1066–1069. https://doi.org/10.1016/s0006-3223(98)00160-7

American Psychiatric Association. (2013). *Diagnostic and statistical manual of mental disorders* (5th ed.). https://doi.org/10.1176/appi.books.9780890425596

Anisman, H., & Zacharko, R. M. (1982). Depression: The predisposing influence of stress. *Behavioral and Brain Sciences, 5*(1), 89–137. https://doi.org/10.1017/S0140525X00010633

Asberg, M., Träskman, L., & Thorén, P. (1976). 5-HIAA in the cerebrospinal fluid. A biochemical suicide predictor?. *Archives of General Psychiatry, 33*(10), 1193–1197. https://doi.org/10.1001/archpsyc.1976.01770100055005

Auerbach, R. P., Pagliaccio, D., Allison, G. O., Alqueza, K. L., & Alonso, M. F. (2021). Neural correlates associated with suicide and nonsuicidal self-injury in youth. *Biological Psychiatry, 89*(2), 119–133. https://doi.org/10.1016/j.biopsych.2020.06.002

Baune, B. T., Fuhr, M., Air, T., & Hering, C. (2014). Neuropsychological functioning in adolescents and young adults with major depressive disorder–A review. *Psychiatry Research, 218*(3), 261–271. https://doi.org/10.1016/j.psychres.2014.04.052

Belujon, P., & Grace, A. A. (2014). Restoring mood balance in depression: Ketamine reverses deficit in dopamine-dependent synaptic plasticity. *Biological Psychiatry, 76*(12), 927–936. https://doi.org/10.1016/j.biopsych.2014.04.014

Bergamini, G., Sigrist, H., Ferger, B., Singewald, N., Seifritz, E., & Pryce, C. R. (2016). Depletion of nucleus accumbens dopamine leads to impaired reward and aversion processing in mice: Relevance to motivation pathologies. *Neuropharmacology, 109*, 306–319. https://doi.org/10.1016/j.neuropharm.2016.03.048

Berman, R. M., Narasimhan, M., Miller, H. L., Anand, A., Cappiello, A., Oren, D. A., Heninger, G. R., & Charney, D. S. (1999). Transient depressive relapse induced by catecholamine depletion: Potential phenotypic vulnerability marker? *Archives of General Psychiatry, 56*(5), 395–403. https://doi.org/10.1001/archpsyc.56.5.395

Blasco-Fontecilla, H., Fernández-Fernández, R., Colino, L., Fajardo, L., Perteguer-Barrio, R., & de Leon, J. (2016). The addictive model of self-harming (non-suicidal and suicidal) behavior. *Frontiers in Psychiatry, 7*, 8. https://doi.org/10.3389/fpsyt.2016.00008

Brown, K. M., & Tracy, D. K. (2013). Lithium: The pharmacodynamic actions of the amazing ion. *Therapeutic Advances in Psychopharmacology, 3*(3), 163–176. https://doi.org/10.1177/2045125312471963

Caetano, S. C., Fonseca, M., Olvera, R. L., Nicoletti, M., Hatch, J. P., Stanley, J. A., Hunter, K., Lafer, B., Pliszka, S. R., & Soares, J. C. (2005). Proton spectroscopy study of the left dorsolateral prefrontal cortex in pediatric depressed patients. *Neuroscience Letters, 384*(3), 321–326. https://doi.org/10.1016/j.neulet.2005.04.099

Chandley, M. J., & Ordway, G. A. (2012). Chapter 3: Noradrenergic dysfunction in depression and suicide. In Y. Dwivedi (ed.), *The neurobiological basis of suicide*. CRC Press/Taylor & Francis. https://www.ncbi.nlm.nih.gov/books/NBK107205/

Chen, X., Chen, H., Liu, J., Tang, H., Zhou, J., Liu, P., Tian, Y., Wang, X., Lu, F., & Zhou, J. (2023). Functional connectivity alterations in reward-related circuits associated with non-suicidal self-injury behaviors in drug-naïve adolescents with depression. *Journal of Psychiatric Research, 163*, 270–277. https://doi.org/10.1016/j.jpsychires.2023.05.068

Clark, L., Stokes, P. R., Wu, K., Michalczuk, R., Benecke, A., Watson, B. J., Egerton, A., Piccini, P., Nutt, D. J., Bowden-Jones, H., & Lingford-Hughes, A. R. (2012). Striatal dopamine D_2/D_3 receptor binding in pathological gambling is correlated with mood-related impulsivity. *NeuroImage, 63*(1), 40–46. https://doi.org/10.1016/j.neuroimage.2012.06.067

Conway, B. R. (2018). The organization and operation of inferior temporal cortex. *Annual Review of Vision Science, 4*, 381–402. https://doi.org/10.1146/annurev-vision-091517-034202

Cooper, J. A., Arulpragasam, A. R., & Treadway, M. T. (2018). Anhedonia in depression: Biological mechanisms and computational models. *Current Opinion in Behavioral Sciences, 22*, 128–135. https://doi.org/10.1016/j.cobeha.2018.01.024

Cottingham, C., & Wang, Q. (2012). α2 adrenergic receptor dysregulation in depressive disorders: Implications for the neurobiology of depression and antidepressant therapy. *Neuroscience and Biobehavioral Reviews, 36*(10), 2214–2225. https://doi.org/10.1016/j.neubiorev.2012.07.011

Culpepper, L. (2015). Impact of untreated major depressive disorder on cognition and daily function. *The Journal of Clinical Psychiatry, 76*(7), Article e901. https://doi.org/10.4088/JCP.13086tx4c

Cyders, M. A., Dzemidzic, M., Eiler, W. J., Coskunpinar, A., Karyadi, K. A., & Kareken, D. A. (2015). Negative urgency mediates the relationship between amygdala and orbitofrontal cortex activation to negative emotional stimuli and general risk-taking. *Cerebral Cortex, 25*(11), 4094–4102. https://doi.org/10.1093/cercor/bhu123

Cyders, M. A., & Smith, G. T. (2008). Emotion-based dispositions to rash action: Positive and negative urgency. *Psychological Bulletin, 134*(6), 807–828. https://doi.org/10.1037/a0013341

Cyrino, L. A. R., Delwing-de Lima, D., Ullmann, O. M., & Maia, T. P. (2021). Concepts of neuroinflammation and their relationship with impaired mitochondrial functions in bipolar disorder. *Frontiers in Behavioral Neuroscience, 15*, Article 609487. https://doi.org/10.3389/fnbeh.2021.609487

Damri, O., & Agam, G. (2024). Lithium, inflammation and neuroinflammation with emphasis on bipolar disorder-A narrative review. *International Journal of Molecular Sciences, 25*(24), 13277. https://doi.org/10.3390/ijms252413277

de Kwaasteniet, B. P., Pinto, C., Ruhé, H. G., van Wingen, G. A., Booij, J., & Denys, D. (2014). Striatal dopamine D2/3 receptor availability in treatment resistant depression. *PLoS ONE, 9*(11), Article e113612. https://doi.org/10.1371/journal.pone.0113612

Dhir, A., & Kulkarni, S. K. (2007). Effect of addition of yohimbine (alpha-2-receptor antagonist) to the antidepressant activity of fluoxetine or venlafaxine in the mouse forced swim test. *Pharmacology, 80*(4), 239–243. https://doi.org/10.1159/000104877

Dionisie, V., Filip, G. A., Manea, M. C., Manea, M., & Riga, S. (2021). The anti-inflammatory role of SSRI and SNRI in the treatment of depression: A review of human and rodent research studies. *Inflammopharmacology, 29*(1), 75–90. https://doi.org/10.1007/s10787-020-00777-5

Dogan, K. H., Unaldi, M., & Demirci, S. (2016). Evaluation of postmortem cerebrospinal fluid S100B protein and serotonin levels: Comparison of suicidal versus nonsuicidal deaths in Konya, Turkey. *Journal of Forensic Sciences, 61*(5), 1285–1291. https://doi.org/10.1111/1556-4029.13124

Doricchi, F. (2022). The functions of the temporal-parietal junction. *Handbook of Clinical Neurology, 187*, 161–177. https://doi.org/10.1016/B978-0-12-823493-8.00020-1

Draganov, M., Vives-Gilabert, Y., de Diego-Adeliño, J., Vicent-Gil, M., Puigdemont, D., & Portella, M. J. (2020). Glutamatergic and GABA-ergic abnormalities in First-episode depression. A 1-year follow-up 1H-MR spectroscopic study. *Journal of Affective Disorders, 266*, 572–577. https://doi.org/10.1016/j.jad.2020.01.138

Drevets, W. C., Price, J. L., & Furey, M. L. (2008). Brain structural and functional abnormalities in mood disorders: Implications for neurocircuitry models of depression. *Brain Structure & Function, 213*(1–2), 93–118. https://doi.org/10.1007/s00429-008-0189-x

Duman, R. S., Sanacora, G., & Krystal, J. H. (2019). Altered connectivity in depression: GABA and glutamate neurotransmitter deficits and reversal by novel treatments. *Neuron, 102*(1), 75–90. https://doi.org/10.1016/j.neuron.2019.03.013

Erhardt, S., Lim, C. K., Linderholm, K. R., Janelidze, S., Lindqvist, D., Samuelsson, M., Lundberg, K., Postolache, T. T., Träskman-Bendz, L., Guillemin, G. J., & Brundin, L. (2013). Connecting inflammation with glutamate agonism in suicidality. *Neuropsychopharmacology, 38*(5), 743–752. https://doi.org/10.1038/npp.2012.248

Fee, C., Banasr, M., & Sibille, E. (2017). Somatostatin-positive gamma-aminobutyric acid interneuron deficits in depression: Cortical microcircuit and therapeutic perspectives. *Biological Psychiatry, 82*(8), 549–559. https://doi.org/10.1016/j.biopsych.2017.05.024

Feyissa, A. M., Chandran, A., Stockmeier, C. A., & Karolewicz, B. (2009). Reduced levels of NR2A and NR2B subunits of NMDA receptor and PSD-95 in the prefrontal cortex in major depression. *Progress in Neuro-Psychopharmacology & Biological Psychiatry, 33*(1), 70–75. https://doi.org/10.1016/j.pnpbp.2008.10.005

Flint, J., & Kendler, K. S. (2014). The genetics of major depression. *Neuron, 81*(3), 484–503. https://doi.org/10.1016/j.neuron.2014.01.027

Fogaça, M. V., Wu, M., Li, C., Li, X. Y., Picciotto, M. R., & Duman, R. S. (2021). Inhibition of GABA interneurons in the mPFC is sufficient and necessary for rapid antidepressant responses. *Molecular Psychiatry, 26*(7), 3277–3291. https://doi.org/10.1038/s41380-020-00916-y

Fox, H. H., & Gibas, J. T. (1953). Synthetic tuberculostats. VII. Monoalkyl derivatives of isonicotinylhydrazine. *The Journal of Organic Chemistry, 18*(8), 994–1002.

Freedman, R. R., Embury, J., Migály, P., Keegan, D., Pandey, G. N., Javaid, J. I., & Davis, J. M. (1990). Stress-induced desensitization of alpha 2-adrenergic receptors in human platelets. *Psychosomatic Medicine, 52*(6), 624–630. https://doi.org/10.1097/00006842-199011000-00004

Frodl, T., Bokde, A. L., Scheuerecker, J., Lisiecka, D., Schoepf, V., Hampel, H., Möller, H. J., Brückmann, H., Wiesmann, M., & Meisenzahl, E. (2010). Functional connectivity bias of the orbitofrontal cortex in drug-free patients with major depression. *Biological Psychiatry, 67*(2), 161–167. https://doi.org/10.1016/j.biopsych.2009.08.022

Frye, M. A., Tsai, G. E., Huggins, T., Coyle, J. T., & Post, R. M. (2007). Low cerebrospinal fluid glutamate and glycine in refractory affective disorder. *Biological Psychiatry, 61*(2), 162–166. https://doi.org/10.1016/j.biopsych.2006.01.024

Gambarana, C., Ghiglieri, O., & Graziella de Montis, M. (1995). Desensitization of the D1 dopamine receptors in rats reproduces a model of escape deficit reverted by imipramine, fluoxetine and clomipramine. *Progress in Neuro-Psychopharmacology & Biological Psychiatry, 19*(5), 741–755. https://doi.org/10.1016/0278-5846(95)00128-i

Gao, L., Zhang, Y., Sterling, K., & Song, W. (2022). Brain-derived neurotrophic factor in Alzheimer's disease and its pharmaceutical potential. *Translational Neurodegeneration, 11*(1), 4. https://doi.org/10.1186/s40035-022-00279-0

Garakani, A., Martinez, J. M., Yehuda, R., & Gorman, J. M. (2013). Cerebrospinal fluid levels of glutamate and corticotropin releasing hormone in major depression before and after treatment. *Journal of Affective Disorders, 146*(2), 262–265. https://doi.org/10.1016/j.jad.2012.06.037

Gilbert, K. E., Kalmar, J. H., Womer, F. Y., Markovich, P. J., Pittman, B., Nolen-Hoeksema, S., & Blumberg, H. P. (2011). Impulsivity in adolescent bipolar disorder. *Acta Neuropsychiatrica, 23*(2), 57–61. https://doi.org/10.1111/j.1601-5215.2011.00522.x

Glikmann-Johnston, Y., Saling, M. M., Reutens, D. C., & Stout, J. C. (2015). Hippocampal 5-HT1A receptor and spatial learning and memory. *Frontiers in Pharmacology, 6*, 289. https://doi.org/10.3389/fphar.2015.00289

Głuch-Lutwin, M., Sałaciak, K., Pytka, K., Gawalska, A., Jamrozik, M., Śniecikowska, J., Kołaczkowski, M., Depoortère, R. Y., & Newman-Tancredi, A. (2023). The 5-HT_{1A} receptor biased agonist, NLX-204, shows rapid-acting antidepressant-like properties and neurochemical changes in two mouse models of depression. *Behavioural Brain Research, 438*, Article 114207. https://doi.org/10.1016/j.bbr.2022.114207

Gong, H., Su, W. J., Deng, S. L., Luo, J., Du, Z. L., Luo, Y., Lv, K. Y., Zhu, D. M., & Fan, X. T. (2025). Anti-inflammatory interventions for the treatment and prevention of depression among older adults: A systematic review and meta-analysis. *Translational Psychiatry, 15*(1), 114. https://doi.org/10.1038/s41398-025-03317-3

Gören, J. L. (2016). Brain-derived neurotrophic factor and schizophrenia. *The Mental Health Clinician, 6*(6), 285–288. https://doi.org/10.9740/mhc.2016.11.285

Grieve, S. M., Korgaonkar, M. S., Koslow, S. H., Gordon, E., & Williams, L. M. (2013). Widespread reductions in gray matter volume in depression. *NeuroImage. Clinical, 3*, 332–339. https://doi.org/10.1016/j.nicl.2013.08.016

Guo, Z., Cui, Y., Qiu, R., Bu, L., Yang, T., Li, Y., & Zhu, X. (2024). The association of impulsivity with depression and anxiety symptoms: A transdiagnostic network analysis and replication. *Journal of Affective Disorders, 359*, 100–108. https://doi.org/10.1016/j.jad.2024.05.076

Gvion, Y., Levi-Belz, Y., Hadlaczky, G., & Apter, A. (2015). On the role of impulsivity and decision-making in suicidal behavior. *World Journal of Psychiatry, 5*(3), 255–259. https://doi.org/10.5498/wjp.v5.i3.255

Ha, T. H., Kim, J. S., Chang, J. S., Oh, S. H., Her, J. Y., Cho, H. S., Park, T. S., Shin, S. Y., & Ha, K. (2012). Verbal and visual memory impairments in bipolar I and II disorder. *Psychiatry Investigation, 9*(4), 339–346. https://doi.org/10.4306/pi.2012.9.4.339

Hakim, A. (2022). Perspectives on the complex links between depression and dementia. *Frontiers in Aging Neuroscience, 14*, Article 821866. https://doi.org/10.3389/fnagi.2022.821866

Hamon, M., & Blier, P. (2013). Monoamine neurocircuitry in depression and strategies for new treatments. *Progress in Neuro-Psychopharmacology & Biological Psychiatry, 45*, 54–63. https://doi.org/10.1016/j.pnpbp.2013.04.009

Hasler, G., Fromm, S., Carlson, P. J., Luckenbaugh, D. A., Waldeck, T., Geraci, M., Roiser, J. P., Neumeister, A., Meyers, N., Charney, D. S., & Drevets, W. C. (2008). Neural response to catecholamine depletion in unmedicated subjects with major depressive disorder in remission and healthy subjects. *Archives of General Psychiatry, 65*(5), 521–531. https://doi.org/10.1001/archpsyc.65.5.521

He, X., Huang, P., Xu, X., Yu, Q., Huang, H., Yang, P., & Yang, B. (2025). Impulsivity and non-suicidal self-injury in adolescents: A systematic review and meta-analysis of longitudinal studies. *Frontiers in Psychiatry, 16*, 1586922. https://doi.org/10.3389/fpsyt.2025.1586922

Herwig, U., Kaffenberger, T., Jäncke, L., & Brühl, A. B. (2010). Self-related awareness and emotion regulation. *NeuroImage, 50*(2), 734–741. https://doi.org/10.1016/j.neuroimage.2009.12.089

Hillhouse, T. M., & Porter, J. H. (2015). A brief history of the development of antidepressant drugs: From monoamines to glutamate. *Experimental and Clinical Psychopharmacology, 23*(1), 1–21. https://doi.org/10.1037/a0038550

Hranilovic, D., Bucan, M., & Wang, Y. (2008). Emotional response in dopamine D2L receptor-deficient mice. *Behavioural Brain Research, 195*(2), 246–250. https://doi.org/10.1016/j.bbr.2008.09.007

Huang, M. H., Kuan, Y. H., Chan, Y. E., Mao, W. C., & Su, T. P. (2024). Poor subjective sleep quality and trait impulsivity in patients with bipolar disorder. *CNS Spectrums, 29*(4), 289–295. https://doi.org/10.1017/S109285292400021X

Huang, H., Rong, B., Chen, C., Wan, Q., Liu, Z., Zhou, Y., Wang, G., & Wang, H. (2023). Common and distinct functional connectivity of the orbitofrontal cortex in depression and schizophrenia. *Brain Sciences, 13*(7), 997. https://doi.org/10.3390/brainsci13070997

Hui, Y. P., Zhang, Q. J., Zhang, L., Chen, L., Guo, Y., Qiao, H. F., Wang, Y., & Liu, J. (2014). Activation of prelimbic 5-HT1A receptors produces antidepressant-like effects in a unilateral rat model of Parkinson's disease. *Neuroscience, 268*, 265–275. https://doi.org/10.1016/j.neuroscience.2014.03.033

Inoshita, M., Umehara, H., Watanabe, S. Y., Nakataki, M., Kinoshita, M., Tomioka, Y., Tajima, A., Numata, S., & Ohmori, T. (2018). Elevated peripheral blood glutamate levels in major depressive disorder. *Neuropsychiatric Disease and Treatment, 14*, 945–953. https://doi.org/10.2147/NDT.S159855

Jha, M. K., & Mathew, S. J. (2023). Pharmacotherapies for treatment-resistant depression: How antipsychotics fit in the rapidly evolving therapeutic landscape. *The American Journal of Psychiatry, 180*(3), 190–199. https://doi.org/10.1176/appi.ajp.20230025

Jiang, X., Xing, G., Yang, C., Verma, A., Zhang, L., & Li, H. (2009). Stress impairs 5-HT2A receptor-mediated serotonergic facilitation of GABA release in juvenile rat basolateral amygdala. *Neuropsychopharmacology, 34*(2), 410–423. https://doi.org/10.1038/npp.2008.71

Jolly, A. E., Raymont, V., Cole, J. H., Whittington, A., Scott, G., De Simoni, S., Searle, G., Gunn, R. N., & Sharp, D. J. (2019). Dopamine D2/D3 receptor abnormalities after traumatic brain injury and their relationship to post-traumatic depression. *NeuroImage. Clinical, 24*, Article 101950. https://doi.org/10.1016/j.nicl.2019.101950

Kalin, N. H. (2020). The critical relationship between anxiety and depression. *The American Journal of Psychiatry, 177*(5), 365–367. https://doi.org/10.1176/appi.ajp.2020.20030305

Kapur, S., & Seeman, P. (2002). NMDA receptor antagonists ketamine and PCP have direct effects on the dopamine D(2) and serotonin 5-HT(2)receptors-implications for models of schizophrenia. *Molecular Psychiatry, 7*(8), 837–844. https://doi.org/10.1038/sj.mp.4001093

Keller, A. S., Leikauf, J. E., Holt-Gosselin, B., Staveland, B. R., & Williams, L. M. (2019). Paying attention to attention in depression. *Translational Psychiatry, 9*(1), 279. https://doi.org/10.1038/s41398-019-0616-1

Kessler, R. C., & Wang, P. S. (2008). The descriptive epidemiology of commonly occurring mental disorders in the United States. *Annual Review of Public Health, 29*, 115–129. https://doi.org/10.1146/annurev.publhealth.29.020907.090847

Kim, Y. H., Jang, J., Kang, N., Jeong, J. H., Kim, J., Ahn, Y. M., Kim, Y. S., & Kim, S. H. (2025). Revealing differential psychotic symptoms in schizophrenia and bipolar I disorder by manifold learning and network analyses. *Translational Psychiatry, 15*(1), 194. https://doi.org/10.1038/s41398-025-03403-6

Kishi, T., Ikuta, T., Sakuma, K., Okuya, M., Hatano, M., Matsuda, Y., & Iwata, N. (2023). Antidepressants for the treatment of adults with major depressive disorder in the maintenance phase: A systematic review and network meta-analysis. *Molecular Psychiatry, 28*(1), 402–409. https://doi.org/10.1038/s41380-022-01824-z

Kitching D. (2015). Depression in dementia. *Australian Prescriber, 38*(6), 209–2011. https://doi.org/10.18773/austprescr.2015.071

Koenig, J., Höper, S., van der Venne, P., Mürner-Lavanchy, I., Resch, F., & Kaess, M. (2021). Resting state prefrontal cortex oxygenation in adolescent non-suicidal self-injury—A near-infrared spectroscopy study. *NeuroImage. Clinical, 31*, Article 102704. https://doi.org/10.1016/j.nicl.2021.102704

Kolasa, M., Nikiforuk, A., Korlatowicz, A., Solich, J., Potasiewicz, A., Dziedzicka-Wasylewska, M., Bugno, R., Hogendorf, A., Bojarski, A., & Faron-Górecka, A. (2025). Unraveling psilocybin's therapeutic potential: Behavioral and neuroplasticity insights in Wistar-Kyoto and Wistar male rat models of treatment-resistant depression. *Psychopharmacology (Berl), 242*(7), 1607–1625. https://doi.org/10.1007/s00213-024-06644-3

Kolata, S. M., Nakao, K., Jeevakumar, V., Farmer-Alroth, E. L., Fujita, Y., Bartley, A. F., Jiang, S. Z., Rompala, G. R., Sorge, R. E., Jimenez, D. V., Martinowich, K., Mateo, Y., Hashimoto, K., Dobrunz, L. E., & Nakazawa, K. (2018). Neuropsychiatric phenotypes produced by GABA reduction in mouse cortex and hippocampus. *Neuropsychopharmacology, 43*(6), 1445–1456. https://doi.org/10.1038/npp.2017.296

Koob, G. F. (2022). Anhedonia, hyperkatifeia, and negative reinforcement in substance use disorders. *Current Topics in Behavioral Neurosciences, 58*, 147–165. https://doi.org/10.1007/7854_2021_288

Kriesche, D., Woll, C. F. J., Tschentscher, N., Engel, R. R., & Karch, S. (2023). Neurocognitive deficits in depression: A systematic review of cognitive impairment in the acute and remitted state. *European Archives of Psychiatry and Clinical Neuroscience, 273*(5), 1105–1128. https://doi.org/10.1007/s00406-022-01479-5

Kuniishi, H., Ichisaka, S., Matsuda, S., Futora, E., Harada, R., & Hata, Y. (2017). Chronic inactivation of the orbitofrontal cortex increases anxiety-like behavior and impulsive aggression, but decreases depression-like behavior in rats. *Frontiers in Behavioral Neuroscience, 10*, 250. https://doi.org/10.3389/fnbeh.2016.00250

Lacerda, A. L., Keshavan, M. S., Hardan, A. Y., Yorbik, O., Brambilla, P., Sassi, R. B., Nicoletti, M., Mallinger, A. G., Frank, E., Kupfer, D. J., & Soares, J. C. (2004). Anatomic evaluation of the orbitofrontal cortex in major depressive disorder. *Biological Psychiatry, 55*(4), 353–358. https://doi.org/10.1016/j.biopsych.2003.08.021

Lambert, G., Johansson, M., Agren, H., & Friberg, P. (2000). Reduced brain norepinephrine and dopamine release in treatment-refractory depressive illness: Evidence in support of the catecholamine hypothesis of mood disorders. *Archives of General Psychiatry, 57*(8), 787–793. https://doi.org/10.1001/archpsyc.57.8.787

Leenaerts, N., Ceccarini, J., Sunaert, S., & Vrieze, E. (2024). The relation between stress-induced dopamine release in the ventromedial prefrontal cortex, fronto-striatal functional connectivity, and negative urgency: A multimodal investigation using [^{18}F]Fallypride PET, MRI and experience sampling. *Behavioural Brain Research, 471*, Article 115138. https://doi.org/10.1016/j.bbr.2024.115138

Lengvenyte, A., Belzeaux, R., Olié, E., Hamzeh-Cognasse, H., Sénèque, M., Strumila, R., Cognasse, F., & Courtet, P. (2023). Associations of potential plasma biomarkers with suicide attempt history, current suicidal ideation and subsequent suicidal events in patients with depression: A discovery study. *Brain, Behavior, and Immunity, 114*, 242–254. https://doi.org/10.1016/j.bbi.2023.08.025

Levine, J., Panchalingam, K., Rapoport, A., Gershon, S., McClure, R. J., & Pettegrew, J. W. (2000). Increased cerebrospinal fluid glutamine levels in depressed patients. *Biological Psychiatry, 47*(7), 586–593. https://doi.org/10.1016/s0006-3223(99)00284-x

Lima, I. M. M., Peckham, A. D., & Johnson, S. L. (2018). Cognitive deficits in bipolar disorders: Implications for emotion. *Clinical Psychology Review, 59*, 126–136. https://doi.org/10.1016/j.cpr.2017.11.006

Lin, I. M., Yu, H. E., Yeh, Y. C., Huang, M. F., Wu, K. T., Ke, C. K., Lin, P. Y., & Yen, C. F. (2021). Prefrontal lobe and posterior cingulate cortex activations in patients with major depressive disorder by using standardized weighted low-resolution electromagnetic tomography. *Journal of Personalized Medicine, 11*(11), 1054. https://doi.org/10.3390/jpm11111054

Lintel, H., Corpuz, T., Paracha, S. U., & Grossberg, G. T. (2021). Mood disorders and anxiety in Parkinson's disease: Current concepts. *Journal of Geriatric Psychiatry and Neurology, 34*(4), 280–288. https://doi.org/10.1177/08919887211018267

Liu, M. Y., Yin, C. Y., Zhu, L. J., Zhu, X. H., Xu, C., Luo, C. X., Chen, H., Zhu, D. Y., & Zhou, Q. G. (2018). Sucrose preference test for measurement of stress-induced anhedonia in mice. *Nature Protocols, 13*(7), 1686–1698. https://doi.org/10.1038/s41596-018-0011-z

Liu, X., Zhang, Y., Chen, J., Xie, M., Pan, L., Hommel, B., Yang, Y., Zhu, X., Wang, K., & Zhang, W. (2025). Altered brain structure and function correlate with non-suicidal self-injury in children and adolescents with transdiagnostic psychiatric disorders. *Journal of Psychiatric Research, 184*, 17–26. https://doi.org/10.1016/j.jpsychires.2025.02.051

Liza, R. G., Yaunin, Y., Yanwirasti, & Rita, R. S. (2024). Association between opioid receptor kappa 1 gene polymorphism, kappa opioid receptor level and dynorphin level with non-suicidal self-injury. *Pakistan Journal of Biological Sciences, 27*(11), 518–525. https://doi.org/10.3923/pjbs.2024.518.525

Lohoff, F. W. (2010). Overview of the genetics of major depressive disorder. *Current Psychiatry Reports, 12*(6), 539–546. https://doi.org/10.1007/s11920-010-0150-6

Lu, X., Zhang, Y., Zhong, S., Lai, S., Yan, S., Song, X., Zhong, Q., Ye, S., Chen, Y., Lai, J., & Jia, Y. (2024a). Cognitive impairment in major depressive disorder with non-suicidal self-injury: Association with the functional connectivity of frontotemporal cortex. *Journal of Psychiatric Research, 177*, 219–227. https://doi.org/10.1016/j.jpsychires.2024.07.008

Lu, J., Zhao, X., Wei, X., & He, G. (2024b). Risky decision-making in major depressive disorder: A three-level meta-analysis. *International Journal of Clinical and Health Psychology, 24*(1), Article 100417. https://doi.org/10.1016/j.ijchp.2023.100417

Luscher, B., Maguire, J. L., Rudolph, U., & Sibille, E. (2023). $GABA_A$ receptors as targets for treating affective and cognitive symptoms of depression. *Trends in Pharmacological Sciences, 44*(9), 586–600. https://doi.org/10.1016/j.tips.2023.06.009

Malejko, K., Hafner, S., Brown, R. C., Plener, P. L., Grön, G., Graf, H., & Abler, B. (2022). Neural signatures of error processing in depressed adolescents with comorbid non-suicidal self-injury (NSSI). *Biomedicines, 10*(12), 3188. https://doi.org/10.3390/biomedicines10123188

McCloskey, M. S., Ben-Zeev, D., Lee, R., Berman, M. E., & Coccaro, E. F. (2009). Acute tryptophan depletion and self-injurious behavior in aggressive patients and healthy volunteers. *Psychopharmacology (Berl), 203*(1), 53–61. https://doi.org/10.1007/s00213-008-1374-6

Merz, E. C., He, X., Noble, K. G., & Pediatric Imaging, Neurocognition, and Genetics Study. (2018). Anxiety, depression, impulsivity, and brain structure in children and adolescents. *NeuroImage. Clinical, 20*, 243–251. https://doi.org/10.1016/j.nicl.2018.07.020

Mikulska, J., Juszczyk, G., Gawrońska-Grzywacz, M., & Herbet, M. (2021). HPA axis in the pathomechanism of depression and schizophrenia: New therapeutic strategies based on its participation. *Brain Sciences, 11*(10), 1298. https://doi.org/10.3390/brainsci11101298

Miller, H. L., Delgado, P. L., Salomon, R. M., Berman, R., Krystal, J. H., Heninger, G. R., & Charney, D. S. (1996). Clinical and biochemical effects of catecholamine depletion on antidepressant-induced remission of depression. *Archives of General Psychiatry, 53*(2), 117–128. https://doi.org/10.1001/archpsyc.1996.01830020031005

Miller, W. R., Seligman, M. E., & Kurlander, H. M. (1975). Learned helplessness, depression, and anxiety. *The Journal of Nervous and Mental Disease, 161*(5), 347–357.

Miranda, L. (2025). Antidepressant and anxiolytic effects of activating 5HT2A receptors in the anterior cingulate cortex and the theoretical mechanisms underlying them—A scoping review of available literature. *Brain Research, 1846*, Article 149226. https://doi.org/10.1016/j.brainres.2024.149226

Mizuno, Y., Ashok, A. H., Bhat, B. B., Jauhar, S., & Howes, O. D. (2023). Dopamine in major depressive disorder: A systematic review and meta-analysis of in vivo imaging studies. *Journal of Psychopharmacology, 37*(11), 1058–1069. https://doi.org/10.1177/02698811231200881

Moret, C., & Briley, M. (2011). The importance of norepinephrine in depression. *Neuropsychiatric Disease and Treatment, 7*(Suppl 1), 9–13. https://doi.org/10.2147/NDT.S19619

Mueller, K., & Nyhan, W. L. (1982). Pharmacologic control of pemoline induced self-injurious behavior in rats. *Pharmacology, Biochemistry, and Behavior, 16*(6), 957–963. https://doi.org/10.1016/0091-3057(82)90052-1

Mulder, M. J., Boekel, W., Ratcliff, R., & Forstmann, B. U. (2014). Cortico-subthalamic connection predicts individual differences in value-driven choice bias. *Brain Structure & Function, 219*(4), 1239–1249. https://doi.org/10.1007/s00429-013-0561-3

Muranishi, M., Inokawa, H., Yamada, H., Ueda, Y., Matsumoto, N., Nakagawa, M., & Kimura, M. (2011). Inactivation of the putamen selectively impairs reward history-based action selection. *Experimental Brain Research, 209*(2), 235–246. https://doi.org/10.1007/s00221-011-2545-y

Nagayama, H., Maeda, T., Uchiyama, T., Hashimoto, M., Nomoto, N., Kano, O., Takahashi, T., Terashi, H., Hamada, S., Hasegawa, T., Hatano, T., Takahashi, T., Baba, Y., Sengoku, R., Watanabe, H., Inoue, M., Kadowaki, T., Kaneko, S., Shimura, H., Kubo, S. I., … Young Japanese Expert Group for Parkinson's Disease and Movement Disorders: YJ-EXPANDS. (2017). Anhedonia and its correlation with clinical aspects in Parkinson's disease. *Journal of the Neurological Sciences, 372*, 403–407. https://doi.org/10.1016/j.jns.2016.10.051

Narita, K., Suda, M., Takei, Y., Aoyama, Y., Majima, T., Kameyama, M., Kosaka, H., Amanuma, M., Fukuda, M., & Mikuni, M. (2011). Volume reduction of ventromedial prefrontal cortex in bipolar II patients with rapid cycling: A voxel-based morphometric study. *Progress in Neuro-Psychopharmacology & Biological Psychiatry, 35*(2), 439–445. https://doi.org/10.1016/j.pnpbp.2010.11.030

Nery, F. G., Chen, H. H., Hatch, J. P., Nicoletti, M. A., Brambilla, P., Sassi, R. B., Mallinger, A. G., Keshavan, M. S., & Soares, J. C. (2009). Orbitofrontal cortex gray matter volumes in bipolar disorder patients: A region-of-interest MRI study. *Bipolar Disorders, 11*(2), 145–153. https://doi.org/10.1111/j.1399-5618.2009.00662.x

Nierenberg, A. A., Agustini, B., Köhler-Forsberg, O., Cusin, C., Katz, D., Sylvia, L. G., Peters, A., & Berk, M. (2023). Diagnosis and treatment of bipolar disorder: A review. *JAMA, 330*(14), 1370–1380. https://doi.org/10.1001/jama.2023.18588

Nishimura, M., Sato, K., Okada, T., Yoshiya, I., Schloss, P., Shimada, S., & Tohyama, M. (1998). Ketamine inhibits monoamine transporters expressed in human embryonic kidney 293 cells. *Anesthesiology, 88*(3), 768–774. https://doi.org/10.1097/00000542-199803000-00029

Niu, M., Wang, Y., Jia, Y., Wang, J., Zhong, S., Lin, J., Sun, Y., Zhao, L., Liu, X., Huang, L., & Huang, R. (2017). Common and specific abnormalities in cortical thickness in patients with major depressive and bipolar disorders. *eBioMedicine, 16*, 162–171. https://doi.org/10.1016/j.ebiom.2017.01.010

Nixon, N. L., Liddle, P. F., Worwood, G., Liotti, M., & Nixon, E. (2013). Prefrontal cortex function in remitted major depressive disorder. *Psychological Medicine, 43*(6), 1219–1230. https://doi.org/10.1017/S0033291712002164

Nudmamud-Thanoi, S., & Reynolds, G. P. (2004). The NR1 subunit of the glutamate/NMDA receptor in the superior temporal cortex in schizophrenia and affective disorders. *Neuroscience Letters, 372*(1–2), 173–177. https://doi.org/10.1016/j.neulet.2004.09.035

Ogawa, S., Koga, N., Hattori, K., Matsuo, J., Ota, M., Hori, H., Sasayama, D., Teraishi, T., Ishida, I., Yoshida, F., Yoshida, S., Noda, T., Higuchi, T., & Kunugi, H. (2018). Plasma amino acid profile in major depressive disorder: Analyses in two independent case-control sample sets. *Journal of Psychiatric Research, 96*, 23–32. https://doi.org/10.1016/j.jpsychires.2017.09.014

Osuch, E., Ford, K., Wrath, A., Bartha, R., & Neufeld, R. (2014). Functional MRI of pain application in youth who engaged in repetitive non-suicidal self-injury vs. psychiatric controls. *Psychiatry Research, 223*(2), 104–112. https://doi.org/10.1016/j.pscychresns.2014.05.003

Pang, X., Wu, D., Wang, H., Zhang, J., Yu, Y., Zhao, Y., Li, Q., Ni, L., Wang, K., Zhang, D., & Tian, Y. (2024). Cortical morphological alterations in adolescents with major depression and non-suicidal self-injury. *NeuroImage. Clinical, 44*, Article 103701. https://doi.org/10.1016/j.nicl.2024.103701

Peckham, A. D., Jordan, H., Silverman, A., Jarvi Steele, S., Björgvinsson, T., & Beard, C. (2020). From urges to action: Negative urgency and nonsuicidal self-injury in an acute transdiagnostic sample. *Archives of Suicide Research, 24*(3), 367–383. https://doi.org/10.1080/13811118.2019.1625831

Peciña, M., Sikora, M., Avery, E. T., Heffernan, J., Peciña, S., Mickey, B. J., & Zubieta, J. K. (2017). Striatal dopamine D2/3 receptor-mediated neurotransmission in major depression: Implications for anhedonia, anxiety and treatment response. *European Neuropsychopharmacology, 27*(10), 977–986. https://doi.org/10.1016/j.euroneuro.2017.08.427

Petty, F., Kramer, G., & Moeller, M. (1994). Does learned helplessness induction by haloperidol involve serotonin mediation? *Pharmacology, Biochemistry, and Behavior, 48*(3), 671–676. https://doi.org/10.1016/0091-3057(94)90330-1

Pizzagalli, D. A., Berretta, S., Wooten, D., Goer, F., Pilobello, K. T., Kumar, P., Murray, L., Beltzer, M., Boyer-Boiteau, A., Alpert, N., El Fakhri, G., Mechawar, N., Vitaliano, G., Turecki, G., & Normandin, M. (2019). Assessment of striatal dopamine transporter binding in individuals with major depressive disorder: In vivo positron emission tomography and postmortem evidence. *JAMA Psychiatry, 76*(8), 854–861. https://doi.org/10.1001/jamapsychiatry.2019.0801

Plener, P. L., Bubalo, N., Fladung, A. K., Ludolph, A. G., & Lulé, D. (2012). Prone to excitement: Adolescent females with Non-suicidal self-injury (NSSI) show altered cortical pattern to emotional and NSS-related material. *Psychiatry Research, 203*(2–3), 146–152. https://doi.org/10.1016/j.pscychresns.2011.12.012

Poewe W. (2008). Non-motor symptoms in Parkinson's disease. *European Journal of Neurology, 15*(Suppl 1), 14–20. https://doi.org/10.1111/j.1468-1331.2008.02056.x

Pothion, S., Bizot, J. C., Trovero, F., & Belzung, C. (2004). Strain differences in sucrose preference and in the consequences of unpredictable chronic mild stress. *Behavioural Brain Research, 155*(1), 135–146. https://doi.org/10.1016/j.bbr.2004.04.008

Pryce, C. R., Azzinnari, D., Spinelli, S., Seifritz, E., Tegethoff, M., & Meinlschmidt, G. (2011). Helplessness: A systematic translational review of theory and evidence for its relevance to understanding and treating depression. *Pharmacology & Therapeutics, 132*(3), 242–267. https://doi.org/10.1016/j.pharmthera.2011.06.006

Pulcu, E., Trotter, P. D., Thomas, E. J., McFarquhar, M., Juhasz, G., Sahakian, B. J., Deakin, J. F., Zahn, R., Anderson, I. M., & Elliott, R. (2014). Temporal discounting in major depressive disorder. *Psychological Medicine, 44*(9), 1825–1834. https://doi.org/10.1017/S0033291713002584

Qian, X., Zhong, Z., Lu, S., & Zhang, Y. (2023). Repeated reserpine treatment induces depressive-like behaviors accompanied with hippocampal impairment and synapse deficit in mice. *Brain Research, 1819*, Article 148541. https://doi.org/10.1016/j.brainres.2023.148541

Querry, M., Botzung, A., Cretin, B., Demuynck, C., Muller, C., Ravier, A., Schorr, B., Mondino, M., Sanna, L., de Sousa, P. L., Philippi, N., & Blanc, F. (2024). Neuroanatomical substrates of depression in dementia with Lewy bodies and Alzheimer's disease. *GeroScience, 46*(6), 5725–5744. https://doi.org/10.1007/s11357-024-01190-4

Rai, S., Griffiths, K. R., Breukelaar, I. A., Barreiros, A. R., Chen, W., Boyce, P., Hazell, P., Foster, S. L., Malhi, G. S., Harris, A. W. F., & Korgaonkar, M. S. (2021). Default-mode and fronto-parietal network connectivity during rest distinguishes asymptomatic patients with bipolar disorder and major depressive disorder. *Translational Psychiatry, 11*(1), 547. https://doi.org/10.1038/s41398-021-01660-9

Ranganath, C. (2006). Working memory for visual objects: Complementary roles of inferior temporal, medial temporal, and prefrontal cortex. *Neuroscience, 139*(1), 277–289. https://doi.org/10.1016/j.neuroscience.2005.06.092

Rao, V. R., Sellers, K. K., Wallace, D. L., Lee, M. B., Bijanzadeh, M., Sani, O. G., Yang, Y., Shanechi, M. M., Dawes, H. E., & Chang, E. F. (2018). Direct electrical stimulation of lateral orbitofrontal cortex acutely improves mood in individuals with symptoms of depression. *Current Biology, 28*(24), 3893-3902.e4. https://doi.org/10.1016/j.cub.2018.10.026

Rehbein, M. A., Kroker, T., Winker, C., Ziehfreund, L., Reschke, A., Bölte, J., Wyczesany, M., Roesmann, K., Wessing, I., & Junghöfer, M. (2023). Non-invasive stimulation reveals ventromedial prefrontal cortex function in reward prediction and reward processing. *Frontiers in Neuroscience, 17*, 1219029. https://doi.org/10.3389/fnins.2023.1219029

Remus, J. L., Stewart, L. T., Camp, R. M., Novak, C. M., & Johnson, J. D. (2015). Interaction of metabolic stress with chronic mild stress in altering brain cytokines and sucrose preference. *Behavioral Neuroscience, 129*(3), 321–330. https://doi.org/10.1037/bne0000056

Réus, G. Z., Manosso, L. M., Quevedo, J., & Carvalho, A. F. (2023). Major depressive disorder as a neuro-immune disorder: Origin, mechanisms, and therapeutic opportunities. *Neuroscience and Biobehavioral Reviews, 155*, Article 105425. https://doi.org/10.1016/j.neubiorev.2023.105425

Rinaldi, R., Lefebvre, L., Joachim, A., & Rossignol, M. (2020). Decision-making of patients with major depressive disorder in the framework of action control. *Cognitive Neuropsychiatry, 25*(1), 71–83. https://doi.org/10.1080/13546805.2019.1685481

Ritter, C., Buchmann, A., Müller, S. T., Volleberg, M., Haynes, M., Ghisleni, C., Noeske, R., Tuura, R., & Hasler, G. (2022). Evaluation of prefrontal γ-aminobutyric acid and glutamate levels in individuals with major depressive disorder using proton magnetic resonance spectroscopy. *JAMA Psychiatry, 79*(12), 1209–1216. https://doi.org/10.1001/jamapsychiatry.2022.3384

Rolls, E. T. (2007). The representation of information about faces in the temporal and frontal lobes. *Neuropsychologia, 45*(1), 124–143. https://doi.org/10.1016/j.neuropsychologia.2006.04.019

Rolls, E. T. (2023). Emotion, motivation, decision-making, the orbitofrontal cortex, anterior cingulate cortex, and the amygdala. *Brain Structure & Function, 228*, 1201–1257. https://doi.org/10.1007/s00429-023-02644-9

Rolls, E. T., Cheng, W., & Feng, J. (2020). The orbitofrontal cortex: Reward, emotion and depression. *Brain Communications, 2*(2), fcaa196. https://doi.org/10.1093/braincomms/fcaa196

Rosenblat, J. D., Kakar, R., Berk, M., Kessing, L. V., Vinberg, M., Baune, B. T., Mansur, R. B., Brietzke, E., Goldstein, B. I., & McIntyre, R. S. (2016). Anti-inflammatory agents in the treatment of bipolar depression: A systematic review and meta-analysis. *Bipolar Disorders, 18*(2), 89–101. https://doi.org/10.1111/bdi.12373

Rosso, G., Porceddu, G., Portaluppi, C., Garrone, C., Di Salvo, G., & Maina, G. (2025). Exploring cognitive symptoms in patients with unipolar and bipolar major depression: A comparative evaluation of subjective and objective performance. *Psychiatry Research, 347*, Article 116422. https://doi.org/10.1016/j.psychres.2025.116422

Rybakowski, J. K. (2023). Application of antipsychotic drugs in mood disorders. *Brain Sciences, 13*(3), 414. https://doi.org/10.3390/brainsci13030414

Sato, K. (2021). Why is lithium effective in alleviating bipolar disorder? *Medical Hypotheses, 147*, Article 110484. https://doi.org/10.1016/j.mehy.2021.110484

Schilbach, L., Bzdok, D., Timmermans, B., Fox, P. T., Laird, A. R., Vogeley, K., & Eickhoff, S. B. (2012). Introspective minds: Using ALE meta-analyses to study commonalities in the neural correlates of emotional processing, social & unconstrained cognition. *PLoS ONE, 7*(2), Article e30920. https://doi.org/10.1371/journal.pone.0030920

Schmaal, L., Hibar, D. P., Sämann, P. G., Hall, G. B., Baune, B. T., Jahanshad, N., Cheung, J. W., van Erp, T. G. M., Bos, D., Ikram, M. A., Vernooij, M. W., Niessen, W. J., Tiemeier, H., Hofman, A., Wittfeld, K., Grabe, H. J., Janowitz, D., Bülow, R., Selonke, M., … Veltman, D. J. (2017). Cortical abnormalities in adults and adolescents with major depression based on brain scans from 20 cohorts worldwide in the ENIGMA Major Depressive Disorder Working Group. *Molecular Psychiatry, 22*(6), 900–909. https://doi.org/10.1038/mp.2016.60

Sekssaoui, M., Bockaert, J., Marin, P., & Bécamel, C. (2024). Antidepressant-like effects of psychedelics in a chronic despair mouse model: Is the 5-HT_{2A} receptor the unique player? *Neuropsychopharmacology, 49*(4), 747–756. https://doi.org/10.1038/s41386-024-01794-6

Seligman, M. E., & Maier, S. F. (1967). Failure to escape traumatic shock. *Journal of Experimental Psychology, 74*(1), 1–9. https://doi.org/10.1037/h0024514

Sengupta, A., Bocchio, M., Bannerman, D. M., Sharp, T., & Capogna, M. (2017). Control of amygdala circuits by 5-HT neurons via 5-HT and glutamate cotransmission. *The Journal of Neuroscience, 37*(7), 1785–1796. https://doi.org/10.1523/JNEUROSCI.2238-16.2016

Shah, P. J., Ogilvie, A. D., Goodwin, G. M., & Ebmeier, K. P. (1997). Clinical and psychometric correlates of dopamine D2 binding in depression. *Psychological Medicine, 27*(6), 1247–1256. https://doi.org/10.1017/s0033291797005382

Shirayama, Y., Takahashi, M., Osone, F., Hara, A., & Okubo, T. (2017). Myo-inositol, glutamate, and glutamine in the prefrontal cortex, hippocampus, and amygdala in major depression. *Biological Psychiatry. Cognitive Neuroscience and Neuroimaging, 2*(2), 196–204. https://doi.org/10.1016/j.bpsc.2016.11.006

Song, Y., Cao, H., Zuo, C., Gu, Z., Huang, Y., Miao, J., Fu, Y., Guo, Y., Jiang, Y., & Wang, F. (2023). Mitochondrial dysfunction: A fatal blow in depression. *Biomedicine & Pharmacotherapy, 167*, Article 115652. https://doi.org/10.1016/j.biopha.2023.115652

Steele, J. S., Bertocci, M., Eckstrand, K., Chase, H. W., Stiffler, R., Aslam, H., Lockovich, J., Bebko, G., & Phillips, M. L. (2021). A specific neural substrate predicting current and future impulsivity in young adults. *Molecular Psychiatry, 26*(9), 4919–4930. https://doi.org/10.1038/s41380-021-01017-0

Stodgell, C. J., Loupe, P. S., Schroeder, S. R., & Tessel, R. E. (1998). Cross-sensitization between footshock stress and apomorphine on self-injurious behavior and neostriatal catecholamines in a rat model of Lesch-Nyhan syndrome. *Brain Research, 783*(1), 10–18. https://doi.org/10.1016/s0006-8993(97)01128-1

Sun, Y., Drevets, W., Turecki, G., & Li, Q. S. (2020). The relationship between plasma serotonin and kynurenine pathway metabolite levels and the treatment response to escitalopram and desvenlafaxine. *Brain, Behavior, and Immunity, 87*, 404–412. https://doi.org/10.1016/j.bbi.2020.01.011

Szałach, ŁP., Lisowska, K. A., Cubała, W. J., Barbuti, M., & Perugi, G. (2023). The immunomodulatory effect of lithium as a mechanism of action in bipolar disorder. *Frontiers in Neuroscience, 17*, 1213766. https://doi.org/10.3389/fnins.2023.1213766

Tafet, G. E., & Nemeroff, C. B. (2016). The links between stress and depression: Psychoneuroendocrinological, genetic, and environmental interactions. *The Journal of Neuropsychiatry and Clinical Neurosciences, 28*(2), 77–88. https://doi.org/10.1176/appi.neuropsych.15030053

Takahashi, T., Oono, H., Inoue, T., Boku, S., Kako, Y., Kitaichi, Y., Kusumi, I., Masui, T., Nakagawa, S., Suzuki, K., Tanaka, T., Koyama, T., & Radford, M. H. (2008). Depressive patients are more impulsive and inconsistent in intertemporal choice behavior for monetary gain and loss than healthy subjects–an analysis based on Tsallis' statistics. *Neuro Endocrinology Letters, 29*(3), 351–358.

Teh, W. L., Lee, S. Y. L., Chandwani, N., Johnson, S. L., Subramaniam, M., & Ho, R. C. M. (2025). Neurocognitive correlates of emotion-based urgency in DSM-5 bipolar disorder: The moderating role of diagnosis status and positive affect in a functional near-infrared spectroscopy (fNIRS) experimental study. *Journal of Affective Disorders, 387*, Article 119534. https://doi.org/10.1016/j.jad.2025.119534

Tsai, S. J. (2017). Role of neurotrophic factors in attention deficit hyperactivity disorder. *Cytokine & Growth Factor Reviews, 34*, 35–41. https://doi.org/10.1016/j.cytogfr.2016.11.003

Uliana, D. L., Gomes, F. V., & Grace, A. A. (2020). Prelimbic medial prefrontal cortex disruption during adolescence increases susceptibility to helpless behavior in adult rats. *European Neuropsychopharmacology, 35*, 111–125. https://doi.org/10.1016/j.euroneuro.2020.04.004

Um, M., Hummer, T. A., & Cyders, M. A. (2020). Relationship of negative urgency to cingulo-insular and cortico-striatal resting state functional connectivity in tobacco use. *Brain Imaging and Behavior, 14*(5), 1921–1932. https://doi.org/10.1007/s11682-019-00136-1

Umehara, H., Numata, S., Watanabe, S. Y., Hatakeyama, Y., Kinoshita, M., Tomioka, Y., Nakahara, K., Nikawa, T., & Ohmori, T. (2017). Altered KYN/TRP, Gln/Glu, and Met/methionine sulfoxide ratios in the blood plasma of medication-free patients with major depressive disorder. *Scientific Reports, 7*(1), 4855. https://doi.org/10.1038/s41598-017-05121-6

Varga, Z., Csabai, D., Miseta, A., Wiborg, O., & Czéh, B. (2017). Chronic stress affects the number of GABAergic neurons in the orbitofrontal cortex of rats. *Behavioural Brain Research, 316*, 104–114. https://doi.org/10.1016/j.bbr.2016.08.030

Vinarcsik, L., Smoller, C., & Grossberg, G. (2025). Use of psychedelic agents in older adults with treatment-resistant major depressive disorder: What the evidence shows. *Drugs & Aging*. https://doi.org/10.1007/s40266-025-01221-5.Advanceonlinepublication.doi: 10.1007/s40266-025-01221-5

Visser, J. E., Bär, P. R., & Jinnah, H. A. (2000). Lesch-Nyhan disease and the basal ganglia. *Brain Research. Brain Research Reviews, 32*(2–3), 449–475. https://doi.org/10.1016/s0165-0173(99)00094-6

Wang, Y., Chen, G., Zhong, S., Jia, Y., Xia, L., Lai, S., Zhao, L., Huang, L., & Liu, T. (2018). Association between resting-state brain functional connectivity and cortisol levels in unmedicated major depressive disorder. *Journal of Psychiatric Research, 105*, 55–62. https://doi.org/10.1016/j.jpsychires.2018.08.025

Wang, K., He, Q., Zhu, X., Hu, Y., Yao, Y., Hommel, B., Beste, C., Liu, J., Yang, Y., & Zhang, W. (2022). Smaller putamen volumes are associated with greater problems in external emotional regulation in depressed adolescents with nonsuicidal self-injury. *Journal of Psychiatric Research, 155*, 338–346. https://doi.org/10.1016/j.jpsychires.2022.09.014

Wang, H., Lesh, T. A., Maddock, R. J., Fassbender, C., & Carter, C. S. (2020). Delay discounting abnormalities are seen in first-episode schizophrenia but not in bipolar disorder. *Schizophrenia Research, 216*, 200–206. https://doi.org/10.1016/j.schres.2019.11.063

Wang, M., Li, H., Zhang, W., Zhang, L., Wang, S., Jia, M., Jia, L., Zhang, Y., Gao, H., Zhang, X., & Yin, Z. (2025). Saikosaponin A alleviates depressive-like behavior induced by reserpine in mice by regulating gut microflora and inflammatory responses. *PLoS ONE, 20*(2), Article e0311207. https://doi.org/10.1371/journal.pone.0311207

Westlund Schreiner, M., Klimes-Dougan, B., Mueller, B. A., Eberly, L. E., Reigstad, K. M., Carstedt, P. A., Thomas, K. M., Hunt, R. H., Lim, K. O., & Cullen, K. R. (2017). Multi-modal neuroimaging of adolescents with non-suicidal self-injury: Amygdala functional connectivity. *Journal of Affective Disorders, 221*, 47–55. https://doi.org/10.1016/j.jad.2017.06.004

Willner, P., Towell, A., Sampson, D., Sophokleous, S., & Muscat, R. (1987). Reduction of sucrose preference by chronic unpredictable mild stress, and its restoration by a tricyclic antidepressant. *Psychopharmacology (Berl), 93*(3), 358–364. https://doi.org/10.1007/BF00187257

Wingert, A. M., Agnorelli, C., Peill, J., Reed, S., Nutt, D. J., & Erritzoe, D. (2025). Serotonergic psychedelics for depression: A comprehensive overview. *International Review of Neurobiology, 181*, 271–304. https://doi.org/10.1016/bs.irn.2025.04.009

Won, G. H., Bae, S., Kim, H. K., & Choi, T. Y. (2023). Subcortical volume analysis in non-suicidal self-injury adolescents: A pilot study. *Psychiatry Research. Neuroimaging, 331*, Article 111617. https://doi.org/10.1016/j.pscychresns.2023.111617

World Health Organization. (2019). *International statistical classification of diseases and related health problems* (11th ed.). https://icd.who.int/

World Health Organization. (2022). *Mental disorders*. Retrieved December 29, 2024, from https://www.who.int/news-room/fact-sheets/detail/mental-disorders

Wu, M., Minkowicz, S., Dumrongprechachan, V., Hamilton, P., Xiao, L., & Kozorovitskiy, Y. (2021). Attenuated dopamine signaling after aversive learning is restored by ketamine to rescue escape actions. *eLife, 10*, e64041. https://doi.org/10.7554/eLife.64041

Wu, S., Yin, Y., & Du, L. (2022). Blood-brain barrier dysfunction in the pathogenesis of major depressive disorder. *Cellular and Molecular Neurobiology, 42*(8), 2571–2591. https://doi.org/10.1007/s10571-021-01153-9

Yan, C. G., Chen, X., Li, L., Castellanos, F. X., Bai, T. J., Bo, Q. J., Cao, J., Chen, G. M., Chen, N. X., Chen, W., Cheng, C., Cheng, Y. Q., Cui, X. L., Duan, J., Fang, Y. R., Gong, Q. Y., Guo, W. B., Hou, Z. H., Hu, L., … Zang, Y. F. (2019). Reduced default mode network functional connectivity in patients with recurrent major depressive disorder. *Proceedings of the National Academy of Sciences of the United States of America, 116*(18), 9078–9083. https://doi.org/10.1073/pnas.1900390116

Yang, Y., Cui, Q., Lu, F., Pang, Y., Chen, Y., Tang, Q., Li, D., Lei, T., He, Z., Hu, S., Deng, J., & Chen, H. (2021). Default mode network subsystem alterations in bipolar disorder during major depressive episode. *Journal of Affective Disorders, 281*, 856–864. https://doi.org/10.1016/j.jad.2020.11.049

Yang, Y., Li, X., Cui, Y., Liu, K., Qu, H., Lu, Y., Li, W., Zhang, L., Zhang, Y., Song, J., & Lv, L. (2022). Reduced gray matter volume in orbitofrontal cortex across schizophrenia, major depressive disorder, and bipolar disorder: A comparative imaging study. *Frontiers in Neuroscience, 16*, Article 919272. https://doi.org/10.3389/fnins.2022.919272

Yang, T., Nie, Z., Shu, H., Kuang, Y., Chen, X., Cheng, J., Yu, S., & Liu, H. (2020). The role of BDNF on neural plasticity in depression. *Frontiers in Cellular Neuroscience, 14*, 82. https://doi.org/10.3389/fncel.2020.00082

Yates, J. R., Darna, M., Gipson, C. D., Dwoskin, L. P., & Bardo, M. T. (2015). Dissociable roles of dopamine and serotonin transporter function in a rat model of negative urgency. *Behavioural Brain Research, 291*, 201–208. https://doi.org/10.1016/j.bbr.2015.05.023

Zahid, Z., McMahon, L., & Lynch, M. (2022). Neural activity across the dorsolateral prefrontal cortex and risk for suicidal ideation and self-injury. *Archives of Suicide Research, 26*(1), 187–207. https://doi.org/10.1080/13811118.2020.1779154

Zhang, P., Huang, H., Gao, X., Jiang, J., Xi, C., Wu, L., Fu, Y., Lai, J., & Hu, S. (2021). Involvement of kynurenine metabolism in bipolar disorder: An updated review. *Frontiers in Psychiatry, 12*, Article 677039. https://doi.org/10.3389/fpsyt.2021.677039

Zhang, L., Qin, K., Pan, N., Xu, H., & Gong, Q. (2025). Shared and distinct patterns of default mode network dysfunction in major depressive disorder and bipolar disorder: A comparative meta-analysis. *Journal of Affective Disorders, 368*, 23–32. https://doi.org/10.1016/j.jad.2024.09.021

Zhao, W., Li, Q., Ma, Y., Wang, Z., Fan, B., Zhai, X., Hu, M., Wang, Q., Zhang, M., Zhang, C., Qin, Y., Sha, S., Gan, Z., Ye, F., Xia, Y., Zhang, G., Yang, L., Zou, S., Xu, Z., … Zhang, H. (2021). Behaviors related to psychiatric disorders and pain perception in C57BL/6J mice during different phases of estrous cycle. *Frontiers in Neuroscience,15*, 650793. https://doi.org/10.3389/fnins.2021.650793

Zhou, J., Cao, X., Mar, A. C., Ding, Y. Q., Wang, X., Li, Q., & Li, L. (2014). Activation of post-synaptic 5-HT1A receptors improve stress adaptation. *Psychopharmacology (Berl), 231*(10), 2067–2075. https://doi.org/10.1007/s00213-013-3350-z

Zhou, H. X., Chen, X., Shen, Y. Q., Li, L., Chen, N. X., Zhu, Z. C., Castellanos, F. X., & Yan, C. G. (2020). Rumination and the default mode network: Meta-analysis of brain imaging studies and implications for depression. *NeuroImage, 206*, Article 116287. https://doi.org/10.1016/j.neuroimage.2019.116287

Zhou, Z., Gao, Y., Feng, R., Zhuo, L., Bao, W., Liang, K., Qiu, H., Cao, L., Tang, M., Li, H., Zhang, L., Huang, G., & Huang, X. (2023). Aberrant intrinsic hippocampal and orbitofrontal connectivity in drug-naive adolescent patients with major depressive disorder. *European Child & Adolescent Psychiatry, 32*(11), 2363–2374. https://doi.org/10.1007/s00787-022-02086-4

Zhu, C., Zhang, T., Li, Q., Chen, X., & Wang, K. (2023). Depression and anxiety during the COVID-19 pandemic: Epidemiology, mechanism, and treatment. *Neuroscience Bulletin, 39*(4), 675–684. https://doi.org/10.1007/s12264-022-00970-2

Chapter 10
Maladaptive Motor Control

1 Introduction

In the introduction to Chap. 3, I discussed my neighbor who lived with schizophrenia. One thing I did not mention is that I could predict when my neighbor would have an episode, screaming for help in the middle of the night, convinced someone was in their apartment trying to harm them. On most days, I noticed that this person would have jerky motions that resembled a spasm. In particular, their shoulders would rise and then roll forward. They also could not stand still, looking like they were squirming in place. However, on days in which my neighbor had an episode, they lacked emotional responses, spoke in a very monotone fashion, and remained perfectly still when I saw them, resembling a statue.

This chapter will focus on motor disturbances. In addition to hallucinations, individuals with schizophrenia display unusual movements, whether uncontrollable movements or a paucity of movement. As you will learn in this chapter, there are numerous conditions characterized by aberrant motor control. To understand the causes underlying motor disturbances, you will first learn about the neural circuits that control movement. Movement-related disorders can result from damage to a specific part of the brain or from damage to motor neurons. In some cases, certain medications can lead to motor impairments.

2 Learning Objectives

By the end of this chapter, you should be able to …

1. Explain the neuromechanisms that control movement.
2. Discuss how aberrations to specific neurocircuits lead to movement-related disorders.
3. Describe how certain drugs can cause motor disturbances in individuals.

J. Yates, *Neuroscience of Maladaptation*,
https://doi.org/10.1007/978-3-032-15602-0_10

3 How Movement Normally Works

We take for granted the complexity of motor control. Think of everything our species can do. We can grip a pencil, alternate between sprinting and skipping, lift weights, play video games, strum a guitar, drive a vehicle, and so on. There are some actions we can perform with little conscious effort, like tying a shoe or driving a vehicle. Other actions require tremendous mental effort, especially as we learn to perform a task (think of how you felt when first learning to drive a vehicle). Given the complexity of movement, there are multiple brain circuits involved in different aspects of motor actions.

3.1 Spinal Cord and Cranial Nerves

In Chaps. 3 and 4, you learned about the spinal cord's role in transmitting somatosensory information to the brain. Motor information travels from the brain to the peripheral nervous system (PNS) through the descending pathway of the spinal cord, exiting through the ventral root of the spinal cord. Motor neurons located in the spinal cord are often called **lower motor neurons**.

Lower motor neurons also include several cranial nerves that have motor functions. You first learned about these nerves in Chap. 3. Out of 12 total cranial nerves, nine of them have motor functions. Cranial nerves III (oculomotor nerve), IV (trochlear nerve), and VI (abducens nerve) control different eye muscles. Cranial nerve V (trigeminal nerve) controls muscles of the jaw, and cranial nerve VII (facial nerve) controls muscles of the face and allows us to raise our eyebrows and close our eyelids. Cranial nerves IX (glossopharyngeal nerve) and XII (hypoglossal nerve) control pharyngeal and tongue muscles, respectively. Cranial nerve X (vagus nerve) controls muscles of the heart, the lungs, and the gastrointestinal tract. Cranial nerve XI (accessory nerve) controls the sternocleidomastoid and the trapezius muscles.

3.2 Cortico-Basal Ganglia-Thalamo-Cortical (CBGTC) Loop

You were introduced to the cortico-basal ganglia-thalamo-cortical (CBGTC) loop in Chap. 5 when you learned about attentional deficits. In addition to sensory gating, I have alluded to the role of the basal ganglia in motor control (see the previous chapter for an example). I need to go into more detail as to how the CBGTC loop is organized. There are three distinct pathways in this loop: the direct, the indirect, and the hyperdirect (Rocha et al., 2023) (see Fig. 1).

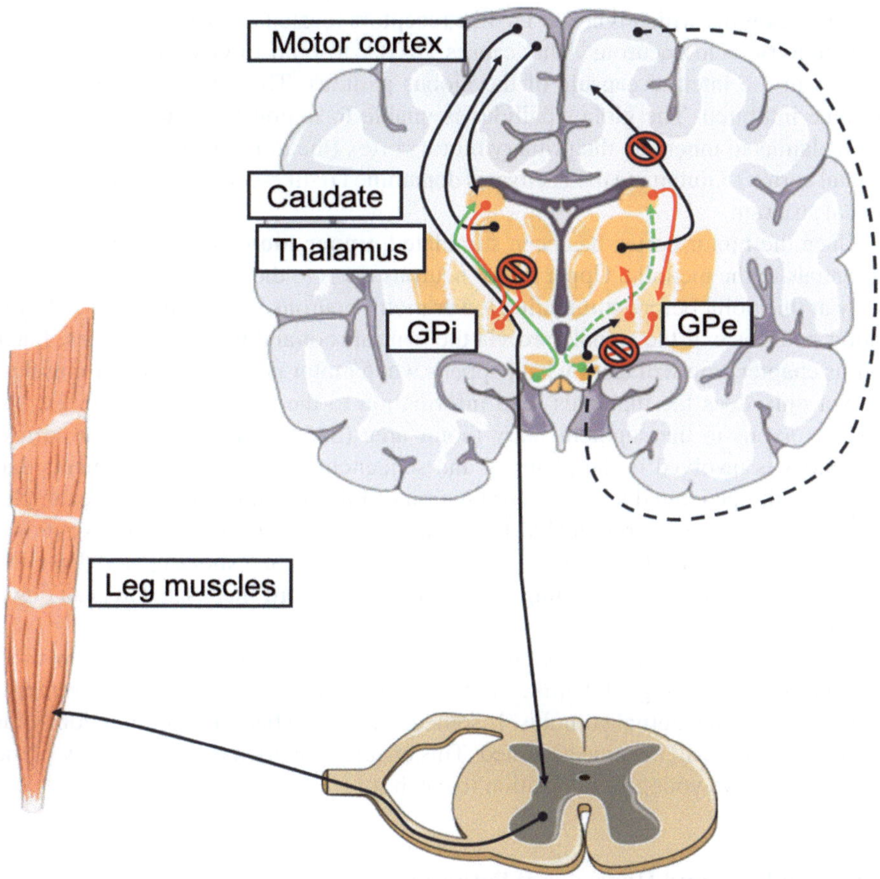

Fig. 1 Representation of the direct, the indirect, and the hyperdirect pathways of the basal ganglia. The direct pathway is presented on the left side of the image. The indirect and the hyperdirect pathways are presented on the right side of the image. Both the direct and the indirect pathways begin in the substantia nigra. The green arrows show dopaminergic projections, with the solid line indicating that dopamine binds to dopamine D_1-like receptors and the dotted line indicating that dopamine binds to D_2-like receptors. Red arrows indicate GABAergic (gamma-aminobutyric acid) projection, and black arrows denote glutamatergic projections. A prohibition sign on an arrow indicates that a projection has become inhibited. The dashed black line represents glutamatergic projections originating from the motor cortex that initiate the hyperdirect pathway. Notice that once the hyperdirect pathway has been initiated, it follows the same steps as the indirect pathway. Image created by the author using Servier Medical Art (https://smart.servier.com/; CC BY 4.0, https://creativecommons.org/licenses/by/4.0/) and Microsoft PowerPoint

3.2.1 Direct Pathway

The direct pathway stimulates purposeful movement. First, dopaminergic neurons originating from the substantia nigra release dopamine into the dorsal striatum of the basal ganglia, which binds to D_1-like receptors located on GABAergic (gamma-aminobutyric acid) neurons. This causes GABA to bind to GABAergic receptors located in the internal capsule of the globus pallidus. These GABAergic neurons become inhibited. The globus pallidus is unable to inhibit the thalamus, allowing the thalamus to innervate the primary motor cortex (the most posterior region of the frontal cortex). Glutamate then activates dopamine D_1-like-containing neurons in the dorsal striatum.

Once the motor cortex has been stimulated, **upper motor neurons** send action potentials to the medulla. Upper motor neurons cross to the contralateral side of the body at this point before stimulating lower motor neurons. As detailed above, lower motor neurons exit the spinal cord via the ventral root and travel to muscles. Later in this chapter, you will learn what happens when motor neurons reach our muscles.

Not only does the thalamus send information to the primary motor cortex, but it sends axons to the supplementary motor area (SMA) and the premotor cortex. The SMA is involved in the planning and sequencing of complex movements that are internally generated and the coordination of bilateral movements. For example, playing a song on an instrument like the guitar involves sequences of complex actions. One hand/wrist is used to strum the strings as the other hand slides up and down the fretboard, with the fingers needing to move to create chord shapes and to push down individual strings.

The premotor cortex is important for selecting a motor plan that will allow us to accomplish a specific goal. Unlike the SMA, the premotor cortex plans movements in response to an external stimulus. If someone throws a ball in your direction, your premotor cortex will become activated. This enables you to map the trajectory of the ball and to move your arm in position to catch it.

3.2.2 Indirect and Hyperdirect Pathways

The indirect pathway also involves dopaminergic transmission from the substantia nigra to the dorsal striatum. However, dopamine binds to inhibitory dopamine D_2-like receptors. GABA then binds to GABAergic receptors in the external capsule of the globus pallidus, preventing the inhibition of the subthalamic nucleus (STN). Glutamate binds to N-methyl-D-aspartate (NMDA) receptors in the internal capsule of the globus pallidus, causing GABA release to inhibit the thalamus. As the thalamus sends glutamatergic neurons to motor cortices, inhibition of the thalamus leads to insufficient stimulation of the primary motor cortex, preventing movement.

The hyperdirect pathway involves glutamatergic projections from the cerebral cortex to the STN. This pathway allows the cerebral cortex to bypass the basal ganglia. Stimulation of the STN causes activity within the internal capsule of the globus pallidus to increase, leading to increased inhibition of the thalamus and the primary motor cortex. The hyperdirect pathway acts in a similar fashion as the indirect pathway; that is, its primary purpose is to inhibit motor actions.

Think of it this way. The indirect pathway prevents new movements from occurring that will interfere with an ongoing action. Imagine you are eating a salad. Your direct pathway will stimulate the actions of grabbing the fork, moving the fork to the salad, and pulling food toward your mouth. Simultaneously, your indirect pathway is going to prevent movements like jerking your hand to the side. Now, imagine as your arm moves toward your mouth, you notice that someone's hair has gotten into your food. Your indirect pathway will kick in, stopping the movement of your hand before you place the food in your mouth.

Comprehension Check 1 Based on the descriptions of the direct, indirect, and hyperdirect pathways, how might the simultaneous activation of both the direct and indirect pathways contribute to the nuanced control of movement, beyond simply initiating or inhibiting actions?

The simultaneous activation of the direct and indirect pathways allows for a fine-tuning of movement. While the direct pathway facilitates the initiation and execution of a desired movement by disinhibiting the thalamus, the indirect pathway, acting concurrently, could modulate this excitation. By increasing inhibition of the thalamus to a certain degree, the indirect pathway might help to control the amplitude, velocity, and smoothness of the movement generated by the direct pathway. This balance between excitation and inhibition could ensure that movements are not just "on" or "off", but rather executed with the appropriate level of force and precision, and that extraneous movements are suppressed.

3.2.3 Parietal Cortex

Have you ever woken up and noticed that your arm has "fallen asleep"? What is it like trying to move your arm? You may notice that guiding your arm becomes much more difficult. I remember accidentally smacking my wife in the face in the middle of the night after my arm had fallen asleep! Just as sensation and perception are intertwined, so are sensation and movement. You learned about the parietal cortex's role in sensation in Chap. 3. I do not think it's a coincidence that the primary motor cortex, the most posterior region of the frontal cortex, is located just anterior to the primary somatosensory cortex, the most anterior region of the parietal cortex.

One important function of the parietal cortex is to integrate sensory information with motor plans (Freedman & Ibos, 2018). Because the parietal cortex is involved in spatial awareness, this region of the brain enables us to map our movements in relation to where objects are located. If you want to reach for a cup on the kitchen counter, your parietal cortex encodes the distance of the cup relative to your current position. This helps you determine how far you should extend your arm. If your arm does not extend far enough, you will fail to grasp the cup. If your arm extends too far, you risk knocking your cup over.

3.3 Cerebellum

The cerebellum is a powerhouse. Despite making up approximately 10% of the brain's volume, it contains more neurons than the rest of the brain combined! Concerning motor control, there is sufficient evidence that the cerebellum acts to modify motor signals to ensure that our movements are smooth and coordinated (Hull, 2020).

One critical function of the cerebellum is motor learning (De Zeeuw & Ten Brinke, 2015). Just as you learn course content by reading a textbook, attending lecture, and testing yourself, you have to learn to coordinate your movements to accomplish specific tasks. When learning to drive, you learn how to insert a key into the ignition (or press the start button for keyless cars), press the brake pedal down, and then shift the car from park to reverse or drive. You also learn how to control turn signals, windshield wipers, and internal controls like temperature. Like your basal ganglia, the cerebellum helps develop muscle memory, often referred to as procedural memory (see Chap. 6).

In Chap. 6, you learned about Pavlovian conditioning (also known as classical conditioning), in which an organism comes to learn the association between two stimuli. In Pavlov's seminal study, he found that dogs will salivate when they hear a metronome that has been previously associated with food. You also learned briefly about the role of Pavlovian conditioning in the addiction process and in posttraumatic stress disorder (PTSD). There is one form of Pavlovian conditioning that has been highly linked to cerebellar function: the *conditioned eyeblink response.* Blowing air into someone's eye will cause them to blink. This eyeblink is a natural response to prevent potential pathogens from entering the eye. Next, a researcher can play a tone before blowing air into the person's eye. On its own, the tone does not elicit a natural response from the individual. If the tone is presented with the puff of air consistently, the tone, even when presented on its own, can elicit an eyeblink response.

Figure 2 shows the neural pathways involved in the conditioned eyeblink response. Information regarding the puff of air travels through the trigeminal nerve to the trigeminal nuclei in the brainstem. Neurons from the trigeminal nuclei project to the inferior olivary nucleus of the medulla, with neurons from this nucleus traveling to the cerebellum. Information regarding the tone travels to the cochlear nucleus in the brainstem and then to the pontine nucleus, located in the pons. The pontine nucleus sends axons to the cerebellum. Notice that sensory information about the puff of air and the sound of the tone travels to the cerebellum. From the cerebellum, motor information travels to the red nucleus. Activation of the red nucleus causes cranial nerve VII to activate muscles responsible for closing our eyes. When the tone and the puff of air are presented enough times, presentation of the tone, on its own, is sufficient to activate the motor pathway from the cerebellum to cranial nerve VII. This is an example of a Hebbian synapse that you learned about in Chap. 6.

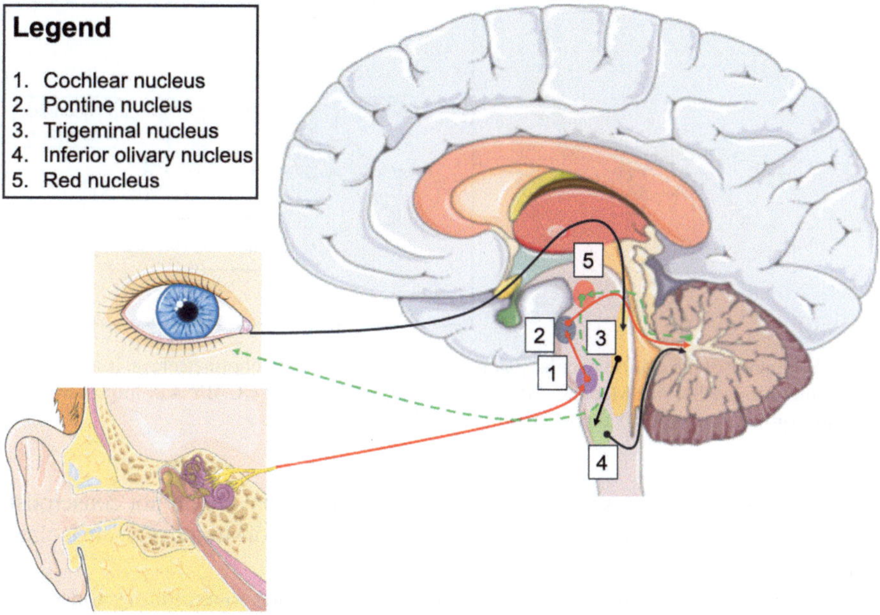

Fig. 2 Schematic showing the neurobiological basis of the conditioned eyeblink response. See text for a full description. Image created by the author using Servier Medical Art (https://smart.servier.com/; CC BY 4.0, https://creativecommons.org/licenses/by/4.0/) and Microsoft PowerPoint

Comprehension Check 2 How might disruptions in the communication or balance of activity between the basal ganglia pathways and the cerebellum manifest in an individual's motor control?

If the basal ganglia and the cerebellum cannot effectively communicate with one another, individuals may be able to initiate movements, but these movements are not performed smoothly. As you will learn later in this chapter, dysregulation of the cerebellum can cause individuals to engage in movements that appear jerky. On the flip side, individuals may perform movements that are smooth, but the initiation of these movements is delayed, and the movement occurs slowly due to dysregulation of the basal ganglia. The important thing to know is that the basal ganglia and the cerebellum work in tandem to ensure that we can complete complex motor actions seamlessly.

3.4 Neuromuscular Junction

Motor neurons exiting the ventral root of the spinal cord eventually form a synapse with a *skeletal* muscle fiber. This synapse is called the **neuromuscular junction** (Fig. 3). Skeletal muscles allow us to consciously move our body. These muscles are

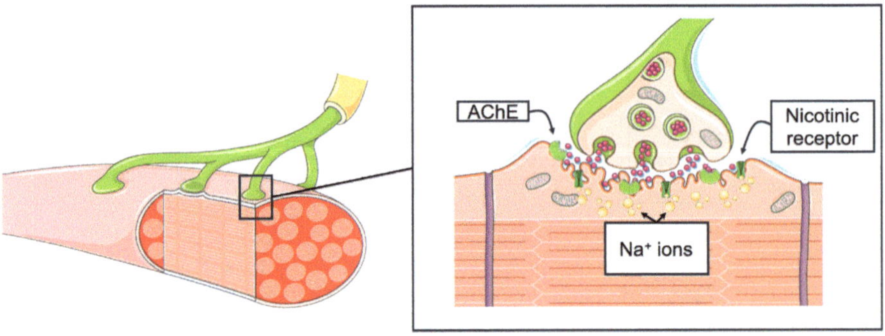

Fig. 3 Schematic of the neuromuscular junction. *Note* AChE = acetylcholinesterase. Image created by the author using Servier Medical Art (https://smart.servier.com/; CC BY 4.0, https://creativecommons.org/licenses/by/4.0/) and Microsoft PowerPoint

different from *smooth* muscle, which is found in organs that we do not consciously control (e.g., stomach), and *cardiac* muscle, which is a type of smooth muscle isolated to the heart.

Acetylcholine is released from motor neurons and binds to excitatory ionotropic nicotinic receptors. Activation of these receptors allows sodium (Na^+) to enter skeletal muscles, causing an action potential to propagate the length of the muscle. This allows more Na^+, as well as calcium (Ca^{++}), to enter the muscle. This ultimately leads to a muscle contraction.

You may be familiar with the nerve agent Sarin, which is an *acetylcholinesterase inhibitor*. Normally, excess acetylcholine is metabolized by acetylcholinesterase. Nerve agents do not allow for the breakdown of acetylcholine (Abou-Donia et al., 2016). As acetylcholine is an important neurotransmitter in the brain, Sarin exposure leads to numerous symptoms that are unrelated to motor impairment such as confusion and drowsiness. Directly related to this chapter, excess acetylcholine levels in the neuromuscular junction lead to excessive muscle contractions, causing tremors, muscle cramps, and respiratory failure. Because acetylcholine stimulates smooth muscle, organs like the lung become overstimulated, preventing an individual from breathing. This is what often leads to death.

Conversely, there are toxins that prevent the release of acetylcholine. Bungarotoxin is found in certain snake venom and acts as an antagonist at nicotinic receptors (Vijayaraghavan et al., 1992). Without acetylcholine to stimulate muscle fibers, individuals become paralyzed. Eventually, paralysis extends to the lungs, leading to death.

4 Conditions Characterized by Maladaptive Movements

There are several medical conditions characterized by maladaptive motor control. First, **dystonia** is a movement disorder in which individuals experience sustained muscle contractions. This leads to twisting and repetitive movements or abnormal postures. The exact cause for dystonia is not fully understood. Originally, the hypothesis was that damage to the basal ganglia leads to symptoms like dystonia. However, there is now evidence suggesting that dysfunction of the CBGTC loop, not just the basal ganglia, underlies dystonia. That is, there could be damage to a different brain region of this network, or messages between regions like the basal ganglia and the cerebral cortex become dysregulated (Neychev et al., 2011). A couple of disorders associated with a dysfunctional basal ganglia system will be detailed below.

4.1 Parkinson's Disease

First described by James Parkinson in 1817 (Parkinson, 2002), Parkinson's disease is classified as a neurodegenerative disease. This means that neurons in the central nervous system (CNS) die, with corresponding symptoms progressively getting worse over time. Parkinson's disease results from death of dopaminergic neurons originating from the substantia nigra. Because dopamine no longer stimulates dopamine D_1-like and dopamine D_2-like receptors in the direct and the indirect pathways of the CBGTC loop, individuals experience a wide range of motor impairments.

The most tell-tale sign of Parkinson's disease is a tremor, which is particularly noticeable in an individual's hands. Remember, the indirect pathway is responsible for inhibiting actions that interfere with normal actions. A trembling hand can make the most routine tasks, like eating with a utensil or tying one's shoes, incredibly difficult. Disruption of the direct pathway manifests as **bradykinesia**, a slowness in movement. Individuals with Parkinson's disease often have difficulty with their gait and their balance, which makes them susceptible to falls.

We know that Parkinson's disease results from the death of dopaminergic neurons, but researchers are still trying to understand *why* these neurons die. In Chap. 6, you learned about dementia with Lewy bodies. Lewy bodies are comprised of abnormal clumps of alpha-synuclein. Individuals with Parkinson's disease also have a buildup of Lewy bodies. Another possible cause of neuronal death in the substantia nigra is mitochondrial damage. Mitochondria are colloquially referred to as the "powerhouse" of a cell, including neurons. This organelle is responsible for generating energy for a cell by oxidizing sugars and ketones into adenosine triphosphate (ATP). If mitochondria do not function properly, cells do not receive the necessary ATP to survive. The question becomes, "what causes mitochondria in the substantia nigra to become damaged?".

Ca^{++}, as you already know by now, is crucial for numerous processes. An influx of Ca^{++} into the presynaptic neuron enables vesicles to release neurotransmitters into the synapse. In the postsynaptic neuron, Ca^{++} signaling pathways are involved in processes like long-term potentiation (LTP) and long-term depression (LTD). You also know that too much Ca^{++} is toxic to cells. Neurons in the substantia nigra are especially sensitive to elevated Ca^{++} levels, with excess Ca^{++} damaging mitochondria. To understand why, I need to describe intra-substantia nigra neurons in more detail.

Up until now, I have discussed that neurons generate an action potential when they are sufficiently stimulated by adjacent neurons. However, this is not always true. In the substantia nigra, dopaminergic neurons generate slow, rhythmic action potentials in the absence of other neuronal input. As such, these neurons act like a pacemaker, with the hypothesis being that this activity allows for better control of dopaminergic levels in regions like the basal ganglia. Neurons in the substantia nigra contain a special type of Ca^{++} ion channel that stays open even when the neuron is in a hyperpolarized state. While this mechanism is important for enabling neurons to generate action potentials at a consistent rate, it leaves them vulnerable to excitotoxicity (Surmeier et al., 2011) (see Fig. 4).

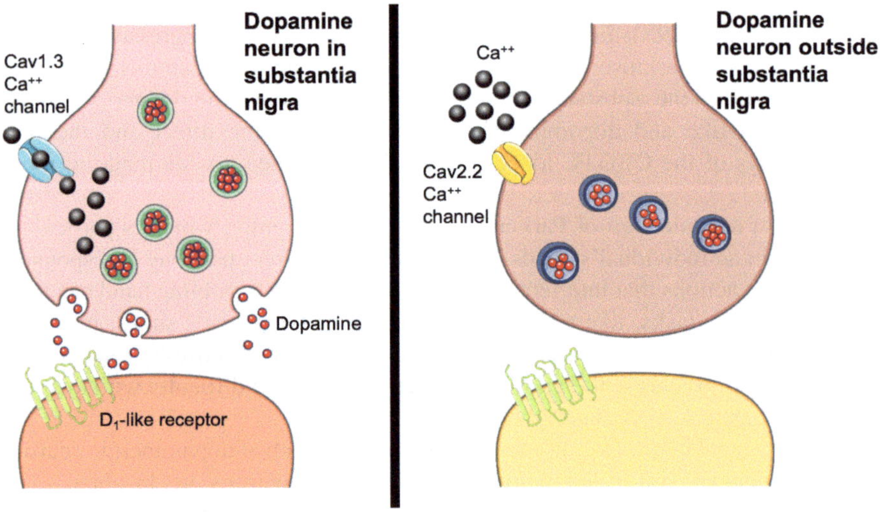

Fig. 4 A comparison of calcium (Ca^{++}) channels found on dopaminergic neurons in the substantia nigra and on dopaminergic neurons outside the substantia nigra. Dopaminergic neurons in the substantia nigra contain an L-type (Ca^{++}) channel known as the Cav1.3 Ca^{2+} channel. This ion channel stays open even the neuron is hyperpolarized, allowing for dopaminergic neurons to release dopamine in a pacemaker-like fashion. Dopaminergic neurons outside of the substantia nigra contain different types of (Ca^{++}) channels. The image above shows a N-type (Ca^{++}) channel known as the Cav2.2 Ca^{2+} channel. This channel only opens when the neuron has been sufficiently depolarized. Image created by the author using Servier Medical Art (https://smart.servier.com/; CC BY 4.0, https://creativecommons.org/licenses/by/4.0/) and Microsoft PowerPoint

Additional evidence supporting the hypothesis that mitochondrial damage underlies Parkinson's disease comes from case reports of young adults developing Parkinsonian-like symptoms in San Jose in 1982. Individuals thought they were taking an analog of the drug meperidine (Demerol), a synthetic opioid used for pain management. Unfortunately, this drug was contaminated with 1-methyl-4-phenyl-1,2,3,6-tetrahydropyridine (MPTP), which gets converted into 1-methyl-4-phenylpyridinium (MPP^+). MPP^+ is a neurotoxin that disrupts the ability of mitochondria to generate ATP; this causes cells to die. MPTP and MPP^+ are now used experimentally to induce Parkinsonian-like symptoms in animals in an effort to find novel treatments for Parkinson's disease (see Pérez-Barrón et al., 2021 for an example). The Experiment Spotlight below details one recent example of how rodent models can be used to identify putative treatments for Parkinson's disease.

Experiment Spotlight: NIR-II Light Therapy Improves Cognitive Performance in MPTP Induced Parkinson's Disease Rat Models: A Preliminary Experimental Study

The full article can be accessed with this link: https://www.sciencedirect.com/science/article/pii/S2405844024088315

I chose this study because it bridges what you learned in Chap. 6 with what you have learned thus far in the present chapter. While Parkinson's disease is primarily characterized by motor impairments, individuals can develop a form of dementia. This occurs because dopamine is no longer able to modulate activity in the frontal cortex; this leads to significant impairments in frontal cortical activity. Recall from Chap. 6 that the frontal cortex is important for memory processes. In the present study, Zhang et al. (2024) injected MPTP into male rats for 7 days. This treatment regimen produces Parkinsonian-like symptoms, as you will read about below. The researchers wanted to determine if *transcranial photobiological modulation* improves MPTP-induced Parkinsonian-like symptoms.

Transcranial photobiological modulation is a technique that utilizes light-emitting diodes or lasers that are directed at a specific part of the brain. Commonly, red-to-near-infrared light waves are used, but this approach is not efficient as most of the light energy decreases as it must pass through the scalp, the skull, the meninges, and other tissue before reaching the target destination. As such, the researchers used light with a longer wavelength compared to what is commonly used (~1070 nm wavelength vs. 700–900 nm). For 10 h per day for 14 days, rats were placed in a box and were exposed to infrared light.

Rats were tested in the Morris water maze to measure learning and memory. They were also subjected to the rotor rod test, which measures motor balance. In this test, rats are placed on a rod that rotates. Researchers record how long the rat can stay on the rod before falling. Behavioral testing occurred before and after light therapy. Rats were euthanized, and immunohistochemistry was performed to determine the number of immune-positive cells in the hippocampus. Specifically, the researchers determined which cells contained Ca^{++}-binding adaptor molecule 1 (IBA1), glial fibrillary acidic protein (GFAP), and apolipoprotein E (APOE). Recall from Chap. 6 that IBA1 is often found in microglia and that APOE is highly linked to conditions like Alzheimer's disease. GFAP is highly expressed in astrocytes, and APOE is often produced by astrocytes.

MPTP-exposed rats took longer to find the platform in the Morris water maze, and they had more difficulty staying on the rotor rod. Exposure to infrared light improved performance in both the Morris water maze and the rotor rod test. IBA1-, GFAP-, and APOE-positive cells were higher in MPTP-treated rats compared to control rats, and infrared light treatment decreased the number of immune-positive cells.

Questions to Consider:

1. Based on the results of this study and the results from the Experiment Spotlight in Chap. 6, how do microglia contribute to conditions like Alzheimer's disease and Parkinson's disease?
2. In addition to the hippocampus, what other brain regions could have been examined? Why?
3. The present study showed that infrared light decreases the number of immune-positive cells, which corresponded with improved learning and motor balance. Based on these results, what other conditions may benefit from infrared light therapy?

Comprehension Check 3 How does the loss of dopamine signaling contribute to the seemingly paradoxical symptoms of both bradykinesia and tremor observed in Parkinson's disease?

Because there is less dopamine to stimulate both dopamine D_1-like and D_2-like receptors, both the direct and the indirect pathways of the basal ganglia system are impaired. Impairment of the indirect pathway gives rise to the characteristic tremor observed in Parkinson's disease. Because the direct pathway no longer functions effectively, individuals have difficulty initiating purposeful movement, such as reaching out to grab an item.

4.2 Huntington's Disease

I first mentioned Huntington's disease in Chap. 6 when discussing dementia. Recall that Huntington's disease is a genetic disorder. Specifically, individuals receive too many copies of the nucleotide sequence cytosine, adenine, and guanine in the *huntingtin* gene. Normally, this sequence repeats approximately 10–26 times. In Huntington's disease, this sequence repeats 40 or more times. Widespread cell death first occurs in the basal ganglia before spreading to other parts of the brain like the frontal cortex. As such, individuals with Huntington's disease experience motor impairments and deficits in implicit memory. Eventually, individuals will experience other memory deficits that are similar to those with Alzheimer's disease.

As the basal ganglia are further damaged, individuals with Huntington's disease will experience symptoms that are similar to those with Parkinson's disease, including bradykinesia, difficulty with balance and coordination, rigidity, and involuntary movements. However, the involuntary movements observed in individuals with Huntington's disease differ from those with Parkinson's disease. Huntington's disease causes muscles to contract uncontrollably, leading to *writhing* (i.e., contortions of the body). This difference occurs because Huntington's disease causes widespread damage throughout the basal ganglia as opposed to the substantia nigra. Although the direct and the indirect pathways become dysregulated in Parkinson's disease, the widespread damage to the basal ganglia causes severe impairments in both pathways for individuals with Huntington's disease.

4.3 Catatonia

Catatonia is a complex neuropsychiatric syndrome characterized by abnormalities in motor behavior. There are 12 symptoms associated with catatonia, with individuals needing to display at least three to be formally diagnosed with this condition (American Psychiatric Association, 2013). The symptoms that can be present in an individual fall on a spectrum, with paucity of movement on one end of the spectrum and excessive movements on the opposite end of the spectrum.

At one extreme end of the spectrum, individuals can appear frozen, maintaining posture for extended periods of time, even if it is uncomfortable. This is known as *catalepsy*. Directly related to catalepsy is *waxy flexibility*. An individual will maintain a posture that is imposed on them by another person. For example, if an examiner crosses the individual's arms, the individual will maintain that position. Individuals can also display little movement and will not respond to external stimuli (*stupor*), and they can engage in *mutism* (lack of speaking).

On the other end of the spectrum, individuals with catatonia can display exaggerated movements (*mannerism*), restlessness/excessive motor activity (*agitation*), and/or repetitive behaviors that serve no purpose (**stereotypy**). You will learn more about stereotypies in the next chapter. Individuals may also contort their face, known as *grimacing*. Finally, individuals may begin to mimic others' actions (*echopraxia*) and/or speech (*echolalia*).

Because *hypokinetic* symptoms of catatonia (e.g., catalepsy, stupor, etc.) are more common than *hyperkinetic symptoms* (e.g., mannerism, stereotypies, etc.), research has focused on examining the neurobiological correlates of hypokinetic catatonia. One hypothesis is that hypokinetic symptoms stem from a hyperactive SMA (Walther et al., 2019). I described the contribution of the SMA to the direct pathway earlier in this chapter. However, the SMA has a direct connection with the STN, a key structure of the indirect and the hyperdirect pathways. If the SMA is hyperactive, activity within the internal capsule of the globus pallidus increases too much, leading to increased inhibition of the thalamus and the primary motor cortex. This can make initiating movement difficult, leading to a hypokinetic state. Indeed, increased blood flow to regions like the SMA and the primary motor cortex are observed in individuals with catatonia (Edinoff et al., 2021).

Beyond the SMA, there is evidence that hyperactivity in other brain regions that make up the CBGTC loop, as well as hyperactivity in the parietal cortex, underlie hypokinetic symptoms of catatonia. Specifically, a hyperactive glutamatergic system in the basal ganglia and a corresponding decrease in $GABA_A$ receptor activity in the right orbitofrontal cortex (OFC) and the right parietal cortex are implicated in catatonic behaviors (Edinoff et al., 2021).

Comprehension Check 4 Given that research on hypokinetic catatonia suggests hyperactivity within the SMA and other components of the CBGTC loop leading to increased thalamic inhibition, how might one reconcile the existence of hyperkinetic symptoms of catatonia? What alternative or additional neural mechanisms might be involved in the expression of agitation, stereotypy, or echolalia?

Going back to the direct and the indirect pathways of the basal ganglia system, the hypokinetic symptoms associated with catatonia appear to be driven by a hyperactive hyperdirect pathway (e.g., the SMA directly stimulates the STN, leading to inhibition of the thalamus and primary motor cortex). Although the hyperdirect pathway is overactive, individuals with hyperkinetic symptoms of catatonia may have deficiencies in the direct and/or indirect pathways. A hyperactive direct pathway and hypoactive indirect pathway would lead to individuals engaging in actions like stereotypies.

4.4 Cerebellar-Based Motor Impairments

Damage to the cerebellum is associated with several motor-related impairments (Manto, 2018). First, **ataxia** is defined by a lack of coordination of voluntary muscle movements, often resulting in difficulties with balance, walking, speech, and fine motor skills. Directly related to ataxia is *dysdiadochokinesia*, in which individuals have difficulty performing rapid, alternating movements. Something as simple as tapping one's feet in an alternating fashion becomes challenging.

Individuals with cerebellar damage can experience *decomposition of movement*. Instead of executing movements as one seamless action, the individual must perform individual actions in a sequence. For example, imagine you want to grab a cup of coffee. This is an action that requires little conscious thought. Your arm seemingly guides itself to the cup. As your arm moves to the cup, your fingers will adjust so you can easily grab the cup. When you're in position, you will either grip the handle with a couple of fingers and your thumb, or you'll place your hand around the cup itself. Someone experiencing decomposition of movement will first need to move their shoulder. Then, they will either need to bend or straighten their elbow to get their hand closer to the cup. After they move their hand to the cup, the individual will then move their fingers. Notice, instead of one fluid motion, an individual with cerebellar damage will need to perform these actions sequentially.

Another potential movement disorder that can arise from cerebellar damage is *intention tremor*. Unlike Parkinson's disease, individuals do not show a tremor at rest. However, a tremor appears as an individual tries to complete a desired action. For example, if an individual wants to grab a cup of coffee, their arm will move toward the cup in a fluid motion. As their hand gets closer to the cup of coffee, their hand will begin to shake uncontrollably. With intention tremor, the shaking gets worse as the individual gets closer to completing the desired action.

Earlier, I discussed the importance of the cerebellum to motor learning. Evidence to support such learning comes from individuals with damage to this region and from animal studies in which the cerebellum is lesioned. In one study (Martin et al., 1996), individuals were asked to throw a ball of clay at a target. Baseline data were collected, meaning the individual threw the clay ball at the target several times. This allowed the researchers to determine how well each individual could hit the target before the experiment began. After the baseline phase, participants were asked to wear a pair of prism glasses while they attempted to hit the target with the clay ball. These glasses distort the location of the target, making it harder to hit. Finally, participants threw the clay ball at the target after removing the prism glasses.

Normally, when individuals attempt to hit the target while wearing prism glasses, they are not accurate. However, they eventually learn to adjust their aim to increase their accuracy. When the glasses are removed, participants have to readjust their aim again. This is not difficult to perform, unless you have cerebellar damage. Individuals with lesions to part of the cerebellum had difficulty adjusting their aim while wearing prism glasses. Even after the glasses were removed, these individuals still had difficulty hitting the target.

Cerebellar damage also impairs the conditioned eye blink response. Animals such as rabbits and rats can be conditioned to blink when they hear a tone. A puff of air is not commonly used in animals, as researchers often utilize periorbital shock; that is, an electrode is implanted into one eyelid, and an electric shock causes the eyelid to close. Disrupting cerebellar function severely impairs the conditioned eye blink response in animals (Attwell et al., 1999; Horiuchi & Kawahara, 2010; McCormick et al., 1982).

4.5 Cerebral Palsy

Cerebral palsy is a broad umbrella term that encompasses conditions that affect movement and posture. Individuals with cerebral palsy have symptoms that overlap with other conditions that have been detailed above, including dystonia, ataxia, and tremors. Cerebral palsy also leads to *spasticity* (muscle stiffness) and *hypertonia* (excessive muscle tone), but individuals can also experience *hypotonia* (loss of muscle tone).

Classified as a neurodevelopmental disorder, the most common causes of cerebral palsy are premature birth or brain damage that occurs during prenatal development, although up to 10% of cerebral palsy cases occur after birth, stemming from an infection or from head trauma (Colver et al., 2014). There is increasing speculation that genetics contribute to cerebral palsy (Friedman et al., 2022; Lewis et al., 2021), but more wide-scale studies are needed to isolate the specific genetic changes that occur that lead to cerebral palsy.

The symptoms of cerebral palsy depend on where brain damage occurs. As already covered, dystonia and ataxia result from damage to the CBGTC loop and thecerebellum, respectively. Spasticity occurs when upper motor neurons cannot be sufficiently inhibited. This causes these neurons to become hyperactive, overstimulating muscles. This is somewhat similar to the overexcitation that occurs in the neuromuscular junction following inhibition of acetylcholinesterase.

4.6 *Muscular Dystrophy*

Like cerebral palsy, the term muscular dystrophy refers to a group of related disorders. These disorders lead to a progressive weakening of skeletal muscles, but smooth muscle can be affected in some types of muscular dystrophy. While cerebral palsy is most commonly caused by brain damage, muscular dystrophy results from genetic mutations (Hoang & Dowdy, 2024). For simplicity, I will highlight just a few muscular dystrophies. Duchenne muscular dystrophy and Becker muscular dystrophy are linked to a mutation in the *DMD* gene, which encodes the protein dystrophin. This protein is critical for protecting the **sarcolemma**, the cell membrane that surrounds a muscle fiber, when our muscles contract and relax. In Duchenne muscular dystrophy and Becker muscular dystrophy, dystrophin is completely or partially missing. The sarcolemma is at increased risk of rupturing. Over time, **fibrosis** occurs, and the resulting scar tissue does not allow muscle fibers to function properly (Fig. 5).

In a different form of muscular dystrophy, Emery-Dreifuss muscular dystrophy, a genetic mutation to the *EMD* gene affects a protein called emerin. Emerin acts like an anchor, allowing the cell nucleus to stay attached to the cytoskeleton of the cell. If emerin function is disrupted, the cell nucleus becomes deformed (Fig. 5). As the nucleus contains the cell's genetic information and controls the cell's function, damage to the nucleus is catastrophic.

4.7 *Paralysis*

Many of you know thatspinal cord injuries can lead to paralysis. Depending on where the spinal injury occurs and the severity of the injury, motor neurons in the descending pathway of the spinal cord will not be able to transmit information to neurons that exit the ventral root. This prevents neurons from stimulating muscles at

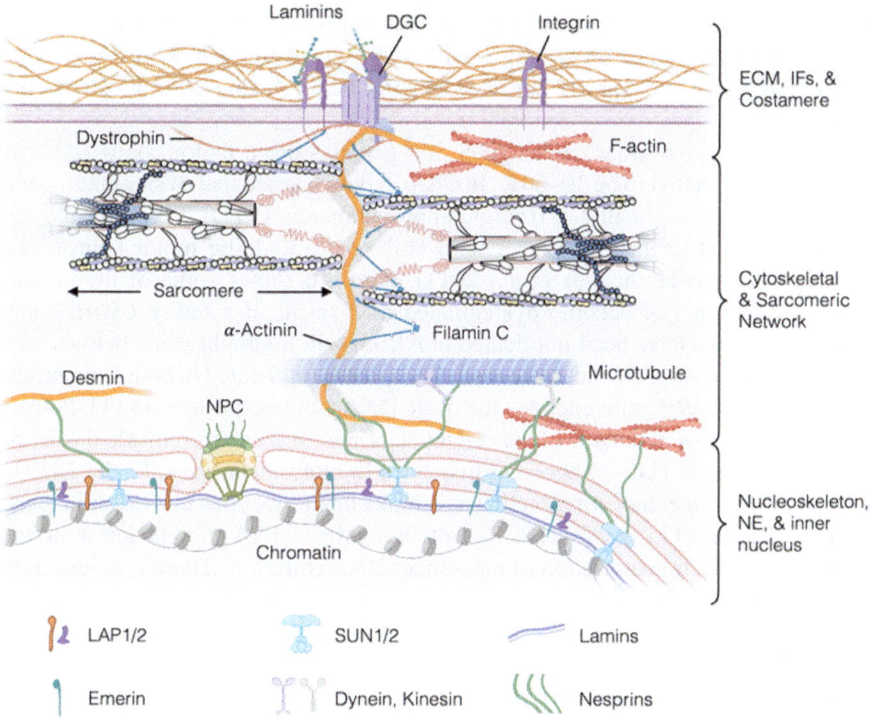

Fig. 5 Schematic of the internal structure of a muscle (note, this image depicts a cardiac muscle). *Note* The location of dystrophin and emerin. Dystrophin is located just below the muscle and acts like a "shock absorber", protecting the muscle as it stretches and contracts. Emerin anchors the cell nucleus to the cytoskeleton of the wall. Dysfunction of either protein can lead to muscular dystrophy. *Note* DGC = dystrophin-glycoprotein complex, ECM = extracelluar matrix, IF = interstitial fluid, LAP1/2 = lamina-associated polypeptides 1 and 2, NE = nuclear envelope, NPC = nuclear pore complex, SUN1/2 = Sad1/UNC-84 domain-containing proteins 1 and 2. *Source* Fig. 1 of Zhang et al. (2023), https://doi.org/10.3389/fphys.2023.1126111. CC BY 4.0 (https://creativecommons.org/licenses/by/4.0/), via Frontiers

the neuromuscular junction. In addition to spinal injuries, there are other conditions that lead to partial or complete paralysis. A couple of disorders will be detailed below.

4.7.1 Amyotrophic Lateral Sclerosis (ALS)

Often colloquially referred to as Lou Gehrig's disease in the United States (for the former New York Yankees baseball player who died from this disease in 1941), amyotrophic lateral sclerosis (ALS) is a neurodegenerative disease like Parkinson's disease. Unlike Parkinson's disease, in which dopaminergic neurons in the substantia nigra are damaged, ALS leads to a progressive loss of both upper and lower motor neurons. Eventually, motor neurons that control respiration stop functioning. Individuals with ALS die, on average, within 3 years, although there are cases in which individuals can survive for decades after being diagnosed (e.g., Dr. Stephen Hawking).

We do not know what causes ALS, but genetics are a major contributor, with heritability estimated to be 30–60%. In Chap. 6, you learned that dysregulation of the *C9orf72* gene is implicated in frontotemporal dementia (FTD). Likewise, the same excessive repeat in the *C9orf72* gene is believed to underlie motor neuron death in ALS (DeJesus-Hernandez et al., 2011). Figure 6 shows some of the potential mechanisms that can become dysregulated as a result of a faulty *C9orf72* gene. Additional genes have been implicated in ALS. I will highlight some below.

As with the *C9orf72* gene, there is one gene that is implicated in both dementia and ALS. The *TARDBP* gene encodes the TAR DNA-binding protein 43 (TDP-43). In FTD, TDP-43 moves outside of the cell nucleus and forms clumps inside the neuron. ALS may occur if TDP-43 becomes displaced in motor neurons. Like the *TARDBP* gene, the *FUS* gene makes a protein that is important for actions that occur in the cell nucleus. The fused in sarcoma (FUS) protein regulates RNA (ribonucleic acid) transcription and the transportation of messenger RNA (mRNA). There is evidence that FUS can become displaced from the cell nucleus, causing clumps to form in motor neurons.

Finally, I want to mention the *SOD1* gene, which encodes the enzyme superoxide dismutase 1 (SOD1). This enzyme is responsible for breaking down *superoxide radicals*. You have already learned about free radicals, unstable molecules that strip electrons away from nearby molecules. Superoxide radicals are a type of reactive oxygen species (ROS). Although we need oxygen to survive, superoxide radicals and other ROS are problematic as they are very unstable. Because these chemicals lack an electron, they attempt to pull electrons from other molecules, which causes extensive intracellular damage. One hypothesis is that a genetic mutation to the *SOD1* gene prevents the breakdown of superoxide radicals in motor neurons, leading to their death.

4.7.2 Guillain-Barré Syndrome

As an autoimmune disease, Guillain-Barré syndrome results from the immune system attacking the peripheral nerves. I could have discussed this condition in Chaps. 3 and/or 4 as individuals experience abnormal sensations like tingling and/or pain as the peripheral nerves break down. I chose to discuss Guillain-Barré syndrome in this chapter as the primary symptom is rapid onset of muscle weakness, which can lead to partial paralysis. Although Guillain-Barré syndrome is rare (approximately 1–2 cases per 100,000 people), one common finding is that individuals that develop this condition experienced an *antecedent event* several weeks before the emergence of symptoms (Shahrizaila et al., 2021). Many individuals that develop Guillain-Barré syndrome recently had an upper respiratory tract infection. Several viruses have been linked to Guillain-Barré syndrome, including Zika, Epstein-Barr, influenza

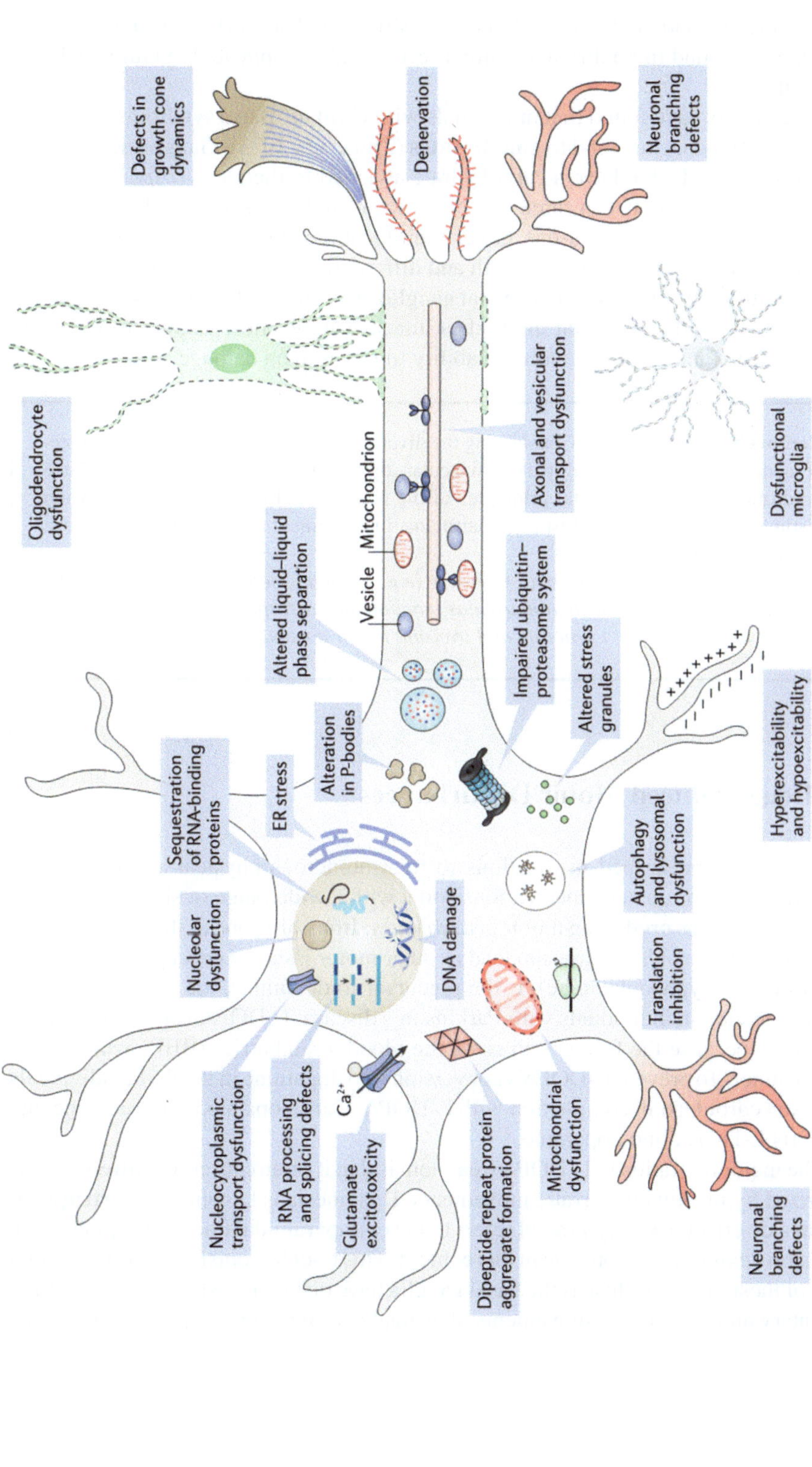

Fig. 6 A schematic detailing how dysfunction of the *C9orf72* gene can cause aberrations of several cellular pathways, which can lead to death of motor neurons. *Source* Fig. 4 of Balendra and Isaacs (2018), https://doi.org/10.1038/s41582-018-0047-2, with permission from Springer Nature

A, hepatitis E, and SARS-CoV-2 (i.e., COVID-19). The bacteria *Campylobacter jejuni*, often found in poultry and animal feces, is also implicated in Guillain-Barré syndrome.

Researchers have found that individuals with Guillain-Barré syndrome have high levels of antibodies that bind to **gangliosides** (Zhu et al., 2023). Gangliosides belong to a class of lipids that have a carbohydrate attached to them (*glycosphingolipids*). Gangliosides have between three and eight sugar molecules and are located in the bilipid cell membrane of neurons. Given their location, gangliosides are proposed to play important roles in cellular growth and differentiation, including axon regeneration. Additionally, there is evidence that gangliosides are involved in axon potentials. One major hypothesis is that antibodies attack gangliosides, leading to damaged myelin sheath and a corresponding inability to repair axon damage.

Comprehension Check 5 Considering the diverse origins and specific cellular targets of conditions like cerebral palsy, muscular dystrophy, ALS, and Guillain-Barré syndrome, what fundamental principles of the nervous and muscular systems, when disrupted, commonly manifest as impairments in movement and posture across these seemingly disparate disorders?

From the brain's motor control centers (e.g., primary motor cortex) down to the peripheral nerves and the neuromuscular junction, any disruption along this pathway will impair the ability to generate and transmit motor commands effectively.

5 Drug-Induced Motor Disturbances

Often, we are prescribed medications to treat numerous physical and psychiatric conditions. Unfortunately, medications come with a wide range of side effects, with some medications producing motor disturbances. Ironically, one of the primary treatments for Parkinson's disease can lead to other motor issues. *Levodopa*, or L-DOPA (l-3,4-dihydroxyphenylalanine), is the precursor to dopamine, the neurotransmitter that is depleted in individuals with Parkinson's disease. L-DOPA is given instead of dopamine because the former can cross the blood–brainbarrier (BBB) whereas the latter cannot. To prevent L-DOPA's conversion into dopamine in the PNS, individuals are given carbidopa in conjunction with L-DOPA. Carbidopa blocks the enzyme that converts L-DOPA into dopamine.

The major limitation to L-DOPA treatment is that it often causes **dyskinesia**, characterized by involuntary, erratic movements. Dyskinesia is known as an **extrapyramidal** side effect. Extrapyramidal refers to the extrapyramidal tract in thespinal cord. The term "extra" means outside in this context. This tract lies outside of the pyramidal tract of thespinal cord. Just as the basal ganglia have different systems for controlling voluntary and involuntary movements, thespinal cord includes the pyramidal and the extrapyramidal tracts. The pyramidal tract regulates voluntary movements while the extrapyramidal tract controls involuntary movements.

If L-DOPA is used to treat Parkinson's disease, why does it cause motor disturbances? At first, L-DOPA is effective at attenuating an individual's symptoms. Because L-DOPA is taken orally several times per day, this leads to varying dopamine levels throughout the day. When an individual wakes up in the morning, dopamine levels are low. When they take L-DOPA, dopamine levels increase. Over a few hours, dopamine levels will decrease. This cycle will repeat several times. When the individual takes L-DOPA, excess dopamine can overstimulate D_1-like receptors in the basal ganglia. This causes the individual to initiate a movement that cannot be sufficiently inhibited by the indirect pathway of the basal ganglia.

In Chap. 3, I discussed antipsychotic medications for the treatment of schizophrenia. The typical (i.e., first-generation) antipsychotic medications act primarily as dopamine D_2-like receptor antagonists. Because dopamine cannot bind to dopamine D_2-like receptors, the indirect pathway of the basal ganglia becomes impaired. One major side effect of typical antipsychotic medications is **tardive dyskinesia**. As you have already learned, individuals with dyskinesia experience involuntary movements. Tardive dyskinesia primarily affects the muscles of the face, the mouth, and the tongue. If you watch someone with tardive dyskinesia, you will notice that their tongue often sticks out of their mouth and will move on its own. The term "tardive" is derived from the French word "tardif", meaning "late". Involuntary movements of the face/tongue appear later than other extrapyramidal side effects.

Additional extrapyramidal side effect associated with typical antipsychotic drugs are dystonia and **akathisia**, a feeling of inner restlessness and an irresistible urge to move, often accompanied by an inability to sit still. In fact, individuals prescribed typical antipsychotic medications often develop Parkinsonian-like symptoms. My grandfather was one such individual. He was diagnosed with schizophrenia after serving in World War II and was prescribed antipsychotic medications. Later in life, he was diagnosed with Parkinson's disease.

Typical antipsychotic medications are not the only drug class known to produce extrapyramidal side effects. Anxiety and mood disorders are often treated with medications that increase norepinephrine and/or serotonin levels. These medications can cause motor disturbances, most likely because the serotonergic system influences dopamine levels in the basal ganglia (Di Matteo et al., 2008).

6 Chapter Summary

This chapter explored the intricate mechanisms of motor control and the various conditions that arise when these mechanisms are compromised. You were reintroduced to the spinal cord/cranial nerves, the CBGTC loop, the parietal cortex, and the cerebellum, learning how these structures regulate our movements. You also learned how the direct, the indirect, and the hyperdirect pathways of the basal ganglia initiate, modulate, and inhibit movement. This chapter introduced the neuromuscular junction, where motor neurons synapse with skeletal muscle fibers to initiate a muscle contraction. Considering the interplay of various neural circuits and neurotransmitter systems in the control of movement, numerous movement-related disorders can arise following disruption to just one part of our vast interconnected motor system. Dysregulation of the CBGTC loop leads to conditions like Parkinson's disease, Huntington's disease, and catatonia. Damage to the CBGTC loop or the cerebellum can cause different forms of cerebral palsy and/or lead to deficits in motor learning. Some conditions like ALS and Guillain-Barré syndrome result from damage to motor neurons, resulting in partial or complete paralysis. The next chapter will continue to focus on motor dysfunction, but attention will shift to repetitious behaviors that become problematic.

Answers to Experiment Spotlight Questions:

1. As immune cells, microglia are involved in inflammation. In Chap. 6, you learned that proliferating microglia are ineffective at removing excess beta amyloid, which may be a significant contributor to Alzheimer's disease. Likewise, excess neuroinflammation may exacerbate the motor and cognitive impairments associated with Parkinson's disease.
2. While the hippocampus is important for memory, examination of immune-positive cells in other regions such as the basal ganglia and the frontal cortex would have enriched the present dataset. The researchers would have been able to determine if neuroinflammation is isolated to the hippocampus or if widespread inflammation in the mesocortical and the nigrostriatal pathways is observed.
3. Individuals with other neurodegenerative diseases such as Alzheimer's disease may benefit from infrared light therapy.

Glossary

Akathisia movement disorder characterized by a feeling of inner restlessness and an irresistible urge to move, often accompanied by an inability to sit still.

Ataxia a neurological condition characterized by a lack of coordination of voluntary muscle movements, often resulting in difficulties with balance, walking, speech, and fine motor skills.

Bradykinesia slowness of movement.

Catatonia neuropsychiatric syndrome characterized by abnormalities in motor behavior, ranging from immobility or stupor to excessive, purposeless activity.

Dyskinesia movement disorder characterized by involuntary, erratic movements.

Dystonia movement disorder characterized by sustained muscle contractions, often causing twisting and repetitive movements or abnormal postures.

Extrapyramidal refers to a specific network of nerve fibers that lies outside of the pyramidal tract; this network regulates involuntary movements, keeping them under control.

Fibrosis thickening and scarring of tissue, often caused by damage.

Ganglioside a complex type of sugar-containing lipid found in cell membranes, particularly in nerve tissue.

Neuromuscular junction the synapse where a motor neuron communicates with a skeletal muscle fiber, enabling nerve impulses to trigger muscle contraction.

Stereotypy frequent repetition of the same, typically purposeless movement, gesture, posture, or vocal sounds or utterances.

Tardive dyskinesia movement disorder characterized by involuntary and repetitive movements, typically affecting the face, mouth, and tongue.

Lower motor neuron includes motor neurons located in the spinal cord and the ventral root, as well as cranial nerves that have motor functions.

Upper motor neuron motor neuron located in the cerebral cortex and the brainstem.

References

Abou-Donia, M. B., Siracuse, B., Gupta, N., & Sobel Sokol, A. (2016). Sarin (GB, O-isopropyl methylphosphonofluoridate) neurotoxicity: Critical review. *Critical Reviews in Toxicology, 46*(10), 845–875. https://doi.org/10.1080/10408444.2016.1220916

American Psychiatric Association. (2013). *Diagnostic and statistical manual of mental disorders* (5th ed.). https://doi.org/10.1176/appi.books.9780890425596

Attwell, P. J., Rahman, S., Ivarsson, M., & Yeo, C. H. (1999). Cerebellar cortical AMPA-kainate receptor blockade prevents performance of classically conditioned nictitating membrane responses. *The Journal of Neuroscience, 19*(24), RC45. https://doi.org/10.1523/JNEUROSCI.19-24-j0003.1999

Balendra, R., & Isaacs, A. M. (2018). C9orf72-mediated ALS and FTD: Multiple pathways to disease. *Nature Reviews. Neurology, 14*(9), 544–558. https://doi.org/10.1038/s41582-018-0047-2

Colver, A., Fairhurst, C., & Pharoah, P. O. (2014). Cerebral palsy. *Lancet, 383*(9924), 1240–1249. https://doi.org/10.1016/S0140-6736(13)61835-8

DeJesus-Hernandez, M., Mackenzie, I. R., Boeve, B. F., Boxer, A. L., Baker, M., Rutherford, N. J., Nicholson, A. M., Finch, N. A., Flynn, H., Adamson, J., Kouri, N., Wojtas, A., Sengdy, P., Hsiung, G. Y., Karydas, A., Seeley, W. W., Josephs, K. A., Coppola, G., Geschwind, D. H., … Rademakers, R. (2011). Expanded GGGGCC hexanucleotide repeat in noncoding region of C9ORF72 causes chromosome 9p-linked FTD and ALS. *Neuron, 72*(2), 245–256. https://doi.org/10.1016/j.neuron.2011.09.011

De Zeeuw, C. I., & Ten Brinke, M. M. (2015). Motor learning and the cerebellum. *Cold Spring Harbor Perspectives in Biology, 7*(9), Article a021683. https://doi.org/10.1101/cshperspect.a021683

Di Matteo, V., Pierucci, M., Esposito, E., Crescimanno, G., Benigno, A., & Di Giovanni, G. (2008). Serotonin modulation of the basal ganglia circuitry: Therapeutic implication for Parkinson's disease and other motor disorders. *Progress in Brain Research, 172*, 423–463. https://doi.org/10.1016/S0079-6123(08)00921-7

Edinoff, A. N., Kaufman, S. E., Hollier, J. W., Virgen, C. G., Karam, C. A., Malone, G. W., Cornett, E. M., Kaye, A. M., & Kaye, A. D. (2021). Catatonia: Clinical overview of the diagnosis, treatment, and clinical challenges. *Neurology International, 13*(4), 570–586. https://doi.org/10.3390/neurolint13040057

Freedman, D. J., & Ibos, G. (2018). An integrative framework for sensory, motor, and cognitive functions of the posterior parietal cortex. *Neuron, 97*(6), 1219–1234. https://doi.org/10.1016/j.neuron.2018.01.044

Friedman, J. M., van Essen, P., & van Karnebeek, C. D. M. (2022). Cerebral palsy and related neuromotor disorders: Overview of genetic and genomic studies. *Molecular Genetics and Metabolism, 137*(4), 399–419. https://doi.org/10.1016/j.ymgme.2021.11.001

Hoang, T., & Dowdy, R. A. E. (2024). A review of muscular dystrophies. *Anesthesia Progress, 71*(1), 44–52. https://doi.org/10.2344/673191

Horiuchi, T., & Kawahara, S. (2010). Effects of ipsilateral cerebellum ablation on acquisition and retention of classically conditioned eyeblink responses in rats. *Neuroscience Letters, 472*(2), 148–152. https://doi.org/10.1016/j.neulet.2010.01.076

Hull, C. (2020). Prediction signals in the cerebellum: Beyond supervised motor learning. *eLife, 9*, e54073. https://doi.org/10.7554/eLife.54073

Lewis, S. A., Shetty, S., Wilson, B. A., Huang, A. J., Jin, S. C., Smithers-Sheedy, H., Fahey, M. C., & Kruer, M. C. (2021). Insights from genetic studies of cerebral palsy. *Frontiers in Neurology, 11*, Article 625428. https://doi.org/10.3389/fneur.2020.625428

Manto, M. (2018). Cerebellar motor syndrome from children to the elderly. *Handbook of Clinical Neurology, 154*, 151–166. https://doi.org/10.1016/B978-0-444-63956-1.00009-6

Martin, T. A., Keating, J. G., Goodkin, H. P., Bastian, A. J., & Thach, W. T. (1996). Throwing while looking through prisms. I. Focal olivocerebellar lesions impair adaptation. *Brain, 119*(Pt 4), 1183–1198. https://doi.org/10.1093/brain/119.4.1183

McCormick, D. A., Clark, G. A., Lavond, D. G., & Thompson, R. F. (1982). Initial localization of the memory trace for a basic form of learning. *Proceedings of the National Academy of Sciences of the United States of America, 79*(8), 2731–2735. https://doi.org/10.1073/pnas.79.8.2731

Neychev, V. K., Gross, R. E., Lehéricy, S., Hess, E. J., & Jinnah, H. A. (2011). The functional neuroanatomy of dystonia. *Neurobiology of Disease, 42*(2), 185–201. https://doi.org/10.1016/j.nbd.2011.01.026

Pang, W., & Hu, F. (2021). Cellular and physiological functions of C9ORF72 and implications for ALS/FTD. *Journal of Neurochemistry, 157*(3), 334–350. https://doi.org/10.1111/jnc.15255

Parkinson, J. (2002). An essay on the shaking palsy. 1817. *The Journal of Neuropsychiatry and Clinical Neurosciences, 14*(2), 223–222. https://doi.org/10.1176/jnp.14.2.223

Pérez-Barrón, G., Montes, S., Aguirre-Vidal, Y., Santiago, M., Gallardo, E., Espartero, J. L., Ríos, C., & Monroy-Noyola, A. (2021). Antioxidant effect of hydroxytyrosol, hydroxytyrosol acetate and nitrohydroxytyrosol in a rat MPP^+ model of Parkinson's disease. *Neurochemical Research, 46*(11), 2923–2935. https://doi.org/10.1007/s11064-021-03379-x

Rocha, G. S., Freire, M. A. M., Britto, A. M., Paiva, K. M., Oliveira, R. F., Fonseca, I. A. T., Araújo, D. P., Oliveira, L. C., Guzen, F. P., Morais, P. L. A. G., & Cavalcanti, J. R. L. P. (2023). Basal ganglia for beginners: The basic concepts you need to know and their role in movement control. *Frontiers in Systems Neuroscience, 17*, 1242929. https://doi.org/10.3389/fnsys.2023.1242929

Shahrizaila, N., Lehmann, H. C., & Kuwabara, S. (2021). Guillain-Barré syndrome. *Lancet, 397*(10280), 1214–1228. https://doi.org/10.1016/S0140-6736(21)00517-1

Smeyers, J., Banchi, E. G., & Latouche, M. (2021). C9ORF72: What it is, what it does, and why it matters. *Frontiers in Cellular Neuroscience, 15*, Article 661447. https://doi.org/10.3389/fncel.2021.661447

Surmeier, D. J., Guzman, J. N., Sanchez-Padilla, J., & Schumacker, P. T. (2011). The role of calcium and mitochondrial oxidant stress in the loss of substantia nigra pars compacta dopaminergic neurons in Parkinson's disease. *Neuroscience, 198*, 221–231. https://doi.org/10.1016/j.neuroscience.2011.08.045

Vijayaraghavan, S., Pugh, P. C., Zhang, Z. W., Rathouz, M. M., & Berg, D. K. (1992). Nicotinic receptors that bind alpha-bungarotoxin on neurons raise intracellular free Ca2+. *Neuron, 8*(2), 353–362. https://doi.org/10.1016/0896-6273(92)90301-s

Walther, S., Stegmayer, K., Wilson, J. E., & Heckers, S. (2019). Structure and neural mechanisms of catatonia. *The Lancet. Psychiatry, 6*(7), 610–619. https://doi.org/10.1016/S2215-0366(18)30474-7

Zhang, B., Powers, J. D., McCulloch, A. D., & Chi, N. C. (2023). Nuclear mechanosignaling in striated muscle diseases. *Frontiers in Physiology, 14*, 1126111. https://doi.org/10.3389/fphys.2023.1126111

Zhang, J., Zhu, Q., Shi, X., Huang, Y., Yan, L., Zhang, G., Pei, L., Liu, J., Han, X., & Zhu, X. (2024). NIR-II light therapy improves cognitive performance in MPTP induced Parkinson's disease rat models: A preliminary experimental study. *Heliyon, 10*(12), Article e32800. https://doi.org/10.1016/j.heliyon.2024.e32800

Zhu, W., Li, K., Cui, T., & Yan, Y. (2023). Detection of anti-ganglioside antibodies in Guillain-Barré syndrome. *Annals of Translational Medicine, 11*(7), 289. https://doi.org/10.21037/atm-20-2285

Chapter 11
Maladaptive Habits

1 Introduction

My alarm goes off at 6:30 am. I roll to my side and turn off the alarm. I walk downstairs and turn my attention to the Keurig in our kitchen. I pop a K-cup in the machine and wait for my coffee to brew. As the water in the Keurig heats up, I begin preparing breakfast. Once my Keurig has finished dispensing my coffee, I sit down at the bar area in the kitchen and enjoy my breakfast. I then feed our fish and then go upstairs to ensure that my wife is awake. Between 7:00 and 7:15, I iron my clothes if I need to dress nicely for work, or I make sure the jeans and T-shirt I am wearing are not too wrinkled. At this point, I take my daily medicine, put deodorant on, brush my teeth, and put my clothes on for the day. By 7:30, I am out the front door. I follow this routine most days during the week.

We are creatures of habit. I imagine most of you reading this book follow some routine in the morning, as well as during the rest of the day. During the week, you likely awaken at the same time. Some individuals will eat breakfast each morning while others skip this meal. Some individuals shower each morning while others opt to shower at night. How do you get to school/work each day? Do you take the same route, or do you take a different route each day? More often than not, you take the same route, assuming there is not an accident or construction that has caused a road to be closed.

We typically follow a routine at school, and many jobs require the same action(s) to be performed repeatedly. From an evolutionary perspective, habits are advantageous because they can be performed seemingly automatically, allowing us to focus on other goals. For example, our morning routine can be performed with little conscious thought. This enables us to devote more mental resources to things like remembering information that will be on an upcoming exam or to work on a presentation we must give at work.

Despite the advantage that habits provide us, we can engage in "bad" habits. That is, there are repetitious actions certain individuals engage in that greatly interfere with

J. Yates, *Neuroscience of Maladaptation*,
https://doi.org/10.1007/978-3-032-15602-0_11

daily functioning. This is the focus of the present chapter. This chapter will begin with a comparison of *goal-directed* behavior and habit formation. You will then learn about a couple of psychiatric conditions in which individuals engage in maladaptive habits that are detrimental to their well-being. This chapter will primarily focus on dysfunction of circuits you have already learned about in this book: the cortico-basal ganglia-thalamo-cortical (CBGTC) loop and the mesocorticolimbic pathway.

2 Learning Objectives

By the end of this chapter, you should be able to ...

1. Differentiate goal-directed behavior from habits, and differentiate the various forms of maladaptive habits.
2. Identify psychiatric conditions characterized, at least in part, by maladaptive habits.
3. Explain how dysregulation of the mesocorticolimbic pathway and the cortico-basal ganglia-thalamo-cortical (CBGTC) loop leads to maladaptive habit formation.

3 Goal-Directed Behavior and Habit Formation

In the previous chapter, you learned how the CBGTC loop controls the initiation and the inhibition of movement. Importantly, our movements often serve a purpose. Some movements are critical for our survival, such as chewing on food or running away from a potential threat. Other movements are not necessary for our survival, but we do them because we *enjoy* doing them. Playing an instrument, pressing buttons on a video game controller, or painting on a canvas are often *positively reinforcing* (see Chap. 1) and are often *intrinsically reinforcing*. This means that we perform these actions because we want to do so. Conversely, many people go to work because they get paid to do so. Even if someone hates their job, they continue going to work because they need to get paid. Here, going to work is *extrinsically reinforcing*. The behavior continues because of some external factor (e.g., your employer).

Whether something is intrinsically or extrinsically reinforcing, we perform complex motor actions in order to accomplish a goal. That goal can be earning money or something as simple as having fun. All animals engage in some form of goal-directed behavior, such as eating, fighting/fleeing from potential predators, and copulating. Non-human mammals even engage in social play.

At a neurobiological level, dopamine is critical for goal-directed behavior because it is involved in reinforcement learning. When we perform an action that is reinforcing (e.g., eating when we are hungry), dopaminergic neurons originating from the ventral tegmental area (VTA) release dopamine in the nucleus accumbens (NAc). This is the mesolimbicdopamine pathway that you have already encountered in this book (see

Chap. 5). Over time, the VTA will stop responding to reinforcers. Instead, the VTA becomes active when we encounter cues that signal the availability of the reinforcer. For example, if we are hungry, our VTA will become active when we smell or see the food. This is our body's way of saying that it knows something good is about to happen if we approach this stimulus.

To this point, you have only heard about the NAc as a single structure. However, the NAc is subdivided into two major regions: the NAc core (inner region) and the NAc shell (outer region). The core and the shell regions have different, but complementary, functions. The shell encodes the *hedonic* aspects of stimuli; that is, the pleasure we experience from some item or activity. I will use myself as an example. I enjoy eating tiramisu. In fact, tiramisu is probably my favorite dessert. When I eat tiramisu, there is high activity in the NAc shell. In addition to hedonia, the NAc shell is important for attributing **incentive salience** to stimuli that become paired with things that we like (think of Pavlovian conditioning). Because I like eating tiramisu, I have come to associate the appearance and the smell of tiramisu with its flavor.

The NAc core is closely connected with our basal ganglia, particularly the dorsal striatum (remember, the NAc is located in the ventral striatum). As such, one primary function of the NAc core is to establish motor plans that allow us to engage in an activity that we find reinforcing. When I see tiramisu, I know exactly what steps need to be taken to eat it. In fact, I do not even have to exert much conscious effort. Even if you are not a fan of tiramisu, I imagine there is something that you enjoy eating. You do not have to sit there and say to yourself "Step 1: grab the utensil. Step 2: move the utensil toward the food…" You just do it. This is habit learning, something the basal ganglia control. In short, the NAc core and the basal ganglia work together to create habits that enable us to complete target behaviors throughout the day.

Habits are also important because they allow us to focus on planning long-term goals. Think of it this way: If I were to ask you what your goals are, how would you answer? Some of you may provide long-term goals like getting a good job after graduating from college, moving to a specific city, getting married, having children, and/or writing a book, just to name a few examples. Some of you may include these long-term goals, but you may also provide more short-term goals like passing an upcoming exam, completing a term paper, or getting through a performance review at work. Interestingly, I expect that many individuals do not include goals that we accomplish each day like getting out of bed, eating something during the day, picking out the right clothes to wear, getting to school/work on time, and so on. That is because many of these goals become part of your daily routine, and you exert little conscious effort to plan and execute these tasks.

In addition to the NAc, the VTA sends dopaminergic neurons to the prefrontal cortex (PFC). The PFC regulates the motivational aspects of a reinforcer. While reinforcement is important for a species' survival, we do not always need certain reinforcers. For example, if you have recently eaten and have satisfied your hunger, there is no advantage to eating dessert. It is most likely going to be reinforcing, but it may not be in your best interest to consume a piece of chocolate cake immediately after eating dinner. In a sense, the PFC helps keep habitual actions in check, ensuring that we are accomplishing a specific goal.

Returning to what you learned in Chap. 6, the basal ganglia are important for procedural memory, which includes habit formation. You have already learned that dopamine originating from the substantia nigra stimulates dopamine receptors that make up the direct and the indirect pathways of movement. Dopaminergic activity in the dorsal striatum is also critical for learning the association between behaviors and outcomes (Yin et al., 2005). As we learn to perform a task, the dorsomedial striatum (DMS) is active. The DMS receives inputs from the PFC and is involved in the planning of motor sequences that enable us to achieve a goal. In primates, the DMS largely corresponds to the caudate nucleus. Over time, activity shifts away from the DMS to the dorsolateral striatum (DLS), which receives inputs from sensorimotor cortices. The DLS is largely confined to the putamen in primates. See Fig. 1 for the DMS and the DLS.

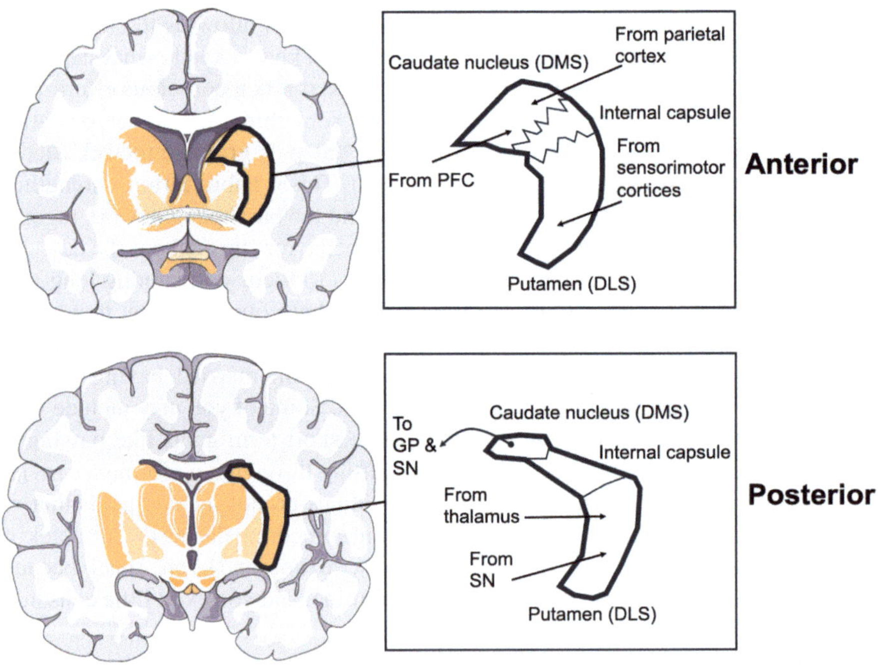

Fig. 1 A schematic of the dorsal striatum. The top and bottom images show the posterior and the anterior regions of the dorsal striatum, respectively. In primates, the caudate nucleus corresponds primarily to the dorsomedial striatum (DMS) while the putamen makes up the dorsolateral striatum (DLS). The internal capsule is a white matter tract (i.e., collection of myelinated axons) that separates the caudate nucleus and the putamen. The insets show the major inputs and outputs of the DMS and the DLS. *Note* GP = globus pallidus, PFC = prefrontal cortex, SN = substantia nigra. Image created by the author using Servier Medical Art (https://smart.servier.com/; CC BY 4.0, https://creativecommons.org/licenses/by/4.0/) and Microsoft PowerPoint

Comprehension Check 1 How does the cooperation between the NAc core and the basal ganglia facilitate the development of habits, and why is this process advantageous for achieving long-term goals?

The cooperation between the NAc core and the basal ganglia facilitates the development of habits by streamlining the execution of reinforcing activities into automatic behaviors. The NAc core works in conjunction with the dorsal striatum, a key component of the basal ganglia, to establish motor plans that allow individuals to perform habitual actions with minimal conscious effort. For instance, once a behavior, such as eating a favorite food, becomes habitual, individuals no longer need to consciously plan each step of the action. Instead, the basal ganglia ensure the behavior is learned and executed efficiently.

This process is advantageous for achieving long-term goals because it frees up cognitive resources for more complex tasks. By automating routine behaviors, such as waking up, eating, or commuting, individuals can focus their attention and planning on broader, long-term objectives, like career advancement or personal development. The habitual execution of daily tasks serves as a stable foundation, enabling people to prioritize and achieve their overarching life goals without being bogged down by repetitive, low-level decision-making processes.

4 Maladaptive Habits

Each day, there are multiple reinforcers that compete with one another. As much as someone may enjoy sleeping in, they may need to wake up early to go to work. Individuals that are interested in getting back into shape are faced with the competing reinforcers of eating a sweet treat and maintaining their fitness goals. Part of goal-directed behavior is knowing when we should engage in specific activities. If one is not scheduled for work, sleeping in is perfectly fine. If an individual is consistent with their workouts and meal plans, indulging in a sweet treat or having a "cheat meal" is healthy. Problems arise when individuals engage in maladaptive habits that interfere with normal daily functioning. Before discussing the neurobiology of maladaptive habits, I need to discuss the major types of habits that are problematic. I will then discuss a couple of psychiatric disorders that are characterized by repetitious actions that become maladaptive.

First, **perseveration** is the *involuntary* and persistent repetition of a thought, word, or action, often beyond a desired point. Perseveration is observed in multiple conditions you have read about in this book already, including schizophrenia and various forms of dementia. In the previous chapter, I mentioned stereotypies, a form of repetitious movements. There is an important distinction between perseveration and stereotypies. A perseverant behavior is one that enables an individual to accomplish a goal but continues to occur even *after* the goal has been met. In the case of an individual with schizophrenia, they may have the goal of discussing a specific topic with another individual. Even after the conversation has shifted to a different topic, the individual will continue to attempt to discuss the original topic. Stereotypies are repetitive behaviors that do not seemingly enable an individual to accomplish a goal. A common stereotypy observed in individuals with schizophrenia is rocking back and forth.

In contrast to perseveration and stereotypies, **compulsivity** is the tendency to engage in repetitive behaviors driven by an irresistible urge. Compulsive behaviors are performed to accomplish some goal, but they are problematic because they are performed excessively. An individual with schizophrenia may engage in repetitive hand washing because they have a fear of germs. Hand washing accomplishes the goal of removing potentially harmful bacteria/germs, but compulsive hand washing can be detrimental. If an individual excessively washes their hands, they can damage their skin. Compulsive behaviors also interfere with other tasks that need to be performed during the day.

Compulsivity is a defining feature of several psychiatric disorders. *The Diagnostic and Statistical Manual of Mental Disorders* (*DSM-5*) (American Psychiatric Association, 2013) and the *International Classification of Diseases (ICD-11)* (World Health Organization, 2022) have an entire section devoted to disorders characterized by compulsions. These disorders will now be discussed.

4.1 Obsessive–Compulsive and Related Disorders

Howard Hughes (1905–1976) was one of the wealthiest individuals in the world during his lifetime. In addition to producing 15 films and directing two films between 1927 and 1957, he was an aerospace engineer, a pilot, and a successful businessman. Sadly, Hughes is more known for his struggles with **obsessive–compulsive disorder (OCD)**, a mental disorder characterized by persistent, unwanted thoughts (*obsessions*) that lead to compulsions. Hughes' primary obsession was germs/contamination. This obsession drove behaviors that appeared strange to others, such as wearing tissue boxes on his feet and burning his clothing if he got too close to someone that was sick. He even had his staff follow strict routines to prevent potential contamination, such as writing a manual on how to open a can of peaches (Dittmann, 2005) and how to retrieve his hearing aid. To retrieve his hearing aid, staff had to:

- Use six to eight tissues to turn the knob on the bathroom door.
- Then use six to eight new tissues to open the bathroom cabinet and remove an unused bar of soap.
- Then wash hands.
- Use at least fifteen tissues to open the door to the cabinet containing the hearing aid.
- Remove the sealed envelope containing the hearing aid with two hands using another fifteen tissues in both hands. (OCD UK, n.d.)

Hughes' OCD became so severe that he often refused to leave his hotel room. In 1967, he famously bought the Desert Inn Hotel so he could continue staying in the penthouse suite on the ninth floor. Hughes eventually moved out of the Desert Inn Hotel, but he spent the rest of his life living in hotels. When he died, Hughes weighed just 90 pounds despite standing at 6 feet and 4 inches; his hair, beard, and fingernails/toenails were also considerably long as Hughes refused to trim his hair/nails during the latter years of his life.

In addition to compulsive hand washing, individuals living with OCD commonly engage in compulsive checking (e.g., repeatedly checking the oven to ensure it is off), ordering/arranging (e.g., putting items in a specific order), and counting (e.g., counting the number of steps taken while walking). Some of these compulsions may seem like they are performed in the absence of a goal, but they are often completed to alleviate stress. This is a form of negative reinforcement. Up until now, you have primarily learned about positive reinforcement; that is, we continue to engage in certain behaviors because we know we will receive something pleasant (e.g., we go to work because we know we will be paid). With negative reinforcement, we continue doing something because the action removes something aversive from our environment. The prototypic example I like to give is taking medicine. Taking something like acetaminophen (Tylenol) to alleviate a headache is not pleasurable *per se*. If acetaminophen is effective at reducing pain, you are more likely to use acetaminophen in the future when you have a headache because you know that this medication will remove something unpleasant (pain). Negative reinforcement will be a common theme as you read through this chapter.

In addition to OCD, the *DSM-5* and the *ICD-11* identify several OCD-related disorders. Like OCD, these disorders involve obsessions and compulsions that negatively impact an individual's life. Individuals with **body dysmorphic disorder (BDD)** obsess over their appearance, fixating on perceived flaws that are non-existent or barely noticeable by others. For example, someone with BDD may think that their nose is too big. This obsession leads to compulsive actions such as mirror checking (i.e., looking at oneself), excessive grooming, and reassurance seeking (Singh & Veale, 2019). Obsession over one's appearance can also lead to cosmetic surgery, with one study reporting that approximately 19% of individuals requesting cosmetic surgery have BDD (Salari et al., 2022).

There are two additional OCD-related disorders that are body-focused. **Trichotillomania** and **excoriation disorder** (also known as dermatillomania) involve compulsive hair pulling and skin picking, respectively. Unlike OCD and BDD, individuals living with trichotillomania or excoriation disorder do not experience obsessive thoughts. Instead, individuals seemingly engage in these actions without thinking about them. However, the behaviors tend to occur as a way to alleviate anxiety. These behaviors are maladaptive as individuals with trichotillomania needlessly experience hair loss, and individuals that engage in excoriation can develop lesions that can then become infected (Grant & Chamberlain, 2021).

The television series *Hoarders* shed light, albeit in a very sensationalized fashion, on **hoarding disorder**, a condition characterized by the obsessive fear of discarding items and the compulsion to keep all possessions. This behavior poses significant safety and health risks for the inhabitant and for visitors. Piles of items can fall, causing serious injury to an individual; they also act as significant fire hazards. Debris can attract vermin that can transmit certain infections.

4.2 *Autism Spectrum Disorder (ASD)*

Like attention-deficit/hyperactivity disorder (ADHD), **autism spectrum disorder (ASD)** is a neurodevelopment disorder. As a spectrum disorder, vast individual differences are observed in individuals diagnosed with autism. Some individuals diagnosed with autism have mild symptoms and can be highly independent and successful. Other individuals can have learning impairments and will need to receive care throughout their life. One defining feature of ASD is performance of limited/repetitive behaviors; individuals with ASD often become fixated on a certain object or activity (Jacob et al., 2009). You will learn more about ASD in Chap. 15 when I discuss aberrant social interactions, a defining feature of this condition.

4.3 *Addiction as a Disorder of Compulsivity*

In Chap. 3, you learned about one of my neighbors that lived with schizophrenia. In the apartment next to this individual lived a middle-aged couple (previously divorced but had reconciled) and their adolescent son. As my wife and I were not from Kentucky, we did not have family or friends in Lexington. This couple became surrogate parents in a sense. My wife and I spent considerable time in their apartment with another neighbor who lived on the first floor, playing Xbox, eating grilled foods, and drinking alcoholic beverages. The more time I spent with my neighbors, the more I noticed how much alcohol my neighbor "Travis" drank during the evening. Travis liked to drink Keystone Light, and he perpetually had a large Taco Bell cup filled with it. I later learned that Travis bought a case of Keystone Light every single day after work and consumed all 24 beers in a single night, often opening the first beer on his way home from work.

During one of our regular visits, Travis joked that he was a "functioning alcoholic". On the surface, Travis seemed to be a highly functioning adult. He had a good job as a machinist and did not seem to have any issues going to work. But beneath the surface, Travis' long-term alcohol use was more nefarious. After my wife and I moved away from Lexington, we heard that Travis had gotten into a physical altercation with his son, which caused his separation from his ex-wife (for a second time). More recently, I learned that Travis' ex-wife had moved back in with him to help care for him as his physical health has deteriorated.

Travis' story is not an isolated one, as there are approximately 29.5 million Americans that have had an alcohol use disorder during the past year (as of 2022) (Substance Abuse & Mental Health Services Administration, 2023). In Chap. 6, I provided a general definition of addiction and its categorization as a relapsing disorder. However, addiction is a complex biomedical disorder that involves more than relapse. Compulsion is another defining trait of addiction (Volkow & Fowler, 2000). *The DSM-5* states that individuals need to meet two or more of the following 11 criteria to be diagnosed with a **substance use disorder**:

1. The substance is often taken in larger amounts or over a longer period than was intended.
2. There is a persistent desire or unsuccessful efforts to cut down or control use of the substance.
3. A great deal of time is spent in activities necessary to obtain the substance, use the substance, or recover from its effects.
4. Craving, or a strong desire or urge to use the substance.
5. Recurrent use of the substance resulting in a failure to fulfill major role obligations at work, school, or home.
6. Continued use of the substance despite having persistent or recurrent social or interpersonal problems caused or exacerbated by the effects of its use.
7. Important social, occupational, or recreational activities are given up or reduced because of use of the substance.
8. Recurrent use of the substance in situations in which it is physically hazardous.
9. Use of the substance is continued despite knowledge of having a persistent or recurrent physical or psychological problem that is likely to have been caused or exacerbated by the substance.
10. [Development of] [t]olerance…
11. [Emergence of] [w]ithdrawal [symptoms upon cessation of substance use]… (American Psychiatric Association, 2013, pp. 577–578).

Notice that the term compulsion is not directly used, but points 2–9 tap into compulsivity. Despite negative consequences that can result from substance use, individuals continue to engage in this behavior. Initially, substance use tends to be positively reinforcing (there are some exceptions as individuals may not enjoy the first alcoholic beverage that they consume). Over time, individuals develop tolerance to the hedonic effects of the drug, which forces them to increase how much they use the substance (e.g., drinking more beer in one sitting or switching from beer to liquor). At this point, using the substance is not pleasurable. Instead, if the individual stops using the substance, they experience unpleasant withdrawal symptoms (in fact, severe alcohol withdrawal can be fatal). At this point, substance use is negatively reinforcing.

Similar to substance use disorders, behavioral addictions include compulsivity as a defining trait. Gambling disorder is the only behavioral addiction officially recognized by the *DSM-5*. Although an individual needs to meet two of 11 criteria to be diagnosed with a substance use disorder, at least four of nine criteria need to be met during a 12-month period before an individual can be diagnosed with a gambling disorder. The criteria are listed below:

1. Needs to gamble with increasing amounts of money in order to achieve the desired excitement.

2. Is restless or irritable when attempting to cut down or stop gambling.
3. Has made repeated unsuccessful efforts to control, cut back, or stop gambling.
4. Is often preoccupied with gambling…
5. Often gambles when feeling distressed…
6. After losing money gambling, often returns another day to get even ("chasing" one's losses).
7. Lies to conceal the extent of involvement with gambling.
8. Has jeopardized or lost a significant relationship, job, or education or career opportunity because of gambling.
9. Relies on others to provide money to relieve desperate financial situations caused by gambling (American Psychiatric Association, 2013, p. 585).

Comprehension Check 2 Which of the following points above indicates that gambling disorder is marked by compulsivity?

Points 2–4 and 8 are related to compulsive gambling behavior. Like substance use disorders, individuals have difficulty reducing their gambling (Point 3) and experience negative emotions when trying to reduce their gambling (Point 2). Individuals also become overly focused on gambling (Point 4) to the point that it interferes with normal daily functioning (Point 8).

In addition to recognizing gambling disorder, the *ICD-11* considers gaming disorder a behavioral addiction. As with substance use disorders and gambling disorder, gaming disorder is characterized by preoccupation with, impaired control over, and continuation of gaming despite negative consequences.

The *ICD-11* also recognizes "compulsive sexual behaviour disorder". Although classified as an impulse-control disorder (you'll learn more about these disorders in the next chapter), compulsive sexual behavior disorder, as the name implies, is characterized by repetitive urges to engage in sexual activities. The symptoms are highly similar to substance use disorders and behavioral addictions:

1. Sexual activities become a central focus of one's life, causing the individual to neglect other aspects of their life.
2. Unsuccessful efforts to reduce repetitive sexual behavior.
3. Continued sexual activity despite adverse outcomes and/or deriving little or no satisfaction from such activity.

5 Neurocircuitry of Maladaptive Habits

Given its role in habit formation, there is considerable interest in understanding how dysfunction of the basal ganglia, particularly the dorsal striatum, leads to maladaptive habits (Robbins et al., 2024). I will begin with a discussion of how aberrant signaling in the dorsal striatum contributes to maladaptive habit formation. I will then discuss

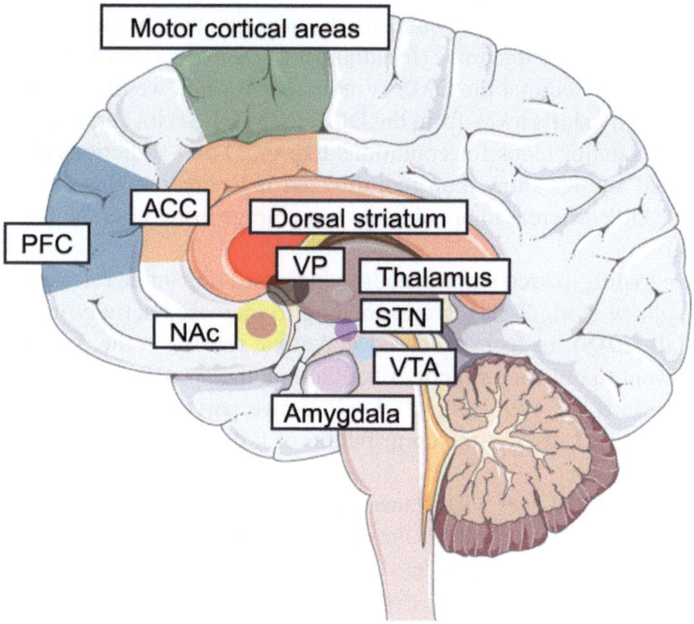

Fig. 2 Schematic showing the brain regions implicated in maladaptive habit formation. *Note* ACC = anterior cingulate cortex; NAc = nucleus accumbens; PFC = prefrontal cortex; STN = subthalamic nucleus; VP = ventral pallidum; VTA = ventral tegmental area. Image created by the author using Servier Medical Art (https://smart.servier.com/; CC BY 4.0, https://creativecommons.org/licenses/by/4.0/) and Microsoft PowerPoint

the contribution of additional circuits to such behavior. Figure 2 shows the brain regions heavily implicated in compulsivity.

5.1 Dysfunction of the Dorsal Striatum

A hyperactive basal ganglia system is commonly observed in individuals that display compulsive behaviors (Huey et al., 2008; Koob & Volkow, 2010). More specifically, a hyperactive dopaminergic system in the basal ganglia is proposed to be a major mediator of conditions like OCD (Jalal et al., 2023). Recall the direct, the indirect, and the hyperdirect pathways of the basal ganglia. While the direct pathway stimulates purposeful movements, the indirect pathway inhibits actions that interfere with purposeful movement, and the hyperdirect pathway stops ongoing actions. Traditional models have proposed that compulsivity results from an imbalance between the direct and the indirect pathways (see Huey et al., 2008; Yager et al., 2015 for discussions). The direct pathway becomes hyperactive, stimulating compulsive behavior.

A hyperactive direct pathway accounts for the finding that individuals with OCD and ASD have a hyperactive thalamus (Friedlander & Desrocher, 2006; Jin et al., 2024).

Earlier, I mentioned that the DMS is more active when we first learn a behavior. Over time, activity shifts away from the DMS to the DLS. Not surprisingly, the DLS is considered a major locus for conditions like OCD and addictions (Lipton et al., 2019). Animal studies show that goal-directed behavior remains intact following lesions to the DLS whereas habitual behaviors emerge faster after the DMS has been inactivated.

Exposure to drug-paired stimuli leads to increased dopamine release in the dorsal striatum (Volkow et al., 2007), corresponding to increased activity in the thalamus (Huang et al., 2018). In fact, increased dopamine release in the dorsal striatum is directly relevant to one major theory of addiction: the *incentive sensitization theory* (Robinson & Berridge, 1993) (Fig. 3). I will discuss part of the theory here, covering the rest of it when I discuss the contribution of a dysregulated mesocorticolimbic pathway to addiction.

When an individual uses a substance such as cocaine or heroin for the first time, there is a spike in dopamine levels in the mesocorticolimbic pathway. As this behavior occurs more frequently, there is less dopaminergic activity in this pathway. This

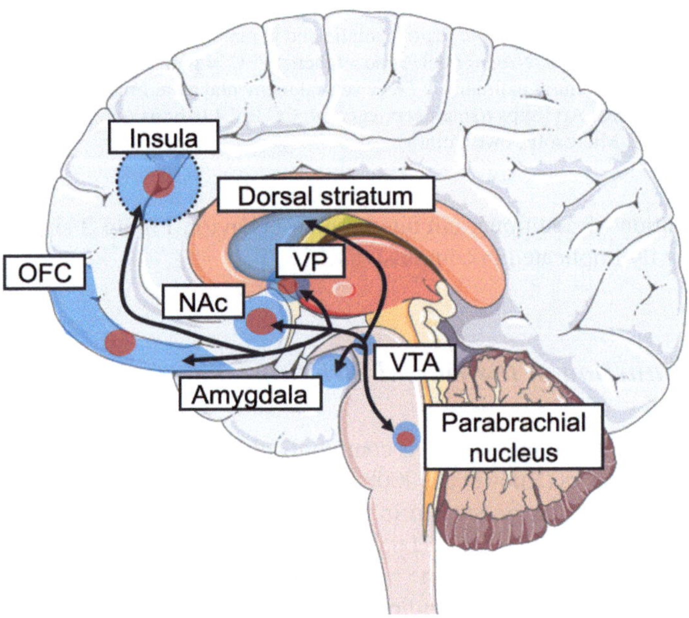

Fig. 3 The neural circuitry associated with drug "liking" (red circles) and drug "wanting" (regions in blue). Dopaminergic neurons originating from the ventral tegmental area (VTA) project to multiple brain regions. Over time, the wanting system sensitizes (i.e., becomes stronger) whereas drug liking decreases. *Note* NAc = nucleus accumbens; OFC = orbitofrontal cortex; VP = ventral pallidum. Image created by the author using Servier Medical Art (https://smart.servier.com/; CC BY 4.0, https://creativecommons.org/licenses/by/4.0/) and Microsoft PowerPoint

corresponds to the decreased pleasure individuals experience following substance use (i.e., tolerance). However, due to Pavlovian conditioning, stimuli become paired with the drug-taking experience. For example, if an individual always uses heroin in their apartment bathroom, this room and any specific drug paraphernalia that are used to take heroin become associated with heroin itself. The individual will begin to attribute incentive salience to these stimuli. When the individual encounters drug-paired stimuli, this leads to increased activity in the hippocampus and the PFC. The dorsal striatum then becomes highly activated due to excitatory inputs from the PFC. This can trigger reflexive-like behaviors that are directly related to substance use and can lead to intense drug craving.

Certain drugs, particularly psychostimulants like cocaine and methamphetamine, produce stereotypies. A common stereotypy associated with long-term substance use is excoriation. Psychostimulants also produce stereotypies in animals, such as excess head bobbing/swaying, sniffing, and biting. Drug-induced stereotypies sensitize over time (i.e., get worse), most likely due to overactivity of dopaminergic signaling in the dorsal striatum. Interestingly, animal research has demonstrated that administration of either dopamine D_1-like or D_2-like receptor agonists produces stereotypies (Charntikov et al., 2011), meaning that drug-induced stereotypies can result from an imbalance of D_1-like and D_2-like receptor activation. There is recent evidence that a hyperactive indirect pathway is involved in stereotypic grooming. Read the Experiment Spotlight below to learn more.

Experiment Spotlight: Hyperactivity of Indirect Pathway-Projecting Spiny Projection Neurons Promotes Compulsive Behavior

The full article can be accessed with this link: https://www.nature.com/articles/s41467-024-48331-z

The present study (Piantadosi et al., 2024) focused on the contribution of D_1-like and D_2-like neurons in the dorsal striatum (referred to as the central striatum in this paper) in compulsive grooming behavior in mice. The researchers used a genetic knockout mouse model that targets the SAPAP3 protein. SAPAP3 stands for synapse-associated protein 90/postsynaptic density protein 95-associated protein 3 (that's a mouthful). The previous chapter's Experiment Spotlight focused on the postsynaptic density (PSD). The image of the PSD that was included in the previous chapter depicted several scaffolding proteins. One of these proteins was labeled as GKAP (guanylate kinase-associated protein). SAPAP3 and GKAP are synonyms. In short, these mice lack an important scaffolding protein, which interferes with normal neuronal communication.

In the first experiment, SAPAP3-knockout mice and wildtype mice received dorsal striatal injections of GCaMP6m, which allowed the researchers to measure intracellular calcium (Ca^{++}) levels. This is used to infer neuronal activity (i.e., increased intracellular Ca^{++} concentrations are interpreted as increased neuronal activity). SAPAP3-knockout mice have greater Ca^{++} levels in the dorsal striatum when grooming compared to wildtype mice, an effect that is not observed when grooming is not occurring.

To isolate which neuron types are involved in compulsive grooming, the researchers used different types of transgenic mice. SAPAP3-knockout mice and wildtype controls were bred with one of two types of transgenic mice (either D1-Cre or A2a-Cre). This is important because it allows the researchers to isolate D_1-like neurons from D_2-like neurons. While Ca^{++} levels did not increase in D_1-like neurons during compulsive grooming, they did significantly increase in D_2-like neurons.

Next, the researchers performed an optogenetic experiment to determine if direct inhibition of D_2-like neurons projecting from the dorsal striatum to the globus pallidus reduces compulsive grooming behavior. Because the dorsal striatum has multiple projections, the researchers had to infuse a *retrograde virus* into the globus pallidus. This virus gets taken back up into the presynaptic neuron and travels to the dorsal striatum. The researchers also injected a virus containing a light-sensitive ion channel into the dorsal striatum. Optogenetic inhibition of the striatopallidal pathway reduced grooming behavior.

Finally, the researchers found that administering a medication that has been used to treat OCD reduces hyperactivity of D_2-like neurons in the striatopallidal pathway.

Questions to Consider:

1. How does the SAPAP3-knockout model contribute to our understanding of compulsive behaviors observed in disorders like OCD? Provide specific examples from the highlighted experiments.
2. Why is it significant that Ca^{++} levels increased specifically in D_2-like neurons during compulsive grooming in SAPAP3-knockout mice? What might this suggest about the role of the indirect pathway in compulsive behaviors?
3. Discuss the importance of optogenetic inhibition in demonstrating the causal role of the striatopallidal pathway in compulsive grooming. How might this approach guide future treatments for OCD?

In fact, one common method for inducing compulsive-like behaviors in rodents is to repeatedly administer a drug called *quinpirole* (Stuchlik et al., 2016). This drug acts as a mixed dopamine D_2 and D_3 receptor agonist. Repeated administration of quinpirole leads to a reduction in dopamine D_2-like receptors in the striatum (Chen et al., 1993). Because there are fewer dopamine D_2-like receptors, certain motor actions cannot be readily inhibited, resulting in habitual actions.

5.2 Mesocorticolimbic Pathway and Cortio-Striatal Dysfunction

One common feature of disorders characterized by compulsivity is that individuals do not engage in repetitive behaviors because they enjoy doing so. In OCD and other compulsive-like disorders, individuals often engage in repetitive actions to alleviate anxiety (i.e., the behavior is negatively reinforcing). Returning to the incentive sensitization theory of addiction, when an individual uses a substance for the first time, they "like" the experience (i.e., it is positively reinforcing). Over time, substance use becomes less positively reinforcing. Recall, the VTA responds less to reinforcers as we experience them more often. If the person stops using the substance, they experience unpleasant withdrawal symptoms. Additionally, when the individual sees drug-paired stimuli, there is increased VTA activity, causing increased activation of the NAc. The individual is highly motivated to use the substance to escape withdrawal symptoms. Now, substance use is negatively reinforcing. The individual no longer uses the substance because they like it. Instead, use continues because the individual "wants" to keep using. Think of wanting as a form of *needing* to do something.

As I mentioned earlier, the PFC is important for controlling when we should seek out certain reinforcers. The PFC has a direct connection with the NAc, modulating dopaminergic activity in this region (Hill et al., 2018). Going back to my love for tiramisu—if I see a slice of tiramisu cake, activity between the VTA and the NAc increases, and I will be highly tempted to eat the cake. However, activity in the PFC also increases. If I have already eaten a piece or two, the PFC will act to "override" the activity occurring in the mesolimbic pathway. Just because an activity is reinforcing, this does not mean we should do it constantly.

Dysfunction of the PFC is strongly linked to the development of maladaptive repetitive actions. Damage to the PFC, especially the orbitofrontal cortex (OFC), leads to increased perseveration (Bari & Robbins, 2013). This should not be surprising as I mentioned earlier in this chapter that perseveration is commonly observed in individuals with dementia; recall from Chap. 6 that the various forms of dementia are often associated with damage to frontal cortical areas. The OFC is involved in making predictions about the consequences of one's actions and for updating the association between stimuli and reinforcers. For example, we can train a rodent to respond on one lever to earn a food pellet. The rodent will learn the association between the lever and the food and will increase the frequency at which it responds on the lever. We can then switch which lever is associated with food. After the rodent responds on the original lever a few times, it should learn that these responses are futile. If the rodent presses the other lever and earns food, it will quickly learn to begin pressing this new lever. When the OFC is damaged in these animals, they show increased perseveration; that is, they continue to respond on the original lever even though it no longer leads to reinforcement.

In contrast to conditions like dementia, individuals with OCD have a *hyperactive* OFC (Maia et al., 2008). While these individuals have increased functional connectivity between the OFC and the dorsal striatum, they have decreased functional connectivity between the NAc and the OFC and between the NAc and the medial PFC (mPFC) (Chen et al., 2021; Liu et al., 2022). Imagine an individual with OCD that engages in compulsive hand washing that is related to an obsession of germs. When this individual washes their hands, there is insufficient activation of the NAc and parts of the PFC. This action is negatively reinforcing, but the problem is that the individual does not receive the full reinforcement that they should. Therefore, the individual quickly begins experiencing negative feelings that they are "contaminated" and need to wash their hands again. Because the dorsal striatum is overactive, this leads to a vicious cycle of compulsive hand washing.

Not all forms of compulsivity are associated with decreased functional connectivity between the NAc and the PFC. Individuals with heroin use disorder show *increased* functional connectivity between the NAc and the OFC (Ma et al., 2010), and craving for alcohol is associated with increased functional connectivity between these two brain regions (Bracht et al., 2021). Remember, the NAc shell is important for assigning incentive salience to stimuli that become paired with a reinforcer. Individuals with an addiction attribute high incentive salience to drug-paired stimuli. As such, when someone with heroin use disorder encounters a syringe, there is increased activity of the mesolimbic pathway, which coincides with intense craving to use heroin. Furthermore, even though functional connectivity between the NAc and the OFC is higher in individuals with a substance use disorder, they have a hypoactive PFC compared to healthy controls (Crews & Boettiger, 2009). Essentially, when an individual encounters a cue that has been consistently paired with substance use, there is a concomitant increase in activity of the NAc and the PFC. However, the PFC is not sufficiently activated to exert control over the NAc, which contributes to compulsive-like drug seeking.

Comprehension Check 3 How do the differing roles of the dorsal striatum and the PFC contribute to the development of compulsive behaviors, and why might dysfunction of these areas lead to contrasting outcomes in disorders such as OCD and substance use disorders?

The dorsal striatum, particularly when overactive, contributes to the repetitive nature of compulsive actions. For instance, in individuals with OCD, hyperactivity in part of the PFC, particularly the OFC, leads to heightened functional connectivity with the dorsal striatum, perpetuating cycles of compulsive behaviors such as hand washing. This overactivity essentially overrides normal decision-making and reinforcement processes associated with the mesocorticolimbic pathway, trapping the individual in a loop of negative reinforcement.

Hypoactivity of the PFC, on the other hand, contributes to the etiology of substance use disorders. The PFC is responsible for overriding excessive mesolimbic pathway activity, regulating responses to drug-paired stimuli. In substance use disorders, while the NAc and the OFC show increased functional connectivity, the PFC fails to exert sufficient control. This dysfunction allows the mesolimbic pathway to dominate, driving compulsive drug-seeking behaviors.

Beyond dopamine, dysregulation of several neurotransmitter systems in the mesocorticolimbic pathway is proposed to underlie compulsive behaviors. First, low serotonergic levels are believed to contribute to conditions like OCD as drugs that increase serotonin levels improve symptoms of OCD (Fineberg et al., 2012). Also, depleting prefrontal serotonin levels increases perseveration in animals (Clarke et al., 2004). Although drugs that increase serotonin levels ameliorate symptoms of obsessive–compulsive-like disorders, research has not consistently found alterations in the serotonergic system in individuals with OCD (Goodman et al., 2021).

Despite the equivocal findings concerning serotonin and OCD, there is evidence that exposure to drugs like cocaine causes alterations to the serotoninergic system in the OFC. Specifically, 5-HT_{2A} receptor mRNA expression significantly increases following cocaine administration in rats while 5-HT_{1A} receptor mRNA expression is unchanged (Wright et al., 2017). 5-HT_{1A} receptors are inhibitory whereas 5-HT_{2A} receptors are excitatory, causing increased glutamate release (I will discuss glutamate's contribution to maladaptive habits below). These results provide an account for the increased functional connectivity observed between the OFC and the NAc observed in individuals with a substance use disorder.

Aberrant glutamatergic signaling in the mesocorticolimbic pathway is also observed in disorders characterized by compulsivity (Goodman et al., 2021), and glutamatergic dysfunction in cortico-striatal circuits is observed in individuals with OCD; indeed, one hypothesis is that medications that elevate serotonin levels ameliorate symptoms of OCD by regulating glutamate levels in the brain (Jalal et al., 2023).

The contribution of glutamate to addiction has gained considerable traction since the 2000s. The **glutamate homeostasis hypothesis** of addiction posits that a loss in glutamate homeostasis in the mesocorticolimbic pathway leads to compulsive drug seeking (Kalivas, 2009). In Chap. 6, I discussed the importance of astrocytes to learning and memory. Glutamate transporters (GLTs), which remove excess glutamate from the synapse, are located on astrocytes. GLTs limit the amount of glutamate that can bind to mGluRs located on both presynaptic and postsynaptic neurons. The glutamate that is left binds to AMPA (α-amino-3-hydroxy-5-methyl-4-isoxazolepropionic acid) and NMDA (N-methyl-D-asparate) receptors that are located on the postsynaptic neuron. In addition to GLTs, astrocytes contain the *cysteine/glutamate transporter*, which trades cysteine, a non-essential amino acid, for glutamate.

GLTs and the cysteine/glutamate transporter are important as they help keep glutamate levels in check. Too much excitation can lead to neuronal damage, so multiple mechanisms exist to prevent excess glutamate levels in the synapse. With repeated substance use, glutamate homeostasis becomes dysregulated. If an individual stops using a substance, glutamate levels increase substantially as the mechanisms used to regulate glutamate homeostasis have become impaired. Increased glutamate leads to enhanced stimulation of ionotropic glutamate receptors, resulting in enhanced long-term potentiation (LTP). When individuals are exposed to drug-paired stimuli, even more glutamate is released, and increased LTP occurs. This, in turn, further strengthens the relationship between the drug-paired stimulus and the substance.

Comprehension Check 4 Even though serotoninergic alterations in conditions like OCD are widely mixed, mediations that increase serotonin levels ameliorate symptoms of OCD. What are a couple of potential mechanisms that can account for symptom improvement after increasing serotonin levels?

Serotonin receptors are located on dopaminergic, GABAergic (gamma-aminobutyric acid), and glutamatergic neurons, meaning that an influx of serotonin can modulate activity of these neurons. For example, presynaptic serotonin receptors located on glutamatergic neurons can reduce the further release of glutamate, thus controlling overexcitation that is observed in obsessive–compulsive-like disorders.

5.3 *Limbic Dysfunction*

The NAc is not the only limbic structure involved in maladaptive habit formation. The amygdala and the anterior cingulate cortex (ACC) also drive the switch from goal-directed behavior to compulsivity.

5.3.1 Amygdala

The amygdala, as you learned in Chap. 6, is involved in processing emotionally salient memories and for fear conditioning. The amygdala also allows us to learn the association between stimuli and rewards (Murray et al., 2009). The basolateral nucleus of the amygdala (BLA) sends information to the DMS while the central nucleus of the amygdala (CeA) projects neurons to the DLS.

Let's return to the example of a rodent learning to press a lever to earn food. When a light above the lever turns on, this indicates that responding on the lever will lead to food. The rodent is going to learn a couple of things. First, pressing the lever results in something positive. Second, pressing the lever only works when the light is on. The BLA is important for encoding information about the association between the light and reinforcement. Over time, the light becomes a *conditioned reinforcer*. That is, animals will eventually press the lever to gain access to the light's illumination even when no food is available. In real life, the prototypic conditioned reinforcer is money. On its own, money has no value. It has value because we associate it with things like food and entertainment.

The CeA is important for transmitting information to the DLS about the relationship between pressing the lever and receipt of food. Imagine if the rodent pressed the lever and then received a foot shock. The rodent would need to update its understanding about what happens when it presses the lever.

Individuals with OCD have decreased volume of the left BLA and increased volume of the right CeA, and they show aberrant functional connectivity between amygdalar subregions and other parts of the brain (Cao et al., 2022b). Specifically, greater activity is observed between the CeA and other regions such as the supplementary motor area (SMA), the cingulate cortex, and the superior temporal gyrus (STG); however, in healthy controls, the BLA shows greater connectivity with these regions compared to the CeA.

Expanding on the results above, individuals with early-onset OCD have increased functional connectivity between the BLA and some regions like the cerebellum, but they show decreased functional connectivity between the BLA and the OFC;

conversely, late-onset OCD is associated with decreased functional connectivity between the BLA and the striatum and increased functional connectivity between the CeA and other parts of the brain (Cao et al., 2022a). Therefore, too much activity in either nucleus of the amygdala is implicated in conditions like OCD, but as the disorder progresses, there appears to be a shift away from BLA activity to CeA activity, mirroring the transition from DMS control of goal-directed behavior to DLS control of habitual actions. Animal research supports this hypothesis as optogenetic stimulation of the BLA-DMS pathway increases compulsive-like behavior in animals (Lee et al., 2024), and hyperactive functional connectivity between the BLA and the mPFC is linked to compulsive-like behavior in mice (Sun et al., 2019).

5.3.2 Anterior Cingulate Cortex (ACC)

You have already learned that the ACC processes the emotional component of pain (Chap. 4) and is involved in attentional processing (Chap. 5). The ACC can be further divided into the dorsal ACC (dACC) and the rostral ACC (rACC). The dACC is involved in goal-directed behavior. More specifically, the dACC helps process reward information, allocates mental resources to accomplish specific tasks, and engages in error processing (e.g., adjusting future behavior if a previous action resulted in a negative outcome). The rACC is also important for reward processing, but it is more involved in processing and regulating emotions. Dysfunction of both dACC and rACC leads to problematic habits.

A hyperactive dACC is observed in individuals with OCD (McGovern & Sheth, 2017), with hyperconnectivity between the ACC and the dorsal striatum being observed in these individuals (Liu et al., 2022). Furthermore, individuals with OCD and ASD have increased glutamate in the ACC (Naaijen et al., 2017). As argued by McGovern and Sheth (2017), a hyperactive dACC leads to "mis-specification" of control signals. They provide an example related to exposure to a contaminant in the environment. Normally, when we are exposed to a contaminant or a potential contaminant, the dACC becomes active, generating a feeling of aversion. The dACC then relays information to other parts of the brain such as the dorsolateral PFC (dlC), the dorsal striatum, and motor cortical areas to generate actions aimed at removing the contaminant (e.g., hand washing, showering, etc.). Once these behaviors occur, the dACC returns to its baseline activity levels. Yet, in OCD and obsessive–compulsive-like disorders, this return to baseline does not occur. The dACC continues to send signals to other brain regions, which further drives compulsive behaviors.

Altered ACC is also implicated in addiction-associated compulsivity. Individuals with a substance use disorder have lower expression of dopamine D_2-like receptors in the dorsal striatum, which leads to lower baseline activity of the ACC and other prefrontal cortical areas (Volkow et al., 2019), which may account for other deficits in cognitive control beyond compulsivity. Although baseline activity levels are lower in regions like the ACC, individuals experiencing drug craving show hyperactivity of the ACC and the ventral striatum, with the magnitude of activity being directly correlated with craving (Zhao et al., 2020).

5.4 Ventral Pallidum (VP)

I mentioned the ventral pallidum (VP) in Chap. 8. Making up the bottom portion of the internal capsule of the globus pallidus, the VP is unique in that it is part of the basal ganglia but has functions that more closely resemble the NAc. In fact, the VP, along with the NAc and part of the amygdala, are major constituents of the basal forebrain. The VP receives information from brain regions such as the mPFC, the VTA, the NAc, the subthalamic nucleus (STN), and the BLA, and it is involved in different types of unconscious and conscious motor responses (Root et al., 2015). Specifically, the VP is inhibited by the NAc via GABAergic inputs but receives glutamatergic inputs from the mPFC, the STN, and the BLA; the VTA projects dopaminergic neurons to the VP.

Blocking intra-VP GABAergic neurons induces compulsive-like behaviors in primates (Saga et al., 2022) and produces stereotypies in rodents; conversely, directly inhibiting the VP blocks relapse-like behavior in animals, thus indicating decreased compulsive drug seeking (Root et al., 2015). These results suggest that hyperactivity of the VP is involved in the development of compulsive-like behaviors. One potential reason is that the VP sends both glutamatergic and GABAergic neurons to the VTA, meaning that hyperactivity of the VP can cause dysregulation of dopaminergic signals emanating from the VTA.

Comprehension Check 5 How do interactions between limbic structures and the VP drive compulsive-like behaviors?

Interactions between limbic structures and the VP play a significant role in driving compulsive-like behaviors by influencing the balance of excitatory and inhibitory signals within key neural circuits. The VP is uniquely positioned, receiving input from regions such as the amygdala and the NAc, which are heavily involved in emotional processing and reward. Hyperactivity in the VP, often due to dysregulated signaling from these limbic regions, can disrupt the normal flow of dopaminergic, glutamatergic, and GABAergic inputs. This hyperactivity leads to an imbalance in the VP's output to the VTA, altering the dopaminergic signaling crucial for reward-based learning and decision-making. Such disruptions result in heightened compulsivity, as the VP's excessive or dysregulated activity reinforces repetitive, habitual actions, bypassing goal-directed behavior. In short, the VP's role as a hub for integrating signals from limbic structures directly impacts motor responses and reward pathways, underscoring its critical involvement in the manifestation of compulsive-like behaviors.

6 Chapter Summary

Habits streamline daily tasks, enabling us to focus on long-term goals, but maladaptive habit formation is a defining feature of several psychiatric disorders like OCD and addiction. The present chapter delved into the neural mechanisms underlying maladaptive habit formation, including dysfunction in dopamine signaling in the CBGTC loop and the mesocorticolimbic pathway, loss of glutamate homeostasis, and altered activity in regions like the dorsal striatum, prefrontal cortical areas, the amygdala, and the VP. In short, hyperactivity of the CBGTC loop and limbic regions can lead to increased compulsivity.

Answers to Experiment Spotlight Questions

1. By knocking out the SAPAP3 protein, which is vital for maintaining the integrity of the PSD, animals display repetitive, compulsive grooming behaviors similar to OCD. The experiments showed that SAPAP3-knockout mice displayed elevated neuronal activity (measured through Ca^{++} levels) in the dorsal striatum specifically during grooming, revealing the hyperactivity of this brain region as a key factor in compulsive actions. This model not only sheds light on the role of specific neural pathways but also provides a platform for testing therapies aimed at reducing such behaviors.
2. The finding that Ca^{++} levels increased specifically in D_2-like receptor-containing neurons during compulsive grooming in SAPAP3 knockout mice is significant because it points to hyperactivity in the indirect pathway of the dorsal striatum. This finding suggests that compulsive behaviors may be driven by dysregulation in the balance of signaling between the direct and the indirect pathways. The hyperactivity in D_2-like neurons implicates the indirect pathway as a critical contributor to the persistence of compulsive grooming behaviors, providing a target for future therapeutic interventions.
3. Optogenetic inhibition of the striatopallidal pathway was pivotal in establishing the causal relationship between this pathway and compulsive grooming behaviors. By using a retrograde virus to trace projections from the globus pallidus to the dorsal striatum and introducing a light-sensitive ion channel into these neurons, researchers were able to selectively inhibit D_2-like neurons. The observed reduction in grooming behavior following this inhibition confirmed the functional role of this pathway in driving compulsive actions. This approach highlights the utility of optogenetics in isolating and manipulating specific neural circuits, offering a precise method to test interventions that could mitigate compulsive behaviors associated with OCD.

   Collectively, the results of this study indicate that reducing hyperactivity in D_2-like neurons or directly targeting the striatopallidal pathway may alleviate symptoms of OCD. The success of optogenetic techniques in modulating neural activity also opens the door for advanced neuromodulation therapies, such as transcranial magnetic stimulation, to be used with individuals diagnosed with a condition like OCD.

Glossary

Autism spectrum disorder (ASD) a neurodevelopmental disorder characterized by persistent challenges with social communication and interaction, and restricted, repetitive patterns of behavior, interests, or activities.

Body dysmorphic disorder (BDD) mental health condition characterized by an obsessive preoccupation with perceived flaws in one's appearance, often leading to significant distress and the emergence of compulsive actions such as viewing oneself in the mirror or engaging in grooming behaviors.

Compulsivity tendency to engage in repetitive behaviors driven by an irresistible urge, often leading to negative consequences.

Excoriation disorder mental health condition characterized by compulsive skin picking.

Glutamate homeostasis hypothesis the hypothesis that disruptions in the tightly regulated balance of glutamate levels in the brain, particularly between synaptic and extrasynaptic spaces, contribute to the pathophysiology of various neurological and psychiatric disorders, including addiction.

Goal-directed behavior purposeful action taken to achieve a specific objective or desired outcome.

Hoarding disorder mental health condition characterized by the persistent difficulty of parting with possessions, often leading to excessive accumulation of items that are not needed or for which there is no space.

Incentive salience a form of attention that motivates us to approach some object or perceived outcome.

Obsessive–compulsive disorder (OCD) mental health condition characterized by persistent, unwanted thoughts that lead to repetitive behaviors.

Perseveration involuntary and persistent repetition of a thought, word, or action, often beyond a desired point.

Substance use disorder a medical condition characterized by a problematic pattern of using a substance (such as alcohol, drugs, or medications) that leads to clinically significant impairment or distress, despite harmful consequences.

Trichotillomania mental health condition characterized by the compulsive urge to pull one's hair out.

References

American Psychiatric Association. (2013). *Diagnostic and statistical manual of mental disorders* (5th ed.). https://doi.org/10.1176/appi.books.9780890425596

Bari, A., & Robbins, T. W. (2013). Inhibition and impulsivity: Behavioral and neural basis of response control. *Progress in Neurobiology, 108*, 44–79. https://doi.org/10.1016/j.pneurobio.2013.06.005

Bracht, T., Soravia, L., Moggi, F., Stein, M., Grieder, M., Federspiel, A., Tschümperlin, R., Batschelet, H. M., Wiest, R., & Denier, N. (2021). The role of the orbitofrontal cortex and the nucleus accumbens for craving in alcohol use disorder. *Translational Psychiatry, 11*(1), 267. https://doi.org/10.1038/s41398-021-01384-w

Cao, L., Li, H., Hu, X., Liu, J., Gao, Y., Liang, K., Zhang, L., Hu, X., Bu, X., Lu, L., Wang, Y., Tang, S., Gong, Q., & Huang, X. (2022a). Distinct alterations of amygdala subregional functional connectivity in early- and late-onset obsessive-compulsive disorder. *Journal of Affective Disorders, 298*(Pt A), 421–430. https://doi.org/10.1016/j.jad.2021.11.005

Cao, L., Li, H., Liu, J., Jiang, J., Li, B., Li, X., Zhang, S., Gao, Y., Liang, K., Hu, X., Bao, W., Qiu, H., Lu, L., Zhang, L., Hu, X., Gong, Q., & Huang, X. (2022b). Disorganized functional architecture of amygdala subregional networks in obsessive-compulsive disorder. *Communications Biology, 5*(1), 1184. https://doi.org/10.1038/s42003-022-04115-z

Charntikov, S., Der-Ghazarian, T., Herbert, M. S., Horn, L. R., Widarma, C. B., Gutierrez, A., Varela, F. A., & McDougall, S. A. (2011). Importance of D1 and D2 receptors in the dorsal caudate-putamen for the locomotor activity and stereotyped behaviors of preweanling rats. *Neuroscience, 183*, 121–133. https://doi.org/10.1016/j.neuroscience.2011.03.037

Chen, J. F., Aloyo, V. J., & Weiss, B. (1993). Continuous treatment with the D2 dopamine receptor agonist quinpirole decreases D2 dopamine receptors, D2 dopamine receptor messenger RNA and proenkephalin messenger RNA, and increases mu opioid receptors in mouse striatum. *Neuroscience, 54*(3), 669–680. https://doi.org/10.1016/0306-4522(93)90238-b

Chen, Y., Ou, Y., Lv, D., Ma, J., Zhan, C., Yang, R., Jia, C., Shang, T., Sun, L., Wang, Y., Sun, Z., Zhang, G., Wang, X., Guo, W., & Li, P. (2021). Decreased nucleus accumbens connectivity at rest in medication-free patients with obsessive-compulsive disorder. *Neural Plasticity, 2021*, 9966378. https://doi.org/10.1155/2021/9966378

Clarke, H. F., Dalley, J. W., Crofts, H. S., Robbins, T. W., & Roberts, A. C. (2004). Cognitive inflexibility after prefrontal serotonin depletion. *Science, 304*(5672), 878–880. https://doi.org/10.1126/science.1094987

Crews, F. T., & Boettiger, C. A. (2009). Impulsivity, frontal lobes and risk for addiction. *Pharmacology, Biochemistry, and Behavior, 93*(3), 237–247. https://doi.org/10.1016/j.pbb.2009.04.018

Dittmann, M. (2005, July 1). Hughes's germ phobia revealed in psychological autopsy. *Monitor on Psychology, 36*(7). https://www.apa.org/monitor/julaug05/hughes

Fineberg, N. A., Brown, A., Reghunandanan, S., & Pampaloni, I. (2012). Evidence-based pharmacotherapy of obsessive-compulsive disorder. *The International Journal of Neuropsychopharmacology, 15*(8), 1173–1191. https://doi.org/10.1017/S1461145711001829

Friedlander, L., & Desrocher, M. (2006). Neuroimaging studies of obsessive-compulsive disorder in adults and children. *Clinical Psychology Review, 26*(1), 32–49. https://doi.org/10.1016/j.cpr.2005.06.010

Goodman, W. K., Storch, E. A., & Sheth, S. A. (2021). Harmonizing the neurobiology and treatment of obsessive-compulsive disorder. *The American Journal of Psychiatry, 178*(1), 17–29. https://doi.org/10.1176/appi.ajp.2020.20111601

Grant, J. E., & Chamberlain, S. R. (2021). Trichotillomania and skin-picking disorder: An update. *Focus, 19*(4), 405–412. https://doi.org/10.1176/appi.focus.20210013

Hill, D. F., Parent, K. L., Atcherley, C. W., Cowen, S. L., & Heien, M. L. (2018). Differential release of dopamine in the nucleus accumbens evoked by low-versus high-frequency medial prefrontal cortex stimulation. *Brain Stimulation, 11*(2), 426–434. https://doi.org/10.1016/j.brs.2017.11.010

Huang, A. S., Mitchell, J. A., Haber, S. N., Alia-Klein, N., & Goldstein, R. Z. (2018). The thalamus in drug addiction: From rodents to humans. *Philosophical Transactions of the Royal Society of London. Series B, Biological Sciences, 373*(1742), 20170028. https://doi.org/10.1098/rstb.2017.0028

Huey, E. D., Zahn, R., Krueger, F., Moll, J., Kapogiannis, D., Wassermann, E. M., & Grafman, J. (2008). A psychological and neuroanatomical model of obsessive-compulsive disorder. *The Journal of Neuropsychiatry and Clinical Neurosciences, 20*(4), 390–408. https://doi.org/10.1176/jnp.2008.20.4.390

Jacob, S., Landeros-Weisenberger, A., & Leckman, J. F. (2009). Autism spectrum and obsessive-compulsive disorders: OC behaviors, phenotypes and genetics. *Autism Research, 2*(6), 293–311. https://doi.org/10.1002/aur.108

Jalal, B., Chamberlain, S. R., & Sahakian, B. J. (2023). Obsessive-compulsive disorder: Etiology, neuropathology, and cognitive dysfunction. *Brain and Behavior, 13*(6), Article e3000. https://doi.org/10.1002/brb3.3000

Jin, X., Zhang, K., Lu, B., Li, X., Yan, C. G., Du, Y., Liu, Y., Lu, J., Luo, X., Gao, X., & Liu, J. (2024). Shared atypical spontaneous brain activity pattern in early onset schizophrenia and autism spectrum disorders: Evidence from cortical surface-based analysis. *European Child & Adolescent Psychiatry, 33*(7), 2387–2396. https://doi.org/10.1007/s00787-023-02333-2

Kalivas, P. W. (2009). The glutamate homeostasis hypothesis of addiction. *Nature Reviews. Neuroscience, 10*(8), 561–572. https://doi.org/10.1038/nrn2515

Koob, G. F., & Volkow, N. D. (2010). Neurocircuitry of addiction. *Neuropsychopharmacology, 35*(1), 217–238. https://doi.org/10.1038/npp.2009.110

Lee, I. B., Lee, E., Han, N. E., Slavuj, M., Hwang, J. W., Lee, A., Sun, T., Jeong, Y., Baik, J. H., Park, J. Y., Choi, S. Y., Kwag, J., & Yoon, B. J. (2024). Persistent enhancement of basolateral amygdala-dorsomedial striatum synapses causes compulsive-like behaviors in mice. *Nature Communications, 15*(1), 219. https://doi.org/10.1038/s41467-023-44322-8

Lipton, D. M., Gonzales, B. J., & Citri, A. (2019). Dorsal striatal circuits for habits, compulsions and addictions. *Frontiers in Systems Neuroscience, 13*, 28. https://doi.org/10.3389/fnsys.2019.00028

Liu, J., Cao, L., Li, H., Gao, Y., Bu, X., Liang, K., Bao, W., Zhang, S., Qiu, H., Li, X., Hu, X., Lu, L., Zhang, L., Hu, X., Huang, X., & Gong, Q. (2022). Abnormal resting-state functional connectivity in patients with obsessive-compulsive disorder: A systematic review and meta-analysis. *Neuroscience and Biobehavioral Reviews, 135*, Article 104574. https://doi.org/10.1016/j.neubiorev.2022.104574

Ma, N., Liu, Y., Li, N., Wang, C. X., Zhang, H., Jiang, X. F., Xu, H. S., Fu, X. M., Hu, X., & Zhang, D. R. (2010). Addiction related alteration in resting-state brain connectivity. *NeuroImage, 49*(1), 738–744. https://doi.org/10.1016/j.neuroimage.2009.08.037

Maia, T. V., Cooney, R. E., & Peterson, B. S. (2008). The neural bases of obsessive-compulsive disorder in children and adults. *Development and Psychopathology, 20*(4), 1251–1283. https://doi.org/10.1017/S0954579408000606

McGovern, R. A., & Sheth, S. A. (2017). Role of the dorsal anterior cingulate cortex in obsessive-compulsive disorder: Converging evidence from cognitive neuroscience and psychiatric neurosurgery. *Journal of Neurosurgery, 126*(1), 132–147. https://doi.org/10.3171/2016.1.JNS15601

Murray, E. A., Izquierdo, A., & Malkova, L. (2009). Amygdala function in positive reinforcement: Contributions from studies of nonhuman primates. In P. J. Whalen & E. A. Phelps (Eds.), *The human amygdala* (pp. 82–104). The Guilford Press.

Naaijen, J., Zwiers, M. P., Amiri, H., Williams, S. C. R., Durston, S., Oranje, B., Brandeis, D., Boecker-Schlier, R., Ruf, M., Wolf, I., Banaschewski, T., Glennon, J. C., Franke, B., Buitelaar, J. K., & Lythgoe, D. J. (2017). Fronto-striatal glutamate in autism spectrum disorder and obsessive compulsive disorder. *Neuropsychopharmacology, 42*(12), 2456–2465. https://doi.org/10.1038/npp.2016.260

OCD UK. (n.d.). *Howard Hughes*. https://www.ocduk.org/ocd/history-of-ocd/howard-hughes/

Piantadosi, S. C., Manning, E. E., Chamberlain, B. L., Hyde, J., LaPalombara, Z., Bannon, N. M., Pierson, J. L., K Namboodiri, V. M., & Ahmari, S. E. (2024). Hyperactivity of indirect pathway-projecting spiny projection neurons promotes compulsive behavior. *Nature Communications, 15*(1), 4434. https://doi.org/10.1038/s41467-024-48331-z

Robbins, T. W., Banca, P., & Belin, D. (2024). From compulsivity to compulsion: The neural basis of compulsive disorders. *Nature Reviews. Neuroscience, 25*(5), 313–333. https://doi.org/10.1038/s41583-024-00807-z

Robinson, T. E., & Berridge, K. C. (1993). The neural basis of drug craving: An incentive-sensitization theory of addiction. *Brain Research. Brain Research Reviews, 18*(3), 247–291. https://doi.org/10.1016/0165-0173(93)90013-p

Root, D. H., Melendez, R. I., Zaborszky, L., & Napier, T. C. (2015). The ventral pallidum: Subregion-specific functional anatomy and roles in motivated behaviors. *Progress in Neurobiology, 130*, 29–70. https://doi.org/10.1016/j.pneurobio.2015.03.005

Saga, Y., Galineau, L., & Tremblay, L. (2022). Impulsive and compulsive behaviors can be induced by opposite GABAergic dysfunctions inside the primate ventral pallidum. *Frontiers in Systems Neuroscience, 16*, 1009626. https://doi.org/10.3389/fnsys.2022.1009626

Salari, N., Kazeminia, M., Heydari, M., Darvishi, N., Ghasemi, H., Shohaimi, S., & Mohammadi, M. (2022). Body dysmorphic disorder in individuals requesting cosmetic surgery: A systematic review and meta-analysis. *Journal of Plastic, Reconstructive & Aesthetic Durgery, 75*(7), 2325–2336. https://doi.org/10.1016/j.bjps.2022.04.098

Singh, A. R., & Veale, D. (2019). Understanding and treating body dysmorphic disorder. *Indian Journal of Psychiatry, 61*(Suppl 1), S131–S135. https://doi.org/10.4103/psychiatry.IndianJPsychiatry_528_1

Substance Abuse and Mental Health Services Administration. (2023). *Key substance use and mental health indicators in the United States: Results from the 2022 National Survey on Drug Use and Health* (HHS Publication No. PEP23-07-01-006, NSDUH Series H-58). Center for Behavioral Health Statistics and Quality, Substance Abuse and Mental Health Services Administration. https://www.samhsa.gov/data/report/2022-nsduh-annual-national-report

Sun, T., Song, Z., Tian, Y., Tian, W., Zhu, C., Ji, G., Luo, Y., Chen, S., Wang, L., Mao, Y., Xie, W., Zhong, H., Zhao, F., Luo, M. H., Tao, W., Wang, H., Li, J., Li, J., Zhou, J., … Zhang, Z. (2019). Basolateral amygdala input to the medial prefrontal cortex controls obsessive-compulsive disorder-like checking behavior. *Proceedings of the National Academy of Sciences of the United States of America, 116*(9), 3799–3804. https://doi.org/10.1073/pnas.1814292116

Stuchlik, A., Radostová, D., Hatalova, H., Vales, K., Nekovarova, T., Koprivova, J., Svoboda, J., & Horacek, J. (2016). Validity of quinpirole sensitization rat model of OCD: Linking evidence from animal and clinical studies. *Frontiers in Behavioral Neuroscience, 10*, 209. https://doi.org/10.3389/fnbeh.2016.00209

Volkow, N. D., & Fowler, J. S. (2000). Addiction, a disease of compulsion and drive: Involvement of the orbitofrontal cortex. *Cerebral Cortex, 10*(3), 318–325. https://doi.org/10.1093/cercor/10.3.318

Volkow, N. D., Fowler, J. S., Wang, G. J., Swanson, J. M., & Telang, F. (2007). Dopamine in drug abuse and addiction: Results of imaging studies and treatment implications. *Archives of Neurology, 64*(11), 1575–1579. https://doi.org/10.1001/archneur.64.11.1575

Volkow, N. D., Michaelides, M., & Baler, R. (2019). The neuroscience of drug reward and addiction. *Physiological Reviews, 99*(4), 2115–2140. https://doi.org/10.1152/physrev.00014.2018

World Health Organization. (2022). *ICD-11: International classification of diseases* (11th revision). https://icd.who.int/

Wright, A. M., Zapata, A., Baumann, M. H., Elmore, J. S., Hoffman, A. F., & Lupica, C. R. (2017). Enduring loss of serotonergic control of orbitofrontal cortex function following contingent and noncontingent cocaine exposure. *Cerebral Cortex, 27*(12), 5463–5476. https://doi.org/10.1093/cercor/bhw312

Yager, L. M., Garcia, A. F., Wunsch, A. M., & Ferguson, S. M. (2015). The ins and outs of the striatum: Role in drug addiction. *Neuroscience, 301*, 529–541. https://doi.org/10.1016/j.neuroscience.2015.06.033

Yin, H. H., Knowlton, B. J., & Balleine, B. W. (2005). Blockade of NMDA receptors in the dorsomedial striatum prevents action-outcome learning in instrumental conditioning. *The European Journal of Neuroscience, 22*(2), 505–512. https://doi.org/10.1111/j.1460-9568.2005.04219.x

Zhao, Y., Sallie, S. N., Cui, H., Zeng, N., Du, J., Yuan, T., Li, D., De Ridder, D., & Zhang, C. (2020). Anterior cingulate cortex in addiction: New insights for neuromodulation. *Neuromodulation*. https://doi.org/10.1111/ner.13291.Advanceonlinepublication.doi:10.1111/ner.13291

Chapter 12
Motor Impulsivity

1 Introduction

On one school night, my wife and I asked our oldest how school was that day. He gave us a brief review of what he did that day. He then complained about one of his classmates. To paraphrase what my son said, "They're so annoying. They're constantly in everyone's business, and they never stop talking". You may have gone to school with a kid like this or currently know someone at work that engages in these behaviors. As a professor, I have taught students that have a difficult time refraining from speaking out or blurting out answers to every question I pose. As frustrating as it can be to share a classroom with individuals like this, one needs to understand that these individuals are not always intentionally trying to be disruptive. I am going to use this chapter to explain one aspect of impulsivity known as **motor impulsivity**. Motor impulsivity is not a unitary construct. An individual with high motor impulsivity can have difficulty inhibiting specific actions (**action restraint**) and/or have difficulty stopping an action that is ongoing (**action cancellation**).

In the previous chapter, you learned about compulsivity. While compulsivity and impulsivity sound like similar terms, there is a key difference between these constructs. Compulsive behaviors are typically performed to alleviate some form of anxiety (e.g., repetitive hand washing to alleviate fear of germs; continued substance use to escape withdrawal symptoms). In the case of impulsivity, individuals spontaneously perform an action with little forethought (e.g., blurting out an answer). Motor impulsivity is observed in various psychiatric conditions. As such, I will begin this chapter by discussing psychiatric conditions characterized by motor impulsivity before diving into the neurocircuitry involved in this form of impulsivity.

J. Yates, *Neuroscience of Maladaptation*,
https://doi.org/10.1007/978-3-032-15602-0_12

2 Learning Objectives

By the end of this chapter, you should be able to …

1. Differentiate action restraint from action cancellation, and identify the ways in which these behaviors can be measured.
2. Describe the disorders that are characterized, at least in part, by motor impulsivity.
3. Explain how the cortico-basal ganglia-thalamo-cortical (CBGTC) loop and the mesocorticolimbic pathway mediate motor impulsivity.
4. Discuss the neurochemical bases of motor impulsivity.

3 Disorders Characterized by Motor Impulsivity

Some disorders you have already learned about are partially characterized by impulsive behaviors. Individuals with schizophrenia, dementia, Parkinson's disease, and substance use disorders/behavioral addictions often display motor impulsivity (Grant et al., 2010; Keszycki et al., 2019; Kozak et al., 2019; Ouzir, 2013; Zhang et al., 2021). For now, I will focus on two groups of disorders in which motor impulsivity is a defining feature.

3.1 Hyperactive/Impulsive Subtype of ADHD

In Chap. 5, you learned about attention-deficit/hyperactivity disorder (ADHD), specifically, the inattentive subtype. The hyperactive/impulsive subtype is characterized by increased movement, restlessness, and impulsivity. Like the inattentive subtype, the *Diagnostic and Statistical Manual of Mental Disorders (DSM-5)* (American Psychiatric Association, 2013) requires that a child under the age of 17 meet at least six of the following criteria over the span of 6 months to be diagnosed with the hyperactive/impulsive subtype:

a. Often fidgets with or taps hands or feet or squirms in seat.
b. Often leaves seat in situations when remaining seated is expected (e.g., leaves his or her place in the classroom, in the office or other workplace, or in other situations that require remaining in place).
c. Often runs about or climbs in situations where it is inappropriate. (**Note:** In adolescents or adults, may be limited to feeling restless).
d. Often unable to play or engage in leisure activities quietly.
e. Is often "on the go", acting as if "driven by a motor" (e.g., is unable to be or uncomfortable being still for extended time, as in restaurants, meetings; may be experienced by others as being restless or difficult to keep up with).
f. Often talks excessively.

g. Often blurts out an answer before a question has been completed (e.g., completes people's sentences; cannot wait for turn in conversation).
h. Often has difficulty waiting his or her turn (e.g., while waiting in line).
i. Often interrupts or intrudes on others (e.g., butts into conversations, games or activities; may start using people's things without asking or receiving permission; for adolescents and adults, may intrude into or take over what others are doing) (p. 60).

The *International Classification of Diseases (ICD-11)* (World Health Organization, 2022) uses similar criteria for the hyperactive/impulsive subtype of ADHD.

3.2 Impulse-Control Disorders

The *DSM-5* and the *ICD-11* have categories for **impulse-control disorders**. These disorders will be briefly detailed below.

Oppositional Defiant Disorder (ODD) is typically associated with children and is characterized by persistent negative, defiant, and hostile behavior toward authority figures. Importantly, individuals have difficulty regulating their emotional responses, leading to hostile behaviors. *Conduct disorder*, like ODD, is typically observed in children, as well as adolescents. This disorder is characterized by a persistent pattern of aggressive, antisocial behavior that violates the rights of others or societal norms. This includes actions such as bullying, stealing, lying, and physical violence.

Intermittent explosive disorder is characterized by recurrent episodes of aggressive outbursts that are disproportionate to the situation that elicits the response. For example, someone may scream at their child for something minor like leaving a toy on the floor. Aggression will be a major topic covered in Chap. 15.

The last two additional impulse-control disorders resemble compulsive disorders as they involve uncontrollable urges. In *pyromania*, individuals have an urge to set fires, and *kleptomania* is a condition in which individuals have an urge to steal. The *ICD-11* also includes compulsive sexual behavior disorder as an impulse-control disorder as opposed to as a compulsive disorder.

4 Measuring Motor Impulsivity

The hyperactive/impulsive subtype of ADHD includes deficits in action restraint (e.g., blurting out answers) and action cancellation (e.g., has difficulty stopping something like hand tapping). Even in individuals without ADHD, motor impulsivity can be measured experimentally in both humans and animals (see Fig. 1).

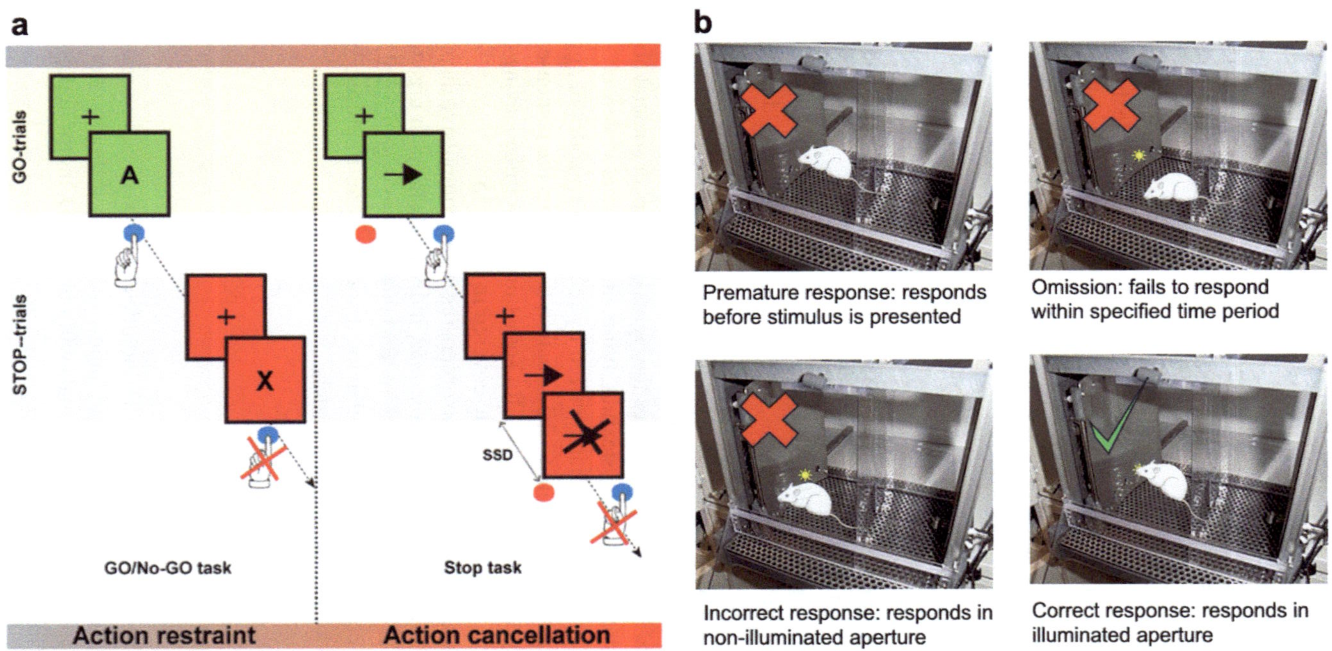

Fig. 1 (**a**) Schematic of the go/no task and the stop signal reaction time task (SSRTT) used to measure action restraint and action cancellation, respectively. *Source* Figure 1 of van Velzen et al. (2014), https://doi.org/. CC BY 3.0 (https://creativecommons.org/licenses/by-sa/3.0/deed.en), via Frontiers. (**b**) Schematic of the five-choice serial reaction time task (5-CSRTT) often used in rodent subjects to measure aspects of attention and motor impulsivity. Different trial outcomes are represented in the image. Image of the operant chamber comes from https://commons.wikimedia.org/wiki/File:5CSRTT_for_mice.jpg from Bd008. CC BY-SA 4.0 (https://creativecommons.org/licenses/by-sa/4.0/deed.en), via Wikimedia Commons. The rest of panel (**b**) was created using images from Servier Medical Art (https://smart.servier.com/; CC BY 4.0, https://creativecommons.org/licenses/by/4.0/) and Microsoft PowerPoint

4.1 Action Restraint

Go/no-go tasks are often used to measure action restraint in humans (Fig. 1a, left panel). Participants are required to respond on a computer key when they are presented with a specific stimulus (e.g., green rectangle) and to refrain from responding when a different stimulus is presented (e.g., blue rectangle). In one version of the go/no-go task, the orientation of the rectangle can be either vertical or horizontal. Vertical rectangles indicate that a "go" stimulus will be presented 80% of the time. Conversely, horizontal rectangles indicate that a "no-go" stimulus will appear 80% of the time (Fillmore et al., 2003).

Animals can also be tested in go/no-go tasks. When I was a graduate student, we trained rats in one variant of a go/no-go task in which they learned to respond for food reinforcement when a cue light above the lever was illuminated. When the cue light was extinguished, responses on the lever did not lead to reinforcement (Yates et al., 2016).

In Chap. 5, you learned about the five-choice serial reaction time task (5-CSRTT), which is used to measure attention (Carli et al., 1983). One major advantage of the 5-CSRTT is that it can be used to measure motor impulsivity as well. Recall that the 5-CSRTT is often tested in a special operant conditioning chamber that consists of five apertures on one wall. When each trial begins, the rodent must wait for a light to be illuminated in one of the apertures before emitting a response. A premature response (i.e., a response before one of the lights is illuminated) is interpreted as a failure in action restraint (Fig. 1b).

Comprehension Check 1 Look at Fig. 1b. You will see that I have included four potential outcomes of an individual trial of the 5-CSRTT. Animals can respond prematurely (i.e., respond before an aperture is illuminated), which is a measure of motor impulsivity. They can fail to respond within a specified time frame (omission) or respond in the wrong aperture (measure of inattention). Finally, a correct response leads to reinforcement. Another behavior that can be measured is perseveration (see previous chapter). How do you think perseveration is measured in the 5-CSRTT?

After the animal has successfully responded and received its food, there is a "time-out" period in which the animal must wait before the next trial begins. Perseveration occurs when the animal continues to respond in the aperture during this time-out period.

4.2 Action Cancellation

Stop signal reaction time tasks (SSRTTs) are commonly used to measure action cancellation (Fig. 1a, right panel). Participants are required to inhibit responses they have already initiated when presented with a cue (Logan et al., 1984). In one animal version of the SSRTT, rodents learn to press one lever, which results in presentation of a second lever. Responding on the second lever results in food delivery; however, on 20% of trials, a tone is played, indicating that lever responses should stop. If the rodent cancels its responding on the second lever, it will receive the food reinforcer (Eagle & Robbins, 2003).

5 Neuromechanisms of Motor Impulsivity

As this chapter focuses on a type of impulsivity characterized by impulsive motor actions, you should not be surprised to learn that the cortico-basal ganglia-thalamo-cortical (CBGTC) loop will be heavily featured. By now, you should be quite familiar with the direct, the indirect, and the hyperdirect pathways of this circuit and how dysregulation of the CBGTC loop leads to attentional deficits (Chap. 5), movement disorders (Chap. 10), and maladaptive habit formation (Chap. 11). This section will begin by exploring how dysregulation of the CBGTC loop can lead to motor impulsivity. You will then learn about the mesocorticolimbic pathway before learning about the contribution of specific neurotransmitter systems to motor impulsivity.

5.1 Cortico-Basal Ganglia-Thalamo-Cortical (CBGTC) Loop

Like compulsivity, a hyperactive dorsal striatum (caudate nucleus and putamen) is implicated in motor impulsivity. The caudate nucleus shows greater activity during "go" trials relative to "no-go" trials in a go/no-go task (Le et al., 2020). This is not problematic *per se* as the increased activity observed in the caudate nucleus should occur as the direct pathway of the CBGTC loop is important for stimulating purposeful movement (e.g., responding appropriately to cues in our environment). However, increased dorsomedial striatal (DMS) activity is associated with impaired inhibition in an SSRTT (Bryden et al., 2012).

Recall that the DMS is more active when we first learn a task, with activation eventually shifting to the dorsolateral striatum (DLS) as the behavior becomes habitual. This finding indicates that a hyperactive DMS can interfere with goal-directed behavior by triggering behavioral responses that are not needed or appropriate for a given situation. Furthermore, increased white matter connectivity between the dorsal striatum and the supplementary motor area (SMA) is associated with enhanced motor impulsivity in a go/no-go task (Hampton et al., 2017). This indicates that there are more neuronal connections between these regions. Given that the SMA is important for selecting which movements to initiate, hyperactivity of the DMS can overstimulate the SMA, leading to an impulsive action.

The frontal cortex, especially the prefrontal cortex (PFC), is critical for impulse control (Bari & Robbins, 2013). Essentially, the PFC prevents inappropriate actions by modulating activity of the basal ganglia. We know that damage to the PFC leads to increased motor impulsivity in both humans and animals (e.g., Humby et al., 2013), and chemogenetic inhibition of PFC neurons that project to the dorsal striatum increases premature responses in the 5-CSRTT (Terra et al., 2020). Conversely, optogenetic stimulation of prefrontal cortical neurons that project to the subthalamic nucleus (STN) improves response inhibition in animals performing a go/no-go task (Li et al., 2020).

Comprehension Check 2 What role does the DMS play in motor impulsivity, and how can its hyperactivity affect behavior?

The DMS plays a significant role in motor impulsivity by influencing goal-directed behaviors. Its hyperactivity can disrupt appropriate responses by triggering unnecessary or unsuitable behaviors for a given context, leading to impaired inhibition and increased impulsive actions.

5.2 *Salience Network*

The frontal cortex is part of another network of brain structures implicated in motor impulsivity: the salience network (also known as the *cingulo-opercular network*). The salience network, as its name suggests, is important for detecting and evaluating highly salient stimuli. This network also helps assign value to both internal and external stimuli, which we use when performing specific behaviors. In Chap. 5, you learned that the anterior cingulate cortex (ACC) and the anterior insula make up part of this network. The salience network also is composed of the medial superior frontal cortex and the frontal operculum (which covers the insula). Regions in the salience network, like the CBGTC loop, become active when individuals commit impulsive actions (Dambacher et al., 2015; Zhukovsky et al., 2020). Additionally, direct stimulation of the ventral region of the ACC leads to an impulsive-like "getting up" from bed response in patients (Caruana et al., 2018). If the salience network is hyperactive, individuals may assign high value to random stimuli that may have very little importance. Increased activation of the salience network is also observed in individuals with attention deficits (Chap. 5), which may account for the attentional and hyperactive/impulsive symptoms displayed by individuals with the combined subtype of ADHD. The Experiment Spotlight below focuses on recent research examining components of the salience network to motor impulsivity.

Experiment Spotlight: Neural Circuit Mapping of Waiting Impulsivity and Proactive Inhibition with Convergent Evidence from fMRI and TMS
The full article can be accessed with this link: https://doi.org/10.1016/j.cortex.2024.12.019

In the study by Baek et al. (2025), healthy adults completed three procedures aimed at measuring both action restraint and *proactive inhibition* across two experiments. A one-choice serial reaction time task (1-CSRTT) (Experiment 1) and a 2-CSRTT (Experiment 2) were used to measure action restraint, and a proactive inhibition test was used to measure proactive inhibition.

In the CSRTTs, participants looked at a computer screen and had to hold the left button down with their dominant index finger. On the computer screen was an empty square (or two squares in the 2-CSRTT); participants were instructed to release the left button and press the right button when they saw a circle appear in the square. A premature response occurred when participants responded before the circle appeared. On 75% of trials, the target appeared after 2 s, but on 25% of trials, the target's appearance was delayed (either 2.5 s or 3 s).

The proactive inhibition test was somewhat similar to the CSRTTs. Baseline trials (n = 60) began with presentation of a white cross on the screen. Participants had to respond like the 1-CSRTT when a green circle appeared in a square. On 40 trials, a red cross appeared, indicating that either a "go" response (30 trials) or a "no-go" response (10 trials) would occur. During no-go trials, a red "X" appeared inside the square. Participants had to keep their finger pressed down on the left button instead of responding on the right button.

In Experiment 1, participants completed the 1-CSRTT and the proactive inhibition test while inside an MRI (magnetic resonance imaging) machine; this enabled the researchers to perform functional magnetic resonance imaging (fMRI; i.e., examine brain activation) while participants completed these tasks. In Experiment 2, participants completed the 2-CSRTT and the proactive inhibition test following continuous theta burst stimulation (cTBS) above the left inferior frontal gyrus (LIFG). The LIFG contains the operculum. This type of stimulation causes inhibition of the LIFG, thus allowing the researchers to determine a causal role for this region in motor impulsivity.

The fMRI results showed that there was overlapping activity of the pre-SMA and the insula when participants completed delayed cue trials of the 1-CSRTT and the proactive inhibition test. Specific to the 1-CSRTT, premature responses were negatively correlated with activity in the dorsomedial PFC (dmPFC) and the LIFG for delayed trials compared to non-delayed trials. cTBS to the left LIFG significantly increased premature responses.

Questions to Consider:

1. Why do you think the researchers used a 1-CSRTT/2-CSRTT instead of a 5-CSRTT that is often used in rodents or a 4-CSRTT often used in human studies?
2. How are the proactive inhibition and the go/no-go task similar and different from one another?
3. How might the current findings inform the development of more targeted and effective interventions for impulse-control disorders and disorders characterized by impaired impulse control, such as ADHD or addictions?

5.3 Mesocorticolimbic Pathway and Connected Structures

The nucleus accumbens (NAc), while primarily associated with reinforcement learning, is involved in motor actions. While the shell subregion is important for attributing incentive salience to stimuli paired with rewards, the NAc core helps establish motor plans that enable an organism to acquire reinforcers (see previous chapter). One seemingly paradoxical finding is that direct stimulation of the NAc shell increases motor impulsivity (Sesia et al., 2010). Let's further discuss what the NAc shell does to understand why stimulating this region can induce motor impulsivity.

In addition to attributing incentive salience to environmental cues, the NAc shell integrates information from brain regions such as the PFC, the amygdala, and the hippocampus; the NAc shell uses this information to influence behaviors. I mentioned in the last chapter that a major output of the NAc is the ventral pallidum (VP). The NAc shell sends GABAergic (gamma-aminobutyric acid) neurons to the VP, which inhibits GABAergic neurons (Root et al., 2015). In a sense, the NAc shell's influence over the VP exerts control over motor actions. Hyperactivity of the NAc shell causes too much inhibition of the VP, preventing the VP from influencing activity in the ventral tegmental area (VTA), which sends dopaminergic projections to the NAc and the PFC.

Indeed, stimulating the VP improves motor impulsivity observed in individuals with Parkinson's disease (Moussawi et al., 2022). Whereas direct infusion of a $GABA_A$ antagonist into VP leads to compulsive-like behaviors, a $GABA_A$ receptor agonist increases motor impulsivity (Saga et al., 2022), thus further highlighting the idea that excess inhibition of the VP drives motor impulsivity.

In Chap. 7, I mentioned that the lateral habenula receives inputs from the PFC and the globus pallidus. More specifically, this part of the brain receives information from the medial PFC (mPFC) and the VP, as well as the NAc (Fig. 2). Directly related to this chapter, the habenula acts as a brake for motor actions. Damage to the habenula leads to hyperactivity and motor impulsivity in animals. This occurs because the habenula is important for regulating dopaminergic activity emanating from the VTA (Hikosaka, 2010). Optogenetic stimulation of the pathway between the VP and the NAc shell to the lateral habenula increases motor impulsivity as these regions project GABAergic neurons to the habenula (Hwang et al., 2024). Like the VP, too much inhibition of the lateral habenula can increase motor impulsivity in individuals.

Like the lateral habenula, impairing the medial habenula exacerbates motor impulsivity. Specifically, disrupting cholinergic projections from the medial habenula to the interpeduncular nucleus (IPN) increases motor impulsivity (Kobayashi et al., 2013). This most likely occurs because the IPN has a direct connection with the VTA, thus controlling dopaminergic activity (Kami et al., 2018).

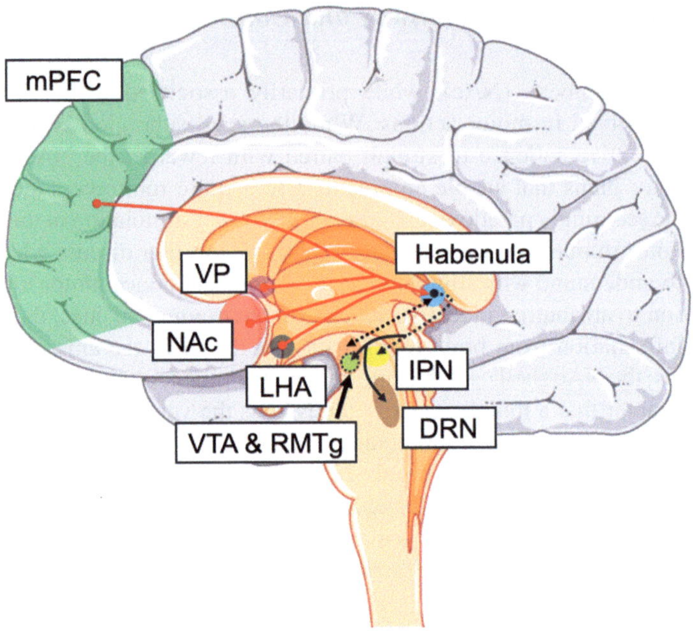

Fig. 2 Schematic showing some of the inputs and outputs of the habenula. Red solid lines indicate afferents to the habenula. Solid black lines indicate efferent projections from the lateral habenula while dotted black lines indicate efferent projections from the medial habenula. *Note* DRN = dorsal raphe nuclei; IPN = interpeduncular nucleus; LHA = lateral hypothalamic area; mPFC = medial prefrontal cortex; NAc = nucleus accumbens; RMTg = rostromedial tegmental nucleus (*Note* This is denoted as dashed green circle within VTA); VP = ventral pallidum; VTA = ventral tegmental area. Image created by the author using Servier Medical Art (https://smart.servier.com/; CC BY 4.0, https://creativecommons.org/licenses/by/4.0/) and Microsoft PowerPoint

Comprehension Check 3 How does hyperactivity of the NAc shell lead to increased motor impulsivity, and what role does the VP play in this process?

Hyperactivity of the NAc shell leads to increased motor impulsivity by causing excessive inhibition of the VP. The NAc shell sends GABAergic neurons to the VP, which are inhibitory. When the NAc shell is overactive, it releases too much GABA into the VP, thereby preventing the VP from effectively influencing activity in the VTA. Because the VP, when functioning normally, is involved in improving motor impulsivity, its excessive inhibition by the NAc shell results in a lack of control over motor actions, thus contributing to impulsivity.

5.4 Catecholamines and Motor Impulsivity

The catecholamines dopamine and norepinephrine have long been linked to impulsivity as medications that are effective at treating individuals with the hyperactive/impulsive subtype of ADHD (e.g., amphetamine, methylphenidate, atomoxetine) work by increasing catecholamine levels. More specifically, ADHD medications primarily exert their effects by targeting norepinephrine and/or dopamine transporters. These results suggest that motor impulsivity may be partially influenced by hypoactive dopamine and norepinephrine systems. Indeed, depleting dopamine levels in the mPFC increases motor impulsivity in a DRL (differential reinforcement of low rates of responding) task (Sokolowski & Salamone, 1994) (note, in a DRL schedule, organisms receive reinforcement if they can wait for a specified period of time before responding; a premature response resets the timer), and dopamine receptor agonist treatment improves motor impulsivity in individuals with Parkinson's disease, which is linked to increased cerebral blood flow in frontal cortical areas like the dorsolateral PFC (dlPFC) and the pre-SMA (Trujillo et al., 2019).

Interestingly, lesions to noradrenergic neurons in the NAc shell do not affect motor impulsivity, but they prevent atomoxetine-induced decreases in motor impulsivity (Benn & Robinson, 2017). Related to this finding is that direct infusion of atomoxetine into the NAc shell, but not the NAc core or the PFC, decreases premature responses (Economidou et al., 2012). If noradrenergic neurons are damaged in the NAc shell, this prevents atomoxetine from blocking norepinephrine transporters to increase synaptic levels of norepinephrine.

While hypoactive dopaminergic and noradrenergic systems are implicated in the etiology of ADHD, this is not quite the case. First, drugs that increase catecholamine levels can exacerbate impulsivity in non-clinical populations. For example, cocaine administration increases motor impulsivity in animals (Zapata & Lupica, 2021). Whereas lesions to noradrenergic neurons in the NAc shell prevent atomoxetine from ameliorating impulsivity, lesions to such neurons in the PFC potentiate amphetamine-induced increases in premature responses (Benn & Robinson, 2017). The PFC contains each type of noradrenergic receptor, but the alpha-2 (α_2) receptor, which is inhibitory, is known to regulate dopamine release in this region (Devoto et al., 2019). When amphetamine is administered to denervated animals, dopamine levels increase significantly without a concomitant increase in norepinephrine. Because there is not enough norepinephrine to stimulate α_2 receptors located on dopaminergic neurons, dopamine levels increase too much, which can cause downstream hyperactivity of the CBGTC loop.

Additional animal research has provided evidence that excess prefrontal cortical dopamine levels exacerbate motor impulsivity. Decreased dopamine transporter function in the orbitofrontal cortex (OFC) is associated with motor impulsivity (Yates et al., 2016). Lower dopamine transporter function indicates that excess dopamine is not effectively removed from the synapse, leading to increased dopamine concentrations. Increased dopamine can overstimulate the direct pathway of the CBGTC loop.

How can hypoactive and hyperactive dopamine levels promote motor impulsivity? Think of the direct and the indirect pathways of the CBGTC loop. As with motor control, a balance between dopamine D_1-like receptor and D_2-like receptor activation is important for regulating when behaviors should occur. Greater availability of dopamine D_1-like and D_2-like receptors is associated with better SSRTT performance (Robertson et al., 2015). With that said, let's look at how catecholamine receptors regulate motor impulsivity.

Research indicates that dopamine and norepinephrine levels are important for stimulating receptors that lead to neuronal inhibition (e.g., D_2-like receptors). Systemic administration (i.e., injecting something into the circulatory system—e.g., subcutaneous or intravenous injection) of dopamine D_2-like agonists or noradrenergic α_2 receptor agonists decreases motor impulsivity (Fernando et al., 2012). Conversely, administration of an α_2 receptor antagonist increases motor impulsivity in the 5-CSRTT, an effect that is reversed following administration of α_1 or beta-1 (β_1) receptor antagonists (Adams et al., 2017). Dopamine D_2/D_3 availability in the inferior frontal cortex (IFC) and the striatum is positively correlated with response inhibition in a stop signal task (Pfeifer et al., 2022). Likewise, dopamine D_2 mRNA expression in the prelimbic region of the mPFC is negatively associated with motor impulsivity as assessed in a DRL task (Simon et al., 2013).

While stimulation of D_2-like and α_2 receptors is important for regulating motor actions, D_1-like receptors promote motor impulsivity. In the NAc shell, higher dopamine D_1 receptor mRNA expression is associated with motor impulsivity (Simon et al., 2013). Relatedly, blocking D_1-like receptors in several brain regions decreases motor impulsivity, including OFC, mPFC, NAc, and dorsal hippocampus (Cheng & Liao, 2017; Winstanley et al., 2010). Collectively, these results suggest that excess catecholamine levels, particularly dopamine, can be problematic as they bind to receptors that lead to overexcitation, thus promoting the initiation of movements that may be unnecessary.

Comprehension Check 4 Explain the seemingly contradictory roles of dopamine and norepinephrine in motor impulsivity, particularly in the context of ADHD treatment versus non-clinical populations.

Essentially, an optimal balance of dopamine and norepinephrine is crucial for regulating motor actions. While insufficient levels contribute to impulsivity in ADHD, excessive levels, particularly of dopamine, can also promote it by binding to receptors that lead to overexcitation and unnecessary movement initiation. This points to an inverted U-shaped function, where both too little and too much catecholamine activity can impair behavioral control.

In non-clinical populations, excessive catecholamine levels can exacerbate impulsivity.

5.4.1 Serotonin and Motor Impulsivity

Serotonin, for the most part, has an inhibitory effect on motor impulsivity. Serotonin levels can be reduced in individuals by having them consume a "cocktail" of amino acids that does not include tryptophan, the precursor to serotonin. This cocktail causes tryptophan levels to be depleted, thus lowering serotonin levels. Tryptophan depletion leads to increased motor impulsivity (Worbe et al., 2014). Likewise, depleting serotonin levels in the rodent brain increases impulsive action (Winstanley et al., 2004) while increasing serotonin levels by inhibiting the serotonin transporter reduces motor impulsivity (Baarendse & Vanderschuren, 2012; Tsutsui-Kimura et al., 2009). Inhibiting serotonin transporters also improves SSRTT performance in individuals with Parkinson's disease, with this effect correlating with prefrontal cortical activation (Ye et al., 2014).

Despite having 14 receptor subtypes, most research has focused on the contribution of 5-HT_1 and 5-HT_2 receptors to motor impulsivity. Above, I mentioned that serotonin tends to have an inhibitory effect on motor impulsivity. Indeed, stimulation of 5-$HT_{1A/1B}$ receptors increases motor impulsivity (Korte et al., 2017), although this effect may be driven more by the 5-HT_{1A} receptor as mice lacking the 5-HT_{1B} receptor are more impulsive (Desrochers & Nautiyal, 2022). Conversely, 5-HT_{1A} receptor antagonists decrease motor impulsivity (Ohmura et al., 2013). 5-HT_{1A} receptors are inhibitory and are found in multiple regions of the brain. However, in the raphe nuclei (where serotoninergic neurons originate), 5-HT_{1A} receptors are primarily found on presynaptic neurons and act to inhibit serotonin release. Therefore, systemic administration of 5-HT_{1A} receptor agonists causes widespread inhibition of serotonin levels in the brain.

The 5-HT_{2A} receptor is excitatory, located on postsynaptic neurons. Theoretically, administration of 5-HT_{2A} receptor agonists should have the opposite effect as 5-HT_{1A} receptor agonists. Yet, direct administration of a 5-HT_{2A} agonist into OFC increases motor impulsivity (Wischhof et al., 2011) whereas blocking these receptors decreases motor impulsivity (Paterson et al., 2012; Sholler et al., 2019). These results are consistent with the finding that high impulsive rats have greater 5-HT_{2A} receptor binding and protein expression in the mPFC (Fink et al., 2015). This is where things get more complicated. In addition to the 5-HT_{2A} receptor, there is a type of 5-HT_2 receptor known as the 5-HT_{2C}. Like the 5-HT_{2A} receptor, the 5-HT_{2C} receptor is excitatory, but these two receptor subtypes often have opposing actions, albeit in a complex manner. While 5-HT_{2A} receptor agonists promote impulsive behavior, both 5-HT_{2C} receptor agonists and antagonists decrease motor impulsivity (Fletcher et al., 2023; Higgins et al., 2016a; Humby et al., 2013).

Motor impulsivity may result from an imbalance of 5-HT_{2A} and 5-HT_{2C} receptors in the mPFC; specifically, impulsive rats have a greater proportion of 5-HT_{2A} receptors relative to 5-HT_{2C} receptors (Anastasio et al., 2015). Remember that serotonin often acts as a major modulator of other neurotransmitter systems. 5-HT_{2A} receptor activation promotes dopamine release in the PFC (Pehek et al., 2006) while *blocking* 5-HT_{2C} receptors has the same effect (Bonaccorso et al., 2002). Opposing effects on motor impulsivity have been observed following intra-NAc administration of 5-HT_{2A} and 5-HT_{2C} receptor antagonists (Robinson et al., 2008). Why?

We need to consider the types of neurons that contain these receptors. The 5-HT_{2A} receptor is largely located on excitatory neurons. Stimulating these receptors, then, promotes increased dopamine release in regions like the PFC. 5-HT_{2C} receptors, on the other hand, are located largely on inhibitory neurons, but they are also found on excitatory neurons. This may account for why 5-HT_{2C} receptor agonists and antagonists can decrease motor impulsivity. Returning to the previous paragraph, if animals have a disproportionate number of 5-HT_{2A} receptors relative to 5-HT_{2C} receptors, overexcitation occurs in areas like the PFC, which sets off a chain reaction in the CBGTC loop, promoting the initiation of movements. Let's continue discussing the role of overexcitation to impulsivity below.

Comprehension Check 5 Considering the complex effects of serotonin on motor impulsivity, what are the primary effects of manipulating 5-HT_1 and 5-HT_2 receptor subtypes, and how do their locations on different neuron types contribute to their varied outcomes on impulsive behavior?

Stimulation of 5-HT_{1A} receptors, which are presynaptic autoreceptors in the raphe nuclei, increases impulsivity by inhibiting overall serotonin release. Conversely, the 5-HT_{2A} receptor, an excitatory postsynaptic receptor, increases motor impulsivity when stimulated, likely by promoting dopamine release and leading to overexcitation of the CBGTC loop. The 5-HT_{2C} receptor, also excitatory but located on both inhibitory and excitatory neurons, presents a more complex picture, as both agonists and antagonists can decrease motor impulsivity. This complexity, combined with the observation that impulsive animals have a higher ratio of 5-HT_{2A} to 5-HT_{2C} receptors in the mPFC, suggests that an imbalance favoring 5-HT_{2A}-mediated excitation, or an overall overexcitation due to inappropriate receptor activation, can lead to uncontrolled motor actions and impulsivity.

5.5 Glutamate and Motor Impulsivity

Although the PFC is important for regulating impulsive actions, hyperactivity of the PFC can lead to increased motor impulsivity (Yates, 2024), as I just discussed. The PFC projects glutamatergic neurons to both the dorsal striatum and the cerebellum. The cerebellum sends glutamatergic neurons to the thalamus. While increased striatal activity can lead to increased inhibition of the thalamus, the increased excitatory inputs from the cerebellum can lead to slight overexcitation of the thalamus. The net result is an inhibited primary motor cortex and premotor cortex. However, the SMA becomes disinhibited. Decreased activity of the premotor and the primary motor cortices prevents the inhibition of motor signals through the spinal cord, leading to increased motor activity (see Fig. 3).

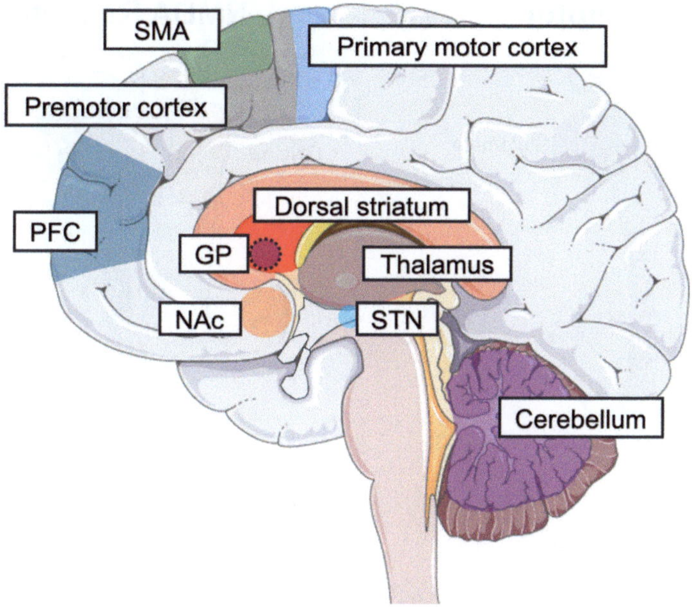

Fig. 3 Schematic showing the brain regions implicated in motor impulsivity. *Note* GP = globus pallidus; NAc = nucleus accumbens; PFC = prefrontal cortex; SMA = supplementary motor area; STN = subthalamic nucleus. Image created by the author using Servier Medical Art (https://smart.servier.com/; CC BY 4.0, https://creativecommons.org/licenses/by/4.0/) and Microsoft PowerPoint

If excessive glutamate is associated with motor impulsivity, controlling glutamate release should be a viable treatment option for impulse-control disorders. Excitatory amino acid transporters (EAATs) remove excess glutamate from the synapse, which should blunt motor impulsivity. Indeed, dopamine-depleted rats that receive repeated treatments of an EAAT2 activator show less motor impulsivity (Das et al., 2025).

Let's return to the glutamatergic receptor subtypes that you have learned about in this book. Understanding the receptor subtypes is important for better understanding the glutamatergic basis of motor impulsivity. There are several classes of metabotropic glutamate receptors: Group I (G_q—stimulatory), Group II (G_i—inhibitory), and Group III (G_i—inhibitory). The major ionotropic glutamate receptors are N-methyl-D-aspartate (NMDA) and α-amino-3-hydroxy-5-methyl-4-isoxazolepropionic acid (AMPA). I need to go into more detail about the composition of the ionotropic glutamate receptors (Fig. 4). Both NMDA and AMPA receptors are composed of four subunits that surround an ion channel. NMDA receptors contain two GluN1 subunits, which is where the *co-agonist* glycine binds. There are two GluN2 subunits; this is where glutamate binds to the receptor. GluN2 subunits can be further subdivided into GluN2A, GluN2B, GluN2C, and GluN2D, with the GluN2A and GluN2B subunits being most common.

Theoretically, blocking excitatory glutamate receptors (ionotropic and Group I mGluRs) and stimulating inhibitory glutamate receptors (Group II and Group III)

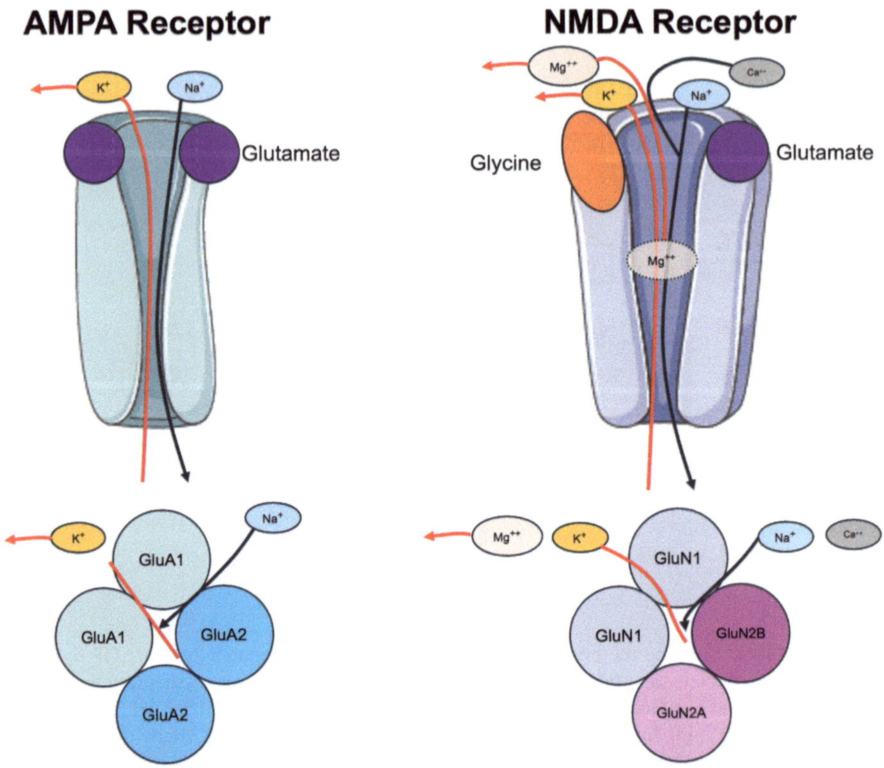

Fig. 4 Schematic of AMPA and NMDA receptor structure. Image created by the author using Servier Medical Art (https://smart.servier.com/; CC BY 4.0, https://creativecommons.org/licenses/by/4.0/) and Microsoft PowerPoint

should decrease motor impulsivity, but this is not quite the case. For example, an AMPA receptor positive allosteric modulator (i.e., increases how well AMPA receptors work) reduces motor impulsivity (Davis-Reyes et al., 2021). Another seemingly paradoxical finding is that blocking GluN2B-containing NMDA receptors increases motor impulsivity (Higgins et al., 2016b). An important consideration is AMPA and NMDA receptors are located on both glutamatergic and GABAergic neurons, especially in the PFC (similar to the 5-HT_{2C} receptor). The findings described above may be related to stimulation/inhibition of ionotropic glutamate receptors located on GABAergic neurons. For example, blocking GluN2B-containing NMDA receptors on GABAergic neurons may prevent the PFC from exerting control over motor control centers. Excess glutamate could lead to a situation in which glutamatergic receptors on both excitatory and inhibitory neurons are activated. Given that glutamatergic neurons outnumber GABAergic neurons by a proportion of 4:1 (Sultan & Shi, 2018), too much glutamate can lead to "runaway" excitation of regions like the striatum and the cerebellum.

Research shows that motor impulsivity is associated with an imbalance of NMDA receptor types. High impulsive animals have much higher concentrations of the GluN2B subunit compared to GluN1 and GluN2B subunit (Davis-Reyes et al., 2019). This finding is interesting because the GluN2B subunit is highly expressed in immature neurons; as neurons mature, GluN2B expression decreases, being replaced by GluN2A-containing NMDA receptors. In fact, the *GRIN2B* gene is associated with ADHD (Dorval et al., 2007).

Rats that display increased motor impulsivity have upregulated GluN1 protein expression in the dorsal striatum (Chuang et al., 2021), indicating increased NMDA receptor expression in this region. Increased stimulation of ionotropic glutamate receptors in the dorsal striatum could lead to hyperactivity in the direct pathway of the CBGTC loop, leading to increased motor actions. Conversely, administration of an $mGluR_4$ positive allosteric modulator (note, $mGluR_4$ is inhibitory) increases motor impulsivity, possibly by inhibiting dopamine D_2-like receptors in the indirect pathway of the CBGTC loop (Isherwood et al., 2017). There is direct evidence that $mGluR_4$ activation in the STN potentiates motor impulsivity as silencing the *GRM4* gene in this region blunts impulsivity of animals classified as "high impulsive" (Piszczek et al., 2022).

Returning to the previous chapter, one hypothesis is that serotonin mediates compulsive behavior by modulating glutamatergic activity. Serotonin appears to influence impulsive behavior through a similar mechanism. As I discussed already, 5-HT_{2A} receptors are located on excitatory (i.e., glutamatergic) neurons. 5-HT_{2A}-induced motor impulsivity is blocked by an $mGluR_{2/3}$ agonist (Wischhof & Koch, 2012). $mGluR_2$ and $mGluR_3$ are part of the Group II family of metabotropic receptors and often act as autoreceptors that regulate the release of glutamate. As such, administration of a drug that activates these autoreceptors can offset the glutamatergic activity generated by 5-HT_{2A} receptors.

I do want to introduce a protein that has not been covered to this point. The *LPHN3* gene encodes a G protein-coupled receptor known as *latrophilin 3 (LPHN3)*. LPHN3 is involved in cell adhesion and neuron guidance, particularly in the development of glutamatergic synapses. You have already learned about the postsynaptic density (PSD); LPHN3 is an important component of the PSD. Alterations to this gene have been implicated in ADHD, and mice lacking this gene are more hyperactive and have elevated dopamine and serotonin levels in the dorsal striatum (Wallis et al., 2012). Research has even shown that zebrafish lacking the *LPHN3* gene leads to a hyperactive/impulsive phenotype (Lange et al., 2012).

Reducing glutamatergic synapses is problematic because glutamate regulates the release of inhibitory neurotransmitters like GABA. Without as much glutamate input, GABAergic neurons become less effective at inhibiting release of other neurotransmitters like dopamine in striatal regions. The end result is hyperactivity of the CBGTC loop and the mesocorticolimbic pathway.

Comprehension Check 6 How can hyperactivity of the PFC ultimately lead to increased motor impulsivity, considering its projections and the resulting effects on the primary motor cortex, premotor cortex, and SMA?

Hyperactivity of the PFC can paradoxically lead to increased motor impulsivity due to a complex interplay with other brain regions. The PFC sends excitatory glutamatergic neurons to both the dorsal striatum and the cerebellum. While the increased striatal activity from these inputs can cause more inhibition of the thalamus, the cerebellum's increased excitatory projections to the thalamus result in a slight overall overexcitation of the thalamus. The net effect on the motor cortices is that the primary motor cortex and the premotor cortex become inhibited. However, critically, the SMA becomes disinhibited. This decreased activity in the primary and the premotor cortices then prevents the normal inhibition of motor signals traveling through the spinal cord, ultimately leading to an increase in uninhibited motor activity and thus, motor impulsivity.

5.6 Cannabinoid and Opioid Contributions to Motor Impulsivity

The endocannabinoid and the endogenous opioid systems serve to inhibit activity of neurons. Cannabinoid receptors are located on presynaptic terminal buttons and inhibit the release of neurotransmitters into the synapse. Opioid receptors are located on postsynaptic dendrites and hyperpolarize the postsynaptic neuron, thus preventing the neuron from generating an action potential and releasing neurotransmitters. Drugs that you are familiar with target cannabinoid or opioid receptors; cannabis is a partial agonist at cannabinoid receptors while opiates (i.e., derived from the opium poppy plant like morphine) and exogenous opioids such as oxycodone and fentanyl are agonists at opioid receptors.

When drugs like cannabis and opioids are taken at high levels, inhibition of the central nervous system (CNS) occurs. At acute levels, cannabinoids and opioids can increase motor activity and potentiate motor impulsivity. Acute cannabis administration increases aspects of motor impulsivity in humans, particularly action cancellation (Rinehart & Spencer, 2021), and animals injected with tetrahydrocannabinol (THC) (the main psychoactive compound in cannabis) exhibit increased motor impulsivity (Zapata & Lupica, 2021). Individuals with opioid use disorder show enhanced motor impulsivity (Baldacchino et al., 2015), and repeated administration of morphine increases impulsivity in animals (Maguire et al., 2016). However, genetic deletion of the gene that encodes the mu opioid receptor leads to low levels of motor impulsivity (Olmstead et al., 2009).

Overstimulation of cannabinoid and opioid receptors can lead to excess inhibition of GABAergic neurons that regulate motor control. For example, the striatum contains numerous GABAergic medium spiny neurons (MSNs) (Fig. 5). In the

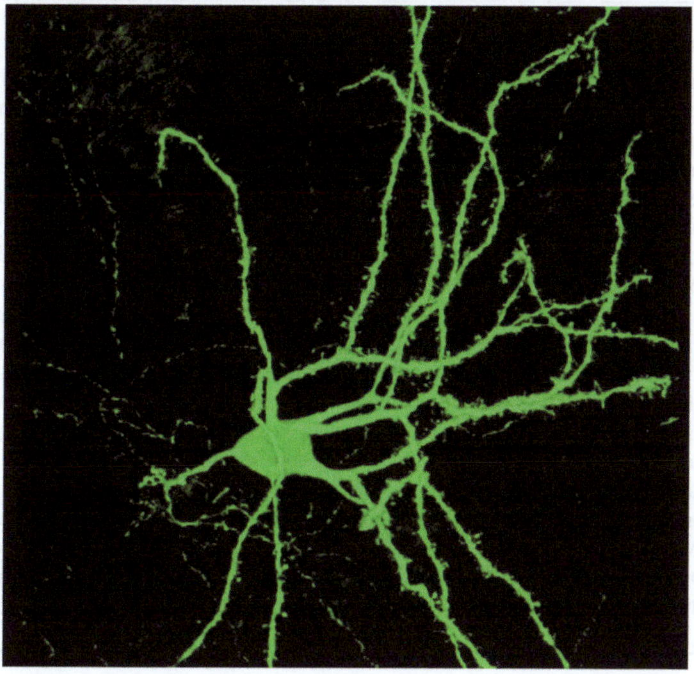

Fig. 5 Image of a medium spiny neuron (MSN). *Source* https://commons.wikimedia.org/wiki/File:Confocal_image_of_spiny_neuron_-_1.jpg from CFCF. Image is in the public domain as it comes from the National Institutes of Health

CBGTC loop, MSNs contain either dopamine D_1-like or D_2-like receptors. By inhibiting MSNs, the direct and the indirect pathways of the CBGTC loop become dysregulated, which can contribute to increased motor impulsivity.

6 Chapter Summary

The present chapter explored the intricacies of motor impulsivity, a critical component of impulsive behavior characterized by challenges in action restraint and action cancellation. You learned about the neurocircuitry underlying motor impulsivity, emphasizing the roles of the CBGTC loop and the mesocorticolimbic pathway. You also learned that too little or too much of specific neurotransmitters (e.g., dopamine and glutamate) can cause increases in motor impulsivity, highlighting the complex interplay of different neurotransmitter systems (e.g., serotonin-mediated control of glutamate release). These findings help explain why medications used to alleviate impulsivity in individuals with ADHD can potentiate motor impulsivity in non-clinical populations.

Answers to Experiment Spotlight Questions

1. When individuals are in an MRI machine, they can barely move as too much movement can interfere with the fMRI results. If you have ever gotten an MRI, you were most likely told to remain as still as possible. If the researchers used something like the 4-CSRTT or the 5-CSRTT, participants would need to press one of multiple buttons on a touchscreen to select the correct square, which could cause erroneous readings to occur, thus confounding the results of the study.
2. These tasks are similar in that individuals need to respond when a specific stimulus is presented and refrain from responding when a different cue is presented. In go/no-go tasks, the no-go signal is presented without prior warning. In the response inhibition task, the red cross tells participants that a no-go signal may appear. This allows researchers to measure proactive response inhibition. In traditional go/no-go tasks, researchers measure *reactive* response inhibition. In the text, I mentioned one version of the go/no-go task that uses vertical and horizontal rectangles to signal the percentage that a no-go cue will appear. This *cued* go/no-go task works in a similar manner as the current proactive inhibition task.
3. Non-invasive brain stimulation techniques like transcranial magnetic stimulation (TMS) (note, cTBS is a form of TMS) or transcranial direct current stimulation (tDCS) applied to the LIFG may be useful in attenuating impulsivity observed in individuals with an impulse-control disorder.

Glossary

Action cancellation the process of stopping or overriding an ongoing motor action, often in response to new information or a change in goals.

Action restraint the proactive prevention of a motor response or the deliberate inhibition of an intended action.

Impulse-control disorder a mental health condition characterized by significant difficulty resisting impulses to perform acts that are harmful to oneself or others.

Motor impulsivity difficulty in suppressing or inhibiting unwanted or inappropriate actions.

References

Adams, W. K., Barrus, M. M., Zeeb, F. D., Cocker, P. J., Benoit, J., & Winstanley, C. A. (2017). Dissociable effects of systemic and orbitofrontal administration of adrenoceptor antagonists on yohimbine-induced motor impulsivity. *Behavioural Brain Research, 328*, 19–27. https://doi.org/10.1016/j.bbr.2017.03.034

American Psychiatric Association. (2013). *Diagnostic and statistical manual of mental disorders* (5th ed.). https://doi.org/10.1176/appi.books.9780890425596

Anastasio, N. C., Stutz, S. J., Fink, L. H., Swinford-Jackson, S. E., Sears, R. M., DiLeone, R. J., Rice, K. C., Moeller, F. G., & Cunningham, K. A. (2015). Serotonin (5-HT) 5-HT2A receptor (5-HT2AR):5-HT2CR Imbalance in medial prefrontal cortex associates with motor impulsivity. *ACS Chemical Neuroscience, 6*(7), 1248–1258. https://doi.org/10.1021/acschemneuro.5b00094

Baek, K., Skandali, N., Sallie, S. N., Sonkusare, S., Mandali, A., Ritou, V., Casero, V., & Voon, V. (2025). Neural circuit mapping of waiting impulsivity and proactive inhibition with convergent evidence from fMRI and TMS. *Cortex, 184*, 194–208. https://doi.org/10.1016/j.cortex.2024.12.019

Baarendse, P. J., & Vanderschuren, L. J. (2012). Dissociable effects of monoamine reuptake inhibitors on distinct forms of impulsive behavior in rats. *Psychopharmacology, 219*(2), 313–326. https://doi.org/10.1007/s00213-011-2576-x

Baldacchino, A., Balfour, D. J., & Matthews, K. (2015). Impulsivity and opioid drugs: Differential effects of heroin, methadone and prescribed analgesic medication. *Psychological Medicine, 45*(6), 1167–1179. https://doi.org/10.1017/S0033291714002189

Bari, A., & Robbins, T. W. (2013). Inhibition and impulsivity: Behavioral and neural basis of response control. *Progress in Neurobiology, 108*, 44–79. https://doi.org/10.1016/j.pneurobio.2013.06.005

Benn, A., & Robinson, E. S. (2017). Differential roles for cortical versus sub-cortical noradrenaline and modulation of impulsivity in the rat. *Psychopharmacology, 234*(2), 255–266. https://doi.org/10.1007/s00213-016-4458-8

Bonaccorso, S., Meltzer, H. Y., Li, Z., Dai, J., Alboszta, A. R., & Ichikawa, J. (2002). SR46349-B, a 5-HT(2A/2C) receptor antagonist, potentiates haloperidol-induced dopamine release in rat medial prefrontal cortex and nucleus accumbens. *Neuropsychopharmacology, 27*(3), 430–441. https://doi.org/10.1016/S0893-133X(02)00311-1

Bryden, D. W., Burton, A. C., Kashtelyan, V., Barnett, B. R., & Roesch, M. R. (2012). Response inhibition signals and miscoding of direction in dorsomedial striatum. *Frontiers in Integrative Neuroscience, 6*, 69. https://doi.org/10.3389/fnint.2012.00069

Carli, M., Robbins, T. W., Evenden, J. L., & Everitt, B. J. (1983). Effects of lesions to ascending noradrenergic neurones on performance of a 5-choice serial reaction task in rats; implications for theories of dorsal noradrenergic bundle function based on selective attention and arousal. *Behavioural Brain Research, 9*(3), 361–380. https://doi.org/10.1016/0166-4328(83)90138-9

Caruana, F., Gerbella, M., Avanzini, P., Gozzo, F., Pelliccia, V., Mai, R., Abdollahi, R. O., Cardinale, F., Sartori, I., Lo Russo, G., & Rizzolatti, G. (2018). Motor and emotional behaviours elicited by electrical stimulation of the human cingulate cortex. *Brain, 141*(10), 3035–3051. https://doi.org/10.1093/brain/awy219

Cheng, R. K., & Liao, R. M. (2017). Regional differences in dopamine receptor blockade affect timing impulsivity that is altered by d-amphetamine on differential reinforcement of low-rate responding (DRL) behavior in rats. *Behavioural Brain Research, 331*, 177–187. https://doi.org/10.1016/j.bbr.2017.05.020

Chuang, C. Y., Tsai, S. Y., Chen, S. F., Yang, Y. H., Chao, C. C., Yen, N. S., & Liao, R. M. (2021). Neurobiological changes in striatal glutamate are associated with trait impulsivity of differential reinforcement of low-rate-response behavior in male rats. *Neurobiology of Learning and Memory, 177*, Article 107352. https://doi.org/10.1016/j.nlm.2020.107352

Dambacher, F., Sack, A. T., Lobbestael, J., Arntz, A., Brugman, S., & Schuhmann, T. (2015). Out of control: Evidence for anterior insula involvement in motor impulsivity and reactive aggression. *Social Cognitive and Affective Neuroscience, 10*(4), 508–516. https://doi.org/10.1093/scan/nsu077

Das, S., Mccloskey, K., Nepal, B., & Kortagere, S. (2025). EAAT2 activation regulates glutamate excitotoxicity and reduces impulsivity in a rodent model of Parkinson's disease. *Molecular Neurobiology, 62*(5), 5787–5803. https://doi.org/10.1007/s12035-024-04644-0

Davis-Reyes, B. D., Campbell, V. M., Land, M. A., Chapman, H. L., Stafford, S. J., & Anastasio, N. C. (2019). Profile of cortical N-methyl-D-aspartate receptor subunit expression associates with inherent motor impulsivity in rats. *Biochemical Pharmacology, 168*, 204–213. https://doi.org/10.1016/j.bcp.2019.07.007

Davis-Reyes, B. D., Smith, A. E., Xu, J., Cunningham, K. A., Zhou, J., & Anastasio, N. C. (2021). Subanesthetic ketamine with an AMPAkine attenuates motor impulsivity in rats. *Behavioural Pharmacology, 32*(4), 335–344. https://doi.org/10.1097/FBP.0000000000000623

Desrochers, S. S., & Nautiyal, K. M. (2022). Serotonin 1B receptor effects on response inhibition are independent of inhibitory learning. *Neurobiology of Learning and Memory, 187*, Article 107574. https://doi.org/10.1016/j.nlm.2021.107574

Devoto, P., Flore, G., Saba, P., Scheggi, S., Mulas, G., Gambarana, C., Spiga, S., & Gessa, G. L. (2019). Noradrenergic terminals are the primary source of α_2-adrenoceptor mediated dopamine release in the medial prefrontal cortex. *Progress in Neuro-Psychopharmacology & Biological Psychiatry, 90*, 97–103. https://doi.org/10.1016/j.pnpbp.2018.11.015

Dorval, K. M., Wigg, K. G., Crosbie, J., Tannock, R., Kennedy, J. L., Ickowicz, A., Pathare, T., Malone, M., Schachar, R., & Barr, C. L. (2007). Association of the glutamate receptor subunit gene GRIN2B with attention-deficit/hyperactivity disorder. *Genes, Brain, and Behavior, 6*(5), 444–452. https://doi.org/10.1111/j.1601-183X.2006.00273.x

Eagle, D. M., & Robbins, T. W. (2003). Inhibitory control in rats performing a stop-signal reaction-time task: Effects of lesions of the medial striatum and d-amphetamine. *Behavioral Neuroscience, 117*(6), 1302–1317. https://doi.org/10.1037/0735-7044.117.6.1302

Economidou, D., Theobald, D. E., Robbins, T. W., Everitt, B. J., & Dalley, J. W. (2012). Norepinephrine and dopamine modulate impulsivity on the five-choice serial reaction time task through opponent actions in the shell and core sub-regions of the nucleus accumbens. *Neuropsychopharmacology, 37*(9), 2057–2066. https://doi.org/10.1038/npp.2012.53

Fernando, A. B., Economidou, D., Theobald, D. E., Zou, M. F., Newman, A. H., Spoelder, M., Caprioli, D., Moreno, M., Hipólito, L., Aspinall, A. T., Robbins, T. W., & Dalley, J. W. (2012). Modulation of high impulsivity and attentional performance in rats by selective direct and indirect dopaminergic and noradrenergic receptor agonists. *Psychopharmacology, 219*(2), 341–352. https://doi.org/10.1007/s00213-011-2408-z

Fillmore, M. T., Rush, C. R., & Marczinski, C. A. (2003). Effects of d-amphetamine on behavioral control in stimulant abusers: The role of prepotent response tendencies. *Drug and Alcohol Dependence, 71*(2), 143–152. https://doi.org/10.1016/s0376-8716(03)00089-9

Fink, L. H., Anastasio, N. C., Fox, R. G., Rice, K. C., Moeller, F. G., & Cunningham, K. A. (2015). Individual differences in impulsive action reflect variation in the cortical serotonin 5-HT2A receptor system. *Neuropsychopharmacology, 40*(8), 1957–1968. https://doi.org/10.1038/npp.2015.46

Fletcher, P. J., Rahbarnia, A., Li, Z., Ji, X., Higgins, G. A., Funk, D., & Lê, A. D. (2023). Effects of 5-HT_{2C} receptor stimulation in male mice on behaviour and Fos expression: Feeding, reward and impulsivity. *Behavioural Brain Research, 447*, Article 114438. https://doi.org/10.1016/j.bbr.2023.114438

Grant, J. E., Potenza, M. N., Weinstein, A., & Gorelick, D. A. (2010). Introduction to behavioral addictions. *The American Journal of Drug and Alcohol Abuse, 36*(5), 233–241. https://doi.org/10.3109/00952990.2010.491884

Hampton, W. H., Alm, K. H., Venkatraman, V., Nugiel, T., & Olson, I. R. (2017). Dissociable frontostriatal white matter connectivity underlies reward and motor impulsivity. *NeuroImage, 150*, 336–343. https://doi.org/10.1016/j.neuroimage.2017.02.021

Higgins, G. A., Silenieks, L. B., Altherr, E. B., MacMillan, C., Fletcher, P. J., & Pratt, W. E. (2016a). Lorcaserin and CP-809101 reduce motor impulsivity and reinstatement of food seeking behavior in male rats: Implications for understanding the anti-obesity property of 5-HT2C receptor agonists. *Psychopharmacology, 233*(14), 2841–2856. https://doi.org/10.1007/s00213-016-4329-3

Higgins, G. A., Silenieks, L. B., MacMillan, C., Sevo, J., Zeeb, F. D., & Thevarkunnel, S. (2016b). Enhanced attention and impulsive action following NMDA receptor GluN2B-selective antagonist pretreatment. *Behavioural Brain Research, 311*, 1–14. https://doi.org/10.1016/j.bbr.2016.05.025

Hikosaka, O. (2010). The habenula: From stress evasion to value-based decision-making. *Nature Reviews. Neuroscience, 11*(7), 503–513. https://doi.org/10.1038/nrn2866

Humby, T., Eddy, J. B., Good, M. A., Reichelt, A. C., & Wilkinson, L. S. (2013). A novel translational assay of response inhibition and impulsivity: Effects of prefrontal cortex lesions, drugs used in ADHD, and serotonin 2C receptor antagonism. *Neuropsychopharmacology, 38*(11), 2150–2159. https://doi.org/10.1038/npp.2013.112

Hwang, E. K., Zapata, A., Hu, V., Hoffman, A. F., Wang, H. L., Liu, B., Morales, M., & Lupica, C. R. (2024). Basal forebrain-lateral habenula inputs and control of impulsive behavior. *Neuropsychopharmacology, 49*(13), 2060–2068. https://doi.org/10.1038/s41386-024-01963-7

Isherwood, S. N., Robbins, T. W., Nicholson, J. R., Dalley, J. W., & Pekcec, A. (2017). Selective and interactive effects of D_2 receptor antagonism and positive allosteric mGluR4 modulation on waiting impulsivity. *Neuropharmacology, 123*, 249–260. https://doi.org/10.1016/j.neuropharm.2017.05.006

Kami, K., Tajima, F., & Senba, E. (2018). Activation of mesolimbic reward system via laterodorsal tegmental nucleus and hypothalamus in exercise-induced hypoalgesia. *Scientific Reports, 8*(1), 11540. https://doi.org/10.1038/s41598-018-29915-4

Keszycki, R. M., Fisher, D. W., & Dong, H. (2019). The hyperactivity-impulsivity-irritiability-disinhibition-aggression-agitation domain in Alzheimer's disease: Current management and future directions. *Frontiers in Pharmacology, 10*, 1109. https://doi.org/10.3389/fphar.2019.01109

Kobayashi, Y., Sano, Y., Vannoni, E., Goto, H., Suzuki, H., Oba, A., Kawasaki, H., Kanba, S., Lipp, H. P., Murphy, N. P., Wolfer, D. P., & Itohara, S. (2013). Genetic dissection of medial habenula-interpeduncular nucleus pathway function in mice. *Frontiers in Behavioral Neuroscience, 7*, 17. https://doi.org/10.3389/fnbeh.2013.00017

Korte, S. M., Prins, J., Van den Bergh, F. S., Oosting, R. S., Dupree, R., Korte-Bouws, G. A., Westphal, K. G., Olivier, B., Denys, D. A., Garland, A., & Güntürkün, O. (2017). The 5-$HT_{1A/1B}$-receptor agonist eltoprazine increases both catecholamine release in the prefrontal cortex and dopamine release in the nucleus accumbens and decreases motivation for reward and "waiting" impulsivity, but increases "stopping" impulsivity. *European Journal of Pharmacology, 794*, 257–269. https://doi.org/10.1016/j.ejphar.2016.11.024

Kozak, K., Lucatch, A. M., Lowe, D. J. E., Balodis, I. M., MacKillop, J., & George, T. P. (2019). The neurobiology of impulsivity and substance use disorders: Implications for treatment. *Annals of the New York Academy of Sciences, 1451*(1), 71–91. https://doi.org/10.1111/nyas.13977

Lange, M., Norton, W., Coolen, M., Chaminade, M., Merker, S., Proft, F., Schmitt, A., Vernier, P., Lesch, K. P., & Bally-Cuif, L. (2012). The ADHD-susceptibility gene lphn3.1 modulates dopaminergic neuron formation and locomotor activity during zebrafish development. *Molecular Psychiatry, 17*(9), 946–954. https://doi.org/10.1038/mp.2012.29

Le, T. M., Zhang, S., Zhornitsky, S., Wang, W., & Li, C. R. (2020). Neural correlates of reward-directed action and inhibition of action. *Cortex, 123*, 42–56. https://doi.org/10.1016/j.cortex.2019.10.007

Li, B., Nguyen, T. P., Ma, C., & Dan, Y. (2020). Inhibition of impulsive action by projection-defined prefrontal pyramidal neurons. *Proceedings of the National Academy of Sciences of the United States of America, 117*(29), 17278–17287. https://doi.org/10.1073/pnas.2000523117

Logan, G. D., Cowan, W. B., & Davis, K. A. (1984). On the ability to inhibit simple and choice reaction time responses: A model and a method. *Journal of Experimental Psychology. Human Perception and Performance, 10*(2), 276–291. https://doi.org/10.1037/0096-1523.10.2.276

Maguire, D. R., Henson, C., & France, C. P. (2016). Daily morphine administration increases impulsivity in rats responding under a 5-choice serial reaction time task. *British Journal of Pharmacology, 173*(8), 1350–1362. https://doi.org/10.1111/bph.13434

Moussawi, K., Kim, M. J., Baybayan, S., Wood, M., & Mills, K. A. (2022). Deep brain stimulation effect on anterior pallidum reduces motor impulsivity in Parkinson's disease. *Brain Stimulation, 15*(1), 23–31. https://doi.org/10.1016/j.brs.2021.11.006

Ohmura, Y., Kumamoto, H., Tsutsui-Kimura, I., Minami, M., Izumi, T., Yoshida, T., & Yoshioka, M. (2013). Tandospirone suppresses impulsive action by possible blockade of the 5-HT1A receptor. *Journal of Pharmacological Sciences, 122*(2), 84–92. https://doi.org/10.1254/jphs.12264fp

Olmstead, M. C., Ouagazzal, A. M., & Kieffer, B. L. (2009). Mu and delta opioid receptors oppositely regulate motor impulsivity in the signaled nose poke task. *PLoS ONE, 4*(2), Article e4410. https://doi.org/10.1371/journal.pone.0004410

Ouzir, M. (2013). Impulsivity in schizophrenia: A comprehensive update. *Aggression and Violent Behavior, 18*(2), 247–254. https://doi.org/10.1016/j.avb.2012.11.014

Paterson, N. E., Wetzler, C., Hackett, A., & Hanania, T. (2012). Impulsive action and impulsive choice are mediated by distinct neuropharmacological substrates in rat. *The International Journal of Neuropsychopharmacology, 15*(10), 1473–1487. https://doi.org/10.1017/S1461145711001635

Pehek, E. A., Nocjar, C., Roth, B. L., Byrd, T. A., & Mabrouk, O. S. (2006). Evidence for the preferential involvement of 5-HT2A serotonin receptors in stress- and drug-induced dopamine release in the rat medial prefrontal cortex. *Neuropsychopharmacology, 31*(2), 265–277. https://doi.org/10.1038/sj.npp.1300819

Pfeifer, P., Sebastian, A., Buchholz, H. G., Kaller, C. P., Gründer, G., Fehr, C., Schreckenberger, M., & Tüscher, O. (2022). Prefrontal and striatal dopamine D_2/D_3 receptors correlate with fMRI BOLD activation during stopping. *Brain Imaging and Behavior, 16*(1), 186–198. https://doi.org/10.1007/s11682-021-00491-y

Piszczek, L., Constantinescu, A., Kargl, D., Lazovic, J., Pekcec, A., Nicholson, J. R., & Haubensak, W. (2022). Dissociation of impulsive traits by subthalamic metabotropic glutamate receptor 4. *eLife, 11*, e62123. https://doi.org/10.7554/eLife.62123

Rinehart, L., & Spencer, S. (2021). Which came first: Cannabis use or deficits in impulse control? *Progress in Neuro-Psychopharmacology & Biological Psychiatry, 106*, Article 110066. https://doi.org/10.1016/j.pnpbp.2020.110066

Robertson, C. L., Ishibashi, K., Mandelkern, M. A., Brown, A. K., Ghahremani, D. G., Sabb, F., Bilder, R., Cannon, T., Borg, J., & London, E. D. (2015). Striatal D1- and D2-type dopamine receptors are linked to motor response inhibition in human subjects. *The Journal of Neuroscience, 35*(15), 5990–5997. https://doi.org/10.1523/JNEUROSCI.4850-14.2015

Robinson, E. S., Dalley, J. W., Theobald, D. E., Glennon, J. C., Pezze, M. A., Murphy, E. R., & Robbins, T. W. (2008). Opposing roles for 5-HT2A and 5-HT2C receptors in the nucleus accumbens on inhibitory response control in the 5-choice serial reaction time task. *Neuropsychopharmacology, 33*(10), 2398–2406. https://doi.org/10.1038/sj.npp.1301636

Root, D. H., Melendez, R. I., Zaborszky, L., & Napier, T. C. (2015). The ventral pallidum: Subregion-specific functional anatomy and roles in motivated behaviors. *Progress in Neurobiology, 130*, 29–70. https://doi.org/10.1016/j.pneurobio.2015.03.005

Saga, Y., Galineau, L., & Tremblay, L. (2022). Impulsive and compulsive behaviors can be induced by opposite GABAergic dysfunctions inside the primate ventral pallidum. *Frontiers in Systems Neuroscience, 16*, 1009626. https://doi.org/10.3389/fnsys.2022.1009626

Sesia, T., Bulthuis, V., Tan, S., Lim, L. W., Vlamings, R., Blokland, A., Steinbusch, H. W., Sharp, T., Visser-Vandewalle, V., & Temel, Y. (2010). Deep brain stimulation of the nucleus accumbens shell increases impulsive behavior and tissue levels of dopamine and serotonin. *Experimental Neurology, 225*(2), 302–309. https://doi.org/10.1016/j.expneurol.2010.06.022

Sholler, D. J., Stutz, S. J., Fox, R. G., Boone, E. L., Wang, Q., Rice, K. C., Moeller, F. G., Anastasio, N. C., & Cunningham, K. A. (2019). The 5-HT_{2A} receptor (5-HT_{2A}R) regulates impulsive action and cocaine cue reactivity in male Sprague-Dawley rats. *The Journal of Pharmacology and Experimental Therapeutics, 368*(1), 41–49. https://doi.org/10.1124/jpet.118.251199

Simon, N. W., Beas, B. S., Montgomery, K. S., Haberman, R. P., Bizon, J. L., & Setlow, B. (2013). Prefrontal cortical-striatal dopamine receptor mRNA expression predicts distinct forms of impulsivity. *The European Journal of Neuroscience, 37*(11), 1779–1788. https://doi.org/10.1111/ejn.12191

Sokolowski, J. D., & Salamone, J. D. (1994). Effects of dopamine depletions in the medial prefrontal cortex on DRL performance and motor activity in the rat. *Brain Research, 642*(1–2), 20–28. https://doi.org/10.1016/0006-8993(94)90901-6

Sultan, K. T., & Shi, S. H. (2018). Generation of diverse cortical inhibitory interneurons. *Wiley Interdisciplinary Reviews. Developmental Biology*, *7*(2). https://doi.org/10.1002/wdev.306.

Terra, H., Bruinsma, B., de Kloet, S. F., van der Roest, M., Pattij, T., & Mansvelder, H. D. (2020). Prefrontal cortical projection neurons targeting dorsomedial striatum control behavioral inhibition. *Current Biology, 30*(21), 4188-4200.e5. https://doi.org/10.1016/j.cub.2020.08.031

Trujillo, P., van Wouwe, N. C., Lin, Y. C., Stark, A. J., Petersen, K. J., Kang, H., Zald, D. H., Donahue, M. J., & Claassen, D. O. (2019). Dopamine effects on frontal cortical blood flow and motor inhibition in Parkinson's disease. *Cortex, 115*, 99–111. https://doi.org/10.1016/j.cortex.2019.01.016

Tsutsui-Kimura, I., Ohmura, Y., Izumi, T., Yamaguchi, T., Yoshida, T., & Yoshioka, M. (2009). The effects of serotonin and/or noradrenaline reuptake inhibitors on impulsive-like action assessed by the three-choice serial reaction time task: A simple and valid model of impulsive action using rats. *Behavioural Pharmacology, 20*(5–6), 474–483. https://doi.org/10.1097/FBP.0b013e3283305e65

van Velzen, L. S., Vriend, C., de Wit, S. J., & van den Heuvel, O. A. (2014). Response inhibition and interference control in obsessive-compulsive spectrum disorders. *Frontiers in Human Neuroscience, 8*, 419. https://doi.org/10.3389/fnhum.2014.00419

Wallis, D., Hill, D. S., Mendez, I. A., Abbott, L. C., Finnell, R. H., Wellman, P. J., & Setlow, B. (2012). Initial characterization of mice null for Lphn3, a gene implicated in ADHD and addiction. *Brain Research, 1463*, 85–92. https://doi.org/10.1016/j.brainres.2012.04.053

Winstanley, C. A., Dalley, J. W., Theobald, D. E., & Robbins, T. W. (2004). Fractionating impulsivity: Contrasting effects of central 5-HT depletion on different measures of impulsive behavior. *Neuropsychopharmacology, 29*(7), 1331–1343. https://doi.org/10.1038/sj.npp.1300434

Winstanley, C. A., Zeeb, F. D., Bedard, A., Fu, K., Lai, B., Steele, C., & Wong, A. C. (2010). Dopaminergic modulation of the orbitofrontal cortex affects attention, motivation and impulsive responding in rats performing the five-choice serial reaction time task. *Behavioural Brain Research, 210*(2), 263–272. https://doi.org/10.1016/j.bbr.2010.02.044

Wischhof, L., Hollensteiner, K. J., & Koch, M. (2011). Impulsive behaviour in rats induced by intracortical DOI infusions is antagonized by co-administration of an mGlu2/3 receptor agonist. *Behavioural Pharmacology, 22*(8), 805–813. https://doi.org/10.1097/FBP.0b013e32834d6279

Wischhof, L., & Koch, M. (2012). Pre-treatment with the mGlu2/3 receptor agonist LY379268 attenuates DOI-induced impulsive responding and regional c-Fos protein expression. *Psychopharmacology, 219*(2), 387–400. https://doi.org/10.1007/s00213-011-2441-y

Worbe, Y., Savulich, G., Voon, V., Fernandez-Egea, E., & Robbins, T. W. (2014). Serotonin depletion induces 'waiting impulsivity' on the human four-choice serial reaction time task: Cross-species translational significance. *Neuropsychopharmacology, 39*(6), 1519–1526. https://doi.org/10.1038/npp.2013.351

World Health Organization. (2022). *ICD-11: International classification of diseases* (11th revision). https://icd.who.int/

Yates, J. R. (2024). Aberrant glutamatergic systems underlying impulsive behaviors: Insights from clinical and preclinical research. *Progress in Neuro-Psychopharmacology & Biological Psychiatry, 135*, Article 111107. https://doi.org/10.1016/j.pnpbp.2024.111107

Yates, J. R., Darna, M., Beckmann, J. S., Dwoskin, L. P., & Bardo, M. T. (2016). Individual differences in impulsive action and dopamine transporter function in rat orbitofrontal cortex. *Neuroscience, 313*, 122–129. https://doi.org/10.1016/j.neuroscience.2015.11.033

Ye, Z., Altena, E., Nombela, C., Housden, C. R., Maxwell, H., Rittman, T., Huddleston, C., Rae, C. L., Regenthal, R., Sahakian, B. J., Barker, R. A., Robbins, T. W., & Rowe, J. B. (2014). Selective serotonin reuptake inhibition modulates response inhibition in Parkinson's disease. *Brain, 137*(Pt 4), 1145–1155. https://doi.org/10.1093/brain/awu032

Zapata, A., & Lupica, C. R. (2021). Lateral habenula cannabinoid CB1 receptor involvement in drug-associated impulsive behavior. *Neuropharmacology, 192*, Article 108604. https://doi.org/10.1016/j.neuropharm.2021.108604

Zhang, J. F., Wang, X. X., Feng, Y., Fekete, R., Jankovic, J., & Wu, Y. C. (2021). Impulse control disorders in Parkinson's disease: Epidemiology, pathogenesis and therapeutic strategies. *Frontiers in Psychiatry, 12*, Article 635494. https://doi.org/10.3389/fpsyt.2021.635494

Zhukovsky, P., Morein-Zamir, S., Meng, C., Dalley, J. W., & Ersche, K. D. (2020). Network failures: When incentives trigger impulsive responses. *Human Brain Mapping, 41*(8), 2216–2228. https://doi.org/10.1002/hbm.24941

Chapter 13
Maladaptive Eating

1 Introduction

During my lunch break, I often watch YouTube videos. This is my go-to method for decompressing after delivering lectures or performing administrative work as a department chair. One day, a thumbnail caught my attention. There was an image of a young man sitting behind a giant plate filled with enough food to feed a small party. I had stumbled on a *mukbang* video, in which an individual records themself eating large amounts of food. Mukbang is a Korean portmanteau, combining the words meongneun ("eating") and bangsong ("broadcast"). Out of curiosity, I played the video. I spent a few minutes watching this individual consume approximately 4 days' worth of food. I was simultaneously impressed and disturbed.

On average, humans can survive up to 3 weeks without food. Other species can survive without food for even longer periods of time. The microscopic tardigrade (better known as the water bear) can survive up to 30 years without food! My undergraduate research mentor I mentioned in the introduction of Chap. 2 currently studies tardigrades.

Not only is food necessary for survival, but it is a major part of our culture (Feldman & Goodman, 2021), and many of our daily routines revolve around eating (Jastran et al., 2009). We tend to eat our meals around the same time each day, and we typically cycle through the same meals throughout the week. Although crucial for our survival, there are several ways eating can become dysregulated. In this chapter, you will learn how some individuals severely limit their caloric intake while others consume too many calories. Other individuals will consume items that have no nutritional value and can cause physical harm.

J. Yates, *Neuroscience of Maladaptation*,
https://doi.org/10.1007/978-3-032-15602-0_13

2 Learning Objectives

By the end of this chapter, you should be able to …

1. Describe the mechanisms that control feeding behavior.
2. Explain the neuromechanisms associated with eating disorders.
3. Discuss the neurobiological underpinnings of obesity.

3 Neurobiological Mechanisms Controlling Feeding Behavior

Consuming food provides us with the nutrients and the energy needed for our bodies to function. We need a lot of energy to keep vital physiological processes functioning, such as thermal regulation, breathing, blood circulation, repairing and growing cells, maintaining membrane potentials, and regulating neurotransmitter/hormone levels. This is known as **basal metabolism**. There are different equations that can be used to determine one's basal metabolic rate, with the revised Harris-Benedict equation (Harris & Benedict, 1918) most commonly used (Roza & Shizgal, 1984). If you are male, use the following formula:

$$\left[(\text{Weight (in kg)} * 13.397) + (\text{Height (in cm)} * 4.799) - (\text{current age} * 5.677)\right] + 88.362$$

The formula for females is the same, but the values differ:

$$\left[(\text{Weight (in kg)} * 9.247) + (\text{Height (in cm)} * 3.098) - (\text{current age} * 4.330)\right] + 447.593$$

Using the formula above, my current basal metabolism requires approximately 1,752 cal per day. Caloric intake needs will vary widely across individuals. A body builder trying to gain muscle mass will need to consume more calories than someone that is actively trying to lose weight. To gain weight, caloric intake needs to surpass the number of calories burned from basal metabolism and from activities like weight training.

Our diet is important because nine of the 20 amino acids that make up proteins come from food sources only: histidine, isoleucine, leucine, lysine, methionine, phenylalanine, threonine, tryptophan, and valine. You have already learned about phenylalanine and tryptophan as they are precursors to the catecholamine dopamine and the indolamine serotonin, respectively. Certain foods also provide important fatty acids and vitamins that our bodies cannot produce. For example, we get vitamin A from dark green leafy vegetables (e.g., spinach), orange-yellow vegetables and fruits (e.g., sweet potato, cantaloupe), and eggs.

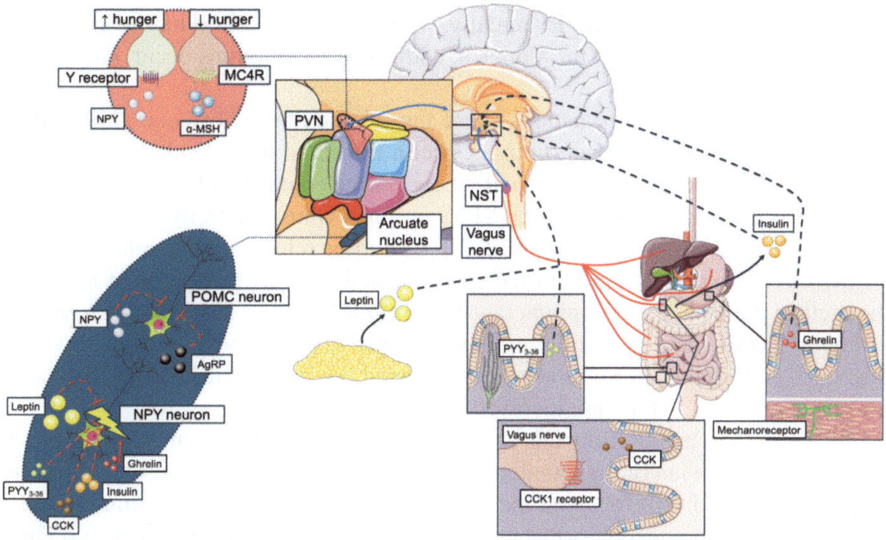

Fig. 1 A schematic detailing the mechanisms that control hunger, satiation, and satiety. Hunger is largely stimulated by release of ghrelin from the stomach, which stimulates neuropeptide Y (NPY) neurons in the arcuate nucleus. NPY neurons release NPY and agouti-related peptide (AgRP), which inhibit proopiomelanocortin (POMC) neurons. NPY travels to the paraventricular nucleus (PVN) and stimulates Y receptors. Satiation is controlled by stimulation of mechanoreceptors in the stomach and the release of cholecystokinin (CCK) from the duodenum. CCK binds to CCK1 receptor located on the vagus nerve. PYY_{3-36} released from the ileum and the colon also contributes to satiation. CCK and PYY_{3-36} inhibit NPY neurons, allowing for POMC neurons to release α-melanocyte-stimulating hormone (α-MSH), which stimulates melanocortin type-4 receptors (MC4Rs). Leptin, released from adipose tissue, and insulin, produced by the pancreas, contribute to satiety as these hormones also inhibit NPY neurons. *Note* NST = nucleus of the solitary tract. Image created by the author using Servier Medical Art (https://smart.servier.com/; CC BY 4.0, https://creativecommons.org/licenses/by/4.0/) and Microsoft PowerPoint

There are numerous mechanisms that control homeostatic regulation of food intake (Cifuentes & Acosta, 2022). These mechanisms control three different processes related to feeding: hunger, satiation, and satiety (see Fig. 1).

3.1 *Hunger*

When our body needs more energy, we become hungry. I am sure you have felt the uncomfortable sensation of hunger pangs; some of you may also feel light-headed or weak when hungry. There is even a term to signify the irritability some people experience when they are hungry—"hangry". Why do we experience hunger pangs? First, at some point after we have eaten, the stomach and the small intestine begin to contract in short bursts. These contractions allow for gastric emptying (i.e., removal of content from the stomach). Eventually, when most of the food has been digested and has been removed from the stomach, additional contractions occur, particularly in the stomach.

Continued stomach contractions cause the release of the hormone *ghrelin* from endocrine cells in the stomach lining. Ghrelin travels to the hypothalamus, specifically the arcuate nucleus. This part of the brain contains *NPY neurons*, which release *neuropeptide Y* (NPY; hence NPY neurons). Ghrelin directly stimulates NPY neurons. In addition to releasing NPY, NPY neurons release *agouti-related peptide* (AgRP). NPY and AgRP bind to receptors located on a different type of neuron in the arcuate nucleus, the proopiomelanocortin (POMC) neuron. NPY and AgRP act to inhibit POMC neurons. You previously learned about POMC in Chap. 4 when I discussed endorphins. In addition to being cleaved into endorphins, POMC is cleaved into melanocortins, which are involved in feeding behavior.

NPY neurons travel from the arcuate nucleus to two different regions of the hypothalamus, the lateral hypothalamic area (LHA) and the paraventricular nucleus (PVN). For now, I will focus on the PVN. This part of the hypothalamus contains neurons that express receptors that are selective for NPY (called Y receptors). There are six known Y receptor types, Y1-Y6. Stimulation of Y1 and Y5 receptors in the PVN has **orexigenic** effects (Beck, 2006) (note, do not confuse the term orexigenic with orexinergic, which is used to describe the orexin system; see Chaps. 8 and 14). Neurons from the PVN travel to the nucleus of the solitary tract (NST) in the medulla. This part of the brain integrates incoming signals from the brain and the peripheral nervous system (PNS) concerning hunger and satiety.

One thing you most likely have noticed during the course of your life is that eating can be quite enjoyable. I imagine there are some foods that you enjoy more than others. Our mesocorticolimbic pathway controls hedonic responses to food (Kenny, 2011). Many foods act as reinforcers, increasing dopamine neurotransmission. When we encounter stimuli that signal the availability of a preferred food, dopaminergic activity in the ventral tegmental area (VTA) increases, leading to increased dopamine release in the nucleus accumbens (NAc) and the prefrontal cortex (PFC).

Endocannabinoids are also important mediators of hunger. One common effect of cannabis is the "munchies", in which appetite increases, particularly for palatable foods that are high in salt, sugar, and/or fat. When individuals consume highly palatable food, both ghrelin and endocannabinoid levels increase in the PNS (Monteleone et al., 2012). Ghrelin also directly interacts with the endocannabinoid system in the VTA, modulating food reward (Edwards et al., 2025).

3.2 Satiation

How do we know we have consumed enough food? Satiation is a process that occurs while we eat, acting to control our meal size and as a signal to stop eating. There are several mechanisms that control satiation. First, as we eat, our stomach stretches. At a certain point, special *mechanoreceptors* are activated, causing the vagus nerve (cranial nerve X) to send a signal to the NST. The vagus nerve is also stimulated by the peptide hormone *cholecystokinin (CCK)*, which is released from the duodenum. From the NST, neurons travel to the hypothalamus to suppress hunger signals.

Furthermore, the cells in the ileum (last part of the small intestine) and the colon release a different peptide hormone called *$PYY_{3\text{-}36}$*. $PYY_{3\text{-}36}$ travels to the arcuate nucleus where it acts in the opposite manner as ghrelin. Instead of stimulating NPY neurons, $PYY_{3\text{-}36}$ inhibits them. This causes disinhibition of POMC neurons, allowing them to send messages to the LHA; appetite-suppressing signals are then transmitted back to the NST. POMC neurons also send axons to the PVN, where they release *α-melanocyte-stimulating hormone (α-MSH)*. This peptide hormone binds to G protein-coupled melanocortin type-4 receptors (MC4Rs). This activates appetite-suppressing neurons in this region.

3.3 Satiety

Satiety is the feeling of fullness after we have finished eating. Notice that satiation and satiety sound very similar. Think of satiation as a process that leads to satiety. When we consume more calories than we burn, we accumulate lipids, fatty acids, and glycerol. Lipids are what we colloquially refer to as "fat". Lipids are stored in adipose tissue, a collection of fat cells. We have two major types of adipose tissue: white and brown. When we eat more calories than we burn, white adipose tissue stores this excess energy. This is what leads to weight gain. Conversely, brown adipose tissue burns energy.

As higher concentrations of lipids are stored in fat cells, these cells release the peptide hormone *leptin*. Leptin, like $PYY_{3\text{-}36}$, inhibits NPY neurons in the arcuate nucleus, thus inhibiting appetite. Fat is beneficial because our body can use it for energy when we experience long periods of food deprivation. Specifically, fat can be metabolized into *glucose*, a type of sugar, and *ketones*, a type of acid, which our body can then use as fuel.

While fat can be converted into glucose, the primary source of this simple sugar is carbohydrates ("carbs"). Carbohydrates are found in foods like grains, fruits, and vegetables. Our body, especially our brain, needs glucose for energy. Due to the importance of glucose, we have a built-in mechanism for storing excess glucose for later use. The pancreas produces the critical hormone *insulin*. Insulin converts glucose into glycogen, a more complex carbohydrate, which is then stored in different parts of the body, particularly the liver and the skeletal muscles. Another pancreatic hormone, *glucagon*, converts glycogen back into glucose when the body needs more energy. Insulin also inhibits NPY neurons, helping suppress appetite.

Diabetes mellitus (often simplified to diabetes) is a serious medical condition in which the pancreas fails to produce insulin (Type I diabetes) or the body no longer uses insulin properly (Type II diabetes). As a result, glucose levels rise to dangerous levels. Because glucose is no longer converted into glycogen, the body must rely on lipids for energy. The excess glucose is excreted through urine, which is where the term diabetes mellitus originates. Diabetes is Greek for "siphon" or "to pass through", and mellitus is Latin for "honeyed" or "sweet"—thus, "passing honey".

As individuals with diabetes cannot store glycogen for future use, it relies heavily on ketones for energy. The excess buildup of ketones leads to *ketoacidosis*. Remember, ketones are a type of acid; excess ketone levels cause the blood to become too acidic. If untreated, ketoacidosis can lead to organ failure and death. One of my former students was hospitalized because of ketoacidosis (fortunately, she was ok after treatment).

Luckily, diabetes can be managed. Individuals can monitor their blood sugar levels, either by testing a small blood sample with a glucose meter or by implanting a continuous glucose monitor under the skin. To control glucose levels, individuals self-inject insulin. Unlike diabetes, the eating disorders that will be discussed for the rest of this chapter are much more difficult to treat as they involve psychological and sociocultural factors beyond aberrant neurobiological processes.

Comprehension Check 1 Explain the difference between satiation and satiety; how do the hormones $PYY_{3\text{-}36}$ and leptin act on NPY neurons to regulate these states?

Satiation is the process that occurs during a meal, signaling that you have consumed enough food and should stop eating. It is primarily a mechanical process driven by the stretching of the stomach. In contrast, satiety is the prolonged feeling of fullness after a meal has ended. It is a long-term signal that helps regulate the time between meals and is controlled by hormones released in response to nutrient absorption. $PYY_{3\text{-}36}$ *is released from the intestines after food consumption and inhibits NPY neurons. This inhibition reduces the hunger signal and helps trigger satiation. Leptin is produced by fat cells as they store more energy; as such, leptin is a long-term signal of energy sufficiency. Leptin also inhibits NPY neurons, contributing to the feeling of satiety and suppressing appetite over a longer period.*

4 Eating Disorders

Eating disorders are serious mental health disorders characterized by abnormal or disturbed eating habits that significantly impact physical and mental health and can be life threatening. Both the *Diagnostic and Statistical Manual of Mental Disorders (DSM-5)* (American Psychiatric Association, 2013) and the *International Classification of Diseases (ICD-11)* (World Health Organization, 2019) include sections on eating disorders. This section will cover four specific eating disorders. As you read this section, you will notice that these eating disorders include maladaptive behaviors that have already been covered in this book (e.g., stress, anxiety, and compulsivity). Because these disorders involve vastly different types of behaviors, the subsections below will describe each eating disorder, as well as altered brain morphology/function and aberrant neurotransmission associated with each of these disorders.

4.1 Anorexia Nervosa

Perhaps the most well-known eating disorder is **anorexia nervosa** (often shortened to anorexia). Individuals with anorexia nervosa have a fear of gaining weight and have a distorted self-body image. Individuals are driven to significantly restrict their caloric intake, resulting in severe low body weight. Because individuals with anorexia nervosa do not consume enough calories to get the nutrients they need, they often experience multiple health complications. Specifically, anorexia nervosa is associated with cardiovascular problems like bradycardia (i.e., slow heartbeat) and heart failure; gastrointestinal issues such as constipation, abdominal pain, heartburn, and vomiting; muscle weakness; decreased bone density, which increases the risk of fractures; hormone dysregulation; and amenorrhea (i.e., loss of menstrual period) (Puckett et al., 2021).

4.1.1 Animal Models of Anorexia Nervosa

As anorexia nervosa can be life threatening, there have been attempts to model this psychiatric condition in animals (Kim, 2012). For simplicity, only a few of these models will be discussed below.

At face value, the simplest way to model anorexia nervosa is to greatly restrict how much food an animal can consume each day. Mild food restriction is commonly used in animal research to motivate subjects to learn an operant response. If one wants to train a rat to press a lever to earn a food pellet, food restriction can help the rat acquire this task at a faster rate. To study anorexia nervosa, researchers can severely limit how much food an animal receives each day. For example, in one study, control mice were allowed to consume as much food as they wanted each day whereas experimental mice were given just 2 g of food each day (Idrobo et al., 1987). Adult mice typically eat between 4 and 5 g of food per day. While food-deprived animals gain significantly less weight than controls, food deprivation does not fully model the food avoidance that is observed in individuals with anorexia nervosa.

In the activity-based anorexia model, animals are first food restricted such that they are given access to food for a limited duration each day (e.g., 1 h). Rodents are also given access to a running wheel. Because rodents enjoy running on a wheel, they will often consume significantly less food compared to control animals that do not have access to a running wheel (Avraham et al., 2001). In extreme cases, rodents have died due to weight loss (Routtenberg & Kuznesof, 1967). This model faces the same major challenge that was mentioned above. Individuals with anorexia nervosa do not experience periods in which they cannot access food. These individuals can acquire food whenever they want, but they do not do so.

Anorexia nervosa can be studied in animals using neuroscience techniques, such as damaging parts of the brain, directly altering neurotransmission, or altering gene expression. Some of this work will be described in the subsection below.

4.1.2 Neuromechanisms Underlying Anorexia Nervosa

As noted by Kaye et al. (2010), studying the neurobiology of anorexia nervosa is challenging because brain alterations observed in these individuals may be the result of malnutrition only; that is, determining if neurobiological differences observed in those with anorexia nervosa are the cause or the result of food restriction is difficult. Despite this methodological challenge, this section will describe some of the potential neuromechanisms involved in anorexia nervosa.

Earlier in this chapter, I mentioned that NPY neurons travel to both the PVN and the LHA. If the LHA is damaged, rodents will eat significantly less food than controls (Lytle & Campbell, 1975; Winn et al., 1990). This finding is consistent with hypothalamic hypoactivity that is commonly observed in individuals with anorexia nervosa (Holsen et al., 2012) and in cancer patients experiencing anorexia (Molfino et al., 2017).

The hypothalamus is not the only brain region that appears to become dysregulated in individuals with anorexia nervosa. When these individuals see food, they show less activation of visual processing areas of the brain and in the precuneus (Datta et al., 2025). Recall the precuneus is an important component of the default mode network (DMN) and is involved in self-referential thought. However, increased activity of the PFC is observed in individuals with anorexia nervosa (Pauligk et al., 2025), and activity of the anterior cingulate cortex (ACC) and the orbitofrontal cortex (OFC) increases in individuals with anorexia nervosa, but not control individuals, following presentation of food images (Uher et al., 2003). Additionally, reduced functional connectivity between the OFC and the hypothalamus and between the dorsolateral PFC (dlPFC) and the dorsal striatum is observed in these individuals (Frank et al., 2016; Muratore et al., 2024). One hypothesis is that anorexia nervosa is characterized by "overcontrol", a form of excessive self-control that enables individuals to avoid eating for extended periods of time.

Altered prefrontal cortical control of feeding seems to be driven, at least partially, by dysfunctional dopaminergic signaling. In animals, optogenetic inhibition of D_1-like-containing neurons decreases feeding (Land et al., 2014). In fact, alterations to the mesocorticolimbic pathway, which relies heavily on dopaminergic activity, is altered in individuals with anorexia nervosa. Overall, hypoactivation of the ventral striatum and the ACC is observed in individuals with anorexia nervosa completing tasks that involve taste (Datta et al., 2025). Even when individuals complete a task involving monetary rewards, individuals with anorexia nervosa show differential activation of the ventral striatum, indicating that these individuals have difficulty identifying the emotional significance of stimuli (Wagner et al., 2007). In the ventral striatum, individuals recovering from anorexia nervosa have increased dopamine $D_{2/3}$

receptor binding (Frank et al., 2005), which could act to inhibit further dopamine release in the NAc.

Another major alteration observed in individuals with anorexia nervosa is *astrogliosis* in the shell region of the NAc and in the VTA (Kawakami et al., 2022). Astrogliosis is an abnormal increase in the number of astrocytes, typically due to damage to neurons. In other words, neuronal connections between the VTA and the NAc are disrupted in individuals with anorexia nervosa. Given the shell region is involved in processing the hedonic aspects of stimuli, damage to this region may explain why individuals with anorexia nervosa show little motivation to eat.

I need to note that studies have shown both increased and decreased activation of the mesocorticolimbic pathway when individuals are presented with images of food (Steinglass et al., 2019), making interpretation of the above findings somewhat difficult. With that said, women with anorexia nervosa show enhanced ventral striatal activation when viewing images of underweight individuals, whereas control women show greater activation when viewing normal weight individuals (Fladung et al., 2010). This finding indicates that women with anorexia nervosa find being underweight reinforcing.

One system that is heavily affected by anorexia nervosa is the hypothalamic–pituitary–adrenal (HPA) axis (see Chaps. 8 and 9). The HPA axis is hyperactive in individuals with anorexia nervosa, and cortisol levels are directly correlated with hedonic appetite (Lawson et al., 2013). Importantly, this hyperactivity may be the consequence of chronic starvation as opposed to being a cause of anorexia nervosa (Bou Khalil et al., 2017). In animals, activating the HPA axis via stress leads to **hypophagia** and significant weight loss (Jiang & Eiden, 2016).

Concerning taste, the anterior insula is closely related to the reward pathway as this structure processes taste of food as well as the rewarding properties of food. When sucrose water is administered, control individuals show greater activation of the anterior insula compared to individuals in recovery for anorexia nervosa; additionally, for control individuals, but not those in recovery for anorexia nervosa, there is a positive correlation between self-reported rating of pleasantness and activation of the anterior insula, as well as the putamen and the ACC (Wagner et al., 2008).

Comprehension Check 2 As you read in the passage above, there is difficulty determining if observed neurobiological changes are a cause of anorexia nervosa or simply a result of prolonged starvation associated with this disorder. How can researchers better differentiate between neurobiological changes that may be a cause of the disorder versus those that are simply a consequence of starvation?

Researchers can address this by studying individuals who are in recovery from anorexia nervosa, as their brains are no longer in a state of starvation. As described above, animal studies can be conducted in which specific pathways are targeted to better elucidate the mechanisms that control feeding behavior.

4.2 Bulimia Nervosa and Binge Eating Disorder

Like anorexia nervous, individuals with **bulimia nervosa** are concerned about weight gain and self-image. Instead of restricting food intake, individuals will engage in *binge eating*, in which they consume more food than what is typical for them, followed by an inappropriate compensatory behavior to prevent weight gain. Such behaviors include self-induced vomiting, misuse of laxatives or diuretics, and/or excessive exercising. These compensatory behaviors can have serious health effects on an individual. Self-induced vomiting can lead to dental problems like periodontal disease, damage to the throat and gastrointestinal tract, dehydration, and cardiovascular complications. Laxative misuse can also cause damage to the gastrointestinal tract and can lead to hypovolemia (i.e., too much loss of bodily fluids) (Mehler & Rylander, 2015).

In contrast to bulimia nervosa, individuals with **binge eating disorder** do not engage in compensatory behaviors to prevent weight gain following a binge. Instead, their behavior resembles those of substance use disorders and behavioral addictions (Gearhardt et al., 2011; Schreiber et al., 2013). The major complication resulting from binge eating disorder is **obesity**, the excessive accumulation of body fat. Approximately 70% of individuals with binge eating disorder are considered obese (Wassenaar et al., 2019). Obesity will be discussed in more detail later in this chapter.

What is the difference between being overweight and being obese? Think of obesity as a more severe form of being overweight. The body mass index (BMI) is often used to calculate an individual's body fat, using the formula weight (in kilograms)/(height (in meters))^2. A BMI under 18.5 is considered underweight, a BMI between 25 and 30 is considered overweight, and a BMI above 30 is considered obese. BMI is not a perfect system for classifying someone as overweight (e.g., body builders can have a BMI above 25). Despite this criticism, BMI provides a simple way of determining if someone is obese.

While obesity is linked to diseases like Type 2 diabetes and cardiovascular disease, individuals with binge eating disorder are at greater risk for developing these conditions compared to individuals with obesity that do not engage in binge eating. Individuals with binge eating disorder are also at risk for developing autoimmune disorders, liver disease, gastrointestinal issues, urinary incontinence, amenorrhea, decreased bone density, and malnutrition (Wassenaar et al., 2019).

4.2.1 Intermittent Food Access to Model Binge Eating Behavior in Animals

Modeling bulimia nervosa in animals is extremely challenging as rodents cannot vomit and do not have access to laxatives. Binge eating can be modeled, however. Instead of restricting access to all food sources as in the food restriction model described above, animals can receive unlimited access to normal food while receiving intermittent access to highly palatable foods like sucrose. In one study, rats were given limited access to a sucrose-rich, chocolate-flavored diet for either 0 h, 2 h, 4 h, or 8 h each day. Rats given just 2 h of access to the highly palatable food showed the greatest binge eating-like behavior (Kreisler et al., 2018).

4.2.2 Neuromechanisms Underlying Bulimia Nervosa and Binge Eating Disorder

Like individuals with anorexia nervosa, aberrant activity is observed in the DMN in individuals with bulimia nervosa. When given a cognitive task, the DMN fails to deactivate in these individuals (Seitz et al., 2016). This may account for why individuals with bulimia nervosa become preoccupied with body image and food intake. Relative to healthy controls, individuals with bulimia nervosa show deactivation of brain structures in frontal-striatal regions (e.g., PFC, striatum, ACC) when performing a conflict-resolution task (Wu et al., 2024). In bulimia nervosa, there is a conflict between food consumption and weight gain avoidance. Alternating binge eating and purging may reflect a deficit in conflict resolution in these individuals (Marsh et al., 2011). Similar to bulimia nervosa, individuals with binge eating disorder have a hypoactive PFC (Wu et al., 2024), and pharmacological inactivation of the PFC increases binge eating in rodents (Sinclair et al., 2019). Insufficient activity in parts of the PFC can lead to poor executive functioning, thus preventing individuals with bulimia nervosa and binge eating disorder from inhibiting themselves from overeating. As you have already learned, hypoactivity of the PFC is observed in individuals with substance use disorder and behavioral addictions (see Chap. 11).

Like addiction, the dopaminergic system in the NAc is hyperresponsive to food-predictive cues (Bello & Hajnal, 2010; Broft et al., 2011). Animal studies show that food-restricted rodents, as well as rodents given intermittent access to sucrose, have greater dopamine release in the NAc when they are given sucrose (Avena et al., 2008; Bocarsly et al., 2014). Relatedly, individuals with binge eating disorder show hyperactivity of the ACC (Wu et al., 2024), which is important for assigning value to rewards, and they show increased activation of the ACC and the insula when exposed to food cues compared to non-food cues (Aviram-Friedman et al., 2018). There is also increased functional connectivity between the dorsal ACC (dACC) and the OFC in individuals with bulimia nervosa. Food may be weighed more heavily in these individuals, accounting for addictive-like eating patterns. The Experiment Spotlight reports how increased activation of regions like the ACC can be used to predict the development of binge eating disorder.

Experiment Spotlight: Neural Risk Factors that Predict the Future Onset of Binge Eating or Compensatory Weight Control Behaviors: A Prospective 4-Year fMRI Study The full article can be accessed with this link: https://pmc.ncbi.nlm.nih.gov/articles/PMC12055025/

Stice et al. (2025) recruited 88 adolescent females (mean age of 14.5 years) to complete an fMRI (functional magnetic resonance imaging) study. Participants performed several tasks while in the MRI machine:

(1) Model image paradigm—participants viewed images of 20 thin models and 20 average-weight models and were asked to think about the model's attractiveness.
(2) Food image paradigm—participants viewed 20 images of high-calorie binge foods and 20 images of glasses of water and were asked to imagine eating each food or drinking the glass of water.
(3) Food receipt paradigm—participants viewed 30 images of milkshakes and 30 images of water (with images randomized) before receiving access to a chocolate milkshake or a tasteless solution. For example, if a participant saw an image of a milkshake, they would receive a delivery of chocolate milkshake they could taste.
(4) Food go/no-go paradigm—participants were presented with images of desserts and vegetables. In one version of the test, participants were instructed to press a button when they saw an image of a vegetable and to withhold from responding when they saw an image of a dessert. Participants completed the test a second time and were given the opposite instructions.
(5) Negative mood induction paradigm—participants viewed statements that were associated with a negative mood or with a neutral mood (20 statements each). Participants were instructed to think about a time in which they felt like the statement they read.

Additional data were collected outside of the MRI machine. Participants completed the Eating Disorder Diagnostic Interview (EDDI), which assesses symptoms of eating disorders. They also completed the Ideal-Body Stereotype Scale-Revised, which measures pursuit of the *thin ideal* (concept of an ideally thin female). Participants assessed the valuation of high-calorie foods and thin models, and they completed the Food Craving Inventory; this scale allows one to measure craving for high-calorie foods. Participants complete other measures that I will not explain here. Finally, participants' biological parents completed a version of the EDDI. Participants were evaluated 4 years later to determine if any of the measures predicted the onset of binge eating disorder.

Out of the 88 participants, five engaged in binge eating, and 17 engaged in some form of compensatory weight control. Participants that developed eating-disordered behaviors showed greater activity of the right dACC when viewing images of thin models compared to average-weight models. These individuals also showed less activation of the right caudate nucleus when viewing images of milkshakes relative to water. No other group differences were observed. Several factors were shown to predict the onset of disordered eating: emotionality, BMI, overvaluation of weight or shape, feeling fat, and parental history of eating disorder symptoms.

Questions to Consider

(1) Given that fMRI measures correlational neural activation (i.e., does not directly measure neuronal activation in response to stimuli), how can this experiment be modified to allow one to draw causal inferences about differential neural responses in individuals with and without disordered eating behaviors?
(2) How would you interpret the finding that individuals with disordered eating show less activation of the caudate nucleus when viewing cues that predict milkshake

delivery? This result seems counterintuitive as individuals that engage in binge eating/bulimia-like behavior show hyperactivity of food-related cues.

(3) This study used female participants only. Why do you believe this was the case?

While a hyperactive is HPA axis associated with anorexia nervosa, the HPA axis is implicated in excessive eating. Normally, acute stress inhibits eating as this action is not as important to us during a "fight-or-flight" situation (e.g., escaping a potential threat). However, stress can promote eating. You may even be familiar with the term "stress eating". Overactivation of the HPA axis can directly affect systems that regulate feeding behavior. Stimulation of glucocorticoid receptors in the mesocorticolimbic pathway is proposed to sensitize us to appetitive stimuli such as drugs and palatable food (Berner et al., 2019). In Chap. 8, I mentioned Cushing syndrome, a condition marked by excess cortisol levels. Individuals with Cushing syndrome experience increased food craving and are at increased risk for obesity (Geer et al., 2016).

Comprehension Check 3 Explain the similarities between bulimia nervosa/binge eating disorder and substance use disorders, focusing on the roles of the mesocorticolimbic pathway, the PFC, and the HPA axis.

The mesocorticolimbic pathway, which heavily relies on dopaminergic signaling, is involved in reward and motivation. In both eating disorders and addiction, this system becomes hyperresponsive to cues associated with food/drugs. This hypersensitivity to cues can drive compulsive behaviors, making individuals more likely to engage in binge eating/substance use. The PFC is critical for executive functions, including impulse control and decision-making. Like substance use disorders, individuals with bulimia nervosa and binge eating disorder show a hypoactive PFC. The HPA axis regulates the body's stress response. Overactivation of the HPA axis is proposed to sensitize the mesocorticolimbic reward system to appetitive stimuli, including highly palatable foods and drugs.

4.3 Pica

So far, you have learned about conditions in which individuals highly limit their eating (anorexia nervosa) or engage in binge eating (bulimia nervosa and binge eating disorder). There is a condition in which individuals perform neither of these actions, instead opting to consume one or more nonnutritive, non-food substances. This is known as **pica**, the Latin word for magpie, a type of bird that is known to have an indiscriminate appetite. To receive a diagnosis of pica, consumption of the non-food substance must occur consistently across a 1-month period and must be severe enough to warrant clinical attention. This consumption must also be "developmentally inappropriate…and not part of a culturally supported or normative practice" (American Psychiatric Association, 2013, p. 330).

Pica is commonly observed in young children and individuals with an intellectual disability. For those reading this book, you most likely know some of the reasons why eating non-food items is problematic. There are numerous health complications that arise from pica (Williams et al., 2022). If an individual consumes something toxic (e.g., lead), this can lead to death. There is also the risk that the individual chokes while trying to swallow something that is not easily broken down by our saliva. Other items can cause physical damage to teeth and to the gastrointestinal tract. Because individuals with pica do not consume enough real food, they can become malnourished.

Pica has been argued to be a form of obsessive–compulsive disorder (OCD) as individuals engage in compulsive-like consumption of non-food items to alleviate anxiety (Rose et al., 2000). Given the rarity of pica, few studies have examined the neurobiological bases of this disorder. Interestingly, one risk factor for pica is repeated blood donations due to resulting iron deficiency (Liu et al., 2022). Excess NPY may also contribute to pica. In one study, administration of NPY increased consumption of *kaolin* (a type of clay mineral) in rats (Madden et al., 1999).

Damage to part of the middle temporal gyrus (MTG) is associated with pica and semantic memory deficits (Funayama et al., 2017). One proposal is that pica occurs because individuals have more difficulty recognizing objects. As discussed by Funayama et al. (2017), individuals with lesions to the posterior MTG examine unrecognizable objects as infants do, by using their hands and their mouth. This then promotes the individual to place the object in their mouth, increasing the chance the individual eats the object.

5 Obesity

Although not classified as an eating disorder, obesity is a serious health condition. From 1990 until 2022, worldwide obesity rates more than doubled, with adolescent obesity rates quadrupling during this period. Almost 1 in 8 people live with obesity (World Health Organization, 2024). Each year, excess weight contributes to between 2.7 and 5.3 million deaths, with approximately 60% of these deaths being attributed to obesity (GBD 2015 Obesity Collaborators et al., 2017).

One account for the rising obesity rates is that we have not been able to evolve quickly enough to adapt to decreasing food scarcity. When our species was still in the "hunter-gatherer" period, being able to store fat was advantageous. During periods of famine, individuals could use fat as an energy source. Some have criticized this hypothesis, arguing that it does not fully account for why obesity rates have only recently spiked (Speakman, 2013). One important consideration is the increased sugar consumption that occurred from the 1970s through the 1990s (Faruque et al., 2019).

5.1 Animal Models of Obesity

The simplest way to study obesity in animals is to provide them with high-fat diets (Hariri & Thibault, 2010). To provide an example, control rats can be fed a standard diet that consists of 67% carbohydrates, 20% protein, and 13% fats (2.9 kilocalories/gram) whereas experimental rats are given a high-fat diet that consists of 35% carbohydrates, 20% protein, and 45% fats (4.73 kilocalories/gram) (Fattiny et al., 2025). Rodents maintained on a high-fat diet gain significantly more weight than rodents given a standard diet.

Another method for studying obesity is to use rodents with genetic mutations to leptin-related genes. The ob/ob mouse (ob is short for obese) has a recessive mutation of the leptin gene, preventing it from producing leptin. The db/db mouse (db is short for diabetic) has a mutation of the *Leprdb* gene, which encodes for the leptin receptor; this prevents leptin signaling. Like the db/db mouse, the fa/fa (fa is short for fatty) Zuker rat has a genetic mutation of the leptin receptor. Limiting leptin signaling causes these rodents to severely overeat.

5.2 Neuromechanisms Linked to Obesity

At the beginning of this chapter, you learned about the contribution of the hypothalamus and several hormones/peptides to feeding behavior. This section will begin by discussing how dysfunction of the hypothalamus contributes to obesity.

5.2.1 Hypothalamic Dysfunction

Damage to the medial portion of the hypothalamus, particularly the ventromedial nucleus (VMH), leads to **hyperphagia**; that is, individuals have an insatiable appetite and will engage in excessive eating (Pinkney et al., 2002). This finding is supported by preclinical research showing obesity-induced lesions to the VMH (Reisin et al., 1987); however, lesions to the VMH do not necessarily cause weight gain via hyperphagia (Balkan et al., 1991). Instead, damage to this part of the brain may impair other processes like metabolism, which contributes to excessive weight gain.

One potential cause for hypothalamic dysfunction is neuroinflammation (Miller & Spencer, 2014). A high-fat diet produces an immune response in the hypothalamus, increasing microglia activation and cytokine levels. This immune response can effective "re-wire" the hypothalamus. Let's continue discussing how inflammation outside of the hypothalamus can promote obesity.

5.2.2 Metaflammation

Obesity is considered a metabolic disease as metabolism is greatly impaired. Specifically, individuals with obesity experience **metaflammation**, a chronic, low-grade inflammatory state primarily induced by metabolic disturbances. Individuals with obesity have increased macrophage expression in *visceral fat* (i.e., fat that surrounds internal organs), which may be attributed to elevated basal **lipolysis** (Patel et al., 2013). Lipolysis is the breakdown of triglycerides (type of lipid) into fatty acids and glycerol. Because individuals with obesity have more adipose tissue, fatty acids accumulate in the bloodstream, and certain fatty acids are linked to inflammation (Fritsche, 2015). Increased fatty acid levels promote the release of cytokines, which interfere with the insulin signaling pathway; this process is hypothesized to contribute to increased lipolysis, which further exacerbates insulin resistance that is often associated with obesity (Schleh et al., 2023).

In short, metaflammation leads to a vicious cycle (see Fig. 2). As visceral fat accumulates, there is increased lipolysis, leading to elevated fatty acid levels. This promotes the release of cytokines, thus attracting more macrophages. Macrophage activity then contributes to lipolysis.

5.2.3 Peptides and Neurotransmitters Implicated in Obesity

Earlier in this chapter, you learned about various peptides that control feeding behavior. As expected, dysregulation of these peptides is associated with obesity. In addition to feeding-related peptides, a couple of neurotransmitters will be discussed.

Leptin

Leptin is highly involved in satiation; as such, disruption of leptin signaling is linked to obesity as individuals feel hunger soon after eating. Leptin resistance is a major contributing factor to obesity (Obradovic et al., 2021). Individuals with obesity often have *increased* mRNA expression of the *ob* gene, which encodes for leptin (think of the ob/ob animal model described earlier), and plasma leptin levels are positively correlated with body weight. This means that individuals with obesity produce high leptin levels, but their body is unable to use it effectively to stimulate satiation. Leptin resistance may occur because leptin is not sufficiently released from white adipose cells and/or leptin receptor expression is downregulated.

Ghrelin, Agouti-Related Peptide (AgRP), and Neuropeptide Y (NPY)

Like leptin, ghrelin resistance is linked to obesity (Cui et al., 2017). This is counterintuitive as ghrelin helps stimulate appetite. Therefore, resistance to ghrelin should lead to a situation in which individuals feel less hunger. Ghrelin is hypothesized to

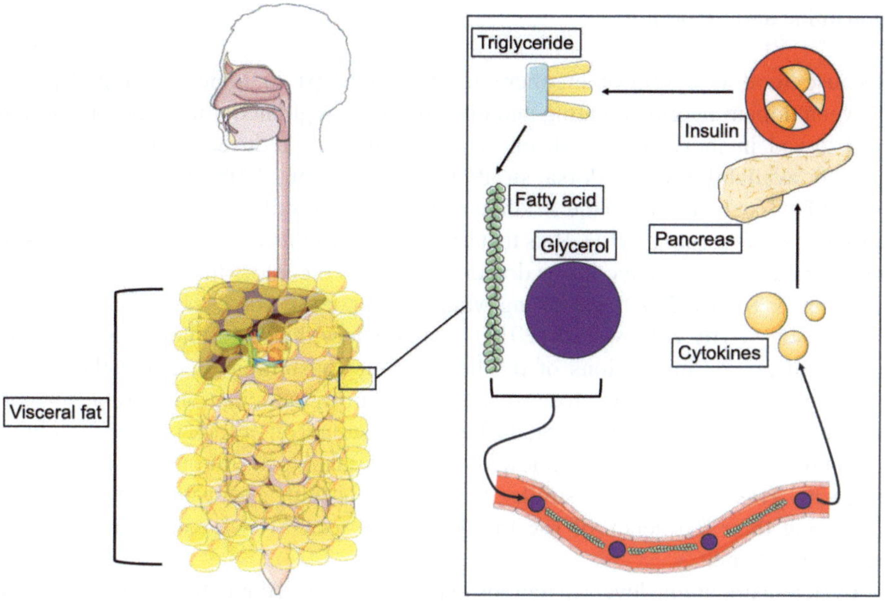

Fig. 2 A depiction of how metaflammation can promote obesity. As visceral fat increases, fat cells are more likely to undergo lipolysis. The accumulation of fatty acids in the bloodstream triggers the release of cytokines, which are involved in immune responses, including inflammation. Increased inflammation of the pancreas impairs insulin synthesis. Because insulin is an important satiety signal, individuals are more likely to overeat, leading to further increases in adipose tissue. This increases the probability of lipolysis and further release of cytokines. Image created by the author using Servier Medical Art (https://smart.servier.com/; CC BY 4.0, https://creativecommons.org/licenses/by/4.0/) and Microsoft PowerPoint

regulate our bodyweight set-point, which is our average-weight range. If an individual gains excessive weight, their set-point will increase substantially. As they gain weight, ghrelin levels increase, perhaps to counteract obesity-induced ghrelin resistance. When this individual begins to lose weight, the ghrelin system becomes hyperactive, promoting the individual to consume more food to return to their new set-point (Zigman et al., 2016). Essentially, a vicious cycle emerges in which individuals see rebound weight gain following weight-loss interventions.

AgRP and NPY levels are significantly elevated in obese animals (Bi & Moran, 2016; Dunbar et al., 2005). Hyperactive AgRP and NPY systems contribute to obesity by promoting hyperphagia, decreasing energy expenditure, and driving metabolic changes that lead to energy conservation (e.g., fat storage), with hypersensitivity of these systems appearing to be driven by leptin resistance (Luo et al., 2011; Zhang et al., 2011).

Dopamine and Serotonin

Because food is a reinforcer, there is much interest in understanding the link between dopaminergic signaling and obesity. A dysregulated dopaminergic system is observed in individuals with obesity, as they have reduced dopamine D_2-like receptor availability in the dorsal striatum (Volkow et al., 2008). Decreased expression of D_2-like receptors in the striatum is linked to compulsive food consumption in animals (Volkow et al., 2011). Rats that are allowed to consume excessive amounts of highly palatable food have lower dopamine D_2-like receptors in the dorsal striatum (Johnson & Kenny, 2010), mirroring what is observed in obese individuals.

As discussed by Volkow et al. (2011), individuals with obesity show less activation of reward processing regions of the brain when consuming food, but they show enhanced activation in somatosensory regions during anticipation of food receipt. Because less dopamine is released during the consumption of food, these individuals eat more to compensate for the low dopamine signaling in the mesocorticolimbic pathway. This is somewhat similar to individuals with a substance use disorder increasing their frequency of use to compensate for lower dopamine release that occurs over time.

Deficient dopamine levels in regions like the PFC can account for loss of inhibitory control overeating (Volkow et al., 2011). Indeed, individuals with obesity often show increased motor impulsivity (Lavagnino et al., 2016) and impulsive decision making (Amlung et al., 2016). Additionally, hypoactivity of the PFC is linked to compulsivity (see Chap. 11), and individuals with binge eating disorder (which often leads to obesity) show enhanced impulsivity and compulsivity (Carr et al., 2021).

Serotoninergic dysfunction is also heavily implicated in weight gain (van Galen et al., 2021). Normally, serotoninergic activity in the hypothalamus serves to promote satiety. Indeed, serotonergic drugs like selective serotonin reuptake inhibitors (SSRIs) decrease appetite in humans and promote hypophagia in animals when administered acutely (Halford et al., 2007). Postmortem studies identify lower serotonin levels in the cerebrospinal fluid (CSF) of individuals with obesity, and obese animals have lower serotonin release in the hypothalamus.

Hypoactive dopaminergic and serotonergic levels explain why antipsychotic medications often cause weight gain. Atypical antipsychotics block both dopamine D_2-like and serotonin 5-HT_{2A} receptors. Weight gain is also a major side effect of selective serotonin reuptake inhibitors (SSRIs). This seems paradoxical as SSRIs increase serotonin levels. By continuously blocking the serotonin transporter, there is a compensatory downregulation of serotonin 5-HT_{2A} receptors. The end result is the same as atypical antipsychotics—there is insufficient stimulation of serotonergic receptors that regulate satiety, leading to increased food consumption and weight gain.

Endocannabinoids

When discussing hunger, I mentioned the "munchies" one can experience after using cannabinoids. The endocannabinoid system is an important regulator of hunger, with excess endocannabinoid levels linked to obesity (Akurati & Hanlon, 2024). Cannabinoid receptors are found throughout the PNS and the central nervous system (CNS). In the hypothalamus, endocannabinoids act to stimulate orexigenic signaling pathways. Endocannabinoids also regulate the reinforcing effects of food as cannabinoid receptors are found in the mesocorticolimbic pathway; stimulation of these receptors causes inhibition of GABAergic (gamma-aminobutyric acid) neurons that normally inhibit dopamine release (Dörnyei et al., 2023).

Given the role of endocannabinoids in hunger, efforts have been made to use cannabinoid receptor antagonists as treatments for obesity. In the early 2000s, the CB_1 **inverse agonist** rimonabant showed promise as an anti-obesity drug. However, it was never approved in the United States as rimonabant also caused serious side effects, especially depressed mood; rimonabant was withdrawn from European countries in 2008 due to concerns over its side effect profile (Sam et al., 2011). Today, novel CB receptor antagonists with minimal side effect profiles have shown promise as treatments for obesity. Specifically, CB receptor antagonists have been developed that do not cross the blood–brainbarrier (BBB) into the CNS. PNS-selective CB receptor antagonists reduce body weight in animals (Gammal et al., 2025; Iyer et al., 2022).

Comprehension Check 4 Both leptin resistance and ghrelin resistance are related to obesity. Explain how this seemingly paradoxical finding contributes to a "vicious cycle" of weight gain and makes it difficult for individuals to sustain weight loss.

As individuals gain weight, their body produces high levels of leptin, a hormone that normally signals satiety and inhibits appetite. However, in obesity, the body becomes resistant to this signal. This means that despite high levels of the "fullness" hormone circulating in the bloodstream, the brain doesn't receive the message to stop eating. This dysfunctional signaling leads to continued overconsumption of food and prevents weight loss, as the body's primary long-term appetite-suppressing system is no longer working effectively. Conversely, when an individual attempts to lose weight, the ghrelin system becomes hyperactive, producing even more ghrelin. This surge in ghrelin drives the individual to eat more, trying to return to their previously higher weight "set-point". This mechanism essentially counteracts weight-loss efforts, leading to rebound weight gain.

6 Chapter Summary

In this chapter, you learned about the mechanisms that control feeding behavior. Several hormone peptides such as ghrelin, NPY, and AgRP, along with neurotransmitters such as dopamine and endocannabinoids, stimulate appetite. Other peptides like CCK, $PYY_{3\text{-}36}$, and leptin, as well as the neurotransmitter serotonin, promote satiation and/or satiety. The neurobiological underpinnings of conditions such as anorexia nervosa, bulimia nervosa, binge eating disorder, and pica were covered, highlighting how disruptions in brain function, from impulse-control issues in the PFC to hypothalamic dysfunction, contribute to these disorders. This chapter concluded by discussing how obesity, while not an eating disorder, is also deeply linked to these neurobiological processes, including hypothalamic dysfunction, as well as chronic inflammation.

Answers to Experiment Spotlight

(1) Researchers could use a methodology such as transcranial magnetic stimulation or transcranial direct current stimulation to temporarily suppress or enhance the activity of the dACC or the caudate nucleus in individuals that display disordered eating patterns. If hyperactivity of the dACC is involved in promoting the thin ideal, disrupting activity in this region should decrease preference for thin models.

(2) Keep in mind that neural activity was recorded before individuals developed disordered eating patterns. This study suggests that a *blunted* reward response to food might be a risk factor that precedes the onset of problematic eating. Another consideration is that the caudate nucleus is connected to the NAc and is involved in reward processing. Lower activity in response to palatable food cues could indicate that young adolescents prone to disordered eating have a lower hedonic response to such foods. This could lead them to seek more intense or different forms of reward, which could manifest as binge eating.

(3) Eating disorders, such as anorexia nervosa and bulimia nervosa, are significantly more prevalent in females compared to males. Also, this study examined the thin ideal, which is a sociocultural construct that is highly specific to women.

Glossary

Anorexia nervosa an eating disorder characterized by an intense fear of gaining weight, a distorted body image, and the restriction of food intake, often leading to severe health complications.

Basal metabolism the minimum amount of energy an organism requires to maintain essential functions at rest.

Binge eating disorder an eating disorder characterized by recurrent episodes of uncontrollable overeating.

Bulimia nervosa an eating disorder characterized by binge eating followed by compensatory behaviors, such as vomiting, excessive exercise, or fasting, to prevent weight gain.

Diabetes mellitus a chronic metabolic disease characterized by elevated blood glucose levels, resulting from the body's inability to produce or properly use insulin.

Hyperphagia an abnormally increased and often insatiable appetite for food, frequently leading to excessive eating.

Hypophagia an abnormally decreased appetite for food, frequently leading to starvation.

Inverse agonist a drug that binds to a receptor and produces a pharmacological effect opposite to that of an agonist by decreasing the receptor's baseline activity.

Lipolysis the process by which triglycerides (type of lipid) are metabolized into fatty acids and glycerol.

Metaflammation a chronic, low-grade inflammatory state primarily induced by metabolic disturbances.

Obesity a chronic disease characterized by excessive accumulation of body fat that poses a risk to health.

Orexigenic refers to a substance or process that stimulates appetite or increases food intake.

Pica an eating disorder characterized by the persistent eating of non-food substances.

References

Akurati, S., & Hanlon, E. C. (2024). Beyond the scale: Exploring the endocannabinoid system's impact on obesity. *Current Diabetes Reports, 25*(1), 6. https://doi.org/10.1007/s11892-024-01562-2

American Psychiatric Association. (2013). *Diagnostic and statistical manual of mental disorders* (5th ed.). https://doi.org/10.1176/appi.books.9780890425596

Amlung, M., Petker, T., Jackson, J., Balodis, I., & MacKillop, J. (2016). Steep discounting of delayed monetary and food rewards in obesity: A meta-analysis. *Psychological Medicine, 46*(11), 2423–2434. https://doi.org/10.1017/S0033291716000866

Avena, N. M., Rada, P., & Hoebel, B. G. (2008). Underweight rats have enhanced dopamine release and blunted acetylcholine response in the nucleus accumbens while bingeing on sucrose. *Neuroscience, 156*(4), 865–871. https://doi.org/10.1016/j.neuroscience.2008.08.017

Aviram-Friedman, R., Astbury, N., Ochner, C. N., Contento, I., & Geliebter, A. (2018). Neurobiological evidence for attention bias to food, emotional dysregulation, disinhibition and deficient somatosensory awareness in obesity with binge eating disorder. *Physiology & Behavior, 184*, 122–128. https://doi.org/10.1016/j.physbeh.2017.11.003

Avraham, Y., Hao, S., Mendelson, S., & Berry, E. M. (2001). Tyrosine improves appetite, cognition, and exercise tolerance in activity anorexia. *Medicine and Science in Sports and Exercise, 33*(12), 2104–2110. https://doi.org/10.1097/00005768-200112000-00020

Balkan, B., Steffens, A. B., Bruggink, J. E., & Strubbe, J. H. (1991). Hyperinsulinemia and glucose tolerance in obese rats with lesions of the ventromedial hypothalamus: Dependence on food intake and route of administration. *Metabolism, 40*(10), 1092–1100. https://doi.org/10.1016/0026-0495(91)90136-k

Beck, B. (2006). Neuropeptide Y in normal eating and in genetic and dietary-induced obesity. *Philosophical Transactions of the Royal Society of London. Series B, Biological Sciences, 361*(1471), 1159–1185. https://doi.org/10.1098/rstb.2006.1855

Bello, N. T., & Hajnal, A. (2010). Dopamine and binge eating behaviors. *Pharmacology, Biochemistry, and Behavior, 97*(1), 25–33. https://doi.org/10.1016/j.pbb.2010.04.016

Berner, L. A., Brown, T. A., Lavender, J. M., Lopez, E., Wierenga, C. E., & Kaye, W. H. (2019). Neuroendocrinology of reward in anorexia nervosa and bulimia nervosa: Beyond leptin and ghrelin. *Molecular and Cellular Endocrinology, 497*, Article 110320. https://doi.org/10.1016/j.mce.2018.10.018

Bi, S., & Moran, T. H. (2016). Obesity in the Otsuka Long Evans Tokushima fatty rat: Mechanisms and discoveries. *Frontiers in Nutrition, 3*, 21. https://doi.org/10.3389/fnut.2016.00021

Bocarsly, M. E., Hoebel, B. G., Paredes, D., von Loga, I., Murray, S. M., Wang, M., Arolfo, M. P., Yao, L., Diamond, I., & Avena, N. M. (2014). GS 455534 selectively suppresses binge eating of palatable food and attenuates dopamine release in the accumbens of sugar-bingeing rats. *Behavioural Pharmacology, 25*(2), 147–157. https://doi.org/10.1097/FBP.0000000000000029

Broft, A. I., Berner, L. A., Martinez, D., & Walsh, B. T. (2011). Bulimia nervosa and evidence for striatal dopamine dysregulation: A conceptual review. *Physiology & Behavior, 104*(1), 122–127. https://doi.org/10.1016/j.physbeh.2011.04.028

Carr, M. M., Wiedemann, A. A., Macdonald-Gagnon, G., & Potenza, M. N. (2021). Impulsivity and compulsivity in binge eating disorder: A systematic review of behavioral studies. *Progress in Neuro-Psychopharmacology & Biological Psychiatry, 110*, Article 110318. https://doi.org/10.1016/j.pnpbp.2021.110318

Cifuentes, L., & Acosta, A. (2022). Homeostatic regulation of food intake. *Clinics and Research in Hepatology and Gastroenterology, 46*(2), Article 101794. https://doi.org/10.1016/j.clinre.2021.101794

Cui, H., López, M., & Rahmouni, K. (2017). The cellular and molecular bases of leptin and ghrelin resistance in obesity. *Nature Reviews. Endocrinology, 13*(6), 338–351. https://doi.org/10.1038/nrendo.2016.222

Datta, N., Hughes, A., Modafferi, M., & Klabunde, M. (2025). An FMRI meta-analysis of interoception in eating disorders. *NeuroImage, 305*, Article 120933. https://doi.org/10.1016/j.neuroimage.2024.120933

Dörnyei, G., Vass, Z., Juhász, C. B., Nádasy, G. L., Hunyady, L., & Szekeres, M. (2023). Role of the endocannabinoid system in metabolic control processes and in the pathogenesis of metabolic syndrome: An update. *Biomedicines, 11*(2), 306. https://doi.org/10.3390/biomedicines11020306

Dunbar, J., Lapanowski, K., Barnes, M., & Rafols, J. (2005). Hypothalamic agouti-related protein immunoreactivity in food-restricted, obese, and insulin-treated animals: Evidence for glia cell localization. *Experimental Neurology, 191*(1), 184–192. https://doi.org/10.1016/j.expneurol.2004.09.002

Edwards, A., DeSante, S., Spencer, C. D., Hyland, L., Smith, A., Sankhe, A. S., Szilvásy-Szabó, A., Fekete, C., Hill, M. N., Chee, M. J., & Abizaid, A. (2025). Ghrelin recruits the endocannabinoid system to modulate food reward. *The Journal of Neuroscience, 45*(9), Article e1620242024. https://doi.org/10.1523/JNEUROSCI.1620-24.2024

Faruque, S., Tong, J., Lacmanovic, V., Agbonghae, C., Minaya, D. M., & Czaja, K. (2019). The dose makes the poison: Sugar and obesity in the united states - A review. *Polish Journal of Food and Nutrition Sciences, 69*(3), 219–233. https://doi.org/10.31883/pjfns/110735

Fattiny, S. Z., Binobead, M. A., Yagoub, A. E. A., Alshammari, G. M., Saleh, A., & Abdo Yahya, M. (2025). Therapeutic efficacy of rhoifolin in type 2 diabetes mellitus: Effects on metabolic parameters, hepatic function, and oxidative stress: A dose-dependent study. *Frontiers in Pharmacology, 16*, 1595323. https://doi.org/10.3389/fphar.2025.1595323

Feldman, Z., & Goodman, M. K. (2021). Digital food culture, power and everday life. *European Journal of Cultural Studies, 24*(6), 1227–1242. https://doi.org/10.1177/1367549421105550

Fladung, A. K., Grön, G., Grammer, K., Herrnberger, B., Schilly, E., Grasteit, S., Wolf, R. C., Walter, H., & von Wietersheim, J. (2010). A neural signature of anorexia nervosa in the ventral striatal reward system. *The American Journal of Psychiatry, 167*(2), 206–212. https://doi.org/10.1176/appi.ajp.2009.09010071

Frank, G. K., Bailer, U. F., Henry, S. E., Drevets, W., Meltzer, C. C., Price, J. C., Mathis, C. A., Wagner, A., Hoge, J., Ziolko, S., Barbarich-Marsteller, N., Weissfeld, L., & Kaye, W. H. (2005). Increased dopamine D2/D3 receptor binding after recovery from anorexia nervosa measured by positron emission tomography and [11c]raclopride. *Biological Psychiatry, 58*(11), 908–912. https://doi.org/10.1016/j.biopsych.2005.05.003

Frank, G. K., Shott, M. E., Riederer, J., & Pryor, T. L. (2016). Altered structural and effective connectivity in anorexia and bulimia nervosa in circuits that regulate energy and reward homeostasis. *Translational Psychiatry, 6*(11), Article e932. https://doi.org/10.1038/tp.2016.199

Fritsche, K. L. (2015). The science of fatty acids and inflammation. *Advances in Nutrition, 6*(3), 293S-301S. https://doi.org/10.3945/an.114.006940

Funayama, M., Muramatsu, T., Koreki, A., Kato, M., Mimura, M., & Nakagawa, Y. (2017). Semantic memory deficits are associated with pica in individuals with acquired brain injury. *Behavioural Brain Research, 329*, 172–179. https://doi.org/10.1016/j.bbr.2017.04.054

Gammal, A., Nassar, T., Soae, Y., Freeman, N., Badihi, A., Baraghithy, S., Nemirovski, A., Tam, J., & Benita, S. (2025). Synthesis and pharmacological characterization of novel peripheral cannabinoid-1 receptor blockers based on a tricyclic scaffold. *Journal of Medicinal Chemistry, 68*(9), 9431–9445. https://doi.org/10.1021/acs.jmedchem.4c03132

GBD 2015 Obesity Collaborators, Afshin, A., Forouzanfar, M. H., Reitsma, M. B., Sur, P., Estep, K., Lee, A., Marczak, L., Mokdad, A. H., Moradi-Lakeh, M., Naghavi, M., Salama, J. S., Vos, T., Abate, K. H., Abbafati, C., Ahmed, M. B., Al-Aly, Z., Alkerwi, A., Al-Raddadi, R., Amare, A. T., & Murray, C. J. L. (2017). Health effects of overweight and obesity in 195 countries over 25 years. *The New England Journal of Medicine, 377*(1), 13–27. https://doi.org/10.1056/NEJMoa1614362

Gearhardt, A. N., White, M. A., & Potenza, M. N. (2011). Binge eating disorder and food addiction. *Current Drug Abuse Reviews, 4*(3), 201–207. https://doi.org/10.2174/1874473711104030201

Geer, E. B., Lalazar, Y., Couto, L. M., Cohen, V., Lipton, L. R., Shi, W., Bagiella, E., Conwell, I., Bederson, J., Kostadinov, J., Post, K. D., & Freda, P. U. (2016). A prospective study of appetite and food craving in 30 patients with Cushing's disease. *Pituitary, 19*(2), 117–126. https://doi.org/10.1007/s11102-015-0690-1

Halford, J. C., Harrold, J. A., Boyland, E. J., Lawton, C. L., & Blundell, J. E. (2007). Serotonergic drugs: Effects on appetite expression and use for the treatment of obesity. *Drugs, 67*(1), 27–55. https://doi.org/10.2165/00003495-200767010-00004

Hariri, N., & Thibault, L. (2010). High-fat diet-induced obesity in animal models. *Nutrition Research Reviews, 23*(2), 270–299. https://doi.org/10.1017/S0954422410000168

Harris, J. A., & Benedict, F. G. (1918). A biometric study of human basal metabolism. *Proceedings of the National Academy of Sciences of the United States of America, 4*(12), 370–373. https://doi.org/10.1073/pnas.4.12.370

Holsen, L. M., Lawson, E. A., Blum, J., Ko, E., Makris, N., Fazeli, P. K., Klibanski, A., & Goldstein, J. M. (2012). Food motivation circuitry hypoactivation related to hedonic and nonhedonic aspects of hunger and satiety in women with active anorexia nervosa and weight-restored women with anorexia nervosa. *Journal of Psychiatry & Neuroscience, 37*(5), 322–332. https://doi.org/10.1503/jpn.110156

Idrobo, F., Nandy, K., Mostofsky, D. I., Blatt, L., & Nandy, L. (1987). Dietary restriction: Effects on radial maze learning and lipofuscin pigment deposition in the hippocampus and frontal cortex. *Archives of Gerontology and Geriatrics, 6*(4), 355–362. https://doi.org/10.1016/0167-4943(87)90014-8

Iyer, M. R., Cinar, R., Wood, C. M., Zawatsky, C. N., Coffey, N. J., Kim, K. A., Liu, Z., Katz, A., Abdalla, J., Hassan, S. A., & Lee, Y. S. (2022). Synthesis, biological evaluation, and molecular modeling studies of 3,4-diarylpyrazoline series of compounds as potent, nonbrain penetrant

antagonists of cannabinoid-1 (CB_1R) receptor with reduced lipophilicity. *Journal of Medicinal Chemistry, 65*(3), 2374–2387. https://doi.org/10.1021/acs.jmedchem.1c01836
Jastran, M. M., Bisogni, C. A., Sobal, J., Blake, C., & Devine, C. M. (2009). Eating routines. Embedded, value based, modifiable, and reflective. *Appetite, 52*(1), 127–136. https://doi.org/10.1016/j.appet.2008.09.003
Jiang, S. Z., & Eiden, L. E. (2016). Activation of the HPA axis and depression of feeding behavior induced by restraint stress are separately regulated by PACAPergic neurotransmission in the mouse. *Stress, 19*(4), 374–382. https://doi.org/10.1080/10253890.2016.1174851
Johnson, P. M., & Kenny, P. J. (2010). Dopamine D2 receptors in addiction-like reward dysfunction and compulsive eating in obese rats. *Nature Neuroscience, 13*(5), 635–641. https://doi.org/10.1038/nn.2519
Kawakami, I., Iritani, S., Riku, Y., Umeda, K., Takase, M., Ikeda, K., Niizato, K., Arai, T., Yoshida, M., Oshima, K., & Hasegawa, M. (2022). Neuropathological investigation of patients with prolonged anorexia nervosa. *Psychiatry and Clinical Neurosciences, 76*(5), 187–194. https://doi.org/10.1111/pcn.13340
Kaye, W. H., Wagner, A., Fudge, J. L., & Paulus, M. (2010). Neurocircuitry of eating disorders. In R. Adan & W. Kaye (Eds.), *Behavioral neurobiology of eating disorders. Current topics in behavioral neurosciences*, (Vol. 6). Springer. https://doi.org/10.1007/7854_2010_85
Kenny, P. J. (2011). Common cellular and molecular mechanisms in obesity and drug addiction. *Nature Reviews. Neuroscience, 12*(11), 638–651. https://doi.org/10.1038/nrn3105
Bou Khalil, R., Souaiby, L., & Farès, N. (2017). The importance of the hypothalamo-pituitary-adrenal axis as a therapeutic target in anorexia nervosa. *Physiology & Behavior, 171*, 13–20. https://doi.org/10.1016/j.physbeh.2016.12.035
Kim, S. F. (2012). Animal models of eating disorders. *Neuroscience, 211*, 2–12. https://doi.org/10.1016/j.neuroscience.2012.03.024
Kreisler, A. D., Mattock, M., & Zorrilla, E. P. (2018). The duration of intermittent access to preferred sucrose-rich food affects binge-like intake, fat accumulation, and fasting glucose in male rats. *Appetite, 130*, 59–69. https://doi.org/10.1016/j.appet.2018.07.025
Land, B. B., Narayanan, N. S., Liu, R. J., Gianessi, C. A., Brayton, C. E., Grimaldi, D. M., Sarhan, M., Guarnieri, D. J., Deisseroth, K., Aghajanian, G. K., & DiLeone, R. J. (2014). Medial prefrontal D1 dopamine neurons control food intake. *Nature Neuroscience, 17*(2), 248–253. https://doi.org/10.1038/nn.3625
Lavagnino, L., Arnone, D., Cao, B., Soares, J. C., & Selvaraj, S. (2016). Inhibitory control in obesity and binge eating disorder: A systematic review and meta-analysis of neurocognitive and neuroimaging studies. *Neuroscience and Biobehavioral Reviews, 68*, 714–726. https://doi.org/10.1016/j.neubiorev.2016.06.041
Lawson, E. A., Holsen, L. M., Desanti, R., Santin, M., Meenaghan, E., Herzog, D. B., Goldstein, J. M., & Klibanski, A. (2013). Increased hypothalamic-pituitary-adrenal drive is associated with decreased appetite and hypoactivation of food-motivation neurocircuitry in anorexia nervosa. *European Journal of Endocrinology, 169*(5), 639–647. https://doi.org/10.1530/EJE-13-0433
Liu, H., Burns, R. T., Spencer, B. R., Page, G. P., Mast, A. E., & NHLBI Recipient Epidemiology Donor Evaluation Study (REDS)-III, 2022 Liu, H., Burns, R. T., Spencer, B. R., Page, G. P., Mast, A. E., & NHLBI Recipient Epidemiology Donor Evaluation Study (REDS)-III (2022). Demographic, clinical, and biochemical predictors of pica in high-intensity blood donors. *Transfusion Medicine, 32*(4), 288–292. https://doi.org/10.1111/tme.12890
Luo, N., Marcelin, G., Liu, S. M., Schwartz, G., & Chua, S., Jr. (2011). Neuropeptide Y and agouti-related peptide mediate complementary functions of hyperphagia and reduced energy expenditure in leptin receptor deficiency. *Endocrinology, 152*(3), 883–889. https://doi.org/10.1210/en.2010-1135
Lytle, L. D., & Campbell, B. A. (1975). Effects of lateral hypothalamic lesions on consummatory behavior in developing rats. *Physiology & Behavior, 15*(3), 323–331. https://doi.org/10.1016/0031-9384(75)90100-6

Madden, L. J., Seeley, R. J., & Woods, S. C. (1999). Intraventricular neuropeptide Y decreases need-induced sodium appetite and increases pica in rats. *Behavioral Neuroscience, 113*(4), 826–832.

Marsh, R., Horga, G., Wang, Z., Wang, P., Klahr, K. W., Berner, L. A., Walsh, B. T., & Peterson, B. S. (2011). An FMRI study of self-regulatory control and conflict resolution in adolescents with bulimia nervosa. *The American Journal of Psychiatry, 168*(11), 1210–1220. https://doi.org/10.1176/appi.ajp.2011.11010094

Mehler, P. S., & Rylander, M. (2015). Bulimia Nervosa - Medical complications. *Journal of Eating Disorders, 3*, 12. https://doi.org/10.1186/s40337-015-0044-4

Miller, A. A., & Spencer, S. J. (2014). Obesity and neuroinflammation: A pathway to cognitive impairment. *Brain, Behavior, and Immunity, 42*, 10–21. https://doi.org/10.1016/j.bbi.2014.04.001

Molfino, A., Iannace, A., Colaiacomo, M. C., Farcomeni, A., Emiliani, A., Gualdi, G., Laviano, A., & Rossi Fanelli, F. (2017). Cancer anorexia: Hypothalamic activity and its association with inflammation and appetite-regulating peptides in lung cancer. *Journal of Cachexia, Sarcopenia and Muscle, 8*(1), 40–47. https://doi.org/10.1002/jcsm.12156

Monteleone, P., Piscitelli, F., Scognamiglio, P., Monteleone, A. M., Canestrelli, B., Di Marzo, V., & Maj, M. (2012). Hedonic eating is associated with increased peripheral levels of ghrelin and the endocannabinoid 2-arachidonoyl-glycerol in healthy humans: A pilot study. *The Journal of Clinical Endocrinology and Metabolism, 97*(6), E917–E924. https://doi.org/10.1210/jc.2011-3018

Muratore, A. F., Foerde, K., Lloyd, E. C., Touzeau, C., Uniacke, B., Aw, N., Semanek, D., Wang, Y., Walsh, B. T., Attia, E., Posner, J., & Steinglass, J. E. (2024). Reduced dorsal fronto-striatal connectivity at rest in anorexia nervosa. *Psychological Medicine, 54*(9), 2200–2209. https://doi.org/10.1017/S003329172400031X

Obradovic, M., Sudar-Milovanovic, E., Soskic, S., Essack, M., Arya, S., Stewart, A. J., Gojobori, T., & Isenovic, E. R. (2021). Leptin and obesity: Role and clinical implication. *Frontiers in Endocrinology, 12*, Article 585887. https://doi.org/10.3389/fendo.2021.585887

Patel, P. S., Buras, E. D., & Balasubramanyam, A. (2013). The role of the immune system in obesity and insulin resistance. *Journal of Obesity, 2013*, Article 616193. https://doi.org/10.1155/2013/616193

Pauligk, S., Seidel, M., Ritschel, F., Geisler, D., Doose, A., Boehm, I., Hellerhoff, I., Ludwicki, F., Roessner, V., King, J. A., & Ehrlich, S. (2025). Overcontrol in anorexia nervosa: Elevated prefrontal activity and amygdala connectivity in a working memory task with food distractors. *International Journal of Clinical and Health Psychology, 25*(1), Article 100544. https://doi.org/10.1016/j.ijchp.2025.100544

Pinkney, J., Wilding, J., Williams, G., & MacFarlane, I. (2002). Hypothalamic obesity in humans: What do we know and what can be done? *Obesity Reviews, 3*(1), 27–34. https://doi.org/10.1046/j.1467-789x.2002.00052.x

Puckett, L., Grayeb, D., Khatri, V., Cass, K., & Mehler, P. (2021). A comprehensive review of complications and new findings associated with anorexia nervosa. *Journal of Clinical Medicine, 10*(12), 2555. https://doi.org/10.3390/jcm10122555

Reisin, E., Wilson, J. R., & Frohlich, E. D. (1987). Hypertension and obesity in rats with ventromedial-hypothalamic lesions and low salt intake. *Journal of Hypertension, 5*(2), 173–178. https://doi.org/10.1097/00004872-198704000-00007

Rose, E. A., Porcerelli, J. H., & Neale, A. V. (2000). Pica: Common but commonly missed. *The Journal of the American Board of Family Practice, 13*(5), 353–358.

Routtenberg, A., & Kuznesof, A. W. (1967). Self-starvation of rats living in activity wheels on a restricted feeding schedule. *Journal of Comparative and Physiological Psychology, 64*(3), 414–421. https://doi.org/10.1037/h0025205

Roza, A. M., & Shizgal, H. M. (1984). The Harris Benedict equation reevaluated: Resting energy requirements and the body cell mass. *The American Journal of Clinical Nutrition, 40*(1), 168–182. https://doi.org/10.1093/ajcn/40.1.168

Sam, A. H., Salem, V., & Ghatei, M. A. (2011). Rimonabant: From RIO to ban. *Journal of Obesity, 2011*, Article 432607. https://doi.org/10.1155/2011/432607
Schleh, M. W., Caslin, H. L., Garcia, J. N., Mashayekhi, M., Srivastava, G., Bradley, A. B., & Hasty, A. H. (2023). Metaflammation in obesity and its therapeutic targeting. *Science Translational Medicine, 15*(723), eadf9382. https://doi.org/10.1126/scitranslmed.adf9382
Schreiber, L. R., Odlaug, B. L., & Grant, J. E. (2013). The overlap between binge eating disorder and substance use disorders: Diagnosis and neurobiology. *Journal of Behavioral Addictions, 2*(4), 191–198. https://doi.org/10.1556/JBA.2.2013.015
Seitz, J., Hueck, M., Dahmen, B., Schulte-Rüther, M., Legenbauer, T., Herpertz-Dahlmann, B., & Konrad, K. (2016). Attention network dysfunction in bulimia nervosa - An fMRI study. *PLoS ONE, 11*(9), Article e0161329. https://doi.org/10.1371/journal.pone.0161329
Sinclair, E. B., Klump, K. L., & Sisk, C. L. (2019). Reduced medial prefrontal control of palatable food consumption is associated with binge eating proneness in female rats. *Frontiers in Behavioral Neuroscience, 13*, 252. https://doi.org/10.3389/fnbeh.2019.00252
Speakman, J. R. (2013). Evolutionary perspectives on the obesity epidemic: Adaptive, maladaptive, and neutral viewpoints. *Annual Review of Nutrition, 33*, 289–317. https://doi.org/10.1146/annurev-nutr-071811-150711
Steinglass, J. E., Berner, L. A., & Attia, E. (2019). Cognitive neuroscience of eating disorders. *The Psychiatric Clinics of North America, 42*(1), 75–91. https://doi.org/10.1016/j.psc.2018.10.008
Stice, E., Yokum, S., Gau, J., & Shaw, H. (2025). Neural risk factors that predict the future onset of binge eating or compensatory weight control behaviors: A prospective 4-year fMRI study. *Psychological Medicine, 55*, Article e48. https://doi.org/10.1017/S0033291724003337
Uher, R., Brammer, M. J., Murphy, T., Campbell, I. C., Ng, V. W., Williams, S. C., & Treasure, J. (2003). Recovery and chronicity in anorexia nervosa: Brain activity associated with differential outcomes. *Biological Psychiatry, 54*(9), 934–942. https://doi.org/10.1016/s0006-3223(03)00172-0
van Galen, K. A., Ter Horst, K. W., & Serlie, M. J. (2021). Serotonin, food intake, and obesity. *Obesity Reviews, 22*(7), Article e13210. https://doi.org/10.1111/obr.13210
Volkow, N. D., Wang, G. J., & Baler, R. D. (2011). Reward, dopamine and the control of food intake: Implications for obesity. *Trends in Cognitive Sciences, 15*(1), 37–46. https://doi.org/10.1016/j.tics.2010.11.001
Volkow, N. D., Wang, G. J., Telang, F., Fowler, J. S., Thanos, P. K., Logan, J., Alexoff, D., Ding, Y. S., Wong, C., Ma, Y., & Pradhan, K. (2008). Low dopamine striatal D2 receptors are associated with prefrontal metabolism in obese subjects: Possible contributing factors. *NeuroImage, 42*(4), 1537–1543. https://doi.org/10.1016/j.neuroimage.2008.06.002
Wagner, A., Aizenstein, H., Mazurkewicz, L., Fudge, J., Frank, G. K., Putnam, K., Bailer, U. F., Fischer, L., & Kaye, W. H. (2008). Altered insula response to taste stimuli in individuals recovered from restricting-type anorexia nervosa. *Neuropsychopharmacology, 33*(3), 513–523. https://doi.org/10.1038/sj.npp.1301443
Wagner, A., Aizenstein, H., Venkatraman, V. K., Fudge, J., May, J. C., Mazurkewicz, L., Frank, G. K., Bailer, U. F., Fischer, L., Nguyen, V., Carter, C., Putnam, K., & Kaye, W. H. (2007). Altered reward processing in women recovered from anorexia nervosa. *The American Journal of Psychiatry, 164*(12), 1842–1849. https://doi.org/10.1176/appi.ajp.2007.07040575
Wassenaar, E., Friedman, J., & Mehler, P. S. (2019). Medical complications of binge eating disorder. *The Psychiatric Clinics of North America, 42*(2), 275–286. https://doi.org/10.1016/j.psc.2019.01.010
Williams, F., Gibbs, S., & Addo, A. S. (2022). The assessment and management of pica in people with intellectual disability. *Bjpsych Advances, 28*(6), 383–392. https://doi.org/10.1192/bja.2022.24
Winn, P., Clark, A., Hastings, M., Clark, J., Latimer, M., Rugg, E., & Brownlee, B. (1990). Excitotoxic lesions of the lateral hypothalamus made by N-methyl-d-aspartate in the rat: Behavioural, histological and biochemical analyses. *Experimental Brain Research, 82*(3), 628–636. https://doi.org/10.1007/BF00228804

World Health Organization. (2019). *International statistical classification of diseases and related health problems* (11th ed.). https://icd.who.int/

World Health Organization. (2024, March 1). *Obesity and overweight.* https://www.who.int/news-room/fact-sheets/detail/obesity-and-overweight

Wu, K., Lo, Y. T., Cavaleri, J., Bergosh, M., Ipe, J., Briggs, R. G., Jann, K. B., Murray, S. B., Mason, X. L., Liu, C. Y., & Lee, D. J. (2024). Neuromodulation of eating disorders: A review of underlying neural network activity and neuromodulatory treatments. *Brain Sciences, 14*(3), 200. https://doi.org/10.3390/brainsci14030200

Zhang, L., Bijker, M. S., & Herzog, H. (2011). The neuropeptide Y system: Pathophysiological and therapeutic implications in obesity and cancer. *Pharmacology & Therapeutics, 131*(1), 91–113. https://doi.org/10.1016/j.pharmthera.2011.03.011

Zigman, J. M., Bouret, S. G., & Andrews, Z. B. (2016). Obesity impairs the action of the neuroendocrine ghrelin system. *Trends in Endocrinology and Metabolism, 27*(1), 54–63. https://doi.org/10.1016/j.tem.2015.09.010

Chapter 14
Maladaptive Sleep

1 Introduction

I roll over and glance at my phone. The time shows 4:00 am, but I am wide awake. I try to go back to sleep, but after 30 min, I realize this is not going to happen. I sneak out of bed and go downstairs, trying to avoid the steps that squeak so as not to wake my wife or children. This is the third night in a row that I have awoken at this hour. There have been multiple times in my life in which I have struggled to fall asleep or to stay asleep throughout the night.

I am certainly not the only person that has experienced bouts of **insomnia**. Many of us have experienced the frustration of waking up in the middle of the night, unable to go back to sleep.

The importance of sleep should be obvious given that we spend approximately one third of our lives sleeping. Even though we can experience insomnia at various points, we cannot indefinitely hold off sleep. Individuals can starve themselves to death, but the longest an individual has stayed awake (officially and willingly) is nearly 19 days (set by Robert McDonald in 1986; note, the Guinness World Records no longer tracks this record for good reason). Think of how you feel the next day when you slept fewer hours than normal. You may have difficulty concentrating or may feel more irritable. Inadequate sleep can negatively affect our physical and psychological well-being.

On the other end of the spectrum, there are individuals that experience bouts of extreme sleepiness (known as **hypersomnia**). No matter how much an individual sleeps, they feel fatigued. This is maladaptive as chronic fatigue can impair daily functions such as going to work/school.

This chapter will focus on the neural bases of sleep disorders. Before discussing sleep disorders, you will first learn about our sleep-wake cycle and the stages of sleep. This knowledge will make understanding why certain sleep disorders occur easier.

J. Yates, *Neuroscience of Maladaptation*,
https://doi.org/10.1007/978-3-032-15602-0_14

2 Learning Objectives

By the end of this chapter, you should be able to ...

1. Explain the neuromechanisms that control circadian rhythms and the sleep-wake cycle.
2. Differentiate the stages of sleep.
3. Discuss the neurobiological bases of sleep disorders.

3 Neuromechanisms Controlling the Sleep-Wake Cycle

Before getting into the neuromechanisms that control sleep, you first need to understand **circadian rhythms**. There are certain physiological processes and behaviors that follow a standard schedule during a 24-hour period. For example, our body temperature remains relatively consistent, but it does fluctuate in a rhythmic pattern. Our body temperature is lowest during the early morning and highest in the evening. Certain hormones are released in the morning while others are released at night. Directly relevant to this chapter, our sleep-wake cycle is the prototypic example of a circadian rhythm.

In the previous chapter, several nuclei of the hypothalamus were detailed. Our circadian rhythm is controlled by another hypothalamic nucleus, the suprachiasmatic nucleus (SCN). In humans, light is crucial for our circadian rhythm to function properly. Not only do neurons from the eye transmit information to the thalamus and the occipital cortex, but they travel to the SCN. This enables the SCN to synchronize circadian rhythms. Regarding wakefulness, when light reaches the SCN, this part of the brain sends a signal to the pineal gland, commanding it to stop the production of *melatonin.* As the day progresses, melatonin levels increase, helping promote the sensation of sleepiness. You may have taken melatonin supplements to help alleviate "jet lag", a condition in which our circadian rhythm is disrupted due to travel across several time zones.

There are multiple systems that regulate the sleep-wake cycle (Scammell et al., 2017). The following neurotransmitters contribute to wakefulness: norepinephrine, dopamine, serotonin, acetylcholine, glutamate, orexin, and histamine. You may be most familiar with histamine if you have taken antihistamines to alleviate allergies. In addition to melatonin, gamma-aminobutyric acid (GABA) and *adenosine* help increase sleepiness. Let's now take a look at the neural circuitry that controls the sleep-wake cycle (Fig. 1).

The SCN sends GABAergic neurons to the subparaventricular zone (SPZ) of the hypothalamus in addition to *vasopressin* and *vasoactive intestinal peptide (VIP).* Vasopressin is an important hormone that is better known for its roles in water retention (i.e., preventing dehydration), blood pressure regulation (specifically, causes vasoconstriction), and social behaviors. In fact, you will learn more about vasopressin

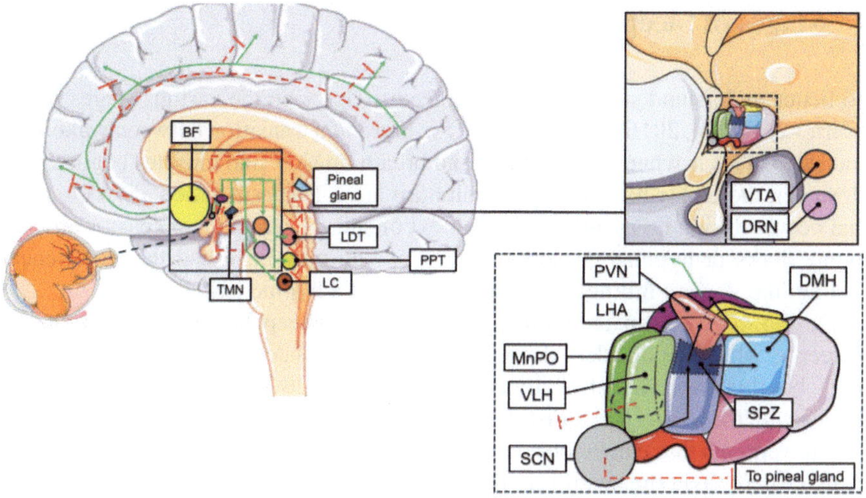

Fig. 1 Schematic detailing the neuromechanisms involved in our sleep-wake cycle. Light entering the eye travels to the suprachiasmatic nucleus (SCN). The SCN projects neurons to the subparaventricular zone (SPZ) of the hypothalamus, which has connections with the dorsomedial nucleus (DMH) and the paraventricular nucleus (PVN) of the hypothalamus. Glutamatergic neurons originating from the DMH activate orexinergic neurons in the lateral hypothalamic area (LHA). Orexinergic neurons travel to multiple brain regions to promote wakefulness, including the locus coeruleus (LC), the pedunculopontine nucleus (PPT), the laterodorsal tegmental nucleus (LDT), the dorsal raphe nuclei (DRN), and the tuberomammillary nucleus (TMN). The SCN also inhibits melatonin production from the pineal gland. As the day progresses, inhibitory neurons originating from the median preoptic nucleus (MnPO) and the ventrolateral hypothalamus (VLH) travel to wake-promoting regions. The basal forebrain (BF) sends both excitatory and inhibitory neurons to cortical areas to mediate the sleep-wake cycle. Image created by the author using Servier Medical Art (https://smart.servier.com/; CC BY 4.0, https://creativecommons.org/licenses/by/4.0/) and Microsoft PowerPoint

in the next chapter. VIP, like vasopressin, has multiple functions beyond sleep regulation. It is also involved in blood pressure regulation, but it acts as a vasodilator. VIP also regulates smooth muscles in the digestive tract and influences digestion. Vasopressin and VIP are important for synchronizing our circadian rhythm.

The SPZ sends axons to the dorsomedial nucleus (DMH) and the paraventricular nucleus (PVN) of the hypothalamus. The DMH then sends glutamatergic projections to the lateral hypothalamic area (LHA) of the hypothalamus, which is where orexinergic neurons originate. Orexin travels to numerous regions of the brain to stimulate the release of neurotransmitters involved in wakefulness.

As the day progresses, the ventrolateral nucleus and the median preoptic nucleus of the hypothalamus (MnPO) send GABAergic signals to wake-promoting regions of the brain. The basal forebrain, although a large cholinergic center, also projects GABAergic neurons to the cerebral cortex. Increased GABA release inhibits wake-promoting centers of the brain, allowing an individual to fall asleep.

4 Stages of Sleep

To better understand sleep disorders, I need to describe the various stages of sleep (Carley & Farabi, 2016) (Fig. 2). *Stage 1* is more of a transition between wakefulness and sleep. This is when our heartbeat and breathing slow and muscles begin to relax. However, there may be a sudden, involuntary muscle spasm called a *myoclonic jerk* that occurs during this stage (in the context of stage 1 sleep, these spasms are called hypnic jerks or hypnagogic jerks). Some of you may have had the experience of feeling like you are falling; this is an example of a hypnic jerk.

During *stage 2* sleep, individuals are considered asleep, but the sleep is light. *Stage 3* sleep is what we consider "deep sleep". You may encounter stage 4 sleep in older texts, but this stage was officially combined with stage 3 in 2007. Finally, *rapid eye movement (REM) sleep* is a distinct stage of sleep that gets its name from the rapid eye movements that can be observed while an individual sleeps. This stage of sleep is also where vivid dreams primarily occur, although dreams can occur during stage 3 sleep. Muscle paralysis also occurs during REM sleep, preventing an individual from acting out their dreams. However, heart rate and breathing increase during this stage.

When we sleep, we cycle through the various stages of sleep multiple times, with each cycle lasting approximately 90–120 min. Even within a cycle, individuals may fluctuate between two stages of sleep. For example, as we fall asleep, we may enter stage 1 sleep before waking for a couple of minutes before transitioning back to stage 1 sleep. We can also oscillate between stage 2 sleep and stage 3 sleep for several minutes. Eventually, we enter REM sleep before transitioning back to stage 1 sleep. We often have brief periods of wakefulness during the night.

The various stages of sleep are characterized by different neuronal patterns of firing. When an electroencephalogram (EEG) is used, we can tell which stage of sleep is occurring based on the pattern of brain activity. While we are awake, high-frequency *beta waves* are observed when we are alert and engaged in some cognitive process. When we are calm, *alpha waves* are more pronounced. Alpha waves have a lower frequency compared to beta waves. As we transition to stage 1 sleep, *theta waves* are observed, which have a lower frequency compared to alpha waves. Stage 2 sleep is also characterized by theta waves, but the brain activity includes high bursts of neuronal firing (called *sleep spindles*) and sharp, high-amplitude waves (*K-complexes*). *Delta waves* are low-frequency waves that dominate stage 3 sleep; as such, stage 3 sleep is also known as slow-wave sleep. REM sleep is characterized by a combination of theta and beta waves, as well as *sawtooth waves*. These waves have a sharp peak, resembling a shark's tooth.

Another characteristic of REM sleep is *PGO waves*. PGO stands for ponto-geniculo-occipital. During REM sleep, neuronal signals originating in the pons (ponto) travel to the lateral geniculate nucleus (LGN) of the thalamus (geniculo) before traveling to the occipital cortex (occipital). This increased activity is hypothesized to underlie the rapid eye movements that are a defining feature of REM sleep. An additional hypothesis argues that PGO waves account for the visual imagery that

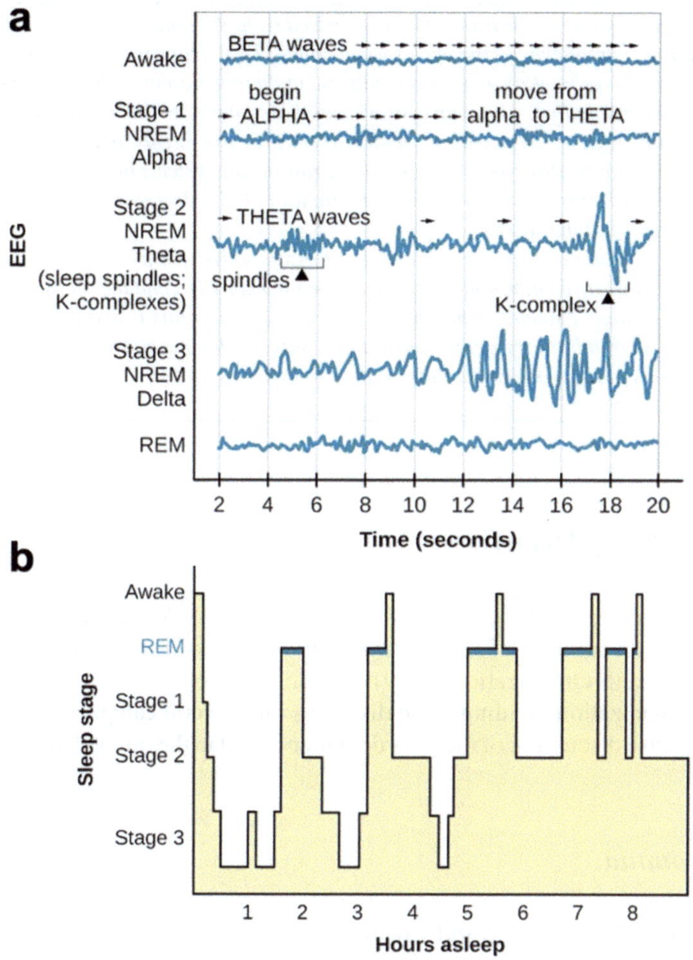

Fig. 2 (**a**) Representation of the neural activity that occurs when we are awake and during each stage of sleep. (**b**) A representative *hypnogram*, which shows how we progress through each stage of sleep during a period of sleep. *Source* Figure 4.8 and Figure 4.12 of Spielman et al. (2020), https://openstax.org/books/psychology-2e/pages/4-3-stages-of-sleep. CC BY 4.0 (https://creativecommons.org/licenses/by/4.0/), via Rice University and OpenStax

comes with dreams. This makes intuitive sense as visual information travels from our eye to the LGN before being relayed to the primary visual cortex in the occipital cortex.

> **Comprehension Check 1** How do the neuronal patterns observed during wakefulness and during each stage of sleep map onto neurochemical processes occurring in the brain?

During wakefulness, the brain is characterized by high-frequency beta waves when alert and lower-frequency alpha waves when calm. These patterns correspond with the activity of wake-promoting neurotransmitters, including norepinephrine, dopamine, serotonin, acetylcholine, glutamate, orexin, and histamine, which are released from various brain regions. When these neurotransmitters are highly active, beta waves are more commonly observed. As these neurotransmitter systems become inhibited by inhibitory neurotransmitters like GABA, neuronal activity decreases, resulting in the lower-frequency alpha waves.

As an individual transitions to sleep, their brain activity shifts to lower-frequency waves due to increased inhibition of brain regions like the LHA and the locus coeruleus. Stage 1 and Stage 2 sleep are marked by theta waves, and this transition is initiated by the release of inhibitory neurotransmitters. Stage 3 sleep is characterized by delta waves. At this point, wake-promoting centers are fully inhibited.

5 Major Sleep Disorders

The *Diagnostic and Statistical Manual of Mental Disorders (DSM-5)* (American Psychiatric Association, 2013) and the *International Classification of Diseases (ICD-11)* (World Health Organization, 2019) include a section on sleep-wake disorders, with each organization dividing these disorders into several categories. This section will begin with a discussion of the disorder I mentioned in the introduction: insomnia.

5.1 Insomnia

Approximately 16% of the world's population experiences chronic insomnia, with prevalence rates being higher in women relative to men (Benjafield et al., 2025). There are numerous factors that can influence insomnia.

First, diet and substance use can cause insomnia. Caffeine is the most consumed drug in the world. This is easy to understand given the ubiquity of caffeinated beverages like coffee, tea, soda, and energy drinks. Chocolate also contains caffeine. Caffeine is a potent adenosine receptor antagonist. By blocking adenosine receptors, adenosine cannot exert its sleep-inducing effects. This is what causes individuals to feel more alert and energized after consuming caffeine. Psychostimulant drugs such as cocaine, methamphetamine, and ADHD (attention-deficit/hyperactivity disorder) medications can increase insomnia by increasing norepinephrine and dopamine levels. Nicotine, which is found in tobacco products, is also associated with insomnia due to its actions on acetylcholine nicotinic receptors.

Second, stress is a major contributor to insomnia. Stress-induced insomnia is something many of us have experienced at some point. Worrying about a major exam or presentation can keep us awake at night. When the hypothalamic-pituitary-adrenal (HPA) axis is activated, adrenocorticotropic hormone (ACTH), cortisol, and noradrenaline levels increase. From Chap. 8, you already know that activation of the HPA axis leads to increased arousal and vigilance. Not surprisingly, stress causes insomnia by activating the HPA axis. Early awakening is attributed to ACTH release, and corticotropin-releasing hormone (CRH) increases noradrenergic activity in the locus coeruleus (Han et al., 2012).

Genetic factors also contribute to insomnia as the heritability of insomnia is estimated to be approximately 40% (Barclay et al., 2021). Gene-wide association studies have identified several genes that are involved with insomnia (Van Someren, 2021). One gene that has received some considerable attention is the *MEIS1* gene, which encodes the homeobox protein Meis1, a transcription factor involved in cell proliferation and differentiation. One study found specific **epigenetic** factors linked to *MEIS1* regulation that can contribute to insomnia (Lam et al., 2022). I have not discussed epigenetics up to this point, but this is a growing research area that has generated much interest. I need to spend some time discussing what epigenetic factors are (see Fig. 3).

Our DNA (deoxyribonucleic acid) wraps itself around a protein called a *histone*. Think of wrapping string around a spool. Epigenetic factors act as "chemical tags", influencing how DNA works without altering its physical structure. One major epigenetic factor is histone modification. Acetylation involves adding an acetyl group to the histone. Because histones are positively charged, the acetyl group "neutralizes" this charge, causing negatively charged DNA to "unwind" from the histone. This causes gene expression to occur. Conversely, histone deacetylases are enzymes that remove acetyl groups from histones. This causes the DNA to wrap itself more tightly around the histone, leading to gene repression.

Another major epigenetic factor is methylation. A methyl group can be added to part of the DNA sequence, which can act to "turn on" or "turn off" a gene. In the study I referenced above, Lam et al. (2022) found altered methylation of parts of the *MEIS1* gene, particularly in inhibitory neurons. Recall, inhibitory neurons are important for promoting sleep as they inhibit brain regions associated with wakefulness. These results indicate that insufficient activation of the *MEIS1* gene in inhibitory neurons can interfere with the onset and/or the duration of sleep.

The *MEIS1* gene is not the only one implicated in insomnia. The Experiment Spotlight below focuses on two additional genes that have recently been reported to contribute to insomnia.

Experiment Spotlight: Exploration of Common Pathogenic Genes between Cerebral Amyloid Angiopathy and Insomnia Based on Bioinformatics and Experimental Validation. The full article can be accessed with this link: https://pmc.ncbi.nlm.nih.gov/articles/PMC12277429/

Li et al. (2025) were interested in determining if cerebral amyloid angiopathy (CAA) and insomnia share common genetic determinants. CAA is considered a cerebral small

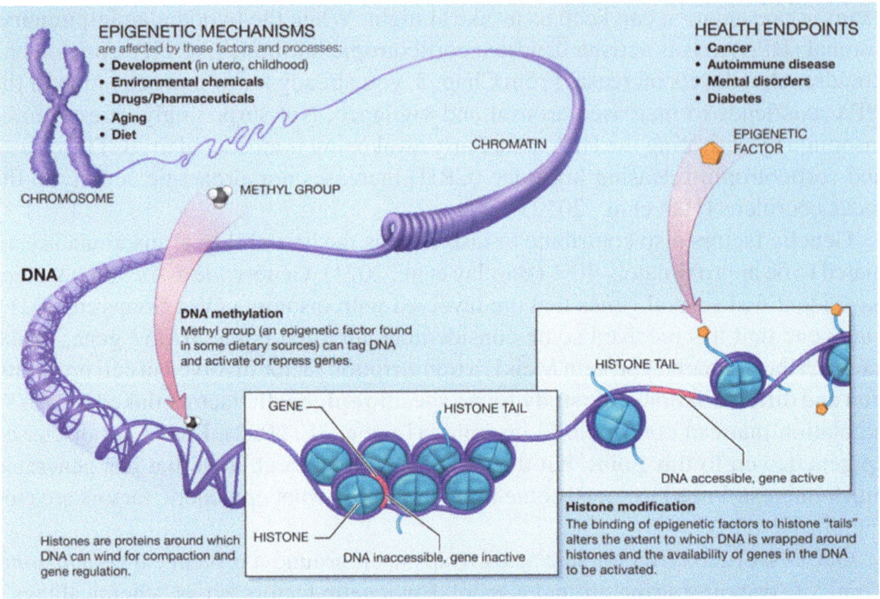

Fig. 3 Image showing two major epigenetic changes that can occur to DNA (deoxyribonucleic acid). First, a methyl group can bind to the DNA, which can activate or repress a gene (turn it on or off). Second, an acetyl group (referred to as an "epigenetic factor" in the image) can be added or removed from a histone tail. This influences how tightly the DNA is wrapped around histones. When the DNA is wrapped around the histone, it is inaccessible. Conversely, when the DNA is unwrapped, it can be used. *Note* Acetylation is not the only epigenetic factor involved in histone modifications. For example, histones can become methylated, phosphorylated, or *ubiquitylated* (ubiquitin is a small protein). *Source* http://commonfund.nih.gov/epigenomics/figure, via the National Institutes of Health. Image is in the public domain

vessel disease; that is, blood vessels in the brain become damaged, which can lead to stroke. In Chap. 6, I briefly discussed vascular dementia. Cerebral small vessel diseases like CAA can lead to vascular dementia. In CAA, beta amyloid protein builds up in blood vessels in the brain, disrupting blood flow (recall the contribution of beta amyloid to dementia from Chap. 6).

In addition to dementia, CAA and other cerebral small vessel diseases are associated with insomnia. As noted by Li et al., the severity of insomnia is linked to cognitive deficits in individuals with conditions like CAA. As such, the overarching goal of the present experiment was to identify differentially expressed genes in those with CAA and insomnia relative to healthy controls.

First, individuals with CAA ($n = 11$) and healthy controls ($n = 11$) were recruited for the study. These individuals provided a blood sample. The researchers then performed RNA (ribonucleic acid) sequencing to determine which genes are actively expressed in these individuals' cells. *Microarray* datasets were pulled from the Gene Expression Omnibus (GEO) database to analyze genetic information for individuals with insomnia and control individuals. I need to explain what a DNA microarray is for this part of the experiment to make sense. When an individual donates something like blood, their DNA can be extracted. Researchers can then denature the DNA strand, separating the DNA into two single-stranded molecules. These single strands are cut into smaller fragments which can then be labeled with something like a fluorescent tag. These fragments are placed on a special chip that contains synthetic DNA. The fragments hybridize (bind) to the DNA on the chip. This enables researchers to examine numerous genes at once. Information related to the microarrays can then be stored in a repository, allowing others to analyze genetic information from specific populations.

Separate experiments were conducted with humans and mice. An additional five individuals with CAA and five healthy controls, as well as five individuals with insomnia and five healthy controls, were recruited and provided blood samples. TgSWDI mice (transgenic mice expressing Swedish, Dutch, and Iowa mutations) were used as they model CAA. These mice accumulate large levels of beta amyloid in the brain and in cerebral blood vessels. C57BL/6 J mice were also used as a control. Quantitative reverse transcription polymerase chain reaction (qRT-PCR) was used to measure RNA expression from peripheral blood mononuclear cells (humans and mice) and from the hippocampus (mice only). Western blot was also performed to measure protein expression from the peripheral blood cells and the hippocampus.

To keep things as simple as possible, I am going to focus on just a few results. Overall, the researchers found 185 genes that were differentially expressed in individuals with CAA and individuals with insomnia. There were two "hub" genes that were identified that are shared across CAA and insomnia: *CBX5* and *POLR1B*. A hub gene is one that is highly connected to other genes. The *CBX5* gene encodes the chromobox protein homolog 5 (often called the heterochromatin protein-1 or HP1). This protein is heavily involved in gene suppression (i.e., prevents certain genes from becoming activated). The *POLR1B* gene encodes the enzyme DNA-directed RNA polymerase 1 subunit RPA2. This enzyme is important for transcription of DNA into RNA. Individuals with CAA and insomnia show less expression of *CBX5* and *POLR1B* genes.

Questions to Consider:

1. How do you think lower expression of *CBX5* and *POLR1B* genes increases the risk of insomnia?
2. If the researchers found 185 genes that are differentially expressed in individuals with CAA and insomnia, why did they primarily focus on the *CBX5* and *POLR1B* genes?
3. What is the next logical step to bridge the gap between the present findings and a practical clinical application?

5.1.1 Fatal Insomnia

There is an extreme (and extremely rare) form of insomnia that has been identified in which individuals lose the ability to sleep entirely. The cause of fatal insomnia is a misfolded protein (i.e., prion) that leads to extensive damage to the thalamus. You learned about prions in Chap. 6 when reading about Creutzfeldt-Jakob disease, a form of dementia. In sporadic fatal insomnia, the cause of prion accumulation is idiopathic. Conversely, in fatal familial insomnia, individuals inherit a gene mutation that leads to prion formation. Regardless of the cause of fatal insomnia, only a few dozen people have been reported to have this condition.

Fatal insomnia is incurable and always leads to death, typically within 2 years from the onset of symptoms. Affected individuals cannot enter stage 3 sleep. Over time, the lack of deep sleep leads to additional symptoms like hallucinations, dysfunction of the autonomic nervous system (e.g., hyperthermia, hypertension, and tachycardia [increased heart rate]), motor impairments (e.g., ataxia, dysarthria), and deficits in attention and memory. Before death, individuals enter a coma (Medori et al., 1992).

Why does destruction of part of the thalamus cause incurable insomnia? Earlier, I mentioned that stage 2 sleep is characterized by sleep spindles and K-complexes. Sleep spindles originate in part of the thalamus called the reticular nucleus. One major hypothesis is that sleep spindles are vital for transitioning us from stage 2 sleep to stage 3 sleep (Jan et al., 2009).

Comprehension Check 2 Compare and contrast insomnia and fatal familial insomnia.

While both conditions are classified as insomnia, they have different causes and symptomology. Insomnia can be caused by various factors, such as lifestyle, stress, and genetics. This condition can lead to difficulty falling asleep and/or staying asleep. Fatal familial insomnia results from prions that destroy the thalamus. Individuals lose the ability to enter slow-wave sleep and die within 2 years.

5.2 Sleep Apnea

Sleep apnea is a breathing-related sleep disorder; individuals with sleep apnea experience repeated interruptions in breathing during sleep. This causes the individual to wake up numerous times throughout the night, leading to excessive daytime sleepiness. The most common form of sleep apnea is obstructive sleep apnea, in which the upper airway collapses during sleep. Men and individuals over the age of 50 are more likely to have obstructive sleep apnea, but the greatest risk factor for developing this condition is obesity (Abbasi et al., 2021).

A second form of sleep apnea results from dysfunction of the central nervous system (CNS) (central sleep apnea). Normally, when we sleep, special chemoreceptors in the *carotid body* (located near where the carotid artery splits into two branches) detect oxygen and carbon dioxide levels. In the CNS, the medulla and the pons detect carbon dioxide levels. When carbon dioxide levels increase, the medulla sends a signal to the lungs to expel carbon dioxide by exhaling and to pull in oxygen by inhaling (Fig. 4).

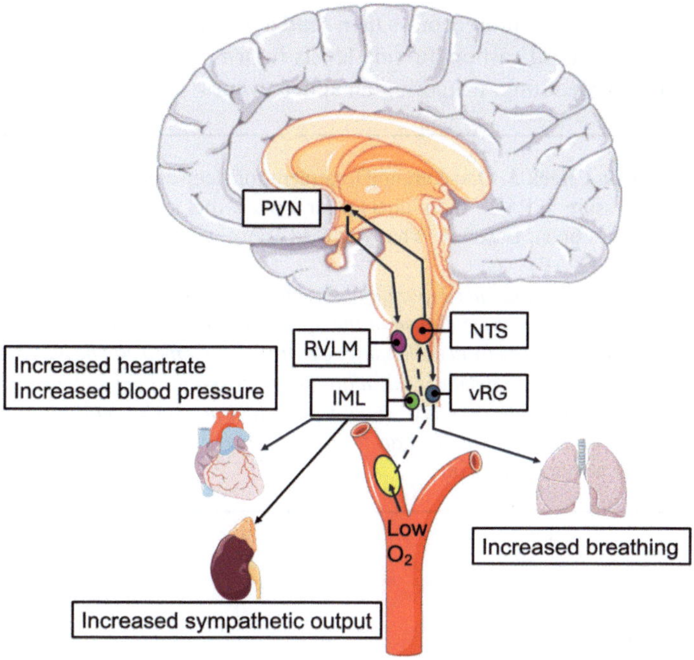

Fig. 4 When chemoreceptors located in the carotid body of the carotid artery detect low oxygen (O_2) levels, a signal is sent to the nucleus of the solitary tract (NTS), which then projects neurons to the ventral respiratory group (vRG). Neurons originating from the vRG serve to increase breathing to pull in more oxygen. Neurons also travel from the NTS to the paraventricular nucleus (PVN). Intra-PVN neurons send axons to the rostral ventrolateral medulla (RVLM), which then relay neurons to the intermediolateral medulla (IML). The IML influences heart rate and adrenal activity. Collectively, these actions increase blood pressure, further allowing for normalization of arterial O_2 levels. Disruption of this system contributes to central sleep apnea. Image created by the author using Servier Medical Art (https://smart.servier.com/; CC BY 4.0, https://creativecommons.org/licenses/by/4.0/) and Microsoft PowerPoint based on Fig. 1 of Iturriaga and Diaz (2025), https://doi.org/10.3390/antiox14060675. CC BY 4.0 (https://creativecommons.org/licenses/by/4.0/), via MDPI

Chemoreceptors also use a negative feedback loop to regulate the amount of carbon dioxide in the blood. This is analogous to how autoreceptors on neurons regulate the further release of neurotransmitters into the synapse. When carbon dioxide levels are sufficiently lowered, chemoreceptors relay messages to the medulla, which then causes breathing to slow. Alterations in chemoreceptor sensitivity and dysregulation of this negative feedback loop are the major contributors to sleep apnea (Roberts et al., 2022). In some individuals, chemoreceptors may be too sensitive to minor increases in carbon dioxide levels. Normally, our body tries to keep carbon dioxide levels to 40 mmHg (millimeters of mercury). For some, a slight increase in carbon dioxide levels to 40.5 mmHg can trigger increased breathing. Respiration rates increase too much, causing carbon dioxide levels to plummet. Due to the low carbon dioxide levels, chemoreceptors then send a message to the medulla that respiration needs to stop until carbon dioxide levels return to 40 mmHg. This causes the individual to stop breathing, triggering an apnea event.

Comprehension Check 3 One common treatment for sleep apnea is continuous positive airway pressure (CPAP). Individuals wear a mask that supplies air, which keeps breathing airways open. Based on what you read above, which form of sleep apnea is better treated by CPAP?

CPAP is more commonly used to treat obstructive sleep apnea. Because obstructive sleep apnea is caused by the collapse of the upper airway, application of continuous airflow prevents the airway from collapsing. Central sleep apnea can be managed with different devices. One such device is the adaptive servo-ventilation machine. Instead of providing fixed pressure like the CPAP machine, the adaptive servo-ventilation machine adjusts the air pressure supplied to an individual. This better regulates oxygen and carbon dioxide levels in the brain.

5.3 Narcolepsy

In the movie *Deuce Bigalow: Male Gigolo*, one character, Carol, provides an inaccurate portrayal of **narcolepsy**. In one scene, Carol abruptly falls asleep while in the act of bowling. Although *sleep attacks* can occur, the most common symptom of narcolepsy is excessive daytime sleepiness. Even if an individual receives adequate sleep, they will experience an overwhelming urge to sleep. Additional symptoms of narcolepsy include cataplexy (loss of muscle control; note, partial cataplexy is more common), sleep paralysis, and **hypnagogic** and/or **hypnopompic hallucinations** (Bassetti et al., 2019). Depending on which symptoms manifest, individuals can be diagnosed with narcolepsy type 1 or narcolepsy type 2. Specifically, narcolepsy type 2 is characterized by the absence of catalepsy.

The primary cause of narcolepsy type 1 is orexin deficiency. As noted by Bassetti et al. (2019), individuals with narcolepsy type 1 have approximately 90% fewer orexinergic neurons than normal. Because there are fewer orexinergic neurons, there is not enough orexin to stimulate other parts of the brain that release wake-promoting neurotransmitters. One hypothesis is that an immune response misidentifies orexinergic neurons as pathogens, leading to their destruction. The human leukocyte antigen (HLA) allele *HLA-DQB1*06: 02* is associated with narcolepsy, although it is more strongly associated with narcolepsy type 1 compared to narcolepsy type 2 (Miyagawa & Tokunaga, 2019). HLAs are proteins found on the surface of cells that regulate the immune system (Medhasi & Chantratita, 2022). Essentially, individuals with this particular allele may have HLAs that are hyperactive, which then target orexinergic neurons.

Diagnosing narcolepsy type 2 is controversial as individuals with this form of narcolepsy have normal orexin levels (Baumann-Vogel et al., 2021). Currently, our understanding of the neurobiological basis of narcolepsy type 2 is severely limited.

Comprehension Check 4 Given the symptoms of narcolepsy, what types of pharmacological treatments do you think are effective at managing this condition?

Stimulant drugs are commonly used to treat narcolepsy. Modafinil (Provigil) is a weak inhibitor of the dopamine transporter, and it increases histamine levels in the brain. These mechanisms help promote wakefulness. Interestingly, another treatment option is sodium oxybate, a depressant. In fact, sodium oxybate is closely related to gamma-hydroxybutyrate, an illicit drug often nicknamed the "date rape drug". This seems counterintuitive, but one important feature of narcolepsy is the feeling of unrest following sleep. By giving someone sodium oxybate, this can improve slow-wave sleep, helping them feel more rested during the day. This can help reduce excessive daytime sleepiness that is frequently observed in individuals with narcolepsy.

5.4 Shift Work Sleep Disorder

Up to this point, I have discussed the sleep-wake cycle for people that are awake during the day and asleep at night. This cycle is not followed by all individuals. There are many individuals that sleep during the day and work at night. My former neighbor is a nurse and worked the night shift for years before transitioning to day shift. Transitioning from day shift to night shift or vice versa is challenging as an individual's circadian rhythm becomes disturbed. There is a sleep disorder associated with the mismatch between one's natural circadian rhythm and their work schedule: **shift work sleep disorder**.

Our SCN acts as an internal clock and uses environmental cues to synchronize its activity with our sleep-wake cycle. As humans are *diurnal* animals (opposite of nocturnal), the SCN relies heavily on sunlight. In fact, sunlight acts as our major *zeitgeber* (German for "time giver"). Earlier in this chapter, I mentioned that light causes our SCN to become active. When individuals switch to evening work, they have difficulty sleeping during the day because light acts as a signal to promote wakefulness.

Shift work sleep disorder leads to several health complications. Shift work sleep disorder is highly associated with obesity (Brum et al., 2020), which is linked to alterations to ghrelin and leptin signaling and increased inflammatory markers (Ko, 2013). Specifically, individuals that sleep during the day have lower leptin levels, and ghrelin levels do not decrease as much following a meal. This causes a hunger signal to persist even after someone has eaten. Additional health problems associated with shift work sleep disorder include hypertension, gastrointestinal problems, stroke, anxiety/mood disorders, and cancer (Jehan et al., 2017).

Comprehension Check 5 How do you think sleep is affected in individuals that live in areas that receive little sunlight for long periods of time (e.g., Alaska, northern parts of Canada, Norwegian countries, etc.)?

Without the consistent presence of sunlight, the SCN would not receive the necessary cue for regulating the circadian rhythm. The absence of a strong zeitgeber could lead to a desynchronized sleep-wake cycle, causing a condition similar to shift work sleep disorder. As a result, individuals might experience difficulty falling asleep, staying asleep, and excessive daytime sleepiness, along with other health complications. To combat this, individuals often use light exposure therapy. By exposing oneself to bright light in the morning, this light can act as a substitute for sunlight, triggering activation of the SCN.

5.5 Parasomnias

A *parasomnia* is a sleep condition in which unwanted events or experiences during sleep or the sleep-wake transition occur. This section will highlight a few parasomnias.

5.5.1 Restless Leg Syndrome

One of the most well-known sleep-related movement disorders is **restless leg syndrome**. Individuals with this condition have an irresistible urge to move their legs during periods of rest. The urge to move one's legs is often accompanied by an uncomfortable sensation. Although primarily associated with the legs, individuals may experience the urge to move other body parts such as the arms, the mouth, and the face.

The cause of restless leg syndrome is unknown, but one leading theory is iron deficiency (Gossard et al., 2021). Iron is a *cofactor* involved in the synthesis of dopamine, which you already know is a neurotransmitter that is critically important for motor regulation. In Chap. 10, you read that levodopa (also known as L-DOPA) is the precursor to dopamine. The precursor to L-DOPA is tyrosine. Tyrosine is converted to L-DOPA by the enzyme tyrosine hydroxylase, which utilizes iron. Without iron, tyrosine hydroxylase cannot convert tyrosine into L-DOPA, meaning that there is not enough L-DOPA to become converted into dopamine. Notice that the difficulty in refraining from moving one's legs is related to deficits of the indirect motor pathway that you previously learned about in Chap. 10.

I mentioned that the *MEIS1* gene is associated with insomnia. This gene is also highly associated with restless leg syndrome. As reviewed by Sarayloo et al. (2019), individuals with a specific haplotype of the *MEIS1* gene have lower *MEIS1* mRNA expression in the thalamus, and *MEIS1* knockout mice show enhanced hyperactivity. Interestingly, individuals with lower *MEIS1* mRNA expression have increased *ferritin* levels and increased expression of the *DMT1* gene. Ferritin is a protein that stores iron, and *DMT1* encodes the divalent metal transporter 1 (DMT1) protein, which is involved in the uptake of iron into cells. Alterations to ferritin and DMT1 levels provide an account for the iron deficiency that is often correlated with restless leg syndrome. The finding that the *MEIS1* gene is implicated in both insomnia and restless leg syndrome is not surprising as restless leg syndrome is highly correlated with insomnia (Bogan, 2006).

5.5.2 REM Sleep Behavior Disorder

Normally, when we are in REM sleep, we experience **atonia** (loss of muscle tone) of our skeletal muscles. Glutamatergic neurons in the sublaterodorsal nucleus of the pons stimulate inhibitory neurons in the ventromedial medulla and the spinal cord; this leads to inhibition of motor neurons. When we are awake, the sublaterodorsal nucleus is inhibited by the ventrolateral region of the periaqueductal gray (PAG), as well as the locus coeruleus and the raphe nuclei (Scammell et al., 2017).

In **REM sleep behavior disorder**, there is a loss of atonia, leading individuals to act out behaviors during sleep. This can be problematic as individuals risk injury to themselves and/or to individuals that share a bed with the individual. Damage to regions such as the sublaterodorsal nucleus (e.g., stroke) can lead to REM sleep behavior disorder, and this condition is observed in individuals with certain neurodegenerative diseases such as Parkinson's disease, dementia with Lewy bodies, and *multiple system atrophy* (Iranzo et al., 2009). Multiple system atrophy, as the name implies, affects various brain systems, primarily the basal ganglia, the cerebellum, and the brainstem.

Individuals with narcolepsy are also at risk of developing REM sleep behavior disorder (Thorpy et al., 2024). Recall from earlier in this chapter that narcolepsy is linked to a deficient orexinergic system. Not only does orexin stimulate wake-promoting neural circuits, but it suppresses REM sleep. Because the orexinergic system is deficient in these individuals, there is a loss of sleep-wake stability. Elements of REM sleep can occur during wakefulness, such as catalepsy; however, aspects of wakefulness can be observed during sleep, such as movement.

Antidepressant medications can produce REM sleep behavior disorder (Dauvilliers et al., 2018), although the percentage of individuals that experience antidepressant-induced REM sleep behavior disorder is relatively low (~6%) (Postuma et al., 2013). As argued by Postuma et al. (2013), antidepressants do not directly lead to symptoms of REM sleep behavior disorder. Instead, these medications accentuate symptoms in those with biobehavioral markers for neurodegenerative diseases. In their study, Postuma et al. found that individuals with antidepressant-induced REM sleep behavior disorder have a higher prevalence of mild cognitive decline, abnormal color vision, motor slowing, and decreased olfaction.

5.5.3 Somnambulism

The term **somnambulism** may not be familiar to many of you reading this, but you most likely are familiar with the term "sleepwalking". In contrast to REM sleep behavior disorder, somnambulism occurs during non-REM sleep, specifically stage 3 sleep, and is often accompanied by impaired consciousness following awakening. However, like REM sleep behavior disorder, somnambulism is associated with various risks to the individual and to others. In one extreme case of reported somnambulism, Kenneth Parks drove 14 miles to his in-laws' house and attacked his father-in-law and mother-in-law. Parks' mother-in-law died from her injuries. Parks was eventually acquitted of second-degree murder and attempted murder after his defense team argued that the attacks occurred while he was asleep (Broughton et al., 1994).

As a slow-wave sleep disorder, sleepwalking typically occurs within the first 3 hours of sleep, when stage 3 sleep is most prevalent. Somnambulism is also associated with *slow-wave sleep fragmentation*, where stage 3 sleep is repeatedly disrupted (Hughes, 2007). When EEG is used to measure cortical activity, those with somnambulism show both delta wave and alpha/beta wave activity (Schenck et al., 1998).

Sleepwalking is a potential side effect of certain medications, particularly sedative-hypnotic drugs. Ironically, zolpidem (Ambien), used to treat insomnia (I was given Ambien to help me battle insomnia when I had mono), can lead to somnambulism (Hoque & Chesson Jr, 2009). Zolpidem is a benzodiazepine-like drug; although it targets $GABA_A$ receptors like traditional benzodiazepines, its chemical structure is different. This leads to a different pharmacological profile, which is why zolpidem is used selectively as a sleep agent as opposed to being used as an anxiolytic or as a muscle relaxant like benzodiazepines. A hypothesis that has been proposed is that zolpidem desensitizes $GABA_A$ receptors, which leads to decreased inhibition of serotonergic neurons (Juszczak, 2011). As you read earlier, serotonin is involved in wakefulness. If GABAergic neurons can no longer effectively inhibit serotonin neurons during slow-wave sleep, serotonin neurons can then stimulate motor neurons, leading to increased motor activity during sleep (Juszczak & Swiergiel, 2005).

Comprehension Check 6 Although restless leg syndrome, REM sleep behavior disorder, and somnambulism all manifest as unwanted or inappropriate motor activity, their underlying neurobiological mechanisms differ significantly. Which neurobiological factors contribute to each of these conditions?

Dopamine is normally associated with wakefulness, but deficient dopamine synthesis caused by iron deficiency appears to be a major determinant of restless leg syndrome. Insufficient stimulation of dopamine D_2*-like receptors in the indirect pathway leads to unplanned movements* (e.g., *moving one's legs*) *that interfere with the onset of sleep. While REM sleep behavior disorder and somnambulism are similar, the former occurs during REM sleep while the latter occurs during stage 3 sleep. In*

REM sleep behavior disorder, signals from the pons to muscles become impaired, potentially due to loss of orexinergic messages. In somnambulism, the GABAergic system is not as effective at inhibiting serotoninergic neurons, which can then stimulate motor control centers of the central nervous system.

6 Chapter Summary

This chapter provided an overview of circadian rhythms, with the sleep-wake cycle being the prototypic example of a circadian rhythm. You learned about the neuromechanisms that promote wakefulness and sleep, as well as the neural signatures associated with each stage of sleep. The remainder of the chapter highlighted various sleep disorders. The conditions insomnia, sleep apnea, narcolepsy, shift work sleep disorder, restless leg syndrome, REM sleep behavior disorder, and somnambulism were discussed. Although some of these conditions have specific causes (e.g., prions lead to fatal familial insomnia, narcolepsy type 1 is caused by loss of orexinergic neurons), other conditions can be caused by multiple factors (e.g., insomnia) or have causes that have yet to be fully elucidated (e.g., narcolepsy type 2).

Answers to Experiment Spotlight Questions

1. Given that HP1 is involved in gene suppression, lower levels of the *CBX5* gene could lead to a situation in which genes that are involved in arousal are not sufficiently silenced, leading to hyperarousal. As the *POLR1B* gene is involved in RNA transcription, lower expression of this gene could impair the production of proteins involved in regulating the sleep-wake cycle.
2. As hub genes, *CBX5* and *POLR1B* form large networks with the other genes that were identified as being differentially expressed in individuals with CAA and insomnia. Hub genes exert considerable control over other genes, thus contributing significantly to biological processes. Differential expression of *CBX5* and *POLR1B* suggests that these genes are not merely altered by CAA or insomnia, instead acting as a major player in the etiology of these conditions.
3. One potential idea is to conduct a larger-scale, longitudinal study. This follow-up research could involve a much bigger cohort of patients with CAA and insomnia, as well as a healthy control group, to confirm the association of *CBX5* and *POLR1B* with conditions like CAA and insomnia. Researchers could measure the expression levels of these genes in patients' blood over time and correlate them with disease progression and symptom severity.

   A different idea is to use in vitro (e.g., cell culture) or in vivo (e.g., mice) studies to manipulate the expression of these specific genes. For example, researchers could selectively target the expression of the *CBX5* or the *POLR1B* gene in mice to determine if they are more likely to develop symptoms related to CAA or insomnia.

This would help establish a stronger causal link between these genes and CAA/insomnia.

Glossary

Atonia a state of complete muscle relaxation or a lack of muscle tone.

Circadian rhythm a natural, internal process that regulates specific physiological processes and behaviors that repeat approximately every 24 hours.

Epigenetics the study of how our environment and behaviors can cause changes that affect the way genes work, without altering the underlying DNA sequence.

Hypersomnia a condition characterized by persistent and excessive daytime sleepiness or prolonged nighttime sleep, even after adequate rest.

Hypnagogic hallucination a vivid, dream-like sensory experience that occurs while falling asleep.

Hypnopompic hallucination a vivid, dream-like sensory experience that occurs while waking up.

Insomnia a sleep disorder characterized by difficulty falling or staying asleep, despite adequate opportunity for sleep.

Narcolepsy a chronic neurological condition characterized by overwhelming daytime sleepiness and sudden attacks of sleep.

REM sleep behavior disorder a sleep disorder in which the muscle paralysis that normally occurs during REM sleep is absent, leading to the acting out of dreams.

Restless leg syndrome a neurological disorder characterized by an irresistible urge to move one's legs, typically accompanied by uncomfortable sensations, particularly during periods of rest or inactivity.

Shift work sleep disorder a circadian rhythm sleep disorder characterized by insomnia and/or excessive sleepiness in individuals whose work schedule conflicts with their body's natural sleep-wake cycle.

Sleep apnea a sleep disorder characterized by repeated interruptions in breathing during sleep, leading to fragmented sleep and daytime fatigue.

Somnambulism a parasomnia characterized by a person getting up and walking or performing other complex behaviors while still in a state of non-REM sleep.

References

Abbasi, A., Gupta, S. S., Sabharwal, N., Meghrajani, V., Sharma, S., Kamholz, S., & Kupfer, Y. (2021). A comprehensive review of obstructive sleep apnea. *Sleep Science, 14*(2), 142–154. https://doi.org/10.5935/1984-0063.20200056

American Psychiatric Association. (2013). *Diagnostic and statistical manual of mental disorders* (5th ed.). https://doi.org/10.1176/appi.books.9780890425596.

Barclay, N. L., Kocevska, D., Bramer, W. M., Van Someren, E. J. W., & Gehrman, P. (2021). The heritability of insomnia: A meta-analysis of twin studies. *Genes, Brain, and Behavior, 20*(4), e12717. https://doi.org/10.1111/gbb.12717

Bassetti, C. L. A., Adamantidis, A., Burdakov, D., Han, F., Gay, S., Kallweit, U., Khatami, R., Koning, F., Kornum, B. R., Lammers, G. J., Liblau, R. S., Luppi, P. H., Mayer, G., Pollmächer, T., Sakurai, T., Sallusto, F., Scammell, T. E., Tafti, M., & Dauvilliers, Y. (2019). Narcolepsy - clinical spectrum, aetiopathophysiology, diagnosis and treatment. *Nature Reviews. Neurology, 15*(9), 519–539. https://doi.org/10.1038/s41582-019-0226-9

Baumann-Vogel, H., Schreckenbauer, L., Valko, P. O., Werth, E., & Baumann, C. R. (2021). Narcolepsy type 2: A rare, yet existing entity. *Journal of Sleep Research, 30*(3), e13203. https://doi.org/10.1111/jsr.13203

Benjafield, A. V., Sert Kuniyoshi, F. H., Malhotra, A., Martin, J. L., Morin, C. M., Maurer, L. F., Cistulli, P. A., Pépin, J. L., Wickwire, E. M., & medXcloud group (2025). Estimation of the global prevalence and burden of insomnia: A systematic literature review-based analysis. Sleep Medicine Reviews, 82, 102121. https://doi.org/10.1016/j.smrv.2025.102121.

Bogan, R. K. (2006). Effects of restless legs syndrome (RLS) on sleep. *Neuropsychiatric Disease and Treatment, 2*(4), 513–519. https://doi.org/10.2147/nedt.2006.2.4.513

Broughton, R., Billings, R., Cartwright, R., Doucette, D., Edmeads, J., Edwardh, M., Ervin, F., Orchard, B., Hill, R., & Turrell, G. (1994). Homicidal somnambulism: A case report. *Sleep, 17*(3), 253–264.

Brum, M. C. B., Dantas Filho, F. F., Schnorr, C. C., Bertoletti, O. A., Bottega, G. B., & da Costa Rodrigues, T. (2020). Night shift work, short sleep and obesity. *Diabetology & Metabolic Syndrome, 12*, 13. https://doi.org/10.1186/s13098-020-0524-9

Carley, D. W., & Farabi, S. S. (2016). Physiology of sleep. *Diabetes Spectrum, 29*(1), 5–9. https://doi.org/10.2337/diaspect.29.1.5

Dauvilliers, Y., Schenck, C. H., Postuma, R. B., Iranzo, A., Luppi, P. H., Plazzi, G., Montplaisir, J., & Boeve, B. (2018). REM sleep behaviour disorder. *Nature Reviews. Disease Primers, 4*(1), 19. https://doi.org/10.1038/s41572-018-0016-5

Gossard, T. R., Trotti, L. M., Videnovic, A., & St Louis, E. K. (2021). Restless legs syndrome: Contemporary diagnosis and treatment. *Neurotherapeutics, 18*(1), 140–155. https://doi.org/10.1007/s13311-021-01019-4

Han, K. S., Kim, L., & Shim, I. (2012). Stress and sleep disorder. *Experimental. Neurobiology, 21*(4), 141–150. https://doi.org/10.5607/en.2012.21.4.141

Hoque, R., & Chesson, A. L., Jr. (2009). Zolpidem-induced sleepwalking, sleep related eating disorder, and sleep driving: Fluorine-18-flourodeoxyglucose positron emission tomography analysis, and a literature review of other unexpected clinical effects of zolpidem. *Journal of Clinical Sleep Medicine, 5*(5), 471–476.

Hughes, J. R. (2007). A review of sleepwalking (somnambulism): The enigma of neurophysiology and polysomnography with differential diagnosis of complex partial seizures. *Epilepsy & Behavior, 11*(4), 483–491. https://doi.org/10.1016/j.yebeh.2007.08.013

Iranzo, A., Santamaria, J., & Tolosa, E. (2009). The clinical and pathophysiological relevance of REM sleep behavior disorder in neurodegenerative diseases. *Sleep Medicine Reviews, 13*(6), 385–401. https://doi.org/10.1016/j.smrv.2008.11.003

Iturriaga, R., & Diaz, H. S. (2025). Reactive oxidative species in carotid body chemoreception: Their role in oxygen sensing and cardiorespiratory alterations induced by chronic intermittent hypoxia. *Antioxidants, 14*(6), 675. https://doi.org/10.3390/antiox14060675

Jan, J. E., Reiter, R. J., Wasdell, M. B., & Bax, M. (2009). The role of the thalamus in sleep, pineal melatonin production, and circadian rhythm sleep disorders. *Journal of Pineal Research, 46*(1), 1–7. https://doi.org/10.1111/j.1600-079X.2008.00628.x

Jehan, S., Zizi, F., Pandi-Perumal, S. R., Myers, A. K., Auguste, E., Jean-Louis, G., & McFarlane, S. I. (2017). Shift work and sleep: Medical implications and management. *Sleep Medicine and Disorders, 1*(2), 00008.

Juszczak, G. R. (2011). Desensitization of GABAergic receptors as a mechanism of zolpidem-induced somnambulism. *Medical Hypotheses, 77*(2), 230–233. https://doi.org/10.1016/j.mehy.2011.04.019

Juszczak, G. R., & Swiergiel, A. H. (2005). Serotonergic hypothesis of sleepwalking. *Medical Hypotheses, 64*(1), 28–32. https://doi.org/10.1016/j.mehy.2004.06.013

Ko, S. B. (2013). Night shift work, sleep quality, and obesity. *Journal of Lifestyle Medicine, 3*(2), 110–116.

Lam, D. D., Antic Nikolic, A., Zhao, C., Mirza-Schreiber, N., Krężel, W., Oexle, K., & Winkelmann, J. (2022). Intronic elements associated with insomnia and restless legs syndrome exhibit cell-type-specific epigenetic features contributing to MEIS1 regulation. *Human Molecular Genetics, 31*(11), 1733–1746. https://doi.org/10.1093/hmg/ddab355

Li, X. Y., Li, K., Wei, Y. H., Yu, W. K., Wu, J. H., Gao, K., He, W. J., Niu, P. P., Zhang, C., Cheng, Y. N., & Li, Y. S. (2025). Exploration of common pathogenic genes between cerebral amyloid angiopathy and insomnia based on bioinformatics and experimental validation. *Scientific Reports, 15*(1), 26365. https://doi.org/10.1038/s41598-025-12553-y

Medhasi, S., & Chantratita, N. (2022). Human leukocyte antigen (HLA) system: Genetics and association with bacterial and viral infections. *Journal of Immunology Research, 2022*, 9710376. https://doi.org/10.1155/2022/9710376

Medori, R., Tritschler, H. J., LeBlanc, A., Villare, F., Manetto, V., Chen, H. Y., Xue, R., Leal, S., Montagna, P., & Cortelli, P. (1992). Fatal familial insomnia, a prion disease with a mutation at codon 178 of the prion protein gene. *The New England Journal of Medicine, 326*(7), 444–449. https://doi.org/10.1056/NEJM199202133260704

Miyagawa, T., & Tokunaga, K. (2019). Genetics of narcolepsy. *Human Genome Variation, 6*, 4. https://doi.org/10.1038/s41439-018-0033-7

Postuma, R. B., Gagnon, J. F., Tuineaig, M., Bertrand, J. A., Latreille, V., Desjardins, C., & Montplaisir, J. Y. (2013). Antidepressants and REM sleep behavior disorder: Isolated side effect or neurodegenerative signal? *Sleep, 36*(11), 1579–1585. https://doi.org/10.5665/sleep.3102

Roberts, E. G., Raphelson, J. R., Orr, J. E., LaBuzetta, J. N., & Malhotra, A. (2022). The pathogenesis of central and complex sleep apnea. *Current Neurology and Neuroscience Reports, 22*(7), 405–412. https://doi.org/10.1007/s11910-022-01199-2

Sarayloo, F., Dion, P. A., & Rouleau, G. A. (2019). *MEIS1* and restless legs syndrome: A comprehensive review. *Frontiers in Neurology, 10*, 935. https://doi.org/10.3389/fneur.2019.00935

Scammell, T. E., Arrigoni, E., & Lipton, J. O. (2017). Neural circuitry of wakefulness and sleep. *Neuron, 93*(4), 747–765. https://doi.org/10.1016/j.neuron.2017.01.014

Schenck, C. H., Pareja, J. A., Patterson, A. L., & Mahowald, M. W. (1998). Analysis of polysomnographic events surrounding 252 slow-wave sleep arousals in thirty-eight adults with injurious sleepwalking and sleep terrors. *Journal of Clinical Neurophysiology, 15*(2), 159–166. https://doi.org/10.1097/00004691-199803000-00010

Spielman, R. M., Jenkins, W. J., & Lovett, M. D. (2020). *Psychology* (2nd ed.) OpenStax. https://openstax.org/books/psychology-2e/pages/4-3-stages-of-sleep

Thorpy, M. J., Siegel, J. M., & Dauvilliers, Y. (2024). REM sleep in narcolepsy. *Sleep Medicine Reviews, 77*, 101976. https://doi.org/10.1016/j.smrv.2024.101976

Van Someren, E. J. W. (2021). Brain mechanisms of insomnia: New perspectives on causes and consequences. *Physiological Reviews, 101*(3), 995–1046. https://doi.org/10.1152/physrev.00046.2019

World Health Organization. (2019). *International statistical classification of diseases and related health problems* (11th ed.). https://icd.who.int/

Chapter 15
Maladaptive Social Behaviors

1 Introduction

My mom likes to tell the story of how I never spoke until I was approximately 6 years of age. As a child, I was shy and became anxious at the thought of speaking to others. I also did not like making eye contact with others. Something about looking into another person's eyes made me uncomfortable. To be honest, making eye contact is still difficult for me, but after years of teaching, I have become more comfortable engaging in this action.

As a mammalian species, we rely heavily on others. Infants need someone to provide nourishment and safety. Many of us go to school with other children and work with others as adults. Yet, there are large individual differences in how much time we like to go out and spend time with others. As a young child, I was extremely **introverted**. Although I was comfortable being around some children that I considered friends, I rarely spoke around them. Now, I am **extroverted**. I do not have difficulty striking up conversations with other individuals.

This brings up an important topic: what is considered maladaptive social behavior? As with many behaviors that have been covered throughout this book, most of the maladaptive social behaviors discussed in the current chapter only become maladaptive when done in excess or are considered maladaptive today due to changes in our social structure over time.

2 Learning Objectives

By the end of this chapter, you should be able to …

1. Discuss the various ways in which communication can become dysregulated, as well as the neurobiological mechanisms accounting for aberrant communication.

J. Yates, *Neuroscience of Maladaptation*,
https://doi.org/10.1007/978-3-032-15602-0_15

2. Compare and contrast the neuromechanisms that lead to social withdrawal and to excessive social bonding.
3. Describe the neurobiological bases of autism spectrum disorder (ASD).
4. Explain the neural correlates of aggression.

3 Dysregulated Communication

Communication is an integral process for numerous species. Humans have various ways of communicating with *conspecifics* (i.e., members of the same species). We can use spoken and written words, as well as nonverbal cues, to convey information to others. Most of us are highly sensitive to facial expressions and can interpret someone's emotional state based on that information alone. With technological advancements, we can easily communicate with individuals from across the world, whether via Zoom, FaceTime, or social media.

Even though communication occurs in other species, we are truly in a class of our own. Our ability to read and to write has enabled us to preserve our history, transmit ideas, and develop new technologies that have advanced societies. However, the ability to communicate is not equal across individuals. As you will learn, there are certain conditions that make the production of speech and written communication difficult or the ability to understand communication difficult. Additionally, despite the great benefits of communication, *homo sapiens* can also use communication for nefarious purposes.

3.1 Aphasias

In 1861, Dr. Pierre Paul Broca (1824–1880), a French surgeon, was assigned a patient named Louis Victor Leborgne. Leborgne was paralyzed on the right side of his body, and he primarily communicated by repeating the French syllable *Tan*, something he had been doing for 21 years. After Leborgne died of gangrene in 1861, Dr. Broca performed an examination of his brain and discovered something fascinating. Extensive damage was observed in part of the left frontal cortex, next to the lateral fissure that separates the frontal cortex from the temporal cortex (Mohammed et al., 2018).

Dr. Broca theorized that this damaged region of the brain is critical for speech production. Patient "Tan", as he was known to most of the world, had a severe form of **Broca's aphasia** (also known as non-fluent aphasia). Aphasias belong to a class of disorders in which language capabilities are disrupted. Aphasia comes from the Greek word *aphatos*, meaning "speechless". In Broca's aphasia, individuals have difficulty producing speech, but their speech is generally comprehensible to others. Unlike Leborgne, individuals with Broca's aphasia can say more than one word; yet, their ability to generate words is diminished. As such, their speech is characterized by short, halting phrases. Speech is not the only aspect of language that is affected. Writing becomes difficult as well. Although language production is challenging for these individuals, they understand what others are saying and show normal reading comprehension. Living with Broca's aphasia can be extremely frustrating as individuals are consciously aware of their condition. Imagine always feeling like you had something "at the tip of your tongue", slowly articulating your thoughts to another individual.

Today, the prevailing hypothesis is that Broca's area is an association area that is critical for integrating language comprehension with motor movements related to articulation (Fedorenko & Blank, 2020). The anterior portion of Broca's area receives inputs from the prefrontal cortex (PFC) and part of the superior temporal gyrus (STG) (which you will learn more about below) (see Fig. 1). These connections allow us to map our thoughts to spoken/written words. Because the PFC is involved in working memory, activation of this area is important for extracting semantic knowledge like using the correct word and using correct sentence syntax (i.e., sentence structure). The posterior portion of Broca's area has connections with the supplementary motor area (SMA) and the somatosensory cortex. To speak, we need to move our tongue. Broca's area takes this somatosensory information and can help modulate activity of the motor cortex to refine tongue movements, as well as vocal cord movements, enabling us to improve articulation. Overall, damage to Broca's area disrupts this integration, accounting for the slowed, labored speech observed in individuals with Broca's aphasia.

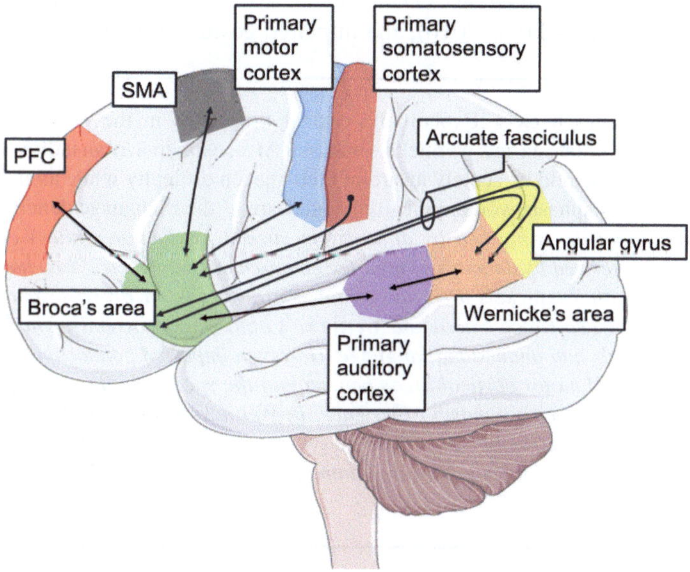

Fig. 1 A simplified schematic showing the connections between Broca's area and Wernicke's area with other parts of the brain. Image created by the author using Servier Medical Art (https://smart.servier.com/; CC BY 4.0, https://creativecommons.org/licenses/by/4.0/) and Microsoft PowerPoint

Inspired by the work of Dr. Broca and other researchers, Dr. Carl Wernicke (1848–1905) studied the neuropsychology of language. He argued that while Broca's area was involved in motor production of speech, language comprehension was subserved by a different part of the brain, located primarily in the STG, extending into part of the parietal cortex. This part of the brain is now often referred to as Wernicke's area (Fig. 1) as we know that damage to this part of the brain leads to **Wernicke's aphasia** (also known as fluent aphasia). In contrast to Broca's aphasia, individuals with Wernicke's aphasia can produce speech, but their speech comes out in the form of a **word salad**. Below is an example of a word salad:

> You know that smoodle pinkered and that I want to get him round and take care of him like you want before. (National Institute on Deafness and Other Communication Diseases, , 2015).

Not only does language come out jumbled, but individuals with Wernicke's aphasia have difficulty understanding what others are saying, have difficulty reading, and are unaware of their language comprehension deficits.

Wernicke's area and Broca's area are directly connected by the *arcuate fasciculus* (Latin for "curved bundle"), a collection (bundle) of nerve fibers that form an arch-shaped (curved) pathway. Another brain region connected to Wernicke's area is the angular gyrus, located in part of the parietal cortex. The angular gyrus is important for visual object recognition and semantic processing of verbal information (Wagner & Rusconi, 2023). Damage to the angular gyrus can lead to *alexia* (inability to read or comprehend written language) (Henderson, 2019), difficulty understanding the meaning of sentences (Price, 2010), and impaired gesture imitation (Seghier, 2024).

Comprehension Check 1 Explain the critical difference in the primary deficits between Broca's aphasia and Wernicke's aphasia. Also, explain *why* an individual with Broca's aphasia would be acutely aware of their speech difficulty while an individual with Wernicke's aphasia would typically be unaware of their language deficits.

Broca's aphasia is marked by difficulty in speech production while Wernicke's area is characterized by deficits in language comprehension. Broca's area has strong connections with Wernicke's area and motor control areas. In Broca's aphasia, the connection between Broca's area and Wernicke's area works correctly, which is why these individuals can understand language. However, impaired connectivity between Broca's area and motor cortical areas can explain the speech production difficulties individuals with Broca's aphasia experience. In Wernicke's aphasia, the connection between Broca's area and motor cortical areas is preserved, which is why these individuals can easily produce speech. However, damage to Wernicke's area disrupts language comprehension.

3.2 *Disorganized Language in Schizophrenia*

Chapter 3 focused on perceptual disturbances, including hallucinations that are characteristic of schizophrenia. Increased activation of Wernicke's area is observed in individuals experiencing auditory hallucinations (Fuentes-Claramonte et al., 2021). While hallucinations are the prototypic symptom of schizophrenia, another major symptom is *disorganized thinking*, in which individuals have difficulty organizing and expressing thoughts in a logical manner. Individuals can either display poverty of speech, somewhat mirroring those with Broca's aphasia, or incomprehensible speech in the form of word salads.

Disorganized speech observed in individuals with schizophrenia is associated with gray matter decreases in several language-processing areas of the brain and reduced white matter connectivity between Wernicke's area and Broca's area. Interestingly, hyperactivity of language-associated regions is observed when individuals with schizophrenia are at rest, but both hyperactivity and hypoactivity are observed when individuals engage in speech production or listen to another individual speak (Cavelti et al., 2018).

Comprehension Check 2 Contrast the underlying causes and mechanisms of poverty of speech in Broca's aphasia and disorganized language in schizophrenia.

Poverty of speech in Broca's aphasia is attributed to dysfunction of the pathway between Broca's area and motor cortical areas, whereas poverty of speech observed in schizophrenia is one component of disorganized speech that is attributed to aberrations in both Broca's area and Wernicke's area.

3.3 *Mutism*

While individuals with Broca's aphasia and schizophrenia display poverty of speech, they communicate orally. **Mutism**, however, is the inability or the unwillingness to speak. When an individual has the capacity to speak but chooses not to, the term *selective* mutism is used. Selective mutism is associated with several psychiatric conditions, most prominently social anxiety disorder. You will learn about a neurodevelopmental disorder later in this chapter that is associated with selective mutism.

Involuntary mutism is most typically caused by neurological conditions. *Cerebellar mutism syndrome* commonly occurs in children undergoing surgery in the posterior fossa, the part of the brain that sits at the base of the skull, containing the brainstem and the cerebellum. Removal of a brain tumor in this area can cause damage to the cerebellum, leading to mutism and other symptoms like ataxia and cognitive impairments (Ashida et al., 2021). By damaging the cerebellum, this region of the brain cannot modulate activity of muscles that control speech production. Luckily, mutism is often temporary.

In *akinetic mutism*, individuals appear to be consciously aware of their surroundings, but they make no effort to move or to speak, with the exception of eye movement. These individuals also do not express any emotions. Although I have included this disorder in this chapter, akinetic mutism is really a disorder related to goal-directed behavior as opposed to a disorder of communication. Akinetic mutism is linked to damage to structures of the mesocorticolimbic pathway, the cortico-basal ganglia-thalamo-cortical (CBGTC) loop, and the anterior cingulate cortex (ACC) (Arnts et al., 2020), which are involved in mediating aspects of motivation.

3.4 Prevarication

Have you ever told someone a lie? If your answer is "no", you are most likely lying. **Prevarication** is a normal part of childhood development. As a father of two boys, I have witnessed both of my children deny doing something even though their mother and I watched them do it. Prevarication is not isolated to telling lies; individuals may try to evade another person's question to avoid having to tell the truth. For example, if I ask my oldest if he pushed his little brother, he can either tell a lie (e.g., "I didn't push him") or he can avoid directly answering my question ("He started it!").

As adults, we often tell *white lies* to avoid hurting another's feelings or coming off as rude. You may tell a friend that their new haircut looks good even if you think it looks terrible. In one study, over 116,000 lies told by 632 participants across a 3-month period were examined (Serota et al., 2022). The good news from this study is that approximately 75% of the participants did not lie much, averaging 0–2 lies per day, with most of these lies being white lies. The most common reasons participants lied were to avoid hurting others and for humor. Only 2% of lies were told to hurt another person. This is not what this section is for. Instead, I am going to discuss individuals that are labeled as *pathological* liars.

Pathological lying appears to be mediated by hyperactivity of prefrontal cortical areas (Abe, 2009). Returning to individuals with Parkinson's disease (see Chaps. 10 and 11), you learned that a hypoactive dopaminergic system affects both the nigrostriatal and the mesocortiolimbic pathways. Interestingly, individuals with Parkinson's disease struggle to make deceptive statements, which may be due to hypoactivity in the PFC. Additional research has shown that transcranial direct current stimulation (tDCS) applied to the PFC facilitates deception. Individuals that have a history of committing fraud have increased white matter volumes in the PFC, particularly the orbitofrontal cortex (OFC), the inferior frontal cortex (IFC), and the middle frontal cortex.

What's interesting is that our PFC is the last region to fully develop as we age. The human PFC reaches maturation by our mid-20 s (Arain et al., 2013). Lying increases throughout childhood, reaching a peak during adolescence, and then decreases during adulthood (Debey et al., 2015). If the PFC is important for executive functions, why does increased activity in this region lead to prevarication? One would think that pathological lying results from impaired inhibitory control. Lying is a complex cognitive-behavioral construct that relies on inhibitory control; individuals need to suppress the truth in order to tell a convincing lie. For individuals that engage in chronic prevarication, increased PFC activation may make this process easier.

Comprehension Check 3 How does hyperactivity of the PFC contribute to prevarication?

Lying is a complex cognitive process that relies on high executive functioning. To tell a successful lie, individuals must inhibit the default response (the truth), formulate a coherent, false narrative, and maintain this narrative against incoming questions.

4 Social Withdrawal

In conditions like social anxiety disorder and agoraphobia, individuals often avoid social interactions to alleviate anxiety. *Social withdrawal* is observed in various neuropsychiatric conditions, including many that have been covered throughout this book like schizophrenia, Alzheimer's disease, and major depressive disorder (MDD).

Withdrawing from social interaction is often adaptive when used short term (Ike et al., 2020). If someone in a household gets sick, they may socially isolate themselves to prevent someone else from getting sick. There are times in which we simply need to get away from someone else, even if we love that person. My wife and I love our sons, but admittedly, getting to spend some time away from them is nice. Long-term social withdrawal is maladaptive because social isolation has detrimental physical and mental health effects.

Concerning physical effects, in older adults, feelings of social isolation are correlated with increased risk of mortality, disability, and dementia (Lyu et al., 2024). Even in younger adults, feelings of loneliness are associated with increased stress, negative affect, decreased cognitive functioning, and increased alcohol use (Vasan et al., 2023). During the COVID-19 pandemic, individuals were asked to utilize social distancing to prevent the spread of the coronavirus. During the pandemic, the prevalence of anxiety and depression increased by 25%, with social isolation being cited as a major source of these mental health conditions (World Health Organization, 2022). Anecdotally, I began taking duloxetine (Cymbalta) for anxiety and depression in 2021 during the height of the pandemic.

Isolation-induced detriments to physical and mental well-being can also be observed in incarcerated individuals that are subjected to solitary confinement (Brinkley-Rubinstein, 2019). In one study, half of individuals subjected to solitary confinement showed signs of anxiety and depression (Reiter et al., 2020). Also, after being released from prison, individuals that were placed in solitary confinement have a higher mortality rate during the first year after release compared to other incarcerated individuals that were never isolated.

4.1 Neuromechanisms of Social Approach and Avoidance

I just discussed the detrimental effects of social isolation, but I need to detail *why* some individuals, even those without a form of social anxiety disorder or agoraphobia, withdraw from social interaction. As detailed by Porcelli et al. (2019), there are several networks related to social interaction. These networks will be discussed below (see Fig. 2).

4.1.1 Perception Network

The *perception network* is important for detecting and processing social stimuli and includes the amygdala, the lateral OFC (lOFC), the ventrolateral temporal pole, the fusiform gyrus, and the superior temporal sulcus (STS). The amygdala is especially important for modulating activity of brain regions involved in recognizing faces like the fusiform face area (FFA) and the occipital face area. The perception network works closely with the salience network you first learned about in Chap. 5.

Salient stimuli activate the dopaminergic system, which helps focus our attention to them. There is an important neuropeptide that modulates dopaminergic activity in response to social stimuli: oxytocin. Often nicknamed the "love hormone", oxytocin is produced by the hypothalamus and enters the bloodstream via the pituitary gland to act as a hormone. In the central nervous system (CNS), oxytocin facilitates dopaminergic responses to social stimuli in the amygdala and the nucleus accumbens (NAc) (Rosenfeld et al., 2011). Oxytocin signaling is modulated by activity of the serotonergic system via 5-HT_{1A} and 5-HT_2 receptors (Bagdy & Kalogeras, 1993;).

Interestingly, individuals with schizophrenia have lower plasma oxytocin levels (Jobst et al., 2014) while individuals with social anxiety disorder have increased oxytocin levels (Hoge et al., 2008). These contradictory findings can be explained by the social salience hypothesis (see Piva & Chang, 2018, for a discussion). Oxytocin does not always act as a prosocial hormone, instead acting to increase the salience of social stimuli. Low oxytocin levels can account for why individuals with schizophrenia often struggle to recognize social interactions (e.g., Cavieres et al., 2023).

Conversely, individuals with social anxiety disorder are more sensitive to social cues, particularly angry faces (Mohlman et al., 2007). Relatedly, social anxiety disorder is linked to thinning of the fusiform gyrus (Syal et al., 2012) and reduced activation of the fusiform gyrus (Binelli et al., 2016). Impaired activation of the fusiform gyrus can lead to difficulty in correctly identifying faces. This provides a potential mechanism for why individuals with social anxiety disorder are more likely to identify neutral faces as being angry (e.g., Peschard & Philippot, 2017).

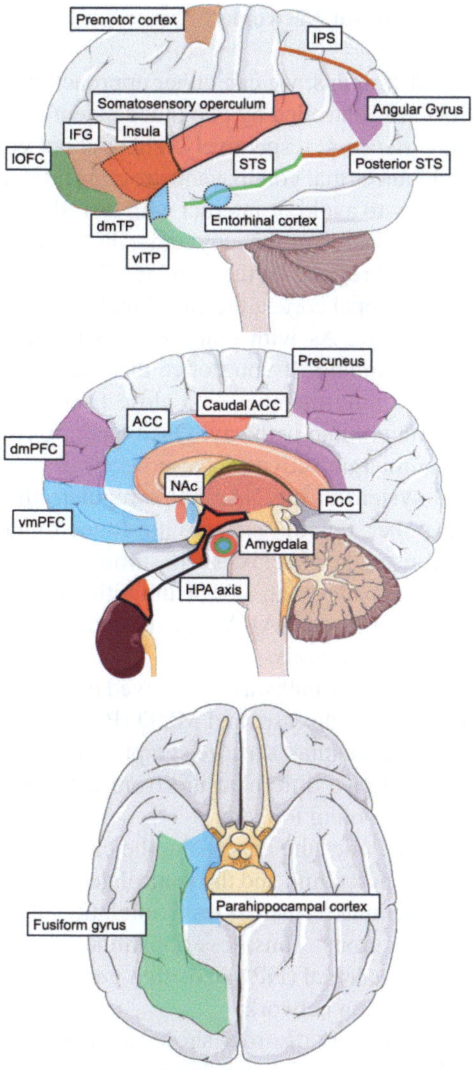

Fig. 2 Brain regions involved in social interaction. Perception network structures are highlighted in green and include the amygdala, the lateral orbitofrontal cortex (lOFC), the ventrolateral temporal pole (vlTP), the fusiform gyrus, and the superior temporal sulcus (STS). Affiliation network structures are highlighted in blue and include the amygdala, the dorsomedial temporal pole (dmTP), the anterior cingulate cortex (ACC), the ventromedial prefrontal cortex (vmPFC), the parahippocampal cortex, the entorhinal cortex, and the ventromedial striatum, particularly the nucleus accumbens (NAc). Aversion network structures are highlighted in red and include the amygdala, the caudal region of the ACC, the insula, the somatosensory operculum, the ventrolateral striatum, and the hypothalamic-pituitary-adrenal (HPA) axis. Mirroring network structures are highlighted in brown and include the inferior frontal gyrus (IFG), the premotor cortex, the posterior STS, and the intraparietal sulcus (IPS). Mentalizing network structures are highlighted in purple and include the dorsomedial PFC (dmPFC), the posterior cingulate cortex (PCC), the precuneus, and the angular gyrus. Image created by the author using Servier Medical Art (https://smart.servier.com/; CC BY 4.0, https://creativecommons.org/licenses/by/4.0/) and Microsoft PowerPoint

4.1.2 Affiliation and Aversion Networks

After recognizing a social stimulus, we can either approach it or avoid it. Approach and avoidance behaviors are controlled by two independent networks: the *affiliation network* and the *aversion network*, respectively. Think about your own experiences. You have friends/acquaintances that you like and are comfortable around. There may be other individuals you try to avoid or interact with as little as possible.

The affiliation network consists of the amygdala, the dorsomedial temporal pole, the rostral and the subgenual regions of the ACC (rACC and sgACC, respectively), the vmPFC, the parahippocampal cortex, the entorhinal cortex, and the ventromedial striatum, particularly the NAc. As with many other stimuli we encounter, social interaction is typically reinforcing. If you notice that the social stimulus you have encountered is a beloved family member or a close friend, dopaminergic activity increases in the mesocorticolimbic pathway. Once again, oxytocin plays a critical role in modulating dopaminergic activity in response to social stimuli. In addition to oxytocin, the opioid system in the NAc helps regulate the hedonic aspects of social interaction. Remember, opioid receptors are found on GABAergic (gamma-aminobutyric acid) neurons in the NAc. As such, stimulation of mu opioid receptors on these GABAergic neurons allows greater disinhibition of dopaminergic neurons. Stimulation of 5-HT_{1B} receptors on GABAergic neurons, which are also inhibitory, helps disinhibit dopaminergic neurons.

An aberrant mesocorticolimbic pathway is observed in individuals with conditions like schizophrenia, Alzheimer's disease, and MDD (Park et al., 2017; Shaikh et al., 2023), accounting for anhedonia that is a core symptom of each of these conditions. Because these individuals derive little pleasure from external stimuli, exposure to social interaction is not reinforcing to them.

The aversion network includes the amygdala, the caudal region of the ACC, the insula, the somatosensory operculum, and the ventrolateral striatum. When we experience unpleasant social encounters, there is less stimulation of mu opioid and 5-HT_{1B} receptors on GABAergic neurons. Thus, less dopamine release in the NAc occurs. The hypothalamic-pituitary-adrenal (HPA) axis that you learned about in Chap. 8 is also a key player of the aversion network.

Individuals with social anxiety disorder show enhanced activation of structures that make up the aversion network (Etkin & Wager, 2007). In a non-clinical sample, greater amygdalar activation in response to fearful and neutral faces is correlated with peak cortisol levels (Roberts et al., 2022). This finding suggests that a hyperactive HPA axis can promote increased amygdala-induced aversion to social cues in individuals with social anxiety disorder.

4.1.3 Mirroring and Mentalizing Networks

There are two additional networks that are important for social interaction: the *mirroring network* and the *mentalizing network*. Composed of the inferior frontal gyrus (IFG), the premotor cortex, the posterior STS, and the intraparietal sulcus (IPS), the mirroring network is important for interpreting the actions and emotions of others; as the name suggests, this network is also involved in imitation of others' behaviors and emotional responses. This network relies heavily on the activation of **mirror neurons** (Jankowiak-Siuda et al., 2011).

The mentalizing network includes the dorsomedial PFC (dmPFC), the posterior cingulate cortex (PCC), the precuneus, and the angular gyrus. This network enables us to use our past experiences and social knowledge to navigate social interactions. There is large overlap between regions of the mentalizing network and the default mode network (DMN) you have already learned about.

Individuals with schizophrenia have decreased connectivity within both the mirroring network and the mentalizing network (Schilbach et al., 2016). The mirror network is also impaired in Alzheimer's disease (Farina et al., 2017). Individuals with social anxiety disorder show decreased prefrontal cortical activation (Binelli et al., 2016). For example, when these individuals are asked to complete a "trust" game, they show less activation of the medial PFC (mPFC) when playing this game with a human partner compared with playing with a computer partner (Sripada et al., 2009). Deficits in prefrontal cortical activity can make navigating social interactions difficult for those with social anxiety disorder; this is directly tied to what was discussed earlier in this section. Because these individuals have difficulty understanding facial cues, this can lead to a feedback loop that reinforces social avoidance. To illustrate: if an individual with social anxiety disorder interprets a neutral face as being angry, they will avoid this individual. By limiting their social interaction, this individual is unable to learn how to better differentiate angry faces from neutral faces, further impairing the perception network and the mentalizing network.

Comprehension Check 4 Social withdrawal is reinforced by a detrimental neurobiological feedback loop in conditions like social anxiety disorder. Using the framework of the social salience hypothesis and the social interaction networks, explain the steps of this feedback loop.

In conditions like social anxiety disorder, individuals exhibit increased oxytocin levels. Under the social salience hypothesis, oxytocin does not inherently promote positive social behavior only; rather, it increases the salience of social stimuli. Because individuals with social anxiety disorder already perceive social situations as threatening, this increased oxytocin intensifies their focus on social cues, particularly negative ones. The heightened salience feeds into a flawed perception network. Social anxiety disorder is linked to structural changes like thinning of the fusiform gyrus, a key area for recognizing and processing faces. This impaired activation leads to the maladaptive tendency to misinterpret neutral faces as angry or threatening. The misinterpretation of a social cue as threatening immediately triggers the aversion network. This powerfully negative experience overwhelms the reinforcing mechanisms of the affiliation network. In a healthy brain, social interaction increases dopamine release in the NAc, promoting future social seeking. In social anxiety disorder, the aversive response is so strong that the potential for pleasure is overridden by fear, leading to the behavioral response of social avoidance. By avoiding the perceived threat, the individual temporarily reduces their anxiety (negative reinforcement). However, this avoidance prevents them from learning to correctly differentiate between neutral and angry faces, further impairing the perception network and mentalizing network. This perpetuates the cycle of avoidance, making chronic social withdrawal a stable, maladaptive pattern.

5 Autism Spectrum Disorder (ASD)

In Chap. 11, you learned about repetitive behaviors that are commonly observed in individuals diagnosed with autism spectrum disorder (ASD). Another major component of ASD is impaired communication and social interaction. According to the *Diagnostic and Statistical Manual of Mental Disorders (DSM-5)* (American Psychiatric Association, 2013), the following social communication and social interaction deficits are defining features of ASD:

1. …social-emotional reciprocity…
2. …nonverbal communicative behaviors…
3. …developing, maintaining, and understanding relationships… (p. 50).

Individuals with ASD may not share interests or emotions with other individuals or engage in normal back-and-forth conversation with others (lack of social-emotional reciprocity); may make little eye contact, have difficulty using gestures, and/or lack facial expressions (deficits in nonverbal communicative behaviors); and may have difficulty adjusting behaviors depending on the social context and/or making friends (deficits in developing, maintaining, and understanding relationships).

5.1 Animal Models of Autism Spectrum Disorder

Both genetic and environmental factors are proposed to contribute to the etiology of ASD. As such, multiple genetic models have been used to study ASD, and preclinical studies often measure how *in utero* or early-life exposure to certain chemicals affect social behaviors later in life (Li et al., 2021). Some of these genetic and environmental models will be discussed below when I cover the neurobiological bases of ASD. I do want to mention one specific animal model.

There is a mouse strain called the BTBR (black and tan brachyury) (Fig. 3). Brachyury is a type of protein; these inbred mice carry a brachyury mutation that affects tail development. BTBR mice also do not have a corpus callosum (Fenlon et al., 2015), which is a large collection of axons that connects the two hemispheres of the brain. These mice also have a much smaller hippocampal commissure, which connects each half of the hippocampus (Dodero et al., 2013). BTBR mice show little

Fig. 3 Two BTBR mice confined in a small chamber. *Source* Figure 1 of Defensor et al. (2011), https://doi.org/10.1016/j.bbr.2010.10.033, with permission from Elsevier

interest in social interaction (e.g., Yang et al., 2012) and show abnormal vocalizations when placed in a social environment (e.g., Scattoni et al., 2011).

5.2 *Neural Correlates of Autism Spectrum Disorder as Related to Impaired Social Communication and Interaction*

I want to begin this section by identifying one piece of misinformation that has spread since the late 1990s—vaccines cause ASD. In 1998, a paper was published by Dr. Andrew Wakefield in *The Lancet* (a prestigious journal) claiming that the MMR (measles, mumps, and rubella) vaccine is highly associated with ASD. This paper was ultimately retracted as serious concerns were raised about the validity and the ethics of this study (Eggertson, 2010). Since Dr. Wakefield's publication, numerous research studies have refuted the idea that vaccines are linked to ASD (e.g., DeStefano et al., 2013; Gabis et al., Gabis et al., 2022; Hviid et al., 2003; Taylor et al., 2014). With that out of the way, let's explore the neural correlates of ASD.

During development, individuals diagnosed with ASD have greater gray matter and white matter volumes early in life, which decreases during adolescence; furthermore, children with ASD have a higher proportion of neurons relative to glia cells (Holanda et al., 2025). One part of the brain that shows abnormal development is the amygdala. Children with ASD have a larger amygdala and greater dendritic spine density in this part of the brain (Beopoulos et al., 2022). Amygdalar growth occurs at a faster rate from 6 months to 24 months of age in children with ASD (Shen et al., 2022). Increased activity of the amygdala is also observed in individuals with ASD, especially when they are asked to direct their visual attention to the eye region of another person (Tottenham et al., 2014), and there is increased negative connectivity between the amygdala and the OFC (Green et al., 2015). Overall, individuals with

ASD may perceive social interaction as a threat due to aberrant amygdalar activity, leading to social withdrawal.

Increased brain volume can be attributed to neuroinflammation, as individuals with ASD have elevated cytokine levels and increased microglia activation (Lv et al., 2025; Vargas et al., 2005). One hypothesis is that exposure to certain chemicals *in utero* or during early life can contribute to neuroinflammation and the development of ASD. *Bisphenol A (BPA)* is commonly found in plastics and epoxy resins and is known to disrupt endocrine processes (Rubin, 2011). As such, BPA is often used in preclinical studies to determine how environmental toxins increase the risk of ASD. To this end, BPA exposure is linked to ASD, at least in boys (Symeonides et al., 2024; Zhang et al., 2025). Some of the evidence linking BPA to ASD in boys will now be discussed.

In male, but not female, rats, *in utero* BPA exposure decreases *SEMA5A* and *SLC9A9* gene expression in the PFC (Kanlayaprasit et al., 2024). The *SEMA5A* gene regulates axon guidance and cell migration, and the *SLC9A9* gene encodes a sodium/hydrogen exchanger. This protein exchanges hydrogen ions for sodium ions; this mechanism keeps the pH of a cell at a consistent level. Dysregulation of the *SEMA5A* gene leads to excessive **neuritogenesis**. *Neurites* are the progenitors of dendrites and axons; these are projections that initially form from a neuron's cell body during development. Excess neurite development can be problematic as connections between neurons can become impaired. Additionally, if the sodium/hydrogen exchanger is impaired, too much hydrogen builds up in neurons, which can damage them. Interestingly, males, but not females, prenatally exposed to BPA show decreased cell death in the mPFC (Drzewiecki et al., 2023), indicating that increased hydrogen buildup in neurons does not necessarily kill these cells; instead, they may alter how these neurons function, further contributing to the symptomatology of ASD.

BPA is also implicated in increased synaptogenesis in the hippocampus of males (Thongkorn et al., 2023). Males exposed to BPA also show reduced neural stem cell expression in the hippocampus, whereas females show BPA-induced increases in neural stem cells (Kasitipradit et al., 2025). Because neural stem cells can differentiate themselves into either neurons or glia cells, females may be less sensitive to BPA exposure as these stem cells can form into glial cells, thus maintaining a balance between these cells and neurons.

A brain region that has received considerable attention is the STS. This part of the brain becomes active when we process language, and it plays an important role in detecting biological motion and directing attention to social cues (Redcay, 2008). Individuals with ASD have reduced cortical thickness in this area, and this part of the brain, along with the amygdala and the fusiform gyrus, is not as active when these individuals are required to process emotional facial expressions (Patil et al., 2024). Reduced activation of regions like the STS and the fusiform gyrus also accounts for general deficits in facial recognition observed in those with ASD (Liu et al., 2025).

While the STS is less active when individuals process emotional facial expressions, this region, as well as the mPFC, is more active at rest in adolescents that display features of ASD. Read the Experiment Spotlight to learn more.

Experiment Spotlight: Resting-State fMRI Activation is Associated with Parent-Reported Phenotypic Features of Autism in Early Adolescence.
The full article can be accessed with this link: https://www.frontiersin.org/journals/child-and-adolescent-psychiatry/articles/10.3389/frcha.2024.1481957/full

The present study (Hickson et al., 2024) included an analysis of 7108 adolescents (53% male) between the ages of 10 and 13 involved in the Adolescent Brain Cognitive Development (ABCD) study. The ABCD study is a long-term study of adolescent brain development in the United States. Data are made publicly available, allowing researchers to perform *secondary analyses.* The ABCD study excludes children/adolescents with a diagnosis of moderate or severe ASD. As such, the researchers had to use three parent-reported questions taken from the *Kiddie Schedule for Affective Disorders and Schizophrenia for DSM-5* (KSADS-COMP) to infer possible phenotypic features of autism (PPFA). The questions were:

1) "In the past two weeks, how often has your child had trouble maintaining eye contact and looking at you or other people when they are talking with your child?"
2) "In the past two weeks, how often have you worked real hard to keep routines and activities the same so your child would not get upset?"
3) "In the past two weeks, how often did your child do unusual body movements like hand flapping, head weaving, body rocking, or body spinning?"

fMRI (functional magnetic resonance imaging) data from the following brain regions were examined: mPFC, PCC, STS, inferior frontal gyrus (IFG), angular gyrus, and temporoparietal junction (TPJ). The mPFC, the PCC, and the angular gyrus are critical components of the DMN. The STS, the IFG, and the TPJ are part of the ventral attention network (VAN). Recall from Chap. 5 that the VAN is closely linked to the salience network.

Results showed that resting-state activity of the mPFC and the STS is positively correlated with PPFA. Conversely, lower resting-state activity of the IFG is associated with PPFA. Another finding was that being assigned male at birth is positively correlated with PPFA.

Question to Consider:

1. How can the results of this study be used to describe the social deficits observed in individuals with ASD?
2. While activity of the mPFC, the STS, and the IFG was correlated with PPFA, no such correlations were observed for the other brain regions examined. Why do you think this is the case?
3. Based on the results of the present study, what is one potential treatment approach for improving social communication in individuals with ASD?

Above, you learned about oxytocin's involvement in social bonding. Vasopressin, which you were introduced to in Chap. 13, is also important for social interaction. The paraventricular nucleus (PVN) and the supraoptic nucleus are smaller in individuals with ASD (Kurth et al., 2011), which is where oxytocin and vasopressin are synthesized. The evidence linking oxytocin and vasopressin to ASD is not clear-cut. Single nucleotide polymorphisms (SNPs) of genes that encode the oxytocin receptor and vasopressin receptors have been linked to ASD (Cataldo et al., 2018). Yet, studies examining oxytocin and vasopressin levels in individuals with ASD are inconsistent, with some studies reporting decreased oxytocin levels, with others reporting no differences in oxytocin or vasopressin levels (see Borie et al., 2021 for a discussion). More work is needed to unravel how social hormones contribute to the etiology of ASD.

Comprehension Check 5 How do abnormalities in structures such as the amygdala, the STS, and the fusiform gyrus contribute to social withdrawal observed in ASD?

Because the amygdala develops faster in individuals with ASD, they may be more prone to respond to social cues as threatening. Conversely, abnormalities in the STS and the fusiform gyrus can make interpreting nonverbal cues difficult. This inability to correctly read cues makes the social world confusing and unpredictable, contributing to difficulty in social-emotional reciprocity and maintaining relationships. This reinforces the initial impulse to withdraw from social interaction.

6 Absence of Empathy

Imagine you learned that a close friend had recently lost their job or had gone through a tough breakup. How would you feel? Would you laugh at your friend's misfortune, or would you feel upset and try to comfort your friend? I think most of you reading this would choose the second option. An important social skill is **empathy**. Most of us have gone through some sort of hardship during our life. Even if we have not gone through the same exact situation as another individual, we can imagine how we would feel if we went through the same thing. Empathy is important because it enables us to build stronger relationships with our peers.

Empathy is composed of several subsystems. I will focus on two such systems. First, the mirror system that you learned about earlier in this chapter is critical for empathy. Second, *theory of mind* is the ability to understand that other individuals have different viewpoints and that they can react differently to situations. Going back to the previous paragraph, theory of mind allows us to imagine how one feels as they go through a situation that we have not personally experienced. Brain regions involved in theory of mind include the mPFC, the STS, the TPJ, and the temporal pole (the most anterior portion of the temporal lobe). One thing to note is that regions like the inferior parietal lobe (IPL) and the mPFC make up part of the DMN. Given that empathy involves "putting yourself in others' shoes" and the DMN is most active when we are engaged in internal thought processes, one should not be surprised to learn that the DMN is important for empathy (Li et al., 2014).

We often experience *empathetic pain* (also known as vicarious pain) when we witness someone else that is in physical and/or emotional pain. Imagine watching someone suffer a significant injury. You may wince as if the injury had occurred to you. Empathetic pain experiences are highly attributed to increased activation of the anterior insula (Gu et al., 2010; Guo et al., 2013).

Some individuals lack empathy, particularly those diagnosed with **antisocial personality disorder** and **narcissistic personality disorder** (American Psychiatric Association, 2013). Individuals with antisocial personality disorder show a disregard for the rights of others and refuse to follow social norms. These individuals frequently lie, engage in impulsive behaviors, and display aggressive tendencies (aggression will be covered later in this chapter). Narcissistic personality disorder is characterized by an inflated sense of self-importance; individuals diagnosed with this personality disorder require admiration from others and have a sense of entitlement.

The term "sociopath" is an antiquated term that was often used to describe someone with antisocial personality disorder. This term is often incorrectly used interchangeably with "psychopath", which is used to describe an individual with **psychopathy**. If you look at the definitions for antisocial personality disorder and psychopathy, you will notice that they seem very similar. Antisocial personality disorder is officially recognized as a mental disorder, with the diagnostic criteria focusing on behaviors. Psychopathy focuses more on personality traits as opposed to observable behaviors. For the sake of this chapter, I will describe the neurobiological changes in individuals that have been officially diagnosed with antisocial personality disorder and in individuals that are classified as meeting criteria for psychopathy.

Dysfunction of the DMN is associated with psychopathy, with emotional detachment being linked to reduced white matter connections between the mPFC and the parietal cortex (Sethi et al., 2015). Even children with *callous-unemotional* traits, a risk for antisocial behaviors, show reduced connectivity within the DMN (Umbach & Tottenham, 2021). In Chap. 5, you learned about networks related to attention. Recall that the dorsal attention network (DAN) works in opposition to the DMN. That is, when we focus our attention on some external stimulus, activity of the DAN increases while activity of the DMN decreases. However, the ability to switch from the DMN to the DAN is impaired in adolescents with high callous-unemotional traits (Winters et al., 2021).

The salience network acts as a "switch" that coordinates the altered activity levels of the DMN and the DAN when individuals begin engaging in cognitively demanding tasks. Impairments in salience network switching are observed in individuals with psychopathy (Anderson et al., 2018; Deming et al., 2023). The inability to effectively switch from a default resting state (DMN) to a more active network (DAN) may provide some support as to why individuals with conditions like antisocial personality disorder are more focused on themselves.

Alterations to the serotoninergic system are implicated in the etiology of antisocial personality traits. SNPs of genes encoding the 5-HT_{1B} and the 5-HT_{2A} receptor are associated with callous-unemotional traits in adolescent boys, with these traits also being correlated with low serum serotonin levels (Moul et al., 2013). Even newborns of individuals with a family history of antisocial personality disorder have lower levels of the serotonin metabolite 5-hydroxyindoleacetic acid (5-HIAA) (Constantino et al., 1997). Furthermore, low hormonal response to fenfluramine, a drug that increases serotonin levels, during childhood is predictive of onset of antisocial personality disorder later in life (Flory et al., 2007). Deficient serotoninergic activity can negatively impact oxytocin, which is crucial for pair bonding. Indeed, adolescents with low oxytocin levels have increased callous-unemotional traits (Dadds et al., 2014; Fragkaki et al., 2019; Levy et al., 2015).

Primary psychopathy, characterized by callousness, manipulativeness, and low anxiety, is related to high tonic and phasic activity of dopaminergic neurons in the mesocorticolimbic pathway (Yildirim & Derksen, 2015). One argument is that individuals with antisocial personality disorder are highly sensitive to external reinforcers; these individuals are not concerned about negative consequences associated with obtaining reinforcement (e.g., hurting another individual).

Comprehension Check 6 How do neurobiological deficits found in individuals with psychopathy, namely, reduced connectivity within the DMN and impaired salience network switching, result in their characteristic emotional detachment and self-focus?

Because the DMN is involved in self-reflection and imagination of others' states, deficient connectivity within this network can drive emotional detachment by making it more difficult to envision the experiences of other individuals. Individuals with psychopathic traits struggle to effectively decrease DMN activity and increase DAN activity when faced with an external, cognitively demanding social stimulus. This impairment locks them into a DMN-dominant, self-focused mode. They remain preoccupied with their own internal thoughts and goals, failing to shift their attention outward to process and respond appropriately to the needs or emotions of others.

7 Excessive Interpersonal Bonding

Have you ever had a friend or a romantic partner that was too "clingy"? This person may have called/texted you at all hours of the day, expecting you to focus your attention on them only. Being expected to devote all your energy to one other individual can be emotionally exhausting and can be scary in certain situations.

There are some psychiatric disorders characterized by excessive interpersonal bonding. Individuals with certain disorders may attach themselves to one individual (i.e., become "clingy") and may become overly dependent on this individual's attention. There are situations in which an individual may overestimate how close they are to an individual, acting like they are best friends with someone when they are mere acquaintances. The disorders detailed below are characterized by more than excessive interpersonal bonding, but dysfunctional interpersonal relationships are a defining symptom.

Borderline personality disorder is perhaps the most well-known psychiatric condition marked by unstable interpersonal relationships. Specifically, individuals with borderline personality disorder have unstable, intense interpersonal relationships (the term "whirlwind romance" is often associated with borderline personality disorder), and they have a fear of abandonment. Other symptoms include unstable self-image, rapid mood changes (similar to bipolar disorder), impulsivity, inappropriate/intense anger, and self-injury (American Psychiatric Association, 2013).

Studies examining the neurobiological bases of excessive interpersonal bonding often use individuals with borderline personality disorder. However, I do want to mention a couple of other disorders characterized by excessive interpersonal bonding. **Dependent personality disorder** is similar to borderline personality disorder, except that individuals with this condition do not experience mood instability or impulsivity. Instead, these individuals become overly dependent on another individual, feeling like they cannot take care of themselves. Like borderline personality disorder, individuals with dependent personality disorder have a fear of abandonment (American Psychiatric Association, 2013). Individuals with **histrionic personality disorder** engage in excessive attention-seeking behavior (i.e., feel like they need to be the center of attention), display superficial emotions that rapidly change, overestimate how close they are to others, and are easily influenced by others. These individuals also display self-dramatization and exaggerated emotional expressions (American Psychiatric Association, 2013).

7.1 Neural Correlates of Excessive Interpersonal Bonding

Earlier, you learned that individuals that engage in social avoidance have lower oxytocin levels. One would assume that excessive interpersonal bonding would be associated with a hyperactive oxytocin system. However, oxytocin levels are lower in individuals with borderline personality disorder (Bertsch et al., 2013; Bocchio Chiavetto et al., 2025; Carrasco et al., 2020; Jobst et al., 2016). Individuals with borderline personality disorder also show lower oxytocin levels following rejection (Jobst et al., 2016), and lower oxytocin levels are correlated with impaired social cognition in these individuals (Galvez-Merlin et al., 2024). As oxytocin reduces activity of the amygdala (Sobota et al., 2015), low oxytocin levels are attributed to a hyperactive amygdala observed in individuals with borderline personality disorder (Donegan et al., 2003; O'Neill & Frodl, 2012; Sicorello & Schmahl, 2021). Additionally, decreased functional connectivity between the OFC and the amygdala is observed in these individuals, which can further account for hyperactivity of the amygdala (New et al., 2007).

Low oxytocin levels and hyperactivity of the amygdala contribute to difficulties in interpersonal relationships. A deficient oxytocin system can make establishing stable interpersonal relationships difficult, and a hyperactive amygdala can make individuals more hypersensitive to interpersonal cues, leading to emotional instability (di Giacomo et al., 2024). Although oxytocin levels are lower in individuals with borderline personality disorder, administration of oxytocin does not improve all aspects of social behavior. Oxytocin reduces social threat avoidance and increases empathy, but it worsens nonverbal behavior, mentalizing, and trust (di Giacomo et al., 2024).

In addition to a hyperactive amygdala, individuals with borderline personality disorder show aberrant DMN connectivity and activity (Kluetsch et al., 2012; Shafie et al., 2023), which is similar to what is observed in those with antisocial personality disorder. Increased dmPFC activation is observed in individuals with borderline personality disorder when they perform a task that involves social inclusion (Malejko et al., 2019). When processing social cues, individuals with borderline personality disorder show increased activation of the STS (Buchheim et al., 2008; Doell et al., 2020). Increased activity of the dmPFC and the STS may reflect a state of hypermentalizing. Individuals with borderline personality disorder need to work harder to decode social signals, often leading to misinterpretations, particularly a bias toward perceiving neutral or ambiguous cues as negative, malevolent, or rejecting. This over-analysis of social information, coupled with other brain differences like heightened amygdala activation, is an important underlying mechanism of dysregulated interpersonal relationships.

As borderline personality disorder is characterized by multiple symptoms such as impulsivity, I want to note that alterations in other brain regions are observed in these individuals, including a smaller hippocampus and ACC (Nunes et al., 2009; O'Neill & Frodl, 2012; Ruocco et al., 2012), thinning of prefrontal cortical areas (Sampedro et al., 2021), and increased ACC activity (Shafie et al., 2023). These changes are similar to what is observed in individuals with MDD (refer back to Chap. 9). In fact, borderline personality disorder and MDD commonly co-occur (Beatson & Rao, 2013).

Comprehension Check 7 While social withdrawal is often associated with high threat perception and increased social anxiety, excessive interpersonal bonding presents a paradoxical pattern of fear combined with hyper-attachment. Explain the neuromechanisms that account for this paradox.

The combination of low oxytocin levels and a hyperactive amygdala directly contributes to the core borderline personality disorder symptoms of relationship intensity and fear of abandonment. Oxytocin is crucial for stable social bonding and trust. Deficits in the oxytocin system make it difficult for individuals with borderline personality disorder to establish and maintain secure, stable interpersonal relationships. Hyperactivity of the amygdala leads to hypersensitivity of emotional and social cues. This heightened state means that minor slights, ambiguous texts, or temporary separations are perceived with excessive intensity as major threats, driving the intense and overwhelming fear of abandonment that dictates individuals' clingy and impulsive relationship behaviors.

8 Aggression

From an evolutionary perspective, aggression can be viewed as an adaptive trait in most species (Buss & Shackelford, 1997). Aggression most likely emerged as a way to deal with certain challenges and to ensure the survival of a species. For example, our earliest ancestors had to hunt and search for food. Because resources were limited, groups of individuals needed to protect the food they had accumulated from other animals, including other tribes. Aggression toward same-sex conspecifics may reflect the need for animals to ensure that the "best" genes are passed on to the next generation. However, as our species evolved, we moved away from living as nomadic hunter-gatherers to living in larger societies with complex laws and regulations governing how we should live. Now, aggression can be problematic as it often violates the norms we have established.

There are several types of aggression: impulsive, premeditated, reactive, and proactive (Babcock et al., 2014). Impulsive and reactive aggression are closely linked as they both involve unplanned aggressive acts that are emotionally driven. The main difference is that impulsive aggressiveness occurs when minimal provocation is present, while reactive aggression stems from perceived threat. Premeditated and proactive aggression are planned ahead of time, but proactive aggression is used to attain some goal (e.g., acquiring money from another individual), whereas premeditated aggression focuses on the act of aggression itself (e.g., torturing another individual).

8.1 Animal Models of Aggression

In Chap. 8, I discussed animal models of stress. In the resident-intruder paradigm, one animal (the intruder) is subjected to repeated bouts of social defeat, which increases stress levels. This paradigm also measures aggressive behavior in the resident animal. Other forms of stress can be used to increase aggression in animals, including social isolation (Takahashi, 2025), overcrowding (Pavlova et al., 2024), and foot shock (Nordman et al., 2020).

Research has also shown that aggressive behavior can be reinforcing to rodents (Golden et al., 2019). In one experiment (Falkner et al., 2016), male mice were first tested in the resident-intruder paradigm. Eventually, a panel containing two apertures was placed in the resident mouse's home cage. Mice were trained to emit a response in one aperture. Completing the response requirement led to the reintroduction of the intruder mouse into the home cage. Approximately 57% of resident mice learned to respond to gain access to the intruder. Mice that were more aggressive during the resident-intruder paradigm were more likely to respond to gain access to the intruder, and the rate of learning was significantly correlated with aggression levels during the resident-intruder paradigm.

In this experiment, neurons in part of the ventral nuclus of the hypothalamus (VMH) became highly active when mice attacked the intruder. As mice learned to emit a nose poke response, neuronal activity in the VMH increased as mice completed the response requirement and while they waited for the introduction of the intruder mouse. This finding is interesting because it mirrors what you learned about reinforcement learning in Chap. 11. That is, dopaminergic neurons initially respond to unexpected reinforcement but then begin responding to cues that signal reinforcement delivery.

8.2 Neural Correlates of Aggression

Returning to the VMH, activation of this region causes glutamatergic signals to the periaqueductal gray (PAG) to increase. The PAG is responsible for organizing motor actions related to aggression. These signals are transmitted through the brainstem and the spinal cord (Hashikawa et al., 2017) (Fig. 4). Optogenetic inhibition of the VMH decreases aggressive behavior while optogenetic stimulation of this region increases aggression (Falkner et al., 2016). Likewise, inhibition of intra-PAG neurons increases the latency of attack behavior in animals (Yu et al., 2021).

The VMH is not the only hypothalamic structure involved in aggression. Inhibiting the pathway between the PVN and the mPFC increases infanticide in rodents (Li

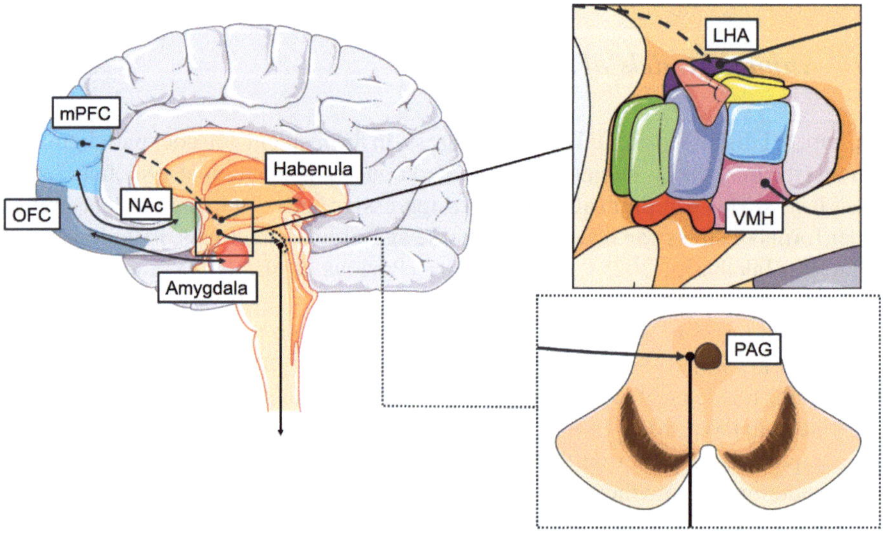

Fig. 4 A simplified schematic of several brain regions implicated in promoting or inhibiting aggressive behavior. See the text for a description of how these regions are involved in aggression. Image created by the author using Servier Medical Art (https://smart.servier.com/; CC BY 4.0, https://creativecommons.org/licenses/by/4.0/) and Microsoft PowerPoint

et al., 2024). Relatedly, GABAergic neurons originating from the prelimbic region of the mPFC inhibit orexinergic neurons in the lateral hypothalamic area (LHA); inhibiting these neurons increases social aggression (Bai et al., 2023). In the previous chapter, you learned about the role of orexin in mediating arousal. Orexinergic activity promotes areas of the brain involved in wakefulness. Because orexin is involved in arousal, increased activity of the orexinergic system can promote aggressive behavior. Research has shown that orexin binds to orexin receptor 2 on GABAergic neurons in the lateral habenula, which corresponds to increased aggression between mice (Flanigan et al., 2020).

Individuals with borderline personality disorder display enhanced amygdalar activity, which coincides with increased reactive aggression (Scott et al., 2017). Individuals with intermittent explosive disorder (see Chap. 12) show increased activation of the amygdala and decreased activation of the OFC when presented with an angry face (Coccaro et al., 2007). Siever (2008) argues that aggression largely stems from a failure of prefrontal cortical structures to inhibit limbic activity. Patients with traumatic brain injury to prefrontal cortical areas display increased aggression (Pardini et al., 2011). Likewise, disruption of prefrontal cortical activity increases aggressive-like behavior in animals (e.g., Kuniishi et al., 2017; Rudebeck et al., 2007; Takahashi et al., 2014). Individuals that display psychopathic traits have decreased functional connectivity between parts of the PFC and regions such as the amygdala, the PCC, the insula, and the hypothalamus, as well as reduced gray matter volumes in these regions and in the STS (Johanson et al., 2020).

As I discussed above, aggressive behavior can be reinforcing to rodents. Although violent offenders with moderate psychopathic traits have a smaller NAc (Boccardi et al., 2013), NAc activity is increased in criminal offenders with psychopathic traits (Hosking et al., 2017), and psychopathic traits predict dopamine release from the NAc (Buckholtz et al., 2010). There is also less connectivity between the ventromedial PFC (vmPFC) and the NAc (Hosking et al., 2017), which can be interpreted as an inability of the PFC to inhibit NAc responses. Collectively, these results indicate that aggression is reinforcing to individuals with psychopathic traits.

Psychopathy is not limited to dopaminergic dysfunction in the ventral striatum. Incarcerated individuals with high psychopathic traits have lower dopamine D_2-like receptor availability in the striatum (Lukkarinen et al., 2024). As these receptors are inhibitory and are critical for inhibiting certain motor actions, lower expression of D_2-like receptors can lead to impulsive aggression. Another hypothesis is that lower D_2-like receptor availability leads to reward deficiency. D_2-like receptors are important for motivation (Eisenegger et al., 2014; Gallo et al., 2018), with disruption of these receptors being linked to anhedonia (Berg et al., 2023). Due to increased anhedonia, individuals with conditions like antisocial personality disorder may need to engage in more extreme forms of behavior to receive reinforcement. Indeed, psychopathy is associated with risky decision making (Snowden et al., 2017) and sensation seeking (van Dongen et al., 2022).

In Chap. 9, you learned that low serotonin levels are correlated with both nonsuicidal self-injury (NSSI) and suicide. Likewise, a hypoactive serotonin system is implicated in impulsive aggression (da Cunha-Bang & Knudsen, 2021). However, a hypoactive serotonergic system does not fully explain increased aggression in individuals. There is much interest in examining how the *low-activity monoamine oxidase A (MAO-A) genotype* influences aggression. Individuals with the nicknamed "warrior gene" are more likely to display aggressive behaviors (Kolla & Bortolato, 2020), especially following provocation (McDermott et al., 2009), and increased aggression observed in these individuals is mediated by negative urgency (see Chap. 9) (Chester et al., 2015). MAO-A knockout mice are more likely to attack conspecifics (Godar et al., 2014).

The low-activity MAO-A genotype leads to *higher* levels of serotonin, dopamine, and norepinephrine. In contrast to impulsive aggression, which is attributed to low serotonin levels (particularly in the PFC), increased serotonin resulting from MAO-A deficiency is more closely linked to reactive aggression. One proposal is that excess serotonin desensitizes serotonergic neurons, making them less sensitive to serotonin release resulting from social stimuli (Godar et al., 2016). In fact, there is evidence that MAO-A knockout mice have desensitized 5-HT_{1A} autoreceptors (Evrard et al., 2002), and preventing the reuptake of serotonin substantially decreases aggressive behavior in these mice (Godar et al., 2014).

Comprehension Check 8 Resolve the apparent contradiction that low serotonin is implicated in impulsive aggression while high serotonin is linked to reactive aggression.

Low serotonin levels in the PFC prevent this region from exerting top-down control over limbic structures, predisposing individuals to engage in impulsive behaviors. However, chronic deficiency in the MAO-A enzyme leads to elevated serotonin levels that can then desensitize 5-HT receptors. This leads to a similar result as hypoactive serotonin levels that are associated with impulsive aggression.

9 Chapter Summary

This chapter explored maladaptive social behaviors, beginning with dysregulated communication like aphasia, disorganized language, mutism, and prevarication, which is linked to dysfunction of frontal and temporal cortical areas. This chapter then discussed social withdrawal, highlighting its detrimental long-term effects and the underlying neural mechanisms of social approach and avoidance, which involve the interplay of the perception, the affiliation, the aversion, the mirroring, and the mentalizing networks. ASD was examined as a condition marked by impaired social communication, with research revealing aberrant amygdala growth and activity, neuroinflammation, and altered connectivity in social processing regions like the STS. This chapter also detailed the absence of empathy, which is tied to dysfunction in the DMN and the salience network, and excessive interpersonal bonding, which is characterized by a hyperactive amygdala and low oxytocin levels. This chapter concluded by discussing aggression, differentiating impulsive, reactive, premeditated, and proactive forms, and linking it to a failure of prefrontal cortical inhibition over limbic structures.

Answers to Experiment Spotlight Questions

1. As the mPFC is an important component of the DMN, individuals with PPFA may engage in too much self-referential thought. As you read about in the section on empathy, the mPFC and the STS are important for theory of mind. Hyperactivity of these regions can lead to hypermentalizing, a phenomenon in which individuals over-interpret social cues and intentions. This can lead to social deficits that are observed in conditions like ASD. As part of the mirroring network, hypoactivity of the IFG may impair the ability of individuals with PPFA to understand and simulate observed actions and movements.
2. As noted by Hickson et al., posterior regions of the brain mature faster than anterior portions of the brain. When discussing prevarication, I mentioned that the PFC is the last region to fully mature. As young adolescents were examined in this study, frontal cortical areas like the mPFC and the IFG are still developing, which may make detecting differences between those with PPFA and without PPFA easier.
3. Techniques such as transcranial magnetic stimulation (TMS) or tDCS could be used to modulate activity of regions like the mPFC, STS, and/or IFG to improve social cognitive processes.

Glossary

Antisocial personality disorder a mental health condition characterized by a persistent pattern of disregard for and violation of the rights of others, often accompanied by a lack of empathy or remorse.

Borderline personality disorder a mental health condition characterized by pervasive instability in mood, interpersonal relationships, self-image, and behavior, often leading to impulsive actions and intense emotional swings.

Broca's aphasia a language disorder characterized by difficulty producing fluent speech, though comprehension is generally preserved.

Communication the process of conveying information or ideas from one person or entity to another through a shared system of symbols, signs, or behaviors.

Dependent personality disorder a mental health condition characterized by an excessive and pervasive need to be taken care of, leading to submissive, clinging behavior and fears of separation.

Empathy the capacity to understand or feel what another person is experiencing from within their frame of reference.

Histrionic personality disorder a mental health condition characterized by a pervasive pattern of excessive emotionality and attention-seeking behavior, often involving dramatic displays, superficial emotions, and a need to be the center of attention.

Mirror neuron a type of brain cell that fires both when an individual performs an action and when they observe someone else performing the same action.

Mutism the inability or unwillingness to speak, which can stem from physical, psychological, or neurological causes.

Narcissistic personality disorder a mental health condition characterized by an excessive sense of self-importance, a pervasive need for admiration, and a significant lack of empathy for others.

Neuritogenesis the process by which new neurites, which are the initial projections that will develop into axons and dendrites, extend from the cell body of a neuron.

Prevarication the act of deviating from or avoiding the truth, often through evasiveness or lying.

Psychopathy a set of personality traits characterized by a persistent pattern of disregard for others, lack of empathy, and antisocial beliefs.

Wernicke's aphasia a language disorder characterized by fluent but nonsensical speech and impaired comprehension.

Word salad a jumbled, incoherent stream of words and phrases that lack logical connection or meaning, often indicative of a mental or neurological disorder.

References

Abe, N. (2009). The neurobiology of deception: Evidence from neuroimaging and loss-of-function studies. *Current Opinion in Neurology, 22*(6), 594–600. https://doi.org/10.1097/WCO.0b013e328332c3cf

American Psychiatric Association. (2013). *Diagnostic and statistical manual of mental disorders* (5th ed.). https://doi.org/10.1176/appi.books.9780890425596.

Anderson, N. E., Maurer, J. M., Steele, V. R., & Kiehl, K. A. (2018). Psychopathic traits associated with abnormal hemodynamic activity in salience and default mode networks during auditory oddball task. *Cognitive, Affective, & Behavioral Neuroscience, 18*(3), 564–580. https://doi.org/10.3758/s13415-018-0588-2

Arain, M., Haque, M., Johal, L., Mathur, P., Nel, W., Rais, A., Sandhu, R., & Sharma, S. (2013). Maturation of the adolescent brain. *Neuropsychiatric Disease and Treatment, 9*, 449–461. https://doi.org/10.2147/NDT.S39776

Arnts, H., van Erp, W. S., Lavrijsen, J. C. M., van Gaal, S., Groenewegen, H. J., & van den Munckhof, P. (2020). On the pathophysiology and treatment of akinetic mutism. *Neuroscience and Biobehavioral Reviews, 112*, 270–278. https://doi.org/10.1016/j.neubiorev.2020.02.006

Ashida, R., Nazar, N., Edwards, R., & Teo, M. (2021). Cerebellar mutism syndrome: An overview of the pathophysiology in relation to the cerebrocerebellar anatomy, risk factors, potential treatments, and outcomes. *World Neurosurgery, 153*, 63–74. https://doi.org/10.1016/j.wneu.2021.06.065

Babcock, J. C., Tharp, A. L. T., Sharp, C., Heppner, W., & Stanford, M. S. (2014). Similarities and differences in impulsive/premeditated and reactive/proactive bimodal classifications of aggression. *Aggression and Violent Behavior, 19*(3), 251–262. https://doi.org/10.1016/j.avb.2014.04.002

Bagdy, G., & Kalogeras, K. T. (1993). Stimulation of 5-HT1A and 5-HT2/5-HT1C receptors induce oxytocin release in the male rat. *Brain Research, 611*(2), 330–332. https://doi.org/10.1016/0006-8993(93)90521-n

Bai, F., Huang, L., Deng, J., Long, Z., Hao, X., Chen, P., Wu, G., Wen, H., Deng, Q., Bao, X., Huang, J., Yang, M., Li, D., Ren, Y., Zhang, M., Xiong, Y., & Li, H. (2023). Prelimbic area to lateral hypothalamus circuit drives social aggression. *iScience, 26*(10), 107718. https://doi.org/10.1016/j.isci.2023.107718

Beatson, J. A., & Rao, S. (2013). Depression and borderline personality disorder. *The Medical Journal of Australia, 199*(6 Suppl), S24–S27. https://doi.org/10.5694/mja12.10474

Berg, M., Riehle, M., Rief, W., & Lincoln, T. (2023). Does partial blockade of dopamine D2 receptors with amisulpride cause anhedonia? An experimental study in healthy volunteers. *Journal of Psychiatric Research, 158*, 409–416. https://doi.org/10.1016/j.jpsychires.2023.01.014

Beopoulos, A., Géa, M., Fasano, A., & Iris, F. (2022). Autism spectrum disorders pathogenesis: Toward a comprehensive model based on neuroanatomic and neurodevelopment considerations. *Frontiers in Neuroscience, 16*, 988735. https://doi.org/10.3389/fnins.2022.988735

Bertsch, K., Schmidinger, I., Neumann, I. D., & Herpertz, S. C. (2013). Reduced plasma oxytocin levels in female patients with borderline personality disorder. *Hormones and Behavior, 63*(3), 424–429. https://doi.org/10.1016/j.yhbeh.2012.11.013

Binelli, C., Muñiz, A., Subira, S., Navines, R., Blanco-Hinojo, L., Perez-Garcia, D., Crippa, J., Farré, M., Pérez-Jurado, L., Pujol, J., & Martin-Santos, R. (2016). Facial emotion processing in patients with social anxiety disorder and Williams-Beuren syndrome: An fMRI study. *Journal of Psychiatry & Neuroscience, 41*(3), 182–191. https://doi.org/10.1503/jpn.140384

Boccardi, M., Bocchetta, M., Aronen, H. J., Repo-Tiihonen, E., Vaurio, O., Thompson, P. M., Tiihonen, J., & Frisoni, G. B. (2013). Atypical nucleus accumbens morphology in psychopathy: Another limbic piece in the puzzle. *International Journal of Law and Psychiatry, 36*(2), 157–167. https://doi.org/10.1016/j.ijlp.2013.01.008

Bocchio Chiavetto, L., Tardito, D., Galbiati, C., Ferrari, C., Lanfredi, M., Pedrini, L., Carcione, A., Riccardi, I., Nicolò, G., Rossi, R., & CLIMAMITHE Study Group. (2025). Reduction of oxytocin plasma levels in borderline personality disorder and normalization induced by psychotherapies. *Psychological Medicine, 55*, e92. https://doi.org/10.1017/S003329172500042X

Borie, A. M., Theofanopoulou, C., & Andari, E. (2021). The promiscuity of the oxytocin-vasopressin systems and their involvement in autism spectrum disorder. *Handbook of Clinical Neurology, 182*, 121–140. https://doi.org/10.1016/B978-0-12-819973-2.00009-5

Brinkley-Rubinstein, L., Sivaraman, J., Rosen, D. L., Cloud, D. H., Junker, G., Proescholdbell, S., Shanahan, M. E., & Ranapurwala, S. I. (2019). Association of restrictive housing during incarceration with mortality after release. *JAMA Network Open, 2*(10), e1912516. https://doi.org/10.1001/jamanetworkopen.2019.12516

Buchheim, A., Erk, S., George, C., Kächele, H., Kircher, T., Martius, P., Pokorny, D., Ruchsow, M., Spitzer, M., & Walter, H. (2008). Neural correlates of attachment trauma in borderline personality disorder: A functional magnetic resonance imaging study. *Psychiatry Research, 163*(3), 223–235. https://doi.org/10.1016/j.pscychresns.2007.07.001

Buckholtz, J. W., Treadway, M. T., Cowan, R. L., Woodward, N. D., Benning, S. D., Li, R., Ansari, M. S., Baldwin, R. M., Schwartzman, A. N., Shelby, E. S., Smith, C. E., Cole, D., Kessler, R. M., & Zald, D. H. (2010). Mesolimbic dopamine reward system hypersensitivity in individuals with psychopathic traits. *Nature Neuroscience, 13*(4), 419–421. https://doi.org/10.1038/nn.2510

Buss, D. M., & Shackelford, T. K. (1997). Human aggression in evolutionary psychological perspective. *Clinical Psychology Review, 17*(6), 605–619. https://doi.org/10.1016/s0272-7358(97)00037-8

Carrasco, J. L., Buenache, E., MacDowell, K. S., De la Vega, I., López-Villatoro, J. M., Moreno, B., Díaz-Marsá, M., & Leza, J. C. (2020). Decreased oxytocin plasma levels and oxytocin receptor expression in borderline personality disorder. *Acta Psychiatrica Scandinavica, 142*(4), 319–325. https://doi.org/10.1111/acps.13222

Cataldo, I., Azhari, A., & Esposito, G. (2018). A review of oxytocin and arginine-vasopressin receptors and their modulation of autism spectrum disorder. *Frontiers in Molecular Neuroscience, 11*, 27. https://doi.org/10.3389/fnmol.2018.00027

Cavelti, M., Kircher, T., Nagels, A., Strik, W., & Homan, P. (2018). Is formal thought disorder in schizophrenia related to structural and functional aberrations in the language network? A systematic review of neuroimaging findings. *Schizophrenia Research, 199*, 2–16. https://doi.org/10.1016/j.schres.2018.02.051

Cavieres, A., Acuña, V., Arancibia, M., & Lopetegui, N. (2023). Differences in social perception in people with schizophrenia and bipolar disorder. *Schizophrenia research. Cognition, 33*, 100286. https://doi.org/10.1016/j.scog.2023.100286

Chester, D. S., DeWall, C. N., Derefinko, K. J., Estus, S., Peters, J. R., Lynam, D. R., & Jiang, Y. (2015). Monoamine oxidase a (MAOA) genotype predicts greater aggression through impulsive reactivity to negative affect. *Behavioural Brain Research, 283*, 97–101. https://doi.org/10.1016/j.bbr.2015.01.034

Coccaro, E. F., McCloskey, M. S., Fitzgerald, D. A., & Phan, K. L. (2007). Amygdala and orbitofrontal reactivity to social threat in individuals with impulsive aggression. *Biological Psychiatry, 62*(2), 168–178. https://doi.org/10.1016/j.biopsych.2006.08.024

Constantino, J. N., Morris, J. A., & Murphy, D. L. (1997). CSF 5-HIAA and family history of antisocial personality disorder in newborns. *The American Journal of Psychiatry, 154*(12), 1771–1773. https://doi.org/10.1176/ajp.154.12.1771

da Cunha-Bang, S., & Knudsen, G. M. (2021). The modulatory role of serotonin on human impulsive aggression. *Biological Psychiatry, 90*(7), 447–457. https://doi.org/10.1016/j.biopsych.2021.05.016

Dadds, M. R., Moul, C., Cauchi, A., Dobson-Stone, C., Hawes, D. J., Brennan, J., & Ebstein, R. E. (2014). Methylation of the oxytocin receptor gene and oxytocin blood levels in the development of psychopathy. *Development and Psychopathology, 26*(1), 33–40. https://doi.org/10.1017/S0954579413000497

Debey, E., De Schryver, M., Logan, G. D., Suchotzki, K., & Verschuere, B. (2015). From junior to senior Pinocchio: A cross-sectional lifespan investigation of deception. *Acta Psychologica, 160*, 58–68. https://doi.org/10.1016/j.actpsy.2015.06.007

Defensor, E. B., Pearson, B. L., Pobbe, R. L., Bolivar, V. J., Blanchard, D. C., & Blanchard, R. J. (2011). A novel social proximity test suggests patterns of social avoidance and gaze aversion-like behavior in BTBR T+ tf/J mice. *Behavioural Brain Research, 217*(2), 302–308. https://doi.org/10.1016/j.bbr.2010.10.033

Deming, P., Cook, C. J., Meyerand, M. E., Kiehl, K. A., Kosson, D. S., & Koenigs, M. (2023). Impaired salience network switching in psychopathy. *Behavioural Brain Research, 452*, 114570. https://doi.org/10.1016/j.bbr.2023.114570

DeStefano, F., Price, C. S., & Weintraub, E. S. (2013). Increasing exposure to antibody-stimulating proteins and polysaccharides in vaccines is not associated with risk of autism. *The Journal of Pediatrics, 163*(2), 561–567. https://doi.org/10.1016/j.jpeds.2013.02.001

di Giacomo, E., Andreini, E., Santambrogio, J., Arcara, A., & Clerici, M. (2024). The interplay between borderline personality disorder and oxytocin: A systematic narrative review on possible contribution and treatment options. *Frontiers in Psychiatry, 15*, 1439615. https://doi.org/10.3389/fpsyt.2024.1439615

Dodero, L., Damiano, M., Galbusera, A., Bifone, A., Tsaftsaris, S. A., Scattoni, M. L., & Gozzi, A. (2013). Neuroimaging evidence of major morpho-anatomical and functional abnormalities in the BTBR T+TF/J mouse model of autism. *PLoS One, 8*(10), e76655. https://doi.org/10.1371/journal.pone.0076655

Doell, K. C., Olié, E., Courtet, P., Corradi-Dell'Acqua, C., Perroud, N., & Schwartz, S. (2020). Atypical processing of social anticipation and feedback in borderline personality disorder. *NeuroImage. Clinical, 25*, 102126. https://doi.org/10.1016/j.nicl.2019.102126

Donegan, N. H., Sanislow, C. A., Blumberg, H. P., Fulbright, R. K., Lacadie, C., Skudlarski, P., Gore, J. C., Olson, I. R., McGlashan, T. H., & Wexler, B. E. (2003). Amygdala hyperreactivity in borderline personality disorder: Implications for emotional dysregulation. *Biological Psychiatry, 54*(11), 1284–1293. https://doi.org/10.1016/s0006-3223(03)00636-x

Drzewiecki, C. M., Brinks, A. S., Sellinger, E. P., Doshi, A. D., Koh, J. Y., & Juraska, J. M. (2023). Brief postnatal exposure to bisphenol a affects apoptosis and gene expression in the medial prefrontal cortex and social behavior in rats with sex specificity. *Neurotoxicology, 94*, 126–134. https://doi.org/10.1016/j.neuro.2022.11.011

Eggertson, L. (2010). Lancet retracts 12-year-old article linking autism to MMR vaccines. *Canadian Medical Association Journal, 182*(4), E199–E200. https://doi.org/10.1503/cmaj.109-3179

Eisenegger, C., Naef, M., Linssen, A., Clark, L., Gandamaneni, P. K., Müller, U., & Robbins, T. W. (2014). Role of dopamine D2 receptors in human reinforcement learning. *Neuropsychopharmacology, 39*(10), 2366–2375. https://doi.org/10.1038/npp.2014.84

Etkin, A., & Wager, T. D. (2007). Functional neuroimaging of anxiety: A meta-analysis of emotional processing in PTSD, social anxiety disorder, and specific phobia. *The American Journal of Psychiatry, 164*(10), 1476–1488. https://doi.org/10.1176/appi.ajp.2007.07030504

Evrard, A., Malagié, I., Laporte, A. M., Boni, C., Hanoun, N., Trillat, A. C., Seif, I., De Maeyer, E., Gardier, A., Hamon, M., & Adrien, J. (2002). Altered regulation of the 5-HT system in the brain of MAO-A knock-out mice. *The European Journal of Neuroscience, 15*(5), 841–851. https://doi.org/10.1046/j.1460-9568.2002.01917.x

Falkner, A. L., Grosenick, L., Davidson, T. J., Deisseroth, K., & Lin, D. (2016). Hypothalamic control of male aggression-seeking behavior. *Nature Neuroscience, 19*(4), 596–604. https://doi.org/10.1038/nn.4264

Farina, E., Baglio, F., Pomati, S., D'Amico, A., Campini, I. C., Di Tella, S., Belloni, G., & Pozzo, T. (2017). The mirror neurons network in aging, mild cognitive impairment, and alzheimer disease: A functional MRI study. *Frontiers in Aging Neuroscience, 9*, 371. https://doi.org/10.3389/fnagi.2017.00371

Fedorenko, E., & Blank, I. A. (2020). Broca's area is not a natural kind. *Trends in Cognitive Sciences, 24*(4), 270–284. https://doi.org/10.1016/j.tics.2020.01.001

Fenlon, L. R., Liu, S., Gobius, I., Kurniawan, N. D., Murphy, S., Moldrich, R. X., & Richards, L. J. (2015). Formation of functional areas in the cerebral cortex is disrupted in a mouse model of autism spectrum disorder. *Neural Development, 10*, 10. https://doi.org/10.1186/s13064-015-0033-y

Flanigan, M. E., Aleyasin, H., Li, L., Burnett, C. J., Chan, K. L., LeClair, K. B., Lucas, E. K., Matikainen-Ankney, B., Durand-de Cuttoli, R., Takahashi, A., Menard, C., Pfau, M. L., Golden, S. A., Bouchard, S., Calipari, E. S., Nestler, E. J., DiLeone, R. J., Yamanaka, A., Huntley, G. W., Clem, R. L., & Russo, S. J. (2020). Orexin signaling in GABAergic lateral habenula neurons modulates aggressive behavior in male mice. *Nature Neuroscience, 23*(5), 638–650. https://doi.org/10.1038/s41593-020-0617-7

Flory, J. D., Newcorn, J. H., Miller, C., Harty, S., & Halperin, J. M. (2007). Serotonergic function in children with attention-deficit hyperactivity disorder: Relationship to later antisocial personality disorder. *The British Journal of Psychiatry, 190*, 410–414. https://doi.org/10.1192/bjp.bp.106.027847

Fragkaki, I., Verhagen, M., van Herwaarden, A. E., & Cima, M. (2019). Daily oxytocin patterns in relation to psychopathy and childhood trauma in residential youth. *Psychoneuroendocrinology, 102*, 105–113. https://doi.org/10.1016/j.psyneuen.2018.11.040

Fuentes-Claramonte, P., Soler-Vidal, J., Salgado-Pineda, P., García-León, M. Á., Ramiro, N., Santo-Angles, A., Llanos Torres, M., Tristany, J., Guerrero-Pedraza, A., Munuera, J., Sarró, S., Salvador, R., Hinzen, W., McKenna, P. J., & Pomarol-Clotet, E. (2021). Auditory hallucinations activate language and verbal short-term memory, but not auditory, brain regions. *Scientific Reports, 11*(1), 18890. https://doi.org/10.1038/s41598-021-98269-1

Gabis, L. V., Attia, O. L., Goldman, M., Barak, N., Tefera, P., Shefer, S., Shaham, M., & Lerman-Sagie, T. (2022). The myth of vaccination and autism spectrum. *European Journal of Paediatric Neurology, 36*, 151–158. https://doi.org/10.1016/j.ejpn.2021.12.011

Gallo, E. F., Meszaros, J., Sherman, J. D., Chohan, M. O., Teboul, E., Choi, C. S., Moore, H., Javitch, J. A., & Kellendonk, C. (2018). Accumbens dopamine D2 receptors increase motivation by decreasing inhibitory transmission to the ventral pallidum. *Nature Communications, 9*(1), 1086. https://doi.org/10.1038/s41467-018-03272-2

Galvez-Merlin, A., López-Villatoro, J. M., de la Higuera-González, P., de la Torre-Luque, A., McDowell, K., Díaz-Marsá, M., Leza, J. C., & Carrasco, J. L. (2024). Decreased oxytocin levels related to social cognition impairment in borderline personality disorder. *Acta Psychiatrica Scandinavica, 149*(6), 458–466. https://doi.org/10.1111/acps.13679

Godar, S. C., Bortolato, M., Castelli, M. P., Casti, A., Casu, A., Chen, K., Ennas, M. G., Tambaro, S., & Shih, J. C. (2014). The aggression and behavioral abnormalities associated with monoamine oxidase a deficiency are rescued by acute inhibition of serotonin reuptake. *Journal of Psychiatric Research, 56*, 1–9. https://doi.org/10.1016/j.jpsychires.2014.04.014

Godar, S. C., Fite, P. J., McFarlin, K. M., & Bortolato, M. (2016). The role of monoamine oxidase a in aggression: Current translational developments and future challenges. *Progress in Neuro-Psychopharmacology & Biological Psychiatry, 69*, 90–100. https://doi.org/10.1016/j.pnpbp.2016.01.001

Golden, S. A., Jin, M., & Shaham, Y. (2019). Animal models of (or for) aggression reward, addiction, and relapse: Behavior and circuits. *The Journal of Neuroscience, 39*(21), 3996–4008. https://doi.org/10.1523/JNEUROSCI.0151-19.2019

Green, S. A., Hernandez, L., Tottenham, N., Krasileva, K., Bookheimer, S. Y., & Dapretto, M. (2015). Neurobiology of sensory overresponsivity in youth with autism spectrum disorders. *JAMA Psychiatry, 72*(8), 778–786. https://doi.org/10.1001/jamapsychiatry.2015.0737

Gu, X., Liu, X., Guise, K. G., Naidich, T. P., Hof, P. R., & Fan, J. (2010). Functional dissociation of the frontoinsular and anterior cingulate cortices in empathy for pain. *The Journal of Neuroscience, 30*(10), 3739–3744. https://doi.org/10.1523/JNEUROSCI.4844-09.2010

Guo, X., Zheng, L., Wang, H., Zhu, L., Li, J., Wang, Q., Dienes, Z., & Yang, Z. (2013). Exposure to violence reduces empathetic responses to other's pain. *Brain and Cognition, 82*(2), 187–191. https://doi.org/10.1016/j.bandc.2013.04.005

Hashikawa, Y., Hashikawa, K., Falkner, A. L., & Lin, D. (2017). Ventromedial hypothalamus and the generation of aggression. *Frontiers in Systems Neuroscience, 11*, 94. https://doi.org/10.3389/fnsys.2017.00094

Henderson, V. W. (2019). Alexia and agraphia from 1861 to 1965. *Frontiers of Neurology and Neuroscience, 44*, 39–52. https://doi.org/10.1159/000494951

Hickson, R., Hebron, L., Muller-Oehring, E. M., Cheu, A., Hernandez, A., Kiss, O., Gombert-Labedens, M., Baker, F. C., & Schulte, T. (2024). Resting-state fMRI activation is associated with parent-reported phenotypic features of autism in early adolescence. *Frontiers in Child and Adolescent Psychiatry, 3*, 1481957. https://doi.org/10.3389/frcha.2024.1481957

Hoge, E. A., Pollack, M. H., Kaufman, R. E., Zak, P. J., & Simon, N. M. (2008). Oxytocin levels in social anxiety disorder. *CNS Neuroscience & Therapeutics, 14*(3), 165–170. https://doi.org/10.1111/j.1755-5949.2008.00051.x

Hosking, J. G., Kastman, E. K., Dorfman, H. M., Samanez-Larkin, G. R., Baskin-Sommers, A., Kiehl, K. A., Newman, J. P., & Buckholtz, J. W. (2017). Disrupted prefrontal regulation of striatal subjective value signals in psychopathy. *Neuron, 95*(1), 221–231.e4. https://doi.org/10.1016/j.neuron.2017.06.030

Holanda, M. V. F., Paiva, E. D. S., de Souza, L. N., Paiva, K. M., Oliveira, R. F., Tavares, É. A. F., Morais, P. L. A. G., de Andrade, A. M., Knackfuss, M. I., & do Nascimento, E. G. C., & Cavalcanti, J. R. L. P. (2025). Neurobiological basis of autism spectrum disorder: Mini review. *Frontiers in Psychology, 16*, 1558081. https://doi.org/10.3389/fpsyg.2025.1558081

Hviid, A., Stellfeld, M., Wohlfahrt, J., & Melbye, M. (2003). Association between thimerosal-containing vaccine and autism. *JAMA, 290*(13), 1763–1766. https://doi.org/10.1001/jama.290.13.1763

Ike, K. G. O., de Boer, S. F., Buwalda, B., & Kas, M. J. H. (2020). Social withdrawal: An initially adaptive behavior that becomes maladaptive when expressed excessively. *Neuroscience and Biobehavioral Reviews, 116*, 251–267. https://doi.org/10.1016/j.neubiorev.2020.06.030

Jankowiak-Siuda, K., Rymarczyk, K., & Grabowska, A. (2011). How we empathize with others: A neurobiological perspective. *Medical Science Monitor, 17*(1), RA18–RA24. https://doi.org/10.12659/msm.881324

Jobst, A., Dehning, S., Ruf, S., Notz, T., Buchheim, A., Henning-Fast, K., Meißner, D., Meyer, S., Bondy, B., Müller, N., & Zill, P. (2014). Oxytocin and vasopressin levels are decreased in the plasma of male schizophrenia patients. *Acta Neuropsychiatrica, 26*(6), 347–355. https://doi.org/10.1017/neu.2014.20

Jobst, A., Padberg, F., Mauer, M. C., Daltrozzo, T., Bauriedl-Schmidt, C., Sabass, L., Sarubin, N., Falkai, P., Renneberg, B., Zill, P., Gander, M., & Buchheim, A. (2016). Lower oxytocin plasma levels in borderline patients with unresolved attachment representations. *Frontiers in Human Neuroscience, 10*, 125. https://doi.org/10.3389/fnhum.2016.00125

Johanson, M., Vaurio, O., Tiihonen, J., & Lähteenvuo, M. (2020). A systematic literature review of neuroimaging of psychopathic traits. *Frontiers in Psychiatry, 10*, 1027. https://doi.org/10.3389/fpsyt.2019.01027

Jørgensen, H., Riis, M., Knigge, U., Kjaer, A., & Warberg, J. (2003). Serotonin receptors involved in vasopressin and oxytocin secretion. *Journal of Neuroendocrinology, 15*(3), 242–249. https://doi.org/10.1046/j.1365-2826.2003.00978.x

Kanlayaprasit, S., Saeliw, T., Thongkorn, S., Panjabud, P., Kasitipradit, K., Lertpeerapan, P., Songsritaya, K., Yuwattana, W., Jantheang, T., Jindatip, D., Hu, V. W., Kikkawa, T., Osumi, N., & Sarachana, T. (2024). Sex-specific impacts of prenatal bisphenol a exposure on genes associated with cortical development, social behaviors, and autism in the offspring's prefrontal cortex. *Biology of Sex Differences, 15*(1), 40. https://doi.org/10.1186/s13293-024-00614-2

Kasitipradit, K., Thongkorn, S., Kanlayaprasit, S., Saeliw, T., Lertpeerapan, P., Panjabud, P., Jindatip, D., Hu, V. W., Kikkawa, T., Osumi, N., & Sarachana, T. (2025). Sex-specific effects of prenatal bisphenol a exposure on transcriptome-interactome profiles of autism candidate genes in neural stem cells from offspring hippocampus. *Scientific Reports, 15*(1), 2882. https://doi.org/10.1038/s41598-025-86392-2

Kluetsch, R. C., Schmahl, C., Niedtfeld, I., Densmore, M., Calhoun, V. D., Daniels, J., Kraus, A., Ludaescher, P., Bohus, M., & Lanius, R. A. (2012). Alterations in default mode network connectivity during pain processing in borderline personality disorder. *Archives of General Psychiatry, 69*(10), 993–1002. https://doi.org/10.1001/archgenpsychiatry.2012.476

Kolla, N. J., & Bortolato, M. (2020). The role of monoamine oxidase a in the neurobiology of aggressive, antisocial, and violent behavior: A tale of mice and men. *Progress in Neurobiology, 194*, 101875. https://doi.org/10.1016/j.pneurobio.2020.101875

Kuniishi, H., Ichisaka, S., Matsuda, S., Futora, E., Harada, R., & Hata, Y. (2017). Chronic inactivation of the orbitofrontal cortex increases anxiety-like behavior and impulsive aggression, but decreases depression-like behavior in rats. *Frontiers in Behavioral Neuroscience, 10*, 250. https://doi.org/10.3389/fnbeh.2016.00250

Kurth, F., Narr, K. L., Woods, R. P., O'Neill, J., Alger, J. R., Caplan, R., McCracken, J. T., Toga, A. W., & Levitt, J. G. (2011). Diminished gray matter within the hypothalamus in autism disorder: A potential link to hormonal effects? *Biological Psychiatry, 70*(3), 278–282. https://doi.org/10.1016/j.biopsych.2011.03.026

Levy, T., Bloch, Y., Bar-Maisels, M., Gat-Yablonski, G., Djalovski, A., Borodkin, K., & Apter, A. (2015). Salivary oxytocin in adolescents with conduct problems and callous-unemotional traits. *European Child & Adolescent Psychiatry, 24*(12), 1543–1551. https://doi.org/10.1007/s00787-015-0765-6

Li, L., Li, Y., Huang, C., Hou, W., Lv, Z., Zhang, L., Qu, Y., Sun, Y., Huang, K., Han, X., He, Z., & Tai, F. (2024). PVN-mPFC OT projections modulate pup-directed pup care or attacking in virgin mandarin voles. *eLife, 13*, RP96543. https://doi.org/10.7554/eLife.96543

Li, W., Mai, X., & Liu, C. (2014). The default mode network and social understanding of others: What do brain connectivity studies tell us. *Frontiers in Human Neuroscience, 8*, 74. https://doi.org/10.3389/fnhum.2014.00074

Li, Z., Zhu, Y. X., Gu, L. J., & Cheng, Y. (2021). Understanding autism spectrum disorders with animal models: Applications, insights, and perspectives. *Zoological Research, 42*(6), 800–824. https://doi.org/10.24272/j.issn.2095-8137.2021.251

Liu, J., Chen, H., Wang, H., & Wang, Z. (2025). Neural correlates of facial recognition deficits in autism spectrum disorder: A comprehensive review. *Frontiers in Psychiatry, 15*, 1464142. https://doi.org/10.3389/fpsyt.2024.1464142

Lukkarinen, L., Tuisku, J., Sun, L., Helin, S., Karlsson, H. K., Venetjoki, N., Salomaa, M., Rautio, P., Hirvonen, J., Lauerma, H., Tiihonen, J., & Nummenmaa, L. (2024). Aberrant type 2 dopamine receptor availability in violent offenders with psychopathy. *NeuroImage, 297*, 120724. https://doi.org/10.1016/j.neuroimage.2024.120724

Lv, T., Wang, M., Kui, L., Wu, J., & Xiao, Y. (2025). Novel inflammatory biomarkers for autism spectrum disorder detected by plasma olink proteomics. *Children, 12*(2), 210. https://doi.org/10.3390/children12020210

Lyu, C., Siu, K., Xu, I., Osman, I., & Zhong, J. (2024). Social isolation changes and long-term outcomes among older adults. *JAMA Network Open, 7*(7), e2424519. https://doi.org/10.1001/jamanetworkopen.2024.24519

Malejko, K., Neff, D., Brown, R. C., Plener, P. L., Bonenberger, M., Abler, B., & Graf, H. (2019). Neural signatures of social inclusion in borderline personality disorder versus non-suicidal self-injury. *Brain Topography, 32*(5), 753–761. https://doi.org/10.1007/s10548-019-00712-0

McDermott, R., Tingley, D., Cowden, J., Frazzetto, G., & Johnson, D. D. (2009). Monoamine oxidase a gene (MAOA) predicts behavioral aggression following provocation. *Proceedings of the National Academy of Sciences of the United States of America, 106*(7), 2118–2123. https://doi.org/10.1073/pnas.0808376106

Mohammed, N., Narayan, V., Patra, D. P., & Nanda, A. (2018). Louis victor Leborgne (tan). *World Neurosurgery, 114*, 121–125. https://doi.org/10.1016/j.wneu.2018.02.021

Mohlman, J., Carmin, C. N., & Price, R. B. (2007). Jumping to interpretations: Social anxiety disorder and the identification of emotional facial expressions. *Behaviour Research and Therapy, 45*(3), 591–599. https://doi.org/10.1016/j.brat.2006.03.007

Moul, C., Dobson-Stone, C., Brennan, J., Hawes, D., & Dadds, M. (2013). An exploration of the serotonin system in antisocial boys with high levels of callous-unemotional traits. *PLoS One, 8*(2), e56619. https://doi.org/10.1371/journal.pone.0056619

National Institute on Deafness and Other Communication Disorders. (2015). Aphasia. *NIH Publication No*, 97–4257. https://www.nidcd.nih.gov/sites/default/files/Documents/health/voice/Aphasia.pdf

New, A. S., Hazlett, E. A., Buchsbaum, M. S., Goodman, M., Mitelman, S. A., Newmark, R., Trisdorfer, R., Haznedar, M. M., Koenigsberg, H. W., Flory, J., & Siever, L. J. (2007). Amygdala-prefrontal disconnection in borderline personality disorder. *Neuropsychopharmacology, 32*(7), 1629–1640. https://doi.org/10.1038/sj.npp.1301283

Nordman, J., Ma, X., & Li, Z. (2020). Traumatic stress induces prolonged aggression increase through synaptic potentiation in the medial amygdala circuits. *eNeuro, 7*(4), ENEURO.0147-20.2020. https://doi.org/10.1523/ENEURO.0147-20.2020

Nunes, P. M., Wenzel, A., Borges, K. T., Porto, C. R., Caminha, R. M., & de Oliveira, I. R. (2009). Volumes of the hippocampus and amygdala in patients with borderline personality disorder: A meta-analysis. *Journal of Personality Disorders, 23*(4), 333–345. https://doi.org/10.1521/pedi.2009.23.4.333

O'Neill, A., & Frodl, T. (2012). Brain structure and function in borderline personality disorder. *Brain Structure & Function, 217*(4), 767–782. https://doi.org/10.1007/s00429-012-0379-4

Pardini, M., Krueger, F., Hodgkinson, C., Raymont, V., Ferrier, C., Goldman, D., Strenziok, M., Guida, S., & Grafman, J. (2011). Prefrontal cortex lesions and MAO-A modulate aggression in penetrating traumatic brain injury. *Neurology, 76*(12), 1038–1045. https://doi.org/10.1212/WNL.0b013e318211c33e

Park, I. H., Lee, B. C., Kim, J. J., Kim, J. I., & Koo, M. S. (2017). Effort-based reinforcement processing and functional connectivity underlying amotivation in medicated patients with depression and schizophrenia. *The Journal of Neuroscience, 37*(16), 4370–4380. https://doi.org/10.1523/JNEUROSCI.2524-16.2017

Patil, M., Iftikhar, N., & Ganti, L. (2024). Neuroimaging insights into autism spectrum disorder: Structural and functional brain. *Health Psychology Research, 12*, 123439. https://doi.org/10.52965/001c.123439

Pavlova, I. V., Broshevitskaya, N. D., Potekhina, A. A., & Shvadchenko, A. M. (2024). The effect of chronic overcrowding on social behavior and expression of neuroinflammation-associated genes in rats. *Biochemistry, 89*(9), 1582–1594. https://doi.org/10.1134/S0006297924090050

Peschard, V., & Philippot, P. (2017). Overestimation of threat from neutral faces and voices in social anxiety. *Journal of Behavior Therapy and Experimental Psychiatry, 57*, 206–211. https://doi.org/10.1016/j.jbtep.2017.06.003

Piva, M., & Chang, S. W. C. (2018). An integrated framework for the role of oxytocin in multistage social decision-making. *American Journal of Primatology, 80*(10), e22735. https://doi.org/10.1002/ajp.22735

Porcelli, S., Van Der Wee, N., van der Werff, S., Aghajani, M., Glennon, J. C., van Heukelum, S., Mogavero, F., Lobo, A., Olivera, F. J., Lobo, E., Posadas, M., Dukart, J., Kozak, R., Arce, E., Ikram, A., Vorstman, J., Bilderbeck, A., Saris, I., Kas, M. J., & Serretti, A. (2019). Social brain, social dysfunction and social withdrawal. *Neuroscience and Biobehavioral Reviews, 97*, 10–33. https://doi.org/10.1016/j.neubiorev.2018.09.012

Price, C. J. (2010). The anatomy of language: A review of 100 fMRI studies published in 2009. *Annals of the New York Academy of Sciences, 1191*, 62–88. https://doi.org/10.1111/j.1749-6632.2010.05444.x

Redcay, E. (2008). The superior temporal sulcus performs a common function for social and speech perception: Implications for the emergence of autism. *Neuroscience and Biobehavioral Reviews, 32*(1), 123–142. https://doi.org/10.1016/j.neubiorev.2007.06.004

Reiter, K., Ventura, J., Lovell, D., Augustine, D., Barragan, M., Blair, T., Chesnut, K., Dashtgard, P., Gonzalez, G., Pifer, N., & Strong, J. (2020). Psychological distress in solitary confinement: Symptoms, severity, and prevalence in the United States, 2017-2018. *American Journal of Public Health, 110*(S1), S56–S62. https://doi.org/10.2105/AJPH.2019.305375

Roberts, A. G., Peckins, M. K., Gard, A. M., Hein, T. C., Hardi, F. A., Mitchell, C., Monk, C. S., Hyde, L. W., & Lopez-Duran, N. L. (2022). Amygdala reactivity during socioemotional processing and cortisol reactivity to a psychosocial stressor. *Psychoneuroendocrinology, 144*, 105855. https://doi.org/10.1016/j.psyneuen.2022.105855

Rosenfeld, A. J., Lieberman, J. A., & Jarskog, L. F. (2011). Oxytocin, dopamine, and the amygdala: A neurofunctional model of social cognitive deficits in schizophrenia. *Schizophrenia Bulletin, 37*(5), 1077–1087. https://doi.org/10.1093/schbul/sbq015

Rubin, B. S. (2011). Bisphenol a: An endocrine disruptor with widespread exposure and multiple effects. *The Journal of Steroid Biochemistry and Molecular Biology, 127*(1–2), 27–34. https://doi.org/10.1016/j.jsbmb.2011.05.002

Rudebeck, P. H., Walton, M. E., Millette, B. H., Shirley, E., Rushworth, M. F., & Bannerman, D. M. (2007). Distinct contributions of frontal areas to emotion and social behaviour in the rat. *The European Journal of Neuroscience, 26*(8), 2315–2326. https://doi.org/10.1111/j.1460-9568.2007.05844.x

Ruocco, A. C., Amirthavasagam, S., & Zakzanis, K. K. (2012). Amygdala and hippocampal volume reductions as candidate endophenotypes for borderline personality disorder: A meta-analysis of magnetic resonance imaging studies. *Psychiatry Research, 201*(3), 245–252. https://doi.org/10.1016/j.pscychresns.2012.02.012

Sampedro, F., Farrés, C. C. I., Soler, J., Elices, M., Schmidt, C., Corripio, I., Domínguez-Clavé, E., Pomarol-Clotet, E., Salvador, R., & Pascual, J. C. (2021). Structural brain abnormalities in borderline personality disorder correlate with clinical severity and predict psychotherapy response. *Brain Imaging and Behavior, 15*(5), 2502–2512. https://doi.org/10.1007/s11682-021-00451-6

Scattoni, M. L., Ricceri, L., & Crawley, J. N. (2011). Unusual repertoire of vocalizations in adult BTBR T+tf/J mice during three types of social encounters. *Genes, Brain, and Behavior, 10*(1), 44–56. https://doi.org/10.1111/j.1601-183X.2010.00623.x

Schilbach, L., Derntl, B., Aleman, A., Caspers, S., Clos, M., Diederen, K. M., Gruber, O., Kogler, L., Liemburg, E. J., Sommer, I. E., Müller, V. I., Cieslik, E. C., & Eickhoff, S. B. (2016). Differential patterns of dysconnectivity in mirror neuron and mentalizing networks in schizophrenia. *Schizophrenia Bulletin, 42*(5), 1135–1148. https://doi.org/10.1093/schbul/sbw015

Scott, L. N., Wright, A. G. C., Beeney, J. E., Lazarus, S. A., Pilkonis, P. A., & Stepp, S. D. (2017). Borderline personality disorder symptoms and aggression: A within-person process model. *Journal of Abnormal Psychology, 126*(4), 429–440. https://doi.org/10.1037/abn0000272

Seghier, M. L. (2024). Symptomatology after damage to the angular gyrus through the lenses of modern lesion-symptom mapping. *Cortex, 179*, 77–90. https://doi.org/10.1016/j.cortex.2024.07.005

Serota, K. B., Levine, T. R., & Docan-Morgan, T. (2022). Unpacking variation in lie prevalence: Prolific liars, bad lie days, or both? *Communication Monographs, 89*(3), 307–331. https://doi.org/10.1080/03637751.2021.1985153

Sethi, A., Gregory, S., Dell'Acqua, F., Periche Thomas, E., Simmons, A., Murphy, D. G., Hodgins, S., Blackwood, N. J., & Craig, M. C. (2015). Emotional detachment in psychopathy: Involvement of dorsal default-mode connections. *Cortex, 62*, 11–19. https://doi.org/10.1016/j.cortex.2014.07.018

Shafie, M., Shahmohamadi, E., Cattarinussi, G., Sanjari Moghaddam, H., Akhondzadeh, S., Sambataro, F., Moltrasio, C., & Delvecchio, G. (2023). Resting-state functional magnetic resonance imaging alterations in borderline personality disorder: A systematic review. *Journal of Affective Disorders, 341*, 335–345. https://doi.org/10.1016/j.jad.2023.09.001

Shaikh, A., Ahmad, F., Teoh, S. L., Kumar, J., & Yahaya, M. F. (2023). Targeting dopamine transporter to ameliorate cognitive deficits in Alzheimer's disease. *Frontiers in Cellular Neuroscience, 17*, 1292858. https://doi.org/10.3389/fncel.2023.1292858

Shen, M. D., Swanson, M. R., Wolff, J. J., Elison, J. T., Girault, J. B., Kim, S. H., Smith, R. G., Graves, M. M., Weisenfeld, L. A. H., Flake, L., MacIntyre, L., Gross, J. L., Burrows, C. A., Fonov, V. S., Collins, D. L., Evans, A. C., Gerig, G., McKinstry, R. C., Pandey, J., St John, T., et al. (2022). Subcortical brain development in autism and fragile X syndrome: Evidence for dynamic, age-and disorder-specific trajectories in infancy. *The American Journal of Psychiatry, 179*(8), 562–572. https://doi.org/10.1176/appi.ajp.21090896

Sicorello, M., & Schmahl, C. (2021). Emotion dysregulation in borderline personality disorder: A fronto-limbic imbalance? *Current Opinion in Psychology, 37*, 114–120. https://doi.org/10.1016/j.copsyc.2020.12.002

Siever, L. J. (2008). Neurobiology of aggression and violence. *The American Journal of Psychiatry, 165*(4), 429–442. https://doi.org/10.1176/appi.ajp.2008.07111774

Snowden, R. J., Smith, C., & Gray, N. S. (2017). Risk taking and the triarchic model of psychopathy. *Journal of Clinical and Experimental Neuropsychology, 39*(10), 988–1001. https://doi.org/10.1080/13803395.2017.1300236

Sobota, R., Mihara, T., Forrest, A., Featherstone, R. E., & Siegel, S. J. (2015). Oxytocin reduces amygdala activity, increases social interactions, and reduces anxiety-like behavior irrespective of NMDAR antagonism. *Behavioral Neuroscience, 129*(4), 389–398. https://doi.org/10.1037/bne0000074

Sripada, C. S., Angstadt, M., Banks, S., Nathan, P. J., Liberzon, I., & Phan, K. L. (2009). Functional neuroimaging of mentalizing during the trust game in social anxiety disorder. *Neuroreport, 20*(11), 984–989. https://doi.org/10.1097/WNR.0b013e32832d0a67

Syal, S., Hattingh, C. J., Fouché, J. P., Spottiswoode, B., Carey, P. D., Lochner, C., & Stein, D. J. (2012). Grey matter abnormalities in social anxiety disorder: A pilot study. *Metabolic Brain Disease, 27*(3), 299–309. https://doi.org/10.1007/s11011-012-9299-5

Symeonides, C., Vacy, K., Thomson, S., Tanner, S., Chua, H. K., Dixit, S., Mansell, T., O'Hely, M., Novakovic, B., Herbstman, J. B., Wang, S., Guo, J., Chia, J., Tran, N. T., Hwang, S. E., Britt, K., Chen, F., Kim, T. H., Reid, C. A., El-Bitar, A., et al. (2024). Male autism spectrum disorder is linked to brain aromatase disruption by prenatal BPA in multimodal investigations and 10HDA ameliorates the related mouse phenotype. *Nature Communications, 15*(1), 6367. https://doi.org/10.1038/s41467-024-48897-8

Takahashi, A. (2025). The role of social isolation stress in escalated aggression in rodent models. *Neuroscience Research, 211*, 75–84. https://doi.org/10.1016/j.neures.2022.07.009

Takahashi, A., Nagayasu, K., Nishitani, N., Kaneko, S., & Koide, T. (2014). Control of intermale aggression by medial prefrontal cortex activation in the mouse. *PLoS One, 9*(4), e94657. https://doi.org/10.1371/journal.pone.0094657

Taylor, L. E., Swerdfeger, A. L., & Eslick, G. D. (2014). Vaccines are not associated with autism: An evidence-based meta-analysis of case-control and cohort studies. *Vaccine, 32*(29), 3623–3629. https://doi.org/10.1016/j.vaccine.2014.04.085

Thongkorn, S., Kanlayaprasit, S., Kasitipradit, K., Lertpeerapan, P., Panjabud, P., Hu, V. W., Jindatip, D., & Sarachana, T. (2023). Investigation of autism-related transcription factors underlying sex differences in the effects of bisphenol a on transcriptome profiles and synaptogenesis in the offspring hippocampus. *Biology of Sex Differences, 14*(1), 8. https://doi.org/10.1186/s13293-023-00496-w

Tottenham, N., Hertzig, M. E., Gillespie-Lynch, K., Gilhooly, T., Millner, A. J., & Casey, B. J. (2014). Elevated amygdala response to faces and gaze aversion in autism spectrum disorder. *Social Cognitive and Affective Neuroscience, 9*(1), 106–117. https://doi.org/10.1093/scan/nst050

Umbach, R. H., & Tottenham, N. (2021). Callous-unemotional traits and reduced default mode network connectivity within a community sample of children. *Development and Psychopathology, 33*(4), 1170–1183. https://doi.org/10.1017/S0954579420000401

van Dongen, J. D. M., de Groot, M., Rassin, E., Hoyle, R. H., & Franken, I. H. A. (2022). Sensation seeking and its relationship with psychopathic traits, impulsivity and aggression: A validation of the Dutch brief sensation seeking scale (BSSS). *Psychiatry, Psychology and Law, 29*(1), 20–32. https://doi.org/10.1080/13218719.2020.1821825

Vargas, D. L., Nascimbene, C., Krishnan, C., Zimmerman, A. W., & Pardo, C. A. (2005). Neuroglial activation and neuroinflammation in the brain of patients with autism. *Annals of Neurology, 57*(1), 67–81. https://doi.org/10.1002/ana.20315

Vasan, S., Eikelis, N., Lim, M. H., & Lambert, E. (2023). Evaluating the impact of loneliness and social isolation on health literacy and health-related factors in young adults. *Frontiers in Psychology, 14*, 996611. https://doi.org/10.3389/fpsyg.2023.996611

Wagner, J., & Rusconi, E. (2023). Causal involvement of the left angular gyrus in higher functions as revealed by transcranial magnetic stimulation: A systematic review. *Brain Structure & Function, 228*(1), 169–196. https://doi.org/10.1007/s00429-022-02576-w

Winters, D. E., Sakai, J. T., & Carter, R. M. (2021). Resting-state network topology characterizing callous-unemotional traits in adolescence. *NeuroImage. Clinical, 32*, 102878. https://doi.org/10.1016/j.nicl.2021.102878

World Health Organization. (2022). COVID-19 pandemic triggers 25% increase in prevalence of anxiety and depression worldwide. https://www.who.int/news/item/02-03-2022-covid-19-pandemic-triggers-25-increase-in-prevalence-of-anxiety-and-depression-worldwide.

Yang, M., Abrams, D. N., Zhang, J. Y., Weber, M. D., Katz, A. M., Clarke, A. M., Silverman, J. L., & Crawley, J. N. (2012). Low sociability in BTBR T+tf/J mice is independent of partner strain. *Physiology & Behavior, 107*(5), 649–662. https://doi.org/10.1016/j.physbeh.2011.12.025

Yildirim, B. O., & Derksen, J. J. (2015). Mesocorticolimbic dopamine functioning in primary psychopathy: A source of within-group heterogeneity. *Psychiatry Research, 229*(3), 633–677. https://doi.org/10.1016/j.psychres.2015.07.005

Yu, H., Xiang, X., Chen, Z., Wang, X., Dai, J., Wang, X., Huang, P., Zhao, Z. D., Shen, W. L., & Li, H. (2021). Periaqueductal gray neurons encode the sequential motor program in hunting behavior of mice. *Nature Communications, 12*(1), 6523. https://doi.org/10.1038/s41467-021-26852-1

Zhang, J., Yuan, M., Liu, Y., Zhong, X., Wu, J., & Chen, W. (2025). Bisphenol a exposure and neurodevelopmental disorders and problems in children under 12 years of age: A systematic review and meta-analysis. *Journal of Hazardous Materials, 490*, 137731. https://doi.org/10.1016/j.jhazmat.2025.137731

Chapter 16
Maladaptive Reproductive Behaviors

1 Introduction

In fifth grade, we learned about the "birds and the bees". I vividly remember our teacher telling us that we better not laugh while watching a sexual education video. During the video, we saw an animation of a penis becoming erect and then ejaculating. I was able to obey my teacher's directive of not laughing until the kid next to me began laughing. Fortunately, most of the classroom began to chuckle or outright laugh during this part of the video. By the time this book is published, my son will have gone through fifth grade and most likely will have learned about sex. I am not sure if I am completely ready to have conversations with him about sex and puberty, but that is my future self's problem.

As embarrassing as discussing sex can be, it plays a vital role in the survival of most species. **Sexual reproduction** allows us to pass on our genes to offspring. Some individuals are unable to pass on their genes through sex due to various reasons. There are some genetic disorders that affect the development of sexual reproductive organs. There are other medical causes that can lead to infertility. Even if individuals do not want children, sexual intercourse is an important component of relationships. However, individuals that are interested in sex can experience physical/psychological difficulties that make intercourse difficult or nearly impossible. There are other conditions that affect one's **libido**. Finally, some individuals engage in sexual behaviors that cause significant distress to themselves and/or physical/psychological harm to others. This last point is critical for determining what maladaptive reproductive behaviors are. If two individuals with low libido are in a fulfilling relationship, this is not maladaptive. For the sake of this chapter, maladaptive reproductive behaviors that will be discussed are as follows:

1. Genetic disorders that affect the development of reproductive organs. As you will learn, these disorders can cause other physical and/or behavioral issues that can negatively impact an individual.

J. Yates, *Neuroscience of Maladaptation*,
https://doi.org/10.1007/978-3-032-15602-0_16

2. Conditions that make sexual intercourse difficult for those that wish to engage in sexual activities.
3. Sexual desires and/or activities that cause significant distress to an individual and/or cause significant physical and/or psychological harm to another individual.

The third part of this chapter carries a ***trigger warning*** as highly sensitive topics will be covered such as sexual aggression.

2 Learning Objectives

By the end of this chapter, you should be able to …

1. Explain the mechanisms governing sexual development, and differentiate the conditions that affect sexual development.
2. Discuss the neuromechanisms that control sexual arousal, and describe the neural correlates of hypoactive sexual desire disorder.
3. Identify the causes of conditions related to sexual intercourse.
4. Compare and contrast the neuromechanisms associated with compulsive sexual behavior, paraphilic disorders, and sexual aggression.

3 Sexual Development

Before discussing maladaptive sexual development, an important distinction needs to be made between the terms *sex* and *gender*. Sex is a biological trait that is determined by our sex chromosomes (X and Y). Females inherit two X chromosomes whereas males inherit one X chromosome and one Y chromosome. Gender is a psychosocial construct that relates to the behaviors, roles, and expectations associated with being male or female. The terms man and woman are used when referring to gender. Distinguishing sex from gender is important because this section will focus on maladaptive sexual development as related to biological sex.

When sperm from a male fertilizes an egg (*ovum*) from a female, a *zygote* is formed (see Fig. 1a). The zygote undergoes several stages of cell division until it becomes a *blastocyst*. The blastocyst attaches itself to the uterine wall, where it becomes an *embryo*. The embryonic period begins during the fifth week of pregnancy and is characterized by the development of major body structures. By the tenth week of pregnancy, the embryo is now a *fetus*. Organs and body structures continue to develop and mature until birth. During the first 6 weeks of prenatal development, the sex organs of an embryo are undifferentiated. That is, embryos have the same internal structures that can develop into either male sex organs or female sex organs. This is also why males have nipples even though they do not need them for nursing an infant. The process of differentiation will be briefly described below (see Fig. 1b).

If an embryo has an XY chromosome pair, the *SRY* gene is transcribed, allowing the SRY protein to bind to DNA (deoxyribonucleic acid) to allow for development

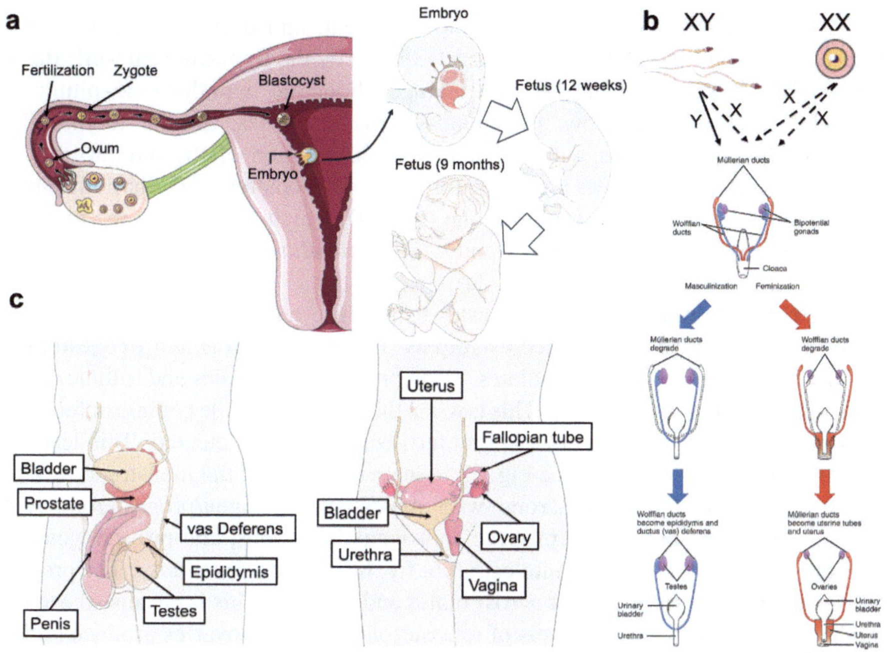

Fig. 1 Stages of embryonic and fetal development (**a**) and sex differentiation (**b**). (**c**) The reproductive system of males (left) and females (right). Image created by the author using Servier Medical Art (https://smart.servier.com/; CC BY 4.0, https://creativecommons.org/licenses/by/4.0/) and Microsoft PowerPoint. The image of sex differentiation depicted in panel (**b**) comes from Figure 28.16 of Betts et al. (2013), https://openstax.org/books/anatomy-and-physiology-2e/pages/28-3-fetal-development. CC BY 4.0 (https://creativecommons.org/licenses/by/4.0/), via Rice University and OpenStax

of male reproductive organs. The undifferentiated **gonads** form into the testes. The testes produce high levels of **androgens**, particularly testosterone and dihydrotestosterone. Testosterone causes the Wolffian ducts (also known as the mesonephric ducts) to form into the epididymis, the Vas deferens, and the seminal vesicles. The testes also produce anti-Müllerian hormone. This prevents the Müllerian ducts (also known as the paramesonephric ducts) from forming into female reproductive organs. By 8–12 weeks, the penis and the scrotum begin to form.

Without a Y chromosome, the undifferentiated gonads form into ovaries. Due to the low testosterone levels, the Wolffian ducts degrade, and the Müllerian ducts begin to form into the fallopian tubes, the uterus, and part of the vagina. External structures such as the vulva, including the clitoris, labia minora, and labia majora also develop in the absence of testosterone. Figure 1c shows the primary reproductive organs of males and females.

Puberty is the period of development in which we become sexually mature. The gonads become fully developed during this time, with the testes and the ovaries able to produce sperm and eggs, respectively. Secondary sexual characteristics also form. These include body hair, breast development, and deepening of the voice. The sexual maturation that occurs is largely driven by the **hypothalamic-pituitary-gonad (HPG) axis**. In Chap. 8, you learned about the hypothalamic–pituitary–adrenal (HPA) axis and how it regulates stress responses. Similar to the HPA axis, the HPG axis begins with hypothalamic stimulation of the pituitary gland, causing it to secrete hormones that travel through the bloodstream. The hypothalamus releases gonadotropin-releasing hormone (GnRH), which then causes the pituitary gland to release *luteinizing hormone* and *follicle-stimulating hormone (FSH).* These hormones stimulate the gonads (Fig. 2a).

Luteinizing hormone causes the testes to produce androgens, primarily testosterone and the ovaries to produce **estrogens** (mainly estradiol) and **progestogens** (mainly progesterone). FSH stimulates sperm production in males and follicle development and ovulation in females. This is when the menstrual cycle begins for females. Menstruation ("period") occurs when unfertilized eggs are detected. This leads to shedding of the uterine wall (see Fig. 2b to learn more about the menstrual cycle).

We often associate testosterone with males and estrogen/progesterone with females. However, both sexes produce androgens, estrogens, and progestogens. In fact, testosterone is converted into estradiol by the enzyme *aromatase*. The proportions of each sex hormone differ across males and females. This is because the testes produce significantly higher levels of testosterone while the ovaries produce higher levels of estrogens and progestogens.

3.1 Conditions Affecting Sexual Development

There are numerous conditions that affect sexual development, with a large number of these conditions resulting from genetic alterations. Some of the genetic disorders discussed below cause individuals to be born as **intersex**. That is, internal reproductive organs and/or external genitalia do not match one's sex chromosomes.

Individuals with *androgen insensitivity syndrome* are genetically male (XY chromosomes) (Delli Paoli et al., 2023). However, due to a mutation of the androgen receptor gene, which is interestingly coded by the X chromosome, individuals do not develop male sex organs because the body is unable to effectively use androgens. Testosterone that is produced cannot bind to androgen receptors. Aromatase then converts the excess testosterone into estradiol. There are varying degrees of androgen insensitivity syndrome. Individuals can develop external genitalia that resemble males, whereas others develop external genitalia that more closely resemble females. Regardless, internal genitalia typically do not form, meaning androgen insensitivity syndrome often leads to infertility.

Somewhat of the reverse of androgen insensitivity syndrome, *congenital adrenal hyperplasia (CAH)* refers to a group of genetic disorders that affect the adrenal glands (Claahsen-van der Grinten et al., 2022). Going back to the HPA axis, you learned that the adrenal glands produce cortisol and noradrenaline/adrenaline. The adrenal

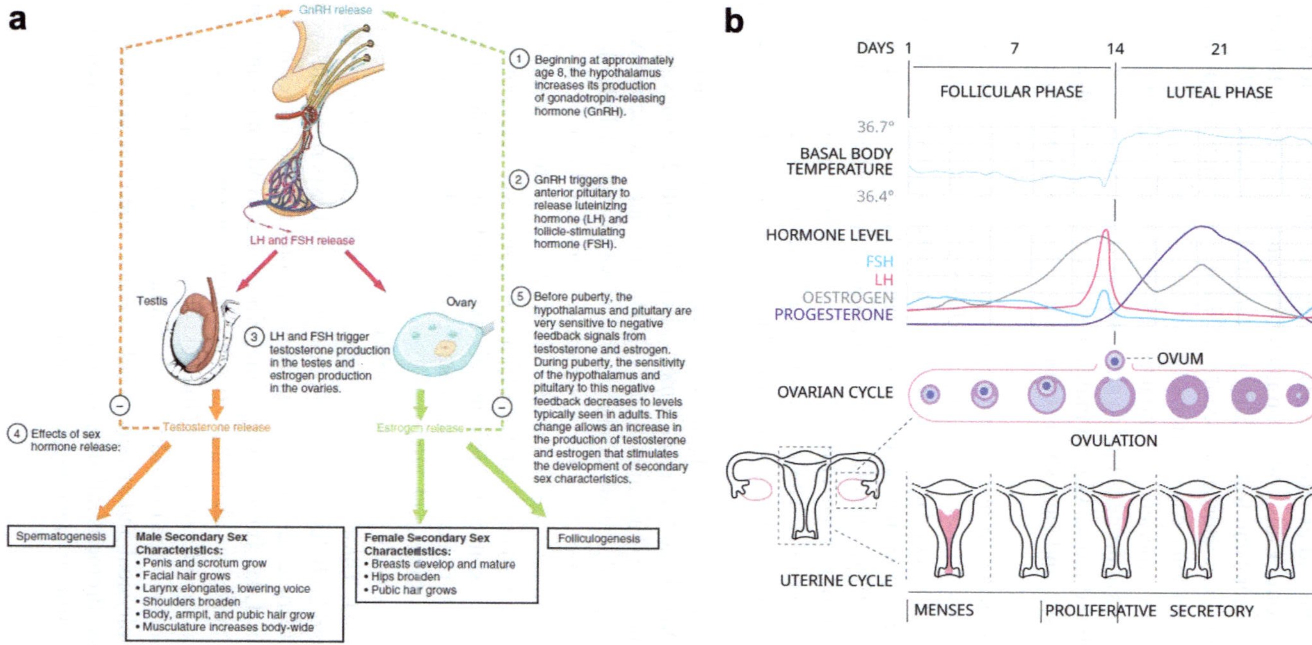

Fig. 2 (**a**) Hormonal influences on sexual development in males and females. *Source* Figure 27.18 of Betts et al. (2013), https://openstax.org/books/anatomy-and-physiology-2e/pages/27-3-development-of-the-male-and-female-reproductive-systems. CC BY 4.0 (https://creativecommons.org/licenses/by/4.0/), via Rice University and OpenStax. (**b**) Depiction of changes that occur throughout the menstrual cycle. *Source* https://commons.wikimedia.org/wiki/File:Menstrual_cycle.svg from Marnanel. CC BY-SA 3.0 (https://creativecommons.org/licenses/by-sa/3.0/deed.en), via Wikimedia Commons. *Note* Oestrogen is an alternative spelling of estrogen

glands also produce androgens, specifically dehydroepiandrosterone (DHEA), dehydroepiandrosterone sulfate (DHEAS), and androstenedione. These androgens are precursors to testosterone and estradiol. CAH causes the adrenal gland to become hyperactive. One symptom of this condition is elevated androgen levels. In biological males, excess androgens can lead to early puberty and short stature. This latter effect occurs because excess androgens accelerate bone maturation and premature fusion of growth plates. In some cases, external genitalia become enlarged, and there are reports of individuals experiencing impaired sexual function and infertility. In biological females, excess androgens lead to ambiguous external genitalia such as an enlarged clitoris and fused labia. As with males, females exposed to high androgen levels reach puberty faster and have short stature; they also can experience fertility issues.

Kallmann syndrome leads to insufficient GnRH production, which affects both males and females (Swee et al., 2021). Puberty is either significantly delayed or does not occur at all. As such, secondary sex characteristics like breasts or an enlarged penis may not develop. Another condition that affects puberty is *Swyer syndrome* (McCann-Crosby et al., 2014). Individuals are genetically male, inheriting both an X chromosome and a Y chromosome. However, scar tissue forms in place of the testes or the ovaries ("streak gonads"). These individuals also develop a vagina, a uterus, and fallopian tubes. Puberty can occur if individuals go on hormone replacement therapy.

There are several genetic conditions in which individuals do not inherit two X chromosomes or one X and one Y chromosome. Individuals with *Turner's syndrome* have a single X chromosome (Huang et al., 2021). The absence of a second X chromosome can lead to multiple health complications. Ovaries do not fully develop, which results in delayed onset of puberty, irregular menstrual periods, and infertility. Individuals with Turner's syndrome often have short stature, a broad chest, a "webbed" neck (extra folds on neck), and heart defects. Females can also inherit three X chromosomes. *Triple X* syndrome does not always lead to noticeable symptoms, but some individuals with triple X syndrome can experience developmental delays and learning disabilities (Otter et al., 2010).

Klinefelter syndrome affects males as they inherit an extra X chromosome (XXY) (Blackburn et al., 2025). These individuals tend to be taller than average height, have reduced facial and body hair, develop breast tissue, and experience delayed or incomplete puberty. Testicles do not fully develop, leading to reduced testosterone and low sperm counts. Like Turner's syndrome, Klinefelter syndrome can lead to infertility. Biological males can also inherit two Y chromosomes and one X chromosome. This is known as *Jacob's syndrome* (Jodarski et al., 2023). Symptoms can vary widely across individuals; some individuals may not have any noticeable symptoms while others display physical and/or behavioral changes. The most common characteristic of Jacob's syndrome is tall stature. Some individuals have delayed speech development or have difficulty reading and writing.

Comprehension Check 1 Considering the distinction between biological sex and psychosocial gender, explain how conditions like androgen insensitivity syndrome and CAH illustrate the complexity of biological sexual development and the potential for a discrepancy between genetic (chromosomal) sex and internal/external reproductive development (phenotype).

These conditions highlight that biological sex is a multi-step process. In androgen insensitivity syndrome, a genetically male individual (XY) develops a female-like external phenotype because their body cannot utilize androgens, thereby illustrating a disconnect between genotype and phenotype due to a receptor issue. In CAH, particularly in a biological female (XX), it leads to the adrenal glands producing excess androgens, causing masculinization of the external genitalia (ambiguous genitalia). Both conditions show that the development of sex organs relies on the presence of appropriate hormones and the ability to respond to them, highlighting how a fault in the hormonal pathway or receptor activity can override the initial chromosomal signal, leading to an intersex variation. These conditions also highlight how someone can be genetically male but have genitalia that resemble females and vice versa.

4 Neuroscience of Sexual Arousal

To better understand the neural processes controlling sexual arousal, researchers present sexual stimuli to participants and measure brain activity. Neuroimaging studies show that numerous brain regions become highly active when individuals are presented sexual stimuli compared to neutral stimuli. One hypothesis states that there are four major components related to sexual arousal, with two additional components involved with inhibition of such arousal (Stoléru et al., 2012). Each of these components and their corresponding neuromechanisms will be detailed below (see Fig. 3).

4.1 Cognitive Component

First, presentation of a sexual stimulus leads to increased activity of brain regions associated with perception and attention, particularly parts of the parietal cortex (inferior parietal lobe and superior parietal lobe; IPL and SPL, respectively) and the primary visual cortex or the primary auditory cortex (if visual stimulus or auditory stimulus, respectively). Next, *appraisal* of the stimulus occurs. The lateral orbitofrontal cortex (lOFC) and part of the inferior temporal cortex (IT) evaluate the stimulus, comparing it to what we have stored in memory concerning sex. For example, when presented with a visual stimulus, the OFC helps determine if the stimulus should elicit arousal (e.g., looking at a coffee cup vs. lingerie).

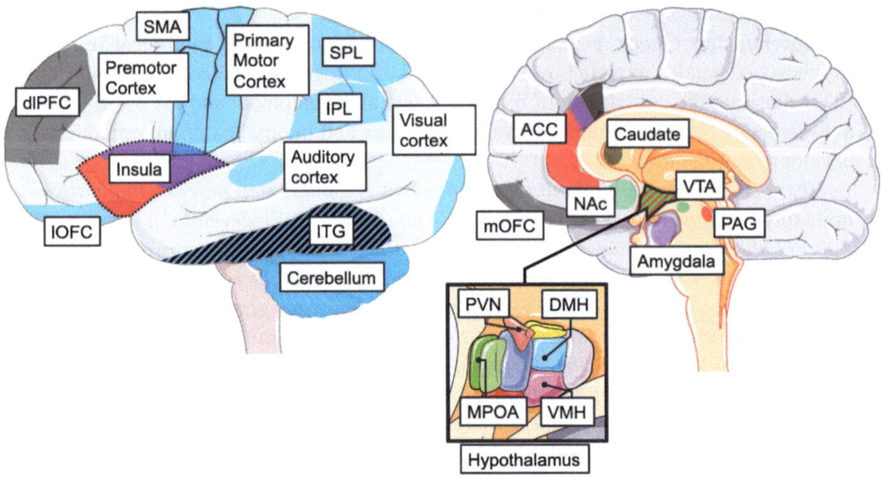

Fig. 3 Brain regions involved in sexual arousal. Cognitive processes associated with sexual arousal are controlled by multiple brain regions highlighted in blue, including primary visual cortex, primary auditory cortex, lOFC (lateral orbitofrontal cortex), inferior temporal gyrus (ITG), IPL (inferior parietal lobe), SPL (superior parietal lobe), SMA (supplementary motor area), premotor cortex, primary motor cortex, and cerebellum. Collectively, these regions are important for attending to sexual stimuli (e.g., visual/auditory cortex, IPL, SPL), appraising sexual stimuli (e.g., lOFC, ITG), and imagining motor actions related to sexual activities (e.g., motor cortices and cerebellum). The hypothalamus, the PAG (periaqueductal gray), the ACC (anterior cingulate cortex), and the anterior insula control autonomic/endocrine processes related to sexual arousal (highlighted in red). Within the hypothalamus, several nuclei play prominent roles in mediating sexual arousal: PVN (paraventricular nucleus), DMH (dorsomedial nucleus), VMH (ventromedial nucleus), and MPAO (medial preoptic area). Brain regions involved in the motivational aspects of sexual arousal (highlighted in green) include the VTA (ventral tegmental area), the NAc (nucleus accumbens), and the hypothalamus. The emotional component is subserved by the amygdala, the posterior insula, and the ACC (highlighted in purple). Finally, inhibition of sexual arousal is largely driven by prefrontal cortical areas such as the dlPFC (dorsolateral prefrontal cortex), the mOFC (medial OFC), and the caudal ACC, as well as the ITG and the caudate nucleus (highlighted in black). Image created by the author using Servier Medical Art (https://smart.servier.com/; CC BY 4.0, https://creativecommons.org/licenses/by/4.0/) and Microsoft PowerPoint

The final aspect of cognitive processing is the motor imagery system. This system consists of the premotor cortex, the supplementary motor area (SMA), the IPL, and the cerebellum. These brain regions are important for motor actions, but they also enable us to imagine such actions. If we deem something to be arousing, we can imagine ourselves performing sexually related actions.

4.2 Autonomic and Endocrine Component

The hypothalamus is a major mediator of sexual behavior. Earlier in this chapter, you learned how the HPG axis regulates sexual maturation. As you have read through this book, you have learned how the hypothalamus contributes to stress responses, feeding behavior, and aggression. In fact, individuals often describe the hypothalamus as controlling the four F's—fighting, fleeing, feeding, and fornicating.

Several hypothalamic nuclei regulate sexual behaviors. The dorsomedial hypothalamic nucleus (DMH), the ventromedial hypothalamic nucleus (VMH), and the paraventricular nucleus (PVN) stimulate parts of the autonomic system that control physiological responses to sexual stimuli (e.g., erection, clitoral enlargement, increased heart rate, and respiration). In particular, oxytocin neurons originating from the PVN travel to the lumbar-sacral region of the spinal cord, which then form synapses with *cavernosal nerves* that control erections (Argiolas & Melis, 2005). These nerves innervate smooth muscles of the *corpora cavernosa*, leading to relaxation of penile muscles. Because these muscles are relaxed, blood can flow into the corpora cavernosa, leading to an erection.

A similar process is observed in females. Oxytocinergic neurons originating from the PVN innervate spinal nerves originating from the sacral region. As in males, stimulating the parasympathetic nervous system leads to smooth muscle relaxation. This causes clitoral engorgement. Also, blood flow to the *vaginal submucosa* (located above the main muscle layers of the vagina) allows *plasma transudate* to pass through the vaginal walls. This is what leads to vaginal lubrication.

The medial preoptic area (MPOA) receives inputs from each sensory system and integrates this information. Neurons from the MPOA travel to the periaqueductal gray (PAG). Neurons from the PAG travel through the brainstem and spinal cord, driving sexual behaviors. In addition to the MPOA, regions like the anterior cingulate cortex (ACC) and the anterior insula transmit integrated sensory information to the brainstem, further helping influence autonomic responses to sexual stimuli (Calabrò et al., 2019). In other words, this region is important for controlling physical actions related to sexual intercourse. For example, lesions to the MPOA decrease copulatory behaviors in rats (Edwards & Einhorn, 1986).

4.3 Motivational Component

Sex is, for the most part, reinforcing. This makes sense as procreation is an important mechanism for ensuring the survival of a species. There is a large network of brain structures that mediate the motivational component of sexual arousal. As you have already learned, initial exposure to a reinforcer like food or sex leads to increased dopaminergic activity in the mesocorticolimbic pathway, which originates in the ventral tegmental area (VTA). Remember, dopamine neurons are important for error signaling. That is, dopamine neurons eventually become highly active when we encounter stimuli that are predictive of sex like nude images or sexually explicit audio. This then causes dopamine levels in the nucleus accumbens (NAc) to increase.

Drugs like cocaine and methamphetamine are proposed to act as aphrodisiacs by directly targeting dopaminergic neurons, leading to increased dopamine in the mesocorticolimbic pathway, thus increasing libido. However, individuals that chronically use stimulant drugs will eventually experience sexual dysfunction (Ghadiagaonkar & Murthy, 2019). This is most likely due to dopamine depletion that occurs with chronic substance use.

The hypothalamus is also implicated in sexual motivation. Above, I cited a study that showed that lesions to the MPOA decrease copulatory behaviors (Edwards & Einhorn, 1986). This study also found that lesions to this region lead to decreased preference for a sexually receptive female. That is, males spend less time around a sexually receptive female compared to a non-receptive female. The MPOA modulates activity of the VTA; pharmacological inhibition of the pathway between the MPOA and the VTA decreases motivation to approach a female without influencing copulatory behaviors (Iyilikci et al., 2017).

In Chap. 11, you learned that the ACC is involved in reward processing, an important component of goal-directed behavior. As such, you should not be shocked to learn that this region is important for sexual motivation. The ACC is activated by sexual stimuli (Long et al., 2020) and is proposed to influence sexual motivation as ACC activity is associated with feelings of sexual desire (Weston, 2012). Activation of the ACC is necessary for "sexual attraction" in animals; that is, when the ACC is inhibited, male rodents spend less time near a sexually receptive female (Wu et al., 2009).

4.4 Emotional Component

The emotional component is closely tied to the motivational component as it is linked to the hedonic quality of sexual arousal. Simply put, sexual arousal feels good. Whereas the dopaminergic system is a major driver of motivation, the endogenous opioid system is important for regulating the hedonic aspects of sex (Pfaus, 2025). Opioids inhibit GABAergic (gamma-aminobutyric acid) neurons that normally inhibit intra-VTA dopaminergic neurons.

In Chap. 11, I discussed the incentive sensitization theory of drug addiction. You learned about the neural correlates of drug wanting and drug liking. Think of wanting as motivation to obtain a reinforcer and liking as the hedonic quality of the reinforcer. What I did not mention in Chap. 11 is that the hedonic "hotspots" associated with substance use are influenced largely by endogenous opioids. Like drugs, dopaminergic activity in the mesocorticolimbic pathway drives our motivation to have sex while opioids lead to a sense of satisfaction during and following sex. Mu opioid receptor availability in regions like the caudate nucleus, the hippocampus, and the ACC is positively associated with sex drive (Nummenmaa et al., 2022), and endogenous opioid release is observed in the hippocampus following orgasm (Jern et al., 2023). Conversely, blocking opioid receptors decreases sexual drive (Quintana et al., 2022).

The amygdala is a critical structure for processing the emotional aspects associated with sex (e.g., pleasure). One major function of the amygdala is to assign emotional significance to stimuli, and the amygdala is highly involved in *sexual conditioning* (Klucken et al., 2016), a form of Pavlovian conditioning. Over time, we associate certain stimuli with sexual intercourse. For example, seeing an image or a video of a nude individual can engender feelings of arousal, just as certain sounds (e.g., moaning, panting) can be sexually arousing.

The posterior insula integrates sensory information from the primary and the secondary somatosensory cortices (e.g., sensation of an erection) and emotional responses (Bogacki-Rychlik et al., 2024). For example, males can awaken with an erect penis (often colloquially referred to as "morning wood"; the technical term is *nocturnal penile tumescence*). The insula integrates the physical sensation with one's current emotional state, which can be one of non-arousal. This process allows the brain to process *why* an erection has occurred.

4.5 Inhibitory Processes

As a species, we have learned that there are times in which we need to suppress sexual arousal. Imagine the potential embarrassment that can come from becoming aroused in the middle of class or at church. One inhibitory process is mediated by parts of the prefrontal cortex (PFC) (particularly medial OFC [mOFC] and dorsolateral PFC [dlPFC]), and the inferior temporal gyrus (ITG); these regions help *devalue* sexual stimuli (i.e., inhibit the appraisal process).

The caudal region of the ACC and the caudate nucleus help suppress mental imagery related to sexual activities and allow us to withhold action related to sex. For example, the ACC is activated when individuals engage in avoidance of sexual stimuli (Kennis et al., 2024). The ACC is important for cognitive control, decision making, and error monitoring. In one study, women were asked to rate how likely they would have sex with a man based on a description. "Low-risk" men were described as having 2–5 previous sexual partners and often used condoms; "high-risk" men were those that had 10–13 previous sexual partners and rarely used condoms. The ACC became more active in response to low-risk men relative to high-risk men (Rupp et al., 2009). Taken together, these results demonstrate that the ACC has dual roles: parts of the ACC help promote sexual motivation and physiological responses related to sex, but other parts are important for sexual decision making, enabling us to consider the potential consequences of engaging in sexual activities (e.g., risk of contracting a sexually transmitted infection, risk of pregnancy, etc.).

There are several neurotransmitters/neuropeptides that are involved in inhibiting sexual arousal. Serotoninergic pathways originating from the raphe nuclei project to multiple brain regions involved in sexual arousal, including cortical areas, limbic structures, and the hypothalamus; serotonergic neurons also make up the descending pathway of the spinal cord. Stimulation of excitatory serotonin receptors like the $5\text{-}HT_{2A}$ receptor on glutamatergic neurons can lead to increased stimulation of GABAergic neurons, which then leads to inhibition of dopamine release (Croft, 2017).

Opioids and endocannabinoids can also act as inhibitors of sexual arousal. This finding is interesting as opioids and endocannabinoids increase dopamine levels in the mesocorticolimbic pathway, and opioids are highly involved in the hedonic qualities of sex. Opioid and cannabinoid receptors are located in various brain regions that mediate sexual arousal. Specifically, stimulation of opioid receptors inhibits the HPG axis (Salata et al., 2022). I mentioned that opioids are released following orgasm; this mechanism helps act as a "signal" to stop sexual behaviors, particularly in males. Like opioids, slight elevations in endocannabinoid levels can facilitate sexual behaviors, but further increases in endocannabinoid signaling inhibit such behavior (Androvicova et al., 2017).

Comprehension Check 2 Differentiate the roles of the hypothalamus and the ACC in sexual arousal.

The hypothalamus is largely involved in controlling physiological responses like erection, clitoral engorgement, and lubrication via the PVN's oxytocin neurons. The hypothalamus is also a key structure that influences the motivation to engage in sexual behaviors by modulating activity of the mesocorticolimbic pathway. In contrast, the ACC is involved in sexual decision-making (e.g., considering the consequences of engaging in sexual behavior), and activation of the caudal region of the ACC is critical for inhibiting sexual imagery. In short, the ACC is critical for ensuring we do not engage in sexual behaviors when doing so is inappropriate.

4.6 Hypoactive Sexual Arousal

Individuals with **hypoactive sexual desire disorder (HSDD)** have low libido, show decreased responsiveness to sexual stimuli, and lack motivation for sexual activity (World Health Organization, 2022). Note, the fifth edition of the *Diagnostic and Statistical Manual of Mental Disorders (DSM-5)* (American Psychiatric Association, 2013) now fractions HSDD into two separate conditions: male hypoactive sexual desire disorder and female sexual interest/arousal disorder. For simplicity, I will use the term HSDD to discuss conditions related to low sexual arousal. Neuroimaging studies, typically with women, show reduced activity in multiple brain regions associated with sexual arousal when presented with sexual stimuli, including the dorsal ACC, the insula, and motor cortical areas; conversely, these individuals show greater activation of frontal cortical areas associated with inhibition of sexual arousal (Kruger & Kneer, 2021).

Hypoactive sexual arousal may stem from dysregulated serotonin, opioid, and/or endocannabinoid systems. As discussed above, these chemicals are largely involved in inhibiting sexual arousal. Not surprisingly, certain drugs that target these systems are linked to sexual dysfunction. Individuals taking selective serotonin reuptake inhibitors (SSRIs) for major depressive disorder (MDD) or anxiety disorders often report low libido, with men also reporting erectile dysfunction (ED) (Atmaca, 2020). Opioid-induced sexual dysfunction is commonly observed in individuals receiving treatment for cancer (Salata et al., 2022), and ED is twice as high in individuals that use cannabis (Pizzol et al., 2019). ED will be discussed in more detail in the next section.

Dysregulation of the mesocorticolimbic dopamine system is also hypothesized to underlie HSDD (Pettigrew & Novick, 2021). This makes sense given dopamine's role in motivation. A hypoactive dopamine system can negatively impact sexual motivation, resulting in low arousal levels associated with HSDD. Evidence to support the dopamine deficiency hypothesis of HSDD comes from the finding that typical antipsychotic medications, which primarily act as dopamine receptor antagonists, are associated with sexual dysfunction (Zhang et al., 2011).

One treatment for HSDD, specifically in younger women, is filbanserin (Addyi). This medication is a 5-HT_{1A} agonist and a 5-HT_{2A} antagonist. Recall that 5-HT_{1A} receptors are inhibitory and often act as autoreceptors to regulate the release of serotonin into the synapse whereas 5-HT_{2A} receptors are located on postsynaptic neurons and are excitatory. As such, filbanserin helps increase sexual arousal by inhibiting the serotonergic system. Modulating serotoninergic activity also indirectly leads to increased dopamine neurotransmission, most likely by disinhibiting intra-VTA dopamine neurons (Invernizzi et al., 2003).

Another potential cause for hypoactive sexual arousal is low testosterone levels. In addition to its involvement in male sexual reproductive development, testosterone activates androgen receptors located in multiple brain regions linked to sexual arousal, including hypothalamus, VTA, amygdala, ACC, and PFC (Nguyen et al., 2022). Androgen receptor expression is observed in the PFC-VTA pathway, and these receptors act as an important modulator of mesocorticolimbic dopamine neurotransmission (Aubele & Kritzer, 2012). In both males and females, low testosterone levels are observed in individuals with low libido (Calabrò et al., 2019). Certain medications that directly block androgen receptors are associated with reduced sexual arousal (de Fátima Viana Santana et al., 2023; Ng et al., 2012). Furthermore, genetic inactivation of androgen receptors decreases sexual motivation in mice (Raskin et al., 2009). To this end, testosterone replacement therapy shows promise in treating HSDD in both men and women (Elliott et al., 2017; Islam et al., 2019).

Comprehension Check 3 Explain how filbanserin (Addyi) and testosterone replacement therapy are proposed to alleviate HSDD; specifically, which component(s) of sexual arousal are affected by filabanserin and testosterone replacement therapy?

By inhibiting the serotonergic system, filbanserin indirectly leads to increased dopamine neurotransmission in the mesocorticolimbic pathway. This action primarily aims to modulate the motivational component of sexual arousal. Additionally, inhibiting serotonin neurotransmission decreases activity of the inhibitory component of sexual arousal. Testosterone replacement therapy allows testosterone to activate androgen receptors located in multiple brain regions linked to arousal (like the hypothalamus, VTA, and ACC). Testosterone's action in the PFC—VTA pathway is a key modulator of mesocorticolimbic dopamine, meaning testosterone replacement therapy also aims to boost the motivational component. Stimulating the hypothalamus is also beneficial for the autonomic/endocrine component of sexual arousal, allowing for the stimulation of physiological processes involved in sexual behaviors.

5 Conditions Affecting Sexual Intercourse

There are situations in which individuals experience arousal, but they have a condition that makes sexual intercourse difficult or impossible.

5.1 Erectile Dysfunction (ED)

Normally, an erection occurs when arteries are dilated, allowing blood to fill the corpora cavernosa and the *corpora spongiosum* of the penis. Disrupting blood flow to the corpora cavernosa and the corpora spongiosum prevents an erection from occurring, even if the individual is sexually aroused. There are numerous causes for ED, such as cardiovascular issues (e.g., high blood pressure, atherosclerosis), neurological conditions (e.g., multiple sclerosis, peripheral nerve damage), hormonal alterations (e.g., low testosterone levels), certain medications, and substances like alcohol. Psychological factors also contribute to ED (Mazzilli, 2022).

In 1989, a group of researchers synthesized sildenafil as a potential treatment for hypertension. When sildenafil was tested in clinical trials, researchers noted that one major side effect was penile erections. This led the drug company Pfizer to market sildenafil as a treatment for ED. While the name sildenafil may not be recognizable, you most likely are familiar with its tradename: Viagra, which was approved by the U.S. Food and Drug Administration in 1998. Sildenafil is an inhibitor of *cyclic guanosine monophosphate (cGMP)-specific phosphodiesterase type 5 (PDE5).* This is a mouthful, so let's break it down.

cGMP is similar to cyclic adenosine monophosphate (cAMP), which you have already learned about (see Chaps. 2 and 6). Both cGMP and cAMP are second messengers that activate protein kinases. In the corpora cavernosa, increased cGMP activity leads to vasodilation, allowing for increased blood flow into the penis. PDE5

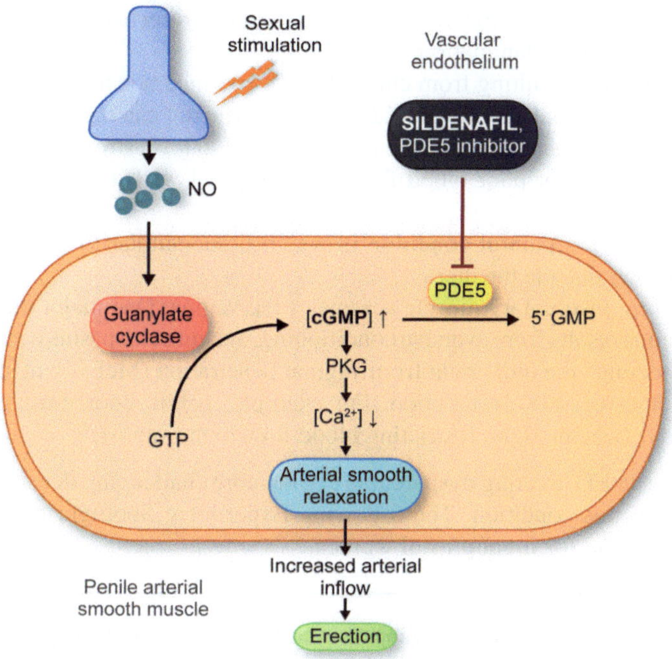

Fig. 4 Mechanism of action of sildenafil. Sexual stimulation causes *nitric oxide* (NO) release in the penis. NO is a gas molecule that relaxes blood vessels to increase blood flow. First, NO binds to guanylate cyclase, which causes guanosine 5′-triphosphate (GTP) to get converted into 3′-5′–cyclic guanosine monophosphate (cGMP). cGMP then activates cGMP-dependent protein kinase (PKG) which acts to decrease calcium (Ca^{++}) levels. This mechanism allows for vasodilation, leading to an erection. Phosphodiesterase-5 (PDE5) metabolizes cGMP into guanosine-5′-monophosphate (5' GMP). https://doi.org/10.3389/fonc.2021.627229. *Source* Figure 1 of Cruz-Burgos et al. (2021), CC BY 4.0 (https://creativecommons.org/licenses/by/4.0/), via Frontiers

metabolizes cGMP. Therefore, sildenafil prevents the breakdown of cGMP, thus improving blood flow to the penis (see Fig. 4).

5.2 *Dyspareunia*

Dyspareunia is a condition in which individuals experience genital pain immediately before, during, and/or immediately after sexual intercourse. This condition is more common in females. Like ED, there are numerous potential causes of dyspareunia (Alimi et al., 2018), including:

1. *Vulvodynia*, a chronic pain condition affecting the vulva (external female genital area). Vulvodynia is idiopathic, although several potential causes have been

proposed, including peripheral/central sensitization (refer back to Chap. 4), injury to the vulva, and inflammation.
2. Perineal trauma resulting from childbirth. The perineum is the area between the genitals and the anus. During childbirth, the perineum can tear.
3. Poor vaginal lubrication, which can be attributed to low estrogen levels (Gandhi et al., 2016) and/or poor blood flow to the vaginal walls (Goncharenko et al., 2019).
4. *Endometriosis*, a painful condition in which tissue similar to the lining of the uterus grows outside the uterus.
5. *Vaginismus*, involuntary muscle spasms of the vagina. One major theory is that vaginismus results from aversive conditioning, and muscle spasms occur as a way of "protecting" the individual from vaginal penetration (McEvoy et al., 2021).
6. Pelvic floor muscle dysfunction. For example, pelvic floor muscles that are overactive become tight, restricting blood flow to the vagina.

In contrast to ED, treating dyspareunia can be more challenging due to the various root causes of this condition. The following Experiment Spotlight highlights one potential treatment for **menopausal** women.

Experiment Spotlight: Effect of Zinc-Containing Pessaries on Menopausal Genitourinary Syndrome, A Double-Blind Randomized Trial
The full article can be accessed with this link: https://pmc.ncbi.nlm.nih.gov/articles/PMC12492771/.

Genitourinary syndrome of menopause (GSM) is a condition that affects women during and after menopause. As the name implies, symptoms are isolated to the genital region and to the urinary tract. Specific to this chapter, the main symptoms associated with GSM are vaginal dryness, burning/irritation in the vulva, and thinning/loss of vaginal tissue. These symptoms lead to dyspareunia. During menopause, estrogen levels decrease, which contributes to the symptomology of GSM. Estrogen replacement therapy can be used to treat GSM, but caution is needed when using this approach as estrogen is linked to breast cancer (Kim & Munster, 2025). An alternative treatment is to selectively apply estrogen to the vaginal area. Women can insert a tablet or *pessary* (type of medical device often made of silicone) into the vagina, or they can apply a cream. Localized estrogen replacement therapy is a safer alternative as breast cancer risk decreases.

In the present study, Mousa et al. (2025) wanted to determine the efficacy of a zinc-containing pessary for the treatment of GSM. Zinc is a mineral that contributes to the *vaginal extracellular matrix*, a network of proteins, polysaccharides, and other molecules that provide structural support to the vaginal wall. Zinc also promotes vaginal lubrication. To this end, Mousa et al. first assessed the eligibility requirements of 62 women. To participate in this study, women had to be postmenopausal with signs of vulvovaginal atrophy or GSM. Participants could not have had a history of breast cancer or pelvic organ prolapse; exclusion criteria also included douching within the past 3 months or using local hormone treatment during the past 6 months. Out of the initial 62 women screened, 50 met eligibility criteria and consented to participate in the clinical trial.

Participants were randomly assigned to two groups. For 14 consecutive nights, women inserted one pessary into the vagina before bedtime. For half of the women, the pessary contained zinc sulfate; for the other half, the pessary did not contain zinc sulfate (placebo condition). Before treatment, participants were asked to answer several questions on a 10-point visual analog scale concerning their symptoms, with a 0 indicating the absence of a symptom and a 10 indicating the worst possible symptom. Participants rated the following symptoms: vaginal pain, burning sensation, itching, vaginal dryness, dyspareunia, and *dysuria* (pain during urination). Clinical data were also collected before testing began. The Vaginal Health Index (VHI) measures vaginal elasticity, vaginal hydration, fluid secretion, pH, and epithelia mucosa. A score of 1 indicates a poor health measure, and a score of 5 indicates the best possible health score. A vaginal smear was also collected to evaluate the percentage of superficial, intermediate, and parabasal epithelia cells. Superficial cells are the most mature cell type found in the epithelium; the parabasal cells are the most immature cell type. A high number of parabasal cells indicates low estrogen levels. Visual analog scores and clinical data were collected at the end of the 14-day treatment period. At the end of the study, participants answered questions about their experience using the pessary.

Before treatment began, no differences in visual analog scores or clinical measures were observed between groups. Following treatment, women that received the zinc-containing pessary reported lower vaginal pain, burning sensation, itching, vaginal dryness, dyspareunia, and dysuria. They also showed more vaginal elasticity and hydration, fluid secretion, and epithelia mucosa, along with decreased pH. No difference was observed in the type of cells found in the epithelium. Participants in the treatment group reported more positive experiences using the pessary compared to the placebo group.

Questions to Consider:

1. Double-blind procedures are often considered the "gold standard" when testing the efficacy of a treatment. Why do you think that is the case?
2. Why do you believe that the zinc-containing pessary did not alter the cell type of the epithelium while it improved all other self-report and clinical assessment data?
3. To answer this question, you need to read the article: Why is a non-hormonal treatment particularly significant for menopausal women, and what mechanism of action does the zinc sulfate heptahydrate complex offer that supports its use as a non-hormonal option?

Comprehension Check 4 How can the pathophysiology of ED be used to partially account for the causes of dyspareunia?

In ED, dysregulation of cGMP does not allow sufficient blood flow to the penis. This prevents males from establishing an erection. Similarly, reduced blood flow to the vaginal walls does not allow for sufficient lubrication of the vaginal. Vaginal dryness is a major cause of dyspareunia.

6 Compulsive Sexual Behavior

In Chap. 11, you learned about compulsivity, the tendency to engage in repetitive behaviors driven by an irresistible urge. While the *DSM-5* does not formally recognize compulsive sexual behavior as a psychological condition, compulsive sexual behavior disorder (CSBD) (sometimes referred to hypersexuality or "sex addiction") is included in the *International Classification of Diseases (ICD-11)* as an impulse-control disorder (World Health Organization, 2019). The diagnostic criteria are very similar to those for substance use disorders and behavioral addictions. Individuals diagnosed with CSBD have difficulty controlling sexual impulses/urges, which results in repetitive sexual behaviors that (a) are distressing to the individual, (b) give the individual little to no satisfaction, and (c) can lead to adverse consequences (e.g., marital conflict, financial difficulties, and health complications). Importantly, compulsive sexual behaviors are not isolated to sexual intercourse. Individuals who are diagnosed with CSBD can have difficulty refraining from masturbating and/or watching pornography.

Because CSBD is characterized by habitual sexual behaviors, one would assume that dysfunction of the cortico-basal ganglia-thalamo-cortical (CBGTC) loop is a major contributor to this condition. To this end, individuals that display hypersexual behavior show greater activation of the caudate nucleus, the thalamus, and the dlPFC (Seok & Sohn, 2015). This finding is interesting as the caudate nucleus is involved in withholding actions related to sex, and the dlPFC is important for devaluing sexual stimuli (see above). Increased activity of regions like the caudate nucleus and the dlPFC may reflect a failed attempt to inhibit sexual arousal mediated by other brain regions. For example, decreased functional connectivity between the dlPFC and the amygdala is observed in individuals that engage in compulsive sexual behaviors (Schmidt et al., 2017). Let's shift our focus from the CBGTC loop to other brain regions implicated in CSBD.

Aberrations are also observed in the mesocorticolimbic pathway of those that engage in hypersexual behaviors. The degree of ventral striatum activity is correlated with symptoms of internet pornography addiction (Brand et al., 2016). The NAc may be overly sensitive to erotic stimuli (Gola & Draps, 2018). Similar to the incentive sensitization theory of drug addiction, "wanting" for sex sensitizes over time due to increased activation of the mesocorticolimbic pathway in response to erotic stimuli while "liking" for sex diminishes. Importantly, decoupling of the PFC and the ventral striatum is observed in hypersexual individuals during sexual conditioning (Klucken et al., 2016). That is, even though both the dlPFC and the ventral striatum are activated by sexual stimuli, the PFC is not as effective at inhibiting signals originating from the NAc. This can lead to a loss in inhibitory control that is characteristic of disorders marked by compulsivity.

One piece of evidence linking mesocorticolimbic dopaminergic dysregulation to compulsive sexual behaviors comes from individuals with Parkinson's disease. As you learned in Chap. 10, these individuals are often given levodopa, the precursor to dopamine. Dopamine agonist treatment increases hypersexuality in individuals with

Parkinson's disease (Codling et al., 2015), with one estimate indicating that approximately 7% of individuals receiving dopamine agonist treatment display hypersexuality (Nakum & Cavanna, 2016). Dopaminergic-based treatments lead to enhanced activation of the ventral striatum, as well as in the OFC and the cingulate cortex, in individuals with Parkinson's disease that show increased hypersexuality (Politis et al., 2013).

Even in individuals without Parkinson's disease, enhanced activation of the cingulate cortex is observed when individuals with hypersexuality are presented with erotic stimuli (Voon et al., 2014; Wojciechowski et al., 2025). This is not surprising as the ACC is normally activated by sexual stimuli. You also learned in Chap. 11 that ACC hyperactivity is implicated in obsessive–compulsive disorder and is observed during drug craving. Hyperactivity of the ACC could reflect an ineffective system that is "working overtime" to exert top-down control over autonomic centers that drive sexual behaviors, similar to the hyperactivity observed in the caudate nucleus and the dlPFC.

Another limbic structure related to CBSD is the amygdala. The amygdala becomes highly active when individuals that engage in compulsive sexual behaviors are presented with erotic stimuli (Seok & Sohn, 2015; Voon et al., 2014; Wojciechowski et al., 2025). Furthermore, individuals that engage in compulsive sexual behaviors show increased amygdalar activation when learning to associate neutral stimuli with sexual stimuli (Klucken et al., 2016). Additional evidence linking the amygdala to compulsive sexual behavior comes from Klüver–Bucy syndrome, which results from lesions to the medial temporal lobe, primarily to the amygdala. Individuals with Klüver–Bucy syndrome display several behavioral alterations, with compulsivity being one major change that is observed (Clay et al., 2019). Compulsive behaviors include overeating and hypersexuality. Damage to the amygdala may cause individuals to assign too much emotional significance to stimuli related to activities like eating and sex. Additionally, damage to the amygdala may prevent it from receiving inhibitory signals from the PFC.

Chapter 11 introduced the glutamate homeostasis hypothesis of addiction. As with substance use disorders, loss of glutamate homeostasis may be a mechanism that contributes to compulsive sexual behaviors. In one clinical trial, *N*-acetylcysteine, which is used to restore glutamate homeostasis, improved compulsive sexual behavior symptoms in 5 out of 8 patients (Blum & Grant, 2022). Given the small sample size, more work is needed to determine the effectiveness of *N*-acetylcysteine in this population.

Comprehension Check 5 Explain how the concepts of "wanting" and "liking" in the incentive sensitization theory relate to the observed dysfunctions in the mesocorticolimbic pathway of individuals with CSBD.

Like substance use disorders, the "wanting" for sex sensitizes over time due to repeated exposure to erotic stimuli, leading to heightened pursuit behavior. Conversely, the "liking" for sex diminishes, meaning the actual satisfaction or pleasure derived from the compulsive sexual behaviors decreases. This imbalance is hypothesized to be driven by an oversensitivity of the nucleus accumbens (NAc), a key structure in the ventral striatum and the mesocorticolimbic pathway, to erotic stimuli, thus increasing the incentive salience (wanting) without an equivalent increase in satisfaction (liking).

7 Paraphilias vs. Paraphilic Disorders

A **paraphilia** is a sexual interest or sexual behavior that involves objects, activities, or situations that are not part of normative sexual stimulation. This is what people colloquially refer to as "kinks" or "fetishes". Many paraphilias do not cause harm to the individual or to others. **Paraphilic disorders** are serious psychiatric conditions that are recognized by the American Psychiatric Association and the World Health Organization. To be diagnosed with a paraphilic disorder, an individual must have a sexual interest that causes physical and/or psychological distress in another individual and/or have a sexual interest involving non-consenting individuals. The individual must also feel distress about their sexual interest that goes beyond worrying about what others think about them. To this end, the following are recognized as paraphilic disorders by the American Psychiatric Association:

1. Fetishistic disorder—arousal stems from non-living objects (e.g., wearing a diaper) or body parts (e.g., feet). This is a case where the disorder arises because it causes significant distress in the individual or impairments in daily functioning.
2. Exhibitionist disorder—arousal stems from exposing one's genitals to a non-consenting individual.
3. Frotteuristic disorder—arousal stems from touching or rubbing one's genitals against a non-consenting individual.
4. Pedophilic disorder—arousal stems from prepubescent children; the individual must be at least 16 years old and at least 5 years older than the child. This requirement is important because it ensures that a 14-year-old that is sexually attracted to a 13-year-old is not labeled as a "pedophile". Related to pedophilia is *hebephilia*, sexual interest in pubescent children that are typically 11–14 years of age.
5. Sexual masochism disorder—arousal stems from being humiliated and/or experiencing pain. This is another case in which the disorder arises because it causes significant distress in the individual or impairments in daily functioning.
6. Sexual sadism disorder—arousal stems from humiliating and/or inflicting pain on a non-consenting individual.
7. Transvestic disorder—arousal stems from dressing in clothes of the opposite sex, and the arousal causes significant distress in the individual or impairments in daily functioning.
8. Voyeuristic disorder—arousal stems from observing unsuspecting individuals undress or engage in sexual activities.

Something I need to note is that there is not always high concordance between paraphilic arousal and paraphilic behaviors. For example, in a recent study using over 400 participants, the percentage of individuals that engaged in pedophilia (1.4% men; 2.4% women) or hebephilia (2.8% men; 1.4% women) was significantly lower than the percentage of individuals that expressed arousal toward pubescent/prepubescent children (3.4%–9.0%) (de Roos et al., 2025). In the same study, correlations between sexual arousal and sexual behavior were higher for other paraphilias such as voyeurism, exhibitionism, and transvestic fetishism. Given the extreme harm associated with pedophilia, many studies examining the underlying neurobiology of paraphilic disorders have focused on pedophilia. This research will be discussed below.

Hypothalamic volume is smaller in individuals with pedophilic disorder that commit child sexual offenses compared to those with pedophilic disorder that have never acted on their sexual urges (Storch et al., 2023). In this chapter, you have learned about the role of the hypothalamus in sexual behavior. The hypothalamus also mediates impulse control by regulating activity of the ventral hippocampus (Noble et al., 2019). This means that a smaller hypothalamus may lead to an imbalance in signals that stimulate sexual behaviors and inhibit such signals. Relatedly, individuals without a history of child sexual offenses show greater connectivity between the ACC and the precuneus and enhanced activation of the superior temporal gyrus (STG), an important structure for social cognition, compared to those that have committed such an offense (Schuler et al., 2022). These findings can explain why individuals may be sexually aroused by children without engaging in sexual activities with such individuals.

In the case of pedophilia, structural and functional abnormalities are observed in frontal cortical areas and limbic structures such as the amygdala (Wiebking & Northoff, 2013). When shown erotic images of adults, individuals with pedophilic disorder show attenuated activity in the dlPFC, the hypothalamus, and the mesocorticolimbic pathway (Walter et al., 2007). Lesions to various brain regions are associated with pedophilic disorder, including the OFC, the precuneus, the posterior cingulate cortex (PCC), the ITG, and the fusiform gyrus (Scarpazza et al., 2021). Collectively, these regions are important for impulse control (OFC), self-referential processing (precuneus and PCC; part of the default mode network [DMN]), and object and face recognition (ITG and fusiform gyrus). Many of these regions are involved in social cognition (see previous chapter). As argued by Scarpazza et al. (2021), acquired pedophilic disorder (i.e., resulting from brain damage) may manifest because individuals have a concurrent loss of inhibitory control and difficulty understanding why their behavior violates societal norms.

A hypoactive serotonin system and/or excessive androgen levels provide plausible explanations for paraphilic disorders. Recall that serotonin is important for inhibiting sexual arousal and sexual behavior while androgens stimulate androgen receptors in various brain regions that promote sexual behaviors. SSRIs, which increase serotonin levels, show promise as treatments for pedophilic disorder (as well as exhibitionist disorder), and antiandrogens can decrease paraphilic behaviors (Assumpção et al., 2014; Culos et al., 2024).

Comprehension Check 6 Explain how structural abnormalities in areas like the hypothalamus and frontal cortical regions are hypothesized to contribute to the manifestation of impulsive or harmful paraphilic behaviors.

A smaller hypothalamic volume is theorized to disrupt the balance between signals that stimulate and those that inhibit sexual behaviors, suggesting an impulse-control deficit. Furthermore, structural abnormalities and attenuated activity in frontal cortical areas, like the dlPFC and OFC, along with limbic structures, are hypothesized to contribute to impulsive paraphilic behaviors by causing a concurrent loss of inhibitory control and difficulty with social cognition, such as understanding the violation of societal norms.

8 Sexual Aggression

One particularly concerning form of sexual behavior is **sexual aggression**, in which sexual activity occurs or is attempted without the explicit consent of an individual (the term sexual violence can also be used). Some of the paraphilic disorders described above can be classified as sexual aggression. For example, sexual activity with prepubescent children is classified as child sexual abuse because children legally cannot provide consent. The age of consent ranges from approximately 13 years of age to 18 years of age depending on the country. In the United States, the age of consent ranges from 16 to 18 years of age. Another form of sexual aggression involves the use of coercion, either by pressuring, threatening, or manipulating an individual to engage in sexual activities. An example of sexual coercion is an employer threatening to fire an employee if they refuse sexual solicitations.

Sexual assault and rape are two forms of sexual aggression that have received considerable attention. Whereas sexual violence includes both attempted and committed sexual acts, sexual assault is specific to sexual actions performed on a non-consenting individual. Sexual assault includes unwanted kissing, touching, grabbing, or other forms of non-consensual sexual contact (e.g., rubbing one's genitals against someone). Rape is a specific form of sexual assault in which penetration of the mouth, anus, and/or vagina occurs, whether with a penis or another object.

Sexual aggression can be modeled in animals. In the sexual aggression test, male rats are first exposed to a sexually receptive female. After the first intromission (i.e., insertion of the penis into the vagina), the sexually receptive female is replaced by a non-sexually receptive female. Researchers are able to measure the number of "forced mounts" and the number of aggressive actions committed by male rats on the non-receptive female. Male rats that display elevated baseline levels of anxiety and non-sexual aggression are more likely to demonstrate sexual aggression (de M Oliveira et al., 2022).

Research examining the neurobiological triggers of sexual aggression is limited. In the previous chapter, you learned about the neural correlates of aggression. The hypothalamus is key component of the HPA axis and the HPG axis. Elevated stress is correlated with sexual aggression (Meyer et al., 2024), and social isolation, a known stressor in rodents, increases sexual aggression toward a non-receptive female (Ngala et al., 2025). Interestingly, sexually aggressive men show *decreased* salivary cortisol levels following presentation of a negative affect-inducing video compared to non-aggressive men (Peterson et al., 2014). One potential explanation is that prolonged exposure to stress causes a dysregulated HPA axis. When individuals are exposed to certain stimuli at a later point, they show a blunted stress response. Due to the lower stress response, individuals that perform sexually aggressive acts may not fear the repercussions of their actions.

Comprehension Check 7 How are low salivary cortisol levels proposed to contribute to sexual aggression?

Low cortisol levels are proposed to represent a blunted stress response due to prolonged stress exposure dysregulating the HPA axis; this reduced stress response may diminish fear of repercussions of sexually aggressive actions.

9 Chapter Summary

This chapter provided a foundational understanding of the neuroscience of sexual arousal, detailing the interplay between cognitive, autonomic, endocrine, motivational, and emotional components, alongside the critical role of neural inhibitory processes mediated by the PFC and other structures. You first learned about genetic disorders that lead to issues in fertility (e.g., androgen insensitivity syndrome, Turner's syndrome). You then learned about several maladaptive sexual behaviors, beginning with HSDD, which is marked by hypoactivity in arousal centers and linked to dysregulated serotonin, dopamine, and testosterone pathways. Physiological disorders like ED and dyspareunia can also be influenced by psychological factors. Sexual disorders characterized by excessive or problematic behavior, such as CSBD, paraphilic disorders, and sexual aggression, were detailed. CSBD shares neurobiological features with substance use disorders, specifically sensitization of "wanting", which is attributed to dysregulation of the mesocorticolimbic pathway over time. The distinction between a paraphilia and a harmful paraphilic disorder was established, with pedophilic disorders and sexual aggression being linked to structural and functional abnormalities in frontal cortical areas and limbic structures governing impulse control, social cognition, and emotional processing, with potential treatments targeting serotonin or androgen levels.

Answers to Experiment Spotlight Questions

1. The double-blind randomized trial is the gold standard because it effectively minimizes bias in patient selection and data collection. Randomization ensures that both known and unknown confounding factors (like baseline health, genetics, environmental factors) are equally distributed between the two groups. Blinding the participants, clinicians, and outcome assessors prevents the placebo effect (participant expectation) and observation bias (researcher preference) from skewing the subjective results like the VAS scores. The use of identical-appearing placebo pessaries with the same base material ensures that any difference in outcome is attributable to the active ingredient (zinc) and not the physical act of treatment.
2. The type of cells located in the epithelium are directly influenced by estrogen levels. Zinc treatment does not increase estrogen levels, meaning that it should not have much effect on the cellular composition of the vaginal extracellular matrix. Because all women are postmenopausal with low estrogen levels, one would expect participants in this study to have a higher proportion of immature parabasal cells.
3. A non-hormonal option is significant because many women are either unwilling or medically uncomfortable with the gold standard treatment of vaginal estrogen supplementation. This is particularly relevant for women with a history of breast or gynecological cancer, who are typically excluded from using hormonal treatments due to safety concerns. The article proposes that the zinc sulfate heptahydrate complex acts as a non-hormonal option through two main mechanisms. First, its strong affinity for water molecules helps with adequate hydration of the atrophic epithelium. Second, zinc acts as an anti-irritant by binding to charged regions on proteins to prevent inflammatory interactions with mucous membranes

Glossary

Androgen a class of sex hormones that contributes to the development and maintenance of masculine characteristics.

Dyspareunia a medical condition in which sexual intercourse causes pain.

Estrogen a class of sex hormones that contributes to the development and maintenance of feminine characteristics.

Gonad an organ that produces reproductive cells, specifically testes in males and ovaries in females.

Hypoactive sexual desire disorder (HSDD) a disorder characterized by a persistent or recurrent deficiency or absence of sexual fantasies and/or desire for sexual activity that causes marked distress or interpersonal difficulties.

Hypothalamic-pituitary-gonad (HPG) axis a neuroendocrine system that regulates reproductive function through the interplay of hormones from the hypothalamus, pituitary gland, and gonads.

Intersex variations in sex characteristics, such as chromosomes, hormones, or anatomy, that do not fit typical definitions of male or female.

Libido an individual's overall sexual drive or desire for sexual activity.

Menopause the natural, permanent cessation of menstrual periods, marking the end of a woman's reproductive life and is medically diagnosed after a woman has gone 12 consecutive months without a period.

Paraphilia an atypical sexual interest or behavior that involves objects, activities, or situations that are not part of normative sexual stimulation.

Paraphilic disorder disorder characterized by recurrent, intense sexual arousal to atypical objects, activities, or situations that cause clinically significant distress or impairment to the individual, or involve harm or the potential for harm to others.

Progestogen a class of hormones that contributes to the female reproductive cycle and pregnancy.

Puberty the period of physical changes during which a child's body matures into an adult body capable of sexual reproduction.

Sexual aggression sexual activity that occurs without the explicit consent of the recipient, often involving force, coercion, or manipulation.

Sexual reproduction the process of creating a new organism by combining the genetic material of two parents.

References

Alimi, Y., Iwanaga, J., Oskouian, R. J., Loukas, M., & Tubbs, R. S. (2018). The clinical anatomy of dyspareunia: A review. *Clinical Anatomy, 31*(7), 1013–1017. https://doi.org/10.1002/ca.23250

American Psychiatric Association. (2013). *Diagnostic and statistical manual of mental disorders* (5th ed.). https://doi.org/10.1176/appi.books.9780890425596

Androvicova, R., Horacek, J., Stark, T., Drago, F., & Micale, V. (2017). Endocannabinoid system in sexual motivational processes: Is it a novel therapeutic horizon? *Pharmacological Research, 115*, 200–208. https://doi.org/10.1016/j.phrs.2016.11.021

Argiolas, A., & Melis, M. R. (2005). Central control of penile erection: Role of the paraventricular nucleus of the hypothalamus. *Progress in Neurobiology, 76*(1), 1–21. https://doi.org/10.1016/j.pneurobio.2005.06.002

Assumpção, A. A., Garcia, F. D., Garcia, H. D., Bradford, J. M., & Thibaut, F. (2014). Pharmacologic treatment of paraphilias. *The Psychiatric Clinics of North America, 37*(2), 173–181. https://doi.org/10.1016/j.psc.2014.03.002

Atmaca, M. (2020). Selective serotonin reuptake inhibitor-induced sexual dysfunction: Current management perspectives. *Neuropsychiatric Disease and Treatment, 16*, 1043–1050. https://doi.org/10.2147/NDT.S185757

Aubele, T., & Kritzer, M. F. (2012). Androgen influence on prefrontal dopamine systems in adult male rats: Localization of cognate intracellular receptors in medial prefrontal projections to the ventral tegmental area and effects of gonadectomy and hormone replacement on glutamate-stimulated extracellular dopamine level. *Cerebral Cortex, 22*(8), 1799–1812. https://doi.org/10.1093/cercor/bhr258

Betts, J. G., Young, K. A., Wise, J. A., Johnson, E., Poe, B., Kruse, D. H., Korol, O., Johnson, J. E., Womblem M., & DeSaix, P. (2013). *Anatomy and physiology*. OpenStax. https://openstax.org/details/books/anatomy-and-physiology-2e

Blackburn, J., Ramakrishnan, A., Graham, C., Bambang, K., Sriranglingam, U., & Senniappan, S. (2025). Klinefelter syndrome: A review. *Clinical Endocrinology, 102*(5), 565–573. https://doi.org/10.1111/cen.15200

Blum, A. W., & Grant, J. E. (2022). N-acetylcysteine in the treatment of compulsive sexual behavior disorder: A case series. *Journal of Psychiatric Research, 154*, 203–206. https://doi.org/10.1016/j.jpsychires.2022.07.066

Bogacki-Rychlik, W., Gawęda, K., & Bialy, M. (2024). Neurophysiology of male sexual arousal-Behavioral perspective. *Frontiers in Behavioral Neuroscience, 17*, 1330460. https://doi.org/10.3389/fnbeh.2023.1330460

Brand, M., Snagowski, J., Laier, C., & Maderwald, S. (2016). Ventral striatum activity when watching preferred pornographic pictures is correlated with symptoms of Internet pornography addiction. *NeuroImage, 129*, 224–232. https://doi.org/10.1016/j.neuroimage.2016.01.033

Calabrò, R. S., Cacciola, A., Bruschetta, D., Milardi, D., Quattrini, F., Sciarrone, F., la Rosa, G., Bramanti, P., & Anastasi, G. (2019). Neuroanatomy and function of human sexual behavior: A neglected or unknown issue? *Brain and Behavior, 9*(12), Article e01389. https://doi.org/10.1002/brb3.1389

Claahsen-van der Grinten, H. L., Speiser, P. W., Ahmed, S. F., Arlt, W., Auchus, R. J., Falhammar, H., Flück, C. E., Guasti, L., Huebner, A., Kortmann, B. B. M., Krone, N., Merke, D. P., Miller, W. L., Nordenström, A., Reisch, N., Sandberg, D. E., Stikkelbroeck, N. M. M. L., Touraine, P., Utari, A., … White, P. C. (2022). Congenital adrenal hyperplasia-Current insights in pathophysiology, diagnostics, and management. *Endocrine Reviews, 43*(1), 91–159. https://doi.org/10.1210/endrev/bnab016

Clay, F. J., Kuriakose, A., Lesche, D., Hicks, A. J., Zaman, H., Azizi, E., Ponsford, J. L., Jayaram, M., & Hopwood, M. (2019). Klüver-Bucy syndrome following traumatic brain injury: A systematic synthesis and review of pharmacological treatment from cases in adolescents and adults. *The Journal of Neuropsychiatry and Clinical Neurosciences, 31*(1), 6–16. https://doi.org/10.1176/appi.neuropsych.18050112

Codling, D., Shaw, P., & David, A. S. (2015). Hypersexuality in Parkinson's disease: Systematic review and report of 7 new cases. *Movement Disorders Clinical Practice, 2*(2), 116–126. https://doi.org/10.1002/mdc3.12155

Croft, H. A. (2017). Understanding the role of serotonin in female hypoactive sexual desire disorder and treatment options. *The Journal of Sexual Medicine, 14*(12), 1575–1584. https://doi.org/10.1016/j.jsxm.2017.10.068

Cruz-Burgos, M., Losada-Garcia, A., Cruz-Hernández, C. D., Cortés-Ramírez, S. A., Camacho-Arroyo, I., Gonzalez-Covarrubias, V., Morales-Pacheco, M., Trujillo-Bornios, S. I., & Rodríguez-Dorantes, M. (2021). New approaches in oncology for repositioning drugs: The case of PDE5 inhibitor sildenafil. *Frontiers in Oncology, 11*, Article 627229. https://doi.org/10.3389/fonc.2021.627229

Culos, C., Di Grazia, M., & Meneguzzo, P. (2024). Pharmacological interventions in paraphilic disorders: Systematic review and insights. *Journal of Clinical Medicine, 13*(6), 1524. https://doi.org/10.3390/jcm13061524

de Fátima Viana Santana, F., Lozi, A. A., Gonçalves, R. V., Da Silva, J., & Da Matta, S. L. P. (2023). Comparative effects of finasteride and minoxidil on the male reproductive organs: A systematic review of in vitro and in vivo evidence. *Toxicology and Applied Pharmacology, 478*, 116710. https://doi.org/10.1016/j.taap.2023.116710

de M. Oliveira, V. E., de Jong, T. R., & Neumann, I. D. (2022). Modelling sexual violence in male rats: The sexual aggression test (SxAT). *Translational Psychiatry, 12*(1), 207. https://doi.org/10.1038/s41398-022-01973-3

de Roos, M. S., Longpré, N., & van Dongen, J. D. M. (2025). When kinks come to life: An exploration of paraphilic behaviors and underlying predictors. *Journal of Sex Research, 62*(3), 317–329. https://doi.org/10.1080/00224499.2024.2319242

Delli Paoli, E., Di Chiano, S., Paoli, D., Lenzi, A., Lombardo, F., & Pallotti, F. (2023). Androgen insensitivity syndrome: A review. *Journal of Endocrinological Investigation, 46*(11), 2237–2245. https://doi.org/10.1007/s40618-023-02127-y

Edwards, D. A., & Einhorn, L. C. (1986). Preoptic and midbrain control of sexual motivation. *Physiology & Behavior, 37*(2), 329–335. https://doi.org/10.1016/0031-9384(86)90242-8

Elliott, J., Kelly, S. E., Millar, A. C., Peterson, J., Chen, L., Johnston, A., Kotb, A., Skidmore, B., Bai, Z., Mamdani, M., & Wells, G. A. (2017). Testosterone therapy in hypogonadal men: A systematic review and network meta-analysis. *British Medical Journal Open, 7*(11), Article e015284. https://doi.org/10.1136/bmjopen-2016-015284

Gandhi, J., Chen, A., Dagur, G., Suh, Y., Smith, N., Cali, B., & Khan, S. A. (2016). Genitourinary syndrome of menopause: An overview of clinical manifestations, pathophysiology, etiology, evaluation, and management. *American Journal of Obstetrics and Gynecology, 215*(6), 704–711. https://doi.org/10.1016/j.ajog.2016.07.045

Ghadiagaonkar, D. S., & Murthy, P. (2019). Sexual dysfunction in persons with substance use disorders. *Journal of Psychosexual Health, 1*(2), 117–121. https://doi.org/10.1177/263183181 9849365

Gola, M., & Draps, M. (2018). Ventral striatal reactivity in compulsive sexual behaviors. *Frontiers in Psychiatry, 9*, 546. https://doi.org/10.3389/fpsyt.2018.00546

Goncharenko, V., Bubnov, R., Polivka, J., Jr., Zubor, P., Biringer, K., Bielik, T., Kuhn, W., & Golubnitschaja, O. (2019). Vaginal dryness: Individualised patient profiles, risks and mitigating measures. *The EPMA Journal, 10*(1), 73–79. https://doi.org/10.1007/s13167-019-00164-3

Huang, A. C., Olson, S. B., & Maslen, C. L. (2021). A review of recent developments in Turner syndrome research. *Journal of Cardiovascular Development and Disease, 8*(11), 138. https:// doi.org/10.3390/jcdd8110138

Invernizzi, R. W., Sacchetti, G., Parini, S., Acconcia, S., & Samanin, R. (2003). Flibanserin, a potential antidepressant drug, lowers 5-HT and raises dopamine and noradrenaline in the rat prefrontal cortex dialysate: Role of 5-HT(1A) receptors. *British Journal of Pharmacology, 139*(7), 1281–1288. https://doi.org/10.1038/sj.bjp.0705341

Islam, R. M., Bell, R. J., Green, S., Page, M. J., & Davis, S. R. (2019). Safety and efficacy of testosterone for women: A systematic review and meta-analysis of randomised controlled trial data. *The Lancet. Diabetes & Endocrinology, 7*(10), 754–766. https://doi.org/10.1016/S2213-8587(19)30189-5

Iyilikci, O., Balthazart, J., & Ball, G. F. (2017). Medial preoptic regulation of the ventral tegmental area related to the control of sociosexual behaviors. *eNeuro, 3*(6), ENEURO.0283-16.2016. https://doi.org/10.1523/ENEURO.0283-16.2016

Jern, P., Chen, J., Tuisku, J., Saanijoki, T., Hirvonen, J., Lukkarinen, L., Manninen, S., Helin, S., Putkinen, V., & Nummenmaa, L. (2023). Endogenous opioid release after orgasm in man: A combined PET/functional MRI study. *Journal of Nuclear Medicine, 64*(8), 1310–1313. https:// doi.org/10.2967/jnumed.123.265512

Jodarski, C., Duncan, R., Torres, E., Gore, R., Raznahan, A., & Similuk, M. (2023). Understanding the phenotypic spectrum and family experiences of XYY syndrome: Important considerations for genetic counseling. *Journal of Community Genetics, 14*(1), 17–25. https://doi.org/10.1007/ s12687-022-00630-y

Kennis, M., Dewitte, M., T'Sjoen, G., Stinkens, K., Sack, A. T., & Duecker, F. (2024). The behavioral component of sexual inhibition and its relation with testosterone levels: An fMRI study in transgender and cisgender individuals. *Psychoneuroendocrinology, 163*, Article 106963. https:// doi.org/10.1016/j.psyneuen.2024.106963

Kim, J., & Munster, P. N. (2025). Estrogens and breast cancer. *Annals of Oncology, 36*(2), 134–148. https://doi.org/10.1016/j.annonc.2024.10.824

Klucken, T., Wehrum-Osinsky, S., Schweckendiek, J., Kruse, O., & Stark, R. (2016). Altered appetitive conditioning and neural connectivity in subjects with compulsive sexual behavior. *The Journal of Sexual Medicine, 13*(4), 627–636. https://doi.org/10.1016/j.jsxm.2016.01.013

Kruger, T. H. C., & Kneer, J. (2021). Neurobiological basis of sexual deviance. In L. A. Craig & R. M. Bartels (Eds.), *Sexual deviance: Understanding and managing deviant sexual interests and paraphilic disorders* (pp. 67–88). Wiley Blackwell. https://doi.org/10.1002/978111977140 1.ch5

Long, X., Tian, F., Zhou, Y., Cheng, B., & Jia, Z. (2020). Different neural correlates of sexually preferred and sexually nonpreferred stimuli. *The Journal of Sexual Medicine, 17*(7), 1254–1267. https://doi.org/10.1016/j.jsxm.2020.03.007

Mazzilli, F. (2022). Erectile dysfunction: Causes, diagnosis and treatment: An update. *Journal of Clinical Medicine, 11*(21), 6429. https://doi.org/10.3390/jcm11216429

McCann-Crosby, B., Mansouri, R., Dietrich, J. E., McCullough, L. B., Sutton, V. R., Austin, E. G., Schlomer, B., Roth, D. R., Karaviti, L., Gunn, S., Hicks, M. J., & Macias, C. G. (2014). State of the art review in gonadal dysgenesis: Challenges in diagnosis and management. *International Journal of Pediatric Endocrinology, 2014*(1), 4. https://doi.org/10.1186/1687-9856-2014-4

McEvoy, M., McElvaney, R., & Glover, R. (2021). Understanding vaginismus: A biopsychosocial perspective. *Sexual and Relationship Therapy, 39*(3), 680–701. https://doi.org/10.1080/14681994.2021.2007233

Meyer, S. R., Hardt, S., Brambilla, R., & Stöckl, H. (2024). A systematic review of theories of stress as a predictor of intimate partner violence. *Trauma, Violence & Abuse*, 15248380241297325. Advance online publication. https://doi.org/10.1177/15248380241297325

Mousa, A., Elgarhi, I., Ibrahim, B., Atalla, E., Abdel-Rasheed, M., Nagy, O., & Shaheen, K. (2025). Effect of zinc-containing pessaries on menopausal genitourinary syndrome, a double-blind randomised trial. *BMC Women's Health, 25*(1), 467. https://doi.org/10.1186/s12905-025-04037-y

Nakum, S., & Cavanna, A. E. (2016). The prevalence and clinical characteristics of hypersexuality in patients with Parkinson's disease following dopaminergic therapy: A systematic literature review. *Parkinsonism & Related Disorders, 25*, 10–16. https://doi.org/10.1016/j.parkreldis.2016.02.017

Ng, E., Woo, H. H., Turner, S., Leong, E., Jackson, M., & Spry, N. (2012). The influence of testosterone suppression and recovery on sexual function in men with prostate cancer: Observations from a prospective study in men undergoing intermittent androgen suppression. *The Journal of Urology, 187*(6), 2162–2166. https://doi.org/10.1016/j.juro.2012.01.080

Ngala, M. E., Hemmings, S. M. J., Womersley, J. S., Shabangu, T. W., & Qulu-Appiah, L. (2025). Social isolation induces sexually aggressive behaviour in male Wistar rats. *BMC Neuroscience, 26*(1), 15. https://doi.org/10.1186/s12868-025-00932-0

Nguyen, V., Leonard, A., & Hsieh, T.-C. (2022). Testosterone and sexual desire: A review of the evidence. *Androgens: Clinical Research and Therapeutics, 3*(1), 85–90. https://doi.org/10.1089/andro.2021.003

Noble, E. E., Wang, Z., Liu, C. M., Davis, E. A., Suarez, A. N., Stein, L. M., Tsan, L., Terrill, S. J., Hsu, T. M., Jung, A. H., Raycraft, L. M., Hahn, J. D., Darvas, M., Cortella, A. M., Schier, L. A., Johnson, A. W., Hayes, M. R., Holschneider, D. P., & Kanoski, S. E. (2019). Hypothalamus-hippocampus circuitry regulates impulsivity via melanin-concentrating hormone. *Nature Communications, 10*(1), 4923. https://doi.org/10.1038/s41467-019-12895-y

Nummenmaa, L., Jern, P., Malén, T., Kantonen, T., Pekkarinen, L., Lukkarinen, L., Sun, L., Nuutila, P., & Putkinen, V. (2022). μ-opioid receptor availability is associated with sex drive in human males. *Cognitive, Affective & Behavioral Neuroscience, 22*(2), 281–290. https://doi.org/10.3758/s13415-021-00960-3

Otter, M., Schrander-Stumpel, C. T., & Curfs, L. M. (2010). Triple X syndrome: A review of the literature. *European Journal of Human Genetics, 18*(3), 265–271. https://doi.org/10.1038/ejhg.2009.109

Peterson, Z. D., Janssen, E., Goodrich, D., & Heiman, J. R. (2014). Physiological reactivity in a community sample of sexually aggressive young men: A test of competing hypotheses. *Aggressive Behavior, 40*(2), 152–164. https://doi.org/10.1002/ab.21512

Pettigrew, J. A., & Novick, A. M. (2021). Hypoactive Sexual desire disorder in women: Physiology, assessment, diagnosis, and treatment. *Journal of Midwifery & Women's Health, 66*(6), 740–748. https://doi.org/10.1111/jmwh.13283

Pfaus, J. G. (2025). Orgasms, sexual pleasure, and opioid reward mechanisms. *Sexual Medicine Reviews, 13*(3), 381–393. https://doi.org/10.1093/sxmrev/qeaf023

Pizzol, D., Demurtas, J., Stubbs, B., Soysal, P., Mason, C., Isik, A. T., Solmi, M., Smith, L., & Veronese, N. (2019). Relationship between cannabis use and erectile dysfunction: A Systematic review and meta-analysis. *American Journal of Men's Health, 13*(6), 1557988319892464. https://doi.org/10.1177/1557988319892464

Politis, M., Loane, C., Wu, K., O'Sullivan, S. S., Woodhead, Z., Kiferle, L., Lawrence, A. D., Lees, A. J., & Piccini, P. (2013). Neural response to visual sexual cues in dopamine treatment-linked hypersexuality in Parkinson's disease. *Brain, 136*(Pt 2), 400–411. https://doi.org/10.1093/brain/aws326

Quintana, G. R., Mac Cionnaith, C. E., & Pfaus, J. G. (2022). Behavioral, neural, and molecular mechanisms of conditioned mate preference: The role of opioids and first experiences of sexual reward. *International Journal of Molecular Sciences, 23*(16), 8928. https://doi.org/10.3390/ijms23168928

Raskin, K., de Gendt, K., Duittoz, A., Liere, P., Verhoeven, G., Tronche, F., & Mhaouty-Kodja, S. (2009). Conditional inactivation of androgen receptor gene in the nervous system: Effects on male behavioral and neuroendocrine responses. *The Journal of Neuroscience, 29*(14), 4461–4470. https://doi.org/10.1523/JNEUROSCI.0296-09.2009

Rupp, H. A., James, T. W., Ketterson, E. D., Sengelaub, D. R., Janssen, E., & Heiman, J. R. (2009). The role of the anterior cingulate cortex in women's sexual decision making. *Neuroscience Letters, 449*(1), 42–47. https://doi.org/10.1016/j.neulet.2008.10.083

Salata, B., Kluczna, A., & Dzierżanowski, T. (2022). Opioid-induced sexual dysfunction in cancer patients. *Cancers, 14*(16), 4046. https://doi.org/10.3390/cancers14164046

Scarpazza, C., Finos, L., Genon, S., Masiero, L., Bortolato, E., Cavaliere, C., Pezzaioli, J., Monaro, M., Navarin, N., Battaglia, U., Pietrini, P., Ferracuti, S., Sartori, G., & Camperio Ciani, A. S. (2021). Idiopathic and acquired pedophilia as two distinct disorders: An insight from neuroimaging. *Brain Imaging and Behavior, 15*(5), 2681–2692. https://doi.org/10.1007/s11682-020-00442-z

Schmidt, C., Morris, L. S., Kvamme, T. L., Hall, P., Birchard, T., & Voon, V. (2017). Compulsive sexual behavior: Prefrontal and limbic volume and interactions. *Human Brain Mapping, 38*(3), 1182–1190. https://doi.org/10.1002/hbm.23447

Schuler, M., Mohnke, S., Amelung, T., Beier, K. M., Walter, M., Ponseti, J., Schiffer, B., Kruger, T. H. C., & Walter, H. (2022). Neural processing associated with cognitive empathy in pedophilia and child sexual offending. *Social Cognitive and Affective Neuroscience, 17*(8), 712–722. https://doi.org/10.1093/scan/nsab133

Seok, J. W., & Sohn, J. H. (2015). Neural substrates of sexual desire in individuals with problematic hypersexual behavior. *Frontiers in Behavioral Neuroscience, 9*, 321. https://doi.org/10.3389/fnbeh.2015.00321

Stoléru, S., Fonteille, V., Cornélis, C., Joyal, C., & Moulier, V. (2012). Functional neuroimaging studies of sexual arousal and orgasm in healthy men and women: A review and meta-analysis. *Neuroscience and Biobehavioral Reviews, 36*(6), 1481–1509. https://doi.org/10.1016/j.neubiorev.2012.03.006

Storch, M., Kanthack, M., Amelung, T., Beier, K. M., Krueger, T. H. C., Sinke, C., Walter, H., Walter, M., Schiffer, B., Schindler, S., & Schoenknecht, P. (2023). Hypothalamic volume in pedophilia with or without child sexual offense. *European Archives of Psychiatry and Clinical Neuroscience, 273*(6), 1295–1306. https://doi.org/10.1007/s00406-022-01501-w

Swee, D. S., Quinton, R., & Maggi, R. (2021). Recent advances in understanding and managing Kallmann syndrome. *Faculty Reviews, 10*, 37. https://doi.org/10.12703/r/10-37

Voon, V., Mole, T. B., Banca, P., Porter, L., Morris, L., Mitchell, S., Lapa, T. R., Karr, J., Harrison, N. A., Potenza, M. N., & Irvine, M. (2014). Neural correlates of sexual cue reactivity in individuals with and without compulsive sexual behaviours. *Plos One, 9*(7), Article e102419. https://doi.org/10.1371/journal.pone.0102419

Walter, M., Witzel, J., Wiebking, C., Gubka, U., Rotte, M., Schiltz, K., Bermpohl, F., Tempelmann, C., Bogerts, B., Heinze, H. J., & Northoff, G. (2007). Pedophilia is linked to reduced activation in hypothalamus and lateral prefrontal cortex during visual erotic stimulation. *Biological Psychiatry, 62*(6), 698–701. https://doi.org/10.1016/j.biopsych.2006.10.018

Weston, C. S. (2012). Another major function of the anterior cingulate cortex: The representation of requirements. *Neuroscience and Biobehavioral Reviews, 36*(1), 90–110. https://doi.org/10.1016/j.neubiorev.2011.04.014

Wiebking, C., & Northoff, G. (2013). Neuroimaging in pedophilia. *Current Psychiatry Reports, 15*(4), 351. https://doi.org/10.1007/s11920-013-0351-x

Wojciechowski, J., Draps, M., Kublik, E., Dubiejko, P., Wolak, T., & Gola, M. (2025). Enhanced conditioning and disrupted extinction processes in men struggling with compulsive sexual behaviors. *Journal of Behavioral Addictions, 14*(1), 191–203. https://doi.org/10.1556/2006.2025.00012

World Health Organization. (2019). *International statistical classification of diseases and related health problems* (11th ed.). https://icd.who.int/

World Health Organization. (2022). *ICD-11: International classification of diseases* (11th revision). https://icd.who.int/

Wu, L. J., Kim, S. S., Li, X., Zhang, F., & Zhuo, M. (2009). Sexual attraction enhances glutamate transmission in mammalian anterior cingulate cortex. *Molecular Brain, 2*, 9. https://doi.org/10.1186/1756-6606-2-9

Zhang, X. R., Zhang, Z. J., Zhu, R. X., Yuan, Y. G., Jenkins, T. A., & Reynolds, G. P. (2011). Sexual dysfunction in male schizophrenia: Influence of antipsychotic drugs, prolactin and polymorphisms of the dopamine D2 receptor genes. *Pharmacogenomics, 12*(8), 1127–1136. https://doi.org/10.2217/pgs.11.46

Chapter 17
Using Neuroscience to Treat Maladaptation

1 Introduction

During my senior year of college I took a course about stress. Our final exam for the course was to use *biofeedback* to control our stress levels. Biofeedback is a technique in which individuals learn to control involuntary physiological responses. Our professor wanted us to use stress-reduction techniques we had learned about in conjunction with biometric data to decrease our heart rate. As a student facing "senioritis", I thought the assignment had little utility. However, biofeedback is an important therapeutic technique that is used to treat conditions like chronic pain and anxiety (Frank et al., 2010).

Throughout this book you have learned about the neuroscience of various forms of maladaptation, ranging from aberrant perception to dysfunctional social interactions. In Chap. 2, I discussed neuroscience techniques that are commonly used to study maladaptation. In some chapters, I cited a few examples in which maladaptive behaviors can be manipulated by these techniques. This chapter serves as an epilogue to the book and will go into more detail as to how neuroscience methods can be used to ameliorate maladaptive behaviors. I will also introduce other neuroscience techniques that have not been discussed in detail. This chapter will begin by discussing a specific form of biofeedback.

2 Learning Objectives

By the end of this chapter, you should be able to …

1. Explain how neuroscience techniques can be used to treat different maladaptive processes.
2. Identify the advantages and the disadvantages of the neuroscience techniques used to treat maladaptation.

J. Yates, *Neuroscience of Maladaptation*,
https://doi.org/10.1007/978-3-032-15602-0_17

3 Neurofeedback

Neurofeedback involves using real-time feedback of brain activity to regulate neuronal activity. The electroencephalogram (EEG) is often used to record changes in neuronal activity in cortical areas. Because individuals often do not know how to interpret EEG patterns, the practitioner can present audio, videos, or a combination of the two to an individual so they can monitor their progress during a neurofeedback session. As the individual focuses on altering their brain activity, they can observe changes in the audio/video output presented to them. Figure 1a shows an example of neurofeedback.

I did not go into too much detail when describing how EEGs work in Chap. 2. Cortical brainwaves can be divided into six main types: delta, theta, alpha, sensorimotor rhythm, beta, and gamma. Delta waves are characterized by low-frequency neuronal activity (1–4 Hz [Hz]) and are most commonly observed when an individual is in deep sleep (refer back to Chap. 14). On the other end of the spectrum, gamma waves are characterized by high-frequency neuronal firing (32–100 Hz) and are observed when individuals are actively processing information and solving problems (Marzbani et al., 2016). See Fig. 1b for more details on brainwave frequencies.

Neurofeedback is commonly used to treat symptoms associated with attention-deficit/hyperactivity disorder (ADHD). Historically, the primary strategy was to have individuals enhance beta wave activity while inhibiting theta wave activity; however, there is some debate as to whether the theta/beta wave ratio is a reliable marker of ADHD (Ölçüoğlu, 2025). Indeed, research has shown that specifically enhancing beta activity is more effective at improving response inhibition in individuals with ADHD compared to targeting beta wave and theta wave activity (Bluschke et al., 2022).

One advantage of EEG is that individuals can utilize neurofeedback techniques from home. Individuals allowed to use EEG-based neurofeedback at home during an 8-week period show reductions in pain (Sakel et al., 2025). Likewise, veterans with posttraumatic stress disorder (PTSD) show reductions in pain intensity, PTSD symptoms, and sleep disturbances following 12 weeks of mobile neurofeedback (Elbogen et al., 2021).

Other neuroscience methods beyond EEG can be used in conjunction with neurofeedback techniques. Functional magnetic resonance imaging (fMRI) is useful as individuals can focus on altering neural activity in specific brain regions. To this end, fMRI-based neurofeedback reduces auditory hallucinations by strengthening connectivity between the default mode network (DMN) and the dorsolateral prefrontal cortex (dlPFC) (Zhang et al., 2025). Also, strengthening connectivity between parts of the DMN and the salience network with the posterior cingulate cortex (PCC) improves symptoms of PTSD (Lieberman et al., 2023). Targeted self-neuromodulation of the amygdala ameliorates pain in individuals with fibromyalgia (Or-Borichev et al., 2025).

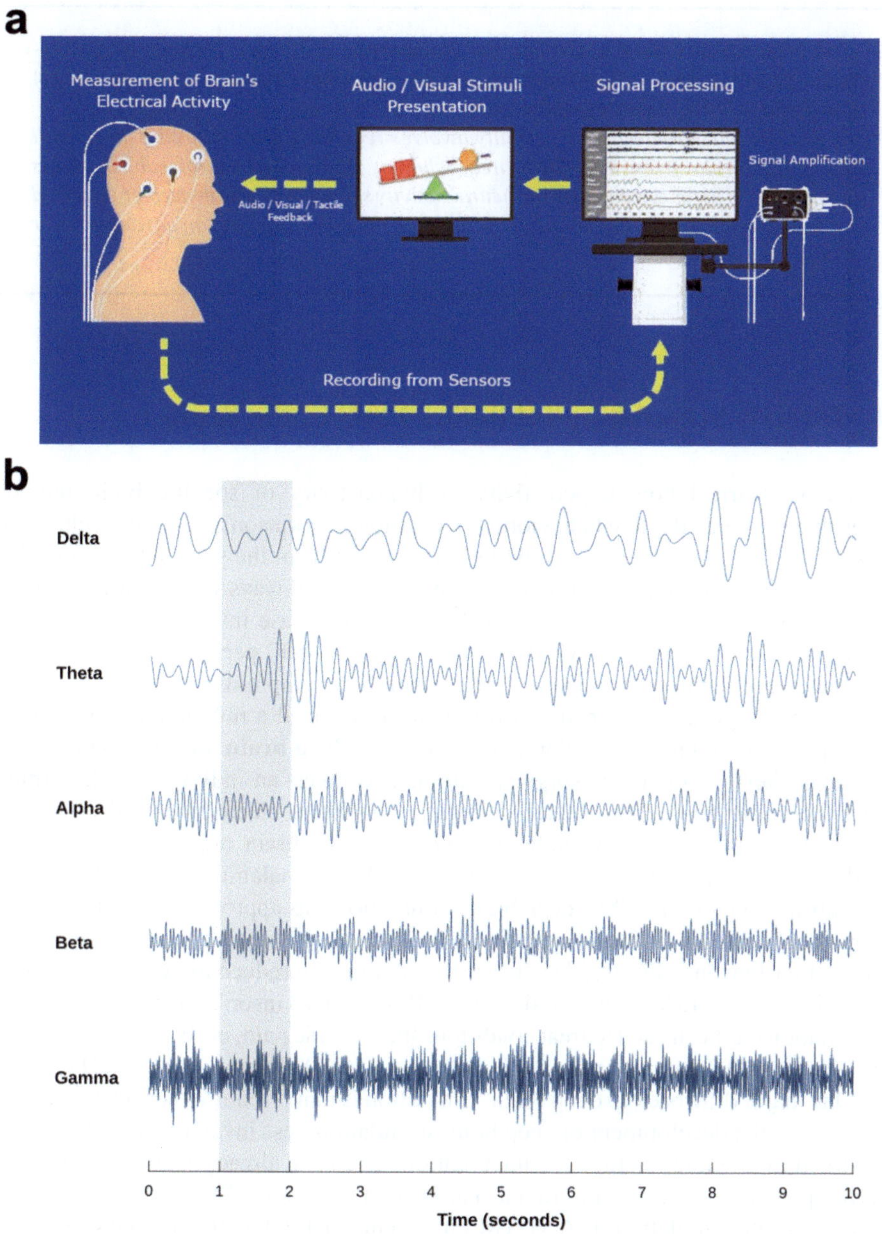

Fig. 1 (**a**) Schematic showing how a neurofeedback session works. *Source* https://commons.wikimedia.org/wiki/File:Neurofeedback_Process_Diagram.png from IsUnsigned. CC BY-SA 4.0 (https://creativecommons.org/licenses/by-sa/4.0/deed.en), via Wikimedia Commons. (**b**) Different brainwave frequencies that are observed. *Source* https://commons.wikimedia.org/wiki/File:EEG_Brainwaves.svg from Laurens R. Krol. CC0 1.0 Universal (https://creativecommons.org/publicdomain/zero/1.0/deed.en), via Wikimedia Commons

Comprehension Check 1 Explain the fundamental difference between how EEG-based neurofeedback is implemented and what an individual is learning to control compared to fMRI-based neurofeedback.

EEG-based neurofeedback typically involves recording and regulating the overall activity, or brainwaves, in cortical areas whereas fMRI-based neurofeedback allows individuals to focus on and alter the neural activity specifically within deep or localized brain regions or strengthen connectivity between networks, which is an advantage over EEG.

4 Directly Stimulating/Inhibiting Parts of the Brain

You have learned how hyperactivity or hypoactivity of specific brain regions contributes to maladaptive behaviors. For example, Parkinson's disease results from destruction of dopaminergic cell bodies originating from the substantia nigra. This prevents innervation of the direct and the indirect pathways of the basal ganglia. Recall that the primary treatments for Parkinson's disease involve pharmacotherapies, particularly L-DOPA, which often lead to "on–off" periods (i.e., individuals fluctuate between improvement and exacerbation of symptoms).

In 1987, a group of French neurosurgeons developed a revolutionary treatment to help treat individuals with Parkinson's disease: **deep brain stimulation** (Fig. 2). A pacemaker-like device is surgically implanted under an individual's skin, often near the collar bone. This device delivers a consistent electrical impulse through an electrode that has been implanted into a specific brain region. In the case of Parkinson's disease, the electrode is placed in the subthalamic nucleus (STN), not the substantia nigra. In 2002, deep brain stimulation was approved as a treatment for Parkinson's disease in the United States. In addition to treating the primary symptoms of Parkinson's disease, deep brain stimulation can reduce motor impulsivity in these individuals (Moussawi et al., 2022). Beyond Parkinson's disease, deep brain stimulation has been used to treat conditions like chronic pain, obsessive–compulsive disorder (OCD), and major depressive disorder (MDD) (Krauss et al., 2021).

One major drawback to deep brain stimulation is that it requires surgical intervention. Since the development of deep brain stimulation, less invasive techniques have emerged as potential therapeutic intervention. You have already learned about such techniques, specifically transcranial magnetic stimulation (TMS) and transcranial direct current stimulation (tDCS). TMS uses a magnet to target neuronal activation, whereas tDCS uses electrodes to directly stimulate or inhibit neuronal populations. TMS and tDCS are effective at improving symptoms related to schizophrenia, MDD, catatonia, OCD, and sleep disorders (Richter et al., 2023; Tortella et al., 2015). Recent work has shown that applying both TMS and tDCS provides additive beneficial effects for those with MDD (Zhou et al., 2024).

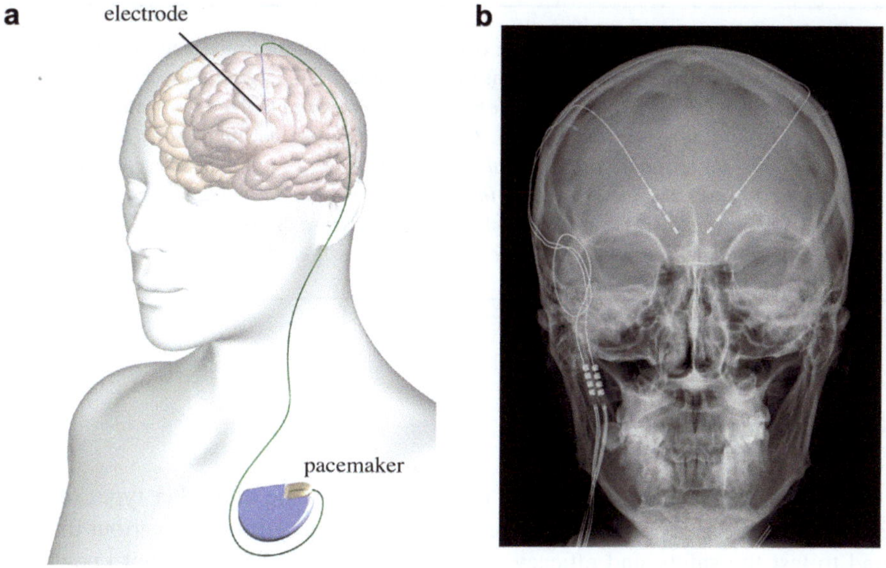

Fig. 2 (**a**) Schematic of deepbrain stimulation. *Source* Figure 1 of Shamir et al. (2014), https://doi.org/10.3389/frym.2014.00012. CC BY 4.0 (https://creativecommons.org/licenses/by/4.0/), via Frontiers. (**b**) Real-life image of a 42-year-old male implanted with electrodes for the treatment of obsessive–compulsive disorder. *Source* https://commons.wikimedia.org/wiki/File:X-ray_of_deep_brain_stimulation_in_OCD,_F.png from Jmarchn. CC BY-SA 3.0 (https://creativecommons.org/licenses/by-sa/3.0/deed.en), via Wikimedia Commons

More recently, **transcranial focused ultrasound (tFUS)** has shown promise as a treatment for schizophrenia, generalized anxiety disorder, MDD, and substance use disorders (Shi & Wu, 2025). As the name implies, tFUS uses sound waves that are focused on a specific region of the brain. When low-intensity ultrasound is used, sound waves create mechanical pressure on ion channels, which can lead to either depolarization or hyperpolarization of neurons. High-intensity ultrasound can be used to damage tissue in a highly localized spot.

Individuals with schizophrenia that receive tFUS to the left dlPFC show improvements in negative symptoms (Zhai et al., 2023). tFUS applied to the left amygdala decreases resting-state functional connectivity between the amygdala and the insula and the hippocampus while increasing functional connectivity between the amygdala and the ventromedial PFC (vmPFC), with these alterations coinciding with decreased self-reported anxiety levels (Chou et al., 2024). Similarly, individuals report improved mood following tFUS applied to the right inferior frontal gyrus (IFG), with mood alterations most likely resulting from decreased functional connectivity between the IFG and other regions, including the orbitofrontal cortex (OFC), the anterior cingulate cortex (ACC), and the entorhinal cortex (Sanguinetti et al., 2020). Finally, a pilot study has shown that targeting the nucleus accumbens (NAc) with tFUS helps reduce drug craving (Mahoney et al., 2023).

Comprehension Check 2 Compare deep brain stimulation to the neuromodulation techniques of tDCS and tFUS in terms of invasiveness and targeting mechanism.

tDCS and tFUS are significantly less invasive compared to deep brain stimulation, which requires surgical intervention. tDCS, like deep brain stimulation, utilizes an electric current to modulate neuronal activity. Instead of implanting an electrode in someone's brain, tDCS uses a device that is placed around an individual's head. tFUS uses focused sound waves to create mechanical pressure on ion channels, leading to depolarization or hyperpolarization of neurons in a targeted region of the brain.

5 Stem Cell Treatments

A **stem cell** is a special type of cell that can differentiate into other types of cells, including neurons. There is much interest in using stem cells to treat various diseases and to test the safety and efficacy of new drugs. Even if you did not know what stem cells do, you most likely have heard of them, whether on the news, social media, and/or in class. There is some opposition to the use of stem cells, particularly embryonic stem cells. While stem cells are found in most types of human tissue, embryonic stem cells are highly versatile as they can differentiate into all cell types in a developing fetus. As such, embryonic stem cells are also known as *pluripotent* stem cells. Contrary to popular belief, embryonic stem cells do not come from abortions. These stem cells come from embryos that are 3–5 days old, most often from *in vitro* fertilization (IVF) clinics. When an individual undergoes IVF, they have excess embryos that can be donated for research purposes.

In addition to some of the ethical concerns associated with stem cell treatment, there are several technical challenges that make stem cell treatment difficult to implement. Because pluripotent stem cells can differentiate into various cell types, there is the risk for *tumorigenicity* (formation of tumors). Another challenge associated with stem cells is donor rejection, which necessitates the use of immunosuppressants (Wei et al., 2024). *Human-induced pluripotent stem cells (hiPSCs)* are a more recent innovation, allowing scientists to "reprogram" adult stem cells into embryonic-like stem cells that can differentiate into a wider range of cell types (Takahashi et al., 2007) (see Fig. 3 for an example). As these cells come from adults, there are fewer ethical concerns compared to the use of embryonic stem cells. Additionally, these cells can be used in the same individual from which the cells were taken originally. This decreases the risk that an individual's immune system rejects hiPSCs.

One promising application of stem cells is the treatment of neurodegenerative diseases (e.g., multiple sclerosis, Alzheimer's disease, Parkinson's disease, amyotrophic lateral sclerosis [ALS], etc.) (Rahimi Darehbagh et al., 2024). Because stem cells can regenerate, the goal is to deliver these cells to brain regions to regrow damaged neurons. The Experiment Spotlight below details how one type of stem cell can be used to treat symptoms associated with multiple sclerosis.

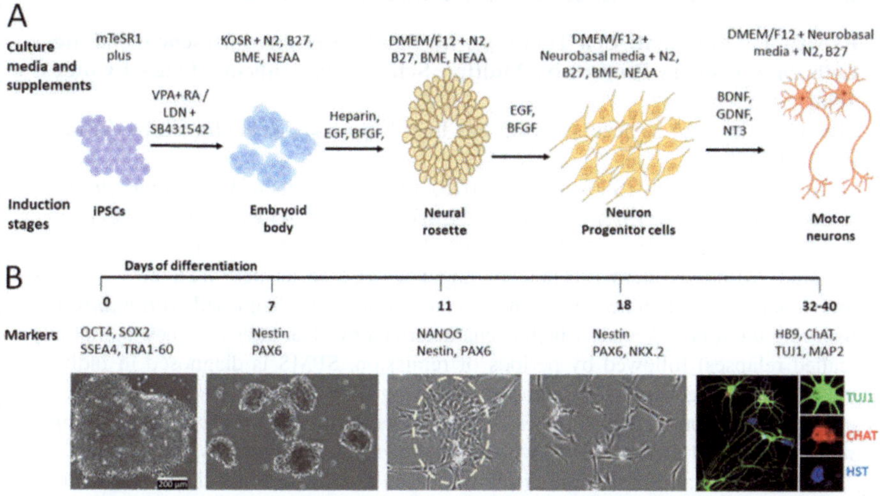

Fig. 3 (**a**) Schematic detailing how human-induced pluripotent stem cells (hiPSCs; abbreviated as iPSCs above) can be used to generate motor neurons. First, cells are derived from a human donor and are exposed to mTeSR1 plus, a type of culture medium. Valproic acid (VPA) and all-trans-retinoic acid (RA) are added, which help the hiPSCs differentiate into neurons. At this point, embryoid bodies form, which are aggregates of pluripotent stem cells. These cells are placed in a medium that contains knockout serum replacement (KOSR) (provides essential nutrients and growth factors), nitrogen (N2), B27 (a supplement that contains lipids and other nutrients), BME (β-mercaptoethanol; helps protect the cells against oxidative stress), and NEAA (non-essential amino acid; adds a mixture of amino acids). Heparin, epidermal growth factor (EGF), and basic fibroblast growth factor (BFGF) are added to help ensure stem cell growth and survival. The embryoid bodies then form into a neural rosette; during early development, the neural rosette forms along the length of the *neural tube*, which gives rise to the brain and the spinal cord during embryonic development. The neural rosette is placed in a medium that contains Dulbecco's modified eagle medium (DMEM) and Ham's nutrient mixture F-12 (simply referred to as F12 above), N2, B27, BME, and NEAA. DMEM and F12 provide the cells glucose, amino acids, vitamins, and nutrients. Once again, EGF and BGFG are added. At this point, neuron progenitor cells begin to form. These cells are placed in the same medium as the neural rosette, but neurobasal medium is added. This medium contains glucose, sodium pyruvate (important energy source), amino acids, vitamins, and inorganic salts. Brain-derived neurotrophic factor (BDNF), glial cell line-derived neurotrophic factor (GDNF), and neurotrophin-3 (NT3) are added to help the progenitor cells differentiate into neurons. The resulting motor neurons are kept in a medium containing DMEM/F12, neurobasal medium, N2, and B27. (**b**) A timeline showing the steps of hiPSC differentiation into motor neurons, as well as real images showing each step of differentiation. OCT4 (octamer-binding transcription factor 4) and SOX2 (SRY [sex determining region Y]-box 2) are transcription factors, and SSEA4 (stage-specific embryonic antigen 4) and TRA1-60 (tumor resistance-associated antigen 1-60) are cell surface antigens associated with human embryonic stem cells. Nestin is a filament protein primarily expressed in developing neural progenitor cells, and PAX6 (paired box 6) is a gene that plays a crucial role in the development of several organs. NANOG is a transcription factor that maintains pluripotency in stem cells; the name comes from Tír na nÓg, a mythological land of eternal youth. NKX.2 (NK2 homeobox) is a transcription factor that is important for cell differentiation and development. HB9 is a transcription factor associated with motor neurons. ChAT is choline acetyltransferase. TUJ1 is the clone name of the monoclonal antibody that recognizes beta-tubulin III, an important protein involved in microtubules. Similar to TUJ1, MAP2 (microtubule-associated protein 2) is important for microtubules. *Source* Figure 1 of Akter and Ding (2022), https://doi.org/10.3390/cells11233796. CC BY 4.0 (https://creativecommons.org/licenses/by/4.0/), via MDPI

Experiment Spotlight: Cell Therapy with Placenta-Derived Mesenchymal Stem Cells for Secondary Progressive Multiple Sclerosis Patients in a Phase 1 Clinical Trial

The full article can be accessed with this link: https://pmc.ncbi.nlm.nih.gov/articles/PMC12062319/

In the present study (Shokati et al., 2025), five individuals with *secondary progressive multiple sclerosis (SPMS)* between the ages of 17 and 45 were recruited to participate in a phase I clinical trial. I first need to define a couple of terms before describing this study. Multiple sclerosis is not a homogenous disease, as there are a few variations individuals can experience. Most individuals are initially diagnosed with *relapsing–remitting multiple sclerosis*, which is characterized by clear attacks of new symptoms (called relapses) followed by periods of remission. SPMS is diagnosed in individuals initially diagnosed with relapsing–remitting multiple sclerosis when symptoms progressively worsen over time. SPMS contrasts with *primary progressive multiple sclerosis*, marked by progressive worsening of symptoms as soon as they appear.

Next, I need to describe the different phases of clinical trials. The present study is considered a phase I clinical trial. This type of clinical trial represents the first time a specific treatment is being tested in a group of individuals. As such, a small sample size is used (hence why a total of five participants was used in this study). The primary purpose of a phase I clinical trial is to determine the safety of the treatment and to identify the dosing that can be used in individuals. If the trial is successful, the treatment can be used with larger samples. At this point, the goal is to determine how effective the treatment is, and safety is continuously assessed. If a treatment passes the third phase of clinical trials, it is approved by a regulatory agency such as the Food and Drug Administration (FDA) in the United States. Even after being approved, the treatment is tested to ensure its safety in a wide range of individuals.

The purpose of this phase I clinical trial was to determine the safety and tolerability of *mesenchymal stem cells (MSCs)* for the treatment of SPMS. MSCs are multipotent stem cells, meaning that they can differentiate into other cell types but are more limited compared to pluripotent stem cells. Typically derived from bone marrow, MSCs can also be extracted from adipose tissue (I will come back to this later), the third molar, amniotic fluid, menstrual blood, and umbilical cord tissue. In fact, this study used MSCs derived from the placenta of individuals undergoing a cesarean section.

Participants received an intravenous infusion of placenta-derived MSCs (3×10^6 cells/kg) across a 15-min period. Participants were closely monitored for 24 h to ensure they did not experience any adverse reactions as a result of the treatment. Safety and tolerability of the treatment continued to be monitored across a 6-month period. Although the primary purpose of a phase I clinical trial is to assess the safety of a treatment, the researchers wanted to measure secondary outcomes related to how well MSC infusion treats symptoms of SPMS.

Cognitive functions were tested with the *minimal assessment of cognitive functions in MS (MACFIMS)*; the following were measured: verbal and memory deficits, processing and motor speed, cognitive flexibility, calculation ability, visuospatial ability, verbal fluency, and executive functions. Psychological assessments were also performed to measure participants' symptoms related to psychoticism, OCD, paranoia, anxiety, hostility, depression, and *somatization* (i.e., physical symptoms that have no underlying physical cause). Fatigue was also measured in participants. The expanded disability status scale (EDSS) was used to assess disability in participants. This scale is specific to individuals with multiple sclerosis and allows one to determine how severe one's symptoms are and to monitor disease progression.

Magnetic resonance imaging (MRI) was used to perform diffusion tensor imaging (DTI) to measure white matter tracts in the brain. fMRI was also used to examine connectivity across brain regions. Finally, blood samples were taken to quantify cytokine and CD-19/CD-20 B-cell marker levels. Cytokines are involved in inflammation, and B cells are white blood cells that produce antibodies. Collectively, cytokines and B cells are important components of our immune system.

No serious adverse events were reported throughout the experiment, although two participants complained of a mild headache following MSC infusion. One month following treatment, EDSS scores decreased in all five participants. For three of the participants, no further decreases in EDSS scores were observed when tested 3 months and 6 months following treatment. For two of the participants, EDSS scores decreased between months 1 and 3. Visuospatial learning and executive functioning improved following treatment, with a corresponding decrease in hostility, fatigue, somatization, and paranoia.

MSC treatment did not cause any brain morphological changes, but fMRI results indicated increased connectivity in the brain, particularly between structures of the DMN in the left hemisphere, the left parietal cortex, and the right parahippocampal formation. DTI results indicated decreased *radial diffusivity*, indicating repaired white matter tracts. The anti-inflammatory cytokine interleukin-10 (IL-10) significantly increased while pro-inflammatory cytokines like tumor necrosis factor-alpha (TNF-α) and IL-6 decreased (with the exception of one participant) following MSC infusion. CD-19/CD-20 B-cell marker levels also decreased following treatment.

Questions to Consider:

1. One thing I did not mention is that all participants were taking rituximab at the time they received the MSC infusion. Rituximab is a monoclonal antibody used to treat autoimmune diseases (like rheumatoid arthritis) and certain cancers like non-Hodgkin's lymphoma and leukemia; however, it is used "off label" for the treatment of multiple sclerosis. How can the use of rituximab act as a confounding variable that makes interpretation of the present results difficult?
2. While the primary purpose of the present study was to determine the safety of placenta-derived MSC treatment, the researchers examined clinically relevant outcomes. What is a major limitation of reporting such clinical data in the present study?
3. Based on the results of the study, provide an explanation as to how placenta-derived MSCs can ameliorate symptoms associated with SPMS.

Another potential use for stem cells is treating obesity (Payab et al., 2018), although most research to date is limited to preclinical studies. Adipose-derived mesenchymal stem cells (ADSCs) are derived from fat tissue. Recall from Chap. 13 that we have white adipose tissue and brown adipose tissue. As brown adipose tissue burns energy, one goal is to use stem cells to convert white adipose tissue into brown adipose tissue. ADSCs are also beneficial because obesity leads to epigenetic and gene expression changes in MSCs associated with inflammation (Xing et al., 2025). This is related to metaflammation I discussed in Chap. 13.

Comprehension Check 3 Distinguish between pluripotent stem cells and hiPSCs, outlining two distinct advantages of using hiPSCs for therapeutic purposes relative to pluripotent stem cells.

Pluripotent stem cells are a type of stem cell found in 3–5-day old embryos, often from in vitro *fertilization clinics. They are the most versatile, as they can differentiate into all cell types in a developing fetus. hiPSCs are adult stem cells that have been "reprogrammed" to be embryonic-like. They can also differentiate into a wider range of cell types. hiPSCs are associated with fewer ethical concerns as they come from adult tissue as opposed to embryonic tissue. hiPSCs can be used in the same individual from which the cells were originally taken. This significantly decreases the risk that the individual's immune system will reject the cells, a major challenge with other stem cell treatments.*

6 Pharmacogenomics

Throughout this book, you have learned about pharmacotherapies that are used to treat a wide range of psychiatric conditions—stimulants for ADHD, serotoninergic drugs for anxiety and mood disorders, L-DOPA for Parkinson's disease, and so on. Treating these disorders is difficult due to the variability in symptom improvement observed across individuals. For example, some children with ADHD respond well to methylphenidate while others show no improvements in their symptoms. An emerging field of study is **pharmacogenomics**, which considers how one's genes influence responses to medications. *Pharmacogenetics* is a closely related term, but it focuses on how single genes impact the pharmacokinetic and pharmacodynamic profiles of drugs.

Returning to methylphenidate, there is evidence that a single nucleotide polymorphism (SNP) of the *SYT1* gene is associated with improvements in inattention and treatment persistence over time (da Silva et al., 2018). This gene encodes the synaptotagmin-1 protein, which is important for synaptic vesicles; this protein detects the presence of calcium (Ca^{++}) in the presynaptic terminal button, which leads to exocytosis. Essentially, individuals may not respond as well to methylphenidate due to alterations in the synaptotagmin-1 protein. If this protein does not work as effectively as normal, vesicles may not readily release dopamine and norepinephrine into the synapse. In that event, methylphenidate, which inhibits transporters for catecholamines, is not going to be beneficial for individuals with ADHD. SNPs of other genes such as the *ADRA2A* (alpha-2_A adrenergic receptor), *COMT* (catechol-O-methyltransferase), and *SLC6A2* (norepinephrine transporter) are associated with methylphenidate responses in children (Myer et al., 2018).

Pharmacogenomic approaches have been studied in relation to anxiety and mood disorders (Kleine Schaars & van Westrhenen, 2023; McGowan, 2020), with attention focused on genes related to monoaminergic systems (primarily serotonin) and the corticotropic releasing hormone receptor (part of our stress system), although genes related to cytochrome P450 enzymes have been examined in relation to mood disorders. Cytochrome P450 enzymes are largely found in the liver and are primarily involved in the metabolism of drugs, and there is evidence that polymorphisms of two specific cytochrome P450 enzymes are related to treatment outcomes for antidepressant medications.

One cytochrome P450 enzyme allele (*CYP2D610*), common in East Asian populations, is associated with improved treatment outcomes for Alzheimer's disease following donepezil treatment (Climacosa et al., 2025). Donepezil is an acetylcholinesterase inhibitor, thus preventing the metabolism of acetylcholine. Variations to the *APOE* gene are also associated with treatment responses to acetylcholinesterase inhibitors and N-methyl-D-aspartate (NMDA) receptor antagonists (Climacosa et al., 2025). Recall from Chap. 6 that the *APOE* gene encodes apolipoprotein E, a protein involved in lipid transport. Individuals with the ε2 and the ε3 alleles respond better to these treatments whereas those with the ε4 allele show poorer outcomes.

Genetics can influence how an individual responds to addictive substances, as well as to treatments for substance use disorders. I will provide just a few examples here. Smoking behaviors, including the age of initiation, trajectory to regular smoking, the number of cigarettes smoked per day, and dependence, are highly associated with the *CYP2A6* gene (encodes a cytochrome P450 enzyme that metabolizes nicotine) and the *CHRNA5-A3-B4* gene cluster (encodes acetylcholine nicotinic receptor subunits) (El-Boraie & Tyndale, 2021). Variations to the *CYP2A6* gene alter how quickly individuals metabolize nicotine. "Slow metabolizers" show less nicotine dependence-like behaviors. While individuals with slow nicotine metabolism show similar treatment responses to nicotine replacement therapy and varenicline, those that metabolize nicotine quickly have superior treatment outcomes following varenicline compared to nicotine patches (Chenoweth & Tyndale, 2017).

Individuals with the cytochrome P450 *CYP3A5*3* allele are at increased risk for fentanyl overdose (Petrovitch et al., 2025), and they are at increased risk of developing an opioid use disorder (Freiermuth et al., 2023). Currently, identifying genetic associations with treatment responses for opioid use disorder is difficult as most studies to this point have relied on small sample sizes that have not been reproduced (Crist et al., 2018). Interestingly, individuals with a single nucleotide polymorphism (SNP) of the *OPRM1 gene*, which encodes the mu opioid receptor, are more likely to report a family history of alcohol use disorder and to experience greater subjective experiences of intoxication, stimulation, sedation, and happiness following alcohol administration; differences in the *OPRM1 gene* also influence naltrexone responses for alcohol use disorder (Kranzler & Edenberg, 2010).

Comprehension Check 4 Describe the mechanism by which genetic polymorphisms of the *SYT1* gene are hypothesized to reduce the efficacy of methylphenidate for some individuals with ADHD.

The SYT1 *gene encodes the synaptotagmin-1 protein, which is critical for synaptic vesicle exocytosis. This protein acts as a* Ca^{++} *sensor in the presynaptic terminal, signaling the vesicles to fuse with the cell membrane and release neurotransmitters like dopamine and norepinephrine into the synapse. A SNP of the* SYT1 *gene leads to an alteration in the synaptotagmin-1 protein, making it less effective. If the protein is impaired, the synaptic vesicles may not readily release dopamine and norepinephrine. If the initial release of these neurotransmitters is compromised due to the* SYT1 *gene variation, there will be less dopamine and norepinephrine in the synapse. Consequently, methylphenidate's intended effect of increasing synaptic catecholamine levels will be diminished, resulting in poorer treatment response for some individuals with ADHD.*

7 Chapter Summary

This chapter explored how various neuroscience techniques are used to treat maladaptive behaviors, beginning with neurofeedback, a form of biofeedback that uses real-time brain activity (often EEG or fMRI) to teach self-regulation of neuronal function. More direct methods include deep brain stimulation, a neurosurgical procedure that implants electrodes to modulate neural activity, and TMS, tDCS, and tFUS, which enable clinicians to excite or inhibit specific brain regions. Stem cell treatments use hiPSCs and MSCs to regenerate damaged neurons or generate new neurons. Finally, pharmacogenomics is a field that examines how an individual's genes affect their response to drugs, and examples of how gene variations influence treatment efficacy for disorders such as ADHD, Alzheimer's disease, and substance use disorders were included.

Answers to Experiment Spotlight Questions

1. The present study reported lower CD-19/CD-20 B-cell marker levels following placenta-derived MSC treatment. However, rituximab is known to decrease CD-20 B-cell marker levels, with these effects lasting 6–9 months. Therefore, the observed positive outcomes reported in this study cannot be definitively attributed to the MSC treatment alone. They may be entirely, or in large part, a residual effect of the pre-existing rituximab therapy.
2. No control group was included in this study. Each participant received the MSC infusion. Related to Question 1, the present study is confounded as participants were already receiving rituximab for their symptoms. Ideally, individuals receiving MSC treatment would be compared to a control group that received a placebo

infusion. This would better enable the researchers to claim that MSC administration leads to symptom improvements.

3. Multiple sclerosis is hypothesized to be caused by an immune response that targets the myelin sheath of neurons. By suppressing the immune system (e.g., reducing pro-inflammatory cytokines and CD-19/CD-20 B-cell marker levels), placenta-derived MSCs can slow the progression of symptoms associated with multiple sclerosis. As radial diffusivity decreased following treatment, these results suggest that suppressing the immune system's activity allows myelination to become repaired. This then enables better functional connectivity across brain regions, which can account for improvements in cognitive and physical symptoms following treatment.

Glossary

Deep brain stimulation a neurosurgical procedure that involves implanting electrodes into specific brain regions to deliver controlled electrical impulses, modulating neural activity for therapeutic benefit.

Neurofeedback a type of biofeedback that uses real-time displays of brain activity, most commonly brainwaves, to teach self-regulation of brain function.

Pharmacogenomics the study of how an individual's genes affect their response to drugs.

Stem cell a type of cell that can differentiate into various cell types and can self-renew.

Transcranial focused ultrasound (tFUS) a non-invasive neurotechnology that uses focused sound waves to target specific brain regions through the intact skull.

References

Bluschke, A., Eggert, E., Friedrich, J., Jamous, R., Prochnow, A., Pscherer, C., Schreiter, M. L., Teufert, B., Roessner, V., & Beste, C. (2022). The effects of different theta and beta neurofeedback training protocols on cognitive control in ADHD. *Journal of Cognitive Enhancement, 6*(4), 463–477. https://doi.org/10.1007/s41465-022-00255-6

Chenoweth, M. J., & Tyndale, R. F. (2017). Pharmacogenetic optimization of smoking cessation treatment. *Trends in Pharmacological Sciences, 38*(1), 55–66. https://doi.org/10.1016/j.tips.2016.09.006

Chou, T., Deckersbach, T., Guerin, B., Sretavan Wong, K., Borron, B. M., Kanabar, A., Hayden, A. N., Long, M. P., Daneshzand, M., Pace-Schott, E. F., & Dougherty, D. D. (2024). Transcranial focused ultrasound of the amygdala modulates fear network activation and connectivity. *Brain Stimulation, 17*(2), 312–320. https://doi.org/10.1016/j.brs.2024.03.004

Climacosa, F. M. M., Ornos, E. D. B., Gapaz, N. C. L. L., Guantia, M. G. R., Cruz, J. M. C., Manalo, R. V. M., Yu, M. L., Qureshi, A. P., Carampel, A. C., Asis, J. L. B., Reyes, J. C. B., Dacasin, A. B., & Anlacan, V. M. M. (2025). The role of genetic polymorphisms on drug response in

Alzheimer's disease: A systematic review. *BMC Medical Genomics, 18*(1), 154. https://doi.org/10.1186/s12920-025-02225-1

Crist, R. C., Clarke, T. K., & Berrettini, W. H. (2018). Pharmacogenetics of opioid use disorder treatment. *CNS Drugs, 32*(4), 305–320. https://doi.org/10.1007/s40263-018-0513-9

da Silva, B. S., Cupertino, R. B., Rovaris, D. L., Schuch, J. B., Kappel, D. B., Müller, D., Bandeira, C. E., Victor, M. M., Karam, R. G., Mota, N. R., Rohde, L. A., Contini, V., Grevet, E. H., & Bau, C. H. D. (2018). Exocytosis-related genes and response to methylphenidate treatment in adults with ADHD. *Molecular Psychiatry, 23*(6), 1446–1452. https://doi.org/10.1038/mp.2017.90

Elbogen, E. B., Alsobrooks, A., Battles, S., Molloy, K., Dennis, P. A., Beckham, J. C., McLean, S. A., Keith, J. R., & Russoniello, C. (2021). Mobile neurofeedback for pain management in veterans with TBI and PTSD. *Pain Medicine, 22*(2), 329–337. https://doi.org/10.1093/pm/pnz269

El-Boraie, A., & Tyndale, R. F. (2021). The role of pharmacogenetics in smoking. *Clinical Pharmacology and Therapeutics, 110*(3), 599–606. https://doi.org/10.1002/cpt.2345

Frank, D. L., Khorshid, L., Kiffer, J. F., Moravec, C. S., & McKee, M. G. (2010). Biofeedback in medicine: Who, when, why and how? *Mental Health in Family Medicine, 7*(2), 85–91.

Freiermuth, C. E., Kisor, D. F., Lambert, J., Braun, R., Frey, J. A., Bachmann, D. J., Bischof, J. J., Lyons, M. S., Pantalon, M. V., Punches, B. E., Ancona, R., & Sprague, J. E. (2023). Genetic variants associated with opioid use disorder. *Clinical Pharmacology and Therapeutics, 113*(5), 1089–1095. https://doi.org/10.1002/cpt.2864

Kleine Schaars, K., & van Westrhenen, R. (2023). Pharmacogenomics and the management of mood disorders—A review. *Journal of Personalized Medicine, 13*(7), 1183. https://doi.org/10.3390/jpm13071183

Kranzler, H. R., & Edenberg, H. J. (2010). Pharmacogenetics of alcohol and alcohol dependence treatment. *Current Pharmaceutical Design, 16*(19), 2141–2148. https://doi.org/10.2174/138161210791516387

Krauss, J. K., Lipsman, N., Aziz, T., Boutet, A., Brown, P., Chang, J. W., Davidson, B., Grill, W. M., Hariz, M. I., Horn, A., Schulder, M., Mammis, A., Tass, P. A., Volkmann, J., & Lozano, A. M. (2021). Technology of deep brain stimulation: Current status and future directions. *Nature Reviews. Neurology, 17*(2), 75–87. https://doi.org/10.1038/s41582-020-00426-z

Lieberman, J. M., Rabellino, D., Densmore, M., Frewen, P. A., Steyrl, D., Scharnowski, F., Théberge, J., Neufeld, R. W. J., Schmahl, C., Jetly, R., Narikuzhy, S., Lanius, R. A., & Nicholson, A. A. (2023). Posterior cingulate cortex targeted real-time fMRI neurofeedback recalibrates functional connectivity with the amygdala, posterior insula, and default-mode network in PTSD. *Brain and Behavior, 13*(3), Article e2883. https://doi.org/10.1002/brb3.2883

Mahoney, J. J., Haut, M. W., Carpenter, J., Ranjan, M., Thompson-Lake, D. G. Y., Marton, J. L., Zheng, W., Berry, J. H., Tirumalai, P., Mears, A., D'Haese, P., Finomore, V. S., Hodder, S. L., & Rezai, A. R. (2023). Low-intensity focused ultrasound targeting the nucleus accumbens as a potential treatment for substance use disorder: Safety and feasibility clinical trial. *Frontiers in Psychiatry, 14*, 1211566. https://doi.org/10.3389/fpsyt.2023.1211566

Marzbani, H., Marateb, H. R., & Mansourian, M. (2016). Neurofeedback: A comprehensive review on system design, methodology and clinical applications. *Basic and Clinical Neuroscience, 7*(2), 143–158. https://doi.org/10.15412/J.BCN.03070208

McGowan, O. O. (2020). Pharmacogenetics of anxiety disorders. *Neuroscience Letters, 726*, Article 134443. https://doi.org/10.1016/j.neulet.2019.134443

Moussawi, K., Kim, M. J., Baybayan, S., Wood, M., & Mills, K. A. (2022). Deep brain stimulation effect on anterior pallidum reduces motor impulsivity in Parkinson's disease. *Brain Stimulation, 15*(1), 23–31. https://doi.org/10.1016/j.brs.2021.11.006

Myer, N. M., Boland, J. R., & Faraone, S. V. (2018). Pharmacogenetics predictors of methylphenidate efficacy in childhood ADHD. *Molecular Psychiatry, 23*(9), 1929–1936. https://doi.org/10.1038/mp.2017.234

Ölçüoğlu, R. (2025). Neurofeedback for ADHD: Exploring the role of quantitative EEG and brain-wave modulation. *Brain and Behavior, 15*(8), Article e70714. https://doi.org/10.1002/brb3.70714

Or-Borichev, A., Lerner, Y., Hamrani, Y., Gurevitch, G., Mor, N., Doron, M., Sarna, N., Ablin, J. N., Hendler, T., & Sharon, H. (2025). Targeted limbic self-neuromodulation for alleviating central sensitization symptoms in fibromyalgia. *BMC Medicine, 23*(1), 304. https://doi.org/10.1186/s12916-025-04138-3

Payab, M., Goodarzi, P., Foroughi Heravani, N., Hadavandkhani, M., Zarei, Z., Falahzadeh, K., Larijani, B., Rahim, F., & Arjmand, B. (2018). Stem cell and obesity: Current state and future perspective. *Advances in Experimental Medicine and Biology, 1089*, 1–22. https://doi.org/10.1007/5584_2018_227

Petrovitch, D., Himes, K. P., Bischof, J. J., Braun, R. S., Brown, J. L., Eleda, I. C., Freiermuth, C. E., Gu, S., Hall, O. T., Johnson, J. A., Kisor, D. F., Lambert, J. W., Lyons, M. S., Maloney, M. V., Punches, B. E., Quarles, E., Littlefield, A. K., & Sprague, J. E. (2025). Scoping review of associations between cytochrome P450 3A4/5 single nucleotide polymorphisms and risk factors for fentanyl overdose. *Pharmacogenomics, 26*(10–12), 391–425. https://doi.org/10.1080/14622416.2025.2562796

Rahimi Darehbagh, R., Seyedoshohadaei, S. A., Ramezani, R., & Rezaei, N. (2024). Stem cell therapies for neurological disorders: Current progress, challenges, and future perspectives. *European Journal of Medical Research, 29*(1), 386. https://doi.org/10.1186/s40001-024-01987-1

Richter, K., Kellner, S., & Licht, C. (2023). RTMS in mental health disorders. *Frontiers in Network Physiology, 3*, Article 943223. https://doi.org/10.3389/fnetp.2023.943223

Sakel, M., Ozolins, C. A., Saunders, K., & Biswas, R. (2025). A home-based EEG neurofeedback treatment for chronic neuropathic pain—A pilot study. *Frontiers in Pain Research, 6*, 1479914. https://doi.org/10.3389/fpain.2025.1479914

Sanguinetti, J. L., Hameroff, S., Smith, E. E., Sato, T., Daft, C. M. W., Tyler, W. J., & Allen, J. J. B. (2020). Transcranial focused ultrasound to the right prefrontal cortex improves mood and alters functional connectivity in humans. *Frontiers in Human Neuroscience, 14*, 52. https://doi.org/10.3389/fnhum.2020.00052

Shamir, R. R., Noecker, A., & McIntyre, C. C. (2014). Deep brain stimulation. *Frontiers for Young Minds, 2*(12). https://doi.org/10.3389/frym.2014.00012

Shi, Y., & Wu, W. (2025). Advances in transcranial focused ultrasound neuromodulation for mental disorders. *Progress in Neuro-Psychopharmacology & Biological Psychiatry, 136*, Article 111244. https://doi.org/10.1016/j.pnpbp.2024.111244

Shokati, A., Nikbakht, M., Sahraian, M. A., Saeedi, R., Asadollahzadeh, E., Rezaeimanesh, N., Chahardouli, B., Gharaylou, Z., Mousavi, S. A., Ai, J., & Naser Moghadasi, A. (2025). Cell therapy with placenta-derived mesenchymal stem cells for secondary progressive multiple sclerosis patients in a phase 1 clinical trial. *Scientific Reports, 15*(1), 16005. https://doi.org/10.1038/s41598-025-00590-6

Takahashi, K., Tanabe, K., Ohnuki, M., Narita, M., Ichisaka, T., Tomoda, K., & Yamanaka, S. (2007). Induction of pluripotent stem cells from adult human fibroblasts by defined factors. *Cell, 131*(5), 861–872. https://doi.org/10.1016/j.cell.2007.11.019

Tortella, G., Casati, R., Aparicio, L. V., Mantovani, A., Senço, N., D'Urso, G., Brunelin, J., Guarienti, F., Selingardi, P. M., Muszkat, D., Junior, B.deS., Valiengo, L., Moffa, A. H., Simis, M., Borrione, L., & Brunoni, A. R. (2015). Transcranial direct current stimulation in psychiatric disorders. *World Journal of Psychiatry, 5*(1), 88–102. https://doi.org/10.5498/wjp.v5.i1.88

Wei, L., Yan, W., Shah, W., Zhang, Z., Wang, M., Liu, B., Xue, Z., Cao, Y., Hou, X., Zhang, K., Yan, B., & Wang, X. (2024). Advancements and challenges in stem cell transplantation for regenerative medicine. *Heliyon, 10*(16), Article e35836. https://doi.org/10.1016/j.heliyon.2024.e35836

Xing, A. L., Zhu, X., Klomjit, N., Bo, L., Al Saeedi, M., Lerman, A., Eirin, A., & Lerman, L. O. (2025). Human obesity decreases the anti-inflammatory functionality of adipose tissue-derived mesenchymal stem/stomal cells by upregulating IL-1β expression. *Stem Cells*, sxaf058. Advance online publication. https://doi.org/10.1093/stmcls/sxaf058

Zhai, Z., Ren, L., Song, Z., Xiang, Q., Zhuo, K., Zhang, S., Li, X., Zhang, Y., Jiao, X., Tong, S., Sun, J., & Liu, D. (2023). The efficacy of low-intensity transcranial ultrasound stimulation on negative symptoms in schizophrenia: A double-blind, randomized sham-controlled study. *Brain Stimulation, 16*(3), 790–792. https://doi.org/10.1016/j.brs.2023.04.021

Zhang, J., Tusuzian, E., Morfini, F., Bauer, C. C. C., Stone, L., Awad, A., Shinn, A. K., Niznikiewicz, M. A., & Whitfield-Gabrieli, S. (2025). Brain structural and functional neuroimaging features are associated with improved auditory hallucinations in patients with schizophrenia after real-time fMRI neurofeedback. *Depression and Anxiety, 2025*, 2848929. https://doi.org/10.1155/da/2848929

Zhou, D., Li, X., Wei, S., Yu, C., Wang, D., Li, Y., Li, J., Liu, J., Li, S., Zhuang, W., Li, Y., Luo, R., Liu, Z., Liu, J., Xu, Y., Fan, J., Zhu, G., Xu, W., Tang, Y., … Zhang, X. Y. (2024). Transcranial direct current stimulation combined with repetitive transcranial magnetic stimulation for depression: A randomized clinical trial. *JAMA Network Open, 7*(11), Article e2444306. https://doi.org/10.1001/jamanetworkopen.2024.44306

Appendix
What's in a Name?

The purpose of this appendix is to provide readers with the etymology of various terms used throughout this book. In other words, why are the things that you read about in this book called what they are?

Etymology of Brain Regions/Structures

There are terms that denote directionality within the nervous system.

- **Anterior**—Latin for "former", with the Latin term *ante* meaning "before"; indicates that one brain region is located in front of another brain region.
- **Caudal**—derived from the Latin word *caudalis*, which is derived from the Latin word *cauda* ("tail"); indicates that a structure is located toward the back (or tail-end) of a region.
- **Dorsal—**derived from the Latin word *dorsum*, meaning "back"; this term indicates that a structure is located toward an animal's back. Because the nervous system in biped animals bends, brain structures that are dorsal are near the top of the head.
- **Inferior**—derived from the Latin word *inferus*, meaning "low"; this term indicates a brain structure that is located beneath another structure.
- **Inter-—**Latin for "beside", this term indicates a structure that is located beside another structure.
- **Intra-—**Latin for "inside", this term indicates a structure located within another structure.
- **Lateral**—derived from the Latin word *lateralis*, which itself comes from the Latin word *latus* ("side"); indicates that something is located near the outer edge.
- **Medial**—derived from the Latin word *medialis*, which itself comes from the Latin word *medius* ("middle"); indicates that a structure is located in the middle of a specific region.
- **Posterior**—Latin for "following", with the Latin term *post* meaning "after"; this term is often used interchangeably with caudal.

J. Yates, *Neuroscience of Maladaptation*, https://doi.org/10.1007/978-3-032-15602-0

- **Pre**—*pre* is a Latin word that is derived from *prae*, meaning "before in time or place"; this term is often used to indicate brain structures that are located in the anterior portion of a brain region.
- **Rostral**—derived from the Latin word *rostralis*, which is derived from the Latin word *rostrum* ("beak, nose"); this term is used interchangeably with the word anterior.
- **Sub-**—Latin for "under" or "below", this term has a similar meaning as inferior.
- **Superior**—derived from the Latin word *superiōrem*, meaning "higher" or "upper"; this term indicates a brain structure that is located above another structure.
- **Supra-**—Latin for "over" or "above", this term has a similar meaning as superior.
- **Ventral**—derived from the Latin word *venter*, meaning "stomach"; this term indicates that a structure is located toward an animal's stomach. Because the nervous system in biped animals bends, brain structures that are ventral are near the bottom of the head.

The terms above can be combined to provide enhanced specificity (e.g., dorsomedial or ventrolateral). With this knowledge, I will not identify each brain region below as you can use the information above to determine the meaning of many brain regions covered in this book.

The major divisions of the brain end with -cephalon, which is derived from the Greek word *enkephalos*, meaning "brain".

- **Prosencephalon**—the term *pros* is derived from the Greek word *proso*, meaning "forward". This is the front part of the brain.
  - **Telencephalon**—the term *tel* is derived from the Greek word *telos*, meaning "end"; name comes from the fact that this is evolutionarily the last portion of the brain to develop.

    Arcuate fasciculus—Latin for "curved bundle", this term describes the curved shape these myelinated axons take to transmit information between language-processing brain structures. See the section Etymology of Terms Related to Neurons to learn about the etymology of the terms "myelin" and "axon".

    Basal forebrain—this part of the brain is located at the base of the forebrain.

    - **Nucleus accumbens (NAc)**—derived from the Latin word *accumbent* ("to lie against" or "to recline"), this collection of cell bodies appears to recline on an adjacent brain structure.

    Basal ganglia—the basal ganglia are located in the basal part of the forebrain (that is, they sit at the base of the forebrain). The term ganglia is a misnomer here as ganglia are found in the peripheral nervous system (PNS) only. The correct term would be basal nuclei to reflect that this part of the brain is composed of different collections of cell bodies.

- **Caudate nucleus**—derived from the Latin word *cauda*, this part of the basal ganglia has a tail-like extension.
- **Globus pallidus**—Latin for "pale globe", this part of the brain gets its name due to the high number of myelinated axons located here, causing it to have a pale appearance.
  - **Ventral pallidum (VP)**—indicates that this structure makes up the bottom portion of the globus pallidus; specifically, it makes up the bottom portion of the internal capsule of the globus pallidus.
- **Lentiform nucleus**—derived from the Latin word *lens* or *lentis* ("lens") and the Latin suffix *-form* ("shape" or "form"), the globus pallidus and the putamen look like a lens when viewed together.
- **Putamen**—Latin for "shell", this part of the brain forms an outer shell with another part of the brain known as the lentiform nucleus.
- **Striatum**—derived from the Latin word *striatus* ("striped"), this part of the brain gets its name from the striped appearance of the caudate and the putamen.

Cerebral cortex—derived from the Latin words *cerebrum* ("brain") and *cortex* ("bark"), this part of the brain is somewhat analogous to the bark on a tree—that is, the cerebral cortex is the outermost layer of the brain.

- The four cortices of the cerebral cortex are also referred to as lobes; the term lobe is derived from the Greek word *lobos*, meaning "a small, rounded projection".
- **Gyrus**—Latin for "circle", which was derived from the Greek word *guros* ("ring"); these ridges often have a circular or oval shape.
- **Sulcus**—Latin for "groove", term denotes the grooves in the cerebral cortex.
- **Frontal cortex**—indicates that this is the front part of the cerebral cortex.
  - **Broca's area**—named for Pierre Paul Broca, a French surgeon who is credited with identifying that this part of the brain is associated with speech production.
  - **Frontal eye field (FEF)**—indicates that this region is located in the front portion of the brain; this area of the brain was first described by German neurologist Otfrid Förster, who named it the *frontales Augenfelds* (German for frontal eye field).
  - **Frontal operculum**—the word *operculum* is Latin for "lid" or "cover". This part of the frontal cortex covers the insular cortex (see next entry).
  - **Insular cortex (or insula)**—the word *insula* is Latin for "island". This region gets its name from the fact that it is hidden deep within the cerebral cortex, like a small island surrounded by the ocean.
  - **Motor cortex**—*motor* is Latin for "movement"; this part of the brain is important for initiating movement.

- **Olfactory sulcus**—olfactory is derived from the Latin word *olfactus* ("the sense of smell"); this part of the brain is located in a groove in the ventral portion of the brain that contains the olfactory bulb and the olfactory tract, which transmit information regarding our sense of smell.
- **Orbitofrontal cortex (OFC)**—the term *orbito* refers to the *orbits*, the cavities in which the eyes sit; this part of the brain is located in the frontal cortex just above our eye sockets.
- **Piriform cortex**—derived from the Latin words *pirum* ("pear") and *formis* ("shaped"), this region is pear-shaped.
- **Prefrontal cortex (PFC)**—this is the most anterior portion of the frontal cortex.

    Infralimbic cortex—*infra* is Latin for "below", and limbic refers to the limbic system; this part of the medial PFC (mPFC) is located below the limbic system.
    Prelimbic cortex—combines *pre* ("before") and limbic system; this part of the mPFC is located in front of the limbic system.

- **Supplementary motor area (SMA)**—supplementary is derived from the Latin word *supplementum* ("that which fills up or makes something complete"); this part of the brain acts as a supplement to the motor cortex.

Occipital cortex—the term occipital is derived from the Latin words *ob* ("behind") and *caput* ("head"); indicates that this region is the back part of the cerebral cortex.

- **Cuneus**—Latin for "wedge"; when viewed from the side, this part of the brain is wedge-shaped.
- **Lingual gyrus**—derived from the Latin word *lingua*, meaning "tongue"; this region is shaped like a tongue.
- **Occipital face area**—part of the occipital cortex involved in face perception.
- **Optic radiations**—the English term *radiations* refers to the spreading of rays; the neurons projecting from the thalamus to the primary visual cortex resemble rays spreading out. Note, the optic radiations begin in the thalamus and travel through part of the temporal cortex before reaching the occipital cortex. See below for the etymology of the thalamus, the visual cortex, and the temporal cortex.
- **Posterior fossa**—*fossa* is Latin for "ditch" or "trench"; this name is used to describe the cranial space at the back of the skull.
- **Visual cortex**—visual is derived from the Latin word *visualis* ("of sight"); this part of the brain is largely responsible for processing visual information.

Parietal cortex—the term parietal is derived from the Latin word *paries*, meaning "wall"; this part of the brain is located beneath the parietal bone of the skull. The parietal bones are located on the sides of the head, effectively creating a wall around the brain.

- **Angular gyrus**—name is derived from its position relative to the angular sulcus, which gets its name from its corner-like shape as it curves upward. Angular is derived from the Latin word *angulāris* ("having corners or angles").
- **Somatosensory cortex**—somatosensory is derived from the Greek word *sōma* ("body") and the Latin word *sensorium* ("pertaining to sensation"). This part of the brain processes information regarding sensation.
- **Somatosensory operculum**—indicates that this part of the somatosensory cortex covers the insula.
- **Supramarginal (SM) gyrus**—marginal comes from the Greek word *margo* ("edge"). This part of the brain sits just above (*supra* is Latin for above) the lateral sulcus, which acts as a boundary between the temporal cortex and the frontal/parietal cortices.

Temporal cortex—the term temporal is derived from the Latin word *tempus* ("time"). This part of the brain is located beneath the temporal bone of the skull, near the temples. The name temporal was chosen because males often first get gray hairs in the temporal portion of the head, signifying the passage of time.

- **Anterior temporal face area**—located in the front portion of the temporal cortex, this area of the brain processes information about faces.
- **Auditory cortex**—auditory is derived from the Latin word *auditorious* ("pertaining to hearing"); this part of the brain primarily processes auditory information.
- **Inferior temporal cortex (IT)**—indicates that this part of the brain is located at the bottom portion of the temporal cortex.
  - **Fusiform gyrus**—derived from the Latin word fusus ("spindle"), this part of the brain has a spindle-like shape. Note, this part of the brain expands into the occipital lobe.

    Fusiform face area (FFA)—part of the fusiform gyrus that helps process facial features.
- **Temporal pole**—in neuroanatomy, a pole is a pointed/rounded end of a cerebral hemisphere; thus, the temporal pole is the most anterior portion of the temporal cortex. Pole is derived from the Latin *polus* and the Greek *polos* ("axis of rotation").
- **Wernicke's area**—named for Carl Wernicke, a German neurologist who is credited with identifying the brain region associated with language comprehension.

Temporoparietal junction (TPJ)—part of the brain where the temporal and the parietal cortices meet.
Corpus callosum—Latin for "hard body"; name comes from the fact that the corpus callosum is a thick ("hard") bundle of myelinated axons.
Limbic—derived from the Latin word *limbus* ("border" or "edge"), this region acts as a border between the cerebral cortex and subcortical brain structures.

- **Amygdala**—derived from the Greek word *amygdalē* and the Latin word *amygdalum* (both meaning "almond"), this structure is almond-shaped.
  - **Basolateral nucleus (BLA)**—the term "basolateral" indicates that this structure is located on the lateral edge of the base of the amygdala.
  - **Central nucleus (CeA)**—indicates that this structure is located in the middle portion of the amygdala.
- **Cingulate cortex**—derived from the Latin word *cingulum* ("belt"), the cingulate cortex looks like a belt that wraps around the corpus callosum.
  - **Subgenual anterior cingulate cortex (sgACC)—***sub* is Latin for "under", and *genu* is Latin for "knee"; this part of the ACC is located below where the anterior portion of the corpus callosum bends.
- **Hippocampus**—derived from the Greek word *hippokampos* ("sea-horse"), this brain structure resembles a seahorse.
  - **Entorhinal cortex**—derived from the Greek term *entos* ("within" or "inside") and the Latin term *rhinal* ("nose"), this structure is located within the olfactory sulcus.
  - **Ventral subiculum**—*subiculum* is Latin for "support". Because this region is located in the ventral portion of the hippocampus, the subiculum resembles a foundation that provides structural support for the hippocampus.
- **Parahippocampal cortex—***para* is Greek for "beside"; this brain structure sits next to the hippocampus (see entry above).

– **Diencephalon**—*die* is derived from the Greek words *dia*, meaning "through"; this part of the brain connects the cerebral cortex to other parts of the nervous system.

Epithalamus—*epi* is Greek for "over". This structure is located above the thalamus.

- **Habenula**—derived from the Latin word *habena* ("rein"), *habenula* translates to "little rein"; this structure's elongated shape resembles a rein.
- **Pineal gland**—derived from the Latin word *pinea* ("pine cone"), the pineal gland resembles a pine cone.

Hypothalamus—this structure is located just below (*hypo*) the thalamus.

- **Arcuate nucleus**—derived from the Latin word *arcuatus* ("arc" or "bow"), this collection of cell bodies forms an arch-like shape.
- **Median preoptic nucleus (MnPO)**—indicates that this nucleus is located near the midline position of the brain in front of the optic chiasm.
- **Paraventricular nucleus (PVN)**—*para* is Greek for "beside"; this part of the hypothalamus is located beside one of the ventricles of the brain.
- **Tuberomammillary nucleus (TMN)**—tuber comes from the Latin term *tuber cinerum* ("ashen swelling"), which is separate part of the hypothalamus noted for its grayish, ash-colored appearance. The term mammillary references the mammillary bodies. Taken together, this collection of cell bodies sits near the tuber cinerum and the mammillary bodies.

Optic chiasm—derived from the Greek words *optikos* ("of the eye") and *chiasma* ("crossing"), this is where the optic nerve crosses from one hemisphere of the brain to the other, creating an X shape.

Pituitary gland—derived from the Latin word *pituita* ("phlegm" or "mucus"), this structure received its name because it was mistakenly believed to produce mucus.

Thalamus—derived from the Greek word *thalamos* ("inner chamber"); this structure was originally believed to act as a reservoir ("chamber") for vital spirits.

- **Geniculate nucleus**—geniculate is derived from the Latin word *geniculatus* ("small knee"); this collection of cell bodies resembles a bent knee, and it is located toward the outer edge of the thalamus.
- **Pulvinar**—Latin for "cushioned couch" or "pillow", this part of the thalamus resembles a pillow.

Mesencephalon—*mes* is derived from the Greek words *mesos*, meaning "middle"; this is the middle part of the brain.

- **Tectum**—derived from the Latin word tēctum ("roof"), this region acts as a "roof" for the other structures in the midbrain.
  - Colliculus—Latin for "little hill", the colliculi look like little hills.
- **Tegmentum**—derived from the Latin word tegumentum ("covering"), this part of the brain is "covered" by the tectum.
  - **Interpeduncular nucleus (IPN)**—collection of cell bodies that lies between the cerebral peduncles (*peduncular*). Peduncular is Latin for *pedunculus* ("little foot" or "footstalk").
  - **Periaqueductal gray (PAG)**—*peri* is Greek for "around", and aqueductal refers to the cerebral aqueduct. This is a collection of cell bodies (gray matter) that surrounds the cerebral aqueduct.
  - **Red nucleus**—this collection of cell bodies appears redish due to their high iron concentration.

- **Substantia nigra**—this is a Latin term for "black substance". This part of the brain has a black hue due to its high melanin concentration; melanin also influences skin color.

Rhombencephalon—*rhomb* is derived from the Greek words *rhombos*, meaning "rhombus"; this part of the brain gets its name from the fact that the structures of this region are located near the rhombus-shaped fourth ventricle.

- **Metencephalon**—*met* is derived from the Greek word *meta*, meaning "beyond" or "after"; this region gets its name from the fact that it develops later during embryonic development.
  - **Cerebellum**—the Latin word cerebrum is combined with the Latin suffix ellum ("little"); this structure looks like a little cerebral cortex.
  - **Pons**—Latin for "bridge"; this structure acts like a bridge, as it is one region where axons from the left hemisphere cross over to the right hemisphere and vice versa.

    Pedunculopontine tegmental nucleus—this part of the brain is located in the tegmentum of the pons near the cerebral peduncle.

- **Myelencephalon**—*myel* is derived from the Greek word myelós, meaning "marrow"; this part of the brain consists of marrow, which is a soft, fatty tissue.
  - **Locus coeruleus**—Latin for "blue spot", this part of the brain has a blueish hue.
  - **Medulla oblongota**—the words medulla and oblongota are Latin for "marrow" and "elongated", respectively. As the primary structure of the myelencephalon, the medulla oblongota is made of marrow.

    Nucleus of the solitary tract (NST)—derived from the Latin expression *tractus solitarus*, this collection of cell bodies received its name because its axons are surrounded by gray matter, making them appear to run alone in the brainstem.
    Ventral respiratory group—this term describes the location and the function of this nucleus; it is located in the ventral portion of the medulla and helps regulate breathing.

  - **Cuneate nucleus**—derived from the Latin word *cuneus* ("wedge"), this wedge-shaped area of the medulla receives sensory information from the upper half of the body.
  - **Gracile nucleus**—derived from the Latin word *gracilis* ("slender" or "thin"), this thin-shaped nucleus in the medulla that receives sensory information from the lower half of the body.
  - **Raphe nuclei**—derived from the Greek word *rhaphē*, meaning "seam"; these cell bodies are located in the middle of the brainstem,

thus resembling a seam or suture. Note, the raphe nuclei are also located in the metencephalon and mesencephalon.

Ventricle—adopted from the Latin word *ventriculus* ("little belly"), this term is used to denote a cavity in an organ. In addition to the brain, the heart contains the left and the right ventricles.
Brainstem—name comes from the fact that this part of the brain acts in a similar fashion as a plant stem, which connects a plant to its roots; the brain stem connects the brain to the spinal cord.

- **Olivary nucleus**—derived from the Latin word *olīvārius* ("of belonging to olives"), this structure resembles an olive.
- **Parabrachial nucleus**—combines the Greek word *para* with the Latin word *brachium* ("arm"). The brachium references the *brachium conjunctivum* ("connecting arm"), which is a structure known as the superior cerebellar peduncle. The parabrachial nucleus sits beside the superior cerebellar peduncle.

Spinal cord—the word spine comes from the Latin word *spina* ("thorn", "backbone"); cord is derived from the French word *corde* ("rope"). The spinal cord resembles a rope and is located toward our back.

- **Cervical**—derived from the Latin word *cervix* ("neck"), the spinal nerves associated with this part of the spine support the neck and back portion of the head.
- **Thoracic**—derived from the Greek word *thōraks* ("breastplate" or "chest"), the spinal nerves associated with this part of the spine support the chest and abdominal region.
- **Lumbar**—derived from the Latin word *lumbus* ("loin"), the spinal nerves associated with this part of the spine transmit sensory information from the groin (loin) to the brain; these nerves also partially support bladder and bowel control, and control sensory and motor information traveling from/going to the lower half of the body.
- **Sacral**—derived from the Latin word *sacrum* ("holy bone"), the triangular-shaped bone near the base of the spine is called the sacrum. An ancient belief was that this bone was important for resurrection.
- **Coccygeal**—derived from the Greek word *kokkyx* ("cuckoo"), the coccyx ("tailbone"—attached to the bottom of the sacrum) resembles the beak of a cuckoo bird.

Etymology of Important Chemistry Terms

I want to provide a section on common chemistry terms that are used. These are terms you will encounter in the sections below. First, I need to define some elements of the periodic table that you will see several times below. I have also included other elements that you read about in the book. You will notice that a lot of chemical elements end with -gen and -ium. The term -gen comes from the Greek term *-genēs* ("formation") to signify that many elements are used to create something else. The term -ium is used in chemistry to denote a metal.

- **H** = hydrogen; derived from the Greek word *hydor* ("water"), this element received its name because when it is burned in oxygen, it produces water.
- **Li** = lithium; derived from the Greek word *líthos* ("stone"); name comes from the fact this element was discovered in a mineral.
- **C** = carbon; derived from the Latin word *carbo* ("coal"), this element is a major constituent of coal.
- **N** = nitrogen; derived from the Greek word *nitron* and the Latin word *nitrum*, which refers to saltpeter (a type of mineral that contains nitrogen).
- **O** = oxygen; derived from the Greek word *oxys* ("sharp" or "acid"), this element was initially believed to be necessary for producing acids.
- **Na** = sodium; the Latin term *natrium* is the reason why sodium is abbreviated as Na. Natrium refers to natron, a naturally occurring salt.
- **Mg** = magnesium; named for the Greek district of Magnesia, which is known for its minerals.
- **P** = phosphorous; derived from the Greek word *phosphoros* ("light-bearing"); this element glows in the dark.
- **S** = sulfur; most likely originates from the Latin word *sulpur*, which most likely relates to "burning".
- **Cl** = chloride; derived from the Greek word *khlōrós* ("pale green"), chlorine gas has a greenish hue.
- **K** = potassium; the Latin term *kalium* is the reason why potassium is abbreviated as K. Kalium is derived from an Arabic term meaning "ash from burnt plants". Potassium can be extracted from the ash of burnt wood and plants.
- **Ca** = calcium; derived from the Latin word *calx* ("limestone"), this element was initially isolated from limestone.

Many of the things you read about in this book are *lipids* or *proteins*. Here is the etymology of these terms:

- **Lipid**—derived from the Greek word *lipos* ("fat"). Lipids are fats.
- **Protein**—derived from the Greek word *proteios* ("primary"), this term signifies that proteins are the building blocks of life.

Now, I need to cover the etymology of several molecular groups that make up things like lipids and proteins.

- **Acetyl**—acetyl is derived from the Latin word *acetum* ("vinegar"); vinegar contains acetic acid, which is where the acetyl group (CH_3CO) comes from.
  - **Acetylation**—indicates the addition of an acetyl group to a molecule.

- **Amide**—derived from ammonia (NH_3), an amide is a compound in which the nitrogen atom is attached to a *carbonyl* group ($C = O$; carbon double bounded to oxygen).
- **Amine**—derived from ammonia (NH_3), an amine is a compound in which the nitrogen atom is attached to two hydrogen atoms (NH_2).
- **Amino acid**—the building blocks of proteins receive their name because they contain an amine that is connected to carboxylic acid (see entry for Carboxyl below).
- **Benzene**—derived from the term "gum benzoin", which is a type of resin (a sticky substance that can transform from a liquid or semi-solid state to a hard, durable substance when exposed to something like air, heat, or light); a benzene is a chemical compound that contains a six-carbon ring structure.
- **Carbonyl**—derived from the element carbon with the suffix -yl, which signifies a compound that has had a single hydrogen atom removed; the chemical structure is $C = O$ (carbon double bonded to oxygen).
- **Carboxyl**—this is a carbonyl group that is attached to a hydroxyl group (see entry for Hydroxy below).
- **Catechol**—derived from catechu, an extract from the acacia tree. In chemistry, a catechol group consists of a benzene ring with two hydroxyl groups (OH—oxygen and hydrogen atom).
- **Cyclic**—derived from the Greek word *kyklikos* ("circle"), this term indicates a chemical compound that contains a ring of atoms (e.g., benzenes are cyclic compounds).
- **Hydroxyl**—combines hydrogen and oxygen; this molecule has a chemical formula of OH.
  - **Alcohol**—in chemistry, an alcohol contains one hydroxyl group attached to a carbon atom. For example, ethanol (which humans consume), has the chemical formula C_2H_5OH.
- **Methyl**—derived from the Greek words *methy* ("wine") and *hylē* ("wood"), the first methyl alcohol was made by distilling wood. A methyl group has the chemical formula CH_3.
  - **Methylation**—indicates the addition of a methyl group to a molecule.
- **Nitro**—derived from the Greek word *nitron* and Latin word *nitrum*, which refers to saltpeter (a type of mineral that contains nitrogen); the chemical formula is NO_2.
  - **Nitration**—indicates the addition of a nitro group to a molecule.
- **Phenyl**—derived from the Greek word *phaino* ("shining"), the term phenyl is used because the first phenyl compounds were isolated from the oily residue of coal gas used for lighting. A phenyl compound contains a benzene ring that is missing one hydrogen atom.
- **Phosphate**—a phosphate is the salt/ester form of phosphoric acid and is often abbreviated as PO_4 (phosphorous atom bonded to four oxygen atoms).

– **Phosphorylation**—indicates the addition of a phosphate group to a molecule.

The following are additional chemistry terms that will help you further understand some of the terms included below:

- **-ate**—this suffix is often used to denote the salt or *ester* form of a compound; esters are formed when an acid is combined with alcohol.
- **Di**—derived from the Greek word *dís* ("double").
- **-ic**—this suffix is often used to denote a chemical that is acidic (pH below 7).
- **-ide**—indicates a binary compound (i.e., contains exactly two different elements—H_2O is dihydrogen monoxide).
- **-idine**—indicates a molecule that contains a nitrogen-containing ring in its structure.
- **-ine**—this suffix is often used to denote a chemical that is basic (pH above 7).
- **Mono**—Greek word meaning "single".
- **-ol**—indicates that a compound is an alcohol.
- **-ous**—this suffix is also often used to denote a chemical that is acidic.
- **Tri**—prefix derived from both Latin and Greek meaning "three".

Because the things you will read about below are encoded by genes, I want to include a few terms related to genetics:

- **Deoxyribonucleic acid (DNA)**—name describes its chemical structure; when you read about depolarization, you learned that de- means "removal". By now, you know oxy is related to oxygen. *Ribose* is a rearrangement of the word *arabinose*, which means "gum arabic"; this is a sugar first isolated from the acacia tree. In short, ribose is a sugar with the chemical formula $C_5H_{10}O_5$; as such, you will often hear ribose described as a 5-carbon sugar. Taken together, deoxyribo means that one oxygen atom has been removed from the ribose backbone: $C_5H_{10}O_4$. The term nucleic refers to the fact that DNA is located in the cell's nucleus. DNA is an acid because it contains acidic phosphate groups in its structure.

  – DNA is composed of four bases (hence why they end with -ine):

    Adenine—derived from the Greek word *adenon*, which means "gland"; adenine was discovered in a glandular extract.
    Cytosine—cytosine received its name because it was first isolated from the thymus cells (cyto) of calves.
    Guanine—derived from the Spanish word *guano*, which refers to bat droppings; guanine was originally isolated from guano.
    Thymine—derived from the Greek word *thymos* ("thyme" or "spirit"), this base was discovered in the thymus gland.

  – When the base is combined with the sugar molecule, it becomes a **nucleoside**.

    Adenosine—derived from the German word *adenosin*, this term combines the words adenine and ribose.
    Cytidine—combines cytosine with the suffix -idine.
    Guanosine—combines the words guanine and ribose.

Thymidine—conbines thymine with the suffix -idine.

- When the nucleoside is combined with a phosphate group, it becomes a **nucleotide**.
- **Histone**—derived from the Greek word *histós* ("loom" or "web"), histones help organize DNA into a web-like structure.

- **Ribonucleic acid (RNA)**—name describes the chemical structure of the ribose backbone, which is $C_5H_{10}O_5$.
  - The bases of RNA are nearly identical to those of DNA, except uracil replaces thymine. Uracil's name comes from urea and acetic acid. This base was discovered in urea, a major component of urine.

Etymology of Terms Related to Neurons

Neuron is Greek for "sinew", "tendon", or "cord"; the neuron received its name because it acts as the sinew that holds the nervous system together.

Major structures of the neuron:

- **Axon**—derived from the Greek word *áxōn* ("axis" or "axel"), the axon acts as a type of axel for neurons by transmitting electrical signals down its length; the axel of a vehicle transmits power from the engine to the tires.
  - **Microtubule**—micro is derived from the Greek word *mīkrós* ("small"). Tubule means small tube. These are small, tube-like structures. Although found in other types of cells, they play a major role in maintaining the axon's structural integrity and transporting certain proteins down the length of the axon.
  - **Nerve**—derived from the Latin word *nervus* ("sinew", "tendon", or "cord"), nerves received their name because the collection of axons resembled tendons.
  - **Tract**—derived from the Latin word *trahere* ("to pull" or "to drag"), this word evolved to mean "drawn-out" (*tractus*), particularly in the context of land ("drawn-out map"). Tracts are collections of axons in the central nervous system that pull information from one area of the brain to another.

- **Cell adhesion molecule**—name describes the function of this protein; it helps anchor neurons to each other.
  - **Cadherin**—portmanteau of calcium, dependent, and adherent (as in adhesion); this molecule requires the presence of calcium to function.
  - **Integrin**—derived from the word "integrate", these proteins are important for cell adhesion.
  - **Laminin**—derived from the Latin word *lamina* ("thin piece"), these proteins form a thin layer in the extracellular matrix, holding cells together.
  - **Neurexin**—a combination of the words neuron and rexin, which is derived from the Latin word *rex* ("king"); black widow venom ("king of neurotoxins") binds to neurexin.

- **Neuroligin—**combines neuro with the Latin word *ligare* ("to bind").
- **Sialyated plasma membrane protein—**sial comes from the Greek word *síalon* ("saliva"), and -ylated refers to siliac acid (which has the same origin as sial). Siliac acids were first discovered in saliva. This term means that these proteins are found in the cell membrane and are activated by siliac acids.

- **Dendrite**—derived from the Greek word *dendron* ("tree"), dendrites resemble tree branches.
- **Ganglion**—Greek for "knot" or "swelling under the skin", the term is used to describe the appearance of cell bodies clustered together in the PNS.
- **Myelin sheath**—myelin is derived from the Greek word *myelos* ("marrow") as Dr. Rudolp Virchow, who coined the term, believed that the myelin sheath resembled bone marrow. A sheath is a cover for a blade or sword; the myelin sheath acts a cover for axons.
  - **Node of Ranvier**—named for Dr. Louis-Antoine Ranvier, who first described the gaps in myelin sheath. The word node comes from the Latin word *nōdus* ("knot") and is used to describe the constrictions in the myelin sheath that lead to gaps.
- **Neurite—**combines the word neuro (for neuron) with -ite; term denotes any projection stemming from a neuron, particularly developing neurons.
- **Nucleus**—nucleus is a Latin term that means "kernel" or "nut"; just as a nut is the central part of a fruit, the nucleus is the central part of the cell body. The term nucleus is also used to indicate a collection of cell bodies in the central nervous system because these cell bodies are compacted like a kernel.
- **Receptor**—derived from the Latin word *recapere* ("to take back" or "to receive"), receptors "receive" something like a neurotransmitter.
  - **Agonist**—derived from the Greek word *agōnistēs* ("contestant" or "combatant"), this term is used to describe the action of binding to a receptor as a "contest".
  - **Antagonist**—derived from the Greek word *antagonistēs* ("opponent" or "competitor"), the antagonist acts as a competitor to the agonist as they both try to occupy the same binding site of a receptor.
  - **Autoreceptor**—auto is Greek for "self"; indicates a receptor located on a neuron that reacts to a chemical released from the same neuron.
  - **Ionotropic receptor**—derived from the Greek words *iono* ("ion") and *trópos* ("turn"), the term ionotropic describes how the receptor protein changes shape ("turns") when a ligand binds to it, causing it to open and allow ions to pass through.
  - **Metabotropic receptor**—metabolic is an English term derived from the Greek word *metabolikós* ("changeable") which refers to chemical changes that occur inside a living organism. Metabotropic receptors are dependent on metabolic processes to function properly.

- **Synapse**—derived from the Greek word *synapsis* ("conjunction" or "point of contact"), which itself was formed from the Greek words *syn-* ("together") and *heptein* ("to clasp"). The word describes that the synapse is the gap between neurons that communicate with one another.
  - **Alpha-synuclein—**alpha is the first letter of the Greek alphabet and was chosen to indicate that this is the first synuclein discovered. Synuclein is a portmanteau of synapse and nucleus. This protein was discovered in synapses and was originally proposed to be located in or near the nucleus of neurons.

    Lewy body—named for Dr. Fritz Heinrich Lewy, who discovered these clumps of proteins.
  - **Postsynaptic density (PSD)—**the term describes the thick appearance of this part of the postsynaptic neuron when viewed under an electron microscope.

    Actin—derived from the Greek word *aktís* ("ray" or "beam"), this protein's filaments have a ray-like pattern.

    - **Activity-related cytoskeleton-associated (ARC) protein**—the term "activity-related" indicates that expression of this protein is triggered by activity of the neuron. The term "cytoskeleton-associated" means that this protein binds to the actin cytoskeleton.

    Guanylate kinase-associated protein (GKAP)—see the section Etymology of Important Chemistry terms above for the entry on guanine, and see the section Etymology of Enzymes to learn about the origin of the term kinase. This term indicates that this protein resembles the enzyme guanylate enzyme.

    Homer protein—named in honor of the Greek poet Homer. Homer's *Odyssey* details a long journey. Likewise, the journey to clone this protein was a long one for scientists.

    Postsynaptic density 95 (PSD-95)—this protein is found in the PSD and has a weight of 95 kilodaltons (kDa).

    SRC homology 3 and multiple ankyrin repeat domains (SHANK)— Src is a family of protein tyrosine kinases (see Etymology of Enzymes to learn about the origin of the term kinase and Terms Related to Chemical Messengers to learn about tyrosine). SRC is short for sarcoma, a type of cancer. Homology comes from the Greek words *homo* ("same") and *logos* ("relation"); this term indicates a shared characteristic or sequence similarity used to identify the protein. The number 3 indicates this was the third conserved sequences identified in the SRC family. Ankyrin is derived from the Greek word *ankyra* ("anchor") as this protein is important for maintaining the structural stability of cells. The terms multiple and repeat indicate that there are multiple ankyrin repeats. The term domain indicates sites of the SRC homology 3 protein and the ankyrin proteins that act as docking sites for other proteins.

Synaptic Ras GTPase-activating protein 1 (SynGAP)—Ras is short for rat sarcoma. GTP stands for guanosine triphosphate (see Etymology of Important Chemistry terms to learn about the terms guanosine, tri, and phosphate). The 1 indicates this was the first protein of this family identified. In short, this protein helps turn off Ras signaling.

Transmembrane AMPA receptor regulatory protein (TARP)—transmembrane indicates this protein spans the cell membrane. As the name suggests, this protein regulates AMPA receptors; it modifies the receptor's function, assembly, and trafficking.

- **Terminal button**—the word terminal is derived from the Latin word *terminalis* ("of or belonging to an end, boundary, or limit"), and the word button is derived from the French word *bouton* ("bud" or "knob"). The terminal button is the end of the neuron, and it looks like a small bud.
- **Transporter**—the word transport is derived from the Latin word *transportare* ("to carry across"); the suffix -er conveys that someone or something is carrying an object from one place to another. The transporter "carries" neurotransmitters from the synapse to the terminal button.
- **Vesicle**—derived from the Latin word *vesicula* ("little bladder"), the term vesicle is used to describe its sac-like appearance.
  - **Synaptotagmin-1 protein—**synapto references the synapse while tagmin combines the words "tag" and "protein". This was the first member of the synaptotagmin family discovered. This protein "tags" synaptic vesicles with calcium (Ca^{++}), allowing vesicles to release their contents into the synapse.

The following are important organelles found in cells, including neurons:

- **Endoplasmic reticulum—**endoplasmic is derived from the Greek prefix *endo-* ("within") and the Greek word *plasma* ("something molded"). Endoplasmic refers to the cytoplasm of the cell. *Reticulum* is Latin for "a little net". This structure has a net-work structure.
- **Lysosome—**derived from the Greek words *lysis* ("dissolution") and *soma* ("body"), this organelle's function as a cellular body that breaks things down.
- **Mitochondrion—**derived from the Greek words *mítos* ("thread") and *khondros* ("grain"), this organelle received its name because it looks like thread-like grains under a microscope.
- **Ribosome—**name comes from ribonucleic acid (see section on Etymology of Important Chemistry Terms to learn about the etymology of RNA) and *soma.*

The following help neurons grow and survive:

- **Brain-derived neurotrophic factor (BDNF)**—the expression "brain-derived" indicates that this protein was first isolated from the brain (pig brain in this case). Neurotrophic combines the Greek words *neuro* ("nerve") and *trophic* ("nourishment" or "growth") and indicates something that promotes neuronal growth. Factor is a general term used to denote a substance that has a specific effect. Together, BDNF acts to promote growth and development of neurons.

- **Nerve growth factor (NGF)**—name describes what this protein does; it promotes neuronal growth.
- **Reelin—**term originates from the abnormal reeling gait of the reeler mouse; these mice lack a particular protein which was named after these mice.
  - **Disabled-1 (Dab1)—**this protein binds to and transmits reelin to other targets. The name disabled-1 reflects that reelin "disables" certain cellular activities that are crucial for neuronal development.
- **Tropomyosin receptor kinase—**tropomyosin is derived from the Greek words *trópos* and *myos*, which is the genitive form of *mus* ("muscle"); however, the name tropomyosin receptor kinase comes from the discovery of an oncogene that contained a kinase domain that was fused with part of a tropomyosin gene (*TPM3*).

The following are important proteins that are involved in certain neuronal processes but are not located on all neurons:

- **Chemoreceptor—**chemo is derived from the Greek word *chemeia* ("chemical"); these receptors respond to chemicals.
- **Cone—**these photoreceptors are named due to their cone-like shape.
- **Hair cell—**name comes from the hair-like stereocilia (see entry below) that protrude from the top of the cell.
  - **Stereocilia**—the term *stereo* is Greek for "solid" or "rigid", and the term *cilium* is Latin for "eyelash" or "hair"; these are rigid hair-like structures.
- **Latrophilin 3 receptor—**latro references the scientific name of the black widow (*Latrodectus*). Philin is derived from the Greek word *philos* ("loving" or "affinity for"). The venom of the black widow (latrotoxin) binds to these receptors. This is the third receptor of this family to be discovered.
- **Low-density lipoprotein (LDL) receptor family**—name describes the function of these receptors; they bind to low-density lipoproteins (i.e., proteins that combine with and transport fats).
  - **Apolipoprotein E (APOE)—**apo is Greek for "away from". A lipoprotein is a compound consisting of a fat and a protein. An apolipoprotein is a specialized protein that helps bind lipids to form a lipoprotein. The letter E helps distinguish this apolipoprotein from other apolipoproteins. APOE binds to various LDLs.
- **Mechanoreceptor—**mechano is derived from the Greek word *mēkhanē* ("machine" or "device"); the literal translation is "machine that receives". These receptors respond to mechanical stimuli like pressure and vibration.
- **Neurokinin receptor 1**—derived from the words neuro and *kinein* ("to move"), this protein received its name because it is involved in contraction of smooth muscles. This was the first receptor of this family that was discovered.
- **Neurotrophic tyrosine kinase receptor type 1**—see entry for BDNF above for the etymology of neurotrophic, and the entries tyrosine and kinase (below) for their etymology. This was the first receptor of this family that was discovered.

Essentially, these receptors are activated by chemicals that stimulate neuronal growth after tyrosine has been added to the receptor.
- **Nociceptor**—this term combines the Latin word *nocere* ("to harm") with the word receptor; this term describes how these receptors respond to painful stimuli.
- **Rod—**these photoreceptors are named due to their rod-like shape.
- **Transient receptor potential (TRP) ion channel**—name comes from a mutation observed in fruit flies that show a *transient* (i.e., short-lived) response to light; activation of TRP allows ions to flow inside the cell. Today, we know that TRPs belong to a superfamily of ion channels that respond to various kinds of stimuli like temperature and chemicals.

The following terms are related to neuronal processes:

- **Action potential**—the first part of this term refers to the action that takes place inside the neuron. The second part of this term refers to the electrical potential difference across the cell membrane.
  - **Depolarization—**derived from the Latin prefix *de-* ("reversal" or "removal") and the Latin word *polaris* ("polar"), this term indicates that the electrical potential difference across the cell membrane is being removed (that is, the number of positively charged and negatively charged ions inside the neuron is similar to the number of positively and negatively charged ions outside the neuron).
  - **Hyperpolarization**—the Greek prefix *hyper-* means "over" or "above"; this term denotes that the electrical potential difference across the cell membrane has become more extreme.
  - **Repolarization**—the English prefix *-re* means "again" or "back"; this term indicates that the neuron is becoming polarized again.
  - **Saltatory conduction**—derived from the Latin words *saltare* ("to hop") and *conductio* ("a leading" or "bringing together"), this term is used to describe how action potentials "hop" from one node of Raniver to the next in a myelinated axon.
- **Endocytosis**—derived from *endo-* and the Greek word *kytos* ("cell"), and the suffix *-osis* ("process" or "condition"), this term describes the process of something moving back inside a cell (e.g., a transporter removing a neurotransmitter from the synapse and placing it in the terminal button).
- **Exocytosis**—derived from the Greek prefix *exo-* ("out of"), *kytos*, and the suffix *-osis*, this term describes the process of something moving outside of a cell (e.g., neurotransmitters being released from the terminal button to the synapse).

The following terms are related to specific types of neurons:

- **Amacrine cells—**derived from the Greek words *a* ("not"), makrós ("long"), and *inós* ("fiber" or "tendon"), these cells received their name because they did not appear to have long axons like other nerve cells.

- **Bipolar cells—**name is based on its physical appearance; two ("bi") prominent projections (dendrite and axon) that extend away from the cell body (like two poles).
- **Ganglion cells—**these neurons received their name because their axons bunch up to form the optic nerve.
- **Horizontal cells—**name is used to describe how the axons of these cells travel horizontally.
- **Interneuron**—term is used to describe neurons that relay messages between sensory and motor neurons.
- **Medium spiny neuron (MSN)**—name comes from the physical appearance of these cells, which are medium-sized and contain numerous dendritic spines.
- **Purkinje cell**—named in honor of Dr. Jan Evangelista Purkyně, who discovered these cells.
- **Pyramidal neuron**—name comes from the pyramid-shaped cell body of these neurons.

The following section details the etymology of the 12 cranial nerves:

- **Olfactory nerve—**see etymology of olfactory sulcus; this nerve transmits information regarding smell.
- **Occipital nerve—**see etymology of occipital cortex; this nerve transmits information regarding vision.
- **Oculomotor nerve—**derived from the Latin words *oculus* ("eye") and *motor*, this nerve controls eye muscles.
- **Trochlear nerve—**derived from the Latin word *trochlea* ("pulley"), this nerve innervates the superior oblique muscle of the eye, which passes through the trochlea. The trochlea acts like a pulley system, helping change the muscle's direction.
- **Trigeminal nerve—**geminal comes from the Latin word *geminus* ("twin"); thus, trigeminal literally translates to "three twins". This term is used to denote the fact this nerve has three major divisions that extend to different parts of the face.
- **Abducens nerve—**abducens comes from the Latin word *abducens* ("leading away"); this nerve is responsible for abducting the eye (i.e., moving it laterally).
- **Facial nerve—**term is used as this nerve controls muscles of the face.
- **Vestibulocochlear nerve—**term combines vestibular nerve with cochlear nerve; these nerves transmit information regarding vestibulation and audition (i.e., sound). Vestibular comes from the Latin word *vestibulum* ("entrance" or "porch"); this term is used in reference to the vestibule of the inner ear. *Cochlea* is Latin for "snail shell"; the cochlear nerve originates from the cochlea of the inner ear.
- **Glossopharyngeal nerve—**derived from the Greek words *glōssa* ("tongue") and *pharynx* (pharynx), this term describes the function of the nerve: control muscles of the tongue and the pharynx.
- **Vagus nerve—**Latin for "wandering", this term describes this nerve's extensive and meandering path through the body.

- **Accessory nerve—**derived from the Latin expression *nervus accessories* ("a nerve in addition to"), this nerve received its name because the cranial portion was initially believed to be a branch of the vagus nerve.
- **Hypoglossal nerve—**term literally translates to "under the tongue"; describes the pathway of this nerve from the brain to the tongue.

Etymology of Terms Related to Non-neuronal Cells

- **Astrocyte**—derived from the Greek words *ástron* ("star") and *kytos* ("cell"), this cell receives its name from its star-like shape.
- **Ependymal cell**—the word ependymal is derived from the Greek words *epi* ("over" or "upon") and *enduma* ("garment" or "investment"); this is translated to "over garment" and is used to represent the thin membrane that covers the lining of the ventricles, which is where these cells are located.
- **Glia**—Greek for "glue", these cells were originally proposed to act as the glue that held neurons together.
- **Lymphocyte—**cyte refers to *cyto,* and lymp is derived from the Latin word *lympha* ("water"); lymp was used to describe the clear fluid in animal bodies.
- **Macrophage—**Greek for "large eater"; used to denote white blood cells that "eat" foreign particles.
- **Microglia**—*micro* is Greek for "small"; these are small glia cells.
- **Oligodendrocyte**—derived from the Greek words *oligos* ("few" or "small"), *dendron*, and *kytos*, the term is used to describe the appearance of this cell ("small tree-like").
- **Radial glia**—radial is derived from the Latin word *radius* ("spoke of a wheel", "ray" or "beam of light"). These cells, like astrocytes, have a star-like shape.
- **Schwann cell**—named for Dr. Theodor Schwann, who first described these cells.

Etymology of Terms Related to Chemical Messengers

Fatty acids are the building blocks of fats. The term fatty acid is used because these molecules were first isolated from animal fats. These molecules have a carboxyl group at one end, making them an acid.

- **Arachidonic acid**—arachidonic is derived from the Latin word *arachis* ("peanut"); it received its name due to its chemical structure being similar to arachidic acid, which is found in peanuts.
- **Glyceride—**combination of the Greek word *glukus* and the chemical suffix -ide. Glyceride has a sweet taste.
- **Prostaglandin**—name comes from the original belief that prostaglandins were produced by the prostate gland.

Hormones are chemical messengers produced by the endocrine system. They travel through the bloodstream to target organs and tissues in the body. Hormone is derived from the Greek words *hormaein* ("to excite") and *horme* ("impulse"); the term was chosen to reflect the function of hormones to set in motion physiological processes.

- **α-Melanocyte-stimulating hormone (α-MSH)**—melanocytes are cells that produce melanin, which influences our skin color. This hormone stimulates melanocyte activity. The α is used to denote that this version of MSH is best characterized.
- **Adrenocorticotropic hormone (ACTH)**—adreno means "near the kidneys" (see epinephrine above); cortico refers to the cortex of the adrenal gland, and tropic is derived from the Greek word *trópos*. The name describes what this hormone does; when released, it "turns on" the adrenal gland to produce cortisol.

  – **Melanocortin receptor (MCR)**—melanocortins are derived from the proopiomelanocortin (POMC) peptide (see below for etymology melanocortin and POMC), with ACTH belonging to this family. The book discusses MC2R and MC4R. The numbers indicate the second and the fourth receptor of this family, respectively.

- **Androgen**—derived from the Greek *andrós* ("man"), androgens are sex hormones associated with the male sex.

  – **Androstenedione**—name describes its chemical structure. This is an androgen that contains a double bond in the steroid ring structure ("sten") and two carbonyl groups ("dione").
  – **Dehydroepiandrosterone (DHEA)**—term describes its chemical structure. Dehydro indicates this androgen steroid is missing two hydrogen atoms. Epi is used in chemistry to denote a difference in its configuration.

    Dehydroepiandrosterone sulfate (DHEAS)—etymology is the same as the entry above, except that a sulfate group has been added to the molecule.

  – **Testosterone**—derived from the Latin word *testis* ("testicle") and the Greek word *sterol*, testosterone is a steroid sex hormone produced in the testes.

    Dihydrotestosterone—this is a testosterone molecule that has two hydroxyl groups added to it.

- **Anti-Müllerian hormone**—Müllerian is named for Dr. Johannes Peter Müller; anti- Müllerian hormone prevents the Müllerian ducts from forming into female reproductive organs.
- **Cholecystokinin (CCK)**—derived from the Greek words *chole* ("bile"), *kystis* ("bladder" or "sac"), and *kinein* ("to move"), the literal translation is "to move the bile-sac". CCK stimulates the contraction of the gallbladder, which is the organ that holds and releases bile.
- **Corticosteroid**—term refers to the fact that these hormones are produced from the cortex of the adrenal gland. The term steroid combines sterol (derived from

cholesterol) and -oid, which means "resembling". In short, steroids have a similar chemical structure as cholesterol (composed of four rings).

- **Corticotropin-releasing hormone (CRH)**—The name describes what this hormone does; it causes the release of ACTH (see ACTH above).
- **Cortisol**—combines cortex with -ol. Although -ol is often used to denote alcohols, it is often used to identify steroid hormones such as cortisol. This hormone is produced in the adrenal cortex.
- **Corticosterone**—combines cortex with ster- and -one. The ster- indicates that this hormone is a steroid. The -one indicates that this hormone is also a ketone.
- **Estrogen**—derived from the Greek word *oistros* and the Latin word *oestrus* ("heat"), the term "in heat" refers to female animals that are receptive to copulation. Thus, estrogen literally translates to "heat producing".
  - **Estradiol—**estra has the same etymology as described for estrogen. The "di" references the two hydroxyl groups of estradiol's chemical structure. The "ol" indicates that this chemical is a steroid.
- **Glucagon**—portmanteau of the words glucose and agonist; this hormone acts as a "glucose agonist" as it increases blood glucose levels. Glucose comes from the Greek word *gleûkos* ("sweet wine") as it was first isolated from grape sugar.
- **Glucocorticoid**—term refers to the fact that these hormones are produced from the cortex of the adrenal gland. Gluco refers to glucose; these hormones were initially linked to their role in glucose metabolism. The -oid is used to identify these hormones as steroids.
- **Ghrelin**—derived from the Proto-Indo-European root *ghre* ("to grow") and the suffix -relin, which indicates a releasing substance; this hormone gets its name due to its role in stimulating growth hormone secretion.
- **Gonadotropin—***gonad* is Greek for "seed of generation". The term gonadotropin literally translates to "substance that turns or changes the gonads".
  - **Follicle-stimulating hormone (FSH)**—name describes its function to stimulate the ovarian follicles (sacs that contain eggs). Follicle is derived from the Latin word *folliculus* ("little bag").
  - **Luteinizing hormone—***luteus* is Latin for "yellow". The corpus luteum ("yellow body") is a temporary gland in the ovary that forms from the follicle after an egg has been released, and it has a yellow hue. Luteinizing hormone is important for ovulation (when eggs are released into the fallopian tube).
- **Gonadotropin-releasing hormone (GnRH)—**name describes the function of this hormone; it causes the release of gonadotropins (see entry above).
- **Insulin**—gets its name from *insula*, as this hormone is produced in the islets of Langerhans, small clusters of cells within the pancreas.
- **Leptin**—derived from the Greek word *leptos* ("thin"), leptin gets its name as it is primarily associated as a hunger-suppressing chemical.

- **Melanocortin—**melano is derived from the Greek words *melas* ("black" or "dark"), and cortin refers to cortisol. Name reflects what was initially discovered about this group of hormones that is cleaved from POMC (e.g., melanin influences skin pigmentation, and cortisol is involved in our stress response).
- **Melatonin**—derived from the Greek words *melas* and *tonos* ("tone" or "color"), as well as the suffix -in (from serotonin, which is closely related to melatonin), melatonin received its name because it has the ability to lighten the skin color of animals.
- **Progestogen—**the term pro- comes from the Latin word meaning "for" or "supporting". Gest- comes from the Latin word *gestare* ("to carry" or "bear"). Therefore, the term progestogen is used to denote a chemical that is produced to support pregnancy.

  – **Progesterone—**the etymology is largely the same as progestogen. The -one at the end indicates that progesterone is a ketone.

- **Proopiomelanocortin (POMC)—**the term pro- means precursor; opiomelanocortin combines opioid, melano, and cortin to reflect which peptides/ hormones come from this polypeptide (opioid peptides like endorphins and melanocortins).
- **PYY$_{3\text{-}36}$**—stands for peptide YY; the numbers indicate that this peptide is 36 amino acids in length, with the first two amino acids removed.

Neuropeptides are short chains of amino acids that act as chemical messengers. The term combines the words neuro and peptide, which is derived from the Greek word *peptose* ("digestible"). Peptides get their name from the fact that amino acids are broken down during digestion.

- **Agouti-related peptide**—the word agouti comes from Guarani language and refers to a specific hair color pattern that is distinct to the agouti rodent. This peptide is encoded by the agouti gene.
- **Neuropeptide Y (NPY)**—the Y comes from the single-letter code for the amino acid tyrosine, which is abundant in this neuropeptide.
- **Opioid**—this word is the combination of opium, which is derived from the Greek word *ópion* ("juice of a plant"), and the English suffix -oid; opioids are chemicals that resemble opium.

  – Two of the endogenous opioids end with the suffix -**orphin**, which is used to denote them as endogenous *morphine* peptides.

    Dynorphin—dyn is derived from the Greek word *dynamis* ("power"); name describes the powerful analgesic effects of dynorphins.

    - **Kappa receptor—**named for ketocyclazocine, which was the first ligand discovered to bind to this receptor subtype.

    Endorphin—portmanteau of endogenous and morphine.

- **Mu receptor—**named for morphine, which was the first ligand discovered to bind to this receptor subtype.

– **Enkephalin**—derived from *enkephalos* ("brain") and the suffix -in; term was used to reference that these opioids were first discovered in the brain.

Delta receptor—named for the mouse vas deferens tissue, which is where this receptor was initially discovered.

- **Orexin (hypocretin)**—derived from the Greek word *orexis* ("appetite"), orexin received its name by its discoverers because of its role in stimulating appetite. Hypocretin gets its name from *hypo* (for hypothalamus), as this is where it is produced, and from secretin, which is a gut hormone that is structurally similar to hypocretin.
- **Substance P**—P is short for peptide.
- **Vasoactive intestinal peptide (VIP)—**name describes the actions this neuropeptide has on blood vessels; "vasoactive" is used to represent VIP's ability to dilate blood vessels. This peptide was first isolated from the intestine, hence the word "intestinal".

Neurotransmitters are chemical messengers released from neurotransmitters that act to excite or inhibit other neurons. These chemicals act as "transmitters", sending a signal to another neuron. The term transmitter comes from the Latin word *transmittere* ("to send across").

- **Acetylcholine**—combines acetyl (see section on Etymology of Important Chemistry Terms above) with choline, derived from the Greek word *chole* ("bile"). Choline was first isolated from bile. Acetylcholine gets its name because its chemical structure consists of an acetyl group that is attached to choline.

  – **Muscarinic receptor—**name comes from muscarine, an alkaloid found the *Amanita muscaria* mushroom that binds to these receptors.
  – **Nicotinic receptor—**name comes from nicotine, the primary psychoactive compound in tobacco, which shows high affinity for these receptors.

- **Adenosine triphosphate (ATP)**—named for its chemical structure; a chain of three phosphate groups has been added to adenosine.
- **Dopamine**—gets its name from dopa, which refers to dihydroxyphenylalanine, a precursor amino acid (see entry below), and amine (see section on Etymology of Important Chemistry Terms above).

  – **Dihydroxyphenylalanine**—composed of the words di, hydroxy, and phenyl (see section on Etymology of Important Chemistry Terms above), as well as alanine (derived from German word referring to aldehyde, which is where alanine was first synthesized).

- **Endocannabinoid**—derived from the Greek words *endo* and *kánnabis* ("cannabis"), and the English suffix -oid ("having the form of"), endocannabinoids are chemicals produced within the body that resemble cannabis.

- **2-Arachidonoylglycerol (2-AG)**—name denotes its chemical structure. Arachidonoyl refers to arachidonic acid (see subsection Fatty Acids for more details). Glycerol comes from the Greek word *glykys* ("sweet") as this compound has a sweet taste. The 2 indicates that the arachidonoyl group is attached to the second carbon of the glycerol molecule.
- **Anandamide**—derived from the Sanskrit word "ananda" ("bliss") and the suffix amide (see section on Etymology of Important Chemistry Terms above). Because endocannabinoids are involved in mediating pleasure, the term anandamide was used when the first endocannabinoid was discovered.

  N-arachidonoyl phosphatidylethanolamine (NAPE)—N-arachidonoyl refers to arachidonic acid that has been added to a nitrogen atom. Phosphatidyl indicates a phospholipid; a phosphate group is added to the third carbon of the glycerol backbone. Ethanolamine combines ethanol with amine; this structure consists of a two-carbon chain with a hydroxyl group at one end and an amine at the other end ($HOCH_2CH_2NH_2$).

- **Gamma-aminobutyric acid (GABA)**—name is based on its chemical structure. Gamma is the third letter of the Greek alphabet; the amino group is attached to the third carbon atom. Amino is derived from amine (see section on Etymology of Important Chemistry Terms above); amino acids contain both an amine and an acidic carboxyl group. Butyric is derived from the Greek word *boúturos* ("butter") as this chemical was first identified in butter.
- **Glutamate**—Derived from the Latin word *gluten* ("glue"), glutamic acid was first discovered in wheat gluten. Glutamate is the salt form of glutamic acid.

  - **α-Amino-3-hydroxy-5-methyl-4-isoxazolepropionic acid (AMPA) receptor—**AMPA is a chemical that shows high affinity for this receptor subtype; its name describes its chemical structure. Propionic acid makes up the backbone of this chemical (CH_3CH_2COOH). The expression 4-isoxazole indicates a five-membered isoxazole ring (contains three-carbon atoms, one nitrogen atom, and one oxygen atom) attached to the fourth position of the propionic acid backbone. The expression 5-methyl indicates that a methyl group is attached to the fifth position of the isoxazole ring. The expression 3-hydroxy indicates that a hydroxyl group is attached to the third position of the isoxazole ring. The expression α-amino indicates an amino group located on the alpha carbon acid (i.e., first carbon atom after the carboxyl group) of the propionic acid backbone.
  - **N-methyl-D-aspartate (NMDA) receptor—**NMDA is a chemical that shows high affinity for this receptor subtype. Aspartate is derived from aspartic acid, a non-essential amino acid. The expression N-methyl indicates that a methyl group has been attached to a nitrogen atom. The "D" is used to identify this compound as the "d-isomer" (i.e., identifies the form of aspartate).

  Quinolinic acid—derived from quinoline, a nitrogen-containing compound, which is derived from quinine (found in tonic water). Quinine is derived from the Spanish word *quina*, which refers to the bark of the chinchona plant; quinine is extracted from this bark.

- **Glutamine**—the etymology is the same as glutamate; the -amine indicates that the precursor chemical to glutamate is an amine.
- ***N*-acetylcysteine**—name describes its chemical structure. An acetyl group is attached to a nitrogen atom, and cysteine is an amino acid.

- **Glycine**—derived from the Greek word *glykys*; this amino acid neurotransmitter has a sweet taste.
- **Histamine**—combines hist- with -amine; hist is derived from the Greek word *histos* ("tissue"); name comes from the fact that this amine was found in many different body tissues.
- **Serotonin (5-hydroxytryptamine)**—this word is a portmanteau of "serum" and "tonic"; serotonin was discovered in blood serum, and researchers noted its ability to constrict blood vessels. 5-hydroxytryptamine refers to the chemical structure of serotonin. 5-hydroxy refers to a hydroxyl group located at the fifth position of the molecule; tryptamine is derived from its parent amino acid tryptophan (see entry below).

  - **5-hydroxyindoleacetic acid (5-HIAA)**—the 5-hydroxy part is identical to what was described for serotonin above. Indole is derived from the color indigo, which was first isolated from plants in India. In chemistry, an indole group consists of a benzene ring fused to a five-membered nitrogen-containing pyrrole ring (four carbon atoms and one nitrogen atom). Acetic acid refers to the CH_2COOH group that is attached to the indole ring.
  - **Tryptophan**—a German word; tryp refers to trypsin, an enzyme that was used to discover tryptophan, and phan is derived from the Greek word *phaínein* ("to appear"). Tryptophan received its name because trypsin causes a colored compound to appear when proteins are broken down during digestion.

    Kynurenine—German word used to denote that this compound was first discovered in dog urine.

Some chemical messengers act as both a hormone and a neurotransmitter.

- **Epinephrine (adrenaline)**—derived from the Greek words *epí* ("upon") and *nephrós* ("kidney"), epinephrine gets its name from the fact that it is produced by the adrenal glands, which are situated above the kidney. The synonym adrenaline is derived from the Latin words *ad* ("near to") and *renal* ("kidney").
- **Monoamine**—denotes a chemical structure shared by neurotransmitters/hormones in this family (a single amine group).

  - **Catecholamine**—combines the words catechol and amine (see section on Etymology of Important Chemistry Terms above).
  - **Indolamine**—combines the words indole and amine (see 5-hydroxyindoleacetic acid and monoamine above).

- **Tyramine**—combines tyrosine (see entry below) with amine. Tyramine is derived from tyrosine.
- **Tyrosine**—this amino acid comes from the Greek word *tyrós* ("cheese") and the suffix -ine; tyrosine was discovered in cheese.

- **Norepinephrine (noradrenaline)**—comes from combining nor- (short for "normal") and epinephrine (see above for epinephrine's etymology); norepinephrine is the "normal" version of epinephrine as it lacks a methyl group (CH_3). The synonym noradrenaline follows the same logic.
- **Oxytocin**—derived from the Greek word *oxús* ("swift") and *tókos* ("childbirth"), oxytocin gets its name from the fact that it is involved in uterine contractions.
- **Vasopressin**—derived from the Latin word *vas* ("vessel") and the English suffix -pressin ("pressure"), vasopressin received its name for its ability to constrict blood vessels, thus increasing blood pressure.

Second messengers are small molecules that relay signals from the cell's surface to other molecules within the cell. The term is used to indicate that these molecules do not send a messenger until they have been activated by something else (e.g., a neurotransmitter binding to a metabotropic receptor).

- **Ca^{++}/calmodulin-dependent kinase II (CaMKII)**—this is a kinase activated by calcium (Ca^{++}), which then binds to calmodulin; calmodulin is a portmanteau, combining the words calcium, modulate, and protein. Essentially, calcium modulates the activity of this protein, which then leads to phosphorylation of other proteins (like AMPA receptors). The II indicates that this is the second member of the CaMK family to be identified. See Etymology of Enzymes to learn about the origin of the word kinase.
- **Cyclic adenosine monophosphate (cAMP)**—name is used to describe its chemical structure; it contains a circular structure (cyclic), an adenine base and ribose backbone (adenosine), and a single phosphate group.
- **Cyclic guanosine monophosphate (cGMP)** - name is used to describe its chemical structure; it contains a circular structure (cyclic), a guanine base and ribose backbone (guanosine), and a single phosphate group.
- **Diacylglycerol (DAG)**—name describes its chemical structure. Glycerol is a three-carbon alcohol molecule. The expression "diacyl" indicates that two fatty acid chains have been added to the glycerol molecule.
- **Inositol 1,4,5-trisphosphate (IP3)**—name describes its chemical structure. Inositol refers to a sugar-like ring and is derived from the Greek word *inós* as it was found in muscle tissue. 1,4,5-Triphosphate refers to the phosphate groups added to the first, the fourth, and the fifth positions of the inositol ring.
- **Phosphatidylinositol 4,5-bisphosphate (PIP2)**—name describes its chemical structure. Phosphatidyl indicates a phospholipid; phosphoglyceride backbone. 4,5-Biphosphate refers to the two phosphate groups added to the fourth and the fifth position of the inositol ring. See IP3 above for the etymology of inositol.

Etymology of Enzymes

Enzyme comes from the Greek words *en* ("in") and *zymē* ("leaven" or "yeast"). The word was first used to describe the process of fermentation of sugar to alcohol by yeast. Enzymes often end with -ase.

- **Acetylcholinesterase**—term denotes that this enzyme acts on acetylcholine to break it down.
- **Alpha-N-acetylneuraminide alpha-2,8-sialyltransferase**—N-acetylneuraminate means that an acetyl group has been added to the sugar neuraminic acid. The term alpha indicates the orientation of the molecule. Neuraminic comes from the German word *neuraminsäure*, which literally translates to "brain acid". Alpha-2,8 refers to the chemical bond formed by this enzyme. The 2 indicates the bond is between the second carbon of one sugar molecule and the eighth carbon of the next sugar molecule. Sialyl refers to salic acid, which comes from the Greek word *sialon* ("saliva"); this acid was first discovered in salvia. Transferases transfer a functional group from one molecule to another.
  - **Cytidine monophosphate (CMP)-N-acetylneuraminate-poly-alpha-2,8-sialyltransferase**—Monophosphate means the ribose backbone contains a single phosphate group. The term "poly" means "many", indicating that this enzyme builds a long polymer.
- **Aromatase**—aroma refers to "aromatic"; this enzyme increases the number of aromatic carbon rings in testosterone, converting it to estradiol.
- **Catechol-O-methyltransferase (COMT)**—name describes the function of this enzyme; a methyl group (CH_3) is transferred to an oxygen atom (the "O") in a catechol group.
- **Cyclic guanosine monophosphate (cGMP)-specific phosphodiesterase type 5 (PDE5)**—see section on Second Messengers section above for the etymology of cGMP. Phosphodiesterase combines phospho for "phosphorous", di for "two", and esterase to denote an enzyme that hydrolyzes an ester bond. The type 5 denotes the type of enzyme in the PDE family. Together, this enzyme metabolizes cGMP by breaking its phosphate bond.
- **Cyclooxygenase-2 (COX-2)**—cyclo is derived from the Greek word *kyklos* ("circle"); this term is used to describe the ring-like structure of prostaglandins that are formed when COX-2 breaks down arachidonic acid. The term oxygenase is derived from *oxys* and *-genēs*; Essentially, COX-2 uses oxygen to create prostaglandins. The number 2 indicates that this is the second COX enzyme to be discovered.
- **Cytochrome P450 enzymes**—cyto references the fact these enzymes are located in the cell body; chrome comes from the Greek word *khrôma* ("color"), and P450 stands for pigment-450 nm (nm). Essentially, these enzymes are good at absorbing light that has a wavelength of 450 nm (corresponding to the color blue).

- **Glutamic acid decarboxylase (GAD)**—indicates the function of this enzyme; decarboxylation is the removal of a carboxyl group from a molecule. Taken together, this enzyme removes a carboxyl group from glutamate to form GABA.
- **Glutaminase**—name means the enzyme that acts on glutamine to convert it to glutamate.
- **Glycogen synthase kinase-3 (GSK-3)**—this kinase adds a phosphate group to glycogen synthase. Glycogen comes from the Greek words *glykys* and *gignomai* ("I produce"); this substance produces sugar. Synthase comes from the Greek word *syntithenai* ("to put together"). Glycogen synthase is responsible for synthesizing glycogen, and GSK-3 is responsible for modulating activity of glycogen synthase.
- **Hypoxanthine–guanine phosphoribosyltransferase (HGPRT)**—xanthine comes from the Greek word *xanthos* ("yellow") because xanthine has a yellowish color; xanthine has the chemical formula $C_5H_4N_4O_2$. The term hypoxanthine means "lies under xanthine" as this compound is an antecedent to xanthine. Phosphoribosyltransferase indicates that a phosphate group is transferred to ribose.
- **Indoleamine 2,3-dioxygenase (IDO)**—see entry for indolamine above. 2,3-dioxygenase means that two oxygen atoms are added to the double bond between the second and the third positions of the indole ring.
- **Kinase**—derived from the Greek word *kineîn* ("to move" or "to set in motion") and the suffix -ase; kinases move phosphate groups to other molecules to activate or inactive them.
  - **Phosphatidylinositol 3-kinase (PI3K)**—the name describes the chemical structure of this kinase. The term phosphatidyl identifies this kinase has a lipid component, and inositol refers to the sugar-like ring that makes up part of its structure. The 3 identifies that a phosphate group has been added to the third position of the inositol ring.
  - **Protein kinase**—name describes the function of this kinase; protein kinases add phosphate groups to proteins.
- **Monoamine oxidase (MAO)**—name describes how this enzyme functions; this enzyme oxidizes monoamines to break them down (i.e., an oxygen atom is added to a molecule).
- **Phospholipase**—combines the Greek root *phosphor* and word *lipos* to describe that this enzyme breaks down phospholipids (i.e., a kind of fat with phosphorous).
- **Tryptophan hydroxylase**—describes the function of this enzyme; it adds a hydroxyl group to tryptophan to convert it to 5-hydroxytryptophan.
- **Tyrosine hydroxylase**—describes the function of this enzyme; it adds a hydroxyl group to tyrosine to convert it to L-DOPA.

Important Proteins for Cell Integrity

In Chap. 10, you learned about motor disturbances. There are numerous proteins that are referenced in this chapter (especially in Fig. 5) that help ensure the structural integrity of cells.

- **Desmin**—derived from the Greek word *désmē* ("bundle"), this protein helps bundle filaments within muscle cells.
- **Dystrophin**—named for dystrophy as mutations to the gene that encodes this protein are linked to this medical condition; dystrophy is derived from the Greek words *dsy* ("bad") and *trophē* ("nourishment"); originally, the hypothesis was that dystrophy resulted from poor nutrition.
- **Emerin**—this protein is named after Emery-Dreifuss muscular dystrophy; this disease is named for Dr. Alan Emery and Dr. Fritz Dreifuss.
- **Filamin C**—filamin combines "filament" with -amin. This protein helps cross-link actin filaments. The C differentiates this filamin from other types of filamin.
- **Nesprin**—portmanteau of nuclear envelope spectrin repeat. Spectrin comes from specter ("ghost") because this protein was first isolated from the cytoskeleton of red blood cells that had the rest of their contents washed away. Nesprin proteins have multiple spectrin proteins embedded in them and act as anchors that hold the cell nucleus to the cytoplasm.
- **Sad1/UNC-84 domain-containing proteins 1 and 2 (SUN1/2)**—Sad1 comes from the species *Schizosaccharomyces pombe* ("fission yeast"), and UNC-84 comes from the species *Caenorhabditis elegans* (type of nematode, aka—roundworm). The numbers indicate the type of protein. These proteins are important for linking the cell nucleus to the cytoplasm.
- **Sarcolemma**—derived from the Greek words *sarx* ("flesh") and *lemma* ("husk" or "sheath"), this term is used to describe the cell membrane that surrounds muscles cells.
  - **Costamere**—combines the Latin word *costa* ("rib") with the Greek suffix *-mere* ("part"). This protein has a rib-like structure when viewed under an electron microscope. These proteins help maintain the mechanical stability of the sarcolemma.

Etymology of Other Important Proteins and Lipids

This section contains the etymology of additional proteins and lipids covered in this book.

- **Cytokine**—combines *cyto* ("cell") with the Greek word *kinesis* ("movement"). Term describes how these proteins move between and influence cells.

- **Divalent metal transporter 1 (DMT1)**—this transporter transports metal ions with a divalent charge; divalent comes from the Latin words *di* ("two") and *valens* ("powerful"). Examples of ions with a divalent charge covered in this book are calcium (Ca^{++}) and magnesium (Mg^{++}). The 1 indicates that this was the first type of metal transporter discovered.
- **Dynein**—the dyne is a unit of force (derived from the Greek word *dúnamis*, which means "force"); this protein is a force-generating ATPase (breaks down adenosine triphosphate) that helps move things like organelles and vesicles in cells.
- **Ferritin**—*ferrum* is Latin for "iron"; this protein stores iron.
- **Fused in sarcoma (FUS)**—the protein is named after the FUS gene, is named for how it was discovered. The gene was initially identified as a *fusion oncogene*; this means that two separate genes fused together to create a gene that has the potential to cause cancer (oncogene). This event was associated with sarcoma, a type of cancer. Sarcoma is derived from the Greek word *sarkōma* ("fleshy growth").
- **Glycosphingolipid** - combines the Greek word *glukús* ("sweet") with sphingo (which refers to the Sphinx), and lipid. Individuals wanted to reference the Sphinx as these lipids had an enigmatic structure.

  - **Ganglioside**—combines the Greek word *ganglion* with the suffix -oside to denote that this is a glycolipid (that is, a molecule made up of sugar and fat). These molecules were first discovered in cell bodies.

- **Homeobox protein Meis1**—*homeo* is Greek for "alike"; box refers to a highly conserved DNA sequence that is common to these genes. Meis1 is a contraction of *myeloid ecotropic viral integration site 1*. Myeloid combines *myelos* ("marrow") and -oid ("likeness") and denotes its role in bone marrow-related blood cells. Ecotropic is derived from the Greek words *oikos* ("house") and *trópos*, this term means that something is confined to a specific environment. The expression viral integration site refers to the site in the genome in which a virus inserted its genetic information. The number 1 indicates that this was the first such site identified.
- **Huntingtin protein**—named due to its association with Huntington's disease, which is named for George Huntington, who first described the disease.
- **Kinesin**—derived from *kinesis*, this protein has a similar function as dynein (see above).
- **Nuclear pore complex**—nuclear refers to the cell nucleus, and pore refers to a channel opening (derived from the Greek word *poros*, meaning "passage"). A complex is a large collection of proteins. This is a collection of proteins near the passage between the nucleus and the cytoplasm.
- **Prion**—originated from the expression "proteinaceous infectious particle". Dr. Stanley Prusiner took the "pro" from protein and "in" from infectious, but he switched the "o" and the "i" to form prion to make it sound like other particles like proton and electron.
- **TAR (trans-activation response) DNA-binding protein 43 (TDP-43)**—TAR refers to a specific nucleic acid sequence found in the human immunodeficiency virus (HIV) genome. TDP-43 binds to this nuclei acid sequence. The number 43 refers to the size of the protein (43 kilodaltons).

Index

J. Yates, *Neuroscience of Maladaptation*, https://doi.org/10.1007/978-3-032-15602-0

I

K

L

M